THE NAMES OF PLANTS

The Names of Plants is an invaluable reference for botanists and horticulturalists. The first section gives an historical account of the significant changes in the ways by which plants have been known and named. It documents the problems associated with an ever-increasing number of common names of plants, and the resolution of these problems through the introduction of International Codes for both botanical and horticultural nomenclature. It also outlines the rules to be followed when plant breeders name a new species or cultivar of plant.

The second section comprises a glossary of generic and specific plant names, and components of these, from which the reader may interpret the existing names of plants and construct new names. With explanations of the International Codes for both Botanical Nomenclature and Nomenclature for Cultivated Plants, this new edition contains a greatly expanded glossary, which includes the Greek, Latin, or other source of each plant name.

THE NAMES OF PLANTS

FOURTH EDITION

David Gledhill

CAMBRIDGE UNIVERSITY PRESS
Cambridge, New York, Melbourne, Madrid, Cape Town, Singapore, São Paulo

Cambridge University Press
The Edinburgh Building, Cambridge CB2 2RU, UK

Published in the United States of America by Cambridge University Press, New York

www.cambridge.org
Information on this title: www.cambridge.org/9780521866453

First published 1985
Second edition 1989
Third edition 2002
Fourth edition 2008

Printed in the United Kingdom at the University Press, Cambridge

A catalogue record for this publication is available from the British Library

ISBN 978-0-521-86645-3 hardback

ISBN 978-0-521-68553-5 paperback

Contents

Preface to the first edition

Originally entitled *The Naming of Plants and the Meanings of Plant Names*, this book is in two parts. The first part has been written as an account of the way in which the naming of plants has changed with time and why the changes were necessary. It has not been the writer's intention to dwell upon the more fascinating aspects of common names but rather to progress from these to the situation which exists today, in which the botanical and horticultural names of plants must conform to internationally agreed standards. The aim has been to produce an interesting text which is equally as acceptable to the amateur gardener as to the botanist. The temptation to make this a definitive guide to the International Code of Botanical Nomenclature was resisted since others have done this already and with great clarity. A brief comment on synonymous and illegitimate botanical names and a reference to recent attempts to accommodate the various traits and interests in the naming of cultivated plants was added after the first edition.

The book had its origins in a collection of Latin plant names, and their meanings in English, which continued to grow by the year but which could never be complete. Not all plant names have meaningful translations. Some of the botanical literature gives full citation of plant names (and translations of the names, as well as common names). There are, however, many horticultural and botanical publications in which plant names are used in a casual manner, or are mis-spelled, or are given meanings or common names that are neither translations nor common (in the world-wide sense). There is also a tendency that may be part of modern language, to reduce names of garden plants to an abbreviated form (e.g. Rhodo for *Rhododendron*). Literal names such as Vogel's *Napoleona*, for *Napoleona vogelii*, provide only limited information about the plant. The dedication of the genus to Napoleon Bonaparte is not informative. Only by further search of the literature will the reader find that Theodor Vogel was the botanist to the 1841 Niger expedition and that he collected some 150 specimens during a rainy July fortnight in Liberia, One of those specimens, number 45, was a *Napoleona* that was later named for him as the type of the new species by Hooker and Planchon. To have given such information would have made the text very much larger.

The author has compiled a glossary which should serve to translate the more meaningful and descriptive names of plants from anywhere on earth but which will give little information about many of the people and places commemorated in plant names. Their entries do little more than identify the persons for whom the names were raised and their period in history, The author makes no claim that the glossary is all-encompassing or that the meanings he has listed are always the only meanings that have been put upon the various entries. Authors of Latin names have not always explained the meanings of the names they have erected and, consequently, such names may have been given different meanings by subsequent writers.

Preface to the fourth edition

This book is intended for use by botanists, gardeners and others who have an interest in plant names, the manner and rules by which they are formed, their origins and their meanings. The evolution of our current taxonomic system, from its origins in classical Greece to its present situation, is dealt with in the first part. This presents an overview of some major aspects of resolving the earlier unregulated way of naming plants. It goes on to explain how the current system evolved, and the use of Latin as the universal, and often innovative, language for those names. It then treats the naming of cultivated plants, from the wild, produced by hybridization or by sporting, maintained only by vegetative means, in horticulture, agriculture or arboriculture, and perhaps differing only in single small features. These are subject to the botanical rules of nomenclature but also have their own set of international rules for the naming of garden variants. Both Codes (the International Code of Botanical Nomenclature and the International Code of Nomenclature for Cultivated Plants) are explained.

The main body of the book has been considerably enlarged for this edition. It consists of a glossary of over 17,000 names or components of names. Each entry contains an indication of the source from which the name is derived. The components (prefixes or suffixes) are often common to medicine and zoology, as are many of the people commemorated in plant names, and where zoology interposes with botany (e.g. gall insects) the gardener will find these explained. Algae and fungi are not primary components of the glossary but many which are commonly encountered in gardening or forestry are included.

The glossary does not claim to be comprehensive but does provide a tool for discovering the meaning of huge numbers of plant names or constructing names for new plants. The author has included some of the views of other writers on the meanings of certain names but accepts that classicists may rue his non-use of diacritics.

The nature of the problem
A rose: by any name?

Man's highly developed constructive curiosity and his capacity for communication are two of the attributes distinguishing him from all other animals. Man alone has sought to understand the whole living world and things beyond his own environment and to pass his knowledge on to others. Consequently, when he discovers or invents something new he also creates a new word, or words, in order to be able to communicate his discovery or invention to others. There are no rules to govern the manner in which such new words are formed other than those of their acceptance and acceptability. This is equally true of the common, or vulgar or vernacular names of plants. Such names present few problems until communication becomes multilingual and the number of plants named becomes excessive. For example, the diuretic dandelion is easily accommodated in European languages. As the lion's tooth, it becomes Lowenzahn, dent de lion, dente di leone. As piss-abed it becomes pissenlit, piscacane, and piscialetto. When further study reveals that there are more than a thousand different kinds of dandelion throughout Europe, the formulation of common names for these is both difficult and unacceptable.

Common plant names present language at its richest and most imaginative (welcome home husband however drunk you be, for the houseleek or *Sempervivum;* shepherd's weather-glass, for scarlet pimpernel or *Anagallis;* meet her i'th'entry kiss her i'th'buttery, or leap up and kiss me, for *Viola tricolor;* touch me not, for the balsam *Impatiens noli-tangere;* mind your own business, or mother of thousands, for *Soleirolia soleirolii;* blood drop emlets, for *Mimulus luteus*). Local variations in common names are numerous and this is perhaps a reflection of the importance of plants in general conversation, in the kitchen and in herbalism throughout the country in bygone days. An often-quoted example of the multiplicity of vernacular names is that of *Caltha palustris*, for which, in addition to marsh marigold, kingcup and May blobs, there are 90 other local British names (one being dandelion), as well as over 140 German and 60 French vernacular names.

Common plant names have many sources. Some came from antiquity by word of mouth as part of language itself, and the passage of time and changing circumstances have obscured their meanings. Fanciful ideas of a plant's association with animals, ailments and festivities, and observations of plant structures, perfumes, colours, habitats and seasonality have all contributed to their naming. So too have their names in other languages. English plant names have come from Arabic, Persian, Greek, Latin, ancient British, Anglo-Saxon, Norman, Low German, Swedish and Danish. Such names were introduced together with the spices, grains, fruit plants and others which merchants and warring nations introduced to new areas. Foreign names often remained little altered but some were transliterated in such a way as to lose any meaning which they may have had originally.

The element of fanciful association in vernacular plant names often drew upon comparisons with parts of the body and with bodily functions (priest's pintle for *Arum maculatum*, open arse for *Mespilus germanicus* and arse smart for *Polygonum hydropiper*). Some of these persist but no longer strike us as 'vulgar' because they are 'respectably' modified or the associations themselves are no longer familiar to us (*Arum maculatum* is still known as cuckoo pint (cuckoo pintle) and as wake robin). Such was the sensitivity to indelicate names that Britten and Holland, in their *Dictionary of English Plant Names* (1886), wrote 'We have also purposely excluded a few names which though graphic in their construction and meaning, interesting in their antiquity, and even yet in use in certain counties, are scarcely suited for publication in a work intended for general readers'. They nevertheless included the examples above. The cleaning-up of such names was a feature of the Victorian period, during which our common plant names were formalized and reduced in number. Some of the resulting names are prissy (bloody cranesbill, for *Geranium sanguineum*, becomes blood-red cranesbill), some are uninspired (naked ladies or meadow saffron, for *Colchicum autumnale*, becomes autumn crocus) and most are not very informative.

This last point is not of any real importance, because names do not need to have a meaning or be interpretable. Primarily, names are mere ciphers which are easier to use than lengthy descriptions, and yet, when accepted, they can become quite as meaningful. Within limits, it is possible to use one name for a number of different things but, if the limits are exceeded, this may cause great confusion. There are many common plant names which refer to several plants but cause no problem so long as they are used only within their local areas or when they are used to convey only a general idea of the plant's identity. For example, *Wahlenbergia saxicola* in New Zealand, *Phacelia whitlavia* in southern California, USA, *Clitoria ternatea* in West Africa, *Campanula rotundifolia* in Scotland and *Endymion non-scriptus* (formerly *Scilla non-scripta* and now *Hyacinthoides non-scripta*) in England are all commonly called bluebells. In each area, local people will understand others who speak of bluebells but in all the areas except Scotland the song 'The Bluebells of Scotland', heard perhaps on the radio, will conjure up a wrong impression. At least ten different plants are given the common name of cuckoo flower in England, signifying only that they flower in spring at a time when the cuckoo is first heard.

The problem of plant names and of plant naming is that common names need not be formed according to any rule and can change as language, or the user of language, dictates. If our awareness extended only to some thousands of 'kinds' of plants we could manage by giving them numbers but, as our awareness extends, more 'kinds' are recognized and for most purposes we find a need to organize our thoughts about them by giving them names and by forming them into named groups. Then we have to agree with others about the names and the groups, otherwise communication becomes hampered by ambiguity. A completely coded numerical system could be devised but would have little use to the non-specialist, without access to the details of encoding.

Formalized names provide a partial solution to the two opposed problems presented by vernacular names: multiple naming of a single plant and multiple application of a single name. The predominantly two-word structure of such formal names has been adopted in recent historic times in all biological nomenclature, especially in the branch which – thanks to Isidorus Hispalensis (560–636), Archbishop of Seville, whose *Etymologies* was a vast encyclopaedia of ancient learning (or truths) and was studied for 900 years – we now call botany (βοτανη, fodder or plants eaten by cattle). Of necessity, botanical names have been formulated from former common names, but this does not mean that in the translation of botanical names we may expect to find meaningful names in common language. Botanical names, however, do represent a stable system of nomenclature which is usable by people of all nationalities and has relevancy to a system of classification.

Since man became wise, he has domesticated both plants and animals and, for at least the past 300 years, has bred and selected an ever-growing number of 'breeds', 'lines' or 'races' of these. He has also given them names. In this, man has accelerated the processes which, we think, are the processes of natural evolution and has created a different level of artificially sustained, domesticated organisms. The names given by the breeders of the plants of the garden and the crops of agriculture and arboriculture present the same problems as those of vernacular and botanical names. Since the second edition was published (1989), genetic manipulation of the properties of plants has proceeded apace. Not only has the innate genetic material of plants been re-ordered, but alien genetic material, from other organisms, even from other kingdoms, has been introduced to give bizarre results. The products are unnatural and have not faced selection in nature. Indeed some may present problems should they interbreed with natural populations in the future. There is still a divide between the international bodies concerned with botanical and cultivated plant names and the commercial interests that are protected by legislation for trademarking new genetic and transgenic products.

The size of the problem

'Man by his nature desires to know' (Aristotle)

Three centuries before Christ, Aristotle of Stagira (384–322 BC), disciple of Plato, wrote extensively and systematically of all that was then known of the physical and living world. In this monumental task, he laid the foundations of inductive reasoning. When he died, he left his writings and his teaching garden to one of his pupils, Theophrastus of Eresus (*c.* 370–287 BC), who also took over Aristotle's peripatetic school. Theophrastus' writings on mineralogy and plants totalled 22 treatises, of which nine books of *Historia plantarum* contain a collection of contemporary knowledge about plants and eight of *De causis plantarum* are a collection of his own critical observations, a departure from earlier philosophical approaches, and rightly entitle him to be regarded as the father of botany. These works were subsequently translated into Syrian, to Arabic, to Latin and back to Greek. He recognized the distinctions between monocotyledons and dicotyledons, superior and inferior ovaries in flowers, the necessity for pollination and the sexuality of plants but, although he used names for plants of beauty, use or oddity, he did not try to name everything.

To the ancients, as to the people of earlier civilizations of Persia and China, plants were distinguished on the basis of their culinary, medicinal and decorative uses – as well as their supposed supernatural properties. For this reason, plants were given a name as well as a description. Theophrastus wrote of some 500 'kinds' of plant which, considering that material had been brought back from Alexander the Great's campaigns throughout Persia, as far as India, would indicate a considerable lack of discrimination. In Britain, we now recognize more than that number of different 'kinds' of moss.

Four centuries later, about AD 64, Dioscorides Pedanius of Anazarbus, a soldier who wrote in Greek and became a Roman doctor, recorded 600 'kinds' of plants and, in about AD 77, the elder Pliny (Gaius Plinius Secundus (23–79), a victim of Vesuvius' eruption), in his huge compilation of the information contained in the writings of 473 authors, described about a thousand 'kinds'. During the 'Dark Ages', despite the remarkable achievements of such people as Albertus Magnus (1193–1280), who collected plants during extensive journeys in Europe, and the publication of the German *Herbarius* in 1485 by another collector of European plants, Dr Johann von Cube, little progress was made in the study of plants. It was the renewal of critical observation by Renaissance botanists such as Rembert Dodoens (1517–1585), Matthias de l'Obel (1538–1616), Charles de l'Ecluse (1526–1609) and others which resulted in the recognition of some 4,000 'kinds' of plants by the sixteenth century. At this point in history, the renewal of critical study and the beginning of plant collection throughout the known world produced a requirement for a rational system of grouping plants. Up to the sixteenth century, three factors had hindered such classification. The first of these was that the main interested parties were the nobility and apothecaries who conferred on plants great monetary value, either because of their rarity or because of the real or imaginary virtues attributed to them, and regarded them as items to be guarded jealously. Second was the lack of any standardized system of naming plants and, third and perhaps most important, any expression of the idea that living things could have evolved from earlier extinct ancestors and could therefore form groupings of related 'kinds', or lineages, was a direct contradiction of the religious dogma of Divine Creation.

Perhaps the greatest disservice to progress was that caused by the doctrine of signatures, which claimed that God had given to each 'kind' of plant some feature which could indicate the uses to which man could put the plant. Thus, plants with kidney-shaped leaves could be used for treating kidney complaints and were grouped together on this basis. The Swiss doctor, Theophrastus Phillipus Aureolus Bombastus von Hohenheim (1493–1541) had invented properties for many plants under this doctrine. He also considered that man possessed intuitive knowledge of which plants could serve him, and how. He is better known under the Latin name which he assumed, Paracelsus, and the doctrinal book *Dispensatory* is usually attributed to him. The doctrine was also supported by Giambattista Della Porta (1537–1615), who made an interesting extension to it, that the distribution of different 'kinds' of plants had a direct bearing upon the distribution of different kinds of ailment which man suffered in different areas. On this basis, the preference of willows for wet habitats is ordained by God because men who live in wet areas are prone to suffer from rheumatism and, since the bark of *Salix* species gives relief from rheumatic pains (it contains salicylic acid, the analgesic principal of aspirin), the willows are there to serve the needs of man.

In spite of disadvantageous attitudes, renewed critical interest in plants during the sixteenth century led to more discriminating views as to the nature of 'kinds', to searches for new plants from different areas and concern over the problems of naming plants. John Parkinson (1567–1650), a London apothecary, wrote a horticultural landmark with the punning title *Paradisi in sole paradisus terestris* in 1629. This was an encyclopaedia of gardening and of plants then in cultivation and contains a lament by Parkinson that, in their many catalogues, nurserymen 'without consideration of kind or form, or other special note give(th) names so diversely one from the other, that . . . very few can tell what they mean'. This attitude towards common names is still with us but not in so violent a guise as that shown by an unknown author who, in *Science Gossip* of 1868, wrote that vulgar names of plants presented 'a complete language of meaningless nonsense, almost impossible to retain and certainly worse than useless when remembered – a vast vocabulary of names, many of which signify that which is false, and most of which mean nothing at all'.

Names continued to be formed as phrase-names constructed with a starting noun (which was later to become the generic name) followed by a description. So, we find that the creeping buttercup was known by many names, of which Caspar Bauhin (1560–1624) and Christian Mentzel (1622–1701) listed the following:

Caspar Bauhin, *Pinax Theatri Botanici*, 1623
Ranunculus pratensis repens hirsutus var. C. Bauhin
repens fl. luteo simpl. J. Bauhin
repens fol. ex albo variis
repens magnus hirsutus fl. pleno
repens flore pleno
pratensis repens Parkinson
pratensis reptante cauliculo l'Obel
polyanthemos 1 Dodoens
hortensis 1 Dodoens
vinealis Tabernamontana
pratensis etiamque hortensis Gerard

Christianus Mentzelius, *Index Nominum Plantarum Multilinguis* (*Universalis*), 1682
Ranunculus pratensis et arvensis C. Bauhin
rectus acris var. C. Bauhin
rectus fl. simpl. luteo J. Bauhin
rectus fol. pallidioribus hirsutis J. Bauhin
albus fl. simpl. et denso J. Bauhin
pratensis erectus dulcis C. Bauhin
Ranoncole dolce Italian
Grenoillette dorée o doux Gallic
Sewite Woode Crawe foet English
Suss Hanenfuss

Jaskien sodky Polish
Chrysanth. simplex Fuchs
Ranunculus pratensis repens hirsutus var. c C. Bauhin
repens fl. luteo simpl. J. Bauhin
repens fol. ex albo variis Antonius Vallot
repens magnus hirsut. fl. pleno J. B. Tabernamontana
repens fl. pleno J. Bauhin
arvensis echinatus Paulus Ammannus
prat. rad. verticilli modo rotunda C. Bauhin
tuberosus major J. Bauhin
Crus Galli Otto Brunfelsius
Coronopus parvus Batrachion Apuleius Dodonaeus (Dodoens)
Ranunculus prat. parvus fol. trifido C. Bauhin
arvensis annuus fl. minimo luteo Morison
fasciatus Henricus Volgnadius
Ol. Borrich Caspar Bartholino

These were, of course, common or vernacular names with wide currency, and strong candidates for inclusion in lists which were intended to clarify the complicated state of plant naming. Local, vulgar names escaped such listing until much later times, when they were being less used and lexicographers began to collect them, saving most from vanishing for ever.

Great advances were made during the seventeenth century. Robert Morison (1620–1683) published a convenient or artificial system of grouping 'kinds' into groups of increasing size, as a hierarchy. One of his groups we now call the family *Umbelliferae* or, to give it its modern name, *Apiaceae*, and this was the first natural group to be recognized. By natural group we imply that the members of the group share a sufficient number of common features to suggest that they have all evolved from a common ancestral stock. Joseph Pitton de Tournefort (1656–1708) had made a very methodical survey of plants and had assorted 10,000 'kinds' into 69 groups (or genera). The 'kinds' must now be regarded as the basic units of classification called species. Although critical observation of structural and anatomical features led to classification advancing beyond the vague herbal and signature systems, no such advance was made in plant naming until a Swede, of little academic ability when young, we are told, established landmarks in both classification and nomenclature of plants. He was Carl Linnaeus (1707–1778), who classified 7,700 species into 109 genera and gave to each species a binomial name (a name consisting of a generic name-word plus a descriptive epithet, both of Latin form).

It was inevitable that, as man grouped the ever-increasing number of known plants (and he was then principally aware of those from Europe, the Mediterranean and a few from other areas), the constancy of associated morphological features in some groups should suggest that the whole was derived, by evolution, from a common ancestor. Morison's family *Umbelliferae* was a case in point. Also, because the basic unit of any system of classification is the species, and some species were found to be far less constant than others, it was just as inevitable that the nature of the species itself would become a matter of controversy, not least in terms of religious dogma. A point often passed over with insufficient comment is that Linnaeus' endeavours towards a natural system of classification were accompanied by his changing attitude towards Divine Creation. From the 365 aphorisms by which he expressed his views in *Fundamenta botanica* (1736), and expanded in *Critica botanica*, (1737), his early view was that all species were produced by the hand of the Almighty Creator and that 'variations in the outside shell' were the work of 'Nature in a sporty mood'. In such genera as *Thalictrum* and *Clematis*, he later concluded that some species were not original creations and, in *Rosa*, he was drawn to conclude that either some species had blended or that one species had given rise to several others. Later, he invoked hybridization as the process by which species could be created, and attributed to the Almighty the creation of the primeval genera, each with a single species. From his observation of land accretion during trips to Öland and Gotland, in 1741, he accepted a continuous creation of the earth and that Nature was in continuous change (*Oratio de Telluris habitabilis incremento*, 1744). He later accepted that fossil-bed remains could only be explained by a process of continuous creation. In *Genera plantarum* (6th edn, 1764) he attributed to God the creation of the natural orders (our families). Nature produced from these the genera and species, and permanent varieties were produced by hybridization between them. The abnormal varieties of the species so formed were the product of chance.

Linnaeus was well aware of the results which plant hybridizers were obtaining in Holland and it is not surprising that his own knowledge of naturally occurring variants led him towards a covertly expressed belief in evolution. However, that expression, and his listing of varieties under their typical species in *Species plantarum*, where he indicated each with a Greek letter, was still contrary to the dogma of Divine Creation and it would be another century before a substantive declaration of evolutionary theory was to be made, by Charles Darwin (1809–1882).

Darwin's essay on *The Origin of Species by Means of Natural Selection* (1859) was published somewhat reluctantly and in the face of fierce opposition. It was concerned with the major evolutionary changes by which species evolve and was based upon Darwin's own observations on fossils and living creatures. The concept of natural selection, or the survival of any life form being dependent upon its ability to compete successfully for a place in nature, became, and still is, accepted as the major force directing an inevitable process of organic change. Our conception of the mechanisms and the causative factors for the large evolutionary steps, such as the demise of the dinosaurs and of many plant groups now known only as fossils, and the emergence and diversification of the flowering plants during the last 100 million years, is, at best, hazy.

The great age of plant hunting, from the second half of the eighteenth century through most of the nineteenth century, produced a flood of species not previously known. Strange and exotic plants were once prized above gold and caused theft, bribery and murder. Trading in 'paper tulips' by the van Bourse family gave rise to the continental stock exchange – the Bourse. With the invention of the Wardian case by Dr Nathaniel Bagshaw Ward, in 1827, it became possible to transport plants from the farthest corners of the world by sea and without enormous losses. The case was a small glasshouse, which reduced water losses and made it unnecessary to use large quantities of fresh water on the plants during long sea voyages, as well as giving protection from salt spray. In the confusion which resulted from the naming of this flood of plants, and the use of many languages to describe them, it became apparent that there was a need for international agreement on both these matters. Today, we have rules formulated to govern the names of about 300,000 species of plants, which are now generally accepted, and have disposed of a great number of names that have been found invalid.

Our present state of knowledge about the mechanisms of inheritance and change in plants and animals is almost entirely limited to an understanding of the causes of variation within a species. That understanding is based upon the observed behaviour of inherited characters as first recorded in *Pisum* by Gregor Johann Mendel, in 1866. With the technical development of the microscope, Marcello Malpighi (1671), Nehemiah Grew (1641–1712) and others explored the cellular structure of plants and elucidated the mechanism of fertilization. However, the nature of inheritance and variability remained clouded by myth and monsters until Mendel's work was rediscovered at the beginning of the twentieth century. By 1900, Hugo Marie de Vries (1848–1935), Carl Erich Correns (1864– 1933), Erich Tschermak von Seysenegg (1871–1962) and William Bateson (1861–1926) had confirmed that inheritance had a definite, particulate character which is regulated by 'genes'. Walter Stanborough Sutton (1877–1916) was the first person to clarify the manner in which the characters are transmitted from parents to offspring when he described the behaviour of 'chromosomes' during division of the cell nucleus. Chromosomes are thread-like bodies which can be stained in dividing cells so that the sequence of events of their own division can be followed. Along their length, it can be shown, the sites of genetic control, or genes, are situated in an ordered linear sequence. Differences between individuals can now be explained in terms of the different forms, or allelomorphs, in which single genes can exist as a consequence of their mutation. At the level of the gene, we must now consider the mutants and alleles as variants in molecular structure represented by the sequences of bases in the deoxyribonucleic acid. Classification can not yet accommodate the new, genetically modified forms that may only be distinguished in terms of some property resultant upon the insertion of a fragment of DNA.

The concept of a taxonomic species, or grouping of individuals each of which has a close resemblance to the others in every aspect of its morphology, and to which a name can be applied, is not always the most accurate interpretation of the true circumstances in nature. It defines and delimits an entity, but we are constantly discovering that the species is far from being an immutable entity. However, botanists find that plant species may have components which have well-defined, individual ecotypic properties (an ability to live on a distinctive soil type, or an adaptation to flower and fruit in harmony with some agricultural practice) or reproductive barriers caused by differences in chromosome number, etc. The plant breeder produces a steady stream of new varieties of cultivated species by hybridization and selection from the progeny. Genetically modified plants with very specific 'economic' properties are produced by techniques which evade nature's safeguards of incompatibility and hybrid sterility and may or may not have to be repeatedly re-synthesized.

If we consider some of the implications of, and attitudes towards, delimiting plant species and their components, and naming them, it will become easier to understand the need for internationally accepted rules intended to prevent the unnecessary and unacceptable proliferation of names.

Towards a solution to the problem

It is basic to the collector's art to arrange items into groups. Postage stamps can be arranged by country of origin and then on face value, year of issue, design, colour variation or defects. The arranging process always resolves into a hierarchic set of groups. In the plant kingdom we have a descending hierarchy of groups through Divisions, divided into Classes, divided into Orders, divided into Families, divided into Genera, divided into Species. Subsidiary groupings are possible at each level of this hierarchy and are employed to rationalize the uniformity of relationships within the particular group. Thus, a genus may be divided into a mini-hierarchy of subgenera, divided into sections, divided into series in order to assort the components into groupings of close relatives. All such components would, nevertheless, be members of the one genus.

Early systems of classification were much less sophisticated and were based upon few aspects of plant structure, such as those which suggested signatures, and mainly upon ancient herbal and medicinal concepts. Later systems would reflect advances in man's comprehension of plant structure and function, and employ the morphology and anatomy of reproductive structures as defining features. Groupings such as Natural Orders and Genera had no precise limits or absolute parity, one with another; and genera are still very diverse in size, distribution and the extent to which they have been subdivided.

Otto Brunfels (1488–1534) was probably the first person to introduce accurate, objective recording and illustration of plant structure in his *Herbarium* of 1530, and Valerius Cordus (1515– 1544) could have revolutionized botany but for his premature death. His four books of German plants contained detailed accounts of the structure of 446 plants, based upon his own systematic studies on them. Many of the plants were new to science. A fifth book on Italian plants was in compilation when he died. Conrad Gesner (1516–1565) published Cordus' work on German plants in 1561 and the fifth book in 1563.

A primitive suggestion of an evolutionary sequence was contained in Matthias de l'Obel's *Plantarum seu stirpium historia* (1576), in which narrow-leaved plants, followed by broader-leaved, bulbous and rhizomatous plants, followed by herbaceous dicotyledons, followed by shrubs and trees, was regarded as a series of increasing 'perfection'. Andrea Caesalpino (1519–1603) retained the distinction between woody and herbaceous plants but employed more detail of flower, fruit and seed structure in compiling his classes of plants (*De plantis*, 1583). His influence extended to the classifications of Caspar (Gaspard) Bauhin (1550–1624), and his brother Jean Bauhin (1541–1613), who departed from the use of medicinal information and compiled detailed descriptions of some 5,000 plants, to which he gave many two-word names, or binomials. P. R. de Belleval (1558–1632) adopted a binomial system which named each plant with a Latin noun followed by a Greek adjectival epithet. Joachim Jung (1587–1657) feared being accused of heresy, which prevented him from publishing his work. The manuscripts which survived him contain many of the terms which we still use in describing leaf and flower structure and arrangement, and also contain plant names consisting of a noun qualified by an adjective. Robert Morison (1620–1683) used binomials, and John Ray (1627–1705), who introduced the distinction between monocotyledons and dicotyledons, but retained the distinction between flowering herbaceous plants and woody plants, also used binomial names.

Joseph Pitton de Tournefort (1656–1708) placed great emphasis on the floral corolla and upon defining the genus, rather than the species. His 69 generic descriptions are detailed but his species descriptions are dependent upon binomials and illustrations. Herman Boerhaave (1668–1738) combined the systems of Ray and Tournefort, and others, to incorporate morphological, ecological, leaf, floral and fruiting characters, but none of these early advances received popular support. As Michel Adanson (1727–1806) was to realize, some sixty systems of classification had been proposed by the middle of the eighteenth century and none had been free from narrow conceptual restraints. His plea that attention should be focused on 'natural' classification through processes of inductive reasoning, because of the wide range of characteristics then being employed, did not enjoy wide publication and his work was not well regarded when it did become more widely known. His main claim to fame, or notoriety, stems from his use of names which have no meanings.

Before considering the major contributions made by Carl Linnaeus, it should be noted that the names of many plant families and genera were well established at the beginning of the eighteenth century and several people had used simplified, binomial names for species. Indeed, August Quirinus Rivinus (1652–1723) had proposed that no plant should have a name of more than two words.

Carl Linnaeus (1707–1778) was the son of a clergyman, Nils, who had adopted the Latinized family name when he became a student of theology. Carl also went to theological college for a year but then left and became an assistant gardener in Professor Olof Rudbeck's botanic garden at Uppsala. His ability as a collector and arranger soon became evident and, after undertaking tours through Lapland, he began to publish works which are now the starting points for naming plants and animals. In literature he is referred to as Carl or Karl or Carolus Linnaeus, Carl Linné (an abbreviation) and, later in life, as Carl von Linné. His life became one of devotion to the classification and naming of all living things and of teaching others about them. His numerous students played a very important part in the discovery of new plants from many parts of the world. Linnaeus' main contribution to botany was his method of naming plants, in which he combined Bauhin's and Belleval's use of binomials with Tournefort's and Boerhaave's concepts of the genus. His success, where others before him had failed, was due to the early publication of his most popular work, an artificial system of classifying plants. In this he employed the number, structure and disposition of the stamens of the flower to define 23 classes, each subdivided into orders on the basis of the number of parts constituting the pistil, with a 24th class containing those plants which had their reproductive organs hidden to the eye – the orders of which were the ferns, mosses, algae (in which he placed liverworts, lichens and sponges), fungi and palms. This 'sexual system' provided an easy way of grouping plants and of allocating newly discovered plants to a group. Originally designed to accommodate the plants of his home parish, it was elaborated to include first the arctic flora and later the more diverse and exotic plants being discovered in the tropics. It continued in popular use into the nineteenth century despite its limitation of grouping together strange bedfellows: red valerian, tamarind, crocus, iris, galingale sedge and mat grass are all grouped under *Triandria* (three stamens) *Monogynia* (pistil with a single style).

In 1735, Linnaeus published *Systema naturae*, in which he grouped species into genera, genera into orders and orders into classes on the basis of structural similarities. This was an attempt to interpret evolutionary relationships or assemblages of individuals at different levels. It owed much to a collaborator and fellow student of Linnaeus, Peter Artendi (d. 1735), who, before an untimely death, was working on the classification of fishes, reptiles and amphibians, and the *Umbelliferae*. In *Species plantarum*, published in 1753, Linnaeus gave each species a binomial name. The first word of each binomial was the name of the genus to which the species belonged and the second word was a descriptive, or specific epithet. Both words were in Latin or Latin form. Thus, the creeping buttercup he named as *Ranunculus repens*.

It now required that the systematic classification and the binomial nomenclature, which Linnaeus had adopted, should become generally accepted and, largely because of the popularity of his sexual system, this was to be the case. Botany could now contend with the rapidly increasing number of species of plants being collected for scientific enquiry, rather than for medicine or exotic gardening, as in the seventeenth century. For the proper working of such standardized nomenclature, however, it was necessary that the language of plant names should also be standardized. Linnaeus' views on the manner of forming plant names, and the use of Latin for these and for the descriptions of plants and their parts, have given rise directly to modern practice and a Latin vocabulary of great versatility, but which would have been largely incomprehensible in ancient Rome. He applied the same methodical principles to the naming of animals, minerals and diseases and, in doing so, established Latin, which was the *lingua franca* of his day, as the internationally used language of science and medicine.

The rules by which we now name plants depend largely on Linnaeus' writings, but, for the names of plant families, we are much dependent on A. L. de Jussieu's classification in his *De genera plantarum* of 1789. For the name of a species, the correct name is that which was first published since 1753. This establishes Linnaeus' *Species plantarum* (associated with his *Genera plantarum*, 5th edition of 1754 and 6th edition of 1764) as the starting point for the names of species (and their descriptions). Linnaeus' sexual system of classification was very artificial and, although Linnaeus must have been delighted at its popularity, he regarded it as no more than a convenient pigeonholing system. He published some of his views on grouping plant genera into natural orders (our families) in *Philosophia botanica* (1751). Most of his orders were not natural groupings but considerably mixed assemblages. By contrast, Bernard de Jussieu (1699–1777), followed by his nephew Antoine Laurent de Jussieu (1748–1836), searched for improved ways of arranging and grouping plants as natural groups. The characteristics of 100 plant families are given in *De genera plantarum*, and most of these we still recognize.

Augustin Pyrame de Candolle (1778–1841) also sought a natural system, as did his son Alphonse Louise (1806–1893), and he took the evolutionist view that there is an underlying state of symmetry in the floral structure which we can observe today and that, by considering relationships in terms of that symmetry, natural alliances may be recognized. This approach resulted in a great deal of monographic work from which de Candolle formed views on the concept of a core of similarity, or type, for any natural group and the requirement for control in the naming of plants.

Today, technological and scientific advances have made it possible for us to use subcellular, chemical and the minutest of morphological features, and to incorporate as many items of information as are available about a plant in computer-aided assessments of that plant's relationships to others. Biological information has often been found to conflict with the concept of the taxonomic species and there are many plant groups in which the 'species' can best be regarded as a collection of highly variable populations. The gleaning of new evidence necessitates a continuing process of reappraisal of families, genera and species. Such reappraisal may result in subdivision or even splitting of a group into several new ones or, the converse process, in lumping together two or more former groups into one new one. Since the bulk of research is carried out on the individual species, most of the revisions are carried out at or below the rank of species. On occasion, therefore, a revision at the family level will require the transfer of whole genera from one family to another, but it is now more common for a revision at the level of the genus to require the transfer of some, if not all the species from one genus to another. Such revisions are not mischievous but are the necessary process by which newly acquired knowledge is incorporated into a generally accepted framework. It is because we continue to improve the extent of our knowledge of plants that revision of the systems for their classification continues and, consequently, that name changes are inevitable.

The equivalence, certainly in evolutionary terms, of groups of higher rank than of family is a matter of philosophical debate and, even at the family level, we find divergence of views as to whether those with few components are equivalent to those with many components. In recent years the two families of lilies, *Liliaceae* and *Amaryllidaceae*, have been subdivided into the following families – mainly by the elevation of their former Englerian sub-families: *Melianthaceae, Colchicaceae, Asphodelaceae, Hyacinthaceae, Hemerocallidaceae, Agavaceae, Aphyllandraceae, Lomandraceae, Anthericaceae, Xanthorrhoeaceae, Alliaceae, Liliaceae, Dracaenaceae, Asparagaceae, Ruscaceae, Convallariaceae, Trilliaceae, Alteriaceae, Herreriaceae, Philesiaceae, Smilacaceae, Haemadoraceae, Hypoxidaceae, Alstoemeriaceae, Doryanthaceae, Campynemaceae*, and *Amaryllidaceae*.

Because the taxonomic species is the basic unit of any system of classification, we have to assume parity between species; that is to say, we assume that a widespread species is in every way comparable with a rare species which may be restricted in its distribution to a very small area. It is a feature of plants that their diversity – of habit, longevity, mode of reproduction and tolerance of environmental conditions – presents a wide range of biologically different circumstances. For the taxonomic problem of delimiting, defining and naming a species we have to identify a grouping of individuals whose characteristics are sufficiently stable to be defined, in order that a name can be applied to the group and a 'type', or exemplar, can be specified for that name. It is because of this concept of the 'type' that changes have to be made in names of species in the light of new discoveries and that entities below the rank of species have to be recognized. Thus, we speak of a botanical 'sub-species' when part of the species grouping can be distinguished as having a number of features which remain constant and as having a distinctive geographical or ecological distribution. When the degree of departure from the typical material is of a lesser order we may employ the inferior category of 'variety'. The term 'form' is employed to describe a variant which is distinct in a minor way only, such as a single feature difference which might appear sporadically due to genetic mutation or sporting.

The patterns and causes of variation differ from one species to another, and this has long been recognized as a problem in fully reconciling the idea of a taxonomic species with that of a biological system of populations in perpetual evolutionary flux. Below the level of species, agreement about absolute ranking is far from complete and even the rigidity of the infraspecific hierarchy (*subspecies, varietas, subvarietas, forma, subforma*) is now open to question.

It is always a cause of annoyance when a new name has to be given to a plant which is widely known under its superseded old name. Gardeners always complain about such name changes, but there is no novelty in that. On the occasion of Linnaeus being proposed for Fellowship of the Royal Society, Peter Collinson wrote to him in praise of his *Species plantarum* but, at the same time, complained that Linnaeus had introduced new names for so many well-known plants.

The gardener has some cause to be aggrieved by changes in botanical names. Few gardeners show much alacrity in adopting new names, and perusal of gardening books and catalogues shows that horticulture seldom uses botanical names with all the exactitude which they can provide. Horticulture, however, not only agreed to observe the international rules of botanical nomenclature but also formulated its own additional rules for the naming of plants grown under cultivation. It might appear as though the botanist realizes that he is bound by the rules, whereas the horticulturalist does not, but to understand this we must recognize the different facets of horticulture. The rules are of greatest interest and importance to specialist plant breeders and gardeners with a particular interest in a certain plant group. For the domestic gardener it is the growing of beautiful plants which is the motive force behind his activity. Between the two extremes lies every shade of interest and the main emphasis on names is an emphasis on garden names. Roses, cabbages, carnations and leeks are perfectly adequate names for the majority of gardeners but if greater precision is needed, a gardener wishes to know the name of the variety. Consequently, most gardeners are satisfied with a naming system which has no recourse to the botanical rules whatsoever. Not surprisingly, therefore, seed and plant catalogues also avoid botanical names. The specialist plant breeder, however, shows certain similarities to the apothecaries of an earlier age. Like them he guards his art and his plants jealously because they represent the source of his future income and, also like them, he has the desire to understand every aspect of his plants. The apothecaries gave us the first centres of botanical enquiry and the plant breeders of today give us the new varieties which are needed to satisfy our gardening and food-production requirements. The commercial face of plant breeding, however, attaches a powerful monetary significance to the names given to new varieties.

Gardeners occasionally have to resort to botanical names when they discover some cultural problem with a plant which shares the same common name with several different plants. The Guernsey lily, around which has always hung a cloud of mystery, has been offered to the public in the form of *Amaryllis belladonna* L. The true Guernsey lily has the name *Nerine sarniensis* Herb. (but was named *Amaryllis sarniensis* by Linnaeus). The epithet *sarniensis* means 'of Sarnia' or 'of Guernsey', Sarnia being the old name for Guernsey, and is an example of a misapplied geographical epithet, since the plant's native area is S Africa. Some would regard the epithet as indicating the fact that Guernsey was the first place in which the plant was cultivated. This is historically incorrect, however, and does nothing to help the gardener who finds that the Guernsey lily that he has bought does not behave, in culture, as *Nerine sarniensis* is known to behave. This example is one involving a particularly contentious area as to the taxonomic problems of generic boundaries and typification but there are many others in which common and Latin garden name are used for whole assortments of garden plants, ranging from species (*Nepeta mussinii* and *N. cataria* are both catmint) to members of different genera ('japonicas' including *Chaenomeles speciosa* and *Kerria japonica*) to members of different families (*Camellia japonica* is likewise a 'japonica'), and the diversity of 'bluebells' was mentioned earlier.

New varieties, be they timber trees, crop plants or garden flowers, require names, and those names need to be definitive. As with the earlier confusion of botanical names (different names for the same species or the same name for different species), so there can be the same confusion of horticultural names. As will be seen, rules for cultivated plants require that new names have to be established by publication. This gives to the breeder the commercial advantage of being able to supply to the public his new variety under what, initially, amounts to his mark of copyright. In some parts of the world legislation permits exemption from the rules and recommendations otherwise used for the names of cultivated plants.

The rules of botanical nomenclature

The rules which now govern the naming and the names of plants really had their beginnings in the views of Augustin P. de Candolle as he expressed them in his *Théorie élémentaire de la botanique* (1813). There, he advised that plants should have names in Latin (or Latin form but not compounded from different languages), formed according to the rules of Latin grammar and subject to the right of priority for the name given by the discoverer or the first describer. This advice was found inadequate and, in 1862, the International Botanical Congress in London adopted control over agreements on nomenclature. Alphonse Louise de Candolle (1806–1893) drew up four simple '*Lois*', or laws, which were aimed at resolving what threatened to become a chaotic state of plant nomenclature. The Paris International Botanical Congress of 1867 adopted the *Lois*, which were:

1 One plant species shall have no more than one name.
2 No two plant species shall share the same name.
3 If a plant has two names, the name which is valid shall be that which was the earliest one to be published after 1753.
4 The author's name shall be cited, after the name of the plant, in order to establish the sense in which the name is used and its priority over other names.

It can be seen from the above *Lois* that, until the nineteenth century, botanists frequently gave names to plants with little regard either to the previous use of the same name or to names that had already been applied to the same plant. It is because of this aspect that one often encounters the words *sensu* and *non* inserted before the name of an author, although both terms are more commonly used in the sense of taxonomic revision, and indicate that the name is being used 'in the sense of' or 'not in the sense of' that author, respectively.

The use of Latin as the language in which descriptions and diagnoses were written was not universal in the nineteenth century, and many regional languages were used in different parts of the world. A description is an account of the plant's habit, morphology and periodicity whereas a diagnosis is an author's definitive statement of the plant's diagnostic features, and circumscribes the limits outside which plants do not pertain to that named species. A diagnosis often states particular ways in which the species differs from another species of the same genus. Before the adoption of Latin as the accepted language of botanical nomenclature, searching for names already in existence for a particular plant, and confirming their applicability, involved searching through multilingual literature. The requirement to use Latin was written into the rules by the International Botanical Congress in Vienna, in 1905. However, the American Society of Plant Taxonomists produced its own Code in 1947, which became known as the Brittonia edition of the Rules or the Rochester Code, and disregarded this requirement. Not until 1959 was international agreement achieved, and then the requirement to use Latin was made retroactive to 1 January 1935, the year of the Amsterdam meeting of the Congress.

The rules are considered at each International Botanical Congress, formerly held at five-, and more recently at six-, yearly intervals during peacetime. The International Code of Botanical Nomenclature (first published as such in 1952) was formulated at the Stockholm Congress of 1950. In 1930, the matter of determining the priority of specific epithets was the main point at issue. The practice of British botanists had been to regard that epithet which was first published after the plant had been allocated to its correct genus as the correct name. This has been called the

Kew Rule, but it was defeated in favour of the rule that now gives priority to the epithet that was the first to be published from the starting date of 1 May 1753. Epithets which predate the starting point, but which were adopted by Linnaeus, are attributed to Linnaeus (e.g. Bauhin's *Alsine media*, *Ammi majus*, *Anagyris foetida* and *Galium rubrum* and Dodoens' *Angelica sylvestris* are examples of binomials nevertheless credited to Linnaeus).

The 1959 International Botanical Congress in Montreal introduced the requirement under the Code that, for valid publication of a name of a family or any taxon of lower rank, the author of that name should cite a 'type' for the name, and that this requirement should be retrospective to 1 January 1958. The idea of a type goes back to Augustin Pyrame de Candolle and it implies a representative collection of characteristics to which a name applies. The type in botany is a nomenclatural type; it is the type for the name and the name is permanently attached to it or associated with it. For the name of a family, the representative characteristics which that name implies are those embodied in one of its genera, which is called the type genus. In a similar way, the type for the name of a genus is the type species of that genus. For the name of a species or taxon of lower rank, the type is a specimen lodged in an herbarium or, in certain cases, published illustrations. The type need not, nor could it, be representative of the full range of entities to which the name is applied. Just as a genus, although having the features of its parent family, cannot be fully representative of all the genera belonging to that family, no single specimen can be representative of the full range of variety found within a species.

For the name to become the correct name of a plant or plant group, it must satisfy two sets of conditions. First, it must be constructed in accordance with the rules of name formation, which ensures its legitimacy. Second, it must be published in such a way as to make it valid. Publication has to be in printed matter which is distributed to the general public or, at least, to botanical institutions with libraries accessible to botanists generally. Since 1 January 1953, this has excluded publication in newspapers and tradesmen's catalogues. Valid publication also requires the name to be accompanied by a description or diagnosis, an indication of its rank and the nomenclatural type, as required by the rules. This publication requirement, and subsequent citation of the new name followed by the name of its author, ensures that a date can be placed upon the name's publication and that it can, therefore, be properly considered in matters of priority.

The present scope of the Code is expressed in the Principles, which have evolved from the de Candollean *Lois*:

1 Botanical nomenclature is independent of zoological nomenclature. The Code applies equally to names of taxonomic groups treated as plants whether or not these groups were originally so treated.
2 The application of names of taxonomic groups is determined by means of nomenclatural types.
3 The nomenclature of a taxonomic group is based upon priority of publication.
4 Each taxonomic group with a particular circumscription, position and rank can bear only one correct name, the earliest which is in accordance with the rules, except in specified cases.
5 Scientific names of taxonomic groups are treated as Latin regardless of their derivation.
6 The rules of nomenclature are retroactive unless expressly limited.

The detailed rules are contained in the Articles and Recommendations of the Code and mastery of these can only be gained by practical experience. A most lucid summary and comparison with other Codes of biological nomenclature is that of Jeffrey (1978), written for the Systematics Association.

There are still new species of plants to be discovered and an enormous amount of information yet to be sought for long-familiar species. In particular, evidence of a chemical nature, and especially that concerned with proteins, may provide reliable indications of phylogenetic relationships. For modern systematists, the greatest and most persistent problem is our ignorance about the apparently explosive appearance of a diverse array of flowering plants, some 100 million years ago, from one or more unknown ancestors. Modern systems of classification are still frameworks within which the authors arrange assemblages in sequences or clusters to represent their own idiosyncratic interpretation of the known facts. In addition to having no firm record of the early evolutionary pathways of the flowering plants, the systematist also has the major problems of identifying clear-cut boundaries between groups and of assessing the absolute ranking of groups. It is because of these continuing problems that, although the Code extends to taxa of all ranks, most of the rules are concerned with the names and naming of groups from the rank of family downwards.

Before moving on to the question of plant names at the generic and lower ranks, this is a suitable point at which to comment on new names for families which are now starting to appear in books and catalogues, and some explanation in passing may help to dispel any confusion. The splitting of the *Liliaceae* and *Amaryllidaceae* into 27 new families was mentioned on pages 11–12, but the move towards standardization has required other family name changes.

Family names

The names of families are plural adjectives used as nouns and are formed by adding the suffix *-aceae* to the stem, which is the name of an included genus. Thus, the buttercup genus *Ranunculus* gives us the name *Ranunculaceae* for the buttercup family and the water-lily genus *Nymphaea* gives us the name *Nymphaeaceae* for the water lilies. A few family names are conserved, for the reasons given above, which do have generic names as their stem, although one, the *Ebenaceae*, has the name *Ebenus* Kuntze (1891) *non* Linnaeus (1753) as its stem. Kuntze's genus is now called *Maba* but its parent family retains the name *Ebenaceae* even though *Ebenus* L. is the name used for a genus of the pea family. There are eight families for which specific exceptions are provided and which can be referred to either by their longstanding, conserved names or, as is increasingly the case in recent floras and other published works on plants, by their names which are in agreement with the Code. These families and their equivalent names are:

Compositae	or	*Asteraceae* (on the genus *Aster*)
Cruciferae	or	*Brassicaceae* (on the genus *Brassica*)
Gramineae	or	*Poaceae* (on the genus *Poa*)
Guttiferae	or	*Clusiaceae* (on the genus *Clusia*)
Labiatae	or	*Lamiaceae* (on the genus *Lamium*)
Leguminosae	or	*Fabaceae* (on the genus *Faba*)
Palmae	or	*Arecaceae* (on the genus *Areca*)
Umbelliferae	or	*Apiaceae* (on the genus *Apium*)

Some botanists regard the *Leguminosae* as including three subfamilies, but others accept those three components as each having family status. In the latter case, the three families are the *Caesalpiniaceae*, the *Mimosaceae* and the *Papilionaceae*. The last of these family names refers to the resemblance which may be seen in the pea- or bean-flower structure, with its large and colourful sail petal, to a resting butterfly (*Papilionoidea*) and is not based upon the name of a plant genus. If a botanist wishes to retain the three-family concept, the name *Papilionaceae* is conserved against *Leguminosae* with the modern equivalent, *Fabaceae*. Thus, the *Fabaceae* are either the entire aggregation of leguminous plant genera or that part of the aggregate which does not belong in either the *Caesalpiniaceae* or the *Mimosaceae*.

Each family can have only one correct name and that, of course, is the earliest legitimate one, **except in cases of limitation of priority by conservation**. In other words, there is provision in the Code for disregarding the requirement of priority when a special case is proved for a name to be conserved. Conservation of names is intended to avoid disadvantageous name changes, even though the name in question does not meet all the requirements of the Code. Names which have long-standing use and wide acceptability and are used in standard works of literature can be proposed for conservation and, when accepted, need not be discarded in favour of new and more correct names.

Some eastern European publications use *Daucaceae* for the *Apiaceae*, split the *Asteraceae* into *Carduaceae* and *Chicoriaceae* and adopt various views as to the generic basis of family names (e.g. *Oenotheraceae* for *Onagraceae* by insisting that Linnaeus' genus *Oenothera* has prior claim over Miller's genus *Onagra*).

Generic names

The name of a genus is a noun, or word treated as such, and begins with a capital letter. It is singular, may be taken from any source whatever, and may even be composed in an arbitrary manner. The etymology of generic names is, therefore, not always complete and, even though the derivation of some may be discovered, they lack meaning. By way of examples:

Portulaca, from the Latin *porto* (I carry) and *lac* (milk) translates as 'milk-carrier'.
Pittosporum, from the Greek, πιττα (tar) and σπορος (a seed) translates as 'tar-seed'.
Hebe was the goddess of youth and, amongst other things, the daughter of Jupiter. It cannot be translated further.
Petunia is taken from the Brazilian name for tobacco.
Tecoma is taken from a Mexican name.
Linnaea is one of the names which commemorate Linnaeus.
Sibara is an anagram of *Arabis*.
Aa is the name given by Reichenbach to an orchid genus which he separated from *Altensteinia*. It has no meaning and, as others have observed, must always appear first in an alphabetic listing.

The generic names of some Old World plants were taken from Greek mythology by the ancients, or are identical to the names of characters in Greek mythology. The reason for this is not always clear (e.g. *Althaea, Cecropia, Circaea, Melia, Phoenix, Tagetes, Thalia, Endymion, Hebe, Paeonia* and *Paris*). However, some do have reasonable floristic associations (e.g. *Atropa* (the third Fate, who held the scissors to cut the thread of life), *Chloris* (the goddess of flowers), *Iris* (messenger to gods of the rainbow), *Melissa* (apiarist who used the plant to feed the bees). The metamorphoses, that are so common in the mythology, provided direct associations for several names (e.g. *Acanthos* (became an *Acanthus*), *Adonis* (became an *Anemone*), *Ajacis* (became a *Narcissus*), *Daphne* (became a laurel), *Hyacinthus* (became, probably, a *Delphinium*) and *Narcissus* (became a daffodil). The gods, however, deviously changed form to further their machinations.

If all specific names were constructed in the arbitrary manner used by M. Adanson (1727–1806), there would have been no enquiries of the author and this book would not have been written. In fact, the etymology of plant names is a rich store of historical interest and conceals many facets of humanity ranging from the sarcasm of some authors to the humour of others. This is made possible by the wide scope available to authors for formulating names and because, whatever language is the source, names are treated as being in Latin. Imaginative association has produced some names which are very descriptive provided that the reader can spot the association. In the algae, the chrysophyte which twirls like a ballerina has been named *Pavlova gyrans* and, in the fungi, a saprophyte on leaves of *Eucalyptus* which has a wide-mouthed spore-producing structure has been named *Satchmopsis brasiliensis* (for Louis Armstrong (1901–1971), Satchmo, diminutive of satchel-mouth). In zoology, a snake has been given the trivial epithet '*montypythonoides*' (for the TV programme *Monty Python's Flying Circus*) and, in palaeontology, the members of the Beatles pop group have been commemorated in the names of

Table 1

Flower part	Greek	Latin	Former meaning
calyx	κάλυξ	—	various kinds of covering
	κύλίξ	—	cup or goblet
sepal	σκέπη	—	covering
corolla	—	*corolla*	garland or coronet
petal	πέταλον	—	leaf
	—	*petalum*	metal plate
stamen	—	*stamen*	thread, warp, string
	σταμίς, σταμίνος	—	pillar
filament	—	*filamentum*	thread
anther	—	*anthera*	potion of herbs
androecium	ἀνδρ-, οἰκός	—	man, house
stigma	στίγμα	—	tattoo or spot
style	στῦλος	—	pillar or post
	—	*stilus*	pointed writing tool
carpel	καρπό	—	Fruit
gynoecium	γυνή-, οἰκός	—	woman, house
pistil	—	*pistillum*	pestle

ammonites. The large vocabulary of botanical Latin comes mostly from the Greek and Latin of ancient times but, since the ancients had few words which related specifically to plants and their parts, a Latin dictionary is of somewhat limited use in trying to decipher plant diagnoses. By way of examples, Table 1 gives the parts of the flower (Latin *flos*, Greek ανθος) (illustrated in Fig.1) and the classical words from which they are derived, together with their original sense.

The grammar of botanical Latin is very formal and much more simple than that of the classical language itself. A full and most authoritative work on the subject is contained in Stearn's book *Botanical Latin* (1992). Nevertheless, it is necessary to know that in Latin, nouns (such as family and generic names) have gender, number and case and that the words which give some attribute to a noun (as in adjectival specific epithets) must agree with the noun in each of these. Having gender means that all things (the names of which are called nouns) are either masculine or feminine or neuter. In English, we treat almost everything as neuter, referring to nouns as 'it', except animals and most ships and aeroplanes (which are commonly held to be feminine). Gender is explained further below. Number means that things may be single (singular) or multiple (plural). In English we either have different words for the singular and plural (man and men, mouse and mice) or we convert the singular into the plural most commonly by adding an 's' (ship and ships, rat and rats) or more rarely by adding 'es' (box and boxes, fox and foxes) or rarer still by adding 'en' (ox and oxen). In Latin, the difference is expressed by changes in the endings of the words. Case is less easy to understand

Table 2

Case	Singular		Plural	
nominative	*flos*	the flower (subject)	*flores*	the flowers
accusative	*florem*	the flower (object)	*flores*	the flowers
genitive	*floris*	of the flower	*florum*	of the flowers
dative	*flori*	to, for the flower	*floribus*	to, for the flowers
ablative	*flore*	by, with, from the flower	*floribus*	by, with, from the flowers

Table 3

Declension	I	II		III				IV		V
Gender	f	m	n	m/f	n	m/f	n	m	n	f
Singular										
nom	*-a*	*-us(-er)*	*-um*	*	*	*-is(es)*	*-e(l)(r)*	*-us*	*-u*	*-es*
acc	*-am*	*-um*	*-um*	*-em*	*	*-em(im)*	*-e(l)(r)*	*-um*	*-u*	*-em*
gen	*-ae*	*-i*	*-i*	*-is*	*-is*	*-is*	*-is*	*-us*	*-us*	*-ei*
dat	*-ae*	*-o*	*-o*	*-i*	*-i*	*-i*	*-i*	*-ui(u)*	*-ui(u)*	*-ei*
abl	*-a*	*-o*	*-o*	*-e*	*-e*	*-i(e)*	*-i(e)*	*-u*	*-u*	*-e*
Plural										
nom	*-ae*	*-i*	*-a*	*-es*	*-a*	*-es*	*-ia*	*-us*	*-ua*	*-es*
acc	*-as*	*-os*	*-a*	*-es*	*-a*	*-es(is)*	*-ia*	*-us*	*-ua*	*-es*
gen	*-arum*	*-orum*	*-orum*	*-um*	*-um*	*-ium*	*-ium*	*-uum*	*-uum*	*-erum*
dat	*-is*	*-is*	*-is*	*-ibus*	*-ibus*	*-ibus*	*-ibus*	*-ibus*	*-ibus*	*-ebus*
abl	*-is*	*-is*	*-is*	*-ibus*	*-ibus*	*-ibus*	*-ibus*	*-ibus*	*-ibus*	*-ebus*

* Denotes various irregular endings.

but means the significance of the noun to the meaning of the sentence in which it is contained. It is also expressed in the endings of the words. In the sentence 'The flower has charm', the flower is singular, is the subject of the sentence and has what is called the nominative case. In the sentence 'I threw away the flower', I am now the subject and the flower has become the direct object in the accusative case. In the sentence 'I did not like the colour of the flower', I am again the subject, the colour is now the object and the flower has become a possessive noun and has the genitive case. In the sentence 'The flower fell to the ground', the flower is once again the subject (nominative) and the ground has the dative case. If we add 'with a whisper', then whisper takes the ablative case. In other words, case confers on nouns an expression of their meaning in any sentence. This is shown by the ending of the Latin word, which changes with case and number and, in so doing, changes the naked word into part of a sentence (Table 2).

Nouns fall into five groups, or declensions, as determined by their endings (Table 3).

Generic names are treated as singular subjects, taking the nominative case. *Solanum* means 'comforter' and derives from the use of nightshades as herbal sedatives. The gender of generic names is that of the original Greek or Latin noun or, if that was variable, is chosen by the author of the name. There are exceptions to this in which masculine names are treated as feminine, and fewer in which compound names, which ought to be feminine, are treated as masculine. As a general guide, names ending in *-us* are masculine unless they are trees (such as *Fagus, Pinus, Quercus, Sorbus,* which are treated as feminine), names ending in *-a* are feminine and names ending in *-um* are neuter; names ending in *-on* are masculine unless they can also take *-um,* when they are neuter, or the ending is *-dendron* when they are also neuter (*Rhododendron* or *Rhododendrum*); names ending in *-ma* (as in terminations such as *-osma*) are neuter; names ending in *-is* are mostly feminine or masculine treated as feminine (*Orchis*) and those ending in *-e* are neuter; other feminine endings are *-ago, -odes, -oides, -ix* and *-es.*

A recommendation for forming generic names to commemorate men or women is that these should be treated as feminine and formed as follows:

for names ending in a vowel,	terminate with *-a*
for names ending in *-a,*	terminate with *-ea*
for names ending in *-ea,*	do not change
for names ending in a consonant,	add *-ia*
for names ending in *-er,*	add *-a*
for Latinized names ending in *-us,*	change the ending to *-ia*

Generic names which are formed arbitrarily or are derived from vernacular names have their ending selected by the name's author. Clearly, a single epithet can be used to commemorate any number of persons sharing that same surname. For instance, the epithet '*meyeri*' can commemorate anyone called Meyer, in addition to those listed in the glossary.

Species names

The name of a species is a binary combination of the generic name followed by a specific epithet. If the epithet is of two words they must be joined by a hyphen or united into one word. The epithet can be taken from any source whatever and may be constructed in an arbitrary manner. It would be reasonable to expect that the epithet should have a descriptive purpose, and there are many which do, but large numbers either refer to the native area in which the plant grows or commemorate a person (often the discoverer, the introducer into cultivation or a noble personage). The epithet may be adjectival (or descriptive), qualified in various ways with prefixes and suffixes, or a noun.

It will become clear that because descriptive, adjectival epithets must agree with the generic name, the endings must change in gender, case and number; *Dipsacus fullonum* L. has the generic name used by Dioscorides meaning 'dropsy', alluding to the accumulation of water in the leaf-bases, and an epithet which is the masculine genitive plural of *fullo,* a fuller, and which identifies the typical form of this teasel as the one which was used to clean and comb up a 'nap' on cloth. The majority of adjectival epithet endings are as in the first two examples listed in Table 4.

Comparative epithets are informative because they provide us with an indication of how the species contrasts with the general features of other members of the genus (Table 5).

Epithets commemorating people

Specific epithets which are nouns are grammatically independent of the generic name. *Campanula trachelium* is literally 'little bell' (feminine) 'neck' (neuter). When they are derived from the names of people, they can either be retained as nouns in the genitive case (*clusii* is the genitive singular of *Clusius,* the Latinized version of l'Écluse, and gives an epithet with the meaning 'of l'Écluse') or be treated as adjectives and then agreeing in gender with the generic noun (*Sorbus leyana* Wilmott is a tree taking, like many others, the feminine gender despite the masculine ending, and so the epithet which commemorates Augustin Ley also takes the feminine ending). The epithets are formed as follows

Table 4

Masculine	Feminine	Neuter	Example	Meaning
-us	*-a*	*-um*	*hirsutus*	(hairy)
-is	*-is*	*-e*	*brevis*	(short)
-os	*-os*	*-on*	*acaulos* ακαυλος	(stemless)
-er	*-era*	*-erum*	*asper*	(rough)
-er	*-ra*	*-rum*	*scaber*	(rough)
-ax	*-ax*	*-ax*	*fallax*	(false)
-ex	*-ex*	*-ex*	*duplex*	(double)
-ox	*-ox*	*-ox*	*ferox*	(very prickly)
-ans	*-ans*	*-ans*	*reptans*	(creeping)
-ens	*-ens*	*-ens*	*repens*	(creeping)
-or	-or	-or	*tricolor*	(three-coloured)
-oides	-oides	-oides	*bryoides* βρυον-οειδης	(moss-like)

Table 5

Masculine	Feminine	Neuter	Example	Meaning
-us	*-a*	*-um*	*longus*	(long)
-ior	*-ior*	*-ius*		(comparative, longer)
-issimus	*-issima*	*-issimum*		(superlative, longest)
-is	*-is*	*-e*	*gracilis*	(slender)
-ior	*-ior*	*-ius*		(comparative, slenderer)
-limus	*-lima*	*-limum*		(superlative, slenderest)
-er	*-era*	*-erum*	*tener*	(thin)
-erior	*-erior*	*-erius*		(comparative, thinner)
-errimus	*-errima*	*-errimum*		(superlative, thinnest)

to names ending with a vowel (except -a) or *-er* is added
-i when masculine singular
-ae when feminine singular
-orum when masculine plural
-arum when feminine plural

to names ending with *-a* is added
-e when singular
-rum when plural

to names ending with a consonant (except *-er*) is added
-ii when masculine singular
-iae when feminine singular
-iorum when masculine plural
-iarum when feminine plural
or, when used adjectivally

to names ending with a vowel (except *-a*) is added
-anus when masculine
-ana when feminine
-anum when neuter

to names ending with *-a* is added

-nus	when masculine
-na	when feminine
-num	when neuter

to names ending with a consonant is added

-ianus	when masculine
-iana	when feminine
-ianum	when neuter

Geographical epithets

When an epithet is derived from the name of a place, usually to indicate the plant's native area but also, sometimes, to indicate the area or place from which the plant was first known or in which it was produced horticulturally, it is preferably adjectival and takes one of the following endings:

-ensis	(m)	*-ensis*	(f)	*-ense*	(n)
-(a)nus	(m)	*-(a)na*	(f)	*-(a)num*	(n)
-inus	(m)	*-ina*	(f)	*-inum*	(n)
-icus	(m)	*-ica*	(f)	*-icum*	(n)

Geographical epithets are sometimes inaccurate because the author of the name was in error as to the true origin of the plant, or obscure because the ancient classical names are no longer familiar to us. As with epithets which are derived from proper names to commemorate people, or from former generic names or vernacular names which are treated as being Latin, it is now customary to start them with a small initial letter but it remains permissible to give them a capital initial.

Categories below the rank of species

The subdivision of a species group is based upon a concept of infraspecific variation which assumes that, in nature, evolutionary changes are progressive fragmentations of the parent species. Put in another way, a species, or any taxon of lower rank, is a closed grouping whose limits embrace all their lower-ranked variants (subordinate taxa). It will be seen later that a different concept underlies the naming of cultivated plants which does not make such an assumption but recognizes the possibility that cultivars may straddle species, or other, boundaries or overlap each other, or be totally contained, one by another.

The rules by which botanical infraspecific taxa are named specify that the name shall consist of the name of the parent species followed by a term which denotes the rank of the subdivision, and an epithet which is formed in the same ways as specific epithets, including grammatical agreement when adjectival. Such names are subject to the rules of priority and typification. The ranks concerned are *subspecies* (abbreviated to subsp. or ssp.), *varietas* (variety in English, abbreviated to var.), *subvarietas* (subvariety or subvar.), *forma* (form or f.). These form a hierarchy, and further subdivisions are permitted, but the Code does not define the characteristics of any rank within the hierarchy. Consequently, infraspecific classification is subjective.

When a subdivision of a species is named, which does not include the nomenclatural type of the species, it automatically establishes the name of the equivalent subdivision which does contain that type. Such a name is an 'autonym' and has the same epithet as the species itself but is not attributed to an author. This is the only event which permits the repetition of the specific epithet and the only permissible way of indicating that the taxon includes the type for the species name. The same constraints apply to subdivisions of lower ranks. For example, *Veronica hybrida* L. was deemed by E. F. Warburg to be a component of *Veronica spicata* L. and he named it *V. spicata* L. subsp. *hybrida* (L.) E. F. Warburg. This implies the existence of a typical subspecies, the autonym for which is *V. spicata* L. subsp. *spicata*.

It will be seen from the citation of Warburg's new combination that the disappearance of a former Linnaean species can be explained. Retention of the epithet *'hybrida'*, and the indication of Linnaeus being its author (in brackets) shows the benefit of this system in constructing names with historic meanings.

Hybrids

Hybrids are particularly important as cultivated plants but are also a feature of many plant groups in the wild, especially woody perennials such as willows. The rules for the names and naming of hybrids are contained in the Botanical Code but are equally applicable to cultivated plant hybrids.

For the name of a hybrid between parents from two different genera, a name can be constructed from the two generic names, in part or in entirety (but not both in their entirety) as a condensed formula; ×*Mahoberberis* is the name for hybrids between the genera *Mahonia* and *Berberis* (in this case the cross is only bigeneric when *Mahonia*, a name conserved against *Berberis*, is treated as a distinct genus) and ×*Fatshedera* is the name for hybrids between the genera *Fatsia* and *Hedera*. The orchid hybrid between *Gastrochilus bellinus* (Rchb.f.) Kuntze and *Doritis pulcherrima* Lindl. carries the hybrid genus name ×*Gastritis* (it has a cultivar called 'Rumbling Tum'!). Alternatively a formula can be used in which the names of the genera are linked by the sign for hybridity '×': *Mahonia* ×*Berberis* and *Fatsia* ×*Hedera*. Hybrids between parents from three genera are also named either by a formula or by a condensed formula and, in all cases, the condensed formula is treated as a generic name if it is published with a statement of parentage. When published, it becomes the correct generic name for any hybrids between species of the named parental genera. A third alternative is to construct a commemorative name in honour of a notable person and to end it with the termination *-ara*; ×*Sanderara* is the name applied to the orchid hybrids between the genera *Brassia*, *Cochlioda* and *Odontoglossum* and commemorates H. F. C. Sander, the British orchidologist.

A name formulated to define a hybrid between two particular species from different genera can take the form of a species name, and then applies to all hybrids produced subsequently from those parent species; ×*Fatshedera lizei* Guillaumin is the name first given to the hybrid between *Fatsia japonica* (Thunb.) Decne. & Planch. and a cultivar of ivy, *Hedera helix* L. 'Hibernica', raised by Lizé Frères in Nantes, France, but which must include all hybrids between *F. japonica* and *H. helix*. Other examples include ×*Achicodonia*, ×*Achimenantha*, ×*Amarygia*, ×*Celsioverbascum*, ×*Citrofortunella*, ×*Chionoscilla*, ×*Cooperanthes*, ×*Halimocistus*, ×*Ledodendron*, ×*Leucoraoulia*, ×*Lycene*, ×*Osmarea*, ×*Stravinia*, ×*Smithicodonia*, ×*Solidaster* and ×*Venidioarctotis*. Because the parents themselves are variable, the progeny of repeated crosses may be distinctive and warrant cultivarietal naming. They may be named under the Botanical Code (prior to 1982 they would have been referred to as nothomorphs or bastard forms) and also under the International Code of Nomenclature for Cultivated Plants as 'cultivars'; thus, ×*Cupressocyparis leylandii* 'Naylor's Blue'. The hybrid nature of ×*Sanderara* is expressed by classifying it as a 'nothogenus' (bastard genus or, in the special circumstances of orchid nomenclature, grex class) by classifying it as a 'nothospecies' (within a nothogenus). For infraspecific ranks the multiplication sign is not used but the term denoting their rank receives the prefix notho-, or 'n-' (*Mentha* ×*piperita* L. nothosubspecies *pyramidalis* (Ten.) Harley which, as stated earlier, also implies the autonymous *Mentha* ×*piperita* nothosubspecies *piperita*.

Hybrids between species in the same genus are also named by a formula or by a new distinctive epithet; *Digitalis lutea* L. ×*D. purpurea* L. and *Nepeta* ×*faassenii* Bergmans ex Stearn are both correct designations for hybrids. In the example of *Digitalis*, the order in which the parents are presented happens to be the correct order, with the seed parent first. It is permissible to indicate the roles of the parents by including the symbols for female '♀' and male '♂' when this information is known, or otherwise to present the parents in alphabetical order.

The orchid family presents particularly complex problems of nomenclature, requiring its own 'Code' in the form of the *Handbook on Orchid Nomenclature and Registration* (Greatwood, Hunt, Cribb & Stewart, 1993). There are some 20,000 species of orchids and to this have been added a huge range of hybrids, some with eight genera contributing to their parentage, and over 70,000 hybrid swarms, or *gregis* (singular *grex*), with a highly complex ancestral history.

In cases where a hybrid is sterile because the two sets of chromosomes which it has inherited, one from each parent, are sufficiently dissimilar to cause breakdown of the mechanism which ends in the production of gametes, doubling its chromosome complement may produce a new state of sexual fertility and what is, in effect, a new biological species. Many naturally occurring species are thought to have evolved by such changes and man has created others artificially via the same route, some intentionally and some unintentionally from the wild. The bread-wheats *Triticum aestivum* L. are an example of the latter. They are not known in the wild and provide an example of a complex hybrid ancestry but whose name does not need to be designated as hybrid. Even artificially created tetraploids (having, as above, four instead of the normal two sets of chromosomes) need not be designated as hybrid, by inclusion of '×' in the name; *Digitalis mertonensis* Buxton & Darlington is the tetraploid from an infertile hybrid between *D. grandiflora* L. and *D. purpurea* L.

Synonymy and illegitimacy

Since a plant can have only one correct name, which is determined by priority, its other validly published names are synonyms. Inevitably, most plants have been known by two or more names in the past. A synonym may be one which is strictly referable to the same type (a nomenclatural synonym) or one which is referable to another type which is, however, considered to be part of the same taxon (this is a taxonomic synonym). The synonymy for any plant or group of plants is important because it provides a reference list to the history of the classification and descriptive literature on that plant or group of plants.

In the search for the correct name, by priority, there may be names which have to be excluded from consideration because they are regarded as being illegitimate, or not in accordance with the rules.

Names which have the same spelling but are based on different types from that which has priority are illegitimate 'junior homonyms'. Clearly, this prevents the same name being used for different plants. Curiously, this exclusion also applies to the names of those animals which were once regarded as plants, but not to any other animal names.

Published names of taxa which are found to include the type of an existing name are illegitimate because they are 'superfluous'. This prevents unnecessary and unacceptable proliferation of names of no real value.

Names of species in which the epithet exactly repeats the generic name have to be rejected as illegitimate 'tautonyms'. It is interesting to note that there are many plant names which have achieved some pleonastic repetition by using generic names with Greek derivation and epithets with Latin derivation: *Arctostaphylos uva-ursi* (bear-berry, berry of the bear), *Myristica fragrans* (smelling of Myrrh, fragrant), *Orobanche rapum-genistae* (legume strangler, rape of broom), *Zizyphus jujuba* (the Greek and Latin from the Arabic, zizouf); or the reverse of this: *Liquidambar styraciflua* (liquid amber, flowing with storax), *Silaum silaus*; but modern practice is to avoid such constructions. In zoological nomenclature tautonyms are commonplace.

The Code provides a way of reducing unwelcome disturbance to customary usage which would be caused by rigid application of the rule of priority to replace with correct names certain names of families and genera which, although incorrect or problematic are, for various reasons (usually their long usage and wide currency in important literature) agreed to be conserved at a Botanical Congress. These conserved names can be found listed in an appendix to the Code, together with names which are to be rejected because they are taxonomic synonyms used in a sense which does not include the type of the name, or are earlier nomenclatural synonyms based on the same type, or are homonyms or orthographic variants.

The Code also recommends the ways in which names should be spelt or transliterated into Latin form in order to avoid what it refers to as 'orthographic variants'. The variety found amongst botanical names includes differences in spelling which are, however, correct because their authors chose the spellings when they published them and differences which are not correct because they contain any of a range of defects which have become specified in the Code. This is a problem area in horticultural literature, where such variants are commonplace. It is clearly desirable that a plant name should have a single, constant and correct spelling, but this has not been achieved in all fields and reaches its worst condition in the labelling of plants for sale in some nurseries.

The International Code of Nomenclature for Cultivated Plants

There can be no doubt that the diverse approaches to naming garden plants, by common names, by botanical names, by mixtures of botanical and common names, by group names and by fancy names, is no less complex than the former unregulated use of common or vernacular names. The psychology of advertising takes descriptive naming into yet new dimensions. It catches the eye with bargain offers of colourful, vigorous and hardy, large-headed, incurved *Chrysanthemum* cvs. by referring to them as HARDY FOOTBALL MUMS. Perhaps the director whose appointment was headlined 'Football Mum appointed to Sainsburys' hopes that she is also 'hardy'. However, we are not here concerned with such colloquial names or the ethics of mail-order selling techniques but with the regulation of meaningful names under the Code.

In 1952, the Committee for the Nomenclature of Cultivated Plants of the International Botanical Congress and the International Horticultural Congress in London adopted the International Code of Nomenclature for Cultivated Plants. Sometimes known as the **Cultivated Code**, it was first published in 1953 and has been revised several times at irregular intervals since then (Trehane, 1995, Brickell *et al.*, 2004). **This Code formally introduced the term 'cultivar' to encompass all varieties or derivatives of wild plants which are raised under cultivation, and its aim is to 'promote uniformity and fixity in the naming of agricultural, sylvicultural and horticultural cultivars (varieties)'**. The term *culton* (plural *culta*) is also proposed as an equivalent of the botanical term *taxon*.

The Cultivated Code governs the names of all plants which retain their distinctive characters, or combination of distinctive characters, when reproduced sexually (by seed), or vegetatively in cultivation. Because the Code does not have legal status, the commercial interests of plant breeders are guarded by the Council of the International Union for the Protection of New Varieties of Plants (UPOV). In Britain, the Plant Variety Rights Office works with the Government to have UPOV's guidelines implemented. Also, in contrast with the International Code for Botanical Nomenclature, the Cultivated Code faces competition from legislative restraints presented by commercial law in certain countries. Where national and international legislation recognize 'variety' as a legal term and also permit commercial trade designation of plant names, such legislative requirements take precedence over the Rules of the Cultivated Code.

The Cultivated Code accepts the International Rules of Botanical Nomenclature and the retention of the botanical names of those plants which are taken into cultivation from the wild, and has adopted the same starting date for priority (precedence) of publication of cultivar names (*Species plantarum* of 1753). It recognizes only the one category of garden-maintained variant, the cultivar (cv.) or garden variety, which should not be confused with the botanical *varietas*. It recognizes also the supplementary, collective category of the Cultivar Group, intermediate between species and cultivar, for special circumstances explained below. The name of the Cultivar Group is for information and may follow the cultivarietal name, being placed in parentheses: *Solanum tuberosum* 'Desiré' (Maincrop Group) or potato 'Desiré' (Maincrop Group).

Unlike wild plants, cultivated plants are maintained by unnatural treatment and selection pressures by man. A cultivar must have one or more distinctive attributes which separate it from its relatives, and may be:

1 Clones derived asexually from (a) a particular part of a plant, such as a lateral branch to give procumbent offspring, (b) a particular phase of a plant's growth

cycle, as from plants with distinctive juvenile and adult phases, (c) an aberrant growth, such as a gall or witches' broom.

2 Graft chimaeras (which are dealt with below).
3 Plants grown from seed resulting from open pollination, provided that their characteristic attributes remain distinctive.
4 Inbred lines resulting from repeated self-fertilization.
5 Multi-lines, which are closely related inbred lines with the same characteristic attributes.
6 F1 hybrids, which are assemblages of individuals that are re-synthesized only by crossbreeding.
7 Topo-variants, which are repeatedly collected from a specific provenance (equivalent to botanical ecospecies or ecotypes).
8 Assemblages of genetically modified plants.

The cultivar's characteristics determine the application of the name – so genetic diversity may be high and the origins of a single cultivar may be many. If the method of propagating the cultivar is changed and the offspring show new characteristics, they may not be given the name of the parent cultivar. If any of the progeny revert to the parental characteristics, they may carry the parental cultivar name.

Plants grafted onto distinctive rootstocks, such as apples grafted onto Malling dwarfing rootstocks, may be modified as a consequence but it is the scion which determines the cultivar name – not the stock. Plants which have their physical form maintained by cultural techniques, such as bonsai and topiary subjects and fruit trees trained as espaliers etc., do not qualify for separate cultivar naming since their characteristics would be lost or changed by cessation of pruning or by pruning under a new regime.

From this it will be seen that with the single category of cultivar, the hybrid between parents of species rank, or any other rank, has equal status with a 'line' selected within a species, or taxon of any other rank, including another cultivar, and that parity exists only between names, not between biological entities. The creation of a cultivar name does not, therefore, reflect a fragmentation of the parent taxon but does reflect the existence of a group of plants having a particular set of features, without definitive reference to its parents. Features may be concerned with cropping, disease resistance or biochemistry, showing that the Cultivated Code requires a greater flexibility than the Botanical Code. It achieves this by having no limiting requirement for 'typical' cultivars but by regarding cultivars as part of an open system of nomenclature. Clearly, this permits a wide range of applications and differences with the Botanical Code, and these are considered in Styles (1986).

The names of cultivars have had to be 'fancy names' in common language and not in Latin. Fancy names come from any source. They can commemorate anyone, not only persons connected with botany or plants, or they can identify the nursery of their origin, or be descriptive, or be truly fanciful. Those which had Latin garden-variety names were allowed to remain in use. *Nigella damascena* L. has two old varietal names, *alba* and *flore pleno*, and also has a modern cultivar with the fancy name 'Miss Jekyll'. In the glossary, no attempt has been made to include fancy names, but a few of the earlier Latin ones have been included.

In order to be distinguishable, the cultivar names have to be printed in a typeface unlike that of the species name and to be given capital initials. They also have to be placed between single quotation marks. Thus, *Salix caprea* 'Kilmarnock' is a weeping variety of the goat willow and is also part of the older variety *Salix caprea* var. *pendula*. Other examples are *Geranium ibericum* Cav. 'Album' and *Acer davidii* Franchet 'George Forrest'.

Cultivar names can be attached to an unambiguous common name, such as potato 'Duke of York' for *Solanum tuberosum* L. 'Duke of York', or to a generic name such as *Cucurbita* 'Table Queen' for *Cucurbita pepo* L. 'Table Queen', or of course to the botanical name, even when this is below the rank of species; *Rosa sericea* var. *omeiensis* 'Praecox'.

Commercial breeders have produced enormous numbers of cultivars and cultivar names. Some have found popularity and have therefore persisted and remained available to gardeners, but huge numbers have not done so and have been lost or remain only as references in the literature. The popular practice of naming new cultivars for people (friends, growers, popular personalities or royalty) or the nursery originating the new cultivar is a form of flattery. For those honouring people who made some mark upon horticulture during their lifetime it is more likely that we can discover more about the plant bearing their name but, for the vast majority of those disappearing into obscurity, the only record may be the use of their name in a nurseryman's catalogue. Alex Pankhurst (1992) has compiled an interesting collection of commemorative cultivarietal names.

For some extensively bred crops and decorative plants there is a long-standing supplementary category, the Cultivar Group. By naming the Cultivar Group in such plants, a greater degree of accuracy is given to the garden name; such as pea 'Laxton's Progress' (Wrinkle-seeded Group) and *Rosa* 'Albéric Barbier' (Rambler Group) and *Rosa* 'Agnes'(Rugosa Group). However, for some trade purposes a cultivar may be allocated to more than one Cultivar Group, such as potato 'Desiré' (Maincrop Group) but also potato 'Desiré' (Red-skinned Group).

The same cultivar name may not be used twice within a genus, or denomination class, if such duplication would cause ambiguity. Thus, we could never refer to cherries and plums by the generic name, *Prunus*, alone. Consequently, the same fancy name could not be used for a cultivar of a cherry and for a cultivar of a plum. Thus, the former cultivars cherry 'Early Rivers' and plum 'Early Rivers' are now cherry 'Early Rivers' and plum 'Rivers Early Prolific'.

To ensure that a cultivar has only one correct name, the Cultivated Code requires that priority acts and, to achieve this, publication and registration are necessary. To establish a cultivar name, publication has to be in printed matter which is dated and distributed to the public. For the more popular groups of plants, usually genera, there are societies which maintain statutory registers of names, and the plant breeding industry has available to it the Plant Variety Rights Office as a statutory registration body for crop-plant names as trade marks for commercial protection, including patent rights on vegetatively propagated cultivars. Guidance on all these matters are provided as appendices to the Code.

As with botanical names, cultivars can have synonyms. However, it is not permissible to translate the fancy names into other languages using the same alphabet; except that in commerce the name can be translated and used as a trade designation. This produces the confusion that, for example, *Hibiscus syriacus* 'Blue Bird' is just a trade name for *Hibiscus syriacus* 'L'Oiseau Bleu' but will be the one presented at the point of sale. Also, translation is permitted to or from another script and the Code provides guidance for this.

In the case of the names of Cultivar Groups, translation is permitted; since these are of the nature of descriptions that may relate to cultivation. An example provided is the Purple-leaved Group of the beech which is the Purpurblätterige Gruppe in German, the Gruppo con Foglie Purpuree in Italian and the Groupe à Feuilles Pourpres in French.

For the registration of a new cultivar name, it is also recommended that designated standards are established. These may be herbarium specimens deposited in herbaria, or illustrations that can better define colour characteristics, or documentation held at a Patents Office or a Plant Variety Protection Office. In each case, the intention is that they can be used as reference material in determining later proposed names. This brings the Cultivated Code closer to the Botanical Code and is a small step towards the eventual establishment of an all-encompassing Code of Bionomenclature.

When the names of subspecies, varieties and forms are used, it is a growing trend to present the full name without indication of these – particularly in America, but also in our own horticultural literature (Bagust, 2001), as a shorthand cross-reference. Thus, *Narcissus bulbocodium* subsp. *bulbocodium* var. *conspicuus* is written as *Narcissus bulbocodium bulbocodium conspicuus*. This is confusing when the cultivar name has a Latin form since this then has the appearance of a pre-Linneaen phrase name (e.g. *Narcissus albus plenus odoratus* and *Rosa sericea omeiensis praecox*).

Graft chimaeras

One group of plants which is entirely within the province of gardening and the Cultivated Code is that of the graft chimaeras, or graft hybrids. These are plants in which a mosaic of tissues from the two parents (not closely related) in a grafting partnership results in an individual plant upon which shoots resembling each of the parents, and in some cases shoots of intermediate character, are produced in an unpredictable manner. The closest analogy amongst animals is the experimental rodent on which a human ear is being grown, or the human into which a heart valve has been grafted from a genetically manipulated pig.

Unlike sexually produced hybrids, the admixture of the two parents' contributions is not at the level of the nucleus in each and every cell but is more like a marbling of a ground tissue of one parent with streaks of tissue of the other parent. Chimaeras can also result from mutation in a growing point, from which organs are formed composed of normal and mutant tissues – as with genetic forms of variegation. In all cases, three categories may be recognized in terms of the extent of tissue 'marbling', called sectorial, mericlinal and periclinal chimaeras. The chimaeral condition is denoted by the addition sign '+' instead of the multiplication sign '×' used for true hybrids. A chimaera which is still fairly common in Britain is that named +*Laburnocytisus adamii* C. K. Schneider. This was the result of a graft between *Cytisus purpureus* Scop. and *Cytisus laburnum* L., which are now known as *Chamaecytisus purpureus* (Scop.) Link and *Laburnum anagyroides* Medicus, respectively. Although its former name *Cytisus* + *adamii* would not now be correct, the name +*Laburnocytisus* meets the requirement of combining substantial parts of the two parental generic names, and can stand.

Combining generic names for graft chimaeras must not duplicate a composite name for a sexually produced hybrid between the same progenitors. Hybrids between species of *Crataegus* and species of *Mespilus* are ×*Crataemespilus* but the chimaera between the same species of the same genera is +*Crataegomespilus*. As in this example, the same progenitors may yield distinctive chimaeras and these may be given cultivar names: +*Crataegomespilus* 'Dardarii' and +*Crataegomespilus* 'Jules d'Asnières'.

It is interesting to speculate that if cell- and callus-culture techniques could be used to produce chimaeral mixtures to order, it may be possible to create some of the conditions which were to have brought about the early 'green revolutions' of the 1950–2000 period. Protoplast fusion methods failed to combine the culturally and economically desirable features of distant parents, which were to have given multi-crop plants and new nitrogen-fixing plants, because of the irregularities in fusion of both protoplasts and their nuclei. It may be that intact cells would prove easier to admix. However, molecular genetics and genetic manipulation have shown that genetic control systems can be modified in ways which suggest that any aspect of a plant can, potentially, be manipulated to suit man's requirements and novel genetic traits can be inserted into a plant's genome by using DNA implants. The genetically modified (GM) results of such manipulation are the products of commercial undertakings, and may be given cultivar names, but are protected commercially by trade designations.

Glossary

This glossary is for use in finding the meanings of the names of plants. There are many plant names which cannot be interpreted or which yield very uninformative translations. Authors have not always used specific epithets with a single, narrow meaning, so there is a degree of latitude in the translation of many epithets. Equally, the spelling of epithets has not remained constant, for example in the case of geographic names. The variants, from one species to another, are all correct if they were published in accordance with the Code. In certain groups such as garden plants from, say, China, and exotics such as many members of the profuse orchid family, commemorative names have been applied to plants more frequently than in other groups. The reader who wishes to add further significance to such names will find it mostly in literature on plant hunting and hybridization, or monographic works on particular taxa.

The glossary contains many examples of words which are part of botanical terminology as well as being employed as descriptive elements of plant names. Much terminology stems from Greek writing and mythology. It has been given Latin form, either by adoption into the Latin of the Romans, or since the renaissance during the sixteenth century. Words from numerous other languages have also been added to nomenclature by being given Latin form. It is not encouraged to compound languages into a single name or epithet, but these do exist. When the roots are from, say, Latin and Greek, the glossary refers to them as botanical Latin. Where a name or epithet is compounded of a name plus a prefix or suffix, it is regarded as legitimate modern Latinization. Hence we have such joys as *cyranostigmus -a -um*, being compounded from Cyrano de Bergerac and the modern Latin *stigma*, from the Greek στιγμα. Where place names have a classical origin, this will be provided in parentheses. Otherwise, the Latinization of place names may be assumed to be modern Latin.

Glossaries of terminology are often to be found in textbooks and Floras. The sixth edition (1955) of Willis' *Dictionary of Flowering Plants and Ferns* (1931) is a particularly rewarding source of information, and B. D. Jackson's *Glossary of Botanic Terms* (1960) is a first-rate source of classical etymological information.

Generic names in the European flora are mostly of ancient origin. Their meanings, even of those which are not taken from mythological sources, are seldom clear, and many have had their applications changed and are now used as specific epithets. Generic names of plants discovered throughout the world in recent times have mostly been constructed to be descriptive and will yield to translation. The glossary contains the generic names of a wide range of both garden and wild plants and treats them as singular nouns, with capital initials. Orthographic variants have not been sought out but a few are presented and have the version which is generally incorrect between brackets. Listings of generic names can be found in Farr (1979–86) and in Brummitt (1992) as well as, on the Internet at www.ipni.org, www.rbgkew.org.uk/epic, etc.

As an example of how the glossary can be used, we can consider the name *Sarcococca ruscifolia*. This is the name given by Stapf to plants which belong to Lindley's genus *Sarcococca*, of the family *Buxaceae*, the box family. In the glossary we find *sarc-*, *sarco-* meaning fleshy and *-coccus -a -um* meaning 'scarlet-berried', and from this we conclude that *Sarcococca* means fleshy-scarlet-berry, or fleshy-scarlet-berried-one (the generic name being a singular noun) and has the feminine gender. We also find *rusci-* meaning butcher's-broom-like or resembling *Ruscus* and *-folius -a -um* meaning *-leaved*, and we conclude that this species of fleshy-scarlet-berried-one has leaves resembling the prickly leaves (leaf-like branches or cladodes) of

Ruscus. The significance of this generic name lies in the fact that dry fruits are more typical in members of the box family than fleshy ones.

From this example, we see that names can be constructed from adjectives or adjectival nouns to which prefixes or suffixes can be added, thus giving them further qualification. As a general rule, epithets which are formed in this way have an acceptable interpretation when '-ed' is added to the English translation; this would render *ruscifolia* as 'butcher's-broom-leaved'.

Since *Sarcococca* has a feminine ending (*-a*), *ruscifolia* takes the same gender. However, if the generic name had been of the masculine gender the epithet would have become *ruscifolius*, and if of the neuter gender then it would have become *ruscifolium*. For this reason the entries in the glossary are given all three endings which, as pointed out earlier, mostly take the form *-us -a -um* or *-is -is -e*.

Where there is the possibility that a prefix which is listed could lead to the incorrect translation of some epithet, the epithet in question is listed close to the prefix and to an example of an epithet in which the prefix is employed. Examples are:

aer-, meaning air- or mist-, gives *aerius -a -um*, meaning airy or lofty;
aeratus -a -um, however, means bronzed (classically, made of bronze).

caeno-, from the Greek καινος, means new- or fresh-, but
caenosus -a -um is from the Latin *caenum* and means mud or filth.

Examples will be found of words which have several fairly disparate meanings. A few happen to reflect differences in meaning of closely similar Greek and Latin source words, as in the example above, and others reflect what is to be found in literature, in which other authors have suggested meanings of their own. Similarly, variations in spelling are given for some names and these are also to be found in the literature, although not all of them are strictly permissible for nomenclatural purposes. Their inclusion emphasizes the need for uniformity in the ways in which names are constructed and provides a small warning that there are in print many deviant names, some intentional and some accidental. Many of the epithets which may cause confusion are either classical geographic names or terms which retain a meaning closer to that of the classical languages. There are many more such epithets than are listed in this glossary.

Glossary

a, ab away from-, downwards-, very-; (privative) un-, without-
aaronis for the prophet Aaron, Aaron's
Abaca a synonym for *Musa textilis*
abactus -a -um repelling, repulsive, driving away, *abigo, abigere, abegi, abactum*
abayensis -is -e from the environs of Lake Abaya, Ethiopia
abbreviatus -a -um shortened, *ab-brevis* (*abbrevio, abbreviare*)
abchasicus -a -um, abschasicus -a -um from Abkhasia in the Caucasus
abditus -a -um hidden, removed, past participle of *abdo, abdere, abdidi, abditum*
Abelia for Dr Clarke Abel (1780–1826), physician and writer on China
abeliceus -a -um *Abelia*-like
Abeliophyllum *Abelia*-leaved (similarity of foliage)
aberconwayi for Charles Melville McLaren (1913–2003) third Lord Aberconway of Bodnant, former President of the RHS.
aberdeenensis -is -e from Aberdeen, Cape Province, S Africa
Aberia from Mount Aber in Ethiopia, provenance of type species
aberrans deviating from the norm, aberrant, differing, present participle from *aberro, aberrare, aberravi, aberratum*
Abies Rising-one, *abeo* (the ancient Latin name for a tall tree or ship)
abietifolius -a -um *Abies*-leaved, *Abies-folium*
abietinus -a -um fir-tree-like, *Abies*
abietis -is -e of *Abies* (*Adelges abietis* gall aphis on spruce)
-abilis -is -e -manageable, -able, -capable of, *habilis* (preceded by some action)
abjectus -a -um abandoned, cast down, unpleasant, *abicio, abicere, abieci, abiectum*
abnormis -is -e unorthodox, departing from normal in some structure, *abnormis*
Abobra from a Brazilian vernacular name
aboriginorum indigenous, of the original inhabitants, *aborigines, aboriginum*
abortivus -a -um miscarried, with missing or malformed parts, *aborior, aboriri, abortus*
abro-, abros soft, delicate, αβρος
Abroma from the Brazilian vernacular name
Abromeitiella Delicate-*Meitiella*, αβρος-μειων (delicate and very small)
Abronia Delicate, αβρος (the involucre)
Abrophyllum Delicate-leaf, αβρος-φυλλον
Abrotanella *Abrotanum*-like (feminine diminutive)
abrotani-, abrotonoides *Artemisia*-like, αβροτανον-οειδης (from an ancient Greek name, αβροτονον, for several fragrant-leaved plants)
abrotanifolius -a -um wormwood-leaved, botanical Latin from αβροτανον with *folium*
Abrotanum, abrotanum Divine, αβροτος, ancient name for southernwood
abruptus -a -um ending suddenly, blunt-ended, past participle of *abrumpo, abrumpere, abrupi, abruptum*
Abrus Soft, αβρος (the foliage of crab's eyes)
abscissus -a -um cut off, past participle of *abscindo, abscindere, abscidi, abscissum*
absconditus -a -um concealed, hidden, residual, *abscondo, abscondere, abscondi* (*abscondidi, absconditum*)
absimilis -is -e different, un-like, *ab-similis*
Absinthium the old generic name for wormwood, αψινθιον, in the works of Xenophon.
absinthius -a -um from an ancient Greek, αψινθιον, or Syrian name for wormwood
absinthoides wormwood-like, αψινθιον-οειδης

absum different, distant, distinct, *absum, abesse, abui*
absurdus -a -um unmusical, absurd, incongruous, *absurdus*
abundiflorus -a -um flowering copiously, *abunde-flora*
abundus -a -um prolific, abounding, *abundo, abundare, abundavi, abundatum*
Abuta from a West Indian vernacular name
Abutilon the Arabic name for a mallow
abyssicolus -a -um inhabiting ravines or chasms, late Latin, *abyssus-colus*, from Greek, α-βυσσος, without bottom
abyssinicus -a -um of Abyssinia, Abyssinian (now Ethiopia)
ac-, ad-, af-, ag-, al-, an-, ap-, ar-, as-, at- near-, towards-
Acacallis etymology uncertain
Acacia Thorn, ακη, ακις (Dioscorides' name, ακακια)
acaciformis -is -a resembling *Acacia, Acacia-forma*
acadiensis -is -e from Nova Scotia (formerly the French colony, Acadia)
Acaena (*Acena*) Thorny-one, ακαινα (the burr-like seeding heads)
acaenoides resembling-*Acaena*, ακαινα-οειδης
acalycinus -a -um lacking a calyx, privative α-καλυκος
Acalypha Unpleasant-to-touch, α-καλος-αφη, from the Greek name, ακελπε, for a nettle (the hispid leaves)
acanthifolius -a -um thorny-leaved, *Acanthus-folium*
acanthium Dioscorides' name, ακανθιον, for a thistle-like plant
acantho-, acanthus thorny-, spiny-, ακανθα, ακανθος, ακανθο-
Acanthocalyx Thorny-calyx, ακανθα-καλυξ
Acanthocarpus, acanthocarpus -a -um Spiny-fruited-one, ακανθο-καρπος
acanthocomus -a -um thorn-haired, ακανθα-κομη
Acanthogilia Spiny-*Gilia* (≡ *Baja californica*)
acanthoides resembling *Acanthus*, ακανθο-οειδης
Acantholimon Thorny-*Limonium*, ακανθο-*Limon*
Acanthometron Thorn-measure, ακανθο-μετρεω (spiny plankton)
Acanthonema Thorn-threaded, ακανθα-νημα (the processes on the filaments of the two lower stamens)
Acanthopanax Spiny-*Panax*, ακανθο-*Panax* (the prickly nature of the plants)
acanthophysus -a -um having inflated spines, ακανθο-φυσα
Acanthopsis Acanthus-like, ακανθιον-οψις
Acanthospermum Spiny-seed, ακανθο-σπερμα
acanthothamnos *Acanthus*-bush, ακανθο-θαμνος (the thorny nature of *Euphorbia acanthothamnos*)
Acanthus Prickly-one, ακανθιον, in Dioscorides (the Nymph, Acantha, loved by Apollo, was changed into an *Acanthus*) (***Acanthaceae***)
acaulis -is -e, acaulos -os -on lacking an obvious stem, privative α-καυλος, *a-caulis*
accedens approaching, agreeing with, present participle of *accedo, accedere, accessi, accessum*
Accipitrina Hawks', *accipiter* (analogy with *Hieracium*)
accisus -a -um with a small acute apical cleft, emarginate, cut into, *accido, accidere, accidi, accisus*
acclivus -a -um uphill, sloping upwards, inclined, *acclinis; acclino, acclinare, acclinavi, acclinatum*
accolus -a -um neighbour, *accola; accolo, accolere, accolui, accuitum*
accommodatus -a -um adaptable, adjusting, accommodating, *accommodo, accommodare, accommodavi, accommodatum*
accrensis -is -e from Accra, Ghana, W Africa
accrescens growing together, coalescing, accreting, *ac-*(*cresco, crescere, crevi, cretum*)
accumbens becoming adjacent, coming face to face, present participle of *accumbo, accumbere, accubui, accubitum*
-aceae -associates, -aceous, the standardized suffix for family names
Acer Sharp, *acer, acris* (Ovid's name for a maple, either from its use for lances or its leaf-shape) etymologically linked to oak, acorn and acre (***Aceraceae***)

acer, acris, acre sharp-tasted, acid, *acer, acris* (sometimes used as *acris -is -e*)
Aceras Without-a-horn, privative α-κερας (the lip has no spur)
acerbus -a -um harsh-tasted, bitter, sour; troublesome, *acerbus*
acerianus, -a -um of maples, living on *Acer (Gypsonoma* is a lepidopteran gall insect on white poplar, αχερωις)
acerifolius -a -um maple-leaved, *Acer-folium*
aceroides maple-like, *Acer-oides*
acerosus -a -um pointed, needle-like; sharp, bitter, *acer, acris*
acerrimus -a -um most bitter, most sharp, superlative of *acer* (burning taste)
acetabulosus -a -um saucer-shaped, saucer-like, *acetabulum-ulosus*
Acetosella, acetosellus -a -um vinegary, slightly acid, feminine diminutive of *acetum*
acetosus -a -um acid, sour, from *acetum* (sour gives the cognate sorrel)
-aceus -a -um -resembling (preceded by a plant name, Rose-aceous)
achaetus -a -um lacking bristles, α-χαιτη
achatinus -a -um banded with colours, chalcedony-like, αχατης
achelensis -is -e from the Sierra de Achela, Argentina
Achicodonia the composite generic name for hybrids between *Achimenes* and *Eucodonia* (properly ×*Achicodonia*)
Achillea after the Greek warrior Achilleios, who reputedly used it to staunch wounds (sneezewort, yarrow)
Achimenantha the composite generic name for hybrids between *Achimenes* and *Trichantha*
Achimenes etymology uncertain, Magic-plant, αχαεμηνις; Tender-one, α-χειμαινω (cold-hating) (Achaemenes was the reputed founder of the Persian dynasty of 553 to 330 BC)
Achlys for Achlys, the goddess of obscurity
achotensis -is -e from the region of the Achote river, Ecuador
achraceus -a -um *Achras*-like
Achras an old Greek name, αχρας, for a wild pear, αχερδος, used by Linnaeus for *Achras sapota*, the sapodilla or chicle tree (marmalade plum)
achro- lacking-light, pale-, α-χρωμα
achy-, achyro- chaffy-, chaff-like-, αχυρον, αχυρο-, αχυ-
achypodus -a -um scaly-stemmed, chaffy-stemmed, αχυ-ποδος
Achyranthes Chaff-flower, αχυρον-ανθος
Achyrophorus, achyrophorus -a -um Chaff-bearer, αχυρο-φορα (the receptacular scales)
Achyrospermum Chaff-seed, αχυρο-σπερμα
acianthus -a -um with pointed flowers, *acus-anthus*
acicularifolius -a -um with needle-like leaves, *aciculus-folium* (having acicles, diminutive of *acus*)
acicularis -is -e needle-shape, diminutive of *acus, acis, aci-*
aciculatus -a -um, aciculine -a -um finely marked as with needle scratches, diminutive from *acus*
aciculus -a -um sharply pointed, diminutive of *acus* (e.g. leaf-tips)
Acidanthera Pointed-anthers, ακις-ανθερα (the cuspidate anthers)
acidissimus -a -um very sour or sharp tasted, superlative of *acidus*
acidosus -a -um acid, sharp, sour, *acidus*
acidotus -a -um sharp-spined, ακιδωτος
acidus -a -um sour-tasting, *aceo, acere*
acinaceus -a -um full of kernels, *acinus-aceous;* scimitar shaped, *acinices, acinacis*
acinacifolius -a -um with leaves like long sabres or scimitars
acinifolius -a -um *Acinos*-leaved, basil-thyme-leaved, *Acinos-folium*
aciniformis -is -e dagger-shaped, *acinaces-forma* (leaves of Hottentot fig)
Acinos Dioscorides' name, ακινος, for a heavily scented calamint (*Clinopodium* or *Satureja*)
acinos, acinosus -a -um *Acinos*-like, berried
Acioa Pointed, *acus* (the needle-toothed bracts of some species)

Aciphylla Pointed-leaf, ακις-φυλλον
Acmella, acmellus -a -um Of-the-best, or Pointed; vigorous, of the best, ακμη
acmo- pointed-, ακμη (followed by a part of a plant) anvil-shaped-
acmopetalus -a -um pointed-petalled, with petals shaped like the pointed part of an anvil, ακμη-πεταλον
acmosepalus -a -um with pointed sepals, like the pointed part of a blacksmith's anvil, ακμη-σκεπη
Acoelorrhaphe Without-hollow-seam, α-κοιλος-ραφη (ovules separate)
Acokanthera Pointed-anther, ακοκε-ανθερα
aconitifolius -a -um aconite-leaved, *Aconitum-folium*
Aconitum the name of a hill in Pontus, used by Theophrastus for the poisonous aconite, ακονιτον
acoroides resembling *Acorus*, ακορον-οειδης
Acorus Without-pupil, ακορον, Dioscorides' name for an iris (its use in treating cataracts)
acostae without ribs, veinless, *a-(costa, costae)*
acpunctus -a -um spotted above, with spots towards the apex, *ac-(pungo, pugere, pupugi, punctum)*
acr-, acra, acro- summit-, highest-, ακρα, ακρις, ακρο- (followed by noun, e.g. hair, or verb, e.g. fruiting)
acracanthus -a -um spine-tipped, ακρο-ακανθος
Acrachne Apical-chaff, ακρ-αχνη (the racemes radiate more than the length of the axis bearing them)
Acradenia. acradenius -a -um Apical-gland, ακρ-αδην, gland-tipped
acraeus -a -um of windy places, of hilltops, ακραης
acreus -a -um of high places, of the summit, the highest, ακρα
Acridocarpus Locust-fruit, (ακρις,ακριδος)-καρπος
acris sharp-tasted, see *acer* (sometimes used as masculine, see *acer*)
Acritochaete Entangled-hair, α-κριτος-χαιτη (the hispidulous upper glume and lower lemma have entangled awns)
acro- summit-, apex-, ακρα
acrobaticus -a -um walking on points, twining and climbing, ακρο-βεινιν, ακροβατες
Acrobolbus Apical-bulb, ακρα-βολβος (the archegonia are surrounded by minute leaves at the apex of the stem)
Acroceras Apex-horn, ακρα-κερας (the glumes have an excurrent vein at the tip)
acrolepis -is -e scale-tipped, ακρο-λεπις
acrostichoides resembling *Acrostichum*, ακρο-στικτος-οειδης
Acrostichum Upper-spotted, ακρο-στικτος (the sori cover the backs or whole of the upper pinnae)
acrotrichus -a -um hair-tipped, ακρο-τριχος (apical tufts of the leaves)
Actaea Pliny's name from the Greek name, ακτεα, ακταια, for elder (the similarity in shape of the leaves)
actaeifolius -a -um *Actaea*-leaved, *Actaea-folium*
actin-, actino- ray-, light-, splendour-, ακτις, ακτινος, ακτινο- (followed by a part of a plant)
actinacanthus -a -um ray-spined, ακτις-ακανθα
Actinella Little-ray, diminutive form of ακτις, a ray (the capitulum)
actineus -a -um, actinia sea-anemone, with a notable radial structure, rayed, ακτις
Actinidia Rayed, ακτις, ακτινος (refers to the radiate styles) (***Actinidiaceae***)
actinidioides resembling *Actinidia, Actinidia-oides*
Actiniopteris Rayed-fern, ακτινο-πτερυξ (the digitate fronds)
actinius -a -um sea-anemone-like, with radial form, of the beach, ακτη
Actinocarpus Radiate-fruit, ακτινο-καρπος (the spreading ripe carpels of thrum-wort)
actinophyllus -a -um with radiate leaves, ακτινο-φυλλον
Actinotus Rayed, ακτινος (the involucre)
acu- pointed-, acute-, sharp like a needle, *acus*

acuarius -a -um prickly, with small needles, *acus*
acuatus -a -um sharpened, *acuo, acuere, acui, acutum*
aculeatus -a -um having prickles, prickly, thorny, *aculeus, aculei* (cognate with eglantine, aiglentina)
aculeolatus -a -um having small prickles or thorns, comparative from *aculeus*
aculiosus -a -um decidedly prickly, comparative from *aculeus*
acuminatus -a -um with a long, narrow and pointed tip (see Fig. 7c), acuminate, *acumen, acuminis*
acuminosus -a -um with a conspicuous long flat pointed apex, *acumen, acuminis*
acutangulus -a -um with sharp edges, *acutus-angulus*
acutidens sharply toothed, *acutus-(dens, dentis)*
acutiflorus -a -um acute-flowered, with pointed petals, *acutus-(floreo, florere, florui)*
acutifolius -a -um with acute leaves, *acutus-folium*
acutiformis -is -a acute-shaped, *acutus-forma* (leaves)
acutilobus -a -um with acute lobes, *acutus-lobus* (leaves)
acutissimus -a -um very pointed, most pointed, sharpest, superlative of *acutus*
acutus -a -um, acuti- acutely pointed, sharply angled at the top, *acutus*
ad-, as- to-, towards-, near-, at-, compared with-, compared with-, *ad (ads-* often becomes *as-*, e.g. *adscendo* becomes *ascendo*)
adamantinus -a -um from Diamond Lake, Oregon, USA; or from Brazil; diamond-like, steely, αδαμαντινος, *adamas, adamantis; adamanteus, adamantinus*
adamantis -is -e from Diamond Head, Hawaii, *adamas, adamantis*, diamond
adamantius -a -um adamant, unyielding, impenetrable, *adamanteus, adamantinus*
adanensis -is -e from Adana, Turkey
Adansonia for Michel Adanson (1727–1806), French botanist in Senegal, author of *Familles des plantes* (1763), who used anatomy and statistics in his work (baobab tree)
adductus -a -um fused together, *adduco, adducere, adduxi, adductum*
adelo- secret, unseen, obscure, uncertain, unseen, αδηλος
Adelostigma Obscure-stigma, αδηλος-στιγμα
adelphicus -a -um brotherly, coupled, closely related, αδελφος
-adelphus -a -um brotherly, fellow-like, coupled, αδελφος (relating to features of the stamens or androecium)
aden-, adeno- gland-, glandular-, αδην, αδηνος, αδηνο-
adenanthus -a -um glandular-flowered, αδην-ανθος
Adenium from Aden (provenance of one species)
Adenocarpus, adenocarpus -a -um Gland-fruit, αδηνο-καρπος (the glandular pod)
adenocaulis -is -e, adenocaulon with a glandular stem, αδηνο-καυλος
adenochaetus -a -um with long glandular hair, αδηνο-χαιτη
adenogynus -a -um with a glandular ovary, αδηνο-γυνη
Adenophora, adenophorus -a -um Gland-bearing, αδηνο-φορα
adenophyllus -a -um glandular-leaved, αδηνο-φυλον
adenopodos, adenopodus -a -um glandular-stemmed, αδηνο-ποδιον
adenoscepes with a glandular surface, αδηνο-σκεπη
Adenostemma Glandular-crown, αδηνο-στεμμα
Adenostyles Glandular-styles, αδηνο-στυλος (actually the stigmatic arms)
adenosus -a -um glandular, comparative from αδηνος
adenothrix glandular hairy, αδηνο-θριξ
adenotrichus -a -um glandular-hairy, αδηνο-τριχος
adfinis -is -e related by marriage, connected to, *adfinis*
adhaerens clinging to, staying close, present participle from *adhaereo, adhaerere, adhaesi, adhaesum* (cognate with adhere and adhesive)
Adhatoda, adhatoda from the Brazilian vernacular name for *A. cydoniifolia*
adiantifolius -a -um with *Adiantum*-like foliage, *Adiantum-folium*
Adiantum Unwetted, αδιαντος (the old Greek name, αδιαντον, refers to its staying unwetted under water) (***Adiantaceae***)
adiantum-nigrum black-spleenwort, αδιαντον (the lower rachis)

Adina Crowded, αδινος (the flowering head)
adiposus -a -um corpulent, *adeps, adipis*
adjacens adjacent, at the border, lying close to, *adiaceo, adiacere, adiacui* (systematic relationship)
Adlumia for Major John Adlum (1759–1836), American viticulturist
admirabilis -is -e to be admired, admirable, *admiror, admirari, admiratus*
adnascens growing on or with, present participle of *ad-(nascor, nasci, natus)*
adnatus -a -um attached through the whole length, adnate, *ad-(nascor, nasci, natus)* (e.g. anthers)
adoneus -a -um *Adonis*-like, resembling pheasant's eye
Adonis for the Greek God, Adonis, loved by Venus, killed by a boar and from whose blood grew a flower called *Adonium*
Adoxa Without-glory, α-δοξα (its small greenish flowers) (***Adoxaceae***)
adoxoides resembling Adoxa, α-δοξα-οειδης
adpressipilosus -a -um with closely flat-lying indumentum, adpressed hairy, *adpressus-pilosus*
adpressus -a -um pressed together, lying flat against, *ad-(premo, premere, pressi, pressum)* (e.g. hairs on a stem)
adriaticus -a -um from the Adriatic region (*Hadriaticus*)
Adromischus Stout-stemmed, αδρος-μισχος (sturdy, grown up)
adroseus -a -um near *roseus -a -um, ad-roseus*
adscendens (see *ascendens*) curving up from a prostrate base, half-erect, ascending, *ascendo, ascendere, ascendi, ascentum*
adscitus -a -um assumed, acquired, alien, *ad-(scisco, sciscere, scivi, scitum)*
adsimilis -is -e similar to, imitating, comparable with, *adsimulo, adsimulare, adsimulavi, adsimulatum* (see *assimilis*)
adspersus -a -um sprayed, sprinkled, past participle of *aspergo, aspergere, aspersi, aspersum*
adstringens constricted, tightened, binding, *ad-(stringo, stringere, strinxi, strictum)*
adsurgens rising up, arising, *ad-(surgo, surgere, surrecxi, surrectum)*
adulterinus -a -um of adultery, forged, *adulterinus* (intermediate between two other species, suggesting hybridity, as in *Asplenium adulterinum*)
aduncus -a -um hooked, having hooks, *ad-uncus*
adustus -a -um fuliginous, soot-coloured, *ad-(uro, urere, ussi, ustum)* to scorch, cauterize, or inflame
adventus -a -um approach, arrival, *ad-(venio, venire, veni, ventum)* (recent mutant or sport)
advenus -a -um exotic, stranger, foreign, *advena*
adzharicus -a -um from Adzhariya, Georgia, near Turkish border (Batumi)
Aechmea Pointed, αιχμη, αιχμο- (a point, edge or lance)
aegaeus -a -um of the Aegean region
Aegiceras Goat's-horn, αιξ-κερας (the shape of the fruit)
Aegilops a name, αιγιλωψ, used for several plants (αιξ, αιγος goat)
Aegirus ancient Greek name, αιγειρος, for *Populus nigra*
Aegithallos Goat's-shoot, αιξ-θαλλος
Aegle one of the Hesperides of mythology
Aegopodium Goat's-foot, αιγο-ποδιον (the leaf shape)
aegypticus -a -um from Egypt, Egyptian
aelophilous -a -um wind-loving, αελλα-φιλος (plants disseminated by wind)
Aeluropus Creeping-stalk, ειλυω-πους (stoloniferous culms)
aemulans, aemulus -a -um jealous, rivalling, imitating, *aemulus*
aeneus -a -um of bronze, bronzed, *aeneus*
Aeolanthus Wind-flower, αελλω-ανθος (their craggy habitats at altitude; Aeolus was god of the winds)
Aeonium Eternity, αειναος (the Latin name from the Greek αει, αιει, for ever)
aequalis -is -e resembling, equal, like, uniform, *aequalis*
aequi-, aequali- equally-, just as-, *aeque*

aequilateralis -is -e, aequilaterus -a -um equal-sided, *aeque-(latus, lateris)*
aequinoctialis -is -e of the equinox, *aequinoctium* (the flowering time); of equatorial regions
aequinoctiianthus -a -um flowering at about the time of the equinox, *aequinoctium-anthus*
aequitrilobus -a -um with three equal lobes, *aeque-tri-lobus*
aer- air-, mist-, *aer, aeris,* αηρ, ερος
Aerangis Air-vessel, αερ-αγγειον (epiphytic orchids)
Aeranthes Air-flower, αηρ-ανθος (epiphytic orchids)
aeranthos -os -on air-flower, αηρ-ανθος (not ground-rooted)
Aeranthus Air-flower, αηρ-ανθος (rootless epiphytes)
aeratus -a -um bronzed, coppery, *aeratus*
aereus -a -um copper (coloured), *aereus*
Aerides Of the air, αηρ-ειδης (epiphytic)
aerius -a -um lofty, of the air, *aerius*
aeruginascens turning verdigris-coloured, *aerugo, aeruginis*
aerugineus -a -um, aeruginosus -a -um rusty, verdigris-coloured, *aerugo*
Aeschynanthus Shame-flower, αισχυνη-ανθος (the curved corolla tube suggests a bowed head)
aeschyno- reverent-, to be ashamed-, to deform-, αισχος, αισχυνη, αισχυνω
Aeschynomene Deformed-moon, αισχυνω-μηνη (the leaves of *Aeschynomene sensitiva* fold when touched. *Aeschynomene aspera* was the source of pith for pith helmets)
aesculi- horse-chestnut like-, *Aesculus*
aesculifolius -a -um *Aesculus*-leaved, *Aesculus-folium*
Aesculus Linnaeus' name from the Roman name, *aesculus, aesculi*, of the durmast oak. The Turks reputedly used 'conkers', horse chestnuts of *Aesculus hippocastanum*, in a treatment for bruising in horses – now attributed to its aescin content
aestivalis -is -e of summer, *aestivus*
aestivus -a -um developing in the summer, *aestas, aestatis, aestivus*
aestuans heating up, glowing, becoming hot, *aestus*
aestuarius -a -um of tidal waters, of estuaries, *aestuarium*
aethereus -a -um of the sky, aerial, *aether* (epiphytic)
Aethionema etymology uncertain; Unusual-filaments, αηθης-νημα (those of the long stamens are winged and toothed). Other meanings have been proposed
aethiopicus -a -um of Africa, African, of NE Africa (the land of burnt faces, αιθειν-ωψ)
aethiops of uncommon appearance, αηθης-ωψ
Aethiorhiza Unusual-root, αηθης-ριζα
Aethusa Burning-one, αιθω, αιθων (for the shining foliage or its pungency). Gilbert-Carter (1964) notes that αιθουσα meant a sunny vestibule or veranda
aethusifolius -a -um *Aethusa*-leaved, *Aethusa-folium*
aetiolatus -a -um lank and yellowish, etiolated, from early French, étieuler
aetnensis -is -e (aethnensis) from Mount Etna (*Aetna*) Sicily
aetolicus -a -um from *Aetolia*, Greece
-aeus -belonging to (of a place)
Aextoxicon Goat-arrow (poison), αιξ-τοξικον (φαρμακον)
afares from Africa, *afer, afri* (≡ *Quercus castaneaefolia*)
afer, afra, afrum of Africa, more extensive than the Roman *Africa, Africae*
affinis -is -e related, similar to, *ad-finis* (to the border of)
afghanicus -a -um from Afghanistan
aflatunensis -is -e from Aflatun, central Asia
afoliatus -a -um without leaves, *a-(folium, folii)*
Aframomum African-*Amomum* (*Amomum* is occidental from E. Indies to Japan)
africanus -a -um African, from Africa, *Africa, Africae*
Afrocalathea African-*Calathea* (*Calathea* is a New World genus)
Afrofittonia African-*Fittonia*
Afrothismia African-*Thismia* (*Thismia, sensu lato,* is in Indo-Malaya and S America)

Afrotrilepis African-*Trilepis*
Afzelia, afzelianus -a -um for Adam Afzelius (1750–1837), Swedish botanist and agricltural advisor in W Africa *c.*1792
aga-, agatho- good, noble, useful, αγαθος, αγα-
aganniphus -a -um of snow coverings, αγαννιφος (living at altitude)
Agapanthus Love-flower, αγαπη-ανθος
Agapetes Beloved, αγαπητος
agapetus -a -um desirable, love, αγαπη
Agaricus Tungus, αγαρικον; from Agaria, Sarmatia (now Ukraine)
Agarista for Agariste, the daughter of Clisthenes of mythology
Agastache Pleasantly-spiked, αγα-σταχυς
agastus -a -um charming, pleasing, admirable, αγαστος, αγητος
Agathelpis Good-hope, αγα-θηλπις (its natural area on the Cape)
Agathis Ball-of-twine, αγαθις (the appearance of the male strobili)
agatho- strong-, noble-, good-, αγαθος, αγαθο-
agathodaemonis -is -e of the good genius, of the noble deity, αγαθο-δαημων (association with rites), some interpret as the good dragon
Agathophytum Good-plant, αγαθος-φυτον (*vide bonus henricus*)
agathosmus -a -um strong-perfumed, pleasantly-perfumed, αγαθ-οσμη
Agave Admired-one, αγαυος, illustrious (Agave was one of the mythical Amazons) (***Agavaceae***)
agavoides resembling *Agave, Agave-oides*
ageratifolius -a -um *Ageratum*-leaved, *Ageratum-folium*
ageratoides resembling *Ageratum, Ageratum-oides*
Ageratum, ageratus -a -um Un-ageing, α-γεραιος (the flower-heads long retain their colour). Dioscorides' name, αγηρατος, was for several plants
agetus -a -um wonderful, αγητος, αγαστος
agglomeratus -a -um in a close head, congregated together, *ag-(glomero, glomerare, glomeravi, glomeratum)*
agglutinatus -a -um glued or firmly joined together, *ag-(glutino, glutinare, glutinavi, glutinatum)*
aggregatus -a -um clustered together, *ad-(grex, greg)*
agius -a -um from Agen, France (*Aginum*)
aglao- bright-, magnificent-, pompous-, delight-, proud-, αγλαια
Aglaodorum Bright-bag, αγλαια-δορος (the spathe around the inflorescence)
Aglaonema Bright-thread, αγλαια-νημα (possibly the naked male inflorescences)
agnatha without a jaw, α-γναθος
agnatus -a -um related, offspring of the father, *agnatus*
agni- lambs-wool-, *agnus, agni*
agninus of lamb (*Valerianella locusta*, the apothecaries' *lactuca agnina*, was Englished by Gerard to lamb's lettuce)
agnipilus -a -um covered with woolly-hair, *agni-pilus*
agnus-castus lamb of heaven, chaste lamb (*agnus* a lamb, αγνος also means pure, chaste, holy)
-ago -like, a feminine suffix on masculine nouns (*vir*, hero, *virago*, heroine)
agraphis -s -e without-writing, α-γραφω (≡ *non-scriptus*)
agrarius -a -um of the land, growing in fields, *ager*
agrestis -is -e rustic, barbarous, wild on arable land, *agrestis*
agri-, agro- grassy-, grass-like-, field-, meadow-, land, *ager, agri*
agricola farmer, countryman, of the fields, rustic, *agris-colo*
Agrimonia Pliny's transliteration of *argemonia.* (Cataract, αργεμον, from the medicinal use of *Papaver argemone*)
agrimonoides resembling *Agrimonia, Agrimonia-oides*
agrippinus -a -um for Marcus Vipsanius Agrippa (63–12 BC), Roman general
Agrocybe Field-cap, αγρος-κυβη (the meadow habitat of some)
agrophilus, -a -um liking grain fields, αγρος-φιλειν
Agropogon the composite generic name for hybrids between *Agrostis* and *Polypogon*

Agropyron (*Agriopyrum*) Field-wheat, αγρος-πυρος (αγριος-πυρος wild-wheat)
Agrostemma Field-garland, αγρος-στεμμα (Linnaeus' view of its suitability for garlands)
Agrosticrinum Grass-like-lily, αγρωστις-κρινον
Agrostis, agrostis -is -e Field-grass (a name, αγρωστις, used by Theophrastus for a wild, αγριος, grass)
-agrus -a -um -chase, -hunt, -capture, αγρα
ai-, aio- eternally-, always-, αιων, αει-, αι-
aianthus -a -um perpetual-flowering; everlasting-flowered, αει-ανθος
Aichryson Dioscorides' name for an *Aeonium*
Aidia Everlasting, αιδιος
ailanthifolius -a -um with *Ailanthus*-like leaves
ailanthoides resembling *Ailanthus*, *Ailanthus-oides*
Ailanthus Tree-of-heaven, from a Moluccan vernacular name, aylanto
Ainsliaea for Sir Whitelaw Ainslie (1767–1837), of the E India Company, author of *Materia Indica*
aiophyllus -a -um always in leaf, evergreen, αειφυλλος
Aiphanes Abrupt, αιφανες, (the apices of the leaflets) αιφνιδιος, sudden
Aira an old Greek name, αιρα, for a crop weed (perhaps darnel grass)
airioides resembling *Aira*, αιρα-οειδης
aitchisonii for Dr James Edward Tierney Aitchison (1836–98), botanist on the Afghan Delimitation Expedition 1884–5
Aitonia for William Townsend Aiton (1766–1849), Superintendent at Kew, succeeding his father William Aiton (1731–93)
aizoides resembling *Aizoon*, *Aizoon-oides*
aizooides resembling *Aizoon*, *Aizoon-oides*
Aizoon, aizoon Always-alive, αει-ζωος (***Aizoaceae***)
ajacis -is -e for Ajax, son of Telemon, from whose blood grew a hyacinth marked AIA
Ajania from Ajan, E Asia (*Chrysanthemum*)
ajanensis -is -e from Ajan, E Asia
Ajuga Scribonius Largus' corrupted Latin for abortifacient (in Pliny, *abigo*, to drive away)
ajugae of bugle, living on *Ajuga* (*Eriophyes*, acarine gall mite)
ajugi- *Ajuga*-, bugle-
akakiensis -is -e from Akaki, Ethiopia, or Akaki, Cyprus
akakus -a -um harmless, innocent, ακακος
akamantis -is -e from Akamas, Cyprus
akane a Japanese vernacular name
akasimontanus -a -um from Mount Akasi, Honshu, Japan
akbaitalensis -is -e from Akbaytai, Tajikistan
Akebia the Japanese name, akebi
akebioides resembling *Akebia*, *Akebia-oides*
akitensis -is -e from Akita, Honshu, Japan
akoensis -is -e from Ako, Honshu, Japan
aktauensis -is -e from Aktau, Kazakhstan
alabamensis -is -e, alabamicus -a -um from Alabama, USA
alabastrinus -a -um like alabaster or onyx, αλαβαστρος, αλαβαστρον
alacranensis -is -e from Arrecife de Alacran, Yucatan
alacriportanus -a -um from Porto Alegre, Brazil
aladaghensis -is -e from the Ala Dag range of mountains, across Asia Minor
alagebsis-is -e from the Alag river, Mindanao, Philippines
alagoanus -a -um, alagoensis -is -e from the Alagoas region of Brazil
alaicus -a -um from the Alai mountains, Tajikistan
alamosensis -is -e from Mount Alamos, Mexico, or the Los Alamos area of the southern Rocky Mountains
Alangium from an Adansonian name for an Angolan tree, some attribute it to a Malabar vernacular name, alangi (***Alangiaceae***)

alaris -is -e winged, alar, *ala, alae*
alaskanus -a-um from Alaska, N America
alatamaha from the environs of the Alatama river, Georgia, USA
alatauensis -is -e, alatavicus -a -um from the Ala Tau mountains, Turkestan/Russia
Alaternus, alaternus -a -um an old generic name for a buckthorn (≡ *Rhamnus*), resembling buckthorn's fissured bark
alatipes with winged stems, *alatus-pes*
alatocaeruleus -a -um blue-winged, *alatus-caeruleus* (stems)
alatum-planispinum winged and with flat spines, *alatus-planus-spina*
alatus -a -um, alati-, alato- wing-like (fruits), winged (stems with protruding ridges which are wider than thick), alate, *alatus*
alb-, albi-, albo- white-, *albus* (followed by and organ or indumentum suffix)
albanensis -is -a from St Albans (*Verulamium*)
albanicus -a -um from Albania
albanus -a -um from Alba Longa, Caspian area (*Albana*)
albatus -a -um turning or dressed in white, *albatus*
albellus -a -um whitish, diminutive of *albus*
albens whitening, whitish, present participle of *albesco, albescere*
albensis -is -e from the region of the river Elba
Alberta for Albertus Magnus (1193–1280) (*A. magna* is from Natal)
alberti, albertianus -a -um for Francis Albert Augustus Charles Emmanuel, Prince Consort (1819–61)
albertii for Dr Albert Regel (1845–1908), Russian plant collector in Turkestan
albertinus -a -um from the environs of Lake Albert, Uganda
albescens turning white, present participle from *albesco, albescere*
albicans being white, present participle of *albico, albicare*
albicaulis -is -e with white stems, *albi-caulis*
albicomus -a -um white-haired, *albi-(coma, comae)*
albidus -a -um, albido- white, *albidus*
albiflorus -a -um, albiflos white-flowered, *albi-floreus*
albifrons with white foliage, *albi-frondeus*
albionis -is -e from Britain, of uncertain Celtic etymology
albivenis -is -e white-veined, *albi-(vena, venae)*
Albizia (Albizzia) for Filippo degli Albizzi, Italian naturalist
albobrunneus -a -um white and brown, modern Latin *albus-brunneus*
albococcineus -a -um white and red, *albus-(coccineus, coccinus)*
albomaculatus -a -um white-spotted, white-stained, *albus-(maculo, maculare, maculavi, maculatum)*
albomarginatus -a -um white-margined, *albus-(margino, marginare)* (leaves etc)
albonigrus -a -um white and black, *albus-(niger, nigri)*
albonitens brilliant white, *albus-(nitens, nitentis)*
albopictus -a -um white-ornamented, with white markings, *albus-(pingo, pingeri, pinxi, pictum)*
albopilosus -a -um white-pilose, *albus-pilosus*
albopurpurescens white and purple coloured, *albus-purpureus-essentia*
alboroseus -a -um white and red coloured, *albus-roseus*
albosinensis -is -e white from China, *albus-sinensis* (*Betula*)
albostriatus -a -um with white stripes, *albus-(striata, striatae)*
alboviolaceus -a -um white and violet, *albus-violaceus*
albrechtii for Dr M. Albrecht, Russian naval surgeon
Albuca Whiter, *albucus*
albucifolius -a -um with *Albuca*-like leaves
albulus -a -um whitish, *albulus* (diminutive from *albus*)
albus -a -um, albi-, albo- bright, dead-white, *albus*
alcaeoides resembling Alcea, αλκαια-οειδης, *Alcea-oides*
alcalinus -a um alkaline, Latinized Middle English, alkali, from Arabic, al-kali (*Mycena alcalina* smells of ammonia)

Alcea the name, αλκαια, αλκεα, used by Dioscorides
alceifolius -a -um having leaves resembling those of *Alcea, Alcea-folium*
alceus -a -um mallow-like, resembling *Alcea*
Alchemilla from Arabic, al-kimiya, in reference either to its reputed property that dew from its leaves could transmute base metals to gold (alchemy) or to the fringed leaves of some species
alchemilloides resembling *Alchemilla, Alchemilla-oides*
alcicornis -is -e elk-horned, *alces-(cornu, cornus; cornum, corni)*
alcockianus -a -um for Sir Rutherford Alcock (1809–97), consul in China
aldabrensis -is -e, aldabricus -a -um from the Aldabra Archipelago, Indian Ocean
Alectra Unwedded or Illicit, α-λεκτρος (mostly parasitic on grasses)
Alectryon Cock, αλεκτρυων (the indumentum of silky-reddish hairs)
alepensis -is -e, aleppensis -is -e, aleppicus -a -um of Halab (Aleppo), N Syria (see *halepensis*)
Aletris Mealy, αλετρον (*Aletris farinosa* re-emphasizes its floury covering)
aletroides resembling Aletris, αλετρον-οειδης
aleur-, aleuro- mealy-, flowery-, αλευρον (surface texture)
Aleura Mealy, αλευρον (the pileus' surface texture)
aleuriatus -a -um, aleuricus -a -um mealy, floury, αλευρον
Aleurites Floury, αλευρον (the mealy covering of tung oil tree leaves)
aleuropus -a -um with meal-covered stalks, αλευρο-πους
aleutaceus -a -um purse-like, softly leathery, *alutus*
aleuticus -a -um Aleutian, from Aleutian Islands, N Pacific
alexandrae for Queen Alexandra Caroline Mary Charlotte Louisa Julia (1844–1925), wife of Edward VII
alexandrinus -a -um from Alexandria, Egypt, or other of the ancient townships named Alexandria
alfalfa the Spanish name for *Medicago sativa*, from Arabic, al-fasfasah
algarvensis -is -e from the Algarve, S Portugal
algeriensis -is -e from Algeria, N Africa
algidus -a -um of cold habitats, of high mountains, *algidus*
algoensis -is -e from Algoa Bay, Cape Province, S Africa
Alhagi the Mauritanian vernacular name for *Alhagi maurorum*
alicae for Princess Alice Maude Mary of Hesse (1843–78)
aliceara for Mrs Alice Iwanaga of Hawaii, orchid hybridist
aliciae for Miss Alice Pegler, plant collector in Transkei, S Africa
alienus -a -um different from, of others, strange, alien, *alienus*
aligerus -a -um winged, bearing wings, (*ala, alae*)-*gero*
alimaculatus -a -um with spotted wing petals, (*ala, alae*)-(*macula, maculae*)
-alis -is -e -belonging to (a noun), adjectival ending signifying of or belonging to the stem noun, e.g. *seges* a corn-field, *segetalis* of cornfields
Alisma, alisma Dioscorides' name, αλισμα, for a plantain-leaved water plant (***Alismataceae***)
alismifolius -a -um having leaves resembling those of *Alisma, Alisma-folium*
alkanet the name given to the imported dye obtained from *Alkanna tinctoria* (Spanish, alcaneta, diminutive of Arabic, al-henna)
Alkanna from the Arabic, al-henna, for *Lawsonia inermis*, the source of henna
alkekengi a name, αλκικαβον, used by Dioscorides (from Persian, al-kakunadj, or al-kakendj, for a nightshade)
Allamanda for Dr Frederick Allamand, or Jean Allamand who sent seeds of this to Linnaeus, from Brazil
allanto- sausage-, αλλας, αλλαντο-, αλλαντ-
Allantodia Sausage-like, αλλαντ-ωδης (the frond shape, (≡ *Athyrium*)
allantoides resembling a sausage, αλλαντ-οειδης
allantophyllus -a -um with sausage-shaped leaves, αλλαντο-φυλλον
Allardia, allardii for E. J. Allard of Cambridge Botanic Garden *c.*1904
allatus -a -um brought, not native, foreign, *adlatus* (*adfero, adferre, attuli, adlatum*)

alleghanensis -is -e from the Alleghany mountains, N USA
Allexis Different (as distinct from *Rinorea*)
alliaceus -a -um, allioides *Allium*-like, *alium-oides* (smelling of garlic)
Alliaria Garlic-like, *alium* (garlic-smelling)
alliariifolius -a -um *Alliaria*-leaved, *Alliaria-folium*
allionii for Carlo Allioni (1728–1804), author of *Flora Pedemontana*
Allium the ancient Latin name for garlic, *alium* (***Alliaceae***)
allo- several-, different-, other-, αλλος, αλλο-; at random, αλλως, αλλο-
Allocasuarina Different-from-*Casuarina*, botanical Latin from αλλος and *Casuarina*
allochrous -a -um varying in complexion, or changing colour, αλλο-χρως
Allosorus Random-sori, αλλος-σωρος (their shapes vary)
Alloteropsis Alien-looking, αλλοτριο-οψις (the irregular grouping of the spikelets)
alluviorus -a -um occupying alluvial habitats, living where silt is washed up, modern Latin *ad-luvio, ad-luvionis*
almus -a -um bountiful, kindly, nourishing, *almus*
alnatus -a -um *Alnus*-like
alni, alni- *Alnus*-like-, alder-like-, living on *Alnus* (gall midges)
alnicolus -a -um living with alder, *alnus-(colo, colere, colui, cultum)* (saprophytic *Pholiota alnicola*)
alnifolius -a -um *Alnus*-leaved, *Alnus-folium*
alnoides resembling *Alnus, Alnus-oides*
Alnus, alnus the ancient Latin name, *alnus, alni*, for the alder
Alocasia Distinct-from-*Colocasia*, αλλο-καλοκασια
Aloe from the Semitic, alloeh, for the medicinal properties of the dried juice, αλοη (Aloë, of Linnaeus)
aloides *Aloe*-like, *Aloe-oides*
aloifolius -a -um *Aloe*-leaved, *Aloe-folium*
aloinopsis -is -e looking like *Aloe*, αλοη-οψις
Alonsoa for Alonzo Zanoni, Spanish official in Bogotá (mask flowers)
alooides resembling *Aloe*, αλοη-οειδης
alopecuroides resembling *Alopecurus*, αλωπηξ-ουρα-οειδης
Alopecurus Fox-tail, αλωπηξ-ουρα, Theophrastus' name αλωπεκουρος
Aloysia for Queen Maria Louisa of Spain (d. 1819)
alpester -tris -tre of mountains, of the lower Alps, *alpes, alpium, alpinus*
alpicolus -a -um of high mountains, *alpes-(colo, colere, colui, cultum)*
alpigenus -a -um born of mountains, living on mountains, *alpes-genus* (*gigno, gignere, genui, genitum*)
Alpinia for Prosper Alpino (1553–1617), Italian botanist who introduced coffee and bananas to Europe
alpinoarticulatus -a -um alpine form of (*Juncus*) *articulatus*
alpinus -a -um of upland or mountainous regions, alpine, of the high Alps, *alpes*
alsaticus -a -um from Alsace, France
Alseuosmia Good-fragrance-of-the-groves, αλσος-ευοσμη
alseuosmoides resembling *Alseuosmia*, αλσος-ευοσμη-οειδης
alsinastrus -a -um resembling *Alsine*, chickweed-like, *Alsine-astrum*
Alsine, alsine a name, αλσινη, used by Theophrastus for a chickweed-like plant (αλσος a grove)
alsinifolius -a -um with *Alsine*-like leaves, chickweed-leaved, *Alsine-folium*
alsinoides chickweed-like, *Alsine-oides*
alsius -a -um of cold habitats, *alsius* (*algeo, algere, alsi*)
also- leafy glade-, of groves-, αλσος, αλσο-
alsodes of woodland, of sacred groves, αλσος-ωδης
Alsophila, alsophilus -a -um Grove-loving, αλσος-φιλεω
Alstonia for Professor Charles Alston (1685–1760), of Edinburgh
alstonii for Captain E. Alston (fl. 1891), collector of succulents in Ceres, S Africa
Alstroemeria for Baron Claus Alströmer (1736–94), Swedish botanist, friend of Linnaeus (Peruvian lilies) (***Alstroemeriaceae***)

altaclerensis -is -e from Highclere, Hampshire (*Alta Clara*), or High Clere Nursery, Ireland
altaicus -a -um, altaiensis -is -e from the Altai mountains of Central Asia
altamahus -a -um from the Altamaha River, Georgia, USA
alte-, alti-, alto- tall, high, *altus*
alternans alternating, present participle of *alterno, alternare, alternavi, alternatum*
Alternanthera Alternating-stamens, *alter-ananthera* (alternate ones are barren)
alternatus -a -um alternating, *alternatus* (phyllotaxy)
alterni-, alternus -a -um alternating on opposite sides, alternate, every other-, *alter, alternus*
alternifolius -a -um with alternate leaves, *alternus-folium*
Althaea (*Althea*) Healer, αλθαινω, a name, αλθαια, used by Theophrastus
althaeoides resembling *Althaea, Althaea-oides*
alticaulis -is -e having tall stems, *altus-caulis*
alticolus -a -um inhabiting high places, *alti-colo*
altifrons tall-canopied, having high leafy growth, *altus-*(*frondeo, frondere*)
altilis -is -e fat, large, nutritious, nourishing, *alo, alere, alui, altum* (*alitum*)
altis -is -e above, on high, from afar, tall, *alte*
altissimus -a -um the tallest (e.g. species of the genus), superlative of *altus*
altus -a -um tall, high, *altus*
alulatus -a -um with narrow wings, diminutive of *alatus*
alumnus -a -um well nourished, flourishing, fostered, *alumnus*
alutaceus -a -um of the texture of soft leather, *alutus*
alveatus -a -um excavated, hollowed, trough-like, *alveus, alvei*
alveolatus -a -um with shallow pits, honeycombed, alveolar, *alveolus, alveoli*
alvernensis -is -e from the Auvergne, France (*Arverni, Arvernus*)
Alyogyne Not-loosening-ovary, α-λυω-γυνη (indehiscent)
alypum a former synonym for *Globularia*
alyssifolius -a -um with leaves resembling those of *Alyssum, Alyssum-folium*
Alyssoides. alyssoides resembling *Alyssum, Alyssum-oides*
Alyssum Pacifier, α-λυσσα (an ancient Greek name, αλυσσος without-fury)
ama-, am- jointly-, together-, αμα
amabilis -is -e pleasing, likeable, lovely, *amo, amare, amavi, amatum*
amada from the Indian vernacular name for *Curcuma amada*
amadelphus -a -um gregarious, αμ-αδελφος
amagianus -a -um from the Amagi mountain, Kyushu, Japan
Amana Japanese vernacular name for *A. edulis*
Amanita Affectionate, αμανιται, *amans, amantis* (attractive but toxic fly-agaric, death-cap and destroying-angel fungi)
Amanitopsis resembling Amanita, αμανιτ-οψις
amanus -a -um from Amman, Jordan or Akmadagh–Amani mountains, N Syria, or Amanus mountain of S Turkey
amaranthoides resembling *Amaranthus, Amaranthus-oides*
Amaranthus (Amarantus) Unfading, α-μαραινω (Nicander's name, αμαραντον, for the 'everlasting' flowers) (***Amaranthaceae***)
amaranticolor purple, *Amaranthus*-coloured
amarantinus -a -um not fading, αμαρανθινος
amaraliocarpus -a -um with fruits resembling those of *Amaralia,* botanical Latin from *Amaralia* and καρπος
Amarcrinum the composite name for hybrids between *Amaryllis* and *Crinum*
amarellus -a -um bitter tasted, diminutive of *amarus* (the Amarelle cherries are distinguished as red or yellow, with clear juice. Morello cherries are black with coloured juice)
Amarine the composite generic name for hybrids between *Amaryllis* and *Nerine*
amarissimus -a -um most bitter tasted, superlative of *amarus*
amarus -a -um bitter, *amarus* (as in the amaras or bitters of the drinks industry, e.g. *Quassia amara,* cognate with amarella and morello)

Amarygia the composite generic name for hybrids between *Amaryllis* and *Brunsvigia*
Amaryllis the name of a country girl in Virgil's writings (***Amaryllidaceae***)
amatolae of the Amatola mountains of S Africa
Amauriella Indifferent, diminutive from αμαυρος (stemless with short inflorescence)
amauro- feeble, indifferent, gloomy, dark, αμαυρος, αμαυρο-
amaurollepidus -a -um having dark scales, dark-bracted, αμαυρο-λεπιδος
amaurus -a -um dark, without lustre, feeble, indifferent, αμαυρος
amazonicus -a -um from the Amazon basin, S America
amb-, ambi- around-, both-, *ambio, ambire, ambii, ambitum; ambi-*
ambianensis -is -e from Amiens, France (*Ambianum*)
ambigens doubtful, of uncertain relationship, *ambi-(genus, generis)*
ambiguus -a -um of ambiguous relationship, *ambigo, ambigere*
ambleocarpus -a -um with blunt (tipped) carpels, αμβλυς-καρπος
ambly- blunt-, αμβλυς, αμβλυ-
amblyandrus -a -um having blunt anthers on the stamens, αμβλυς-(ανηρ, ανδρος)
amblyanthus -a -um feeble-flowering, blunt-flowered, αμβλυς-ανθος
amblycalyx with a blunt calyx, αμβλυς-καλυξ
amblygonus -a -um blunt-angled, αμβλυς-γωνια
amblyodon, amblyodontus -a -um blunt-toothed, αμβλυς-οδων
Amblyopetalum Blunt-petalled-one, αμβλυς-πεταλον
amblyotis -is -e with blunt, or weak ears, αμβλυς-ωτος (lobes)
amboinensis -is -e (amboynensis), amboinicus -a -um from Ambon (*Amboina*), Indonesia
ambovombensis -is -e from Ambovombe, Madagascar
Ambrosia Elixir-of-life, Dioscorides' name, αμβροσια, for *Ambrosia maritima* (divine food, food of the gods, immortality)
ambrosiacus -a -um *Ambrosia*-like, similar to *Ambrosia*
Ambrosina diminutive of *Ambrosia*
ambrosioides ambrosia-like, αμβροσια-οειδης
amecaensis -is -e from Ameca, Mexico
amecamecanus -a -um from Amecameca, Mexico
Amelanchier a Provençal name, amelancier, for *A. ovalis* (snowy-*Mespilus*)
Amelasorbus the composite generic name for hybrids between *Amelanchier* and *Sorbus*
amelloides resembling *Amellus, Amellus-oides*
Amellus, amellus a name used by Virgil for a blue-flowered composite from the River Mella, near Mantua, Italy
amentaceus -a -um having catkins, of-catkins, *amentum, amenti* (*Ciboria amentacea* grows on fallen alder and willow catkins)
amenti- catkin-, *amentum, amenti* (literally a strap used to impart spin when throwing a javelin)
americanus -a -um from the Americas, American
amesianus -a -um for Frederick Lothrop Ames (1835–93), American orchidologist, or for Professor Oakes Ames (1874–1950) of Harvard Botanic Garden, orchidologist
amethystea, amethystinus -a -um the colour of amethyst gems, violet, αμεθυστος
amethystoglossus -a -um amethyst-tongued, αμεθυστος-γλωσσα (*Cattleya*)
amianthinus -a -um violet on top or upwards, (αμ, ανα)-ιανθινος
Amicia for Jean Baptiste Amici (1786–1863), Italian physicist
amicorum of the Friendly Isles, Tonga (*amicus, amici,* friendly)
amictus -a -um clad, clothed, *amicio, amicire, amictus*
amiculatus -a -um cloaked, mantled, with a cloak, *amiculum, amiculi*
Ammi Sand, a name, αμμη, used by Dioscorides for *Carum copticum* and reapplied by Linnaeus
ammo- sand-, αμμος, αμμη; ψαμμος

Ammobium Sand-dweller, αμμο-βιο
Ammocalamagrostis the compound name for hybrids between *Ammophila* and *Calamagrostis*
Ammocharis Sand-beauty, αμμο-χαριεις (habitat)
ammodendron tree of the sand, αμμο-δενδρον (habitat)
Ammoides, ammoides resembling Ammi, αμμη-οειδης
ammoniacum gum ammoniac, αμμονιακος, of Ammon (ammonia was first noted at the temple of Ammon, Siwa, Egypt, and its modern Latin name was given in the eighteenth century) an old generic name for *Dorema ammoniacum*
Ammophila Sand-lover, αμμος-φιλος
ammophillus -a -um sand-loving, αμμος-φιλος (the habitat)
amnicolus -a -um growing by a river, *amnis-colo*
Amoeba Changing-one, αμοιβη (having no fixed shape)
amoenolens delightfully scented, *amoenus-olens*
amoenulus -a -um quite pleasing or pretty (diminutive of *amoenus*)
amoenus -a -um charming, delightful, pleasing, *amoenus*
Amomum, amomum Purifier, α-μωμος (probably from an Arabic name, the Indian spice plant *Amomum* was used to cure poisoning)
amorginus -a -um from the Greek Amorgos islands
Amorpha Deformed-one, α-μορφη (flowers of the genus of greyish-downy lead plant *Amorpha canescens* lack wing and keel petals)
Amorphophallus Deformed-phallus, αμορφος-φαλλος (the enlarged spadix)
amorphus -a -um, amorpho- deformed (α-μορφη, shapeless, without form
ampelas having the habit of a vine, αμπελος
ampelo- wine-, vine-, grape-, αμπελος
Ampelodesmos (*Ampelodesma*) Wine-cable, αμπελο-δεσμος
ampeloprasum leek of the vineyard, a name, αμπελοπρασσον, in Dioscorides
Ampelopsis Vine-resembling, αμπελο-οψις (***Ampelidaceae*** ≡ ***Vitaceae***)
Ampelopteris Vine-fern, αμπελο-πτερυξ (the scrambling habit)
amphi-, ampho- on-both-sides, in-two-ways-, both-, double-, of-both-kinds-, around-, αμφις, αμφι-, αμφοτερος, αμφο-
amphibius -a -um with a double life, growing both on land and in water, αμφι-βιος
amphibolus -a -um fired at from all sides, ambiguous, doubtful, αμφι-βολος (αμφι-βολια doubt)
amphicarpos with curved pods, αμφι-καρπος
Amphicome Haired-about, αμφι-κομη (the seeds have hair tufts at each end) (≡ *Incarvillea*)
amphidoxa of all-round glory, αμφι-δοξα (seasonal flower and foliage colouring)
amphioxys lancet-like, tapered to each end, sharp all round, αμφι-οξυς
amphoratus -a -um amphora-shaped, αμφι-φορευς
Amphorella Small-wine-jar, diminutive of αμφορευς
amplectens stem-clasping (leaf bases), *amplector, amplecti, amplexus* to embrace or encircle
amplex-, amplexi- loving-, embracing-, *amplexor, amplexare, amplexatus*
amplexans twisting together, surrounding, embracing, αμ-πλεκτος
amplexicaulis -is -e embracing the stem, αμπλεκτος-καυλος (e.g. the base of the leaf, see Fig. 6d)
amplexifolius -a -um leaf-clasping, *amplexus-folium*
ampli- large-, double-, *amplus*
ampliatus -a -um enlarged, *amplio, ampliare, ampliavi, ampliatum*
ampliceps large-headed, *amplus-ceps*, some interpret as clasped head, αμπλι-κεφαλη
amplissimus -a -um very large, the biggest, superlative of *amplus*
amplus -a -um eminent, large, abundant, *amplus*
ampullaceus -a -um, ampullaris -is -e bottle-shaped, flask-shaped, *ampulla*
ampulli- bottle-, *ampulla, ampullae*
Amsinckia for W. Amsinck (1752–1831), of Hamburgh

Amsonia for Charles Amson, eighteenth-century Virginian physician and traveller in America
amurensis -is -e, amuricus -a -um from the region of the Amur river, eastern Siberia
amydros indistinct, dim, αμυδρος
amygdalifolius -a -um almond-leaved, *Amygdalus-folium*
amygdalinus -a -um almond-like, kernel-like, of almonds, αμυγδαλινος
amygdaloides almond-like, αμυγδαλος-οειδης
amygdalopersicus -a -um Persian almond, αμυγδαλος-περσικος
amygdalus the Greek name, αμυγδαλος, for the almond-tree (from Hebrew, megdh-el, sacred-fruit)
amylaceus -a -um starchy, floury, αμυλον
an-, ana- upon-, without-, backwards-, above-, again-, upwards-, up-, ανα-, αν-
Anabasis Without-pedestal, ανα-βασις (has no gynophore)
anabasis -is -e going upwards, climbing, ανα-(βασις, βασεως)
Anacampseros Love-returning, ανα-καμπτω-ερος (a love charm)
Anacamptis Bent-back, ανα-καμπτω (the long spur of the flower)
anacanthus -a -um lacking thorns, αν-ακανθα
Anacardium Heart-above, ανα-καρδια (Linnaeus' name refers to the shape of the false-fruit) (***Anacardiaceae***)
Anacharis Without-charm, ανα-χαρις
anachoretus -a -um not in chorus, growing in seclusion, ανα-(χορος, χορητος)
Anacylus Lacking-a-circle, ανα-κυκλος (the arrangement of the outer florets on the disc)
Anadelphia Without brothers or sisters, αν-αδελφος (racemes lack homogamous pairs of spikelets)
anagallidifolius -a -um *Anagallis*-leaved, *Anagallis-folium*
Anagallis Unpretentious, ανα-αγαλλω, or Delighting, αναγελαω
anagallis-aquatica water-*Anagallis (Veronica), Anagallis-aquaticus*
Anagyris Backward-turned, ανα-γυρος (the curved pods)
anagyroides resembling *Anagyris, Anagyris-oides* curved backwards
Ananas probably from a Tupi-Guarani vernacular name, nana, anana, ananas
ananassus -a -um small-lipped, pineapple-like, *ananas* (the fruiting receptacle)
anandrius -a -um unmanly, lacking stamens, ανανδρος
ananta not-direct, uphill, endless, αν-αντα, αν-αντης
ananthocladus -a -um having non-flowering shoots, αν-ανθο-κλαδος
Anaphalis Greek name for an immortelle, derivation obscure
anaphysemus -a -um turned-back-bladder, ανα-φυσα (the swollen tip of the curved spur), or without a bladder
Anastatica, anastaticus -a -um Resurrection, αναστασις (*Anastatica hierochuntica*, resurrection plant or rose of Jericho)
anastomans intertwining, anastomozing, αναστομαω (forming 'mouths')
anastreptus -a -um twisted-backwards, curved-backwards, ανα-στρεπτος
anatinus -a -um healthy, rewarding, ανα-τινω
anatolicus -a -um from Anatolia, Turkish
anatomicus -a -um skeletal, cut-up, ανα-τεμνειν (leaves)
anceps doubtful, dangerous, two-edged, two-headed, *anceps, ancipitis* (stems flattened to form two edges)
Anchomanes a name used by Dioscorides for another arum (the stems are prickly, αγκυρα-μανια)
anchoriferus -a -um bearing flanges, anchor-like, poor Latin *anchora-fero*, from Greek αγκυρα-φερω
Anchusa Strangler, αγχω, or Close, αγχου (Aristophanes' name, αγχουσα, εγχουσα, formerly for an alkanet, yielding a red dye)
anchusifolius -a -um *Anchusa*-leaved, *Anchusa-folium*
anchusoides *Anchusa*-like, αγχουσα-οειδης
ancistro- fish-hook-, αγκιστρον
Ancistrocheilus Fish-hook-lip, αγκιστρο-χειλος (the deflexed lip)

Ancistrorhynchus Fish-hook-beak, αγκιστρο-ρυγχος (the shape of the pollinarium)
Ancistrophyllum Fish-hook-leaved, αγκιστρο-φυλλον (the leaf rachis terminates in hooked spines), or Quick-changing-leaf, the various leaflets and spines on the rachis, αγχι-στροφο-φυλλον
Ancistus Barbed-one, αγκιστριον
ancylo- hooked-, curved-, αγκυλος, αγκυλο-
ancyrensis-is -e from Ankara (*Ancyra*), Turkey
andalgalensis -is -e from Andalgal, Argentina
andaminus -a -um from the Andaman Islands, SE Bay of Bengal
andegavensis -is -e from Angers in Anjou, France (*Andegava*)
Andersonia for William Anderson (1750–78), botanist on Cook's second and third voyages
andersonii for Thomas Anderson (1832–70), botanist in Bengal, or for J. Anderson (fl. 1909), who collected in the Gold Coast (Ghana), or Messrs Anderson, patrons of botany
andesicolus -a -um from the S American Andes cordillera, *Andes-colo*
andicolus -a -um from the Colombian Andean cordillera, *Andes-colo*
andigitrensis -is -e from the Andigitra mountains, Madagascar
andinus -a -um from the high Chilean Andes
Andira, andina from the Brazilian vernacular name
andongensis -is -e from Andonga, NW Angola
Andrachne the ancient Greek name, ανδραχνε (for an evergreen shrub)
andrachnoides resembling *Andrachne*, ανδραχνε-οειδης
andreanus -a -um for E. F. André (1840–1911) Parisian landscape gardener
andrewsianus -a -um, andrewsii for H. C. Andrews, early nineteenth-century botanical writer
andrieuxii for G. Andrieux, plant collector in Mexico
andro-, -andrus -a -um male, man, stamened-, anthered-, ανηρ, ανδρ, ανδρος, ανδρο-
Androcymbium Male-cup, *andro-cymbium* (the petal-limbs enfold the stamens)
androgynus -a -um with staminate and pistillate flowers on the same head, hermaphrodite, ανηρ-γυνη, *androgynus, androgyni*
Andromeda after Andromeda, the daughter of Cepheus and Cassiope, whom Perseus rescued from the sea monster
Andropogon Bearded-male, ανδρο-πωγων (awned male spikelet)
Androrchis Male-testicle, ανδρ-ορχις, *Orchis mascula* or man orchid
Androsace Man-shield, ανδρο-σακος (the exposed stamens of heterostyled spp.)
androsaceus -a -um *Androsace*-like (the pink and brown pileus of the horse hair fungus)
androsaemifolius -a -um *Androsaemum*-leaved, *Androsaemum-folium*
Androsaemum, androsaemum Man's-blood, ανδρος-αιμα (Dioscorides' name, ανδροσαιμον, for the blood-coloured juice of the berries)
andrus -a -um -stamened, ανηρ, ανδρ-
Andryala etymology uncertain
anegadensis -is -e from Anegada Island, Puerto Rica
Aneilema Without-a-cover, α(ν)-ειλυμα (the absence of an involucre)
aneilematophyllus -a -um having leaves resembling those of *Aneilema*
Anemarrhena Exposed-males, ανεμ-αρρην (the exposed stamens)
Anemia (Aneimia) Naked, ανειμων (the sori have no indusia)
Anemiopsis Naked-looking, ανειμων-οψις (the sparsely leaved stems)
Anemone a name used by Theophrastus. Possibly a corruption of Naaman, a Semitic name for Adonis, from whose blood sprung the crimson-flowered *Anemone coronaria*
Anemonella Little-*Anemone*-resembling, diminutive termination
anemones of or upon *Anemone* (*Urocystis* smut fungus)
anemoniflorus -a -um *Anemone*-flowered
anemonifolius -a -um *Anemone*-leaved, *Anemone-folium*
anemonoides *Anemone*-like, *Amenone-oides*

Anemopaegma Wind-sportive, ανεμος-παιγνημων (παιγνια, sport)
anethiodorus -a -um *Anethum*-scented, smelling of dill, *Anethum-odorus*
Anethum Undesireable, ανεθελητος, an ancient Greek name
anfractifolius -a -um having twisted leaves, *anfractus-folium*
anfractus -a -um, anfractuosus -a -um twisted, twining, bent, winding, *anfractus*
Angelica Angel, the name, *herba angelica*, in Matthaeus Sylvaticus (healing powers, see *Archangelica*)
Angelonia the South American vernacular name, angelon, for one species
angio- urn-, vessel-, enclosed-, (αγγειον, αγγος vessel, receptacle, urn)
angiocarpus -a -um enclosed fruit, αγγειο-καρπος (the perianth segments fuse to the fruit)
Angiopteris Winged-vessel, αγγειον-πτερυξ (the aggregated sporangia)
anglicus -a -um, anglicorum English, of the English, *Anglia, Anglicus*
anglorum of the English people, *Angles*
angolanus -a -um, angolensis -is -e from Angola, W Africa
angraecoides resembling *Angraecum, Angraecum-oides*
Angraecopsis *Angraecum*-like, *Angraecum*-οψις
Angraecum a Malayan name, angurek, for epiphytes
angui-, anguinus -a -um serpentine, *anguis, anguis*, a serpent; eel-like, wavy, *anguilla, anguillae,* an eel
anguiceps snake-headed, *anguis-ceps* (floral structure)
anguifugus -a -um snake-banishing, *angui-fugus*
Anguillulina Minute-eel, diminutive of *anguilla* (an eelworm causing galls on about 330 plant species)
angularis -is -e angular, *angulus, anguli*
angulatus -a -um somewhat angled, *angulatus*
anguligerus -a -um hooked, having hooks, *angulus-gero*
angulosus -a -um having angles, angular, *angulus, anguli*
Anguria a Greek name for a cucumber, αγγυρον
angustatus -a -um somewhat narrow, *angustus, angusti*
angusti-, angustus -a -um narrow, *angustus, angusti-*
angustiflorus -a -um narrow-flowered, *angusti-flora*
angustifolius -a -um narrow-leaved, *angusti-folium*
angustior narrower, comparative of *angustus*
angustisectus -a -um narrowly divided, *angusti-*(*seco, secare, secui, sectum*) (leaves)
angustissimus -a -um the most narrow, superlative of *angustus*
anhweiensis -is -e from Anhui province, China
Anigosanthus, Anigosanthos Open-flower, ανοιγος-ανθος (the expanding inflorescence stalks)
anis-, ani- not equal-, unequal-, dissimilar-, αν-ισος
anisandrus -a -um having unequal stamens, αν-ισος-ανηρ
Anisanthus (Anisantha) Unequal-flower, ανισος-ανθος (flowers vary in their sexuality)
anisatus -a -um Anise-scented, from *Anisum*
Aniseia Different, ανισος
aniso- unequally-, unequal-, uneven-, ανισος, anise-, dill-, ανισον *(anisum)*
Anisochylus Unequal-lipped, ανισος-χειλος
Anisodontia Unequal-toothed, ανισος-οδοντος
anisodorus -a -um Anise-fragrant, *anisum-odoro*
Anisodus Unequal-toothed, ανισος-οδους
Anisopappus Unequal-pappus, *an-iso-pappus*
Anisophyllea Unequal-leaved, ανισο-φυλλον (the pairs of large, maturing, and small, transient, leaves)
anisophyllus -a -um with leaves that are oblique at the base, αν-ισος-φυλλον
Anisopus Unequal-stalked, α(ν)-ισο-πους (the nodal inflorescences)
Anisosorus Differing-sori, ανισο-σορος (some are straight and others lunate)
Anisostichus Unequally lined, ανισος-στιχος

Anisotes Inequality, αν-ισοτης
Anisotome Unequally cut, ανισο-τομη (the divisions of the leaves)
anisum aniseed, ανισον (an old generic name, ανισον, for dill or anise)
ankylo- crooked-, αγκυλος
annae for the Roman goddess *Anna Perenne*
annamensis -is -e from Anam, Vietnam
annectens fastening upon, binding to, present participle of *an-(necto, nectere, nexi; nexui, nexum)*
Annona (Anona) from the Haitian vernacular name, menona (***Annonaceae***)
annosus -a -um long-lived, aged, *annosus* (parasitic and growing for several years)
annotinus -a -um one year old, last year's, *annotinus* (with distinct annual increments)
annularis -is -e, annulatus -a -um ring-shaped, having rings, *annulus, annuli* (markings)
annuus -a -um one year's, annual, *annuus*
ano- upwards-, up-, ανω, towards the top-
Anoda Without-joint, *a-nodus* (pedicel feature distinguishes from *Sida*)
Anodia Impervious, ανοδος
anodontus -a -um with outwards-pointing teeth, ανω-(οδους, οδοντος)
Anogeissus Towards-the-top-tiled, ανω-γεισσον (the scale-like fruiting heads)
Anogramma Towards-the-top-lined, ανω-γραμμη (the sori first mature towards the tips of the pinnae)
anomalus -a -um unlike its allies, out of the ordinary, anomalous, αν-ομαλος
Anomatheca unequal-boxes, ανωμα-θηκη (anther structure) (≡ *Lapeyrousia*)
anomocarpus -a -um having variable fruit shapes, ανωμοιος-καρπος
Anomochloa Lawless-grass, ανομος-χλοη
Anonidium Like-*Annona*
Anoplobatus Unarmed-thornbush, ανοπλος-βατος
Anoplophytum Unarmed-plant, ανοπλος-φυτον (≡ *Tillandsia*)
Anopterus Winged-upwards, ανω-πτερον (the seeds)
anopetalus -a -um erect-petalled, ανω-πεταλον
anoplo- unarmed-, thornless-, ανοπλος
Anoplophytum Unarmed-plant, ανοπλος-φυτον (≡ *Tillandsia*)
Anopyxis Upright-capsule, ανω-πυξις (the fruit is held upright until it dehisces)
anosmus -a -um without fragrance, scentless, αν-οσμη
ansatus -a -um having a handle, *ansa, ansae*
Ansellia for Mr Ansell, collector for RHS Chiswick on the ill-fated Niger Expedition of 1841
anserinifolius -a -um with leaves similar to those of *Potentilla anserina* (which Linnaeus called goose-weed, Gåsört), *anserina-folium*
anserinoides *anserina*-like, *anserina-oides*
anserinus -a -um of goose greens, of the goose, *anser, anseris*
ansiferus -a -um bearing a handle, *ansa-fero* (petiolate or petiolulate); some interpret as sword-bearing
antalyensis -is -e from Antalya, SW Turkey
antanambensis -is -e from Antanamba, Madagascar
antarcticus -a -um from the Antarctic continent, αντ-αρκτικος (opposite to the north)
ante- before-, *ante, antea*
Antennaria, antennaria Feeler, *antenna, antennae* (literally, projecting like a boat's yard-arm, the hairs of the pappus)
antenniferus -a -um bearing antennae, with 'feeler-like' stamens, *antenna-(fero, ferre, tuli, latum)*
anthelminthicus -a -um vermifuge, worm expelling, αντι-(ελμινς, ελμινθος)
anthemi- *Anthemis*-, chamomile-, ανθεμοεις
anthemifolius -a -um having leaves resembling those of *Anthemis*
Anthemis Flowery, ανθος (name, ανθεμις, used by Dioscorides for a plant also called χαμαιμηλον and λευκανθεμον)

anthemoides Anthemis-like, ανθεμις-οειδης
-anthemus -a -um, -anthes -flowered, ανθεμις
Anthericum, anthericum from Theophrastus' name, ανθερικος, used by Linnaeus for an asphodel (St Bernard's lily), originally an ear of corn or a stalk
antherotes brilliant-looking, ανθεω-ωτης
-antherus -a -um -flowering, ανθηρος, ανθηρα (in the botanical sense, -stamens, or -anthered)
-anthes, -anthus -a -um -flowered, ανθος
anthiodorus -a -um fragrant-flowered, *anthus-odorus*
antho- flower-, ανθος, ανθο-
Anthocercis Rayed-flower, ανθος-κερκις
Anthoceros Flower-horn, ανθο-κερας (the conspicuously elongate, dark-brown, bivalved capsules)
Anthocleista Closed-flower, ανθο-κλειστος (the flower-shape)
anthocrene fountain of flowers, ανθο-κρηνη
Antholyza Rage-flower, ανθο-λυσσα (resemblance of flower to animal's maw)
anthomaniacus -a -um frenzied-flowering, ανθος-μανιακος
Anthophorus Flower-bearing, ανθο-φορα (the clustered spikelets fall together)
anthopogon bearded-flowered, ανθος-πωγων (*Rhododendron anthopogon*'s weird flowers)
anthora resembling *Ranunculus thora* in poisonous properties, *an-thora*
Anthospermum Flower-seed, ανθο-σπερμα (dioecious)
anthosphaerus -a -um globe-flowered, ανθο-σφαιρα
Anthostema Floral-crown, ανθο-στεμμα (the heads of flowers)
Anthoxanthum Yellow-flower, ανθος-ξανθος (the spikelets at anthesis)
anthracinus -a -um black-as coal, ανθραξ, ανθρακος
anthracophilus -a -um growing on burnt wood, liking living-coal, (ανθραξ, ανθρακος)-φιλεω
Anthriscus from a Greek name, αθρυσκον, for another umbellifer (chervil)
anthropophagorus -a -um of the man-eaters, ανθρωπο-φαγεω (Cannibal tomato)
anthropophorus -a -um man-bearing, ανθρωπο-φορα (flowers of the man orchid)
Anthurium Flower-tail, ανθ-ουρα (the tail-like spadix)
-anthus -a -um -flowered, ανθος
anthyllidifolius -a -um with *Anthyllis*-like leaves, *Anthyllis-folium*; having leaves tubercled above, αν-τυλη-*folium*
Anthyllis Downy-flower, ανθ-υλλις (the name used by Dioscorides, for the calyx hairs)
anti- against-, opposite-, opposite-to-, for, like-, false-, αντι-, *anti-*
Antiaris Against-association (the Javan vernacular name, antja, for the upas tree, *Antiaris toxicaria* (George Stevens (1736–1800) fabricated the upas tree's reputed ability to cause the death of anyone who sleeps beneath it. Malayan, upas, poison)
anticarius -a -um from the area around Caria, *anti-Caria*; from Antequera, Andalusia, S Spain
Anticharis Lacking-charm, αντι-χαριεις (formerly *Doratanthera*, leathery flowered)
antichorus -a -um distinctive, standing apart from the throng, αντι-χορος
anticus -a -um turned inwards towards the axis, in front, *anticus* (antonym, *posticus*)
Antidesma Against-a-band, αντι-δεσμος
antidysentericus -a -um against dysentery, αντι-δυσ-εντερια (use in medical treatment)
Antigonon Opposite-angled, αντι-γονον (the zig-zag stems of coral vine)
antillarus -a -um from either the Greater or Lesser Antilles, West Indies
antioquiensis -is -e from Antioquia, Colombia
antipodus -a -um from the Antipodes, αντιποδης (from the other side of the world, literally, with the feet opposite)
antipolitanus -a -um from the Antibes (*Antipolis*)

antipyreticus -a -um against fire, *anti-(pyra, pyrae)* (the moss *Fontinalis antipyretica* was packed around chimneys to prevent thatch from igniting); others interpret as against fevers
antiquorum of the ancients, of the former, old-world, *antiquus, anticus*
antiquus -a -um traditional, ancient, *antiquus*
antirrhiniflorus -a -um *Antirrhinum*-flowered, *Antirrhinum-florum*
antirrhinoides resembling *Antirrhinum*, αντι-ρινος-οειδης, *Antirrhinum-oides*
Antirrhinum Nose-like, αντι-(ρις, ρινος) (a name, αντιρρινον, used by Dioscorides)
Antrophyum Upwards-brooding, αν-τρωφη (the creeping rhizomes)
antrorsus -a -um forward or upward facing, *antero-versus*
Anubias for Anubis, son of Nephthytis and Typhonis
anulatus -a -um with rings (*anulus, anuli*) on, ringed, *anulatus*
-anus -a -um suffix to imply -having, -belonging to, -connected with, -from
anvegadensis -is -e see *andegavensis*
anvilensis -is -e from Anvil Creek area on the Seward Peninsula, Alaska, USA
ap- without-, up-, before-, απο; towards-, to-, *ap-*
Apargia Of-neglect, απο-αργια (old meadowland provenance, ≡ *Leontodon*)
aparine a name, απαρινη, used by Theophrastus for goosegrass (clinging, seizing)
apenninus -a -um (appenina) of the Italian Apennines
Apera a meaningless name used by Adanson
aperantus -a -um open-flowered. *aper-anthus*
aperti-, apertus -a -um open, bare, naked, past participle of *aperio, aperire, aperui, apertum*
apetalus -a -um without petals, α-πεταλον
aphaca a name, αφακη, used in Pliny for a lentil-like plant, φακος
aphan-, aphano- unseen-, inconspicuous-, not-seen, α-φανερος, αφαντος, αφαν-
aphanactis resembling the Andean genus *Aphanactis* (*Erigeron aphanactis*)
Aphananthe Inconspicuous-flower, αφαν-ανθος
Aphanes Inconspicuous, unnoticed, αφανης,
Aphelandra Simple-male, αφελης-ανηρ
aphelandraeflorus -a -um with flowers similar to those of *Aphelandra*
aphelandroides resembling *Aphelandra*, αφελης-ανηρ-οειδης, *Aphelandra-oides*
aphthosus -a -um with an ulcerated or tubercled throat, αφθαι (suggesting thrush)
Aphyllanthes, aphyllanthes Leafless-flower, α-φυλλον-ανθος (flowers on rush-like stems); with apetalous flowers,
aphyllus -a -um without leaves, leafless, α-φυλλον (perhaps at flowering time)
apianus -a -um of bees, liked by bees, *apis*
apiatus -a -um bee-like, spotted, *api-atus*
apicatus -a -um with a pointed tip, capped, *apex, apicis*
apiculatus -a -um with a small broad point at the tip, apiculate, diminutive of *apex* (see Fig. 7e) (*apicula*, a small bee)
apifer -era -erum bee-like, bee-bearing, *apis-fero* (flowers of the bee orchid), bee-flowered
apii- parsley-, *Apium-*
apiifolius -a -um celery-leaved, *Apium*-leaved, *Apium-folium*
Apios Pear(-rooted), απιον
Apium a name, *apium*, used in Pliny for celery-like plants (απιον in Dioscorides). Some relate it to the Celtic 'apon', water, as its preferred habitat (***Apiaceae*** ≡ ***Umbelliferae***)
apo- up-, without-, free-, from-, απο
Apocynum Against-dogs, απο-(κυων, κυνος), Dioscorides' name, αποκυνον, for *Cionura oreophila* and redefined by Linnaeus (Dog's-bane *Apocynum venetum* is supposed to be poisonous to dogs) (***Apocynaceae***)
apodectus -a -um acceptable, welcome, απο-δεκτος
apodus -a -um without a foot, stalkless, α-(πους, ποδος)
Aponogeton Without-trouble-neighbour, απονος-γειτων (analogy with *Potamogeton*) (some derive it from the Celtic, apon) (water hawthorn) (***Aponogetonaceae***)

aporo- without means of achieving-, without thoroughfare-, α-πορος
appendiculatus -a -um with appendages, appendaged, *appendicula, appendiculae*
applanatus -a -um flattened out, *ap-planus*
applicatus -a -um enfolded, placed close together, *ap-(plico, plicare, plicavi (plicui), plicatum (plicitum))*
appressus -a -um lying close together, adpressed, *ad-(presso, pressare)*
appropinquatus -a -um near, approaching, *appropinquo, appropinquare, appropinquavi, appropinquatum* (resemblance to another species)
approximans drawn close together but not united, to the nearest, *ad-proximus*
approximatus -a -um near together, *ad-proximus*
apricus -a -um sun-loving, of exposed places, *apricus* (cognate with *praecox*)
aprilis -is -e of April, *Aprilis* (the flowering season)
Aptenia Wingless, α-(πτην, πτηνος) (the capsules)
apterus -a -um without wings, wingless, α-πτερον
Aptosimum Not-falling, α-(πτωμα, πτωσις, πτωσιμος) (persistent capsules)
apulus -a -um from Apulia, S Italy
apus lacking a stalk, α-πους
aquaticus -a -um living in water, *aquaticus*
aquatilis -is -e living under water, *aquatilis*
Aquifolium Thorny-leaved, *acus-folium* (a former generic name for *Ilex*) (***Aquifoliaceae***)
aquifolius -a -um holly-leaved, with pointed leaves, spiny-leaved, *acus-folium*
Aquilegia Eagle, *aquila* (claw-like nectaries) or from medieval German Acheleia, Akelei
aquilegifolius -a -um with *Aquilegia*-like leaves, *Aquilegia-folium*
aquilinus -a -um of eagles, eagle-like (Linnaeus noted that in the obliquely cut rhizome of *Pteridium* the appearance of the vasculature *'refert aliquatenus aquilam imperialis'*, suggests to some extent the imperial eagle)
aquilus -a -um blackish-brown, swarthy, *aquilus*
aquosus -a -um rainy, humid, damp, *aquosus*
arabicus -a -um, arabus -a -um of Arabia, Arabian (*Arabia, Arabiae*)
Arabidopsis *Arabis*-resembling, *Arabis-opsis*
Arabis Arabian, *arabia, arabiae* (derivation obscure)
Arachis ancient Greek name for a leguminous plant, αρακος, *arachus* in Galen (groundnut). Some translate it as α-ραχις, without a branch
arachniferus -a -um cobwebbed, bearing a weft of cobweb-like hairs, αραχνη-φερω
Arachniodes Spider-like, αραχνη-ωδης
arachnites spider-like, αραχνη-ιτης
arachnoides, arachnoideus -a -um cobwebbed, αραχνη-οειδης, covered in a weft of hairs
aragonensis -is -e from Aragon, NE Spain (*Aragonia*)
araiophyllus -a -um slender leaved, αραιο-φυλλον
Araiostegia Thin-cover, αραιος-στεγη (the indusium)
Aralia origin uncertain, could be from French Canadian, aralie (***Araliaceae***)
aralioides resembling *Aralia, Aralia-oides*
araneosus -a -um spider-like, like a cobweb, *arania, araneosus*
aranifer -era -erum spider-bearing, *aranea-fero*
araraticus -a -um from Agri Dagi (Ararat mountains), Turkey
araroba the Brazilian name for the powdery excretion produced by *Andira araroba*
aratophyllus -a -um with plough-like leaves, *aratrum, aratri*, a plough
araucanus -a -um from the name of a tribal area of Chilean Indians in southern Chile
Araucaria from the Chilean name, araucaros, for the tree (***Araucariaceae***)
araucarioides resembling *Araucaria, Araucaria-oides*
Araujia from the Brazilian name for the cruel plant
arborea-grandiflora tree-like and large-flowered, *arbor-grandis- floris*
arborescens becoming or tending to be of tree-like dimensions, *arbor-essentia*

arboreus -a -um tree-like, branched, *arbor; arbos, arboris*
arboricolus -a -um living on trees, *arbori-colo* (symbionts, parasites and saprophytes)
arbortristis -is -e melancholy-tree, *arbor-tristis* (*Oxydendron arbortristis*, the sorrowful tree)
arbor-vitae tree of life, *arbor-(vita, vitae)* (N American equivalence of *Thuja occidentalis* with the Mediterranean *Cupressus* as durable and fragrant, and planted in graveyards)
arbusculus -a -um, arbuscularis -is -e small-tree, shrubby (diminutive of *arbor*)
arbustivus -a -um coppiced, growing with trees, of plantations, *arbustum, arbusti*
arbutifolius -a -um with *Arbutus*-like leaves, *Arbutus-folium*
Arbutus the ancient Latin name, *arbutus*, or Celtic arboise for a rough fruit
arcadensis -is -e from Arcady, Arcadian, from paradise
arcadiensis -is -e Arcadian, from Peloponnese, S Greece
arcanus -a -um of coffins, of boxes, of cages, *arcanus* (use in basketry)
Arceuthobium Juniper-life, αρκευθος-βιοσ (European species is a parasite on *Juniperus*)
Archaefructus Ancient-fruit, *archi-fructus* (Chinese fossilized fruit)
archaeo- ancient, αρχαιος
Archangelica supposedly revealed to Matthaeus Sylvaticus by the archangel as a medicinal plant
arche-, archi- beginning-, original-, primitive-, αρχη-
archeri for S. Archer, who sent plants from Barbados to Kew
archonto- majestic-, ruler-, αρχων, αρχοντος
Archontophoenix Majestic-fig, αρχοντος-φοινιξ (Queen Alexandra's fig)
arct-, arcto- bear-, αρκτος, northern-,
Arctanthemum Northern-flower, αρκτ-ανθεμιον (arctic *Chrysanthemum arctium*)
Arcterica Arctic-*Erica*, αρκτος-ερεικη, *arcticus-Erica*
arcticus -a -um of the Arctic regions, Arctic, αρκτικος, *arcticus*
Arctium Bear-like, αρκτος (a name in Pliny, the shaggy hair)
arctopoides bear's-foot-like, resembling *Arctopus*, αρκτο-ποδος-οειδης
Arctopus Bear's-foot, αρκτο-πους
Arctostaphylos Bear's grapes, αρκτο-σταφυλη (this is the Greek version of *uva-ursi*, giving one of the repetitive botanical binomials, *Arctostaphylos uva-ursi*)
Arctotis Bear's ear, αρκτ-ωτος
Arctous, arctous Boreal-one, αρκτος, or That-of-the-bear, αρκτωος (the black bearberry)
arcturus -a -um bear's-tail-like, αρκτ-ουρα (cognate with *arctium* and *arctous*)
arcuatus -a -um curved, arched, bowed, *arcus*
ardens glowing, fiery, *ardens, ardentis*
ardesiacus -a -um slate-grey, slate-coloured, modern Latin from French, ardoise
ardeus -a -um shining, burning, *ardeo, ardere, arsi, arsum*; from Tivoli (*Ardea Tibur*), Lazio, Italy
Ardisea Pointed, αρδις (the anthers are shaped like spear-heads)
Ardisiandra *Ardisia*-anthers, αρδις-ανδρος
ardoinoi for H. Ardoino (1819–74), of Mentone, author of the flora of the Maritime Alps
ardonensis-is -e from the Ardon river area, Caucasus, SW Russia
ardosiacus -a -um slate-grey, modern Latin from French, ardoise, slate
arduennensis -is -e from the Ardennes, France/Belgium (*Arduenna*)
arduinoi for Pietro Arduino (1726–1805), botanist at Padua
Areca from the Malabar vernacular name, areek (betel nut palm) (***Arecaceae*** ≡ ***Palmae***)
arecina *Areca*-like, *Areca*
Aregelia for E. A. von Regel (1815–92), of St Petersburg Botanic Garden (≡ *Neoregelia*)
Aremonia derived from a Greek plant name, αρεμον, for *Agrimonia*

aren-, areba-, areni-, areno- sand-, of sandy habitats, *harena, harenae, arena, arenae*
Arenaria Sand-dweller, *(h)arena, (h)arenae*
arenarius -a -um, arenosus -a -um growing in sand, of sandy places, *(h)arenosus*
arenastrus -a -um resembling *Arenaria, Arenaria-aster*
arendsii for Georg Adalbert Arends (1863–1952), German nurseryman of Wuppertal–Ronsdorf
Arenga from the Malaysian vernacular name for *Arenga caudata*
arenicolus -a -um sand-dwelling, *(h)arena-colo*
arenivagus -a -um straggling across the sand, *arena-(vagor, vagare, vagatus)*
arenosus -a -um gritty, sandy, *harenosus*
areolatus -a -um with angular spaces or scars, *areola, areolae* (in or on stems or leaves)
arequipensis-is -e from Arequipa region of southern Peru
arequitae from Arequit, Peru
aretioides resembling *Aretia, Aretia-oides* (*Androsace*)
arfakianus -a -um from the Arfak mountains, western New Guinea or Irian Jaya
Argemone a name, αργεμωνη, used by Dioscorides for a poppy-like plant used medicinally as a remedy for cataract
argent-, argente-, argenti- silver-, silvery-, shining-, *argentum, argenti*
argentatus -a -um silvered, silver-plated, *argentatus*
argentauratus -a -um silvery-gilded, *argentum-aureum*
argenteo-, argenteus -a -um, argentus -a -um silvery, of silver, *argenteus*
argenteoguttata guttating silvery drops, *argentum-(guttata, guttatae)*
argentifolius -a -um silvery-leaved, *argentum-folium*
argentissimus -a -um most silvery, purest, superlative of *argentus*
argi- whitened-, clay-, αργης, αργι-, *argilla. argillae, argi-*
argillaceus -a -um growing in clay, whitish, clay-like, of clay, *argilla, argillae*
argillicolus -a -um living on clay soils, *argilla-colo*
argipeplus -a -um white-robed, *argi-peplum*
argo- bright, pure white-, silvery-, αργος
Argocoffopsis Looking-like-silvery-*Coffea*, αργο-*coffea-opsis*
argolicus -a -um from the area of Argolis peninsula, NE Peloponnese, S Greece
argophloius -a -um having shining bark, αργος-φλοιος
argophyllus -a -um silvery-leaved, αργος-φυλλον
Argostemma Bright-crown, αργος-στεμμα
argun, argunensis -is -e from the northern Manchurian republic of Argun
arguti- clear, graceful; sharply saw-toothed, sharp, *argutus, arguti-*
argutifolius -a -um with sharply toothed leaves, *arguti-folium*
argutus -a -um sharply toothed or notched; clear, graceful, *argutus*
Argylia for Archibald Campbell, of Whitton, Middlesex, third Duke of Argyll and plant introducer (*Lycium barbarum* was wrongly labelled as tea, Duke of Argyll's tea-plant)
argyr-, argyro- silvery, silver-, αργυρος, αργυρο-, αργυρ-
argyraceus -a -um resembling silver, αργυρος
argyraeus -a -um silvery-white, αργυρειος
Argyranthemum, argyranthus -a -um Silver-flower, αργυρ-ανθεμιον (formerly included in *Chrysanthemum*)
argyratus -a -um, argyrites silvered, αργυρος
Argyreia, argyreus -a -um Silvery-one, αργυρειος, αργυρεος (with a silvery appearance of the leaves)
argyrellus -a -um silverish, pale silver, diminutive of αργυρος
argyrocalyx silver-calyxed, αργυρο-καλυξ
argyrocarpus -a -um silver-fruited, αργυρο-καρπος
argyrocoleon sheathed in silver, αργυρο-κολεος
argyrocomus -a -um silver-leaved, αργυρο-κομη
Argyroderma Silver-skin, αργυρο-δερμα (the foliage colour)
argyroglochin silver-tipped, silver-pointed, αργυρο-γλωχις

Argyrolobium Silver-podded-one, αργυρο-λοβος
argyrophyllus -a -um silver-leaved, αργυρο-φυλλον
argyrotrichus -a -um silver-haired, αργυρο-τριχος
Argyroxiphium Silver-sword-like, αργυρο-ξιφος-ειδος (for the leaves)
arhizus -a -um lacking roots, rootless, not rooted, α-ριζα
ari- *Arum-*
Aria, aria a name, αρια, used by Theophrastus for a whitebeam
arianus -a -um from Afghanistan, Afghan, after the Alexandrian priest Arius (*c.* 250–336) founder of the creed of Arianism
aridi- withered, meagre, dry, *aridus*
aridicaulis -is -e having dry or withered-looking stems, *aridi-caulis*
aridus -a -um of dry habitats, dry, arid, *aridus*
arietinus -a -um like a ram's head, ram-horned, *aries, arietis, arietinus*
arifolius -a -um *Arum*-leaved, *Arum-folium*
ariifolius -a -um, ariaefolius -a -um having leaves resembling those of *Sorbus aria*
arillatus -a -um with seeds having a partially enveloping funicular expansion or aril, *arillatus*
arilliformis -is -e bag-shaped, *arillus-forma*
aripensis -is -e from the environs of Mount Aripo, Trinidad
-aris -is -e -pertaining to
Arisaema Blood-*Arum*, αρον-αιμα (spathe colour)
Arisarum a name used by Dioscorides
-aristus -a -um -ear of corn, *arista, aristae* (used for awns or awn-like appendages)
aristatus -a -um with a beard, awned, aristate, *arista* (see Fig. 7g)
Aristea Point, *arista, aristae* (the acute leaf tips)
Aristida Beard, *arista* (the barley-like appearance due to conspicuous awns)
aristideus -a -um bristled, like an ear of corn, *arista, aristae*
Aristolochia Best-childbirth, αριστος-λοχος, Theophrastus' name, αριστολοχια, for one species (abortifacient property) (***Aristolochiaceae***)
aristolochioides birthwort-like, resembling *Aristolochia*, αριστο-λοχος-οειδης
aristosus -a -um with a strong beard, heavily awned, comparative of *arista*
Aristotelia for Aristotle of Stagira (384–322 BC), Greek philosopher
aristuliferus -a -um bearing small awns, of noble bearing, *arista-fero*
-arius -a -um -belonging to, -having
ariza from a vernacular name for *Browneia* in Bogotá, Colombia
arizelus -a -um notable, eye-catching, conspicuous, αρι-(ζηλοω, ζηλωτος)
arizonicus -a -um from Arizona, USA
arkansanus -a -um from Arkansas, USA
armandii for Père Armand David (1826–1900), Jesuit missionary and plant collector in China
armatissinus -a -um most protected or armed, superlative of *armatus*
armatus -a -um thorny, armed, *armatus*
armeniacus -a -um Armenian (mistakenly for China), the dull orange colour of *Prunus armeniaca* fruits
armentalis -is -e of the herd, *armentum, armentalis*
armenus -a -um, armeniacus -a -um from Armenia, Armenian
Armeria, armeria ancient Latin name for a *Dianthus*, French, armoires
armiferus -a -um bearing arms, armoured, *armifer, armiferi; (arma, armorum)-fero*
armigerus -a -um arms-bearer, *armiger, armigeri*
Armillaria Braceleted-one, *armilla* a bracelet (the collar round the stipe of honey fungus *Armilleria mellea*)
armillaris -is -e, armillatus -a -um bracelet-like, having a collar, *armilla, armillae*
armoraceus -a -um horse-radish-like, resembling *Armoracia*
Armoracia of uncertain meaning, αρμορακια, used by Columela and Pliny, formerly for a cruciferous plant, possibly the widespread *Raphanus raphanistrum* rather than *Armoracia rusticana* (horse radish)
armoraciifolius -a -um with leaves resembling those of *Armoracia*

armoricensis -is -e from Brittany peninsula, NW France (*Armorica*)
armstrongii for the land agent who discovered *Freesia armstrongii* in Humansdorp, S Africa
Arnebia from an Arabic vernacular name
Arnica Lamb's-skin, αρνακις (from the leaf texture)
arnicoides resembling *Arnica*, αρνακις-οειδης
arnoldianus -a -um of the Arnold Arboretum, Massachusetts, USA
arnoldii for Nicolas Joseph Arnold, Belgian Colonial Administrator
Arnoseris Lamb-succour, αρνος-σερις (Lamb's succory, fragrant, aromatic potherb)
arnotianus -a -um, arnotii either for George A. Walker Arnott (1799–1868), Scottish botanist, or for Hon. David Arnot, Commissioner for Griqualand, *c.* 1867
aroanius -a -um from Aroania, Arcadia, S Greece; or of ploughed fields, of farmed land, *aro, arare, aravi, aratum*
aromaphloius -a -um having aromatic or spicy bark, αρωμα-φλοιος
aromaticus -a -um fragrant, aromatic, αροματικος
Aronia a derivative name from *Aria*
aronioides resembling *Aronia*, *Aronia-oides*
arpadianus -a -um sickle-shaped, *harpe*, αρπη (leaves)
Arrabidaea, arrabidae for Bishop Antonio de Arrabida, editor of *Flora Fluminiensis*, *c.* 1827
arranensis -is -e from the island of Arran, W Scotland
arrectus -a -um raised up, erect, *adrectus, arrectus* (steep)
arrhen-, arrhena- strong-, male-, stamen-, αρρην, αρρενος, αρσην, αρσηνος, αρσεν
Arrhenatherum Male-awn, αρρην-αθερος (the male lower spikelet is long-awned)
arrhizus -a -um without roots, rootless, α-ριζα (the minute, floating *Wolffia* has no roots)
arrhynchus -a -um not beaked, αρ-ρυγχος
arrigens freezing, stiffening up, becoming erect, *ar-*(*rigeo, rigere*)
Arsenococcus Male-berry, αρσενο-κοκκος
Artabotrys Hanging-fruit, αρταω-βοτρυς (the tendrillar structure)
artacarpifolius -a -um having leaves resembling those of *Artocarpus*
Artanema Thread-bearer, αρταω-νημα
Artemisia Dioscorides' name for Artemis (Diana), wife of Mausolus, of Caria, Asia Minor (*Artemisia dracunculus* is tarragon, Arabic, tarkhun)
artemisioides, atremesioides resembling *Artemisia*, *Artemisia-oides*
arthr-, arthro- joint-, jointed-, αρθηρον, αρθηρο-, αρθρον-, αρθρο-, αρθρ-
Arthraxon Jointed-stem, αρθρ-αξων
Arthrocnemum Jointed-thread, αρθρο-κνημη
Arthrolobium Jointed-pod, αρθρο-λοβος
Arthropodium, arthropodius -a -um Jointed-foot, αρθρο-ποδιον (the jointed pedicels)
Arthropteris Jointed-fern, αρθρο-πτερυξ (the rachis of the frond is jointed towards the base)
arthrotrichus -a -um having jointed hairs, αρθρο-τριχος
articulatus -a -um, arto- knuckled, jointed, joint-, articulated-, *articulus, articuli*
artitectus -a -um completely fabricated, αρτιος-τεκτοω, fully roofed, *arti-tectum*
Artocarpus, artocarpus Bread-fruit, αρτος-καρπος (the large, edible composite fruit)
artosquamatus -a -um covered with crumb-like scales, botanical Latin from αρτος and *squamatus*
artus -a -um close, tight, narrow; joint, limb, *artus*
aruanus -a -um from Aru Kep Island, off W New Guinea, Indonesia
Arum a name, αρον, used by Theophrastus (***Araceae***)
Aruncus the name in Pliny
arundarus -a -um of pens, canes, rods, flutes, combs, *(h)arundo, (h)arundinis*
arundinaceus -a -um *Arundo*-like, reed-like, *(h)arundo, (h)arundinis*
Arundinaria, arundinarius -a -um Cane or Reed-like, derived from *Arundo*
Arundinella Little *Arundo* (but may grow to 3.5 m)
Arundo the old Latin name, *harundo*, for a reed or cane, *(h)arundo, (h)arundinis*

arvalis -is -e of arable or cultivated land, *arvus, arvae*
arvaticus -a -um from Arvas, N Spain
arvensis -is -e of the cultivated field, of ploughed fields, *arvus, arvum*
arvernensis -is -e from Auvergne, France (once the region occupied by the *Arverni* Gauls under Vercingetorix)
arvoniensis -is -e from the area around Caernarvon, Wales (from Celtic, arfon)
arvorus -a -um of ploughed fields, *arvus, arvum*
asafoetida stinking-laser, (*laserpicium*) *laser-foetidus* (the gum-resin of *Ferula foetida*), botanical Latin from Persian, aza, mastic, with *foetidus*
asarabacca medieval Latin compounded from ασαρον, *asarum*, and βακχαρις, *baccaris*
asarifolius -a -um *Asarum*-leaved, *Asarum-folium*
Asarina, asarina from the Spanish vernacular name for *Antirrhinum*, having leaves similar to those of *Asarum*
Asarum a name, ασαρον, used by Dioscorides for asarabacca
ascalonicus -a -um from Ashqelon, SW Israel (*Ascalon*)
ascendens obliquely upwards, ascending, *ascendo, ascendere, ascendi, ascensum*
ascendiflorus -a -um flowering up the stem, *ascendo-florum*
-ascens -becoming, -turning to, -tending-towards, -being, ουσια, *essentia*
asclepiadeus -a -um resembling a milkweed, *Asclepias*-like
Asclepias for Aesculapius, mythological god of medicine (milk weeds) (***Asclepiadaceae***)
asco- wine-skin, bag-like-, bag-, ασκος
Ascocoryne Bag-like-club, ασκο-κορυνη (the saprophyte's concave-topped fruiting body)
Ascolepis Bag-scale, ασκο-λεπις (the hypogynous scale encloses the achene in some)
Ascyrum, ascyron not hard, soft, α-σκυρος
asiaticus -a -um from Asia, Asiatic
Asimina (Assimina) from the French-Canadian name, asiminier, used by Adanson
asininus -a -um, asinius -a -um ass-like (eared), loved by donkeys
*aspalathoide*s like a thorny shrub, ασπαλαθος-οειδης
asparaginus -a -um somewhat similar to *Asparagus*
asparagoides *Asparagus*-like, *Asparagus-oides*
Asparagus the Greek name, ασπαραγος, for plants sprouting edible turions from the rootstock (***Asparagaceae***)
asper -era -erum, asperi- rough, *asper, asperi* (the surface texture)
asperatus -a -um rough, *asper*
asperens becoming rough or sharp, present participle of *aspero, asperare, asperavi, asperatum*
aspergilliformis -is -e shaped like a brush, with several fine erect branches, *aspergillum-folius*
Aspergillus Brush, botanical Latin from *aspergillum*, the brush used to sprinkle holy water (for the closely erect branches in the sporulating stage)
asperifolius -s -um rough-leaved, *asperi-folium*
aspermus -a -um without seed, seedless, α-σπερμα
aspernatus -a -um rejected, disdained, despised, *aspernor, aspernare, aspernatus*
asperocarpus -a -um having rough-walled fruit, *asperus-carpus*
asperrimus -a -um with a very rough epidermis, superlative of *asper*
aspersiculus -a -um finely roughened, diminutive from *aspersus*
aspersus -a -um with spattered markings, sprinkled, *aspergo, aspergere, aspersi, aspersum*
asperugineus -a -um somewhat *Asperugo*-like, slightly roughened or uneven, *asper*
Asperugo Rough-one, *asper* with feminine suffix *-ugo*
Asperula Little-rough-one, feminine diminutive of *asper* (woodruff)
asperulatus -a -um somewhat resembling *Asperula*
asperuloides *Asperula*-like, *Asperula-oides*
asperulus -a -um finely roughened, *asper*

asperus -a -um rough, *asper*
Asphodeline *Asphodelus*-like
asphodeloides *Asphodelus*-like. *Asphodelus-oides*
Asphodelus the Latin name, *asphodilus*, in Homer, ασφοδελος, for *Asphodelus ramosus* (silver rod) (***Asphodelaceae***)
Aspidistra Small-shield, ασπιδισεον (the stigmatic head, analogy with *Tupistra*)
Aspidium Shield, ασπιδιον, diminutive of ασπις, ασπιδος (the shape of shield fern's indusium) (***Aspidiaceae***)
Aspidoglossum Shield-tongue, ασπιδος-γλωσσα (the dorsally flattened corolla lobes)
Aspidotis Shield-like-eared, ασπιδος-ωτος (the false indusium)
Aspilia Without-blemish, α-σπιλος
aspleni- *Asplenium-*, spleen-wort-
asplenifolius -a -um *Asplenium*-leaved, *Asplenium-folium*
Aspleniophyllitis the compound name for hybrids between *Asplenium* and *Phyllitis*
Asplenium Without-spleen, α-σπλην Dioscorides' name, ασπληνον, for spleenwort (***Aspleniaceae***)
assa-foetida fetid-mastic, botanical Latin from Persian, azu, and Latin, *foetida*
assamensis -is -e, assamicus -a -um from Assam, India
assimilis -is -e resembling, like, similar to, *adsimulo, adsimulare, adsimulavu, adsimulatum* (another species)
assinboinensis -is -e from the area of Assinboine mountain, Alberta, Canada; or Assinboia, Saskatchewan, Canada
assoanus -a -um from Aswan, Egypt
assurgens, assurgenti- rising upwards, ascending, present participle of *adsurgo, adsurgere, adsurrexi, adsurrectum*
assurgentiflorus -a -um with flowers presenting upwards, *adsurgens-florum*
assyriacus -a -um, assyricus -a -um from northern Iraq (*Assyria*)
Astartea for Astarte, the Syrian equivalent of Venus
Astelia Stemless, α-στηλη (some are epiphytes)
astelifolius -a -um *Astelia*-leaved, *Astelia-folium*
Aster Star, αστηρ, αστερος (***Asteraceae***)
-aster -ra -rum, -istrum -partial similarity, -wild, -inferior, (used as a suffix to the generic name to denote a section, e.g. *Trifoliastrum*); star-, stellate-, αστηρ, αστερος, *astrum, astri*
Asteranthera Star-flowered, αστηρ-ανθος (the disposition of the anthers)
asterias star-like, αστηρ, αστερος
asterictos unsupported, weak, αστηρικτος
asterioides, asterodes, asteroides *Aster*-like, αστερ-ωδης, αστηρ- οειδης
Asteriscus Small-star, αστερισκος
Asteromoea Resembling-*Aster*, αστηρ-ομοιος
Asterophora Star-bearer, αστηρ-φορα (the processes covering the chlamydospores)
asterosporus -a -um with star-shaped spores, αστερος-σπορος (the spores have blunt spines)
Asthenotherum Poor-harvest, ασθενης-θερος (a desert grass)
asthmaticus -a -um of asthma, ασθμα (its medicinal use for shortage of breath)
astictus -a -um immaculate, without blemishes, unspotted, α-στικτος
Astilbe Without-brilliance, α-στιλβη, α-στιλβος (the flowers)
Astilboides, astilboides *Astilbe*-like, α-στιλβω-οειδης
astracanicus -a -um from Astrakhan province, Volga Delta, Russia
astrachinus -a -um having a dark, curly indumentum; from Astrakhan, Russia (simile with fleece of karakul lambs)
Astraeus Star-shaped, *astrum*, the outer wall of spore producing body of the earth-star fungus (*Astraeus* was father of the winds – by which the spores are dispersed)
Astragalus Ankle-bone, αστραγαλος, a Greek name in Pliny for a plant with vertebra-like knotted roots
Astranthium Star-flower, αστηρ-ανθος

Astrantia Masterwort (l'Ecluse's name, from *magistrantia*, Meisterwürz)
astrantioideus -a -um resembling *Astrantia*
astrictus -a -um drawn together, *adstringo, adstringere, adstrinxi, adstrictum*
astringens contracting, becoming drawn in, present participle of *adstringo*
Astripomoea Star-*Ipomoea*, αστηρ-ιψ-ομοιος
astro- star-shaped-, *astrum*, αστρον, αστρο-
Astrocarpus Star-fruit, αστρο-καρπος
astroites star-like, αστρον-ιτικος
Astronium Star-like, αστρον (the flowers)
astrophoros star-bearing, αστερ-φορος
Astrophytum Star-plant, αστρο-φυτον (the morphology of the plant)
astrotrichus -a -um having star-shaped hairs, stellate-hairy, αστρο-τριχος
-astrum somewhat like, wild, inferior, as good as, *ad instar*
asturicus -a -um, asturiensis -is -e from Asturias, NW Spain
astutus -a -um cunning, deceptive, *astutus*
astylus -a -um lacking a distinct style, α-στυλος
asymmetricus -a -um irregular, lacking symmetry, α-συμ-μετριος
Asyneuma derivation uncertain (?α-συν-ευ-μα, relationship to *Phyteuma*)
Asystasia derivation uncertain, α-συστασις, lacking association
atacamicus -a -um from the Atacama desert of Chile
Ataenidia Without-a-small-band, α-ταινια (no spur on the staminode)
Atalantia for Atalanta, the swift-footed huntress of Greek mythology
atalantoides resembling *Atalantia, Atalantia-oides*
atamasco an Amerindian vernacular name for *Zephyranthes atamasco*
atavus -a -um great-great-great-grandfather, of great age, ancient, *atavus, atavi*
ataxacanthus -a -um having irregularly arranged prickles, ατακτος-ακανθα
ater, atra, atrum matt-black, *ater, atris*
Athamanta Athamas-one, *athamanticus -a -um (athemanticus),* of Mount Athamas, Sicily, or for the King Athamas of the Minyans, in mythology
Athanasia Immortal, α-θανασιμος (without death, funerary use of *Tanacetum*)
atheniensis -is -e from Athens (*Athenae*)
athero- bristle-, beard-, αθηρ, αθερος
atherodes bristle-eared, (αθηρ, αθερος)-ωδης
Atherosperma Bearded-seed, αθηρο-σπερμα
athois -is -e, athous -a -um from Mount Athos, NE Greece
athro- crowded, αθροος, αθρο-
athrostachyus -a -um having crowded spikes, αθρο-σταχυς
Athrotaxis Crowded-order, αθροος-ταξις
Athyrium Sporty, αθυρω (sporty in an earlier sense of variability, from the varying structure of ladyfern sori) (***Athyriaceae***)
-aticus -a -um,-atilis -is -e -from (a place)
atlanticus -a -um of the Atlas Mountains (*Atlas, Atlantis*), N Africa, of Atlantic areas (the western limits of the classical Old World)
atlantis -is -e from the Atlas mountains (*Atlas, Atlantis*), Morocco/Algeria/Tunisia, N Africa; for the giant Atlas of mythology
atomarius -a -um small, pigmy, unmoved, indivisible, ατομος, *atomus, atomi*
atomerius -a -um speckled (atomate); having very small parts, ατομ-μερις
Atractogyne Spindle-fruited-one, ατρακτος-γυνη (fusiform fruits of some)
Atractylis Spindles, ατρακτος (the long spines of the outer bracts)
atractyloides resembling *Atractylis, Atractylis-oides*
atramentarius -a -um with black eruptions, *atra-mentagra* (sycosis)
atramentiferus -a -um carrying a black secretion, *atra-mentagra-fero*
atrandrus -a -um having dark stamens, botanical Latin *ater-andrus*
Atraphaxis an ancient Greek name, ατραφαξυς, for *Atriplex* (q.v.)
atratus -a -um blackened, clothed in black, blackish, *ater*
atrebatus -a -um swarthy, blackened, from the Celtic Atrebates tribe of Roman England, South of the Thames and in Gaul – famous as ironworkers

atri-, atro- better-, dark-, black-(a colour), *ater, atra, atrum, atro-*
atrichus -a -um lacking hairs, α-τριχος
Atriplex the name used by Pliny, *ater-plexus,* black and intertwined (from the ancient Greek name, ατραφαξυς)
atriplicifolius -a -um *Atriplex*-leaved, *Atriplex-folium*
atriplicis -is -e of *Atriplex* (aphis)
atrispinus -a -um having black spines or thorns, *atra-spina*
atrocarpus -a -um dark-fruited, botanical Latin *atro-carpus*
atrocaulis -is -e having dark stems, botanical Latin *atro-caulis*
atrocinereus -a -um dark-grey, dark-ashen coloured, *atro-cineris*
atrocintus -a -um being girdled with black, *atro-(cingo, cingere, cinxi, cinctum)*
atrococcus -a -um black-berried, ατηρ-κοκκος
atrocyaneus -a -um dark-blue, ατηρ-κυανος
atrofuscus -a -um dark-swarthy, dark-brown coloured, *atro-fuscus*
atroides somewhat darkened, ατηρ-οειδης
Atropa Inflexible, ατροπος (Atropa, one of three Fates or Μοιραι)
Atropanthe *Atropa*-flowered-one, Ατροπος-ανθερος
atropatanus -a -um from Azarbaijan, N Iran (the area of N Media was given by Alexander to Atropates and became the kingdom of Artopatene)
Atropis, atropis -is -e Keel-less-one, without a keel, α-τροπις
atropurpureus -a -um dark-purple coloured, *atro-purpureus*
atrorubens dark-red coloured, *atro-rubens*
atrosanguineus -a -um the colour of congealed blood, black-blooded, *atro-(sanguis, sanguinis)*
atrosquamosus -a -um having dark scales, *atro-(squama, squamae)*
atrotomentosus -a -um having a dark-hairy tomentum, *atro-(tomentum, tomenti)*
atrovaginatus -a -um with a black sheath, *atro-(vagina, vaginae)*
atroviolaceus -a -um very dark violet coloured, *atro-(viola, violae)*
atrovirens, atroviridis -is -e very dark green, *atro-(viresco, virescere), atro-viridis*
atrox hideous, dreadful, savage, *atrox, atrocis*
atrum black, *ater, atri*
attavirius -a -um from Mount Atáviros, Rhodes, Greece
attenuatus -a -um tapering, drawn out to a point, flimsy, weak
atticus -a -um from around Athens, Greece (*Attica*)
attractus -a -um drawn towards, attractive, past participle of *attraho, attrahere, attraxi, attractum*
-atus -a -um -rendered, -being, -having (prefixed by some observable attribute)
aubertii for Père George Aubert, French missionary in China *c.* 1899
aubretioides resembling *Aubretia, Aubretia-oides*
Aubrieta (Aubretia) for Tournefort's artist friend, Claude Aubriet (1665?–1742)
auct. of authors, used by a writer to indicate a name used in an alternative sense by other authors. See hort., below, and *sensu*, page 14 and below.
auctus -a -um enlarged, augmented, great, past participle of *augeo, augere, auxi, auctum*
Aucuba (Aukuba) Latinized Japanese name, aokiba
aucuparius -a -um bird-catching, of bird catchers, *avis capio* (use of fruit as bait), *aucupor, aucupare, aucupatus*
audax bold, proud, audacious, *audax, audacis*
augescens increasing, past participle of *augesco, augescere*
augurius -a -um of the soothsayers, *augur, auguris, augurius*
augusti-, augustus -a -um stately, noble, tall, majestic, *augustus*
augustifolius -a -um having impressive foliage, majestic-leaved, *augusti-folium*
augustinii for Dr Augustine Henry (1857–1930), plant collector in China and Formosa, Professor of Forestry, Dublin
augustissimus -a -um the most majestic, superlative of *augustus*
aulacanthus -a -um having grooved spines, αυλακο-ακανθος
aulaco- furrowed-, grooved-, αυλαξ, αυλακος, αυλακο-

aulacocarpus -a -um having furrows in the fruit wall, αυλακο-καρπος
aulacospermus -a -um having ridged seed coats, αυλακο-σπερμα
Aulax Furrow, αυλαξ, αυλακος (the furrowed leaves of some)
aulicus -a -um of the court, *aulicus*
aulo- tube-, αυλος, αυλο-
Aulocalyx Tubular-calyx, αυλος-καλυξ
aurantiacus -a -um resembling an orange, orange-coloured, *aurantium*
auraniticus -a -um from Hawran, SW Syria (the Roman province of *Auranitis*)
aurantifolius -a -um with golden leaves, or having *Citrus*-like leaves, *auranti-folium*
aurantius -a -um orange-coloured, the colour of an orange, *aurantium*
aurarius -a -um, aureus -a -um golden, ornamented with gold, *aurum*
auratus -a -um metallic yellow, golden, ornamented with gold, gilt, *auratus*
aureafolius -a -um with golden leaves, *aureus-folium*
aureatus -a -um like gold, golden, *aureus*
aurelianensis -is -e, aureliensis -is -e from Orleans, France (*Aurelianum*)
aurellus -a -um yellowish, diminutive of *aureus*
aureo-, aureus -a -um golden-yellow, *aureum*
Aureoboletus Golden-mushroom, *aureus-boletus*
aureolus -a -um golden-yellow, *aureus*
aureomaculatus -a -um having golden spots, *aureo-(macula, maculae)*
aureonitens shining gold, *aureo-(niteo, nitere)*
aureosulcatus -a -um having golden grooves, *aureo-(sulcus, sulci)* (on the stems)
aurescens turning golden, *aureo-(fio, fiere, factus)*
auricolor having a golden lustre, *aureo-(color, coloris)*
auricomus -a -um with golden hair, *aurum-coma* (golden-hairy leaved)
Auricula, auriculus -a -um Ear, *auricula* (the leaf shape of *Primula auricula* or the 'Jew's-ear' fruiting body of *Auricularia auricula-judae*)
Auricularia Ear-like, *auricula, auriculae*
auriculatissimus -a -um most lobed, superlative of *auriculatus* (the huge lobes on the petiole of *Senecio auriculatissimus*)
auriculatus -a -um, auricularis -is -e lobed like an ear, with lobes, *auris, auricula*
auricula-ursofolius, auriculae-ursifolius from Clusius' name, *auricula ursi*, bear's ear, for the leaves of *Primula auricula*
aurigeranus -a -um from Ariège, France (*Aurigera*)
Aurinia Of-the-breeze, αυρα, αυρη, *aura, aurae* (plants of montane crags)
Auriscalpum Ear-pick, *auris-scalpium* (the tapered stipe is inserted laterally on the somewhat ear-shaped cap)
auritextus -a -um cloth of gold, woven-from-gold, *aureus-(texo, texere, texui, textum)*
auritus -a -um with ears, long-eared, having long ears, *auris* (stipules)
aurorius -a -um orange, like the rising sun, *aurora, aurorae*
aurosus -a -um of day-break, of sunrise, *aurora, aurorae*
australasiae from southern Asia, botanical Latin
australasicus -a -um Australian, South Asiatic, botanical Latin
australiensis -is -e from Australia, Australian
australis -is -e southern, of the South, *australis*
austriacus -a -um from Austria, Austrian
austrinus -a -um from the south, *australis*
austro- southern, *australis, austro-*
austroafricanus -a -um from southern Africa, botanical Latin
austroalpinus -a -um from the southern alps, *austro-alpinus*
Ausrocedrus Southern-cedar, *australis-Cedrus*
austromontanus -a -um from southern mountains, *austro-montanus* (of North America)
auto- self-, alone-, the same-, αυτος, αυτο-
autochthonus -a -um not introduced, indigenous, αυτοχθων
autumnalis -is -e of the autumn, *autumnus, autumni* (flowering or growing)
auxillaris -is -e helpful, aiding, *auxillaris*; increasing, αυξη (vegetatively)

Auxopus Different-stalk, αυξο-πους (the yellowish weak stems of this parasitic plant)
avasmontanus -a -um from Auas Berg mountains of Namibia
Avellana an old name in Pliny, *nux avellana*, for the hazel nut, from Fonte Avellana (*Avella*), in the Italian Apennines
avellanae of hazel, living on *Corylus avellana* (*Eriophyes*, acarine gall mite)
avellanarius living in hazel woods (*Muscardinus avellanarius*, dormouse)
avellaneus -a -um hazel-brown, *Avellana*
avellanidens with tearing teeth, (*avello, avellere, avelli(avulsi), avulsum*)-(*dens, dentis*) (*Agave* leaf-margins)
avellanus -a -um from Avella, Italy (Pliny's name, *nux avellana*, for hazel-nut)
Avellinia Small-oat-like, diminutive from *Avena*
avellinus -a -um hazel-brown, *Avellana*
Avena Nourishment, *avena* (also meant oats, reed and shepherd's pipe)
avenaceus -a -um oat-like, *avena*
avenius -a -um lacking or with obscure veins, *a*-(*vena, venae*)
avenoides resembling *Avena*, oat-like, *Avena-oides*
Avenula Like-a-small-oat, feminine diminutive of *avena*
Averrhoa for Averrhoes, twelfth-century Arabian physician, translator of Aristotle's work
Avicennia, avicennae for Ibn Sina (*Avicenna*) (980–1037), Arabian philosopher and physician
avicenniifolius -a -um having leaves resembling those of the white mangrove, *Avicennia-folium*
avicennioides resembling *Avicennia*, mangrove-like, *Avicennia-oides*
avicularis -is -e of small birds, eaten by small birds, *avis*
aviculus -a -um omen, of small birds, diminutive of *avis*
avisylvanus -a -um of undisturbed woods, botanical Latin, *avis-sylva*
avium of the birds, *avis*
avocado from a Nahuatl name, ahuacatl, for the fruit (cognate with alligator [pear])
axillariflorus -a -um, axilliflorus -a -um with axillary flowers, with flowers produced in the leaf axils, *axilla-florum*
axillaris -is -e in the armpit, arising from the leaf axils, axillary, *axilla, axillae*
Axonopus Axle-stalked, αξων-πους (the spicate racemes radiate around the upper part of the rachis)
Axyris Without-edge, αξυρις (the bland flavour)
ayabacanus -a -um from Ayabaca, NW Peru
Azalea Of-dry-habitats, αζαλεος (etymology uncertain, formerly used by Linnaeus for *Loiseleuria*)
azaleanus -a -um *Azalea*-like
azaleodendron *Azalea* (flowered) tree, αζαλεος-δενδρον
azaleoides resembling *Azalea*, αζαλεος-οειδης
Azana from a Mexican vernacular name
Azanza from a Mexican vernacular name
Azara for J. N. Azara, Spanish patron of botany and other sciences in the early nineteenth century
azarolus the Italian vernacular name, azarolo, for *Crataegus azarolus*
azedarach, azadarachtus -a -um a middle-eastern vernacular name, azaddirakht, for the bead tree, *Melia azadarachta*
Azolla etymology uncertain, possibly from a South American name thought to refer to its inability to survive out of water, or αζο-ολλυμι, to dry-to kill (***Azollaceae***)
azonites ungirdled, (*a-zona, zonae*)
Azorella Without-scales, α-ζοραλεος, feminine diminutive of α-ζωρος not strong or stout
azoricus -a -um from the Azores Islands, mid-Atlantic
aztecorum from the lands of the Aztecs

azureovelatus -a -um blue-clothed, *azureus-(velo, velare, velavi, velatum)*
azureus -a -um sky-blue, Latin *azureus*, from Arabic, al-lazaward, for lapis-lazuli

babadagicus -a -um from the Babatag mountains, Uzbekistan
babae wonderful!, ahh!, *babae* (an interjection of awe)
Babiana Baboon, from the Afrikaans, babianer, for baboon (which feed on the corms)
babingtonii for Charles Cardale Babington (1808–95), Professor of Botany at Cambridge, author of *Manual of British Botany*
babylonicus -a -um from Babylon, *babylon, babylonius*
bacaba a South American vernacular name for the wine palm, *Oenocarpus bacaba*
bacatus -a -um of pearls; berried, *baca, bacae; bacca, baccae*
baccans becoming berried-looking (shining red to purple, berry-like fruits of *Carex baccans*)
baccatus -a -um having berries, *baca, bacca, bacae, baccae* (fruits with fleshy or pulpy coats)
Baccharis an ancient Greek name (doubtful etymology, perhaps Ecstatic, from βακχος, the spicy smell of the roots)
baccifer -era -erum, bacifer -era -erum olive-bearing, bearing berries, *bacca-fero*
bacciformis -is -e berry-shaped, *bacca-forma*
bacillaris -is -e rod-like, staff-like, stick-like (used botanically for very small rod-like entities), *bacillum, bacilli* a lictor's staff
Backhousia, backhousianus -a -um for James Backhouse (1794–1869), nurseryman of York
Bacopa derivation uncertain
bacterio-, -bacterium stick-, staff-, βακτηρια (rod bacteria)
bacteriophilus -a -um bacteria-liking, symbiotic, βακτηρια-φιλος
Bactris Cane, βακτρον (use in making walking sticks)
baculiferus -a -um staff-carrying, with reed-like stems, *baculum-fero*
baculus -a -um stick, staff, *baculum, baculi*
badachschanicus -a -um from Badakshan, Afghanistan
badiocarpus -a -um having chestnut-brown fruits, botanical Latin *badio-carpus*
badius -a -um, badio- reddish-brown, chestnut-coloured, *badius*
Baeckea for Abraham Baeck, friend of Linnaeus and physician
baeo- small-, βαιος, βαιο-
baeocephalus -a -um small-headed, βαιο-κεφαλη (inflorescence)
Baeometra Of-small-measure, βαιο-μετρον (its small stature)
Baeospora Small-spore, βαιο-σπορος (the spores are about 3×1.5 μm)
baeticus -a -um from S Spain, Andalusia (*Baetica*)
baffinensis -is -e from Baffin Island or Baffin Bay, N Canada
Bafutia for its provenance, Bafut-Ngemba, Cameroon, W Africa
bagamoyensis -is -e from Bagamoyo, Tanzania (one-time coastal HQ of the German East Africa Company)
bahamanus -a -um from the Bahamas, Bahamian
bahianus -a -um from Bahia State, E Brazil
baicalensis -is -e, baicalicus -a -um, baikalensis -is -e from the area around Lake Baikal (Baykal), E Siberia
Baikiaea for Dr William Balfour Baikie, surgeon and naturalist who commanded the Niger Expeditions of 1854 and 1857
baileyanus -a -um, baileyi either for Captain F. M. Bailey, who collected in Tibet *c.* 1913, or Major Vernon Bailey, who collected on Mount Wichita, Oklahoma *c.* 1906, or Liberty Hyde Bailey (1858–1954), Professor of Horticulture at Cornell University, USA
Baillonia for H. Baillon (1827–95), French botanist
bainesii for John Thomas Baines (1820–75), student of the aloes of S Africa
Bakerantha for John Gilbert Baker (1834–1920), British botanist and author of *Handbook of the Bromeliaceae*

Bakerisideroxylon Engler's generic name for Baker's *Sideroxylon revolutum* (≡ *Vincentella revoluta)*
Balanites Acorn-having, βαλανος-ιτης (the Greek name, βαλανος, describes the fruit of some species)
balanoideus -a -um resembling an acorn, βαλανος-οειδης, *balanus-oides*
Balanophora Acorn-carrying, βαλανο-φορα (the nut produced by these total parasites of tropical trees) (***Balanophoraceae***)
balansae, balansanus -a -um for Benjamin (Benedict) Balansa (1825–92), French plant collector who botanized in many parts of the tropical world
balanus the ancient name, βαλανος, for an acorn
balata a Guyanese Carib vernacular name for trees producing an edible fruit and the gutta-percha-like latex, balata (e.g. *Mimusops balata*)
balaustinus -a -um pomegranate-fruit coloured, βαλαυστιον
Balbisia, balbisianus -a -um, balbisii for Giovanni Battista Balbis (1765–1831), Professor of Botany at Turin, Italy
balcanicus -a -um, balcanus -a -um from the Balkans, Balkan
balcoous -a -um from a Bengali vernacular name
baldaccii for Antonio Baldacci (1867–1950) of the Bologna Botanic Garden, Italy
Baldellia for Bartolommeo Bartolini-Baldelli, nineteenth-century Italian nobleman
baldemonia a medieval name for *Meum athamanticum*, baldmoney
baldensis -is -e from the area of Mount Baldo, N Italy
baldschuanicus -a -um from Baldschuan (Baldzhuan), Bokhara, Uzbekistan
balearicus -a -um from the Balearic Islands (*Baliares Insulae*), Mediterranean
balfourii for Sir Isaac Bayley Balfour (1853–1922), collector in Socotra, Professor of Botany, Edinburgh
ballatrix dancing, feminine form of late Latin, *ballo, ballare*, from Italian, ballo, a dance
Ballota Dioscorides' Greek name, βαλλωτη, for *Ballota nigra*
balsamae, balsameus -a -um, balsamoides balsam-like, yielding a balsam, βαλσαμον-οειδης
balsamifer -era -erum yielding a balsam, producing a fragrant resin, βαλσαμον-φερω
Balsamina Balsam, βαλσαμον (a former generic name for *Impatiens*) (***Balsaminaceae***)
balsamina, balsamitus -a -um an old generic name, βαλσαμινη, for alecost (*Tanacetum balsamita*)
Balsamorhiza Balsamic-root, βαλσαμον-ριζα (the resinous roots)
balsamus -a -um of balm, βαλσαμον, *balsamum, balsami*
balticus -a -um from the Baltic Sea or surrounding lands
baluchistanicus -a -um from Baluchistan Province of Pakistan
bamboosarus -a -um of bamboos (stem morphology)
bambos from the Malayan vernacular name, mambu
Bambusa from the Malayan vernacular name, mambu
bambusaefolius -a -um with bamboo-like leaves, *Bambusa-folium*
bambusetorum in bamboo-dominated vegetation, of bamboo forests, *Bambusa*
bambusoides resembling a bamboo, *Bambusa*-like, *Bambusa-oides*
banana a W African vernacular name, banam, from Arabic, banana, for a finger
banatus -a -um, banaticus -a -um, bannaticus from the Banat area (parts of Romania, Hungary and Vojvodina, Yugoslavia)
bancanus -a -um from the island of Pulau Banca (Banka, Bangka), Sumatra, Indonesia
bandaensis -is -e from the islands surrounding the Banda Sea, Indonesia
Banisteria, banisteri for Reverend John Baptist Banister (1650–92), English botanist in Virginia
Banksia, banksii, banksianus -a -um for Sir Joseph Banks (1743–1820) one-time President of the Royal Society and patron of the sciences
banksiae for Lady Dorothea Banks, wife of Sir Joseph, *vide supra* (*Rosa banksiae* was sent to England by the Kew collector William Kerr, in 1807, from China)

banksiopsis looking similar to *Rosa banksiae*
banyan Sir Thomas Herbert's name, reflecting its use by Indian traders for a place of worship, vanija, in 1628
baobab the vernacular name recorded by Prospero Alpini in his *De Plantis Aegypti*, 1592
baoulensis -is -e from Baoule, Ivory Coast, W Africa, or from the Baoule tributary of the Niger, Mali
Baphia Dyer, βαφευς (cam-wood, *Baphia nitida*, gives a red dye, it is also used for violin bows)
baphicantus -a -um of the dyers, dyers', dye(-producing), βαφευς, βαφη
Baptisia Dyeing, βαπτω (several yield false indigo)
barbacensis -is -e from the area around Barbacena, Minas Gerais, Brazil
barbadensis -is -e from Barbados Island, West Indies, or the Barbary coast of N Africa
barba-jovis Jupiter's beard, (*barba, barbae*)*-iovis*
Barbarea Lyte's translation of Dodoens' *Herba Sanctae Barbarae*, for St Barbara
barbarus -a -um foreign, from Barbary, *barbaria* (outside Greece, N African coast)
barbatulus -a -um having a short beard, short-awned, diminutive of *barba*
barbatus -a -um of philosophers, with tufts of hair, with a beard, *barba, barbae*
barbellatus -a -um having small barbs, feminine diminutive of *barba*
barberae, barberii for Mrs F. W. Barber (1818–99), who collected in S Africa
barbi-, barbigerus -a -um bearded, *barbiger, barba-gero*
barbinervis -is -e with bristly veins, *barba-vena*
barbinodis -is -e having bearded nodes, *barba-nodus*
barbulatus -a -um having barbules, with small barbs, diminutive of *barba*
barcellensis -is -e from Villa de Barra area (Baracelos), Amazonas, N W Brazil
barcinonensis -is -e from Barcelona, Catalonia, Spain (*Barcinona*)
bargalensis -is -e from Bargal, on the coast at the N E tip of Somalia
Barkhausia for Gottlieb Barkhaus, of Lippe
Barleria for Reverend J. Barlier (Barrelier) (d. 1673), French botanist
Barnadesia for Michael Barnadez, Spanish botanist
barnumae for Mrs Barnum of the American Mission at Kharput, 1887
barometz from a Tartar word, barants, meaning lamb (the woolly fern's rootstock)
Barosma Heavy-odour, βαρυ-οσμη
Barteria, barteri for C. Barter (d. 1859), of the 1857 Niger Expedition
bartlettii for Harley Harris Bartlett (*c.* 1886), American biochemist
bartonianus -a -um, bartonii for Major F. R. Barton, who collected in Papua
bartramianus -a -um either for John Bartram (d. 1777), King's botanist in America, or his son William Bartram (1739–1823), nurseryman of Delaware, Pennsylvania, USA
Bartschella for Dr Paul Bartsch of the United States National Museum
Bartsia for Johann Bartsch (1709–38), Prussian botanist in Surinam
bary- heavy-, deep-, hard-, strong, βαρυς, βαρυ-
baryosmus -a -um heavily scented, βαρυ-οσμη
barystachys heavily branched, having dense spikes, βαρυ-σταχυς
basalis -is -e sessile-, basal-, *basis*
basalticolus -a -um living in areas of basaltic rock, *basaltes-colo*
basalticus -a -um of basaltic soils, *basaltes* (cognate with *basanites*)
-basanus -a -um -testing, βασανος
baselicis -is -e of Basle, Switzerland (*Basilea*)
Basella the Malagar vernacular name (***Basellaceae***)
baselloides like *Basella, Basella-oides*
basi-, -bassos foot, of the base-, from the base-, βασις, *basis*
basidio- short-pedestal-, *basidium*
basilaris -is -e relating to the base, *basis*
Basilicum, basilicus -a -um princely, royal, βασιληις, kingly-herb, βασιλικος-φυτον
basilongus -a -um having a long lower portion, *basis-longus*

basirameus -a -um much branched from the base, *basis-ramus*
basisetus-a -um with a hairy base, *basis-saeta* (stem)
basitonae extended to the base, βασις-τονος
basjoo the Japanese name for fibre from *Musa basjoo*
Bassia for Ferdinando Bassi (1714–74), Italian botanist and Directer of Bologna Botanic Garden, or for George Bass (d. at sea 1803), navigator who commended Botany Bay for settlement
bastardii for T. Bastard (1784–1846), author of the Flora of Maine & Loire, 1809
bastardus -a -um not natural, abnormal, debased, medieval Latin *bastardus*
basuticus -a -um from Lesotho, S Africa (Basutoland until 1966)
bataanensis -is -e from Bataan, Luzon, Philippines
batalinii for A. F. Batalin (1847–96), Botanic Garden, St Petersburg
batatas Haitian name, batata, for sweet potato, *Ipomoea batatas* (cognate via Portugese, patatas, with potato)
bataua from a vernacular name for an oil palm
batavicus -a -um, batavinus -a -um from Jakarta (Batavia), NW Java, Indonesia
Batemannia for James Bateman (1811–97), orchid collector and monographer of *Odontoglossum* etc.
bathy- thick-, deep-, βαθος, βαθυ-
bathyphyllus -a -um densely leaved, thick-leaved, βαθυ-φυλλον
Batis Thorn-bush, βατος (***Bataceae***)
batjanicus -a -um from Bacan Island, Maluka (Batjan, N Molucca) Indonesia
Batrachia section of *Geranium*, (resemblance to *Ranunculus acris)*
batrachioides water-buttercup-like, *Batrachium*-like, βατραχος-οειδης
Batrachium Little-frog, diminutive of βατραχος (Greek name for several *Ranunculus* species)
Batrachospermum Frog-seed, βατραχος-σπερμα (mucilaginous appearance)
batrachospermus -a -um having mucilaginous seed, βατραχος-σπερμα
battandieri for Jules Aime Battandier (1848–1922), of the Algiers Medical School
Batodendron Thorny-tree, βατος-δενδρον (≡ *Vaccinium)*
Batopedina Little-thorny-tangle, βατος-πεδαω
-batus -a -um accessible, passable; -thorn bush, βατος (sectional suffix in *Rubus*)
baudotii for Herr Baudot (fl. 1837), a German amateur botanist
Bauera, bauera, baueri, bauerianus -a -um for H. Gottfried and Franz Bauer, botanists, travellers and illustrators, and Ferdinand Bauer (1760–1826), botanical artist and traveller (***Baueraceae***)
Bauhinia for the Swiss botanists Caspar (Gaspard) Bauhin (1560–1624) and his brother Johann (Jean) (1541–1613)
bauhiniiflorus -a -um having flowers resembling those of *Bauhinia*
baumannii either for Baumann brothers, nurserymen at Bollweiler, Alsace, or Herrn E. H. Baumann of Bolivia, who produced *Begonia baumannii*
Baumea for Baume, Luzon (*Cyperaceae*)
baumeanus -a -um from Baume, Luzon
baurii for Reverend L. R. Baur (1825–89), who collected *Rhodohypoxis* in S Africa
bavaricus -a -um from Bayern State, Germany (Bavaria)
bavosus -a -um from a Mexican vernacular name, bavoso
bay berry, from the Old French, baie (*Laurus nobilis, baccae-lauri*; laurels were awarded to scholars, hence baccalaureate)
baytopiorus -a -um, baytopii for Professor Turhan Baytop (1920–2002), Turkish pharmacist, plant collector and writer
bdellium sticking, leach-like, βδελλα (has a Semitic origin, via βδελλιον, referring to the resin of some *Balsamodendron* (≡ *Commiphora*) species)
-bdolon -smelling, -stench, βδολος
bealei for Thomas Chay Beale (*c.* 1775–1842), Portuguese Consul in Shanghai, who facilitated Robert Fortune's collecting work
beanianus -a -um, beanii for William J. Bean (1863–1947) (*Trees and Shrubs Hardy in the British Isles*)

beatricis for Beatrice Hops, who discovered *Watsonia beatricis* in S Africa (*c.* 1920)
beatus -a -um abundant, prosperous, *beatus*
Beaufortia for Mary Somerset (*c.* 1630–1714), Duchess of Beaufort, patroness of botany
Beaumontia for Lady Diana Beaumont (d. 1831), of Bretton Hall, Yorkshire
beauverdianus -a -um for Gustave Beauverd (1867–1942), of the Boissier Herbarium, Geneva
bebbianus -a -um for Michel Schuck Bebb (1833–95)
bebius -a -um from the Bebisch mountains, Dalmatia (firm, steady, trusty, βεβαιος)
beccabunga from an old German name 'Bachbungen', mouth-smart or streamlet-blocker
beccarianus -a -um, beccarii for Odoardo (Odordo) Beccari (1843–1920), botanist and traveller in Borneo
Beccariophoenix Beccari's date palm, botanical Latin from Beccari and *phoenix*
Beckmannia for Johann Beckmann (1739–1811), professor at Göttingen
bedeguaris -is -e brought by the wind, from Persian, bad awar (supposed cause of the Hymenopteran-induced gall, rose bedeguar, or Robin's pin-cushion)
Bedfordia, bedfordianus -a -um for John Russell (1766–1839), Sixth Duke of Bedford
beesianus -a -um for Bees' nursery of Ness, Cheshire, plant introducers from China and elsewhere
Befaria for Dr Bejar, a Spanish botanist (a Linnaean spelling error)
Begonia for Michel Begon (1638–1710), French Governor of St Dominique and patron of botany (***Begoniaceae***)
begonifolius -a -um, begoniifolius -a -um having *Begonia*-like leaves
begonioides *Begonia*-like, *Begonia-oides*
beharensis -is -e from Behara, Madagascar (felt leaf, *Kalenchoe beharensis*)
behen from the Arabic name for several plants
beissnerianus -a -um, beissneri for Ludwig Beissner of Poppelsdorf (1843–1927), writer on *Coniferae*
Belamcanda from an Asian vernacular name for the leopard lily
belgicus -a -um from Belgium, Belgian, *Belgae, Belgicus*
belinensis -is -e from Belin, Turkey
belizensis -is -e from Belize, NE Central America
belladonna beautiful lady, botanical Latin from Italian, bella donna (the juice of the deadly nightshade was used to beautify by inducing pallid skin and dilated eyes when applied as a decoction)
bellamosus -a -um, bellatus -a -um quite beautiful, *bella*
Bellardia, bellardii for C. A. L. Bellardi (1741–1826), Italian physician and botanist
bellatulus -a -um somewhat beautiful, diminutive of *bellus*
Bellevalia for P. R. de Belleval (1558–1632), early systematist
bellicus -a -um warlike, fierce, armed, *bellicus*
bellidi- *Bellis*-like-, daisy-
bellidiastrus -a -um daisy-flowered, *Bellis*-flowered, *Bellis-astrum*
bellidiflorus -a -um daisy-flowered, *Bellis-florum*
bellidifolius -a -um with daisy-like leaves, *Bellis-folium*
bellidiformis -is -e, daisy-like, *Bellis-forma*
bellidioides, bellidoides daisy-like, *Bellis-oides*
bellinus -a -um neat, pretty, pleasing, comparative of *bellus*
Bellis Pretty, a name, *bellus*, used in Pliny
Bellium Resembling-*Bellis*
bellobatus -a -um beautiful bramble, botanical Latin *bellus-batus*
belloides daisy-like, *Bellis-oides*
bellulus -a -um pretty little one, diminutive of *bellus*
bellus -a -um handsome, beautiful, neat, pretty, choice, *bellus, bella*
belmoreanus -a -um for the Earl of Belmore, Governor of New South Wales, 1868
belo- pointed like a dart or arrow or javelin, frightening, βελονη, βελος, βελο-
Belonophora Arrow-head-bearing, βελονη-φορα (the apex of the connective)

Beloperone Dart-clasp, βελος-περονη (the shape of shrimp plant's connective)
belophyllus -a -um dart-leaf, terrifying-leaf, βελος-φυλλον
beluosus -a -um of monsters, monstrous, *belua, beluae; beluosus*
bemban from a Javanese vernacular name
benedictus -a -um well spoken of, blessed, healing (*Cnicus benedictus* was once used as a cure for gout; herb bennet, or blessed herb, *Geum urbanum*, was prized for its fragrant root, used to make Benedictine liqueur)
bengalensis -is -e, benghalensis, -is -e from Bengal (Benghala), India
benguelensis -is -e from Bunguela, Angola
benguetensis -is -e from Benguet, Luzon, Philippines
Benincasa for Conte Giuseppe Benincasa (d. 1596), Italian botanist
benjamina from an Indian vernacular name, ben-yan, or Gujarati, vaniyo
benjan the Indian vernacular, ben-yan, for weeping fig, *Ficus benjan*
bennettii for A. W. Bennett (1833–1902), British botanist
Bensoniella, bensoniae, bensonii for Colonel Robson Benson (1822–94), who collected in Malabar etc.
Benthamia, Benthamidia, benthamianus -a -um, benthamii for George Bentham (1800–84) author of *Genera Plantarum*, with Sir Joseph Hooker (*Benthamia* ≡ *Cornus*)
Benzoin, benzoin from an Arabic or Semitic name, luban-jawi, signifying Javanese perfume or gum
Berberidopsis resembling *Berberis, Berberis-opsis*
berberifolius -a -um having leaves resembling those of *Berberis*
Berberis Bar-berry, medieval Latin, *barbaris*, from an Arabic name for N Africa (***Berberidaceae***)
Berchemia for M. Berchem, seventeenth-century French botanist
bergamius -a -um from the Turkish name, beg-armydu (beg-armodi), prince's pear, applied to the Bergamot orange (and lemon-mint and Oswego tea)
Bergenia for Karl August von Bergen (1704–60), German physician and botanist of Frankfurt am Oder
bergeri, bergerianus -a -um for Alwin Berger (1871–1931), Curator of the Hanbury Garden at La Mortola and writer on succulents
beringensis -is -e from the region around the Bering Sea (named for Vitus Bering)
Berkheya for Jan Le Francq van Berkhey (1729–1812), Dutch botanist
Berlandiera, berlandieri, berlandierianus -a -um for J. L. Berlandier (d. 1851), Belgian botanist who explored in Texas and Mexico
Berlinia for Andreas Berlin (1746–73) Swedish botanist in W Africa
bermudana, bermudense, bermudiana from Bermuda (however, *Sisyrinchium bermudiana* is endemic in Ireland)
bernalensis -is -e from Bernal, New Mexico
bernardii for Sir Charles Bernard of the India Office, London
berolinensis -is -e from Berlin, Germany (*Berolinum*)
Berteroa for Carlo Guiseppe L. Bertero (1789–1831), Italian physician and traveller who died at sea between Tahiti and Chile
berthelotii for Sabin Berthelot (1784–1880), co-author with P. B. Webb of *Histoire Naturelle des Iles Canaries*
Bertholletia for Claude-Louis Berthollet (1748–1822), French chemist (Brazil nut)
Bertolonia, bertolonae, bertolonii for A. Bertoloni (1755–1869), Italian botanist and writer
Berula the Latin name in Marcellus Empyricus
berylinus -a -um having the colour of beryl, βερυλλος, *beryllus*
Beschorneria for Friedrich Wilhelm Christian Beschorner (1806–73), German botanist
bessarabicus -a -um from Ukraine/Moldovia region (*Bessarabia*)
Bessera, besserianus -a -um for Dr Wilibald Swibert Joseph Gottlieb von Besser (1782–1842), Professor of Botany at Brody, Ukraine
Besseya For Charles Edwin Bessey (1845–1915), American systematic botanist

Beta the Latin name for beet, *beta*
betaceus -a -um beet-like, beetroot coloured, resembling *Beta*
bethlehemensis -is -e from Bethlehem (Palestine, S Africa or America)
bethunianus -a -um for Captain Bethune, who brought *Chirita bethunianus* from Borneo in 1849
betinus -a -um beetroot-purple, *beta*
betle from the Malayan vernacular name, vettila, for the masticatory leaves, betel, of *Piper betle*
betoni- *Betonica*-like-, *Betonica*
Betonica from a name, *Vettonica*, in Pliny for a medicinal plant from Vectones (*Vettones*), Spain
betonicifolius -a -um betony-leaved, *Betonica-folium*
betonicoides *Betonica*-like, *Betonica-oides*
betonicus -a -um betony-like, resembling *Betonica*
Betula Pitch, the name, *betula*, in Pliny (bitumen is distilled from the bark) (***Betulaceae***)
betularus -a um of birches, *Betula* (saprophytes on birch leaves etc.)
betuletarus -a -um associated with birch woodland, *Betula*
betulifolius -a -um with leaves similar to those of a birch, *Betula-folium*
betulinus -a -um, betulus -a -um *Betula*-like, birch-like, living on *Betula* (symbionts, parasites and saprophytes)
betuloides resembling *Betula*, *Betula-oides*
beyrichianus -a -um, beyrichii for Karl Beyrich (b. 1834), Canadian gardener
bholua from a Nepalese vernacular name, bholu-swa, for a *Daphne*
bhotanicua -a -um, bhutanensis -is -e, bhutanicus -a -um from Himalayan Bhutan
bi-, bis- two-, twice-, double-, *bis, bi-*
biacutus -a -um having two points, twice sharply tipped, *bis-acutus*
biafrae, biafranum from the Biafra region of southern Nigeria
bialatus -a -um two-winged, *bis-alatus* (usually the stem)
Biarum a name used by Dioscorides for an *Arum*-like plant
biauritus -a -um two-eared, having two long ears, *bis-auritus*
bicalcaratus -a -um two-spurred, *bis-calcaris*
bicallosus -a -um having two callosities, *bis-callosus*
bicameratus -a -um two-arched, two-chambered, *bis-camera*
bicapsularis -is -e with two small boxes, having two capsules, *bis-capsula*
bicarinatus -a -um having two keels, double-keeled, *bis-carina*
bicaudatus -a -um having two tails, *bis-cauda*
bichlorophyllus -a -um two-green-leaved, botanical Latin from δι-χλωρο-φυλλον (the marked difference in colour of the two surfaces of the leaves)
bicolor, bicoloratus -a -um of two colours, twice-coloured, *bis-*(*coloro, colorare, coloravi, coloratum*)
bicolor-rosea of two shades of red, *bicolor-rosea*
bicornis -is -e, bicornutus -a -um two-horned, *bi-cornu*
bicuspidatus -a -um having two cusps, two-tipped, *bis-*(*cuspis, cuspidis*)
Bidens, bidens Two-teeth, *bi-dens* (the scales at the fruit apex)
bidentatus -a -um double-toothed, having toothed teeth, *bi-dentatus*
bidwillii for J. C. Bidwill (1815–53), Director of Botanic Garden, Sydney
Biebersteinia, biebersteinianus -a -um, biebersteinii for Friedrich August Frieherr Marschall von Bieberstein (1768–1826), author of *Flora Taurico-Caucasica* and other works
biennis -is -e (with a life) of two years, biennial, *bi-annus*
bifarius -a -um in two opposed ranks, two-rowed, in two parts, *bi-fariam* (leaves or flowers)
bifidus -a -um deeply two-cleft, bifid, *bifidus*
bifloriformis -is -e two-flowered form, *bi-florum-forma*
biflorus -a -um two-flowered, *bi-florum*
bifolius -a -um two-leaved, *bi-folium*

biformis -is -e having two shapes (the parental segregates of *Rumex* × *pratensis*)
bifrons having a double garland of leaves, two-boughed, *bi-(frons, frondis)*
bifurcatus -a -um divided into equal limbs, bifurcate, *bi-(furca, furcae)*
Bigelovia, bigelowii for Dr John M. Bigelow (1804–78), American pharmacist and botanist who worked on the Mexican Boundary Mission
bigibbus -a -um having two humps, *bi-(gibbus, gibbi)*
biglandulosus -a -um with two glands, two-glandular, *bi-glandulae* (male florets of *Euphorbia*)
biglumis -is -e with two glumes, *bi-gluma* (*Juncus*)
Bignonia (Bignona) for Abbé Jean Paul Bignon (1662–1743), librarian to Louis XIV (***Bignoniaceae***)
bignoniaceus -a -um, bignonioides *Bignonia*-like, *Bignonia-oides*
biharamulensis -is -e from Biharamulo, at the SW end of Lake Victoria, Tanzania
bijugans, bijugus -a -um two-together, twin yoked, *bi-(-iugis, iugus)* (staminal arrangement or leaves with two pairs of leaflets)
bilamellatus -a -um having two flat ridges, with two lamellae, double-gilled, *bi-lamellatus*
bilateralis -is -e bilateral, zygomorphic, having two (mirror image) sides, *bi-latus*
bilimbi an Indian vernacular name for the cucumber-tree (*Averrhoa bilimbi*)
biliottii for Alfred Biliotti, British Consul and collector at Trabzon, Turkey
-bilis -is -e -able, -capable, *habilis*
Billardiera, billardierei (billardierii) for Jaques Julien Houtou de la Billardière (1755–1834), French botanist
Billbergia for J. G. Billberg (1772–1844), Swedish botanist (angel's tears)
billbergioides resembling *Billbergia, Billbergia-oides*
billotii for Paul Constant Billot (1796–1863), Professor of Botany at Hagenau, Alsace
bilobatus -a -um, bilobus -a -um two-lobed, *bi-lobus* (see Fig. 8a)
bimaculatus -a -um having two conspicuous spots, *bi-macula*
binatus -a -um with two leaflets, bifoliate, paired, *bini, binae, bina, binatus*
binervis -is -e two-veined, *bi-vena*
binervosus -a -um two-veined-ish, two vein patterns, *bi-(nervus, nervi)* (1- to 3-veined)
binnendijkii for S. Binnendijk (1821–83), of the Bogor Botanic Garden, Java
binocularis -is -e two-eyed, marked with two eye-like spots, *bini-(oculus, oculi)*
-bios, bio- life, βιος, βιο- (βιοω, βιωσις, biosis or mode of life)
Biophytum Life-plant, βιο-φυτον (sensitive leaves)
Biorhiza Root-liver, βιο-ριζα (gall midge females lays eggs in rootlets)
bioritsensis -is -e from Biarritz, France
-biosis -living, -mode of life, βιωσις
Biota Of-life or of-bows, βιος (≡ *Platycladus*)
bipartitus -a -um almost completely divided into two, *bi-(partio, partire, partivi, partitum; partior, partiri)*
bipinnatifidus -a -um twice-divided but not to the rachis, *bi-pinnatus-(findo, findere, fidi, fissum)* (leaves)
bipinnatus -a -um twice-pinnate, *bi-pinnatus* (leaves)
bipulvinaris -is -e with two cushions, having a double pulvinus, with two-pulvinate petioles, *bi-(pulvinus, pulvini)*
bipunctatus -a -um two-spotted, *bi-(pungo, pungere, pupugi, punctum)*
birameus -a -um two-branched, *bi-(ramus, rami)* (inflorescence)
birostratus -a -um having two beaks, *bi-(rostrum, rostri)* (spurs or nectaries)
birschelii for F. W. Birschel, who collected plants for Kew in Caracas *c.* 1854
Biscutella, biscutellus -a -um Two-small-trays, *bi-scutella* (the walls of the dehisced fruit)
bisectus -a -um cut into two parts, *bi-(seco, secare, secui, sectum)*
bisepalus -a -um having two sepals, having a double calyx, *bi-sepalum*
biseptus -a -um twice-hedged, partitioned into two, *bi-septum*

biserialis -is -e having two ranks, rows or sequences, *bi-(series, seriem)* (of leaves or floral parts)
biserratus -a -um twice-saw-toothed, double-toothed, *bi-serra* (leaf margin teeth themselves toothed)
bisetaeus -a -um, bisetus -a -um two-bristled, *bi-saeta* (calyx plicae of *Gentiana*)
bispinosus -a -um, bispinus -a -um two-spined, with paired thorns, *bi-(spinus, spini)*
bisporus -a -um having two spores, *bi-spora* (basidia of *Agaricus* typically have four spores)
bisquamatus -a -um two-scaled, *bi-(squama, squamae)* (the calyx lobes)
Bistorta, bistortus -a -um twice twisted, *bis-(torqueo, torquere, torsi, tortum)* the medieval name (for the rhizomes, some say for the inflorescences)
bistortoides resembling bistort, *Bistorta-oides*
bisulcatus -a -um, bisulcus -a -um with two grooves, cloven, *bi-(sulcus, sulci)*
bisumbelatus -a -um twice-umbellate, having an umbel of umbels, *bis-(umbella, umbellae)* (e.g. the compound umbels of many *Apiaceae*)
bisuntinus -a -um from Besançon, France (*Vesontio, Bisuntinus*)
bitchiuensis -is -e, bitchuensis -is -e from Bitchu province, Japan
biternatus -a -um twice ternate, with three lobes each divided into three, *bi-ternatus* (leaves or inflorescences)
bithynicus -a -um from Turkey (*Bithynia, Paphlagonia*, NW Anatolia)
bitorquis -is -e two-necklaced, *bi-torquatus* (the stipe of *Agaricus bitorquis* has two separate sheathing rings)
bituminosus -a -um tarry, clammy, adhesive, smelling of tar, *bitumen, bituminis*
biunciferus -a -um, biuncinatus -a -um bearing two hooks, with paired hooks, *bi-(uncus, unci)-fero* (e.g. the hooked leaves of *Dioncophyllaceae*)
bivalvis -is -e two-valved, botanical Latin *bi-valva* (literally, two folding doors)
bivittatus -a -um with two headbands, two-banded, *bi-(vitta, vittae)* (e.g. *Cryptanthus* leaf stripes)
bivonae for Antonio de Bivona-Bernardi (1774–1837), Sicilian botanist and author of *Sicularum Plantarum* (1806)
Bixa from a S American native name for *Bixa orellana*, the annatto tree (***Bixaceae***)
Blackstonia for John Blackstone (1712–53), English apothecary and botanical writer
Blaeria for Patrick Blair (1666–1728), Scottish surgeon and botanical writer
blagayanus -a -um for Count Blagay, who discovered *Rheum ribes c.* 1837
blandaeformis -is -e, blandiformis -is -e having an attractive shape or appearance, *blandus-forma*
Blandfordia, blandfordii for George Spencer-Churchill (1766–1840), Marquis of Blandford
blandulus -a -um quite pleasing, diminutive of *blandus*
blandus -a -um pleasing, alluring, not harsh, bland, *blandor, blandire, blanditus*
blastophorus -a -um sprouting, suckering, βλαστος-φορα
-blastos, -blastus -a -um -shoot, -sprout, bud, growth, βλαστος
Blattaria, blattarius -a -um an ancient Latin name in Pliny, cockroach-like, *blatta, blattarae*
Blatti an Adansonian name, ≡ *Sonneratia* (***Blattiaceae***)
blechnoides similar to *Blechnum, Blechnum-oides*, βληχνον-οειδης
Blechnum the Greek name, βληχνον, for a fern (***Blechnaceae***)
-blemmus -a -um -suggesting, -looking, βλεμμα
blennius -a -um, blennus -a -um slimy, mucus-like, βλεννος (sticky surface texture)
blepharicarpus -a -um having fringed fruit, βλεφαρο-καρπος
blephariglossus -a -um having a fringed tongue or lip, βλεφαρο-γλωσσα
Blepharis Eyelash, βλεφαρις (the fringed bracts and bracteoles)
blepharistes resembling *Blepharis*, with a pronounced eyelash-like fringe, βλεφαρις
blepharo- fringe-, eyelash-, βλεφαρις, βλεφαρο-
blepharocalyx having a fringed calyx, βλεφαρο-καλυξ
blepharophyllus -a -um with fringed leaves, βλεφαρο-φυλλον

blepharopus fringed with a ring of hairs at the base of the stalk, βλεφαρο-πους
Blepharospermum, blepharospermus -a -um Fringed-seeded-one, βλεφαρο-σπερμα, with fringed seeds (≡ *Calothamnus*)
Bletia for Louis Blet (*c.* 1794), apothecary who kept a garden in Algeciras
Bletilla *Bletia*-like
Blighia for William Bligh (1754–1817), thrice mutinied-against British sailor and author of an account of sailing the South Seas (1792)
blitoides resembling *Blitum*, βλιτον-οειδης (from a plant name used by Greek and Latin writers)
Blitum an ancient name, βλιτον, βλητον, for a kind of spinach, others say either for *Chenopodium* or *Amaranthus* (cognate, blite, formerly for *Chenopodium*, and sea-blite, for *Sueda*)
Bloomeria, bloomeri for H. G. Bloomer (1821–74), pioneer Californian botanist
blossfeldianus -a -um, blossfeldii for Robert Blossfeld, seedsman and cactus dealer of Potsdam, Berlin
Blumea, blumei for Karel Lodewijk Blume (1789–1862), Dutch writer on the E Indies
Blumenbachia for Johann Friedrich Blumenbach, FLS (1752–1840), medical doctor of Gottingen
Blysmus meaning uncertain (βλυζω surge out)
boarius -a -um, bovarius -a -um of the ox, or of oxen, *bos, bovis*
Bobartia for Jacob Bobart (1641–1719), Professor of Botany at Oxford
bocasanus -a -um, bocasensis, bocensis from the Sierra de Bocas, Mexican Panama, or from Las Bocas, Mexico
Bocconia, bocconei, bocconii for Dr Paolo Boccone (1633–1704), Sicilian physician and botanist
bockii for Herr Bock, the German Consul General in Oslo *c.* 1891
bodinieri for Emile Maria Bodinier (1842–1901), French missionary and botanist in China
bodnantensis -is -e from Bodnant gardens, N Wales
Boea for Reverend Dr Bau of Toulon, France
Boehmeria, boehmeri for George Rudolph Boehmer (1723–1803), professor of botany at Württemberg
Boenninghausenia for C. F. von Boenninghausen (1785–1864), German botanist
boeoticus -a -um from Voiotia district (Boeotia), Greece
Boerhaavia, boerhaavii for Herman Boerhaave (1668–1739), Professor at Leiden, early systematist
Boesenbergia for Clara and Walter Boesenberg, in-laws of C. E. Otto Kuntze
bogotensis -is -e from Bogota, S America
bohemicus -a -um Bohemian, from Bohemia, present-day Czech Republic
boissieri, boissii for Pierre Edmond Boissier (1810–85), of Geneva (author of *Flora Orientalis*)
bokharicus -a -um from Bukhara (Buchara, Bockhara), central Uzbekistan
bolanderi for Professor H. N. Bolander (1831–97) of Geneva, plant collector in California and Oregon
bolanicus -a -um from the environs of Mount Bolan, Papua New Guinea
bolaris -is -e dark red, brick-coloured, modern Latin, *bolaris*; netted, βολος (the surface is tessellated with reddish scales)
Bolax Small mound, βωλος (the mounded cushion habit)
Bolbitis With-bulbs, *bulbus* (many bear gemmae)
Bolbitius Occurring-with-*Bolbitis*
bolboflorus -a -um having bulbous flowers, *bulbus-florum*
boldo, boldus -a -um from an Araucarian vernacular name for the fruit of *Peumus boldo*
Boletinus Little-bolete, diminutive of *boletus*
Boletus, boletus Mushroom, *boletus, boleti*; Clod (of earth), βωλος, βωλιτης (for its lumpy shape)

bolivianus -a -um, boliviensis -is -e from Bolivia, S America
Bollea, bolleana for Carl Bolle (1821–1909), Berlin dendrologist
bollwyllerianus -a -um, bollwylleri of Bollwyller, Alsace
Boltonia for James B. Bolton (d. 1799), English botanist
-bolus -a -um -throwing, βολη, βολις, βολιδος
Bolusanthus, Bolusiella, bolusii Bolus'-flower, for Harry Bolus (1834–1911), writer on the flora of S Africa
Bomarea for Jacques Christophe Valmont de Bomare (1731–1807), of Paris
bombaciflorus -a -um silk-flowered, with *Bombax*-like flowers, *bombyx-florum*
Bombax Silk, βομβυξ or *bombyx*, a silkworm (for the kapok covering of the seeds) (***Bombacaceae***)
bombicis -is -e, bombici- of silk, silk- (silkworms, *bombyx, bombycis*, feed on *Morus bombycis*)
bombyciferus -a -um silk-bearing, *bombyx-fero*
bombycinus -a -um silky, with silky hairs, *bombycinus*
bombyliferus -a -um bearing bumble bees, with bee-like flowers, *bombyx-fero*
bombyliflorus -a -um bumble-bee flowered, *bombyx-florum*
bona-nox good night, *bonus-(nox, noctis)* (night-flowering)
bonariensis -is -e from Buenos Aires, Argentina (*Bonaria*)
bonarota, bonarotianus -a -um for Michel Angelo Buonarotti (1475– 1564), of Florence, in whose garden was found *Tulipa bonarotiana*
bondaensis -is -e from Bondi, Australia
bonduc the Arabic vernacular name, bonduq, for a nut
bonensis -is -e from Bon, Chad, C Africa
Bongardia for Heinrich Gustav Bongard (1786–1839), German botanist
boninensis -is -e from Bonin Island, SE Japan
bononiensis -is -e from either Boulogne, France or Bologna, N Italy (both having the Roman name *Bononia*)
Bonplandia, bonplandii for Aimé J. A. Bonpland (1773–1858), authority on the flora of Tropical America
Bonstedtia for Carl Bonstedtia (1866–1953) Gartenoberinspektor at Göttingen
bonus -a -um good, *bonus* (in various senses, see *melior*, comparative, and *optimus*, superlative)
bonus-henricus good King Henry, an apothecaries' name to identify it from *malus henricus*, a poisonous plant (allgood or mercury)
Boophone Ox-killer, βοο-φονη (narcotic property)
boothii for either H. Booth, gardener *c.* 1864, or T. J. Booth, collector in Assam and Bhutan *c.* 1850
boottianus -a -um, boottii for Fr C. M. Boott (1792–1863), American physician and botanist
Borago Shaggy-coat, *burra* with feminine suffix (the leaves) (***Boraginaceae***)
boranensis -is -e of the lands of the Oroma people (Borana) of S Ethiopia
Boraphila North-lover, βορεας-φιλος
Borassus from Linnaeus' name for the spathe of the date palm, βορα, food
borbasii for Vincenz von Borbas (1844–1905), Hungarian Director of the Botanic Garden at Klaustenburg (Cluj-Napoca), Romania
Borbonia, borbonianus -a -um for Gaston Jean Baptiste de Bourbon (1608–60), the Duke of Orleans, third son of Henry IV of France, patron of botany
borbonicus -a -um from Réunion Island (Ile de Bourbon), Indian Ocean
borealis -is -e of the North wind, northern, of the North, βορεας, *boreas, boreae*
Boreava, boreaui for Alexander Boreau (1801–75), Belgian botanist director of the Angers Botanic Garden and author of *Flore du centre de la France*
borinquenus -a -um from Puerto Rico (Amerindian name, Borinquen)
boris-regis for King Boris III of Bulgaria (1894–1943)
borneensis -is -e from Borneo Island
Bornmeullera, bornmuelleri for J. Franz N. Bornmueller (1862–1948), Hanoverian botanist

Boronia for Francesco Boroni (d. 1794), assistant to Humphrey Sibthorp in Greece
Borreria, borreri for William Borrer (1781–1862), British botanist
borszczovii, borszczowii for I. G. Borszczov (Borshchow) (1833–78), Russian explorer of Turkestan
boryanus -a -um, boryi for Baron J. B. M. Bory de St Vincent (1780–1846), French traveller and naturalist
borysthenicus -a -um from the environs of the Dnieper River (Greek, βορυσθενης), Belarus and Ukraine
boschianus -a -um for J. van der Bosch (1807–54), Governor General of the Dutch E Indies
Boscia, boscii for Louis A. G. Bosc (1759–1828), French professor of Agriculture
bosniacus -a -um from Bosnia, E Europe
Boswellia for Dr James Boswell (later Lord Auchinleck) (1740–95), lawyer, diarist and biographer of Samuel Johnson (*Boswellia sacra*, frankincense)
bothrio- minutely pitted-, pitted-, βοθρος
Bothriochloa Pitted-grass, βοθριο-χλοη (the pitted lower glume of some)
botry-, botrys bunched-, panicled-, βοτρυς, βοτρυ-
botrycephalus -a -um having a clustered head of flowers, βοτρυ-κεφαλη
Botrychium Little-bunch, βοτρυχιον,the fertile portion of the frond of moonwort
botryodes, botryoides, botrys resembling a bunch of grapes, grape-like, βοτρυς-ωδης, βοτρυς-οειδης
botryosus -a -um having many flower clusters, βοτρυς
Botrytis, botrytis -is -e Grape-like, βοτρυς (the microscopic appearance of the massed conidia on their conidiophores), or smelling fruity
botrytis -is -e racemose, racemed, bunched, βοτρυς
botuliformis -is -e sausage-shaped, alantoid, *botulus-forma*
botulinus -a -um shaped like small sausages, *botulus* (branch segments)
Bouchea for C. D. Bouche (1809–81)
Bougainvillea for Louis Antoine de Bougainville (1729–1811), French naval officer and circumnavigator (1766–9)
bourgaei, bourgaeanus -a -um, bourgeauianus -a -um for Eugene Bourgeau (1813–77), French traveller and collector
bourgatii for M. Bourgat, who collected in the Pyrenees *c.* 1866
Boussingaultia for Boussingault (1802–87), French chemist who recognized the plant's need for a supply of nitrogenous material (artificial fertilizers)
Bouteloua for Claudio (1774–1842) and Esteban Boutelou (1776–1813)
bovarius -a -um of cattle, *bos, bovis* (vide *boarius*)
bovicornutus -a -um Ox-horned, *bovis-cornu* (the leaf marginal processes)
bovinus -a -um of, or affecting cattle, *bos, bovis*
bowdenii for the collector, Athelstan Bowden, who first sent *Nerine bowdenii* to his mother in Newton Abbot in 1902
Bowenia for Sir George Fergusson Bowen (1821–99), first Governor of Queensland (1859), New Zealand (1863), Victoria (1877)
Bowiea for J. Bowie (1789–1869), who collected in Brazil for Kew
Bowkeria, bowkeri for James Henry Bowker (1822–1900) and his sister Mary Elizabeth, S African botanists
bowringianus -a -um for J. C. Bowring (1821–93), orchidologist of Windsor
boydii for William B. Boyd (1831–1918), rock gardener of Faldonside, Melrose, Scotland
Boykinia for Samuel Boykin (1786–1846), American field botanist of Milledgville, Georgia, USA
brachi-, brachy- short-, βραχυς, βραχυ-
brachialis -is -e about 18 inches in length, arm-like, βραχυς, *bracchium*
brachiatus -a -um branched at about a right-angle, widely branching, *brachialis*
Brachyachne Short-chaff, βραχυ-αχνη (the minute spikelets)
brachyandrus -a -um short stamened, βραχυ-ανδρος
brachyantherus -a -um having short stamens, βραχυς-ανθηρος

brachyanthus -a -um having short (tubular) flowers, βραχυς-ανθος
Brachyaria Armed-*Aira* (the arm-like branches of the inflorescence)
brachyarthrus -a -um short-jointed, having short internodes, βραχυς-αρθρον
brachyatherus -a -um having short beards, shortly-bearded, βραχυς-(αθηρ, αθερος)
brachybotryus -a -um, brachybotrys short-clustered, shortly bunched, βραχυ-βοτρυς
brachycalyx with a short calyx, βραχυς-καλυξ
brachycarpus -a -um short-fruited, βραχυς-καρπος
brachycaulos short-stemmed, βραχυς-καυλος
brachycentrus -a -um having a short spur, βραχυς-κεντρον
brachycerus -a -um short-horned, βραχυς-κερας
Brachychiton Short-tunic, βραχυς-χιτων
Brachycome (*Brachyscome*) Short-hair, βραχυς-κομη
brachycory having a short helmet, short-bracteate, βραχυς-(κορυς, κοροθος)
Brachycorythis Short-helmeted, βραχυς-κοροθος (the short adaxial petals)
Brachyglottis Short-tongue, βραχυς-γλωσσα (short ligulate florets)
brachylobus -a -um with short lobes, βραχυς-λοβος
brachypetalus -a -um with short petals, βραχυς-πεταλον
brachyphyllus -a -um with short leaves, βραχυς-φυλλον
Brachypodium Short-little-foot, βραχυς-ποδιον (pedicels 1–2 mm long)
brachypodus -a -um, brachypus short-stalked, βραχυς-πους
brachyscyphus -a -um shallowly cupped, βραχυς-σκυφος
brachysiphon short-tubed, βραχυς-σιφον (corolla tube shorter than wide)
Brachystachium, brachystachys, brachystachyus -a -um short-spiked, βραχυς-σταχυς
Brachystelma Short-crown, βραχυς-στελμα (the coronna)
brachytrichus -a -um short-haired, βραχυς-τριχος
brachytylus -a -um slightly swollen, short-pegged, with small callosities, βραχυς-τυλος
brachyurus -a -um short-tailed, βραχυς-ουρα (style of *Clematis*)
bracte-, bracteo- bract-, *brattea-, bractea-* (classically gold leaf)
Bracteantha having-bracteate-flowers, *bractea-florum*
bractealis -is -e, bracteatus -a -um with bracts, bracteate, *bractea* (as in the inflorescences of *Hydrangea, Poinsettia* and *Acanthus*)
bracteolaris -is -e having distinct bracteoles, diminutive of *bractea* (*bratteola* classically fine gold leaf)
bracteosus -a -um with large or conspicuous bracts, comparative of *bractea*
bractescens with late-enlarging bracts, becoming conspicuously bracteate, *bractea-essentia*
Bradleia for Richard Bradley (1688–1732), Cambridge professor and horticultural writer
brady- lazy-, slow-, heavy-, βραδυς
bradypus heavy-stalked, with slow stem growth, βραδυς-πους
Brahea for Tycho Brahe (1546–1601), Danish astronomer who wrote *De nova stella* (1573) and thereby questioned divine creation
Brainea for C. J. Braine, merchant who collected ferns in Hong Kong *c*. 1844–52
brandegeeanus -a -um, brandegeei, brandegei for Townsend S. Brandegee (1843–1925), N American collector and author
brandisianus -a -um, brandisii for Sir Dietrich Brandis (1824–1907), dendrologist of Bonn
brasilianus -a -um, brasiliensis -is -e from Brazil, Brazilian
Brassia for William Brass (d. 1783 at sea), orchidologist, collected in W Africa
Brassica Pliny's name, *brassica, brassicae,* for various cabbage-like plants (***Brassicaceae***)
Brassicella diminutive of *Brassica*
brassici- cabbage-, *Brassica-*
brassicifolius -a -um kale or cabbage leaved, *Brassica-folium*
brassicolens smelling of cabbage, *Brassica-*(*oleo, olere, olui*)
Brathys Dioscorides' name, βραθυ, for a *Juniperus* (≡ *Hypericum*)

braun-blanquetii for Josias Braun-Blanquet (1884–1980), pioneer of Life Form Spectra for regional floristic comparisons
braunianus -a -um for Johannes M. Braun (1859–93), German collector in Cameroon, W Africa
braunii for Alexander Braun (1805–77), professor of Botany at Karlsruhe
Bravoa for Leonardo Bravo (b. 1903) and Miguel Bravo (b. 1903), Mexican botanists
Braya for Count Francisci Gabriela von Bray (1765–1831), German botanist
braziliensis -is -e from Brazil, Brazilian
brazzavillensis -is -e from Brazzaville, Congo Republic (named for Pierre Paul François Camille Brazza (1852–1905), French explorer)
Brenania for J. P. M. (Pat) Brenan (1917–85), Director at Kew, collector on the second Cambridge expedition to Nigeria, 1947–48
bretschneideri for Emil Bretschneider (1833–1901), physician to the Russian Legation in Peiping (Peking), China *c.* 1870–80
brevi-, brevis -is -e short-, abbreviated-, *brevis, brevi-*
breviacanthus -a -um with short thorns, *brevis-acanthus*
brevialatus -a -um short, or narrow-winged, *brevis-alatus*
breviarticulatus -a -um having short joints, *brevis-articulus* (internodes)
brevibracteatus -a -um with short bracts, *brevis-bractea*
brevicalcar having short spurs, *brevis-(calcar, calcaris)* (floral)
brevicaudatus -a -um short-tailed, *brevis-(cauda, caudae)*
brevicaulis -is -e short-stemmed, *brevis-caulis*
brevicornu short-horned, *brevis-(cornu, cornus)*
breviculus -a -um somewhat short, diminutive of *brevis*
brevifimbriatus -a -um with a short fringe, *brevis-fimbriae*
breviflorus -a -um short-flowered, *brevis-florum*
brevifolius -a -um short-leaved, *brevis-folium*
brevifrons short-fronds, *brevis-(frons, frondis)*
breviglumis -is -e short-glumed, *brevis-gluma*
brevihamatus -a -um shortly-hooked, *brevis-hamatus* (the tip of a leaf etc)
brevilabris -is -e, brevilabrus -a -um short-lipped, *brevis-(labrum, labri)* (corolla)
breviligulatus -a -um having a short ligule, *brevis-ligula* (various tongue- or strap-shaped appendages)
brevilobis -is -e short-lobed, *brevis-lobus*
breviochreatus -a -um with a short ochrea, *brevis-ocrea* (stipular sheath in *Polygonaceae*)
brevior shorter, smaller, diminutive of *brevis*
brevipedicellatus -a -um short-pedicelled, *brevis-pedicellus* (flower stalk)
brevipedunculatus -a -um short-peduncled, *brevis-pedunculus* (inflorescence stalk)
breviperulatus -a -um with short-protective scales, *brevis-perula* (flower or leaf buds, literally small wallets)
brevipes short-stalked, short-stemmed, *brevis-podus*
brevipilis -is -e, brevipilus -a -um shortly hairy, with short stiff hairs, *brevis-pilus*
breviracemosus -a -um with short racemes, *brevis-racemus*
breviramosus -a -um having short branches, *brevis-(ramus, rami)*
brevirimosus -a -um with short cracks, *brevis-(rima, rimae)* (as on bark)
brevis -is -e of low stature, shallow, short-lived, short, *brevis*
breviscapus -a -um short-stalked, with a short scape, *brevis-(scapus, scapi)* (inflorescence stalk)
brevisectus -a -um cut short, blunt-tipped, *brevis-(seco, secare, secui, sectum)*
breviserratus -a -um with finely serrate margins, *brevis-(serra, serrae)*
brevispinus -a -um with short spines, *brevis-spina*
brevissimus -a -um the shortest, superlative of *brevis*
brevistylus -a -um short-styled, *brevis-stylus*
breweri for S. Brewer (1670–1743), gardener to the Duke of Beaufort at Badminton
Brexia Rain, βρεχις (for the protective leaves) (***Brexiaceae***)

Breynea (*Breynia*), *breynianus -a -um* for Jacob Breyne (1670–1743), collector of exotic plants in Danzig
Bridelia for S. E. von Bridel-Brideri (1761–1828), Swiss botanist
bridgesii for Thomas Bridges (1807–65), collector in S America
Brigandra the composite name for hybrids between *Briggsia* and *Didissandra*
brigantes, brigantiacus -a -um of the Brigantes tribe, *brigantes, briganticus* (of ancient Britain); or from Bregenz, Austria (*Brigantium*); or from Coruña, NW Spain (*Brigantium*)
brigantinus -a -um from the area around Lake Constance, Switzerland/Germany, (*Lacus Brigantinus*)
Briggsia for Munro Briggs Scott (1889–1917), Kew botanist
Brillantaisia for M. Brillant-Marion, who accompanied J. P. M. F. Palisot de Beauvois (1752–1820) in W Africa
briseis for Achilles' slave, Briseis
bristoliensis -is -e from Bristol (*Bristolium*)
britannicus -a -um from Britain, British (*Britannia*)
Briza Nodding, βριζω (an ancient Greek name, for rye, βριζα, Persian brizi, cognate with *oryza*)
briziformis -is -e, brizoides resembling *Briza, Briza-forma*, βριζω-οειδης
Brocchinia for Giovani Battista Brocchi (1772–1826), director of the Brescia botanic garden
broccoli little shoots, from Italian, broccolo, for cauliflower (q.v.), diminutive of *brocco* (*borecole* is from the Dutch, boerenkool, peasant's kale)
Brodiaea for James Brodie (1744–1824), Scottish botanist who discovered *Pyrola uniflora* in Britain
-broma meat, food, βρωμα, βρωμη
Bromelia for O. Bromel (1629–1705), Swedish botanist (***Bromeliaceae***)
bromeliifolius -a -um with *Bromelia*-like leaves
bromelioides *Bromelia*-like, *Bromelia-oides*
bromoides resembling brome grass, *Bromus*-like, βρομα-οειδης, *Bromus-oides*
Bromus Food, βρωμα (the Greek name, βρωμα, for an oat-like edible grass)
bronchialis -is -e throated, of the lungs, βρονχος, βρονχια, windpipe (medicinal use)
brooklynensis -is -e from Brooklyn, New York, USA
Brosimum, brosimus -a -um Edible-one, βρωσις, βρωσιμος
broteri, broteroi for Felix de Avellar Brotero (1744–1828), Professor of Botany at Coimbra, Portugal
-brotus an eating, edible, βρωσις, βρωτυς
broughtonii-aureum Broughton's golden, botanical Latin from Broughton and *aureum* (*Rhododendron*)
Broughtonia for Arthur Broughton, eighteenth-century Bristol physician and botanist, died 1796 in Jamaica
Broussonetia, broussonetii for Pierre Marie August Broussonet (1761–1807), Professor of Botany at Montpellier, France (paper mulberry)
Browallia for John Browall (1707–55), Bishop of Abo and supporter of Linnaeus' sexual system of classification
Brownea for Patrick Browne (1720–90), Irish physician and author of *History of Jamaica*
browneae for Señora Mariana Browne
brownei for Nicholas E. Brown (1849–1934), botanist at Kew
brownii for Robert Brown FRS (1773–1858), English botanist
Browningia for W. E. Browning, director of the Instituto Ingles, Santiago, Chile
Brownleea for Reverend J. Brownlee (1791–1871), missionary in Caffraria, S Africa
Bruckenthalia for S. von Bruckenthal (1721–1803), an Austrian nobleman
brumalis -is -e of the winter solstice, winter-flowering, *bruma, brumae*
Brunella the earlier spelling for *Prunella* (q.v.)
Brunfelsia for Otto Brunfels (1489–1534) who pioneered critical plant illustration
Brunia for Corneille de Bruin, Dutch traveller in the Levant (***Bruniaceae***)

bruniifolius -a -um with close awl-like leaves like *Brunia, Brunia-folium*
brunneifolius -a -um having brown leaves, *brunneus-folium*
brunneocroceus -a -um yellowish-brown, brownish-saffron, *brunneus-croceus*
brunneo-incarnatus -a -um brownish-flesh-coloured, *brunneus-incarnatus*
brunneoviolaceus -a -um brownish-violet, *brunneus-violaceus*
Brunnera for Samuel Brunner (1790–1844), Swiss botanist
brunnescens browning, turning brown, *brunneus-essentia*
brunneus -a -um russet-brown, *bruneus, brunneus*
Brunnichia for M. T. Brunnich, eighteenth-century Danish naturalist
Brunonia, brunonianus -a -um, brunonis Smaethman's name to commemorate Robert Brown (*vide infra*) (***Brunoniaceae***)
brunonianus -a -um, brunonis -is -e for Robert Brown FRS (see *brownii*)
Brunscrinum the composite name for hybrids between *Brunsvigia* and *Crinum*
Brunsdonna the composite name for hybrids between *Amaryllis belladonna* and *Brunsvigia*
Brunsvigia to the honour of the House of Brunswick (Charles William Ferdinand, Duke of Brunswick-Luneburg, was father of George IV's estranged wife, Caroline (1768–1821)
brutius -a -um from Calabria, S Italy (*Brutia*)
Bryanthus, bryanthus -a -um Moss-flower, βρυ-ανθος
bryicolus -a -um of mossy habitats, living with mosses, *bryum-cola*
brymerianus -a -um for W. E. Brymer MP, of Islington House, Dorchester
bryoides moss-like, βρυον-οειδης, *bryum-oides*
bryomorphus -a -um of moss-like form, βρυον-μορφη, *bryum-morphus*
Bryonia Sprouter, βρυω (Dioscorides also called the plant black vine, αμπελος μελαινα)
bryoniifolius -a -um having leaves resembling those of *Bryonia, Bryonia-folium*
bryophorus -a -um bearing epiphyllous bryophytes, moss-bearing, βρυω-φορος (the foliage)
Bryophyllum Leaf-sprouter, βρυω-φυλλον (ability to produce plantlets at leaf margins)
bryotrophis -is -e nourished by mosses, living on mosses, βρυον-τροφη
Bryum Moss, βρυον, *bryum*
bubalinus -a -um, bubulinus -a -um of cattle, of oxen, dull brown, *bubulus*
buboni- of the groin, βουβων
Bubonium a name for a plant used to treat groin swellings, βουβωνος
bucciferus -a -um, buciferus -a -um cheek-bearing, *bucca-fero* (the inner perianth members of *Tigridia*)
buccinatorius -a -um, buccinatus -a -um trumpet-shaped, horn-shaped, trumpeter, *bucinator, bucinatoris*
buccosus -a -um cheeky, distended like a trumpeter's cheek, *bucca, buccae*
bucculentus -a -um inflated, distended, *bucca, buccae*
bucephalophorus -a -um ox-head bearing, *bucephalus-fero* (? the fruiting heads)
bucephalus -a -um bull-headed, βου-κεφαλη (Bucephalus was Alexander the Great's favourite horse)
bucerus -a -um ox-horn-shaped, βου-κερας
Buchanani for Francis Buchanan Hamilton (1762–1829) of Calcutta Botanic Garden, John Buchanan (1819–98), specialist on New Zealand plants, or other Buchanans
bucharicus -a -um from Bokhara (Bukhara or Buchara), Uzbekistan
Buchloe Cow-grass, βοοσ-χλοη
Buchnera, buchneri for Dr Wilhelm Buchner of Nuremberg, alpine botanist
Buchosia for the French botanist P. J. Buc'hoz (1731–1807), author of *Plantes nouvelles découvertes* (1779), who named two species from Chinese art and saddled taxonomy with two inexact epithets (*Buchosia foetida* was raised to express contempt for his work)
bucinalis -is -e, bucinatus -a -um trumpet-shaped, trumpet-like, *bucina, bucinae*

Bucklandia for William Buckland (1784–1856), geology professor at Oxford and Dean of Westminster
Buckleya for S. B. Buckley (1809–84), American botanist
Buda an Adansonian name of no meaning
Buddleia (Buddleja) for Reverend Adam Buddle (*c.* 1660–1715), English vicar and botanist of Farnbridge (***Buddleiaceae***)
buddlejifolius -a -um with *Buddleia*-like foliage, *Buddleia-folium*
buddlejoides, buddleoides resembling *Buddleia*, *Buddleia-oides*
buergerianus -a -um, buergeri for Thomas J. Buertgers (*c.* 1881), collector in Kaiser Wilhelmsland
buffonius -a -um for Georges-Louis Leclerc, Compte de Buffon (1707–88), French evolutionist and author of *Histoire naturelle, générale et particulière*
bufo, bufonius -a -um of the toad, *bufo* (living in damp places)
bugeacensis -is -e from the Bugeac Plain, S. Moldova
Buglossoides bugloss-like, βουγλωσσος-οειδης
buglossus -a -um ox-tongued, βουγλωσσος, (the Greek name for an *Anchusa*, referring to the rough-textured leaves, bugloss is *Lycopsis arvensis*)
bugula the apothecaries' name for bugle, *Ajuga reptans*
bugulifolius -a -um with *Ajuga* or bugle-like leaves, *Bugula- folium*
bukobanus -a -um from Bukoba, Tanzania
bukoensis -is -e from Buka Island, Papua New Guinea
bulbi-,bulbo- bulb-, bulbous-, bulbus, βολβος, βολβο-
bulbifer -era -erum, bulbigerus -era -erum carrying bulbs, *bulbus-fero, bulbus-gero* (often when these take the place of normal flowers)
bulbigenus -a -um arising from bulbs, bulb-borne, *bulbus-genus* (*Drosera*)
bulbilliferus -a -um bearing bulbils, *bulbillus-fero* (small usually aerial bulb-like propagules)
Bulbine Little-bulb, diminutive from βολβος, the Greek name for a bulb
Bulbinella, bulbinellus -a -um diminutive of *Bulbine* (Maori onion, *B. hookeri*)
bulbispermus -a -um producing 'seed bulbs' or offsets of the parent bulb, βολβος-σπερμα
Bulbocastanum, bulbocastanus -a -um Chestnut-brown-bulbed, βολβο-καστανον (the tuber)
bulbocodioides resembling *Bulbocodium*, βολβος-κοδεια-οειδης
Bulbocodium Bulb-headed, βολβος-κοδεια (the inflorescence)
Bulbophyllym Bulb-leafed, βολβος-φυλλον (the pseudobulb is surmounted by the leaf)
Bulbostylis Bulbous-styled, βολβος-στυλος
bulbosus -a -um swollen, having bulbs, bulbous, βολβος
bulgaricus -a -um from Bulgaria, Bulgarian
bullatifolius -a -um having puckered leaves, bullate-leaved, *bulla-folius*
bullatus -a -um puckered, blistered, bullate, with knobbles, *bulla* (also signifies adolescence, wearing the *bulla* of childhood)
bulleyeanus -a -um for A. K. Bulley (1861–1942), rock-plant specialist of Ness, Cheshire
bulliarda for P. Bulliard (1752–93), French mycologist
bullulatus -a -um with small bumps or blisters, diminutive of *bulla, bullae*
bumalda for Josepheus Antonius Bumaldus Ovidio Montalbani (1601–71), botanical writer of Bologna
bumammus -a -um with cow-like teats, having large tubercles, botanical Latin from βου and *mamma*
Bumelia an ancient Greek name for an ash tree
-bundus -a -um suffix implying -having the capacity for, or -copiously
bungadinia a vernacular name, bunga diniyah, for sacred flower
bungeanus -a -um, bungei for A. von Bunge (1813–66), herbalist of Kiev, Ukraine
Bunias, bunias Linnaean generic name from the Greek name, βουνιας, for a kind of turnip; reapplied as an epithet for *Antidesma bunias*

Bunium a name, βουνιον, used by Dioscorides
buphthalmoides *Buphthalmum*-like, βου-οφθαλμος-οειδης
Buphthalmum, buphthalmus Ox-eyed, from medieval Latin, *oculus bovis* (Dioscorides' βουφθαλμον was a yellow ox-eyed composite)
bupleuroides resembling *Bupleurum*, βου-πλευρα-οειδης
Bupleurum Ox-rib, βου-πλευρα, an ancient Greek name, βουπλευρος, used by Nicander
buprestius -a -um with bright metallic colours, like or of jewel beetles, *Buprestidae*
Burbidgea, burbidgei for F. W. Burbidge (1847–1905). Collector in Borneo and Curator of the Botanic Garden, Trinity College Dublin
Burchardia for H. Burchardt, physician and writer on plants
burchardii for Oscar Burchard (*c.* 1864–1949), expert on Canary Island plants
Burchellia for William John Burchell (1781–1863), plant collector in St Helena, Brazil and S Africa
bureaui, bureauii, bureavii for Edouard Bureau (1830–1918), collector in China during 1894–8
bureauoides resembling *Rhododendron bureauii*
burejaeticus -a -um from the Bureya mountains, north of the Amur basin, E Russia
burgundensis -is -e from Burgundy (*Burgundiones*)
burjaticus -a -um from Buryatiya, east of Lake Baikal, E Siberia (Buryatskaya)
burkwoodii for Burkwood nursery, Kingston Hill, Surrey
burmanicus -a -um (birmanicus) Burmese, from Myanmar (Burma)
burmani-linearifolium Burman's linear-leaved, botanical Latin from Burman and *linea-folium*
Burmannia, burmannii either for J. Burmann (1707–79), physician, or Nicolaus Burmann (1733–93), Professor of botany at Amsterdam
burnatii for Emile Burnat (1828–1920), French botanist
burs-, bursa- leathery; pouch-, purse-, βυρσα, *bursa*
bursa-pastoris shepherd's purse, *bursa-(pastor, pastoris)*
Bursaria Resembling-a-pouch, *bursa* (the fruiting capsule)
Bursera, burseri, burserinus -a -um for Joachim Burser (1583–1649), author of *Introductis ad Scientiam Naturalem* (***Burseraceae***)
bursiculatus -a -um formed like a purse, pouch-like, *bursa*
Burtonia for D. Burton (d. 1792), Kew Gardens plant collector
Butea for John Stuart, third Earl of Bute (1713–92), who negotiated the end of the Seven Years' War with France
Butia uncertain etymology
Butomus Ox-cutting, βου-τομος (a name used by Theophrastus, for the sharp-edged leaves or the mouth bleeding caused by the acrid sap) (***Butomaceae***)
butyraceus -a -um greasy, oily, buttery, βουτυρον, *butyrum*
Butyrospermum Butter-seed, βουτυρον-σπερμα (oily seed of shea-butter tree)
butyrosus -a -um greasy, buttery, βουτυρον, *butyrum*
Buxbaumia, buxbaumii for Johann Christian Buxbaum (1683–1730), German botanical writer (the bug-on-a-stick-moss, for its capsule)
buxi-, buxi *Buxus*-, box-, πυξος, of box, living on *Buxus* (symbionts, parasites and saprophytes)
buxifolius -a -um box-leaved, *Buxus-folium*
Buxus an ancient Latin name, *buxus*, used by Virgil, for *B. sempervirens*, πυξος *(**Buxaceae**)*
Byblis for Byblis, the daughter of Melitus (insectivorous aquatic shrubs) (βυβλος, papyrus) (***Byblidaceae***)
byrs-, byrsa- pelt-, hide-, (leather-), βυρσα
Byrsonima Hide-necessity, *bursa-nimius* (murice, important in leather tanning in Brazil)
byssaceus -a -um like fine linen, cobwebbed, βυσσινος
byssinus -a -um textured or coloured like fine linen, βυσσινος

byssitectus -a -um with a covering like fine linen, botanical Latin from βυσσινος and *tectum*
byzantinus -a -um, byzantius -a -um from Istanbul (*Byzantium*, Constantinople), Turkish

cabbage from the medieval French name, caboche, for a head
cabardensis -is -e from Cabar, Croatia
Cabomba from a Guyanese vernacular name for *Cabomba aquatica* (***Cabombaceae***)
cabralensis -is -e from the Brazilian mountain chain named for Pedro Alvares Cabral (1467–1520), Portuguese navigator
cabrerensis -is -e from the environs of Cabrera, Dominican Republic, or for Jeronimo Luis de Cabrera, who founded Argentina's second largest city, Cordoba, or for several other persons bearing the name Cabrera
cabulicus -a -um from Kabul, Afghanistan
cabuya a W Indian vernacular name for a Mauritius hemp-like fibre
cac-, caco- bad-, dying or drying, κακ-, κακο-, καχ-
Cacalia Very-hurtful, κακο-λιαν (name used by Dioscorides)
cacaliifolius -a -um with leaves resembling those of *Cacalia, Cacalia-folium*
cacao Aztec name, kakahuatl, for the cacao tree, *Theobroma cacao* (cognate with the Nahuatl vernacular, xocoatl, cocoa and chocolate)
cacaponensis -is -e from the valley of the Cacapon river (confluent with the Potomac river), western Virginia, USA
cacatuus -a -um brightly coloured, botanical Latin from Malayan, kakatua
Caccinia for Mateo Caccini, seventeenth-century plant introducer of Florence
cacharensis -is -e from the Cachar administrative district of Assam
cachemerianus -a -um, cachemiricus -a -um from Kashmir, W Himalaya (Cachemere)
cachinalensis -is -e from Cachinal, N Chile
Cachrys Parched barley, καχρυς, or Pine-cone-like, *cachrys* (the appearance of the fruit)
cachyridifolius -a -um having strobilar-looking leaves, *cachrys-folius*
cacomorphus -a -um of bad form, ugly-looking, κακο-μορφη
cacti- cactus-like- (originally the Greek κακτος was an Old World spiny plant, not one of the *Cactaceae*)
cacticolus -a -um living with or on cacti, *Cactus-colo*
cactiformis -is -e succulent, cactus-like, *Cactus-forma*
Cactus Linnaeus' name, *Cactus*, derived from the former *Melocactus* (melon thistle) (***Cactaceae***)
cacumenus -a -um, cacuminis -is -e of the point, of the mountain top, *cacumen, cacuminis*
Cadia from the Arabic vernacular name, kadi
cadmeus -a -um for Cadmus, or from the area he established at Thebes (*Cadmea*)
cadmicus -a -um with a metallic appearance, *cadmia*
caduci- falling-, abscising-, *caducus*
caducifolius -a -um having leaves that fall early, caduceus-leaved, *caducus-folium*
caducus -a -um transient, not persisting, caducous, *caducus*
cadens tumbling, cascading, becoming pendulous, *cado, cadere, cecidi, casum*
caeciliae for Cecil J. Brooks (1875–*c.* 1953) who collected plants in Borneo
caecus -a -um blind, obscure, uncertain, dead-ended, (*intestinus-*)*caecus*
caelestis -is -e celestial blue, *caelestis*
caeno-, caenos- new, unheard of, strange, fresh-, recent-, καινος, καινο-
caenosus -a -um muddy, growing on mud, *caenum*
caeruleatus -a -um blued, made blue, blue-tinged, *caeruleus*
caeruleoracemosa having racemes of sky-blue flowers, *caeruleus-racemosus*
caerulescens turning blue, conspicuously blue, *caeruleus-esse*
caeruleus -a -um dark sky-blue, dark sea-green, dusky, *caeruleus*

Caesalpinia for Andrea Caesalpini (1519–1603), Tuscan botanist and physician to Pope Clement VIII (***Caesalpiniaceae***)
caesareus -a -um imperial, Caesar's (*Amanita caesarea*, Caesar's mushroom)
caesi-, caesius -a -um bluish-grey, lavender-coloured, *caesius*
caesiifolius -a -um having lavender-blue leaves, *caesius-folium*
caesiocyaneus -a -um greyish-blue, *caesius-cyaneus*
caesioglaucus -a -um glaucous-blue, with a bluish bloom, *caesius-glaucus*
caesiomurorum of the grey walls, *caesius-*(*murus, muri*) (*Hieraceum*)
caespitellus -a -um somewhat tufted, feminine diminutive from *caespes, caespitis*, a sod
caespiticius -a -um turf-forming, formed into turf, with matted roots, *caespes, caespitis*
caespitosus -a -um growing in tufts, matted, tussock-forming, *caespes, caespitis*
caffensis -is -e from Al Kaf (The Rock), NW Tunisia
caffer -ra -rum, caffrorum from S Africa, of the unbelievers, from Arabic, kaffir, kafir
cagayanensis -is -e from the Cagayan river area or Cagayan Sula Island, Philippines
cainito the W Indian vernacular name for the star apple, *Chrysophyllum cainito*
Caiophora (Cajophora) Burn-carrier, καιω-φορα (the stinging hairs)
cairicus -a -um from Cairo, Egypt (*Cairus*)
cajamarcensis -is -e from the Cajamarca department of N Peru
cajambrensis -is -e from the Cajambre river valley area, Colombia
Cajanus, cajan from the Malay vernacular name, katjang, for pigeon pea
cajennensis -is -e from Cayenne, French Guiana
cajonensis -is -e from either Sierra del Cajon, Argentina, or Cajon Pass, S California, or Cajon Canyon, SW Colorado, USA
cajuputi the Malayan name, kiya putih (white tree), for *Melaleuca cajuputi*, source of cajuput oil
Cakile from an Arabic name
cala- beautiful-, καλος, καλο-
calaba the W Indian vernacular name for the fruit and seed of *Calophyllum calaba*
calabriensis -is -e, calabrus -a -um, calabricus -a -um from Calabria, S Italy
calaburus -a -um W Indian vernacular name for the Jamaican cherry, *Muntingia calabura*
Caladenia Beautifully-glanded, καλος-αδην (the prominent, coloured glands on the labellum)
Caladium from the Indian name, kaladi (for an elephant ears aroid)
Calamagrostis, calamagrostis Reed-grass, the name, καλαμος-αγρωστις, used by Dioscorides
calamarius -a -um reed-like, resembling *Calamus*
calami- *Calamus-*, reed-, καλαμος
calamifolius -a -um reed-leaved, *Calamus-folium* (the phyllodes of broom wattle)
calamiformis -is -e having a reed-like form, *Calamus-forma*
calaminaris -is -e, calaminarius -a -um cadmium-red, growing on the zinc ore, calamine, *cadmia*
Calamintha, calaminthus -a -um Beautiful-mint, καλο-μινθη
calaminthifolius -a -um with leaves similar to *Calamintha*, *Calamintha-folium*
calamistratus -a -um curled, *calamister* (*calamistrum*) *calamistri*
calamitosus -a -um ruinous, causing loss, blighted, dangerous, miserable, *calamitosus*
calamondin the Philippine vernacular Tagalog name, kalamunding, for the fruit of × *Citrofortunella*
Calamovilfa Reed-like-*Vilfa*, botanical Latin from καλαμος and *Vilfa*
Calamus the name, καλαμος, for a reed (from Arabic, kalom) for the reed-like stems of rattan palms
calamusoides resembling a rattan palm, *Calamus-oides*
Calandrina for J. L. Calandrini (1703–58), Genevan botanist
calandrinoides resembling *Calandrina*, *Calandrina-oides*

calandrus -a -um having attractive stamens, καλ-ανηρ
Calanthe Beautiful-flower, καλ-ανθος
calanthoides resembling *Calanthe*, καλ-ανθος-οειδης
calanthus -a -um beautiful-flowered, καλ-ανθος
Calathea Basket-flower, καλαθος (the inflorescence)
calathiformis -is -e shaped like a cup or basket, concave, (*calathus, calathi*)*-forma*
calathinus -a -um basket-shaped, basket-like, *calathus, calathi*
calcaratus -a -um, calcatus -a -um spurred, having a spur, *calcar*
calcareophilus -a -um lime-loving, (*calx, calcis*)*-philus*
calcareus -a -um of lime-rich soils, chalky, *calcarius*
calcar-galli cock's-spur, *calcar, calcaris*
calcatus -a -um trampled on, spurned, *calco, calcare, calcavi, calcatus*; of chalk
calceatus -a -um shod, with shoe- or slipper-like structure, *calceatus*
calcensis -is -e from Calca, one-time Inca capital, Peruvian Andes
Calceolaria Slipper-like, *calceolus*
calceolaris -is -e, calceolatus -a -um shoe-shaped, slipper-shaped, *calceolus*
calceolus -a -um like a small shoe, diminutive of *calceus*
calci- lime-, *calx, calcis, calci-*
calcicolus -a -um living on limy soils, calcicole, *calcis-colo*
calcifugus -a -um disliking lime, avoiding limy soils, *calcis-fugo*
Calcitrapa, calcitrapa Caltrop, Old English, calcatrippe, for plants catching on one's feet (the fruit's resemblance of the spiked ball used to damage the hooves of charging cavalry horses) (≡ *Centaurea pro-parte*)
calcitrapoides *Centaurea*-like, resembling *Calcitrapa, Calcitrapa-oides* (for *Centaurea*)
calculus pebble-like, *calculus, calculi* (habit of some succulents)
caldasensis -is -e from the Caldas department, Central Andes of Colombia
caldasii for Francisco Jose de Caldas (1770–1816), director of Bogota Observatory and collector in S America
Caldcluvia for Alexander Caldcleugh (fl. 1820s–1858), who sent plants to England from Chile
Caldesia for Luigi Caldesi (1821–84), Italian botanist
Caleana for G. Caley (1770–1829), superintendent of St Vincent botanic garden
caledonensis -is -e from the area of the Caledon river, SE Africa, tributary of the Orange river
caledonicus -a -um from Scotland (*Caledonia*), Scottish, of northern Britain
Calendula First-day-of-the-month (Latin *calendae*, associated with paying accounts and settling debts; for pot marigold's long flowering period)
calendulaceus -a -um with golden flower-heads, resembling *Calendula*
calenduli- *Calendula-*, marigold-
calenduliflorus -a -um *Calendula*-like-flowered, *Calendula-florum*
Calepina an Adansonian name perhaps relating to Aleppo
calicaris -is -e like a small cup, *calix, calicis*
calicarpus -a -um fiery-fruit, calyx-fruited , καλυξ-καρπος (persistent calyx)
calicinus -a -um see *calycinus*
calicope having the appearance of a small cup, καλυξ-ωψ
caliculimentus -a -um having a cup-like depression on the lip, diminutive of *calix-mentum*
calidicolus -a -um of heat, inhabiting warm places, (*calidus, caldus*)*-colo*
calidus -a -um fiery, warm, *calidus, caldus*
californicus -a -um from California, USA
caligaris -is -e of dim places, of mists, *caligo, caliginis*
caliginosus -a -um dark, misty, obscure, *caliginosus*
calignis -is -e darkish, of mists, misty, *caligo, caliginis*
calimallianus -a -um from Calimalli, Baja California, Mexico
Calimeris Beautiful-parts, καλος-μερις (≡ *Aster*)
calipensis -is -e from Calipan, Mexico (pincushion cactus, *Coryanthes calipensis*)
Caliphruria Beautiful-prison, καλος-φρουρα (the spathe)

calisaya Andean vernacular for the yellow-bark *Cinchona calisaya*
calistemon resembling *Callistemon*
Calla Beauty, καλλος (a name, *calsa*, used in Pliny)
calleryanus -a -um for J. M. M. Callery (1810–62), missionary and botanist in Korea and China
calli-, callis- beautiful-, καλλος, καλλι-
Calliandra Beautiful-stamens, καλλι-ανηρ (shaving-brush tree)
Callianthemum Beautiful-flower, καλλι-ανθεμος
callianthus -a -um, callianthemus -a -um beautifully flowered, καλλι-ανθος, καλλι-αντηεμις
callibotryon, callibotrys having beautiful clusters, beautifully bunched, καλλι-βοτρυς
Callicarpa, callicarpus -a -um Beautiful-fruit, καλλι-καρπος (purple mulberry's metallic-violet drupes)
callicephalus -a -um having beautiful heads, beautifully headed, καλλι-κεφαλη
Callichilia Beautiful-box, καλλι-χηλος (the fruit)
callichromus -a -um beautifully coloured, καλλι-χρωμα
callichrous -a -um beautifully complexioned, with a beautiful surface, καλλι-(χρως, χρωτος)
callichrysus -a -um of a beautiful golden colour, καλλι-χρυσους
callicomus -a -um beautifully hairy, καλλι-κομη
callidictyon beautifully netted, καλλι-δικτυον (venation)
Calliergon Beautiful-thing, καλλι-εργον
callifolius -a -um *Calla*-leaved, *Calla-folium*
Calligonum Beautiful-joints, καλλος-γονυ (the nodes)
Callilepis Beautiful-scaled-one, καλλος-λεπις
callimischon beautifully pedicelled, καλλι-μισχος
callimorphus -a -um of beautiful form or shape, καλλι-μορφη
calliprinos beautifully early, καλλι-πριν; beautiful *Ilex* (*Prinus* ≡ *Ilex*; or the holly-leaved *Quercus prinus*)
Callipsyche Beautiful-soul, καλλος-ψυχη (Ψυχη may also be portrayed as a butterfly)
Callirhoe for the daughter of the river god, Achelous, and wife of Alcmaeon
Callisia Beauty, καλλος (appearance and violet-fragrance)
callistachys, callistachuus -a -um beautifully spicate, καλλι-σταχυς
callistegioides resembling *Calystegia, Calystegia-oides*
Callistemon, callistemon Beautiful-stamens, καλλι-στεμον (bottle-brush tree)
Callistephus Beautiful-crown, καλλι-στηφος (the flower-heads of China aster)
callisteus -a -um the most beautiful, καλλιστευω
callistus -a -um very beautiful, καλλιστευω
Callitriche Beautiful-hair, καλλι-(θριξ, τριχος) (***Callitrichaceae***)
callitrichis -is -e, callithrix having beautiful hair, καλλι-(θριξ, τριχος)
callitrichoides resembling *Callitriche*, καλλι-τριχος-οειδης
Callitris Beauty, καλλος (the general morphology)
Callitropsis *Callitris*-looking, καλλος-οψις
Callixene Beautiful-stranger, καλλι-ξενος (for the surprise that it graces Magellanica)
callizonus -a -um beautifully girdled, banded or zoned, καλλι-ζωνη (colouration)
callosus -a -um hardened, with a hard outer layer, *callosus*
Calluna Beautifier, καλλυνω (former common use as brushes)
callybotrion fine-racemed, botanical Latin from καλλος and *botryosus*
calo- beautiful-, καλλος, καλος, καλι-, καλο-, καλ-
calobotrys having beautiful clusters, καλο-βοτρυς
calocarpus -a -um beautiful fruit, καλο-καρπος
Calocedrus Beautiful-cedar, καλο-κεδρος
Calocephalus, calocephalus -a -um Beautiful-headed-one, καλο-κεφαλη
Calocera Beautiful-antler, καλο-κερας (the bright yellow fruiting bodies)
Calochilus, calochilus -a -um Beautiful-lip, καλο-χειλος (the labellum)
calochlorus -a -um of beautiful green colour, καλο-χλωρος

Calochortus Beautiful-grass, καλο-χορτος (the grass-like foliage)
calocodon having beautiful bell-shaped flowers, καλο-κωδων
Calocybe Beautiful-cover, καλο-κυβη (the flesh-pink pileus)
Calodendrum (-on), calodendron Beautiful-tree, καλο-δενδρον
calomelanos beautifully dark, καλο-μελανω
Calomeria Beautiful-parts, καλο-μερις
Caloncoba Beautiful-*Oncoba*, botanical Latin from καλλος and *Oncoba*
caloneurus -a -um well, or beautifully veined, καλλος- νευρα
Calonyction Beauty-of-night, καλο-νυξ (night flowering)
Calophaca Beautiful-lentil, καλο-φακη
calophrys with dark margins, καλο-οφρυς
Calophyllum, calophyllus -a -um with beautiful leaves, καλο-φυλλον
calophytus -a -um beautiful plant, καλο-φυτον
Calopogon Beautifully-bearded-one, καλο-πωγων
calopus -a -um beautifully stalked, καλο-πους (the stipe of *Boletus calopus* is shaded red below and has a white reticulation)
calorubrus -a -um beautifully red, botanical Latin from καλος and *rubrum*
Calostemma Beautiful-crown, καλο-στεμμα
calostomus -a -um beautifully throated, καλο-στομα
calostrotus -a -um beautifully covered, beautifully spreading, καλο-στρωτος
Calothamnus Beautiful-shrub, καλο-θαμνος
calothyrsus -a -um beautifully wreathed, καλο-θυρσος
Calotis Beautiful-eared-one, καλο-(ους, ωτος, -ωτικος) (the pappus-scales)
calotrichus -a -um having a beautiful indumentum, καλο-τριχος
Calotropis Beautiful-keel, καλο-τροπις (floral members)
calourus -a -um having beautiful tails, καλο-ουρα
caloxanthus -a -um of a beautiful orange-yellow colour, καλο-ξανθος
calpetanus -a -um from Gibraltar (*Calpe*)
calpi- urn(-shaped), καλπις, καλπι-
calpo- estuarine- (*Calpe* was the Rock or Straits of Gibraltar)
calpodendron estuarine tree, *calpe-dendron*
calpophilus -a -um estuary-loving, estuarine (*Calpe* was the Rock or Straits of Gibraltar)
Calpurnia for Calpurnius, who wrote in the style of Virgil (*Calpurnia* having affinities with *Virgilia*)
Caltha old Latin name, *caltha*, used by Pliny for a marigold, from καλαθος, goblet
calthifolius -a -um with *Caltha*-like leaves, *Caltha-folium*
Calvatia Bald-head-like, *calva, calvae* (appearance of the fruiting body)
calvatus -a -um smooth, bald-headed, hairless on top, *calva, calvae*
calvescens with non-persistent hair, becoming bald, present participle of *calvesco*
calviceps with a hairless head, *calva-caput*
calviflorus -a -um with hairless flowers, *calva-florus*
calvifolius -a -um smooth-leaved, *calva-folium*
calvus -a -um smooth, naked, hairless, bald, *calvus*
caly-, calyc-, calyci-, -calyx calyx-, καλυξ, καλυκος, καλυκο-
calycanthemus -a -um having flowers with conspicuous calyces, καλυξ-ανθεμος
Calycanthus Calyx-flower, καλυξ-ανθος (the undifferentiated tepals of the spiral perianth of allspice) (***Calycanthaceae***)
calycinoides resembling *calycinus* (in *Rubus*)
calycinus -a -um, calycosus -a -um with a persistent or conspicuous calyx, calyx-like, καλυξ
Calycocarpum Cup-fruit, καλυκο-καρπος (the concavity on one side of the stone)
calycogonius -a -um having the ovary attached to the calyx (hypanthium), καλυκο-γυνη
calycoideus -a -um calyx-like, καλυκο-οειδης, *calyx-oides*
Calycolobus Calyx-lobed-one, καλυκο-λοβος (the two enlarging calyx lobes)
Calycoma Leafy-calyx, καλυξ-κομη (foliate sepals)

Calycophyllum Calyx-leaf, καλυκο-φυλλον (the leaf-like expansion of one calyx lobe)
Calycopteris Winged-calyx, καλυκο-πτερον (the conspicuous keels of the calyx tube)
calycosus -a -um with a conspicuous calyx, καλυξ,
Calycotome Severed-calyx, καλυξ-τομη (abscission of the upper part of the calyx before anthesis)
calyculatus -a -um with a calyx-like bracts, resembling a small calyx, *calyculatus*
Calydorea Beautiful-gift, καλος-δορεα
calymmatosepalus -a -um having enveloping sepals, veiled by the sepals, καλυμμα-σκεπη
Calypso for Calypso, mythological nymph
calyptr-, calyptro- hooded-, lidded-, καλυπτρα, καλυπτρο-, καλυπτω, καλυπτος
calyptraeformis -is -e veil-like, καλυπτρα-*forma* (the conical pileus)
Calyptranthes Veiled-flower, καλυπτρα-ανθος (the calyx is calyptrate)
calyptratus -a -um with a cap-like cover over the flowers or fruits, καλυπτρα
Calyptridium Small-covering, diminutive from καλυπτρα
Calyptrocalyx Covering-calyx, καλυπτρα-καλυξ (the leafy-based spadices)
Calyptrochilum Covered-lip, καλυπτρο-χειλος
Calyptrogyne Covered-pistil, καλυπτρα-γυνη
calyptrostele having a sheathed style, καλυπτρο-στηλη
Calystegia (*Calistegia, Calycostegia*) Calyx-cover, καλυξ-στεγη (bearbind's calyx is at first obscured by two large bracts or prophylls)
Calythrix (*Calycothrix, Calytrix*) Calyx-hair, καλυξ-θριξ (the hair-like calyx apices)
camaldulensis -is -e from the Camaldoli gardens near Naples
camanchicus -a -um of the area of the Camanche tribe, USA
Camaridium Arched, καμαρα (the apex of the stigma)
Camarotis Vaulted-lobe, καμαρα-ωτος (the form of the lip)
camarus -a -um, -camarus -a -um arched, chambered, vaulted, καμαρα, καμαρη, *camera, camerae;* a W Indian name, kamara
Camassia a N American Indian name, quamash, for an edible bulb
cambessedesii for Jaques Cambessedes (1799–1863), co-author of *Flora Basilae Meridionalis*
cambodgensis -is -e from Kampuchea, SE Asia (Cambodia or, when in French control, Cambodge)
cambodiensis -is -e from Cambodia
cambogiensis -is -e from Cambodia
cambrensis -is -e, cambricus -a -um Welsh, from Wales (*Cambria*)
Camelina Dwarf-flax, χαμαι-λινον (a name used by l'Obel)
Camellia for Georg Joseph Kamel (Cameli) (1661–1706), Moravian Jesuit botanist, plant illustrator and traveller in Luzon
camelliiflorus -a -um *Camellia*-flowered, *Camellia-florum*
camelopardalis -is -e giraffe-like, with tawny spots, καμηλοπαρδαλις
camelorus -a -um of camels, καμηλος (they feed upon the camel thorn, *Alhagi camelorum*, also known as the manna plant because of the crust of dried honey-dew forming on the leaves overnight)
cameronicus -a -um, cameroonianus -a -um from Cameroon, W Africa
cammarus -a -um lobster, *cammarus* (from a name used by Dioscorides)
Camoensia for the Portuguese poet, Luis Camoens
Campanaea Bell, *campana, campanae* (shape of the corolla)
campanellus -a -um resembling a small bell, diminutive of *campana*
campani- bell-, *campana*
campaniflorus -a -um with bell-like flowers, *campana-florum*
campanilis -is -e bell-tower-like, from Italian (the tall bell-flowered inflorescences)
Campanula Bell-like, diminutive of *campana* (***Campanulaceae***)
campanulae of harebell, living on *Campanula* (coleopteran gall weevil)
campanularius -a -um, campanulatus -a -um, campanulus -a -um bell-shaped, bell-flowered, diminutive of *campana*
campanuloides resembling *Campanula*, *Campanula-oides*

campanus -a -um Capuan, from Campania, Italy
campbelliae, campbellii for Mrs and Dr Archibald Campbell (1805–74) respectively (he travelled with Hooker in the Himalayas)
campechianus -a -um of the Campeche area of the Mexican Yucatan peninsula
campester -tris -tre of the pasture, from flat land, of the plains, *campester*
camphorus -a -um, camphoratus -a -um camphor-like scented, from Arabic, kafur, Sanskrit, Karpura
Camphorosma, camphorosmae Camphor-odour, καμφορ-οσμη (the fragrance)
campicolus -a -um living on plains, of flat areas, *campus-colo*
campinensis -is -e from Campinas, São Paulo, Brazil
camporus -a -um of plains, savanna and open woodland, *campus, campi*
camposii for Dom Pedro del Campo, of Granada, Spain (*c.* 1849)
campto- bending-, curved-, καμπτος
camptocladus -a -um having bent or arching branches, καμπτο-κλαδος
Camptomanesia for Count Rodriguez de Camptomanes, Spanish patron of botany
camptophyllus -a -um having curved leaves, καμπτο-φυλλον
Camptosema curved-standard, καμπτο-σημα (the lateral appendages of the vexillum)
Camptosorus Curved-sorus, καμπτο-σορος
camptotrichus -a -um with curved hairs, καμπτο-(θριξ, τριχος)
Campsidium Resembling-*Campsis*
Campsis Curvature, καμπη (the bent stamens of trumpet creepers)
campto- (kampto-) bent-, καμπη, καμπτω
camptodon hooked teeth, καμπτω-οδοντος
Camptosorus Curved-sorus, καμπτω-σορος
Camptostylus Bent-style, καμπτω-στυλος (the long curved style)
Camptothecium Bent-theca, καμπτω-θηκη (the curved capsule)
campyl-, campylo- bent-, curved-, καμπυλος, καμπυλο-
Campylanthus Curved-flower, καμπυλος-ανθος
campylocalyx with a curved calyx (tube), καμπυλος-καλυξ
campylocarpus -a -um having curved fruits, καμπυλος-καρπος
Campylocentrum Curved-spur, καμπυλος-κεντρον
campylocladus -a -um with flexible or curved branches, καμπυλος-κλαδος
campylogynus -a -um having a curved ovary, καμπυλος-γυνη
Campylopus Curved-stalk, καμπυλος-πους
Campylotropis Curved-keel, καμπυλος-τροπις (the curved, rostrate keel petals)
camschatcensis -is -e from the Kamchatka Peninsula, Siberia
camtschatcensis -is -e, camtschaticus -a -um from the Kamchatka Peninsula, Siberia
camulodunus -a -um from Colchester (*Camulodunum*), Essex, England
canadensis -is -e from northern American continent, from Canada, Canadian
canaliculatus -a -um furrowed, channelled, *canalis*
Cananga from a Tagalog name, alang ilang, *Cananga odorata*, for the tree, the perfume produced from its flowers, and from the macassar or kenanga oil produced
canangoides resembling *Cananga, Cananga-oides*
canariensis -is -e of bird food, from the Canary Isles (*Canaria insula*, dog island, one of the *Insulae fortunatae*)
Canarina from the Canary Islands, *Canaria insula*
canarinus -a -um canary yellowish, resembling *Canarium*
Canarium the Malayan vernacular name, canari, for *Canarium commune*
Canavalia from the Malabar vernacular name, canavali
canbyi for William Marriott Canby (1831–1904), American botanist
cancellatus -a -um cross-banded, chequered, latticed, *cancelli, cancellorum* (a grating, crossbars or barrier, cognate with chancel, and chancellor, the porter who manned the barrier)
cancroides somewhat crab-like, (*cancer, cancri*)*-oides*
candelabrus -a -um candle-tree, like a branched candlestick, *candelabrum*

candelaris -is -e taper or candle-like, with tall tapering stems, *candela, candelae*
candelilius -a -um small candles, diminutive from *candela*
candens dazzling-white, *candens, candentis*
candicans white, hoary-white, with white woolly hair, present participle of *candico*
Candollea, candollei for Augustin Pyramus de Candolle (1778–1841), Professor of Botany at Geneva
candidissimus -a -um the whitest, superlative of *candidus*
candidulus -a -um off-white, whitish, diminutive of *candidus*
candidus -a -um shining-white, *candidus* (*candeo, candere*)
candiae from Crete, Cretan (*Candia*)
candius -a -um from Crete, Cretan (*Candia*)
Canella Little reed, diminutive of *Canna* (for the rolled, peeling bark) (***Canellaceae***)
canephorus -a -um like a basket bearer, bearing a basket-like structure, κανεον-φορα
canescens turning hoary-white, with off-white indumentum, present participle of *canesco, canescere*
caninus -a -um of the dog, sharp-toothed or spined, repellent to dogs, usually implying inferiority, wild or not of cultivation, *canis, canis*
canis-dalmatica white with spots, like a Dalmatian dog
Canistrum Wicker-basket, *canistrum, canistri* (the inflorescence)
Canna, canna Reed, *canna, cannae* (an uncertain comparison for Indian shot) (***Cannaceae***)
cannabifolius -a -um with *Cannabis*-like leaves
cannabinus -a -um *Cannabis*-like (leaves of hemp agrimony, wild hemp, κανναβις αγρια of Dioscorides)
Cannabis Hemp, κανναβις (Dioscorides' name for hemp) (***Cannabaceae***)
cannalinus -a -um somewhat like *Canna*, reed- or pipe-like, *canna*
cannifolius -a -um having leaves resembling those of *Canna*
cano- grey-, white-, hoary-, *canus*
canocarpus -a -um hoary-fruited, *cano-carpus*
canovirens greyish-green, *cano-virens*
-cans -being (e.g. *albicans* being white)
Canscora from a Malabari vernacular name, kansjan-cora, for *Canscora perfoliata*
cantabile having a smooth or flowing habit, Italian for singable
cantabricus -a -um from Cantabria, N Spain
cantabrigiensis -is -e from Cambridge (*Cantabrigia*)
cantaburiensis -is -e from Canterbury (*Cantaburia*) (England, New Zealand or USA)
cantalupensis -is -e from Villa Mandela (Cantalupo), between Tivoli and Arsoli, Italy, where warty melons (Cantaloups) were first raised from seed
canterburyanus -a -um for Alfred Milner (1845–1925), Viscount Canterbury, of Sturry Court, Kent
Cantharellula feminine diminutive of *Cantharellus*
Cantharellus Little-tankard, diminutive of *cantharus* (the fruiting body of the edible chanterelle fungus) (*cantharis*, Spanish-fly beetle)
Canthium from the Malabar vernacular name, cantix
cantianus -a -um from Kent, England (*Cantium*)
cantonensis -is -e, cantoniensis -is -e from Kuangchou, or Guangzhuo, SE China (Canton)
Cantua from a Peruvian vernacular name, cantu, for *C. buxifolia*
cantuariensis -is -e from Canterbury (*Durovernum* or *Cantaburia*)
canus -a -um whitish-grey, white, *caneo, canere, canui*
caoutchouc from the Amerindian vernacular name, cauchuc, for the solidified latex of *Hevea brasiliensis*
capax wide, broad, able, fit, *capax*
capensis -is -e from Cape Colony, or Cape of Good Hope, S Africa
caperatus -a -um of goats, *caper, capri*
caperratus -a -um wrinkled, *caperro, caperrare*

capillaceus -a -um, capillaris -is -e hair-like, very slender, *capillus, capilli*
capillaensis -is -e from either Capilla near Buenos Aires, or Capilla del Monte near Cordoba, Argentina
capillatus -a -um long-haired, *capillatus*
capillifolius -a -um with thread-like leaves, *capillus-folium*
capilliformis -is -e hair-like, thread-like, *capillus-forma*
capillipes with a very slender stalk, *capillus-pes*
capillus-junonis Juno's-hair, *capillus-iunonis* (Jupiter's wife, ≡ Hera)
capillus-veneris Venus' hair, *capillus-(Venus, Veneris)*
capinensis -is -e from Capina, Brazilian Amazonia
capistratus -a -um muzzled, strap-like, haltered, *capistrum, capstri*
capitatus -a -um growing in a head, head-like, *caput, capitis* (inflorescence or stigma)
capitellatus -a -um growing in a small head, small head-like, diminutive of *capitatus*
capitis-york from the Cape York Peninsula, Queensland, Australia, (*caput, capitis*)-York
capitulatus -a -um having a small head, *capitulum*
capituliflorus -a -um having flowers in small heads, diminutive of *capitus*
capituliformis -is -e shaped into small capitulae, *capitulum-forma*
capnodes, capnoides smoke-coloured, καπνος-οειδης
Capnophyllym Smoke-leaf, καπνος-φυλλον
capollin, capulin from a vernacular name for the fruit of *Prunus cerasifera salicifolia*
cappadocicus -a -um, cappadocius -a -um from Cappadocia, Asia Minor (Turkey)
Capparis from the Arabic, kabar, for capers, through Greek, καππαρις (***Capparidaceae***)
capra, capri- of the goat, goat-like (smell), *capra, caprae*
capraeus -a -um of the roe, reddish, *caprea, capreae*
Capraria Of-the-she-goat, *capra, caprae* (can also mean with the smell of arm-pits)
capreolatus -a -um sprawling between supports, twining, winding, tendrilled, *capreolus*
capreolus the roe deer, roebuck-coloured, *capreolus*
capreus -a -um, caprinus -a -um goat, of goats, *capra, caprae*
capricornus -a -um, capricornu of the winter solstice, of capricorn or the goat's horn, *capra-cornu*
Caprifolium Goat-leaf, *caprae-folium* (an old generic name) (***Caprifoliaceae***)
Capriola Uncultivated, *caprificus* (≡ *Cynondon*)
Capsella Little-case, diminutive of *capsa* (the form of the fruit of shepherd's purse)
capsicastrum resembling *Capsicum*, biting, peppery, *Capsicum-astrum*
Capsicum Biter, καπτο (the hot taste of peppers), or Case, *capsa*
capsularis -is -e producing capsules, capsular-fruited, *capsula, capsulae*
capuli, capulinos see *capollin*
caput, capitis head, *caput, capitis* (cognate are the Italian cabochia, French caboche and English cabbage)
caput-avis bird's-head, *caput-avis*
caput-bovis ox-head, *caput-(bos, bovis)*
caput-galli cock's-head, *caput-gallus*
caput-medusae Medusa's head, *caput-Medusa*
caput-viperae adder's-head, snake-headed, serpentine, *caput-(vipera, viperae)*
caracalla for Emperor Caracalla (Marcus Aurelius Severus Antoninus Augustus) (188–217), who gave Roman citizenship to all free people of the then empire, in AD 212; some interpret as spiralled or to do with charcoal
caracalus -a -um lynx-like (*Felis caracala*), having black (ear-like) tufts, Turkish, kara-kulak, black-ears
caracarensis -is -e from the Cara Cara mountains of Bolivia
caracasanus -a -um from Caracas, Venezuela
Caragana the Mongolian name, caragan, for the plant
caraguatatubus -a -um from Caraguatatuba, Brazil

caraguatus -a -um a S American vernacular name for fibres of *Eryngium pandanifolium*, and other plants, or from the Caraguata river of Uruguay, or Caraguatay, Paraguay
Caralluma possibly a Telinga (British India) vernacular name, Car-allum
caramanicus -a -um from Karaman, Turkey, a former principality of Anatolia
carambola a Portuguese vernacular name for the carambola-tree (*Averrhoa carambola*)
caramicus -a -um from Karamea, New Zealand
carana a vernacular name for the balsam resin from species of *Protium*
carandas the Malayan vernacular name, karandang, for Christ's thorn (*Carissa carandas*)
Carapa from a S American vernacular name, caraipa, for *Carapa procera*
carassanus -a -um from Carassa, Minas Gerais, Brazil
carataviensis -is -e from Karatau, Kazakhstan
carbonarius -a -um of charcoal burners, of burnt ground, *carbonarius, carbonari* (*Poliota* habitat)
carchariodontus -a -um having sharp teeth, καρχαρ-οδοντος
Cardamine Dioscorides' name, καρδαμινη, for cress
cardaminifolius -a -um having leaves resembling those of *Cardamine*
cardaminioides resembling *Cardamine*, καρδαμινη-οειδης
Cardaminopsis *Cardamine*-resembling, καρδαμινη-οψις
Cardamon the Greek name, καρδαμον, for garden cress
cardamomum ancient Greek name for the Indian spice, καρδαμωμον
Cardaria Heart-like, καρδια (the fruiting pods)
cardi-, cardio- heart-shaped-, καρδια, καρδιο-
cardiacus -a -um antispasmodic, dyspeptic, of heart conditions, καρδιακος (medicinal use)
Cardiandra Heart-shaped-stamens, καρδια-ανηρ (the shape of the anthers)
cardianthus -a -um with heart-shaped flowers, καρδια-ανθος
cardinalis -is -e deep-scarlet, the colour of the cassock worn by a Cardinal, upon whom the Roman Church hinges (*cardo*, a hinge)
cardiobasis -is -e heart-shaped base, καρδιο-βασις
Cardiocrinum Heart-lily, καρδιο-κρινον (the leaves of giant lilies)
cardiopetalus -a -um with heart-shaped petals, καρδιο-πεταλον
cardiophyllus -a -um with heart-shaped leaves, καρδιο-φυλλον
Cardiospermum Heart-seed, καρδιο-σπερμα (refers to the white, heart-shaped aril on the black seeds)
cardiostictus -a -um with heart-shaped markings, καρδιο-στικτος
cardius -a -um for the heart, καρδια
carduacus -a -um thistle-like, *carduus, cardui; cardus*
carduchorus -a -um of the Kurdes (Carduchoi), from Kurdistan
carduelinus -a -um somewhat thistle-like, *carduus*
cardui- *Carduus*-, thistle-, living on *Carduus* (symbionts, parasites and saprophyts)
Carduncellus, carduncellus -a -um Little-*Cardunculus*
Cardunculus, cardunculus -a -um Thistle-like, small thistle-like, diminutive of *carduus* (cognate with cardoon)
Carduus Thistle, a name, *carduus*, in Virgil; Celtic, ard, for a prickle
Careya for Reverend William Carey (1761–1834) of Serampore, botanist and linguist
careyanus -a -um, careyi for John Carey (1797–1880), plant collector in America
Carex Cutter, κειρο (the sharp leaf margins of many)
caribaceus -a -um, caribaeus -a -um from the Caribbean (*Caribaea*)
Carica From-Caria (mistakenly thought to be the provenance of the Pawpaw, *Carica papaya*) (***Caricaceae***)
carici- sedge-, *carex, caricis*
caricifolius -a -um with sedge-like leaves, *Carex*-leaved, *carici-folius*
caricinus -a -um, caricosus -a -um sedge-like, resembling *Carex*

caricoides resembling Carex, *carex-oides*
caricis -is -e of *Carex*, *caricis*
caricus -a -um, cariensis -is -e from the ancient region of Caria, SW Anatolia, Turkey
cariflorus -a -um having keeled or boat-shaped flowers, *carina-florus*
carinalis -is -e keel-like, *carina*
carinatus -a -um keeled, having a keel-like ridge, *carina, carinae*
carinellus -a -um having a small keel, diminutive of *carina*
carinensis -is -e from Karin on the N coast of Somalia
carinthiacus -a -um from Kärnten (*Carinthia*), S Austria
carinulatus -a -um slightly keeled, diminutive from *carina*
cariocae, cariocanus -a -um from the Carioca mountain range, Brazil
caripensis -is -e from Caripito and environs of Rio Caribe, N Venezuela
Carissa from Sanskrit, krishna-pakphul, for *Carissa carandas*, Christ's thorn or carunda; some derive from Italian for beloved but etymology is unclear
carissimus -a -um most esteemed or revered
carlcephalum for the *Viburnum* hybrid between *V. carlesii* and *V. macrocephalum*
carlesii for William Richard Carles (1848–1929), British Consul in China, who collected *Viburnum carlesii* in W Korea (1889)
Carlina for Charlemagne (742–814), *Carolinus*, his army was supposed to have been cured of the plague with a species of *Carlina*, which the Archangel had revealed to him
carlinus -a -um thistle-like, like *Carlina*, of *Carlina*
carlsruhensis -is -e either from Karlsruhe, SW Germany, or from Carlsruhe, Poland
Carludovica for Charles IV of Spain (1748–1819) and his wife Louisa (1751–1819)
carmanicus -a -um from the region of Kerman, Iran (*Carmannia* of Alexandrian times)
carmesinus -a -um carmine-like, reddish-purple, medieval Latin from Arabic, kirmiz or qirmiz (*vide kermesinus*)
Carmichaelia, carmichaelii for Captain Dugald Carmichael (1772–1827), plant hunter, author of Flora of Tristan da Cunha
carminatus -a -um relieving flatulence, *carmino, carminare, carminavi, carminatum* (*carmen, carminis*, song, prophecy, formula, moral text, cognate with charm)
carmineus -a -um carmine, Arabic kirmiz or qirmiz (*vide kermesinus*)
carnauba Tupi vernacular for the wax palm, *Copernicia cerifera*, and its leaf-wax
Carnegiea for the Scottish philanthropist Andrew Carnegie (1835–1919)
carneolus -a -um fleshy, flesh-coloured, succulent, *carneus*
carneus -a -um, carnicolor the colour of flesh, of flesh, *caro, carnis; carneus*
carniolicus -a -um from Carniola, Slovenia
carnosiflorus -a -um having fleshy flowers, *carnosus-florum*
carnosulus -a -um somewhat fleshy, diminutive of *carnosus*
carnosus -a -um pulpy, fleshy, thick and soft textured, *carnosus*
carnulosus -a -um somewhat fleshy, diminutive of *carnosus*
caroli for Carl Ludwig Ledermann (1875–1958), Swiss curator of Victoria Botanic Garden, Cameroon
carolinae-septentrionalis from North Carolina
carolinianus -a -um, carolinensis -is -e, carolinus -a -um of N or S Carolina, USA
carota the old name, καρωτον, for carrot (*Daucus carota*)
caroviolaceus -a -um fleshy-violet coloured (*caro, caronis*)-*violaceus*
carpathicus -a -um, carpaticus -a -um, carpathus -a -um from the Carpathian mountains between Vienna and Romania
carpathus -a -um from the Carpathian mountains
Carpenteria for William M. Carpenter (1811–1848), Professor at Louisiana
carpetanus -a -um of the *Carpetano* tribe, from the Toledo area of Spain
carpini- hornbeam-like-, *Carpinus-* (followed by a structure, e.g. leaf or fruit)
carpini of hornbeam, living on *Carpinus* (symbionts, parasites and saprophytes)
carpinifolius -a -um hornbeam-leaved, *Carpinus-folium*
carpinoides resembling *Carpinus, Carpinus-oides*

Carpinus, carpinus the ancient Latin name, *carpinus*, for hornbeam, some derive it from Celtic, car-pix, for a wood-headed yoke
carpo-, carpos-, -carpus -a -um (karpo-) fruit-, -fruited, -podded, καρπος, καρπο-, botanical Latin *carpus*
Carpobrotus Edible-fruit, καρπο-βρωτυς (the edible fruiting structure of hottentot fig)
Carpodetus Bound-fruit, καρπο-δετης (external appearance of the putaputawheta fruit)
Carpodinus Top-like-fruit, καρπος-δινευω
Carpolyza Angry-fruit, καρπος-λυσσα (its dehiscence)
carpophilus -a -um fruit liking, καρπος-φιλη (fungus on rotting beech-mast)
carrerensis -is -e from Carrera Island, Trinidad
Carrichtera for Bartholomaeus Carrichter, physician to Maximillian II, author of Kreutterbüch (1575)
Carrierea, carrierei for E. A. Carrière (1816–96), French botanist
carringtoniae for Lady Carrington, wife of the Governor of New South Wales, Sir Charles Robert Wynn Carrington
carsonii for Alexander Carson (1850–96), who collected plants in Tanganyika
cartagensis -is -e from Cartagena, either Spain or Colombia (*Carthago Nova*)
cartagoanus -a -um from Cartago, Costa Rica
carthaginiensis -is -e from Carthage (*Carthago*), Tunis
carthamoides resembling *Carthamus, Carthamus-oides*
Carthamus Painted-one, from Hebrew, qarthami, an orange-red dye (false saffron, Arabic, safra, is made from safflower, *Carthamus tinctorius*)
carthusianorum, carthusianus -a -um of the Grande Chartreuse Monastery of Carthusian Monks, Chartreuse (*Carthusia*), France
cartilagidens having firm but not bony teeth, *cartilago-dens*
cartilagineus -a -um, cartilaginus -a -um cartilage-like, *cartilago, cartilaginis* (texture of some part, e.g. leaf margin)
cartwrightianus -a -um for J. Cartwright, who, as British Consul in Constantinople, discovered *Crocus cartwrightianus* (*c.* 1844)
Carum from Caria, Dioscorides' name, καρως, for caraway
carunculatus -a -um with a prominent caruncle, *carunculus* (seed coat outgrowth, usually obscuring the micropyle, literally a bit of flesh)
carunculiferus -a -um bearing a caruncle, *carunculus-fero*
carvi (*carui*) from Arabic, karwiya, caraway (Pliny derives it from an origin in Caria, Asia Minor)
carvifolius -a -um caraway-leaved, *carui-folium*
Carya Walnut, καρυα (Dion's daughter, Carya, was changed into a walnut tree by Bacchus)
caryo- (karyo-) nut-, clove-, καρυον, καρυο-
Caryocar Nut, καρυον (for the butter-nut) (***Caryocaraceae***)
caryocarpus -a -um having dry indehiscent fruits, καρυο-καρπος
Caryolopha Nut-crest, καρυο-λοφια (the nutlets are borne in a ring)
caryophyllacearus -a -um of chickweeds, living on *Caryophyllaceae (Melampsorella,* basidiomycete witches broom on *Abies* and telutospore phase on *Cerastium* and *Stellaria*)
caryophyllaceus -a -um, caryophylleus -a -um resembling a stitchwort, clove-pink coloured, from Arabic, karanful, for cloves or clove pinks
caryophylloides resembling *Dianthus caryophyllus*, clove-pink-like
Caryophyllus Nut-leaved, καρυοφυλλον (a former generic name for clove tree, Arabic karanful, *Eugenia caryophyllata*); clove-fragrance or colour in other genera has transferred this meaning to the epithet, and given such cognate names as gillyflower (*Dianthus caryophyllus, Orobanche caryphyllacea, Cyperus caryophyllea*)
caryopteridifolius -a -um *Caryopteris*-leaved, *Caryopteris-folium*
Caryopteris Wing-nut, καρυο-πτερυξ (the fruit-body splits into four winged nutlets)
Caryota a name, καρυοτις, used by Dioscorides for a date palm

caryotideus -a -um like a small palm, *Caryota*-like
caryotifolius -a -um having leaves resembling those of fish-tail palms, *Caryota-folium*
caryotoides resembling *Caryota, Caryota-oides*
cascadenis -is -e from the volcanic Cascade mountain range from British Columbia to California, or from waterfalls elsewhere
cascarillus -a -um resembling the purgative *Croton cascarilla* (cascara sagrada, Spanish for sacred cascara, was once a common laxative/purgative, made from the bark of *Rhamnus purschiana*)
cashemirianus -a -um, cashmiriensis -is -e from Kashmir
cashew from the Tupi vernacular name, acaju, for the nuts
Casimiroa for Cardinal Casimiro Gomez de Ortega (1740–1818), Spanish botanist
caspicus -a -um, caspius -a -um of the Caspian area
Cassia a name, κασια, used by Dioscorides from a Hebrew plant name, quetsi'oth, used by Linnaeus for *C. fistula* (medicinal senna)
cassidius -a -um helmet-shaped, *cassis, cassidis*
Cassine, cassine from a N American vernacular name for hottentot cherry (formerly the generic name for *Mauricenia capensis*)
cassinifolius -a -um having leaves resembling those of *Cassine, Cassine-folium*
cassinoides resembling *Cassine, Cassine-oides* (hottentot-cherry-like)
Cassinia for Count A. H. G. de Cassini (1781–1832), French botanist
cassioides resembling *Cassia*, κασια-οειδης
Cassiope, cassiope for the Queen of Ethiopia and mother of Andromeda in Greek mythology; heath-like, *Cassiope*-like
Cassipourea from a vernacular name from Guyana
cassubicus -a -um from Gdansk, Danzig (*Cassubia*), Poland
Castalia Spring-of-the-Muses, at Castalia, Mount Parnassus, Greece (≡ *Nymphaea*)
Castanea old Latin name, *castanea*, for the sweet chestnut, from the Greek καστα (cognate with Kastanis and Chestnut)
castaneaefolius -a -um, castaneifolius -a -um with leaves resembling those of *Castanea*
castaneus -a -um, castanus -a -um chestnut-brown, *castanea, castanae*
castanii-, castani- chestnut-, chestnut-brown-, *castanea*
castaniifolius -a -um with leaves resembling those of the chestnut, *castanea-folius*
castanioides resembling *Castanea, castanea-oides*
castanopsicolus -a -um living on *Castanopsis, Castanopsis-colo* (epiphytic)
Castanopsis Chestnut-like, καστα-οψις, *castanea-opsis*
Castanospermum *Castanea*-seeded, καστα-σπερμα
castellanus -a -um of a fortress, *castellanus*; from Castille, Spain (*Castella*)
castello-paivae for Baron Castello de Paiva, of Portugal
castellorus -a -um of strongholds or high places, *castellum, castelli*
Castilleja for Domingo Castillejo (1744–93), Spanish botanist of Cadiz
castus -a -um chaste, clean, flawless, spotless, pure, *castus*
Casuarina Cassowary-like, from the Malayan vernacular name, pohon kasuari (compares the pendulous branches with the feathers of *Cassuarius cassuarius*) (***Casuarinaceae***)
casuarinoides resembling *Casuarina, Casuarina-oides*
cat-, cata-, cato- below-, outwards-, downwards-, from-, under-, against-, along-
Catabrosa Eating-up, καταβρωσις to swallow (the nibbled appearance of the tip of the lemmas, and also much liked by cattle)
catacanthus -a -um having downwards-pointing thorns, κατ-ακανθος
catacosmus -a -um adorned, κατακοςμεω
catafractus -a -um enclosed, armoured, closed in, mail-clad, καταφρακτος
catalaunicus -a -um from Catalonia, Montserrat, Spain (*Catalaunia*)
catalinae from Catalina, USA
Catalpa, catalpa from an E Indian vernacular name, katuhlpa
catalpifolius -a -um *Catalpa*-leaved, *Catalpa-folius*
catamarcensis -is -e from Caramarca, NW Argentina

Catananche, catananche Driving-force, καταναγκε (of Cupid's dart, used by Greek women in love potions)
cataphractus -a -um enclosed, covered, shut in, covered in armour, καταφρακτος, *cataphractes, cataphractae*
cataphysaemus -a -um having a swollen lower portion, κατα-φυσαω
Catapodium Trivial-stalk, κατα-ποδιον (the spikelets are subsessile)
catappa from a native E Indian name for olive-bark tree
cataractae, cataractarus -a -um, (catarractae, catarractarum) growing near waterfalls, tumbling like a waterfall, καταρακτης
cataractalis -is -e cascading, καταρακτης
cataria of cats, late Latin, *cattus*, old name for catmint (*herba cataria*)
Catasetum Downwards-bristle, κατα-σετα (the two cirri on the column)
catawbiensis -is -e from the area of the North American Indian Catawba tribe, from the Catawba River area, N Carolina, USA
catechu a Tamil vernacular name, caycao or kaku or katti-shu, for the betel (*Acacia catechu*)
catenarius -a -um, catenatus -a -um linked, chain-like, *cateno, catenare, catenavi, catenatum*
catenulatus -a -um somewhat resembling a small chain, somewhat fettered, diminutive from *catenus*
cateriflorus -a -um four-flowered, French quatre-fleur; well-flowered, κατηρης
caterviflorus -a -um having crowded flower-heads, (*caterva, catervae)-florum*
Catesbaea, catesbyi for Mark Catesby (1674–1749), of Suffolk, author of *Natural History of Carolina*
Catha from an Arabic vernacular name, khat, for *Catha edulis* (the leaves are eaten and used to brew a beverage)
Catharanthus Perfect-flower, καθαρος-ανθος
catharticus -a -um cleansing, purging, cathartic, καθαρτης, καθαρτικος
cathartius -a -um cleansing, purifying, καθαρτης; of the territory of the vulture, *Cathartae*
cathayanus -a -um, cathayensis -is -e from China (Cathay)
Cathcartia, cathcartii for John Ferguson Cathcart (1802–51), Judge in Bengal
catholicus -a -um Linnaeus used this to imply of Catholic lands (Spain and Portugal), orthodox, worldwide, universal, καθολικος
catingaensis -is -e from the dry, thorn forests (caatingas) of S and C America
catingicolus -a -um living in the Brazilian caatinga (thorn-scrub woodland or white forest)
Catonia name, *sedi incertis*, for Marcus Porcius Cato (234–149 BC), author of *De agri cultura* (160 BC)
Catopsis Looking-down, κατω-οψις (epiphytic Bromeliads)
catopterus -a -um having wings on the lower part, κατ(ο)-πτερον
Cattleya, cattleyanus -a -um for William Cattley (d. 1832), English plant collector and patron of botany
caucalifolius -a -um with leaves resembling those of *Caucalis, Caucalis-folium*
Caucalis, caucalis old Greek name, καυκαλις, for an umbelliferous plant
caucasiacus -a -um, caucasicus -a -um from the Caucasus, Caucasian
caucasigenus -a -um born in or originating in the Caucasus, *Caucasia-(gigno, gignere, genui, genitum)*
cauda- tail-, *cauda, caudae* (used for any long appendage)
cauda-felis cat's-tailed, *cauda-(feles, felis)*
caudatifolius -a -um with tailed leaves, *caudatus-folius* (apices)
caudatilabellus -a -um with the lip drawn out into a tail, *caudatus-labellum*
caudatus -a -um, caudi- produced into a tail, tailed, *caudatus* (see Fig. 7a)
caudescens developing tails, *cauda-essentia*
caudiculatus -a -um with a thread-like caudicle, diminutive of *cauda* (as the tail-like threads, *caudiculae*, of orchid pollinia)
caul-, caule-, cauli, caulo- stalk-, καυλος, *caula, caulae*

Caulerpa Cabbage-stalk, *caulis* (the stipe)
caulescens becoming distinctly stalked, beginning to stem, *caulis-essentia*
caulialatus -a -um having winged or alate stems, *caula-alatus*
cauliatus -a -um, -caulis -is -e, -caulo, -caulos of the stem, -stemmed, -stalked, καυλος, botanical Latin *-caulis*
caulibarbis -is -e with bearded stalks, *caula-(barba, barbae)* (the beard may consist of rigid hairs or barbs)
caulicolus -a -um living on stems, *caula-colo* (certain fungi)
cauliculatus -a -um with diminutive stalks, diminutive of *caula* (may be sprouts on an old stem or stipes of fungi)
cauliferus -a -um stem-bearing, *caula-fero*
cauliflorus -a -um bearing flowers on the main stem, flowering on the old woody stem, *caula-florum* (cocoa (*Theobroma*) flowers and fruits on old stems and is cauliflorous, but the English cauliflower derives from cole-flower or cabbage-flower)
caulocarpus -a -um fruiting on the stem, καυλος-καρπος
Caulophyllum Stem-leaf, καυλος-φυλλον (the stalk has a single compound leaf at its apex)
caulorapus -a -um stem-turnip, *caulo-rapa* (kohlrabi)
caulorrhizus -a -um having rooting stems, with adventitious roots, καυλος-ριζα
caurinus -a -um of the NW wind (*Caurus* or *Corus*) (a seasonal feature)
causticus -a -um with a caustic taste, burning, καιειν, καυστος, καυστικος, (causing inflammation or having a hot taste)
cauticolus -a -um growing on cliffs, cliff-dwelling, *cautes-colo*
cautleoides resembling *Cautlea, Cautlea-oides*
Cautleya, Cautlea for Major General Sir P. Cautley (1802–71), British naturalist
cautleyoides resembling *Cautlea, Cautlea-oides*
Cavendishia, cavendishii for George Spencer Cavendish (1790–1858), Sixth Duke of Devonshire, at Chatsworth
cavenius -a -um hedging, for cages, *cavea, cavae* (its use as living fences)
cavernicolus -a -um growing in caves, cave-dwelling, *caverna-colo*
cavernosus -a -um full of depressions or holes, *caverna*
cavernus -a -um of caves or cavities, *caverna*
cavipes hollow-stalked, *cavus-pes*
cavus -a -um of caves, excavated, hollow, cavitied, *cavo, cavare, cavavi, cavatum*
cayapensis -is -e from the area of the Cayapa Indians of W Equador
cayenensis -is -e, cayennensis -is -e from Cayenne, French Guyana (Cayenne pepper is named from its Tupi Guarani vernacular name, kyinha, from Cayenne Island)
Cayratia the vernacular name, cay-rat, in Annam for a vine (≡ *Columella*)
cazorlanus -a -um, cazorlensis -is -e from the Cazorla mountains, Andalusia, Spain
ceanothoides resembling *Ceanothus*, κεανοθος-οειδης
Ceanothus Linnaeus' re-use of the ancient Greek name, κεανοθος, used by Theophrastus for another plant
cearensis -is -e from Ceará state, NE Brazil
cebennensis -is -e from the Cevennes (*Cebenna*), France
cebolletus -a -um chive-like, from the French for chives, ciboullete
cecidiophorus -a -um, cecidophorus -a -um bearing galls, κηκις-φορος
ceciliae for the daughter of Sir Frederick A. Weld, once Governor of the Straits Settlements
Cecropia for Cecrops, legendary king and builder of ancient Athens
cecropioides resembling *Cecropia, Cecropia-oides*
cedarbergensis -is -e from Cedarberg, Cape Province, S Africa, or Wisconsin
cedarmontanus -a -um, cedrimontanus -a -um from the Cedarberg mountains, Cape Province, S Africa, or Cedar mountain, Oregon, or Cedarberg, Wisconsin
cedratus -a -um, cedron, cedrum of cedars, resinous, fragrant, κεδρος, *cedrus, cedri* (in the Latin sense, *Juniperus*, perfume and oils are also included)
Cedrela Cedar-like, diminutive of κεδρος (the wood has a similar fragrance)
cedricolus -a -um living on cedar, *Cedrus-colo*

cedroensis -is -e from Cedros Island off Baja California peninsula, Mexico
Cedronella Resembling-cedar, diminutive of κεδρος (fragrance)
cedrorus -a -um of cedars, *Cedrus*
Cedrus the ancient Greek name, κεδρος, for a resinous trees with fragrant wood, Arabic, kedri
Ceiba from a vernacular S American name for silk-cotton tree
ceilanicus -a -um from Sri Lanka (Ceylon, Ceilan)
celans hiding, becoming hidden, present participle of *celo, celare, celavi, celatum*
celastri-, celastrinus -a -um *Celastrus*-like-
Celastrus Theophrastus' name, κηλαστρος, for an evergreen tree, retaining fruit over winter (possibly an *Ilex*) (***Celastraceae***)
celatocaulis -is -e with concealed stems, *celans-caula* (concealed by density of growth)
celatus -a -um hidden, concealed, *celo, celare, celavi, celatum*
celebicus -a -um from the Indonesian island of Sulawesi (Celebes)
celer, celeratus -a -um hastened, growing rapidly, *celero, celerare*
cellulosus -a -um with little rooms, many-celled, *cellula, cellulae* (tube-stalked sori of *Trichomanes cellelosum*)
-cellus -a -um -lesser, -somewhat
Celmisia for Celmisius, the son of the nymph Alciope, in Greek mythology
Celosia Burning (from κελος, for the burnt or dry flowers of some)
celosioides resembling *Celosia, Celosia-oides*
Celsia, celsianua -s -um, celsii for Olaf Celsius (1670–1756), professor at Uppsala University, author of *Hierobotanicon*
celsimontanus -a -um of high mountains, *celsus-montanus*
celsus -a -um haughty, eminent, loft, high, *celsus*
celtibiricus -a -um from Aragon, Spain (*Celtiberia*)
celticus -a -um from Gaul, of the area of the Celtic people, *celtae, celtarum; celticus*
celtidifolius -a -um with leaves resembling those of *Celtis, Celtis-folium*
Celtis ancient Greek name, κηλτις, for a tree with sweet fruit; Linnaeus applied this to the European hackberry
cembra the old Italian name for the arolla or Swiss stone pine; some derive it from German, Zimmer, a room
cembroides, cembrus -a -um resembling *Pinus cembra, cembra-oides*
Cenarrhenes Empty-male, κενος-αρρην (the stamen-like glands)
cenchroides resembling *Cenchrus*, κεγχρος-οειδης
Cenchrus Piercing-one, κεγχρος, (the involucre of sharp, sterile spikelets create burrs that attach to animal fur)
Cenia Empty, κενος (the hollow receptacle of the inflorescence)
cenisius -a -um from Mont Cenis (Monte Cenisío) on the French–Italian border
ceno-, cenose- empty-, fruitless-, κενος, κενο-, κεν-
cenocladus -a -um having empty (leafless) branches, κενο-κλαδος
Cenolophium Hollow-bristles, κενο-λοφια
Centaurea Centaur, Centauros (mythical creature with the body of a horse replacing the hips and legs of a man, the name, κενταυριον, κενταυρειον, used by Hippocrates, *centaureum, centauria* in Pliny; in Ovid the centaur Chiron was cured with this plant of Hercules' arrow wound in the hoof)
centaureoides resembling *Centaurea*, κενταυριον-οειδης
Centaurium, centaurium for the centaur, Chiron, who was fabled to have a wide knowledge of herbs and used this plant medicinally, cognate with *Centaurea*
centi- one hundred-, many-, *centum*
centifolius -a -um many-leaved, *centum-folium*
centilobus -a -um many-lobed, *centum-lobus*
Centopedia, centipedus -a -um Many-stemmed-one, *centum-pes*
Centotheca Prickly-sheath, κεντο-θηκη (reflexed bristles on the upper lemmas)
centra-, centro-, -centrus -a -um, -centron spur-, -spurred, κεντρον, κεντρο-, κεντρ-
Centradenia Spur-gland, κεντρ-αδεν (the anthers have spur-like glands)

centralis -is -e in the middle, central, *centrum* (distributional or systematic position)
Centrantherum Spurred-anthers, κεντρον-ανθερος
centranthifolius -a -um having leaves resembling those of *Centranthus*
centranthoides resembling *Centranthus*, κεντρον-ανθερος-οειδης
Centranthus (Kentranthus) Spur-flower, κεντρον-ανθος (valerians)
centratus -a -um spined, spurred, κεντρον, botanical Latin, *centratus*
centrifugus -a -um developing outwards from the centre, *centrum-fugo*
centro- sharply-pointed-, spur-like-, κεντρον, κεντρο-
centrodes spur-like, κεντρον-ωδης
Centronia Spur, κεντρον (the spurred anthers)
Centropetalum Spurred-petal, κεντρον-πεταλον, (the appendages on the labellum)
Centropogon Centred-beard, κεντρο-πογον (the fringe around the stigma)
Centrosema Spurred-standard, κεντρο-σημα (lateral appendages of the vexillum)
Centunculus Small-patch or Saddle-cloth, *centunculus, centunculi* (Pliny's name, re-used by Dillenius (chaffweed, ≡ *Anagallis minima*)
centunculus saddle-cloth, *centunculus, centunculi* (velvety tan surface of *Naucoria* pileus)
cepa, cepae- the old Latin name, *caepa*, for an onion, onion- (cognate with cive and chive) (*ascalonia caepa*, onion from Ascalon, Palestine, gives us the cognate, shallot)
cepaceus -a -um onion-like, *caepa*
cepaeifolius -a -um onion-leaved, *caepa-folium*
cepaeus -a -um grown in gardens, κηπαιος, from the ancient Greek for a salad plant
Cephaelis Head-like, κεφαλη (the dense corymbose inflorescences)
cephal-, cephalidus -a -um head-, head-like-, κεφαλη, κεφαλ-
Cephalanthera Head-anther, κεφαλ-ανθερα (its position on the column)
Cephalanthus, cephalanthus -a -um Head-flower, κεφαλ-ανθος (flowers are in axillary globose heads)
Cephalaria Head, κεφαλη (the capitate inflorescence)
cephalidus -a -um having a head, κεφαλη
cephallenicus -a -um from Cephalonia (κεφαλλινια), one of the Ionian Islands
Cephalocereus Headed-*Cereus*, botanical Latin, *cephalo-cactus* (the woolly flowering head)
cephalonicus -a -um from Cephalonia (κεφαλλινια), one of the Ionian Islands
Cephalophora, cephalophorus -a -um Head-bearer, κεφαλη-φορα (with capitate inflorescence)
Cephalophyllum Head-of-leaves, κεφαλη-φυλλον
Cephalostigma Headed-stigma, κεφαλη-στιγμα
Cephalotaxus Headed-yew, κεφαλη-ταξος (the globose heads of staminate 'flowers' of plum yews) (***Cephalotaxaceae***)
Cephalotus, cephalotus -a -um, cephalotes Large-flower-headed, with large flower-heads, having a head-like appearance, κεφαλη-ωτης (***Cephalotaceae***)
-cephalus -a -um -headed, κεφαλη, with small heads, κεφαλις
cepifolius -a -um onion-leaved, *caepa-folius*
cepiformis -is -e having the shape of an onion, *cepae-forma*
-ceps -heads, -headed, κεφαλη, *caput, capitis*; from French cèpe, Gascon cep, mushroom
cepulae of or upon *Alium* species (*Urocystis* smut fungus)
ceraceus -a -um waxy, κερος; *cera, cerae; cereus*
ceracophyllus -a -um waxy-leaved, κερος-φυλλον
cerae- waxy-, κερος, *cera, cerae*
Ceramanthe, Ceramanthus Jug-flower, Imprisoning-flower, κεραμος-ανθη(ος) (the form of the corolla)
ceramensis -is -e from Seram island, Indonesia
ceramicus -a -um of the potter, of clay soils, κεραμος (κεραμιχος, κεραμινος, κεραμευς a potter, *Ceramicus, Ceramici* an Athenian cemetery)
-ceras -horned, -podded, κερας, κερατος

ceraseidos resembling (*Prunus*) *cerasus*
ceraseus -a -um cherry-like, *cerasus*
cerasifer -era -erum bearing cherries, *cerasus-fero* (cherry-like fruits)
cerasiformis -is -e cherry-shaped, *cerasus-forma*
cerasinus -a -um cherry-red, *cerasus*
cerasocarpus -a -um cherry-fruited, with cherry-like fruits, *cerasus-carpus*
cerastioides resembling *Cerastium, Cerastium-oides*
Cerastium Horned, κερας (the fruiting capsule's shape)
Cerasus, cerasus from an Asiatic name, kirhas, κερασος, Caucasian (Lucullus imported the sour cherry to Rome from a place later named *Cerasus*, in *Pontus*, Asia)
ceratacanthus -a -um having horned spines, κερατο-ακανθος
ceratites, ceratinus -a -um horn-like, κερατινος (texture or shape)
cerato- horn-shaped-, κερας, κερατος, κερατο-
ceratocarpus -a -um with horn-shaped fruits, κερατο-καρπος
ceratocaulus -a -um with horned stems, κερατο-καυλος
Ceratochloa Horned-grass, κερατο-χλοη (the lemmas are horn-like)
Ceratolobus Horned-pod, κερατο-λοβος (the horned spathe resembles a pod)
Ceratonia Podded-one, κερατιον (the fruit of the locust tree, the seeds of which provided the unit of weight, the carat, of goldsmiths)
Ceratopetalum Antler-petal, κερατο-πεταλον (the antler-like petals of *Ceratopetalum gummiferum*)
ceratophorus -a -um horned, horn-bearing, κερατο-φορα (the corolla spurs)
Ceratophyllum Horn-leaf, κερατο-φυλλον (the stag's horn shape of the leaf) (***Ceratophyllaceae***)
Ceratopteris Horned fern, κερατο-πτερυξ (the appearance caused by the inflexed margins of the fertile fronds of floating stag's-horn fern)
Ceratosanthes Horned-flowered-one, κερατος-ανθος (petal shape)
Ceratostema Horned-stamened-one, κερατο-στεμα
Ceratostigma Horned-stigma, κερατο-στιγμα (the shape of the stigmatic surface)
Ceratotheca Horned-box, κερατο-θηκη (the fruit shape)
Ceratozamia Horned-*Zamia*, κερατο-*zamia* (the cone-scales bear two horns)
ceratus -a -um waxy, *ceratus*
Cerbera (us) Dangerous-one, after *Cerberus*, the three-headed guardian dog of Hades (the plant's poisonous properties)
Cercestis from the Greek name, κερκις, a weaver's shuttle
cercidifolius -a -um *Cercis*-leaved, *Cercis-folius*
Cercidiphyllum, cercidiphyllus -a -um *Cercis*-leaved (***Cercidiphyllaceae***)
Cercidium Shuttle-like, κερκις (the fruits are rod-like or flattened pods)
cerciformis -is -e rod-like or shuttle-like, *cercis-forma*
Cercis The ancient Greek name, κερκις (Judas tree fruit resembles a weaver's shuttle, κερκις, κερκιδος)
Cercocarpus Tail-fruit, κερκιδος-καρπος (the tail-like, persistent, long, plumose style on the fruit)
cerealis -is -e of meal, of corn, for *Ceres*, the goddess of agriculture
cerebriformis -is -e having a brain-like form, convoluted like a walnut, *cerebrum-forma*
cerefolius -a -um pleasing-leaved, χαιρειν-φυλλον (cognate with *caerefolium*, *Chaerophyllum*, and chervil)
cereiferus -a -um, cerifer -era -erum wax-bearing, (κερα, κερος)-φερω
cereoides resembling *Cereus*, *cereus-oides*
cereolus -a -um, cereus -a -um waxen, waxy-yellow, κερα, *cereus*
Cereus, cereus -a -um Torch, *cereus* (wax or wax taper, for the flower-shape)
cereusculus -a -um slightly waxen, diminutive of *cereus*
cerinanthus -a -um having waxy flowers, κερα-ανθος
Cerinthe Wax-flower, κερα-ανθος (≡ *Hieracium cerinthoides*)
cerinthoides resembling *Cerinthe*, honey-wort-like, κερυνθη-οειδης

cerinus -a -um the colour of bees-wax, waxy-yellow, κερα, κερος
Ceriporiopsis Looking-like-porous-wax, κερα-πορος-οψις (texture of fruiting body)
cernuus -a -um drooping, curving forwards, facing downwards, *cernuus*
Ceropegia Fountain-of-wax, κερος-πηγη (appearance of the inflorescence)
Ceroxylon Waxy-wood, κερος-ξυλον (for the waxy exudate on the trunks)
cerris the ancient Latin name, *cerrus*, for turkey oak
cerulatus -a -um waxen, waxy, *cerula, cerulae*
cerusatus -a -um white, appearing to be painted with white lead, *cerussa, cerussae*
cervianus -a -um from Cervia, N Italy; of the hind or stag, *cervus, cervinus* (*Mollugo cervianus*)
cervicarius -a -um, cervicatus -a -um constricted, keeled, *cervix, cervicarius*
cervicornis -is -e, cervicornu curved like a deer's horn, *curvus-cornu*
cervinus -a -um tawny, stag-coloured, *cervus, cervinus* (a hind)
Cespedesia for Juan Maria Cespedes, priest of Santa Fé de Bogota
cespitosus -a -um growing in tufts (see *caespitosus*)
cestroides *Cestrum*-like, κεστρον-οειδης
Cestrum an ancient Greek name, κεστρον, of uncertain etymology
Ceterach etymology dubious, from an Arabic name, chetrak, for a fern, or from the German, Krätze, for an itch (referring to the scurfy epidermis)
cetratus -a -um armed with a targe, *caetratus* (shape of the pileus)
ceylanicus -a -um from Ceylon (Sri Lanka)
chacaoensis -is -e from Chacao, Venezuela, or the area of the Chacao Straits, Chile
chacoanus -a -um from the chaco (annually flooded flatlands of Argentina, extending into Paraguay and Bolivia)
Chaenactis Gaping-ray, χαινω-ακτις (the ray florets have a conspicuous mouth)
chaeno- splitting-, gaping-, χαινω
Chaenomeles Gaping-apple, χαινω-μηλον (Japanese quince)
chaenomeloides resembling *Chaenomeles*, χαινω-μηλον-οειδης
Chaenorhinum, Chaenorrhinum Gaping-nose, χαινωρρινον (analogy with *Antirrhinum*)
Chaenostoma Gaping-mouth, χαινω-στομα (the corolla)
chaero- pleasing-, rejoicing-, χαιρω
chaerophylloides resembling *Chaerophyllum*, χαιρω-φυλλον-οειδης
Chaerophyllum, chaerophyllus -a -um, chaerophyllon Pleasing-leaf, χαιρω-φυλλον (the ornamental foliage)
Chaetacanthus Hair-thorned, χαιτη-ακανθα
Chaetanthera Haired-anther, χαιτη-ανθερα (the anthers have hair-tufts)
chaeto- long hair-like-, χαιτη, χαιτ-
Chaetocalyx Bristly-calyx, χαιτη-καλυξ
chaetocarpus -a -um hairy-fruited, χαιτη-καρπος
Chaetogastra Bristly-belly, χαιτη-γαστηρ (the bristly calyx tube)
chaetomallus -a -um having a main, fleecy-haired, χαιτη-μαλλος
chaetophyllus -a -um fleecy-leaved, χαιτη-φυλλον
chaetorhizus -a -um having hair-like or hair-covered roots, χαιτη-ριζα
chailaricus -a -um of Chailu, Punjab region of Pakistan (salt range area)
chaixii for Abbé Dominique Chaix (1731–1800), French botanist and a collaborator of Villars in producing *Histoire des plantes Dauphinoises*
chalarocephalus -a -um having open or tired-looking heads (of flowers), χαλαρος-κεφαλη
chalcedonicus -a -um from Chalcedonia, Turkish Bosphorus
chalcospermus -a -um having seeds of a coppery appearance, χαλκο-σπερμα
chalepensis -is -e from Aleppo, from the Arabic name (see *halapensis*)
chalusicus -a -um from Chalus, France, or the Chalus river area of the Elbrz mountains, N Iran
chalybaeus -a -um grey, the colour of steel, χαλυψ, χαλυβος
chamae- on-the-ground-, lowly-, low-growing-, prostrate-, false-, χαμαι
Chamaeangis Dwarf-vessel, χαμαι-αγγειον

Chamaebatia Dwarf-bramble, χαμαι-βατος
Chamaebatiaria *Chamaebatia*-like
chamaebuxus like dwarf box, χαμαι-πυξος
chamaecarpus -a -um fruiting upon the ground, χαμαι-καρπος
chamaecerassus like dwarf cherry, χαμαι-κερασος
Chamaecereus Prostrate-*Cereus*
chamaecissus ivy of the ground, *chamaicissus*, χαμαικισσις
chamaecistus (chamaeacistus) like dwarf *Cistus*, χαμαι-*Cistus*
Chamaecladon Short-branched, χαμαι-κλαδος (the peduncle)
chamaecristus -a -um with a small crest, χαμαι-, *crista*
Chamaecyparis Dwarf-cypress, χαμαι-κυπειρος
Chamaecyparissus, chamaecyparissus Pliny's name, from χαμαι-κυπαρισσος, for a ground-hugging cypress
Chamaecystus Dwarf-ivy, χαμαι-(κισσος, κιστος)
Chamaedaphne Ground-laurel, χαμαι-*Daphne*
chamaedendrus -a -um dwarf tree, χαμαι-δενδρον
Chamaedorea Low-gift, χαμαι-δορεα (the accessible fruits)
chamaedoron lowly-gift, gift from the earth, χαμαι-δωρον
chamaedrifolius -a -um chamaedrys-leaved, *chamaedrys-folium*
chamaedryoides resembling *chamaedrys, chamaedrys-oides*
Chamaedrys Ground oak, Theophrastus' name, χαμαιδρυς, for a small oak-leaved plant
chamaeiris dwarf-Iris, χαμαι-*Iris*
chamaejasme dwarf jasmine-like, dwarf jessamine, botanical Latin from χαμαι and yesamin or *Jasminum*
Chamaelaucium, Chamelaucium derivation uncertain (dwarf λαυκανια throat?)
chamaeleus -a -um ground-lion, covering the ground, χαμαιλεον (from the Greek name for the chameleon)
Chamaelirium Dwarf-lily, χαμαι-λειριον
chamaemelifolius -a -um *Chamaemelum*-leaved, *Chamaemelum-folium*
Chamaemelum Ground-apple, χαμαι-μελον (the habit and fragrance), chamomile
chamaemespilus dwarf *Mespilus*, botanical Latin from χαμαι and *Mespilus*
chamaemoly dwarf-magic-herb, χαμαι-μολυ (*Allium*)
Chamaemorus Dwarf-mulberry, botanical Latin from χαμαι and *morus*
Chamaenerion Dwarf-oleander, χαμαι-νηριον Gesner's name for rosebay willow herb
chamaepeplus -a -um ground-robe, clothing the ground, χαμαι-πεπλος
Chamaepericlymenum Dwarf-climbing-plant, χαμαι-περικλυμενον
chamaepeuce ground fir, χαμαι-πευκη
chamaephytus -a -um ground plant, χαμαι-φυτον (perennials with their resting buds at about ground kevel)
chamaepitys Theophrastus' name for a dwarf pine-like plant, χαμαι-πιτυς
Chamaeranthemum dwarf *Eranthemum*, botanical Latin from χαμαι- εραω-ανθεμιον
chamaerops Low-bush, χαμαι-ρωψ (in comparison with the tall, tropical fan-palms)
chamaesulus -a -um quite close to the ground, botanical Latin diminutive from χαμαι
chamaesyce dwarf fig tree, χαμαι-συκη
chamaethomsonii dwarf (*Rhododendron*) *thomsonii*
Chamaethyoides Dwarf-*Thuja*-like
chamaeunus -a -um lying on the ground, χαμαι-ευνης
chamaezelus -a -um jealous of the earth, ground seeking, χαμαι-ζηλοω
Chamagrostis Dwarf-*Agrostis*, χαμ-αγρωστις (≡ *Mibora*)
chamberlaynianus -a -um, chamberlaynii for Reverend Hon Joseph Chamberlayn (1836–1914), orchid grower of Birmingham, England
chamelensis -is -e from Chamela, Mexico
chameleon changing appearance, χαμαιλεον
Chamelum Humble, χαμελος (the dwarf habit)

chamissoi, chamisonis -is -e for Louis Charles Adalbert von Chamisso (1781–1838) poet and botanist on the Romanzof expedition (1815–18)
Chamomilla, chamomilla Apple-of-the-ground, Dioscorides' name, χαμαιμηλον, for a plant smelling of apples
champaca the Hindi vernacular name for *Michelia champa*
championiae, championi for John George Champion (1815–54), who collected in Hong Kong and Ceylon
champlainensis -is -e from the envirous of Lake Champlain, New York and Vermont, USA
chaneti for L. Chanet, who found *Sedum fimbriatum-chaneti*
changuensis -is -e from Changu, Sikkim, India
chantavicus -a -um from Chantada, Spain
chantrieri for Chantrier Frères, nurserymen of Mortefontaine, France
chapadicolus -a -um living in the Chapada do Araripe mountains of northeastern Brazil
chapalensis -is -e from Chapala, Mexico
Chaptalia for M. Chaptal (1756–1831), French chemist
Chara Delight, χαρα (camphor fragrance and morphological appearance); *chara* was a name for an unidentified vegetable
Characias, characias the name in Pliny for a spurge with very caustic latex
charantius -a -um graceful, χαρα, χαρις (the pendent fruits)
charianthus -a -um with elegant flowers, χαρις-ανθος
Charieis Charming, χαριεις
-charis -graceful, -pleasant, χαρις, χαριτος
charitopes with graceful stems, *caritas-pes*
charkoviensis -is -e from Charkov (Kharkov), Ukraine
chartaceus -a -um parchment-like, papery, χαρτης, *charta, chartae*
chartophyllus -a -um with a papery textured leaves, χαρτης-φυλλον
Chascanum Yawning, χασκω (the gaping corolla)
Chasmanthe, chasmanthus -a -um Gaping-flower, χασμα-ανθος
Chasmanthium Gaping-flower, χασμα-ανθινος
chasmophilus -a -um liking crevices, χασμα-φιλεω
chasmophyticus -a -um crevice-living plant, χασμα-φυτον
Chasmopodium Agape-foot, χασμα-ποδος (pedicels are angled to the rachis)
Chassalia Open-mouthed, botanical Latin from χασμα
chathamicus -a -um from the Chatham Islands, S Pacific
Chauliodon With-projecting-teeth, χαυλι-οδων (the lip's callus before the opening of the spur)
chauno- gaping-, χαινω
chaunostachys having open spikes, χαινω-σταχυς
chebulicus -a -um, chebulus -a -um from Kabul, Afghanistan
Cheilanthes Lip-flower, χειλ-ανθος (the false indusium of the frond margin covers the marginal sori)
cheilanthifolius -a -um *Cheilanthes*-leaved, *Cheilanthes-folium*
cheilanthus -a -um with lipped flowers, χειλ-ανθος
cheilo- lip-, lipped-, χειλος, χειλο-, χειλ-
cheilophyllus -a -um having leaves folded along the mid-rib to suggest lips, χειλο-φυλλον
cheima-, chimon- winter, frost, cold, χειμα, χειμων
cheimanthus -a -um winter-flowering, χειμων, χειμωνος, χειμα
cheir- red, from Arabic, kheri
Cheiranthera Hand-flower, χειρ-ανθερα (the finger-like disposition of the stamens)
cheiranthifolius -a -um wallflower-leaved, *Cheiranthus-folium*
Cheiranthus, cheiranthos Red-flower (from an Arabic name, kheyri, for wallflower)
cheiri, cheiri- red-flowered, wallflower-; sleeve, χειρις, χειρι-
Cheiridopsis Sleeve-like, χειρις-οψις (the sheathing leaf remains)
cheirifolius -a -um wallflower-leaved, botanical Latin from χειρις and *folium*

cheiro- hand-, hand-like-, χειρ, χειρο-; bad-, mean-, weak-, χειρων
cheiropetalus -a -um having fingered petals, lobed petals, χειρο-πεταλον
cheirophyllus -a -um having hand-like, lobed leaves, χειρο-φυλλον
Cheirostemon Hand-of-stamens, χειρο-στεμον (the disposition of the anthers on the united filaments)
cheirostyloides resembling *Cheirostylis*, χειρο-στυλος-οειδης
Cheirostylis Hand-column, χειρο-στυλος (alluding to the finger-like appendages and rostellar-lobes)
chelidoniifolius -a -um with leaves like those of *Chelidonium*, *Chelidonium-folius*
chelidonioides resembling *Chelidonium*, χελιδων-οειδης
Chelidonium Swallow-wort, Dioscorides' name, χελιδων, Greek for a swallow (flowering at the time of their migratory arrival)
Chelone Turtle-like, χελωνη (the form of turtle head's corolla)
chelonius -a -um similar to *Chelone*
chelonoides resembling *Chelone*, χελωνη-οειδης
Chelonopsis *Chelone*-like, χελωνη-οψις
chelophyllus -a -um having claw-like leaves, botanical Latin from χελη-φυλλον
chelsiansis -is -e for Bull's nurseries at Chelsea
chenopodii of fat hen, living on *Chenopodium* (*Pegomyia*, leaf miner)
chenopodioides resembling *Chenopodium*, χηνο-ποδιον-οειδης
chenopodiophyllus -a -um with *Chenopodium* like leaves, χηνο-ποδιον-φυλλον
Chenopodium Goose-foot, χηνο-ποδιον (l'Ecluse's name refers to the shape of the leaves) (***Chenopodiaceae***)
chensiensis -is -e from Shensi (Shaanxi), Henan Province, China
cherimola, cherimolia a Peruvian-Spanish name, cherimoya
Cherleria for J. H. Cherler (1570–1610), Swiss physician and son-in-law of Johann Bauhin
chermisinus -a -um red, crimson, see *kermesinus* (as if dyed with spruce gall aphid, *Chermes*)
cherokeensis -is -e from Cherokee, Texas, or Iowa, USA
chersinus -a -um living in dry habitats, χερσος, χερσαιος
chersonensis -is -e from Kherson or Cherson, in the Crimea
chestertonii for Mr Chesterton, collector for Sander nurseries *c.* 1883
Chevalieria for Jean Baptiste Pierre Antoine de Monet Chevalier de Lamarck (1744–1829), French evolutionist
Chevreulia derivation uncertain
chia from the Greek island of Chios
chiapasanus -a -um from the Chiapas mountains, river or state, Mexico
Chiastophyllum Crosswise-leaf , χιαζειν-φυλλον (the phyllotaxy)
chichibuensis -is -e from the Chichibu Basin, Honshu, Japan
Chickrassia etymology uncertain (Chitagong wood)
chihuahuensis -is -e from the city or state of Chihuahua, N Mexico
childsii for Childs, the American nurseryman
chilensis -is -e from Chile, Chilean
Chilianthus, chilianthus -a -um Thousand-flowered, (χιλιας, χιλιοι)-ανθος (*chiliarchus* was an officer in command of 1000 men)
chilinus -a -um from Chile
Chiliotrichum Thousand-haired-one, χιλιοι-(θριξ, τριχος) (the pappus)
chillanensis -is -e from Chillan, Bio-Bio region of central Chile
chilli from the Nahuatl vernacular name, for the fruits of *Capsicum frutescens*
Chilocarpus, chiliocarpus -a -um Thousand-fruited, χιλιοι-καρπος (many-fruited)
chiloensis -is -e from Chiloe Island off Chile
Chilopsis Lip-like, χειλος-οψις (the calyx)
-chilos, -chilus -a -um -lipped, χειλος
chima-, chimon- winter-, χειμα, χειμων, χειμωνος
chimaerus -a -um monstrous, fanciful (the mythological χιμαιρα, Chimera, was a fire-breathing she-goat with lion's head and serpent's tail)

chimanimaniensis -is -e from the area of Chimanimani, on Zimbabwe's border with Mozambique
Chimaphila Winter-love, χειμα-φιλος (wintergreen)
chimboracensis -is -e from Chimborazo mountains of central Equador
Chimonanthus Winter-flower, χειμων-ανθος (winter sweet)
Chimonobambusa Winter-*Bambusa*, botanical Latin from χειμωνος and *Bambusa*
china China-root (*Smilax china*), former gout remedy from China and Japan (China is added to plant names when they were introduced from there, or plants such as pe-tsai and pak-choi are dubbed Chinese cabbage etc)
chinensis -is -e from China, Chinese, see *sinensis*
chino Chinese (see *sino-*)
chio-, chion-, chiono- snow-, χιων, χιωνος, χιονο-
Chiococca Snow-berry, χιων-κοκκος
Chiogenes Borne of snows, χιων-γενεα (the snow-white berries/winter-growing/flowering)
Chionanthus Snow-flower, χιων-ανθος (fringe tree's abundant white flowers)
chionanthus -a -um snow-flowered, flowering in the snow, χιων-ανθος
chionatus -a -um of the snow, χιων
chioneus -a -um snowy, χιων
chionocephalus -a -um with snow-white (flower) heads, χιονο-κεφαλη
Chionochloa Snow-verdure, χιονο-χλοη (Snow tussock grass)
Chionodoxa Glory of the snow, χιονο-δοξα (Boissier's name reflects the very early flowering, during snow-melt)
chionogenes growing in snowy conditions, χιονο-γενεα (the snow-white berries/winter growing/flowering)
Chionographis Snow-brush, χιονο-γραφις (the appearance of the white inflorescence)
Chionohebe Snow-*Hebe*, χιονο-ηβη
Chionophila, chionophilus -a -um Snow-lover, χιων-φιλος (its Rocky Mountain habitat)
chionophyllus -a -um having foliage during the winter snows, χιωνο-φυλλον
Chionoscilla the composite name for hybrids between *Chionodoxa* and *Scilla*
chiriquensis -is -e from Chirique (Chiriqui) volcanic region or province, Panama
Chirita from the Hindustani vernacular name for a *Gentiana* species
chiro- hand, χειρ, χειρος
Chironia, chironius -a -um, chironus -a -um after Chiron the centaur of Greek mythology, who taught Jason and Achilles the medicinal use of plants
chirophyllus -a -um with hand-shaped leaves, χειρ-ανθος
chisanensis -is -e from Chisan river area, USA
-chiton -covering, -protective covering, -tunic, χιτων
chitralensis -is -e from Chitral, NW Pakistan
chitria from a Hindu vernacular name for a *Berberis*
chius -a -um of Khios Island (Chios), Aegean; snow, χιων (flowering in the snow)
-chlainus -a -um -cloaked, -mantled, χλαινα
chlamy-, chlamydo- cloak-, cloaked-, χλαμυς, χλαμυδος, χλαμυδο-, χλαμυδ-
chlamydanthus -a -um shrouded or enveloped flowers, χλαμυδο-ανθος
chlamydiflorus -a -um shrouded or enveloped flowers, botanical Latin from χλαμυδο- and *florum*
Chlamydocardia Cloak-of-hearts, χλαμυδο-καρδια (the large, cordate, obscuring bracts of the inflorescence)
Chlamydomonas Single-cloak, χλαμυδο-μονας (the chloroplast)
chlamydophorus -a -um bearing a cloak or indusium, χλαμυδο-φορα
chlamydophyllus -a -um cloaked with leaves, χλαμυδο-φυλλον
Chlidanthus Luxurious-flower, χλιδη-ανθος
-chloa -verdure, χλοη
Chloachne Pale-green-chaff, χλωρηις-αχνη
chloodes grass-green, with the appearance of young grass, χλοη-ωδης

chloophyllus -a -um having fresh green leaves, χλοη-φυλλον
chlor-, chloro-, chlorus -a -um yellowish-green-, χλωρος, χλωρο-
Chlora an Adansonian name (≡ *Blackatonia*), χλωρος, greenish-yellow
chloracrus -a -um with green tips, green-pointed, χλωρ0ς-ακρος
Chloraea Pale-green, χλωρηις (the flowers of several species)
chloraefolius -a -um with leaves resembling those of *Chloraea*
Chloranthus, chloranthus -a -um green-flowered, χλωρος-ανθος (***Chloranthaceae***)
chlorideus -a -um *Chloris*-like, pale-green, χλωρηις
chlorifolius -a -um with pale green leaves, botanical Latin from χλωρος and *folium*
chlorinus -a -um yellowish-green (Sir Humphry Davy's name, chlorine, for the colour of the gaseous element 17)
Chloris for the earth nymph, Chloris, pursued by Zephyr and changed into Flora, goddess of flowers, χλωρηις, pale green
Chlorocodon Green-bell, χλωρο-κοδον (the flower shape)
chlorodryas green wood nymph, χλωρο-δρυας (ground flora)
Chlorogalum Green-milk, χλωρο-γαλα (the sap)
chloroleucus -a -um greenish-white, silvery-green, χλωρο-λευκη
chloromelas very dark green, χλωρο-μελας
chloropetalus -a -um green-petalled, χλωρο-πεταλον
chlorophorus -a -um green carrying, χλωρο-φορα (produces a green dye)
chlorophyllus -a -um green-leaved, χλωρο-φυλλον
Chlorophytum Green-plant, χλωρο-φυτον (foliage of spider plant)
chlorops with a green eye, χλωρο-ωψ
chlorosarcus -a -um green fleshed, χλωρο-σαρκος (fruit)
chlorosolen with a green tube, χλωρο-σωλην (flower)
Chlorosplenium Green-milt, χλωρο-σπλην
chloroticus -a -um yellowish-green, chlorotic, a diminutive from χλωρος
Chloroxylon, chloroxylon Green-wood, χλωρο-ξυλον (the timber of satin-wood)
chnoodes with a surface covered in down, χνοος-ωδης (χνοος foam or crust)
Choananthus, chaonanthus -a -um Funnel-flower, χοανος-ανθος (the perianth)
chocolatinus -a -um chocolate-brown, the colour of chocolate, from Nahuatla, chocolati, for the food made from *Theobroma* beans
Choiromyces Swine-fungus, χοιρος-μυκης
choisianus -a -um *Choisya*-like
Choisya for Jacques Denis Choisy (1799–1859), Swiss botanist (Mexican orange flower)
choli- bile-like, χολη, χολος
Chomelia for J. B. Chomel, physician to Louis XV and author of *Abgrégé de l'histoire des plantes usuelles* (1712)
chondracnis -is -e rough chaff, with rough bracts, χονδρ-αχνη
chondro- grain-like, rough-, lumpy-, coarse-, cartilage-, χονδρος, χονδρο-, χονδρ-
Chondropetalum a composite name for hybrids between the orchid genera *Chondrorhyncha* and *Zygopetalum*
Chondrorhyncha Cartilaginous-beak, χονδρο-ρυγχος (the rostellum)
chondrospermus -a -um granular (coated) seed, χονδρος-σπερμα
Chondrostereum Solid-cartilage, χονδρος-στερεος (the brackets of the silver-leaf fungus become brittle on maturing)
Chondrosum Grain, χονδρος (≡ *Bouteloua*)
Chonemorpha Funnel-shaped, χονη-μορφη (the flowers)
chontalensis -is -e from SE Mexio, the area of the Chontal tribe of Mayan Indians
Chorda Rope, *chorda, chordae* (*Chorda filum* avoids tautology by being thread-like rope!)
chordatus -a -um cord-like, χορδη, *chorda, chordae*
chordo-, -chordus -a -um string-, slender-elongate-, χορδη, *chorda, chordae*
chordophyllus -a -um having long slender leaves, χορδη-φυλλον
chordorhizus -a -um, chordorrhizus a- u-m slender creeping rooted, with string-like roots, χορδη-ριζα, botanical Latin *chorda-rhiza*

Chordospartium Thread-*Spartium*, χορδη-σπαρτον (the slender shoots)
chori-, -choris separate-, apart-, χωριζο, χωρις, χωρι-
Chorisia for J. Ludwig (Louis) Choris (1795–1828), the artist who circumnavigated the world with Kotzbue
Chorispora Separated-seed, χωρι-σπορος (winged seeds are separated within the fruit)
choristaminius -a -um having distinct and free stamens, χωρι-(σταμις, σταμινος)
Chorizema Dance-with-drink, χωρι-ζεμα (La Billardière is said to have danced for joy at finding the plant and a necessary freshwater spring)
chortophilus -a -um pasture loving, food-loving, χορτος-φιλος
chotalensis -is -e from Chota, Amazonas, N Peru
Christensenia for Dr Carl Christensen (b. 1872), of Copenhagen, author of *Index filicum*
Christiana for Christen Smith (1785–1816), Norwegian plant collector in Congo
christii for Hermann Christ (1833–1933), Swiss rose specialist, of Basle
christyanus -a -um for Thomas Christy FLS, orchidologist of Sydenham, England, who flowered *Catasetum christyanum* in 1882
-chromatus -a -um, chromosus -a -um, -chromus -a -um, -conspicuously-coloured, χρωμα, χρωματο-
chrono- time-, χρονος, χρονο-
chronosemium time-flag, χρονο-σεμειον (section of *Iris* with the standard enlarging and enclosing the fruit)
chroo- coloured-, χρωος. χρωο-
Chroococcus, chroococcus -a -um Coloured-berry, χρωος-κοκκος (the unicellular cocci)
Chroogomphus Colourful-nail (vinaceous, chrome and glowing red fungus)
chroolepis with coloured (yellow) scales, χρωος-λεπις
chroosepalus -a -um with coloured sepals, χρωος-σκεπη
-chrous -a -um body or skin, χροα, χροια, χρως, χρωτος, χρωο-
Chrozophora Stain-bearing, χρωζω-φορα
chrys-, chryso- golden-, χρυσος, χρυςο-, χρυς-
chrysacanthus -a -um with golden thorns, χρυσ-ακανθα
Chrysalidocarpus Pupa-like-fruit, χρυσαλλις-ειδο-καρπος; Golden-looking-fruit, χρυσος-ειδος-καρπος
Chrysanthellum *Chrysanthemum*-like, diminutive of *chrysanthum*
chrysanthemoides *Chrysanthemum*-like, χρυσ-ανθεμιον-οειδης
Chrysanthemum Golden-flower, χρυσ-ανθεμιον (Dioscorides' name for *C. coronarium*), now treated as several genera such as *Ajania, Arctanthemum, Argyranthemum, Dendranthema, Leucanthemum, Leucanthemella, Leucanthemopsis, Nipponanthemum, Pyrethropsis, Rhodanthemum* and *Tanacetum*
chrysantherus -a -um having golden anthers, χρυσ-ανθηρος
chrysanthus -a -um golden-flowered, χρυς-ανθος
chrysellus -a -um somewhat golden, diminutive from χρυσος
chrysenteron golden-entrailed, χρυς-εντερον (the flesh revealed by cracking of the surface of the pilea)
chryseus -a -um golden-yellow, gold, χρυσεος
Chrysobalanus Golden-acorn, χρυσο-βαλανος (the fruit of some is acorn-like) (***Chrysobalanaceae***)
chrysocarpus -a -um golden-fruited, χρυσο-καρπος
chrysocephalus -a -um golden-headed, χρυσο-κεφαλη
Chrysochloa Golden-grass, χρυσο-χλοη
chrysochlorus -a -um bronzed (leaves), golden-green, χρυσο-χλωρος
chrysocodon golden-bell (flowered), χρυσο-κωδων
Chrysocoma, chrysocomus -a -um Golden-hair, χρυσο-κομη (goldilocks' inflorescence)
chrysocrepis -is -e golden-shoe (flowered), χρυσο-κρηπις
chrysocyathus -a -um golden-cupped, χρυσο-κυαθος (the corolla)
chrysodon golden-toothed, χρυς-(οδους, οδοντος) (the golden scales at the margin of the pileus)

chrysodoron presenting gold, golden gift, χρυσο-δωρον
Chrysoglossum Golden-tongue, χρυσο-γλωσσα (the lip)
Chrysogonum, chrysogonus -a -um Golden-joints, with golden knees, χρυσο-γωνια (the nodes)
chrysographes marked with gold lines, as if written upon in gold, χρυσο-γραφις
chrysogyne golden ovary, golden fruited, χρυσο-γυνη
chrysolectus -a -um picked out in yellow, yellow at maturity, χρυσο-λεκτος
chrysolepis -is -e with golden scales, χρυσο-λεπις
chrysoleucus -a -um with gold and yellow, χρυσο-λευκος (flower parts)
chrysomallus -a -um with golden wool, golden-woolly-hairy, χρυσο-μαλλος
chrysomanicus -a -um a riot of gold, abundantly golden, χρυσο-μανικος
chrysonemus -a -um having golden threads, golden filaments, χρυσο-νημα
chrysophaeus -a -um dull golden, swarthy yellow, χρυσο-φαιος
chrysophylloides resembling *Chrysophyllum*, χρυσο-φυλλον-οειδης
Chrysophyllum Golden-leaf, χρυσο-φυλλον
chrysophyllus -a -um with golden leaves, χρυσο-φυλλον
Chrysopogon Golden-bearded, χρυσο-πωγων
chrysops with a golden eye, χρυς-ωψ
chrysopsidis -is -e resembling *Chrysopsis* (former N American generic name)
Chrysopsis Golden-looking, χρυσ-οψις
chrysorrheus -a -um running with gold, χρυσο-ρεω (the sulphur-yellow milky exudate)
chrysosphaerus -a -um with golden globes (flowers or flower-heads)
chrysosplenifolius -a -um having leaves resembling those of *Chrysosplenium*
Chrysosplenium Golden-spleenwort, χρυσο-σπλην (used for diseases of the spleen)
chrysostephanus -a -um gold-crowned, χρυσο-στεφανος
chrysostomus -a -um wiith a golden throat, χρυσο-στομα
Chrysothamnus Golden-shrub, χρυσο-θαμνος (its appearance when in full flower)
Chrysothemis Divine-gold, χρυσο-θεμις (Golden *Anthemis*)
chrysotoxus -a -um golden arching, golden-bowed, χρυσο- τοξον
chrysotrichus -a -um having golden hairs, χρυσο-τριχος
chrysotropis -is -e turning golden-yellow, χρυσο-τροπη
chrysoxylon yellow-wooded, χρυσο-ξυλον
chrysus -a -um golden coloured, χρυσος
-chthon-, chthono- -underground, earth-, χθων, χθονος
chumbyi from Chumbi, Tibet (southern enclave between Sikkim and Bhutan)
chungensis -is -e from the Chung Shan mountains of E Taiwan
chusanus -a -um from Zhoushan (Chusan), China
Chusquea from the Colombian vernacular name
chusua the Nepalese vernacular name, chu-swa
chyllus -a -um from a Himalayan vernacular name for some *Pinus* species
chylo- sappy-, χυλος, χυλο-
chylocaulus -a -um with a succulent stem, χυλο-καυλος
Chysis Pouring, χυσις (the confluent pollen masses)
chytraculius -a -um like a small pot or jug, χυτρα, χυτρος, χυτριδος (turbinate calyx parts at fruiting, of *Calyptranthus*)
Chytroglossa Jug-tongued, χυτρος-γλωσσα (depression on lip)
cibarius -a -um common, edible, food, *cibus, cibi, cibarius* (*Cantharellus cibarius*, chanterelle)
Cibotium Little-box, diminutive of κιβωτος (the sporangia) (κιβοριον is the name for the cup-shaped seed vessel of the water lily, and *ciborium* is the cup-cover or small shrine to cover the sacrament)
cicatricatus -a -um, cicatricosus -a -um marked with scars, *cicatrix, cicatricis* (left by falling structures such as leaves)
Cicendia an Adansonian name with no obvious meaning
cicer, cicerus -a -um the old Latin name, *cicer*, for the chick-pea

Cicerbita Italian name for *Sonchus oleraceus*, from Marcellus Empiricus' name, *Cicharba*, for a thistle
cichoriaceus -a -um chicory-like, resembling *Cichorium*
Cichorium Theophrastus' name, κιχωριον, from Arabic, kesher (cognate with cicoree, chicory, and succory)
ciclus -a -um mangel-wurzel-like, old name Cicla
ciconius -a -um resembling the neck of a stork, *ciconia*
Cicuta the Latin name for *Conium maculatum*
cicutarius -a -um resembling *Cicuta*, with large bi- or tri-pinnate leaves
-cidius -a -um -murder, -cide, -killing, *caedo, caedere, cecidi, caesum*
Cienfuegosia for Bernard Cienfuego, sixteenth-century Spanish botanist
cigarettiferus -a -um cigarette-bearing (the white sheaths of *Cheiridopsis cigarettifera* sheath the lower third of the narrow, greyish-green leaves)
cili- eyelash-, *cilium, cili-* (marginal cilia)
cilianensis -is -e from Ciliani, N Italy
ciliaris -is -e, ciliatus -a -um fringed, *ciliatus* (with hairs extending from an edge)
ciliatifolius -a -um having hair-fringed leaves, *cilium-folius*
ciliatulus -a -um somewhat fringed, diminutive of *ciliatus*
ciliceus -a -um Cilician, from S Turkey
cilicicus -a -um from Cilicia, S Turkey
cilicioides eyelash-like, *cilium-oides* (the fibrils within the cap)
ciliicalyx with a hair-fringed calyx, *cili-calyx*
ciliidens having hair-fringed teeth, *cili-dens*
cilinodus -a -um with finely hairy nodes, *cili-nodus*
ciliolaris -is -e, ciliolatus -a -um with a minutely fringed appearance, diminutive of *ciliatus*
ciliospinosus -a -um with hair-like spines, *cilium-spinosus*
ciliosus -a -um markedly fringed with hairs, comparative of *ciliatus*
cilius -a -um any superficial fine-hair-like feature, fringe, *cilium*
-cillus -a -um -lesser, diminutive suffix
cimex a bug, *cimex, cimicis*
cimicarius -a -um of bugs, bug-like, *cimex, cimicis* (the oily smell)
cimicidius -a -um, cimicinus -a -um of bugs or small insects, *cimes, cimicis*
cimiciferus -a -um bug-bearing, *cimicis-fero* (appearance of small flowers or fruits)
Cimicifuga Bug-repeller, *cimicis-(fugo, fugare, fugavi, fugatum)* (bugbanes)
cimiciodorus -a -um smelling of bugs, *cimicis-(odoro, odorare)*
cinaeus -a -um glaucous, ashen, ash-grey, *cinis, cineris*
Cinchona (*Chinchona*) for the Countess of Chinchon, wife of the Viceroy of Peru. She was cured of fever with the bark, the source of quinine, in 1638, and introduced it to Spain in 1640. In Qhuechua, the medicinal bark, kina kina, became Portugese, quinaquina, and our quinine
cincinalis -is -e, cincinnatus -a -um with crisped hair, *cincinnus*, curly-haired, *cincinnatus*
cincinus -a -um curled, *cincinnus*
cinctulus -a -um with a small girdle, diminutive of *cinctus*
cinctus -a -um, -cinctus -a -um girdled, -edged, *cinctus*
cineoliferus -a -um with moving parts, shaking, κινεω-φερω
cineraceus -a -um, cinerarius -a -um ash-coloured, covered with ash-grey felted hairs, *cinis, cineris*
cinerascens, cinerescens becoming ashen, *cineris-essentia*
Cineraria, cinerarius -a -um Ashen-one, *cinis, cineris* (the foliage colour)
cineraria-oleifolia with ashen olive-like leaves, botanical Latin, *cineris-olea-folium*
cinerariifolius -a -um having leaves resembling those of *Cineraria, Cineraria-folium*
cinereus -a -um ash-grey, *cinis, cineris*
cinerioides ash-like, (*cinis, cineris*)*-oides* (the colour of grey coral fungus)
cingulatus zoned, girdled, *cingulus, cingulum*

cinnabari, cinnabarinus -a -um cinnabar-red, κινναβαρι (the colour of mercury sulphide)
cinnameus -a -um resembling *Cinnamomum*
Cinnamodendron Cinnamon-tree, κινναμωμον-δενδρον
cinnamomeobadius reddish cinnamon-brown, *cinnamomum-badius*
cinnamomeus -a -um reddish-brown, endearing, *cinnamomum* (Ovid)
cinnamomifolius -a -um cinnamon-leaved, *Cinnamomum-folium*
Cinnamomum the Greek name, κινναμωμον, used by Theophrastus, from Hebrew, qinnamon, cinnamon
cinquefolius -a -um five-leaved, poor Latin from French, cinque and *folium*
cintranus -a -um from Sintra, Portugal
cio- erect-, κιων
Cionosicyos Erect-*Sicyos*, κιων-σικυος (erect-cucumber)
Cionura Erect-tailed, κιων-ουρα (*Cionura erecta*)
cipoanus -a -um from Serra do Cipo, Minas Gerais, Brazil
Cipura etymology uncertain
Circaea for the enchantress Circe, of mythology (Pliny's name for a charm plant, used variously by l'Obel)
circaeoides resembling *Circaea, Circaea-oides*
circaezans enchanting, having the property of *Circe*
circassicus -a -um Circassian, modern Latin *circassia*, from Russian, Cherkes, N Caucasus
circinalis -is -e, circinatus -a -um curled round, coiled like a crozier, circinate, *circino, circinare*
circularis -is -e surrounding, growing in crowds, *circulor, ciculari*
circum- around-, about-, *circum*
circumalatus -a -um edged with a wing, *circum-(ala, alae)*
circumplexus -a -um embraced, clasped-around, *circumplector, circumplecti, circumplexus*
circumscissus -a -um cut or opening all round, *circum-(scindo, scindere, scindi, scissum)*
circumseptus -a -um enclosed all round, fenced about, *circum-(septum, septa)*
circumserratus -a -um toothed around, with spines all around, *circum-(serra, serrae)* (the leaf)
circumtextus -a -um woven all round, *circum-(texo, texere, texui, textum)*
cirratus -a -um, cirriferus -a -um curled, having or carrying tendrils, (*cirrus, cirri*)-*fero*
Cirrhaea Tendrilled, *cirrhus* (the elongated rostellum)
cirrhatus -a -um, cirrhiferus -a -um having or bearing tendrils, *cirrhus-fero*
cirrhifolius -a -um with tendril-like leaves, *cirri-folium*
Cirrhopetalum Yellowish-petalled, κιρρο-πεταλον
cirrhosus -a -um yellowish, tawny-coloured, κιρρος
cirsioides resembling *Cirsium*, κιρσιον-οειδης
Cirsium the ancient Greek name, κιρσιον, for a thistle
cis- prefix denoting near or the-same-side-of, *citra-, cis-*
cisalpinus -a -um of the southern Alps, *cis-(alpes, alpium)* (literally on this side of the Italian Alps)
cisplatanus -a -um, cisplatinus -a -um on the southern side of the river Plata or near La Plata, Argentina, botanical Latin, *cis* with Plata
cismontanus -a -um on the Italian side of the mountains, *cis-(mons, montis)*
ciss-, cisso- ivy-, κισσος, κισσ, κισσο-
Cissampelos Ivy-vine, κισσ-αμπελος (the growth is like ivy and the inflorescence like a grape)
cissifolius -a -um ivy-leaved, botanical Latin, κισσος with *folium*
cissoides resembling *Cissus*, κισσος-οειδης
Cissus the ancient Greek name, κισσος, for ivy
cistaceus -a -um *Cistus*-like, resembling a rock rose; box-like, κιστη (flowers)
Cistanche *Cistus*-strangler, κιστυς-αγχω (root parasite)

cistenus -a -um box-like, κιστη, *cisterna, cisternae*
cisti- *Cistus*-like-
cistifolius -a -um with *Cistus*-like leaves, *Cistus-folius*
cistoides resembling *Cistus*, κιστυς-οειδης
Cistus Capsule, κιστυς (rock roses are conspicuous in fruit) (***Cistaceae***)
Citherexylum Fiddlewood, κιθαρα-ξυλον (used for making lyres etc)
citratus -a -um *Citrus*-like, lemon-scented, *Citrus*
citreus -a -um citron-yellow, *Citrus*
citri- citron-like-, *Citrus*
citricolor citron-yellow coloured, *citri-color*
citrifolius -a -um *Citrus* leaved, *Citrus-folium*
citrinellus -a -um slightly citron yellowish, diminutive of *Citrus*
citriniflorus -a -um *Citrus*-flowered, lemon-yellow-flowered, *Citrus-florus*
citrinifolius -a -um *Citrus*-leaved, *Citrus-folius*
citrinus -a -um citron-yellow, *citrinus*
citriodorus -a -um citron-scented, lemon-scented, *citri-odorus*
Citrofortunella the composite name for hybrids between *Citrus* and *Fortunella*
citroides resembling *Citrus medica, Citrus-oides*
Citroncirus the composite name for hybrids between *Citrus medica* and *Poncirus*
citronellus -a -um with the fragrance of citronella oil, diminutive of *Citrus*
Citropsis resembling *Citrus*, κιτρον-οψις
citrulloides resembling *Citrullus, Citrullus-oides*
Citrullus, citrullus -a -um Little-orange, diminutive of *citrus* (the cucurbit's fruit colour)
Citrus from the ancient Latin name, *citrus*, from Greek κιτρον
citus -a -um quick, ephemeral, *cieo, ciere, civi, citum*
civilis -is -e gracious, courteous, *civilis*
clad-, clado- shoot-, branch-, of the branch-, κλαδος, κλαδο-
Cladanthus, cladanthus -a -um flowering on the branches, κλαδος-ανθος (the terminal flowers)
Cladium, cladius -a -um Small-branch, diminutive of κλαδος (the short branches of the compressed panicle)
Cladothamnus Branched-shrub, κλαδο-θαμνος (the much-branched habit)
Cladrastis Fragile-branched, κλαδο-ραστος (the brittle branches)
clandestinus -a -um concealed, hidden, secret, *clandestinus*
clandonensis -is -e from the home of Arthur Simmonds (1892–1968) at West Clandon, Surrey
Claoxylon Brittle-wood, κλαω-ξυλον
claraensis -is -e from Santa Clara, Cuba
clarinervius -a -um clearly or distinctly nerved, *clarus-nerva*
clarkei for Charles Baron Clarke (1832–1906), British student of Indian plants
Clarkia for Captain William Clark (1770–1838), co-leader of the Rocky Mountain expedition of 1803–6, with Meriwether Lewis
claroflavus -a -um of a bright yellow colour, *clarus-flavus*
claroviridis -is -e bright green, *clarus-viridis*
clarus -a -um clear, bright, *clarus, clari-, claro-*
clathratus -a -um latticed, barred, cage-like, *clatratus, clathratus* (*clatri, clathri*)
clausus -a -um shut, closed, enclosed, *claudo, claudere, clausi, clausum*
clava-hercules Hercules' club (thorny shrubs, e.g. devil's walking sticks in *Aralia* or *Xanthoxylum*)
Clavaria Clubbed, *clava-aria* (the clavate fruiting bodies)
Clavariadelphus *Clavaria*'s-brother, botanical Latin from *Clavaria* and αδελφος
clavatus -a -um, clavi-, clavus -a -um clubbed, distally enlarged like a club, *clava*
clavellatus -a -um, clavellinus -a -um with small clubs, diminutive of *clavus*
clavennae for Niccolo Chiavena (d. 1617), Italian apothecary
Claviceps Thick-head, (*clavus, clavi*)-*caput* (the hard distension caused by ergot fruiting bodies)

clavicornis -is -e having club-shaped horns, (*clavus, clavi*)-*cornu*
Clavicorona Club-shaped-crown, (*clavus, clavi*)-*corona* (the fruiting body)
claviculatus -a -um tendrilled, having vine-like tendrils, *clavicula, claviculae*
claviformis -is -e club-shaped, (*clavus, clavi*)-*forma*
clavigerus -a -um club-bearing, (*clavus, clavi*)-*gero*
Clavija for Don José de Viera y Clavijo (1731–1813), Spanish translator of Buffon's works
Clavinodus Knotted-noded-one, (*clavus, clavi*)-*nodus* (*Clavinodus oedogonatum*)
clavipes club-stalked, (*clavus, clavi*)-*pes*
clavularis -is -e, clavulatus -a -um like a small nail, diminutive of *clavus*
Clavulina Little-club, Little-nail, feminine diminutive of *clavus, clavi*
Clavulinopsis Resembling-*Clavulina*
Claytonia, claytonianus -a -um for John Clayton (1686–1773), British physician and botanist in America
Cleistanthus, cleistanthus -a -um Hidden-flower, κλειστο-ανθος (concealed by prominent, hairy bracts)
cleio-, cleisto- shut-, closed-, κλειω, κλειστος, κλειστο-
Cleistocactus Closed-cactus, κλειστος-*cactus* (the flowers barely open)
cleistocarpus -a -um enclosed fruit, κλειστο-καρπος
cleistogamus -a -um closed-marriage, κλειστο-γαμος (self-fertilization occurs before anthesis)
Cleistopholis Closed-scales, κλειστο-φολις (the arrangement of the inner petals)
Cleistostoma Closed-mouth, κλειστο-στομα (entrance to spur is obstructed)
-clema vine-twig, shoot, branch, κλημα
clematideus -a -um like vine twigs, κληματιδος
clematiflorus -a -um *Clematis*-flowered, *Clematis-florus*
Clematis the Greek name, κληματις, for several climbing plants, diminutive of κλημα, a vine shoot
clematitis -is -e with long vine-like branches, brushwood, vine-like, Dioscorides' name αριστολοχεια κληματιτις
Clematoclethra Climbing-*Clethra* (resembles *Clethra* but climbs like *Clematis*)
Clematopsis Resembling-*Clematis*, κληματις-οψις
clemens, clementis -is -e mild, gentle, merciful, *clemens, clementis* (not thorny)
clementinae for the wife of George Forrest, collector in China (the hybrid *Citrus* called a Clementine commemorates Father Clément, who raised it *c.*1900 at Oran, Algeria)
Cleome a name used by Theophrastus (spider flower)
cleomifolius -a -um having leaves resembling those of spider flowers, *Cleome-folium*
Clerodendron (um) Chance-tree, κλερο-δενδρον (early names for Ceylonese species *arbor fortunata* and *arbor infortunata*)
Clethra ancient Greek name, κληθρη, for alder (similarity of the leaves of sweet pepper bush) (***Clethraceae***)
clethroides resembling *Clethra*, κληθρη-οειδης
Cleyera for Andrew Cleyer, seventeenth-century physician working for the Dutch East India Company
Clianthus Glory-flower, κλεος-ανθος (parrot's bill)
clidanthus -a -um wanton, delicate, fine, beautiful, κλιδη-ανθος
Clidemia for Kleidemys, ancient Greek botanist
cliffordiae for Lady de Clifford (d. 1845) of London
Cliffortia for George Clifford (1685–1760), whose Amsterdam garden plants were recorded in Linnaeus' *Hortus Cliffortianus*, 1737
cliffortioides resembling (*Nothofagus*) *cliffortii*
Cliftonia for William Clifton, eighteenth-century lawer and Attorney General of the state of Georgia
cliftonii for J. Talbot Clifton of Lytham Hall, Lancashire
Clino- prostrate-, bed-, κλινη
Clinopodium, clinopdius -a -um Bed-foot, κλινη-ποδιον (Dioscorides' name, κλινοποδιον, for the knob-shaped appearance of the inflorescence)

Clintonia for De Witt Clinton (1769–1828), Governor of New York State and originator of the Erie Canal, writer on American science
clipeatus -a -um armed with a shield, shield-shaped, *clipeatus*
Clitandra Inclined anthers, κλιτυς-ανδρος
Clitopilus Smoothed-down-felt, (the cap of the miller fungus has a kid-leather-like texture)
Clitoria Clitoris, κλειτορις (by analogy with the young legume in the persistent flower-parts)
Clivia for Lady Charlotte Clive, wife of Robert Clive (1725–74) of India (kaffir lilies); for Duchess of Northumberland (d. 1866), *nee* Clive
clivorum of the hills, *clivus, clivi*
cloiphorus -a -um carrying a strong collar, κλοιος-φορεω
Clonostylis branched-style, κλων-στυλος
closterius -a -um spindle-shaped, κλωσοτηρ
closterostyles having a spindle-shaped style, κλωσοτηρ-στυλος *clostrum-stylus*
Clostridium Little-spindle, diminutive of κλωσοτηρ (pathogenic bacteria)
Clowesia for Reverend John Clowes (1777–1846), orchid grower of Manchester
Clusia, clusii, clusiana for Charles de l'Écluse (Carolus Clusius) (1526–1609), Flemish renaissance botanist, author of *Rariorum plantarum historia* (***Clusiaceae*** ≡ ***Guttiferae***)
clusiifolius -a -um having leaves resembling those of *Clusia*
Clutia (Cluytia) for Outgers Cluyt (*Clutius*) (1590–1650), of Leyden
clymenus -a -um from an ancient Greek name (see *periclymenum*)
clypeatus -a -um, clypeolus -a -um having structures resembling circular Roman shields, *clipeus, clipei*
Clypeola (Clipeola) Shield, diminutive of *clipeus, clipei* (the shape of the fruit)
clypeolatus -a -um like a small circular shield, diminutive of *clypeus*
Clytostoma Beautiful-mouth, κλυτος-στομα (the flowers)
cnemidophorus -a -um wearing greaves, with a sheathed stem, κνημιδο-φορος
-cnemis, cnemi-, cnemido- -covering, ancient Greek, κνημις, for a greave or legging
-cnemius -calf-of-the-leg, internodes, ancient Greek, κνημο
cnemo- of wooded valleys, κνημος
-cnemum -the-internode, Theophrastus used κναμα, κνημη, tibia, for the part of the stem between the joints
Cneorum, cneorum of garlands, the Greek name, κνεορον, for an olive-like shrub (***Cneoraceae***)
Cnestis Scraper, κναω (the hair covering of the fruit)
Cnicus the Greek name, κνηκος, of a thistle used in dyeing
co-, col-, com-, con- together-, together with-, firmly-
coacervatus -a -um accumulated, clustered, in clumps, *co-acervatio, co-acervationis*
coactilis -is -e growing densely, crowded, *cogo, cogere, coegi, coactum*
coadenius -a -um with united glands, botanical Latin from *com-* and αδην
coadnatus -a -um, coadunatus -a -um united, held-together, joined into one, *co-(adnascor, adnasci, adnatus)*
coaetaneus -a -um of the same age, ageing together, *co-(aetas, aetatis)* (leaves and flowers both senesce together)
coagulans curdling, from *coagulum* (rennet)
coahuilensis -is -e from the Coahuila area of Mexico
coalifolius -a -um with joined leaves, *(coalesco, coalescere, coalui, coalitum)-folium*
coarctatus -a -um pressed together, bunched, contracted, *coarto, coartare, coartavi, coartatum*
Cobaea (Coboea) for Father B. Cobo (1572–1659) Spanish Jesuit and naturalist in Mexico and Peru (cup and saucer vine)
cobanensis -is -e from Coban, Guatemala
cobbe from the Singhalese vernacular name, kobbae
Coburgia for Prince Leopold of Saxe-Coburg, later King of Belgium
coca the name used by S American Indians of Peru

cocciferus -a -um, coccigerus -a -um scarlet-bearing, *coccum-fero, coccum-gero* (*Quercus coccifera* is host of kermes insect, *Kermes illicis*, from which the dye was prepared)
coccioides resembling (*Crataegus*) *coccinea*
Coccinia Scarlet, κοκκινος, *coccineus* (fruit colour)
coccinelliferus -a -um bearing the cochineal scale insect, *Dactylobius coccus*
coccinellus -a -um light-scarlet, diminutive of *coccineus*
coccineus -a -um, (cochineus) crimson, scarlet, *coccineus* (the dye produced from galls on *Quercus coccifera*)
Coccocypselum Fruit-vase, κοκκος-κυψελη (the fruit)
Coccoloba Berry-pod, κοκκος-λοβος (sea grape is distinguished amongst *Polygalaceae* by having a succulent fruit)
coccoloboides resembling *Coccoloba*, κοκκος-λοβος-οειδης
coccos scarlet-berried, κοκκος
coccospermus -a -um having cochineal-insect-like scarlet seeds, κοκκινος-σπερμα
cocculoides resembling *Cocculus, Cocculus-oides*
Cocculus, cocculus -a -um Small-berry, diminutive of κοκκος (or from *coccum*, for the scarlet fruits)
coccum scarlet, κοκκος, *coccum, cocci*
-coccus -a -um scarlet-berried, κοκκος (in botany the derived Latin suffix, *-coccus*, is used for spherical bodies, *cocci*, of many sorts, as in fruits, algae, fungal spores, bacteria)
cochinchinensis -is -e from Vietnam, Laos or Cambodia (formerly French Cochinchina)
cochlea-, coclea- snail, *cochlea, cochleae;* spoon, *cocleare, coclearis*
Cochlearia Spoon, *cocleare, coclearis*, via German Löffelkraut, *cochlear*, for the shape of horseradish's basal leaves (Dodoens described its use as an antiscorbutic, scurvy-grass)
cochlearifolius -a -um with spoon-shaped leaves, *Cochlearia-folium*
cochlearis -is -e spoon-shaped, *cocleare, coclearis*
cochlearispathus -a -um having a spathes resembling the bowls of a spoon, *cocleare, coclearis*
cochleatus -a -um twisted like a snail-shell, cochleate, *coclea, cocleae; cochlea cochleae*
cochlio-, cochlo- twisted-, spiral-, κοχλιας, κοχλος, *cochlea*
Cochlioda Small-snail, κοχλιας (the callus shape)
cochliodes, cochlioides resembling *Cochlioda*, κοχλιας-ωδης
cocoides *Cocos*-like, coconut-like, *Cocos-oides*
cocoinus -a -um from the Cocos islands, resembling a coconut (smell or colour)
Cocos from the Portuguese, coco, for bogeyman, the features of the end of the coconut's shell
Codiaeum from a Malayan vernacular name, kodiho or codebo
-codium -fleeced, κωδιον, κωας, -headed, κωδειον, poppy-headed, κωδεια
-codon -bell, -mouth, κωδων
Codonantha (*Codonanthe*) Bell-flower, κωδων-ανθος (*Gesneriaceae*)
Codonanthus, codonanthus -a -um Bell-flower, κωδων-ανθος (≡ *Calycobolus, Convolvulaceae*)
Codonatanthus the composite name for hybrids between *Codonanthe* and *Nematanthus*
Codonoprasum Bell-shaped-leek, κωδωνος-πρασον
codonopsifolius -a -um having leaves similar to those of *Codonopsis*
Codonopsis Bell-like, κωδων-οψις (flower shape)
Coelachyrum Hollow-chaff, κοιλο-αχυρον (the hollowed shape of the grain)
coelebo- unmarried, κοιλεβς (pistillate)
coelestinus -a -um, coelestis -is -e, coelestus -a -um sky-blue, heavenly, *caelum, caeli; coelum, coeli*
coeli- sky-blue-, heavenly-, *caelum, caeli; coelum, coeli*
coelicus -a -um heavenly, somewhat blue, *coelum*

coeli-rosa rose of heaven, *coeli-rosa*
coelo- hollow-, κοιλος, κοιλο-
Coelocaryon Hollow-nut, κοιλο-καρυον (the cavity in the seed)
Coeloglossum Hollow-tongue, κοιλο-γλωσσα (the depression at the base of the lip of the flower)
Coelogyne Hollow-woman, κοιλος-γυνη (the hollow style of the pistil)
Coelonema Hollow-threads, κοιλος-νημα
coeloneuron having hollowed veins, impressed veins, κοιλος-νευρα
coelophloeus -a -um having cavitied bark, κοιλο-φλοιος
Coelorhachis Hollow-rachis, κοιλο-ραχις
coelospermus -a -um hollow-seeded, κοιλο-σπερμα
coen-, coenos- common-, κοινος
coenobialis -is -e cloistered, having structures sharing a common investment, κοινοβιον (colonials such as *Volvox*, multinucleate (coenocytic) fungal structures, some fruits)
coenosus -a -um common, polluting, κοινοω
coeris -is -e restrained, blue, *caeruleus, coeruleus*
coerulans, coerulescens turning blue, conspicuously blue, *caeruleus* (*vide caerulescens*)
coeruleus -a -um sky-blue, blue, *caeruleus, coeruleus*
coetaneus -a -um existing together, *co-*(*aetas, aetatis*) (flowering and fruiting)
coetanus -a -um crowded together, κοιτη, κοιτος; *coetus, coitus*
Coffea from the Arabic name, qahwah, for the drink made by infusing the dry seeds
cogens clearly together, *co-*(*ago, agere, egi, actum*)
coggygria the ancient Greek name for *Cotinus*
cognatus -a -um closely related, *cognatus, cognati*
cognobilis -is -e of equal fame or note, *cog-nobilis*
coherens sticking together, *cohaereo, cohaerere, cohaesi, cohaesum*
cohune the Miskito vernacular name for the oil-rich fruit of the palm *Orbigyna cohune*
Coinochlamys Hairy-throughout, κοινος-χλαμυς
Coix the ancient Greek name, κωηξ, for Job's tears grass
Cola from the Mende, W African name, ngolo
colchaguensis -is -e from Colchagua province, Chile
colchiciflora -a -um *Colchicum*-flowered, *Colchicum-florum*
colchicifolius -a -um having leaves similar to those of *Colchicum, Colchicum-folium*
Colchicum Colchis, a Black Sea port, used by Dioscorides as a name, κολχικον, for *Colchicum speciosum* (meadow saffron)
colchicus -a -um from *Colchis, Colchidis*, the Caucasian area once famous for concocting poisons
cole-, colea-, coleo- sheath-, κολεος, κολεο-
Colea for General Sir Lowrey Cole (1772–1842), Governor of Mauritius
coleatus -a -um sheathed, sheath-like, κολεος
colebrookianus -a -um for Henry Thomas Colebrook FRS FLS (1765–1837), Sanskrit scholar and naturalist
Colensoa, colensoi for Reverend William Colenso FRS (1811–1899), student of the New Zealand flora
Coleochloa Grass-sheathed, κολεο-χλοη (the leaf sheath is open on one side, as in grasses)
Coleogyne Sheathed-ovary, κολεο-γυνη
coleoides resembling *Coleus*, κολεος-οειδης
Coleonema Sheathed-filaments, κολεος-νημα (the filaments of the sterile stamens lie in channels in the petals)
coleophyllus -a -um having sheathing leaves, κολεο-φυλλον
coleospermus -a -um with sheathed seeds, κολεο-σπερμα
Coleotrype Sheath-hole, κολεο-τρυπημα (inflorescences pierce the leaf-sheaths)
Coleus Sheath, κολεος (the filaments around the style) (flame nettle)

coliandrus -a -um   coriander-like (see *coriandrum*)
coliformis -is -e   rod-shaped, *colis-forma* (the pillars of *Myriostoma coliformis*)
coll-, -collis -is -e   -necked, *collum, colli*
collariferus -a -um   bearing a collar or sheath, *collare-fero*
collaris -is -e   having a collar, necklace or band, collared, *collare, collaris*
collatatus -a -um   gathered together, *confero, conferre, contuti, conlatum*
Colletia   for Philibert Collet (1643–1718), French botanist and writer (anchor plants have flattened thorn-tipped branches)
colletianus -a -um   resembling *Colletia*
collettii   for Colonel Sir Henry Collett (1836–1901), collector in Shan States
colliculinus -a -um   of low hills, of hummocked land, *colliculus, colliculi* (diminutive of *collis*)
colliniformis -is -e   mound-forming, making small hills, *collis-forma*
collinitus -a -um   besmeared, *collino, collinere, collevi, collitum* (the bands of velar remnants)
Collinsia   for Zaccheus Collins (1764–1831), of Philadelphia Academy of Natural Sciences
Collinsonia   for Peter Collinson (1694–1768), plant introducer and correspondent of Linnaeus
collinus -a -um   of the hills, growing on hills, *collis, collis*
collo-   gluey, sticky, mucilaginous, κολλα
collococcus -a -um   mucilaginous-berried, κολλα-κοκκος
Collomia   Mucilaginous-one, κολλα (the sticky seed coat when wet)
collum-cygni   shaped like a swan's neck, (*collum, colli*)-(*cycnus, cycni*)
colmariensis -is -e   from Colmar, Alsace region of France
colo-, colob-   shortened-, κολοβοω, κολος
Colobanthus   Shortened-flower, κολοβοω-ανθος
colobodes   cut-short, κολοβοω-ωδης
Colocasia   the Greek name, κολοκασια, from the Arabic, kulkas (for taro, the root of *Colocasia antiquorum*); Latin *colocasia* is Egyptian bean, *Caladium*
colocynthis   ancient Greek name, κολοκυνθις, for the bitter apple cucurbit, *Citrullus colocynthis*
coloides   resembling *Cola, Cola-oides*
colombianus -a -um   from Colombia, Colombian
colombinus -a -um   dove-like, *columbus, columbi*
coloneurus -a -um   having short veins, κολος-νευρα
coloniatus -a -um   forming colonies or patches, *colonia, coloniae*
colonus -a -um   forming a mound, humped, *colonus*
coloradensis -is -e   from Colorado, USA
colorans, coloratus -a -um   colouring, coloured, *coloro, colorare, coloravi, coloratum*
colosseus -a -um (*colloseus*)   very large, κολοσσος, *colossus* (literally descriptive of statues)
Colossoma   Large-bodied, κολοσσος-σωμα (*Colossoma macropomum*)
colpodes   hollow-looking, κολπος-ωδης
colpophilus -a -um   bay- or estuary-loving, κολπος-φιλεω
Colquhounia   for Sir Robert Colquhoun (d. 1838), patron of the botanic garden, Calcutta
colubrinus -a -um   wily, snake-like, *colcubrinus* (*colubra, colubrae*, snake)
columbaris -is -e   with some form of collar, collared, *columbar, columbaris*
columbariae, columbarius -a -um   dove-blue, dove-coloured, of doves, *columba, columbae* (*columbarium*, a dove-cote)
columbettus -a -um   young dove, *columba* (colouration)
columbianus -a -um   from the Columbia river or British Columbia
columbinus -a -um   pigeon-like, dove-like, *columbus, columbi; columba, columbae*
Columella   for Lucius Junius Moderatus Columella (b. first century BC), Roman soldier and author of *De re rustica* and *De arboribus*
columellaris -is -e   having or forming small pillars, *columella, columellae*

Columellia as for *Columella*
columnaris -is -e pillar-like, columnar, *columna, columnae*
Columnea, columnae for Fabio Colonna of Naples (1567–1640), publisher of *Phytobasanos*, 1592
columniferus -a -um column-bearing, *columni-fero* (growth habit)
Coluria Deprived, κολουρος (either tail-less seeds or the lysing styles equated to a dying swan)
colurnus -a -um the ancient name, *colurnus*, for Turkish hazel (*Corylus colurna*)
colurnoides resembling (*Corylus*) *colurna*
-colus -a -um -loving, -inhabiting, -adorning, *colo, colere, colui, cultum* (follows a place, plant type or habitat)
Colutea an ancient Greek name, κολουτεα, used by Theophrastus and Dioscorides for a tree (bladder senna)
Coluteocarpus *Colutea*-fruited, κολουτεα-καρπος (similar capsule shape)
colvilei for Sir James Colvile FRS (1810–80), Indian Judge
Colvillea for Sir Charles Colville (1770–1843), Governor of Mauritius
colvillei for James Colville (1746–1822) and James Colville (1777–1832), nurserymen predecessors of Veitch at Chelsea
com- with-, together with-, *com*
coma-aureus -a -um golden-haired, with golden foliage, *coma-aureus*
comans hairy, plumed, leafy, tufted, *comans, comantis*
Comarostaphylis Grape-*Comarum*, κομαρος-σταφυλη (the fruiting clusters)
comarrhenus -a -um having hairy or long hair-like stamens, κομη-αρρενος
Comarum from Theophrastus' name, κομαρος, for the strawberry tree (their similar fruiting structures)
comatus -a -um long-haired, leafy, tufted, *comatus*
comaureus -a -um with golden hair, golden-haired, *coma-aureus*
Combretodendron *Combretum*-like-tree, botanical Latin from *Combretum* and δενδρον
Combretum a name used by Pliny for an undetermined climbing plant (***Combretaceae***)
comedens devouring, *comedo, comesse, comedi, comesum* (spreading cortical saprophyte)
Comesperma Haired-seed, κομη-σπερμα (the hair-tufts on the end of the seeds)
cometes comet-like, κομητης
Commelina for Caspar (1667–1731) and Johann (1629–98) Commelijn, Dutch botanists (***Commelinaceae***)
Commelinidium *Commelina*-like (the foliage)
commemoralis -is -e commemorative, memorable, *commemero, commemorare, commemoravi, commemoratum*
Commiphora Resin-bearer, κομμι-φορα (*Commiphora myrrha* and *C. habyssinica*, myrrh)
commixtus -a -um mixed together, mixed up, *commisceo, commiscere, coomiscui, commixtum*
commodus -a -um opportune, pleasant, *commodus;* just, *commodum*
communis -is -e growing in clumps, gregarious, common, *communis*
commutatus -a -um changed, altered, *commutato, commutatare, commutavi, commutatum* (e.g. from previous inclusion in another species)
comonduensis -is -e from Comondu, Mexico
comophorus -a -um bearing long hair, κομη-φορα
comorensis -is -e from Comoro Islands, off Mozambique, E Africa
comosus -a -um long-haired, shaggy-tufted, with tufts formed from hairs or leaves or flowers, κομη, *coma*
compactus -a -um close-growing, closely packed together, dense, *compingo, compingere, compegi, compactum*
compar comrade, husband, wife; well-matched, equal, *compar, comparis*
complanatus -a -um flattened out upon the ground, *complano, complanare*

complectens becoming entwined and enfolding, present participle of *complector, complecti, complexus*
complectus -a -um, complex, complexus -a -um encircled, embraced, twining, *complector, complecti, complexus*
complicatus -a -um folded back, pleated, *complico, complicare*
compositus -a -um with flowers in a head, *Aster*-flowered, compound, *compono, compnere, composui, compositum*
compressicaulis -is -e having a flattened stems, *compressi-caulis*
compressus -a -um flattened sideways (as in stems), pressed together, *comprimo, comprimere, compressi, compressum*
Comptonia, comptonianus -a -um for Henry Compton (1632–1713), Bishop of Oxford, then Bishop of London
comptus -a -um union; ornamented, with a head-dress, elegant, *como, comere, compsi, comptum*
con- with-, together with-, a form of *com-*
conabilis -is -e with handiness or expertise, difficult, *con-habilis* (culture)
Conanthera Anther-cone, κωνος-ανθερα (before full anthesis the anthers present the appearance of a cone)
concanensis -is -e from the Concan region of India
concatenans, concatenatus -a -um joined together, forming a chain, *con-catenatus*
concavissimus -a -um greatly hollowed out, superlative of *concavus*
concavus -a -um basin-shaped, concave, *concavus* (with a hollow, *con-cavus*)
concentricatrix, concentricus -a -um with concentric markings, *con-centrum* (King Alfred's cake fungus)
conchae-, conchi- shell-, shell-like-, *concha*
conchiferus -a -um bearing shell(-shaped structures), *concha-fero*
conchiflorus -a -um shell-flowered, *concha-florum*
conchifolius -a -um with shell-shaped leaves, *concha-folium*
Conchocelis Shell-concealing, *concha-celo* (red alga)
conchoideus -a -um shell-shaped, shell-like, *concha-oides*
concholobus -a -um shell-lobed, *concha-lobus*
concinnulus -a -um neat, pretty, diminutive of *concinnus*
concinnus -a -um well-proportioned, neat, elegant, harmonious, symmetrical, *concinno, concinnare, concinnavi, concinnatum*
concolor uniformly coloured, coloured similarly, *concolor, concoloris*
concrescens clotting, coalescing, *concresco, concrescer, concrevi, concretum* (the fruiting bodies become conjoined)
concretus -a -um hardened, congealed, grown together, *concresco, concrescere, concrevi, concretum*
concurrens flocking or happening together, *concurro, concurrare, concurri, concursum*
condensatus -a -um crowded together, compacted, *condenso, condensare; condenseo, condensere*
conduplicatus -a -um twice-pleated, double-folded, *conduplico, conduplicare* (e.g. aestivation of *Convolvulus*)
condylatus -a -um having knob-like swellings, κονδυλος
condylobulbo having knobbly bulbs, κονδυλος-βολβος
condylodes knobbly, with knuckle-like bumps, κονδυλος-ωδης
confertiflorus -a -um with dense or crowded flowers, *confertus-florum*
confertifolius -a -um with dense foliage, with crowded leaves, *confertus-folium*
confertissimus -a -um most compact, superlative of *confertus*
confertus -a -um crowded, pressed-together, past participle of *confercio, confercire, confertum*
Conferva Seething or Passionate, *con-(ferveo, fervere, ferbui; fervo, fervere, fervi)*
confervaceus -a -um resembling *Conferva*
confervoides crowded-looking, *Conferva-oides*
confinalis -is -e, confinis -is -e close to, adjoining, akin, *confinis*

confluans flowing-together, *fluo, fluere, fluxi, fluxum*
confluens flowing-together, *fluo, fluere, fluxi, fluxum* (growing in dense tufts)
confluentes from Koblenz, Germany, *Confluentes*
conformis -is -e symmetrical, conforming to type or relationship, *conformatio, conformationis*
confragosus -a -um breaking into pieces, *confringo, confringere, confregi, confractum*
confusus -a -um easily mistaken for another species, disordered, past participle of *confundo, confundere, confundi, confusum*
congestiflorus -a -um with crowded flowers, *congestus-florus*
congestissimus -a -um very densely packed together, superlative of *congestus*
congestus -a -um arranged very close together, crowded, past participle of *congero, congerere, congessi, congestum*
conglobatus -a- um massed into a ball, past participle of *conglobo, conglobare, conglobavi, conglobatum*
conglomeratus -a -um rolled up, crowded together, *conglomero, conglomerare*
congolanus -a -um from the Congo river area, Africa
congregatus -a -um clustered together, *congrego, congregare, congregavi, congregatum*
congruus -a -um agreeable, *congruus*
conicoides, conicoideus -a -um cone-shaped, κωνικος-οειδης, *conicus-oides*
conicus -a -um, -conicus -a -um cone-shaped, conical, *conus*
conifer -era -erum, coniferus -a -um cone-bearing, (*conus, coni*)-*fero*
coniflorus -a -um cone-flowered, *coni-florum*
conii- hemlock-like, resembling *Conium*
coniifolius -a -um hemlock-leaved, *Conium-folium*
conimbrigensis -is -e from Condeixa (*Conimbriga*), Portugal
conio- dust, ashes, covered with dust or ashes, whitewashed; conidia, κονια, κονιω
Coniogramme Sprinkled-lines, κονιω-γραμμη (the sori along the veins)
Coniophora Bearing ash, κονια-φορα (mealy surface)
Conioselinum the name formula for hybrids between *Conium* and *Selinum*
Conium the Greek name, κωνειον, for hemlock plant and poison
conjugens growing together, present participle of *coniugo, coniugare*
conjugialis -is -e, conjugatus -a -um joined together in pairs, conjugate, *coniugo, coniugare* (ovaries of some *Lonicera* species)
conjunctus -a -um joined together, *coniungo, coniungere, coniunxi, coniunctom*
connatus -a -um born at the same time, united, joined, *con-natus*
connectilis -is -e, connexus -a -um joined-up, past participle of *con-*(*necto, nectere, nexi, nectum*) (fern sori)
connivens winking, converging, connivent, *coniveo, conivere, conivi* (*conixi*)
cono- cone-shaped-, κωνος, κωνο-
conocarpodendron cone-fruited-tree, κωνο-καρπο-δενδρον
Conocarpus, conocarpus -a -um Cone-fruit, κωνο-καρπος (the infructescence shape)
Conocephalum, conocephalus -a -um Cone-headed, κωνο-κεφαλη
Conocybe Cone-cap, κωνο-κυβη (the usual shape of the pileus)
conoides, conoideus -a -um cone-like, κωνος-οειδης
conophalloides resembling a cone-shaped phallus, κωνο-φαλλος-οειδης
Conopharyngia With a cone-shaped throat, κωνο-φαρυγξ
Conophora, conophorus -a -um Cone-bearer, κωνο-φορα
Conophytum Cone-plant, κωνο-φυτον (its inverted conical habit)
Conopodium Cone-foot, κωνο-ποδιον (the enlarged base of the styles)
conopseus -a -um (*conopea*) looking like a cloud of gnats or mosquitoes, κωνος-ωψ, κωνος-ωπος
conothelis -is -e having a conical ovary, κωνο-θηλυς
conradinia for an unidentified lady, Conradina
Conringia for Hermann Conring, seventeenth-century German academic of Helmstedt
consanguineus -a -um closely related, of the same blood, *consanguineus*
consimilis -is -e just like, much resembling, *consimilis*

consobrinus -a -um cousin, related, *consobrinus*
consocialis -is -e, consocius -a -um associating, formed into clumps, *consocio, consociare, consociavi, consociatum*
Consolida Whole-maker, *con-solida*, the ancient Latin name from its use in healing medicines (cognate with *Solidago*)
consolidatus -a -um, consolidus -a -um stable, firm, *con-(solido, solidare)*
consors sharing or shared in common, *consors* (cognate with consortium)
conspersus -a -um speckled, scattered, *con-(spergo, spergere, spersi, spersum)*
conspicuus -a -um easily seen, marked, conspicuous, *cospicio, conspicere, conspexi, conspectum*
constans stable, consistent, present participle of *consto, constare, constiti, constatum*
constantinopolitanus -a -um from Istanbul (formerly Constantinople), Turkey
constantissius -a -um uniform, consistent, superlative of *constans*
constantius -a -um steady, consistent, *constans, constantis*
constrictus -a -um narrowed, constricted, drawn together, erect, dense, past participle of *constringo, constringere, constrinxi, constrictum*
contactus -a -um infectious; touching, confined, past participle of *contingo, contingere, contigi, contactum*
contaminans becoming impure (through breeding), defiling, making unclean, present participle of *contamino, contaminare, contaminavi, contaminatum*
contaminatus -a -um defiled, not pure, mixed, past participle of *contamino, contaminare, contaminavi, contaminatum*
contemptus -a -um worthless, despised, past participle of *contemno, contemnere, contempsi, contemptum*
conterminus -a -um closely related, close in habit or appearance, neighbouring, *conterminus*
contextus -a -um woven-together, coherent, *con-(texo, texere, texui, textum)*
contiguus -a -um close and touching, closely related, adjoining, *contiguus*
continentalis -is -e of moderation; continental, of any of the larger land masses, from sixteenth-century Latin, *terra continens*, for a continuous land mass
continuatus -a -um without a break, joined-together, *continuo, continuare, continuavi, continuatum*
continuus -a -um joined, successive, uninterrupted, *continuus*
contortuplicatus -a -um very complicated, *contortuplicatus* (tangled growth habit)
contortus -a -um twisted, bent, intricate, *contorqueo, contorquere, contorsi, contortum*
contra-, contro- against-, *contra*
contractus -a -um drawn together, *con-(traho, trahere, traxi, tractum)*
contrarius -a -um opposite, harmful, *contrarius*
controversus -a -um doubtful, controversial, *controversus*
Convallaria, convallarius -a -um Of-the-valley, *con-vallis* (the natural habitat of lily-of-the-valley) (***Convallariaceae***)
convallarioides resembling *Convallaria, Convallaria-oides*
convalliodorus -a -um lily-of-the-valley-scented, *Convallaria-(odor, odoris)*
convergens coming together, inclined towards, *con-(vergo, vergere)*
conversus -a -um turning towards, turning together, *converto, convertere, converti, conversum*
convexulus -a -um slightly rounded outwards, diminutive of *convexus*
convexus -a -um humped, bulged outwards, convex, *convexus*
convolutus -a -um rolled together, *convolvo, convolvere, convolvi, convolutum*
convolvulaceus -a -um bindweed-like, similar to *Convolvulus*
convolvuloides resembling *Convolvulus, Convolvulus-oides*
Convolvulus, convolvulus Entwined, *convolo* (a name in Pliny) (***Convolvulaceae***)
Conyza a name, κονυζα, used by Theophrastus
conyzae similar to *Inula conyza*
Cookia, cookii for Captain James Cook (1728–79), antipodean explorer who was murdered in Hawaii
Cooperanthes the composite name for hybrids between *Cooperia* and *Zephyranthes*

Cooperia, for Joseph Cooper, nineteenth-century gardener to Earl Fitzwilliam at Wentworth, Yorkshire
cooperi, cooperianus -a -um for either Mr Cooper, orchid grower of London *c.* 1865, or Edward Cooper (b. 1877), writer on orchids, or Thomas Cooper (1815–1913), who collected for W. W. Saunders in S Africa, or Edgar Franklin Cooper (1833–1916)
coopertus -a -um covered over, overwhelmed, *co-operio, co-operire, co-operui, co-opertum*
Copaifera, copaiferus -a -um Copal-bearing, from the Brazilian vernacular name, copaiba, for the balsamic juice of the plant, botanical Latin from Nahuatl, kopalli, and *fero*
copaius -a -um Brazilian vernacular name, copaiba, for the resinous, gummy exudate
copallinus -a -um from a Nahuatl name, kopalli, yielding copal-gum
cophocarpus -a -um basket-fruited, κοφινος-καρπος
Copiapoa from Copiapo, Chile
copiosus -a -um abundant, copious, *copiosus*
copra from the Malayan vernacular name, kappora, for the coconut
Coprinus Of-dung, κοπρος (*Coprinus comatus* is the coprophilous shaggy-cap fungus)
coprophilus -a -um dung-loving, coprophilous, κοπρος-φιλος
Coprosma Dung-smelling, κοπρος οσμη (the odour of the bruised leaves)
copticus -a -um from Coptos, near Thebes, Egypt; of the Copts
Coptis Cutting, κοπτω (the leaves)
coptophyllus -a -um cut-leaved, κοπτω-φυλλον
copulatus -a -um joined, coupled, united, *copulo, copulare, copulavi, copulatum*
coquimbanus -a -um, coquimbensis -is -e from Coquimbo, N Chile
coracanus -a -um grain-like (kurakkan)
coracinus -a -um raven-black, κοραξ, κορακος
coraeensis -is -e from Korea, Korean
corallicolus -a -um inhabiting coral formations, *corallus-(colo, colere, colui, cultum)*
coralliferus -a -um coral-bearing, *corallum-fero*, κοραλλιον-φερω
coralliflorus -a -um having coral-red flowers, *corallinus-florus*
corallinus -a -um coral-red, *corallum*, κοραλλιον
corallioides coral-red, resembling coral, κοραλλιον-οειδης
corallipes having a coralloid stalk, *corallus-(pes, pedis)*
corallodendron coral-tree, κοραλλιον-δενδρον (appearance when in deep-red flower and leafless)
coralloides resembling coral, κοραλλιον-οειδης
Corallorhiza Coral-root, κοραλλιον-ριζα (the rhizomes)
Corallospartium Coral-red-*Spartium*, κοραλλιον-σπαρτον
coranicus -a -um from Corani, Cochabamba, Bolivia (*Ammocharis*)
corazonicus -a -um from Mount Corazon, Ecuador
corbariensis -is -e from Corbières, S France
corbularius -a -um like a small basket, *corbicula, corbiculae* (growth habit; also a bee's pollen basket)
corbulus -a -um like a small basket, *corbula, corbulae* (diminutive of *corbis*)
corchorifolius -a -um with leaves similar to those of *Corchorus*
Corchorus the Greek name for jute; etymology uncertain
corcovadensis -is -e from Mount Corcovados (Hunchback), Rio de Janeiro, Brazil
corcyraeus -a -um, corcyrensis -is -e from the Greek island or department of Corfu (*Corcyra*)
cord-, cordi- heart-shaped, *cor, cordis, cordi-*
cordatus -a -um, cordi- heart-shaped, cordate, *cor* (see Fig. 6e) (literally, wealthy)
Cordia for Henricus Urbanus (Enricus Cordus) (1486–1535), and his son Valerius (1515–44), German botanists
cordiacus -a -um dyspeptic, of the heart, καρδιακος, *cor, cordis*

cordiferus -a -um, cordigerus -a -um bearing hearts, *cordis-fero* or *-gero* (*Serapias cordigera*, heart-flowered orchid)
cordifoliatus -a -um having heart-shaped leaflets, (*cor, cordis*)*-foliatus*
cordifolius -a -um with heart-shaped leaves, *cordi-folium*
cordiformis -is -e heart-shaped, *cordi-forma*
cordilabrus -a -um having a heart-shaped lip, *cordi-labrum*
cordobensis -is -e from any of the Córdobas in Argentine, Colombia or Mexico
cordubensis -is -e from Cordova (*Corduba*) Spain
Cordyceps Rope-stake, *chorda-cippus* (fruit-body form, via Gascon-French, cep, a tree-trunk mushroom)
cordufanus -a -um from Kordofan (Kurdufan), Sudan
Cordyline Club, κορδυλη (some cabbage palms have large club-shaped roots)
coreanus -a -um from Korea, Korean
Corema Broom (Greek name, κορεμα, suggested by the bushy habit)
Coreopsis Bug-like, κορις-οψις (the shape of the fruits)
Corethrogyne Broom-styled, κορεθρον-γυνη (the styles)
coriaceifolius -a -um thick or leathery leaved, *corium-folium*
coriaceus -a -um tough, thick, leathery, like leather, *corium* (the leaves)
coriandrifolius -a -um coriander-leaved, *Coriandrum-folium*
coriandrinus -a -um resembling coriander, like-*Coriandrum*
Coriandrum Theophrastus' name, κοριαννον or κοριανδρον, for *Coriandrum sativum* (coriander has seeds resembling bed-bugs, κορις)
Coriaria Leather, κωρυκος, *corium* (used in tanning) (***Coriariaceae***)
coriarioides resembling *Coriaria, Coriaria-oides*
coriarius -a -um of tanning, leather-like, of the tanner, *coriarius*
corid- *Coris*-like
corifolius -a -um leathery-leaved, *corius-folium*
corii- leathery-, *corium; corius*
coriifolius -a -um, coridifolius -a -um with leaves resembling those of *Coris*
coriophorus -a -um bug-bearing, κορις-φορα (smell of foliage); bearing helmets, κορυς-φορα (flower shape)
Coris a name, κορυς, used by Dioscorides
coritanus -a -um from the East Midlands (home of the *Coritani* tribe of ancient Britons)
corius -a -um leathery, hide-like, *corium, corius, cori-, corii-*
cormiferus -a -um producing corms, κορμος-φερω
corneus -a -um horny, of horn-like texture, *corneus*
corni, corni, -cornis -is -e living on *Cornus* (*Craneiobia*, dipteran gall midge); horned-, horn-bearing-, *Cornus-*
cornicinus -a -um crow- or raven-black, *cornix, cornicis*
corniculatus -a -um having small horn- or spur-like appendages or structures, diminutive of *cornus*
cornifer -era -erum, corniferus -a -um corniger -era -erum, cornigerus -a um horned, horn-bearing, *cornu-*(*fero, ferre, tuli, latum*) or *-cornu-*(*gero, gerere, gessi, gestum*)
cornifolius *Cornus*-leaved, *Cornus-folium*
cornolium medieval name for cornel, *Cornus mas*, (the fruit of cornelian cherry, was *cornolia*)
cornubiensis -is -e, cornubius -a -um from Cornwall (*Cornubia*), Cornish
cornu-cervi shaped like a deer's horn, *cornu-cerva*
Cornucopiae Horn of plenty, *cornu-copiae* (hooded grass)
cornucopiae, cornucopioides shaped like a cornucopia, *cornu-copiae* (cf. *pharmacopoeia*, poison-making, φαρμακον-ποιος)
Cornus Horn, *cornu, cornus; cornum* (the ancient Latin name, *cornum*, for the cornelian cherry, *Cornus mas*) (***Cornaceae***)
-cornus -horned, *cornus*
Cornutia for Jaques Cornutus (1606–51), French traveller in Canada, author of *Historia Plantarum Canadensium*

cornutus -a -um horn-shaped, *cornus*
corocoroensis -is -e from the Corocoro river area of NW Venezuela
Corokia from a New Zealand Maori vernacular name
corollarius -a -um having a corolla, *corolla, corollae*
corolliferus -a -um bearing a corolla, *corolla-fero*
corollinus -a -um with a conspicuous corolla, *corolla, corollae*
coromandelicus -a -um from the Coromandel coastal area, SE India, or the similarly named area in New Zealand
coronans encircling, garlanding, crowning, present participle of *corono, coronare, coronavi, coronatum*
Coronaria Crown-material, *corona* (Latin translation of στεφανωτικη, used in making chaplets, cf. *Stephanotis*)
coronarius -a -um garlanding, forming a crown, *corona, coronae*
corona-sanctistephani St Stephen's crown, botanical Latin from *corona, sanctus* and Stephan
coronatus -a -um crowned, *coronatus*
Coronilla, coronillus -a -um Little-garland, diminutive of *corona* (the flower-heads)
coronopifolius -a -um crowfoot-leaved, *Coronopus-folium*
Coronopus, coronopus Theophrastus' name, κορωνοπους, for crowfoot, κορωνη-πους (leaf-shape)
Correa for Jose Francesco Correa de la Serra (1750–1823), Portuguese botanist
correlatus -a -um with relationship, related, *con-relatus*
corriganus -a -um improved, like a shoelace, slender, *corrigia, corrigiae*
Corrigiola Shoe-thong, diminutive of *corrigia* (the slender stems)
corrugatus -a -um wrinkled, corrugated, *corrugo, corrugare*
corsicanus -a -um, corsicus -a -um from Corsica, Corsican
Cortaderia Cutter, from the Spanish-Argentinian name, cortadera, for *Cortaderia selloana* (refers to the sharp leaf margins of pampas grass)
corticalis -is -e, corticosus -a -um with a notable, pronounced or thick bark, *cortex, corticis*
corticolus -a -um living on tree bark, *cortex-colo*
Cortinarius Vaulted, with a distinct cortina, *cortina, cortini* (the covering between the edge of the pileus and the stalk of a toadstool enclosing the gills)
Cortusa for Jacobi Antonii Cortusi (1513–93), director of the Padua Botanic Garden
cortusifolius -a -um with leaves similar to those of *Cortusa*
cortusioides, cortusoides resembling *Cortusa*
coruscans shaking, quivering, fluttering, flashing, present participle of *corusco, coruscare*
coruscus -a -um tremulous, oscillating, shimmering, glittering, *coruscus*; helmet-shaped (stamens), κορυς; of the NW wind, *Corus*
Corvisatia for Jean Nicolas Corvisat des Marets, physician to Napoleon Bonaparte
Coryanthes Helmet-flower, κορυς-ανθος (the labellum shape)
coryanthus -a -um having helmet-shaped flowers, κορυς-ανθος
coryanus -a -um, coryi for Reginald Radcliffe Cory (1871–1934), benefactor of the Cambridge Botanic Garden
Corybas Ecstatic (from analogy to frenzied dances of the Corybantes, priests of Cybele)
Corydalis (*Corydallis*), *corydalis -is -e* Lark, κορυδαλλις (Durante's name refers to the spur of the flowers); fumitory-like
coryli, coryli- (parasitic on) hazel; hazel-, *Corylus*
corylifolius -a -um with leaves similar to those of *Corylus, Corylus-folium*
corylinus -a -um hazel-like, resembling *Corylus*
Corylopsis Hazel-resembler, botanical Latin from *Corylus* and οψις
Corylus Helmet, *corylus* (the Latin name refers to the concealing nature of hazel's calyx) (***Corylaceae***)
corymbiferus -a -um bearing corymbs, *corymbus, corymbi* (Fig. 2d)

corymbiflorus -a -um with flowers in flat-topped heads, botanical Latin, *corymbiflorum*
Corymbium Corymbose, κορυμβος (the flowering habit)
Corymborchis Flat-headed-orchid, κορυμβος-ορχις
corymbosus -a -um with flowers arranged in corymbs, κορυμβος, with a flat-topped raceme (see Fig. 2d)
corymbulosus -a -um with small corymbs, κορυμβος
corynacanthus -a -um having stout thorns, κορυνη-ακανθα
Corynanthe Club-flower, κορυνη-ανθος (the orbicular appendages of the corolla lobes)
coryne-, coryno- club-, club-like-, κορυνη
Corynephorus Club-bearer, κορυνη-φορα (the clubbed awns)
corynephorus -a -um clubbed, bearing a club, κορυνη-φορα
corynestemon having club-shaped stamens, κορυνη-στεμων
Corynocarpus, corynocarpus -a -um Club-fruited-one, κορυνη-καρπος (the stylar structure) (***Corynocarpaceae***)
corynodes club-shaped, thickened towards the distal end, κορυνη-ωδης
coryophorus -a -um bug-bearing, κορις-φορα (colour markings)
coryph- at the summit-, κορυφη
coryphaeus -a -um leading, the best, *coryphaeus, coryphaei*
coryphoides resembling a talipot palm, *Corypha-oides*
corys- ,-corythis -is -e helmet-, -cucculate, κορυς, κοροθος
coscayanus -a -um from Coscaya, N Chile
Coscinium Sieve, κοσκινον (the timber of window-wood, *C. fenestratum*)
cosmetus -a -um well-apparelled, *cosmeta, cosmetae* (wardrobe master)
cosmoides resembling *Cosmos, Cosmos-oides*
cosmophyllus -a -um having leaves resembling those of *Cosmos*
Cosmos Beautiful, κοσμος (the ornamental flowers)
-cosmus -a -um -beauty, -decoration, κοσμιος
cossus of the goat moth, *Cossus cossus* (the smell of the flesh of goat moth wax cap fungus)
costalis -is -e, costatus -a -um with prominent ribs, with a prominent mid-rib, *costa, costae* (-cost, aromatic herb, *costum, costi*)
costaricanus -a -um, costaricensis -is -e from Costa Rica
costatus -a -um prominently ridged, ribbed, *costa, costae*
Costus a name, *costum*, used in Pliny (κοστος, for an Indian plant with scented roots, possibly from the Arabic, koost) (cognate with cost, as in costmary and alecost)
-costus -a -um -aromatic, *costum, costi*
cosyrensis -is -e from Pantelleria Island, Mediterranean, near Sicily (*Cossyra*)
cotinifolius -a -um having leaves resembling those of *Cotinus, Cotinus-folius*
cotinoides resembling *Cotinus, Cotinus-oides*
Cotinus, cotinus ancient Greek name, κοτινος, for a wild olive
Cotoneaster Wild-quince, *cotonea-aster* (Gesner's name suggests that the leaves of some species are similar to quince, *cotonea* in Pliny)
Cotula, cotulus -a -um Small-cup, κοτυλη, via the Italian vernacular name, cota, for *Anthemis cota* (the leaf arrangement)
cotuliferus -a -um bearing small cup(-like structures), κοτυλη-φερω
Cotyledon, cotyledon Cupped, κοτυληδων (Pliny's name refers to the leaf shape)
cotyledonis -is -e cup-like, κοτυλη-οειδης (leaves)
coulteri for Thomas Coulter (1793–1843), Irish physician and botanist
coum from a Hebrew name for *Cyclamen coum* (coumarin, from Tupi, kumaru, relates to the Tonka Bean's use as a flavouring)
courbaril from a vernacular name for the timber
cous Coan, from the island of *Cos, Cous, Coi*, Aegean Turkey (pearl millet, couscous, derives from Arabic, kuskus, that which has to be pounded)
cowa an Indian vernacular name for the fruit of *Garcinia cowa*

Cowania, cowanii for James Cowan (d. 1823), of London, who introduced plants from Mexico
crabroniferus -a -um bearing hornet(-like flowers), *crabro, crabronis*
cracca name used in Pliny, for a vetch
Crambe ancient Greek name, κραμβη (for a cabbage-like plant)
cranichoides resembling *Cranichis, Cranichis-oides*
crantzii for H. J. von Crantz (1722–99), botanical writer
cranwellia, cranwellae for Lucy Cranwell-Smith (1907–2000), palaeobotanist
Craspedia Fringe, κρασπεδον (the pappus)
craspedodronus -a -um fringed course, κρασπεδον-δρομος
crassi- thick-, fleshy-, *crassus, crassi-*
crassicaulis -is -e thick-, fleshy-stemmed, *crassi-caulos*
crassicollus -a -um thick-necked, *crassi-collum*
crassifolius -a -um with thick, fleshy or leathery leaves, *crassi-folium*
crassinodis -is -e, crassinodus -a -um having swollen nodes, *crassi-nodus*
crassior thicker (than the type), comparative of *crassus*
crassipes thick-stalked, *crassi-pes*
crassirhizomus -a -um with a thick rhizome, *crassi-(rhizoma, rhizomata)*
crassistipulus -a -um with thick stipules, *crassi-(stipula, stipulae)*
crassiusculus -a -um somewhat thick, a little thickened, diminutive of *crassus*
Crassocephalum Thick-headed, botanical Latin, *crassus-cephalum* (the expanded peduncle)
Crassula Succulent-little-plant, feminine diminutive of *crassus* (***Crassulaceae***)
crassus -a -um thick, fleshy, *crassus*
crataegi of hawthorn, living on *Crataegus* (symbionts, parasites and saprophytes)
crataegifolius -a -um hawthorn-leaved, *Crataegus-folium*
crataeginus -a -um hawthorn-like, *Crataegus*
Crataegomespilus the composite name for the chimaera involving *Crataegus* and *Mespilus*
Crataegus Strong, κραταιος (the name, κραταιγος, used by Theophrastus for hawthorn's timber)
Crataemespilus the composite name for hybrids between *Crataegus* and *Mespilus*
crateri-, cratero- strong-, κρατερος, καρτερος; goblet-shaped-, a cup, κρατηρ, κρατηρος, κρατηρο; *crater, crateris*
Crateranthus, crateranthus -a -um Bowl-flower, κρατηρ-ανθος (the shape of the corolla tube)
crateriformis -is -e goblet- or cup-shaped, with a shallow concavity, *crateris-forma*
Craterispermum Saucer-shaped-seeded, κρατηρ-σπερμα
Crataeva for Cratevas (Creteuas) (first century BC), Greek physician and artist to Mythridates VI (the elder Pliny refers to Crateva's books with coloured illustrations – no extant works remain)
cratus -a -um strong, superior, κρατος
Crawfurdia for Sir John Crawfurd (1783–1868), Governor of Singapore
creber -ra -rum, crebri- densely clustered, frequently, *creber, crebri*
crebriflorus -a -um densely flowered, *crebri-florus*
crebrus -a -um frequent, crowded, prolific, *crebro*
Cremanthodium Hanging flower-head, κρεμαστος-ανθωδης
Cremaspora Pendulous-seeded, κρεμαστος-σπορος
cremastogyne with hanging or pendent ovary, κρεμαστος-γυνη
cremastus -a -um with hanging or swinging flowers, κρεμαστος
cremeus -a -um burnt-looking, fiery, singed, *cremo, cremare, cremavi, crematum* (leaf margins)
cremnophilus -a -um liking steep slopes or precipices, κρημνος-φιλεω
cremnophylax cliff-top sentinel, κρημνος-φυλαξ (protected habitat)
crenati-, crenatus -a -um notched, with small rounded teeth, modern Latin, *crena* (the leaf margins, see Fig. 4a)
crenatiflorus -a -um with crenate-lobed flowers, *crenatus-florum*

crenatifolius -a -um, crenifolius -a -um having crenate-margined leaves, *crenatus-folium* (Fig. 4a)
crenophilus -a -um spring-loving, κρηνη-φιλεω
crenulatus -a -um having small rounded and flat teeth around the leaves, diminutive of *crenatus*
Creolophus hairy-fleshed-one, κρεας-λοφος (the scaled surface texture)
creophagus -a -um flesh-eating, κρεας-φαγω
crepidatus -a -um wearing sandals, sandal- or slipper-shaped, *crepidatus*
crepidiformis -is -e slipper-shaped, *crepida-forma*
crepidioides resembling *Crepis*, *Crepis-oides*
Crepis a name, κρηπις, used by Theophrastus (meaning not clear, κρηπι, κρηπιδος is a shoe or enclosing wall)
crepitans creaking, rattling, present participle of *crepo, crepare, crepui, crepitum* (as the seeds in the pod of the sandbox tree, *Hura crepitans*)
crepitatus -a -um clattering or creaking, *crepitus*
Crescentia for Pietro de Crescenzi (1230–1321), of Bologna
Cressa Cretan, *cressa, cressae* (*Cressa cretica* is not a tautonym!)
cretaceus -a -um inhabiting chalky soils, of chalk, *creta*
cretensis -is -e, creticus -a -um from Crete, Cretan (*Creta*)
cretiferus -a -um bearing chalk, *creta-fero* (superficial deposit from chalk glands)
cretus -a -um descended, born, *cretus;* appearing, thriving, increasing, *cresco, crescere, crevi, cretum*
cribratus -a -um sieve-like, *cribrum, cribri*
crinatus -a -um with a tuft of long, fine hairs, *crinatus*
crini- hair-, *crinis, crinis, crini-*
criniferus -a -um, criniger -era -erum carrying a tuft of fine hairs, *crinis-fero, crinis-gero*
criniformis -is -e hair-like, much elongated, *crini-forma*
Crinitaria Long-hair, *crinitus* (the inflorescence)
crinitus -a -um with long soft hairs, *crinitus*
Crinodendron Lily-tree, κρινον, δενδρον (floral similarity)
Crinum Lily, κρινον
crispatulus -a -um slightly waved, *crispus*
crispatus -a -um undulate, closely waved, curled, *crispus*
crispifolius -a -um with wavy-edged leaves, *crispus-folium*
crispipilus -a -um curly-haired, *crispi-pilus*
crispulus -a -um slightly waved, diminutive of *crispus*
crispus -a -um curled, wrinkled, with a waved or curled margin, *crispus*
crista-galli cock's comb, (*crista, cristae*)-*gallus* (the crested bracts)
cristatellus -a -um having a small crest, diminutive of *cristatus*
cristatus -a -um tassel-like at the tips, crested, *cristatus*
cristiflorus -a -um having crested flowers, *crista-florum*
cristus -a -um plumed, crested, *crista*
crithmifolius -a -um having leaves resembling those of *Crithmum*
crithmoides resembling *Crithmum*, *Crithmum-oides*
Crithmum Barley, κριθη (the Greek name, κρηθμον, refers to the similarity of the fruits)
croaticus -a -um from Croatia
crobulus -a -um with a crest or tuft of hair, κρωβυλος
crocatus -a -um citron-yellow, saffron-like, κροκος (used in dyeing, or the orange exudate from *Mycena crocata*)
croceiflorus -a -um *Crocus*-flowered, having saffron-yellow flowers, *croceus-florum*
croceo-caeruleus -a -um yellow and bluish coloured, *croceus-caeruleus*
croceocarpus -a -um having saffron-yellow fruits, *croceus-carpus*
croceofolius -a -um (chrome-)yellow leaved, *croceus-folium* (fruiting body of *Cortinarius croceofolius*)
croceus -a -um saffron-coloured, yellow, *croceus*

crocidatus -a -um felted, with a felt-like surface, κροκυς, κροκυδος
crocifolius -a -um yellow-leaved, *croci-folium*
crocipodius -a -um yellow-stalked, *croci-(pous, podos)* (stipe of yellow cracking bolete)
Crocosmia Saffron-scented, κροκος-οσμη (the dry flowers) (***Crocosmataceae***)
crocosmifolius -a -um with *Crocosmia*-like leaves, *Crocosmia-folium*
crocostomus -a -um yellow-throated, κροκος-στομα (flowers)
crocothyrsos woolly-panicled, saffron-panicled, κροκος-θυρσος
Crocus Thread, κροκος, from the Chaldean name, κροκη a thread (for the stigmas of *Crocus sativus*, from which is produced true saffron, Arabic, zacfaran)
croesus flowing with wealth, for Croesus (d. 546 at Sardus), King of Lydia
Crossandra Fringed-anther, κροσσαι-ανηρ
Crossandrella Resembling-*Crossandra*, feminine diminutive of *Crossandra*
Crossopteryx Fringed-wing, κροσσαι-(πτερυξ, πτερυγος) (the seed)
crossosepalus -a -um having fringed sepals, κροσσαι-σκεπη
Crossosoma Stepped-bodies, κροσσαι-σωμα (the reniforme, arillate seeds) (**Crossosomataceae**)
crotalarioides resembling *Crotalaria*, *Crotalaria-oides*
Crotalaria Rattle, κροταλον, *crotalum* (seeds become loose in the inflated pods of some)
Croton Tick, κροτον (the seeds of some look like ticks)
Crotonogyne Female-*Croton*, κροτον-γυνη
crotonoides *Croton*-like, *Croton-oides*
Crowea for James Crowe (1750–1807), British botanist of Norwich
Crucianella Little-cross, diminutive of *crux* (for the phyllotaxy, ≡ *Phuopsis*)
Cruciata Cross, *crux, crucis* (Dodoens' name refers to the cruciate arrangement of the leaves)
cruciatus -a -um with leaves in alternate pairs at right angles to the pair below; instrument of torture, torture, misfortune, *crucio, cruciare, cruciavi, cruciatum* (fiercely armed with thorns set crosswise)
crucifer -era -erum, crucigerus -a -um cross-bearing, cruciform, *crux-fero* or *-gero*
cruciformis in the form of a cross, *crucis-forma*
crucis cross, *crux, crucis* (the corolla)
crudelis -is -a coarse, cruel, bloody, *crudelis; crudus*
cruentatus -a -um stained with red, bloodied, blood-red, *cruor, cruoris*
cruentus -a -um bloody, blood-red, blood-coloured, *cruentus* (*cruor, cruoris*)
crumenatus -a -um pouched, like a purse, *crumena, crumenae*
crura, cruris legged, leg, shin, *crus, cruris*
crus leg, shin, *crus, cruris*
crus-andrae St Andrew's cross, *crux, crucis*
crus-corvi raven's spur, (*crus, cruris*)-, (*corvus, corvi*)
crus-galli cock's spur, (*crus, cruris*)-*gallus* (thorns)
crus-maltae, crux-maltae Maltese cross (*crux, crucis*)-*melita*
crustaceus -a -um brittle, hard-surfaced, *crusta, crustae*
crustatus -a -um encrusted, having a hard surface, *crusta, crustae*
crustuliniformis -is -e shaped like small pastries, *crustulum, crustuli*
cruzianua -a -um from any place called Santa Cruz, in Bolivia or elsewhere
cryo- cold, frost, ice, κρυος, κρυο-
cryocalyx having a frosted-looking calyx, κρυο-καλυξ
cryophilus -a -um cold-loving, κρυο-φιλεω
Cryophytum Ice-plant, κρυο-φυτον (the appearance of *Cryophytum crystalinum* ≡ *Mesembryanthemum crystalinum*)
Crypsis Covered, κρυψιος, κρυπτος (the short flowering head is embraced by two inflated leaf-sheaths)
crypt-, crypto- obscurely-, covered-, hidden-, κπυπτειν, κρυπτος, κρυπτο-, κρυπτ-; *crypta, crypto-*
cryptandrus -a -um having inconspicuous or concealed stamens, κρυπτ-ανηρ

Cryptantha Hidden-flower, κρυπτ-ανθος (long calyx lobes)
Cryptanthus Hidden-flower, κρυπτ-ανθος (the concealed flowers of earth star)
cryptanthus -a -um having concealed flowers, κρυπτ-ανθος
Cryptbergia the composite name for hybrids between *Cryptanthus* and *Billbergia*
cryptocarpus -a -um having concealed fruits, κρυπτο-καρπος
Cryptocarya Covered-nut, κρυπτο-καρυον (the mace which surrounds Brazilian nutmeg)
cryptocephalus -a -um having protected or concealed flower-heads, κρυπτο-κεφαλη
Cryptocoryne Hidden-club, κρυπτο-κορυνη (the spathe encloses the spadix)
cryptodontus -a -um having obscure teeth, κρυπτ-(οδους, οδοντος)
Cryptogramma (Cryptogramme) Hidden-lines, κρυπτο-γραμμη (the concealed lines of sori)
cryptolanatus -a -um hidden in wool, *crypto-(lana, lanae)*
Cryptolepis Hidden-scaled-one, κρυπτο-λεπις (the coronnal scales within the corolla tube)
Cryptomeria Hidden-parts, κρυπτο-μερος (the inconspicuous male cones of Japanese cedar)
cryptomerioides resembling *Cryptomeria*, κρυπτο-μερος-οειδης
Cryptomonas Hidden-unit, κρυπτο-μονας
cryptophytus -a -um minuscule, obscure or concealed plant, κρυπτο-φυτον
cryptopodius -a -um with a concealed stalk, κρυπτο-(πους, ποδος)
Cryptostegia Hidden-cover, κρυπτο-στεγη (the coronna conceals the anthers)
Cryptotaenia Obscured-ribbons, κρυπτο-ταινια
Cryptothladia Hidden-eunuch, κρυπτο-θλαδιας
cryptus -a -um covered, concealed, *crypta, crypto-*
crystallinus -a -um with a glistening surface, as though covered with crystals, κρυσταλλος, *crystallinus*
Ctenanthe Comb-flower, (κτεις, κτενος)-ανθος (the bracteate flower-head)
Ctenitis, ctenitis Little-comb, κτενιτος
Ctenium Comb, κτενος (the one-sided, awned, spike-like inflorescence)
cteno- comb-, κτεις, κτενος, κτενο-
ctenoglossus -a -um having a much-divided, comb-like labium, κτενο-γλωσσα
ctenoides comb-like-, κτενος-οειδης
Ctenolophon Comb-crest, κτενο-λοφος (the comb-like aril of the seed)
ctenophorus -a -um bearing fimbriate, comb-like structures, κτενο- φερω
-ctonus -a -um, ctono- slaughter, κτονος
cuajonesensis -is -e from Mina Cuajones, Peru
cubeba the Arabic vernacular name, kubaba, for the unripe fruits of *Piper cubeba* (used medicinally and to flavour cigarettes)
cubensis -is -e from Cuba, Cuban
cubicus -a -um cuboid, κυβος
cubili Javanese vernacular name for the nut of *Cubilia cubili*
cubitalis -is -e a cubit tall, *cubitalis* (the length of the forearm plus the hand); forming a cushion, *cubital, cubitalis*
cubitans reclining, lying on a slope, present participle of *cubo, cubare, cubui, cubitum*
Cucubalus a name in Pliny
cuculi of the cuckoo, *cuculus* (flowering about the time the cuckoo arrives, May–June)
cucullaris -is -e, cucullarius -a -um, cucullatus -a -um (*cuccularia*) hooded, hood-like, *cucullus*
cucumerinus -a -um resembling cucumber, cucumber-like, *Cucurbita*
cucumeroides similar to *Cucumis, Cucumis-oides*
Cucumis the name, *cucumis*, used in Pliny for cucumbers grown for Tiberius, etymology not certain
Cucurbita the Latin name, *cucurbita, cucurbitae*, for the bottle-gourd, *Lagenaria siceraria* (***Cucurbitaceae***)

cucurbitinus -a -um melon- or marrow-like, gourd-like, *Cucurbita*
Cudrania from a Malayan vernacular name for the silk-worm thorn, *Cudrania tricuspidata*
cuencamensis -is -e from Cuencamé, Durango, Mexico
cujete a Brazilian vernacular name for the gourds of *Crescentia cujete*
Culcas the Arabic name for *Colocasia antiquorum*
Culcasia from the Arabic vernacular name, kulkas
culiciferus -a -um bearing gnats, (*culex, culicis*)-*fero*; bearing small cups, (κυλιξ, κυλικος)-φερω
culinaris of food, of the kitchen, *culina, culinae*
culmicolus -a -um growing on other plant's stalks, *culmus-colo*
cultoris, cultorus -a -um of gardens, of gardeners, *cultor, cultoris*
cultratus -a -um, cultriformis -is -e shaped like a knife-blade, *culter, cultri*
cultrifolius -a -um having leaves shaped like a knife-blade, *cultri-folius*
cultus -a -um cultivated, grown, past participle of *colo, colere, colui, cultum*
-culus -a -um -lesser
cumbalensis -is -e from Cumbal, Nevado, Colombia
cumberlandensis -is -e from any of the Cumberlands in Australia, Canada, England, USA or Vanatu
cumingianus -a -um, cumingii for Hugh Cuming (1791–1865), collector in Malaya and Philippines
Cuminum Mouse-plant, κυμινον, for the near-tautological *Cuminum cyminum*
cumulatus -a -um piled-up, enlarged, perfect, *cumulo, cumulare, cumulavi, cumulatum*
cumuliflorus -a -um having massed heads of flowers, (*cumulus, cumuli*)-*florus*
cunarius -a -um from the area of the Cuna Indians of Panama
cundinamarcensis -is -e from Cundinamarca, Colombia
-cundus -a -um -dependable, -able
cuneatifolius -a -um, cuneifolius -a -um with wedge-shaped leaves, broader to the apex, *cuneatus-folius, cune-folius*
cuneatus -a -um, cuneiformis -is -e narrow below and wide above, wedge-shaped, cuneate, *cuneus-forma*
cuneiflorus -a -um triangular-flowered, *cune-florus* (isosceles shaped)
Cunila the ancient Latin name for a fragrant herb
Cunninghamia, cunnunghamii for either James Cunningham, discoverer in 1702 of *C. lanceolata* in Chusan, China, or his brother A. Cunningham, botanist in Australia (Chinese firs)
Cunonia for J. C. Cuno (1708–80), Dutch naturalist (***Cunoniaceae***)
cupatiensis -is -e from the Cupati mountains of Colombia
Cuphea Curve, κυφος (cigar flower's capsule shape)
cupidus -a -um desirous, passionate, *cupidus*
cupreatus -a -um bronzed, coppery, late Latin *cuprum* (from *cyprium aes*, Cyprus metal)
cupreiflorus -a -um having bronze-coloured flowers, *cuprum-florum*
cupressifolius -a um cupressoid-leaved, *Cupressus-folium*
cupressiformis -is -e Cypress-like, *Cupressus-forma* (conical habit)
cupressinus -a -um, cupressoides cypress-like, resembling *Cupressus, Cupressus-oides*
Cupressocyparis the composite name for hybrids between *Cupressus* and *Chamaecyparis*
cupressorus -a -um of cypresses, *cupressus, cupressi*
Cupressus Symmetry, κυο-παρισος (the conical shape) (in mythology Apollo turned Cupressos into an evergreen tree, cognate with cypress) (***Cupressaceae***)
cupreus -a -um copper-coloured, coppery, *cuprum*
cuprispinus -a -um having coppery thorns, *cuprus-spina*
Cupularia Cupped, *cupula* (the fused, outer whorl of pappus hairs)
cupularis -is -, cupulatus -a -um forming, with, or subtended, by a cup-like structure, *cupula*

cupuliferus -a -um bearing small cup-like structures, *cupula-fero*
cupuliflorus -a -um having small cup-like flowers, *cupula-florum*
curassavicus -a -um from Curaçao, Leeward Islands, W Indies
curcas ancient Latin name for *Jatropha curcas* (physic nut)
Curculigo Weevil, *curculio* (the beak of the fruit)
curculigoides resembling *Curculigo*, *Curculigo-oides*
Curcuma the Arabic name, kurkum, kunkuma, for turmeric and its saffron-like colour
curiosus -a -um different, requiring thought, *curiosus*
curti-, *curto-*, *curtus -a -um* arched, curved, κυρτος, κυρτο-; shortened-, short, *curtus*, *curti-*
curtipendulus -a -um shortly-pendent, *curti-pendulus*
curtipes short-stalked, *curtus-(pes, pedis)*
curtophyllus -a -um having curved leaves, κυρτος-φυλλον
Curtisia for William Curtis (1746–99), founder of the *Botanical Magazine* and *Flora Londiniensis*
curtisii for Charles Curtis (1853–1928), Assistant Superintendent of Gardens and Forests, Straits Settlements
curtisiliquus -a -um short-podded, botanical Latin, *curti-siliquus*
curtus -a -um as if cut short or broken off, incomplete, short, *curtus*
curuzupensis -is -e from area of the Curuzupe, Paraguay
curvatus -a -um, *curvi-* curved, past participle of *curvo, curvare, curvavi, curvatum*
curvibracteatus -a -um with curved bracts, *curvi-bracteatus*
curvidens with curved teeth, *curvi-dens*
curviflorus -a -um with curved flowers, *curvi-florum*
curvipes with a curved stalk, *curvi-pes*
curviramus -a -um having arched branches, *curvi-(ramus, rami)*
curvisiliquus -a -um having long, curved pods, *curvi-(siliqua, siliquae)*
curvistylus -a -um having a curved style, *curvi-stylus*
curvulus -a -um slightly curved, diminutive of *curvus*
cusickii for William C. Cusick (1842–1922) of Oregon, USA
Cuscuta the name used by Rufinus (thirteenth-century botanist) for dodder, from Arabic, kechout (***Cuscutaceae***)
cuscutiformis -is -e resembling dodder, *Cuscuta-forma*, of a *Cuscuta*-like nature (slender runners with plantlets at their tips)
cuspidatus -a -um, *cuspidi-* cuspidate, abruptly narrowed into a short rigid point, *cuspis, cuspidis*
cuspidifolius -a -um having short leaves with a pointed apex, *(cuspis, cuspidis)-folium*
Cussonia for Pierre Cusson (1727–83), Professor of Botany at Montpellier
Cuthbertia for Alfred Cuthbert (1857–1932), collector of SE American plants
cuticularis -is -e cuticulate or heavy-barked, skin-like, *cuticula, cuticulae*
cutispongeus -a -um spongy-barked (*Polyscias cutispongea* is the sponge-bark tree)
Cuviera for Georges Léopold Chrétien Frédéric Dagobert, Baron Cuvier (1769–1832), French zoological anatomist and systematist
cuzcoensis -is -e from Qosqo (Cusco or Cuzco), one-time Inca capital of SE Peru
cyan-, *cyano-* dark-blue-, corn-flower-blue-, κυανεος, κυανο-, κυαν-
cyanandrus -a -um having blue stamens, κυαν-ανηρ
Cyananthus, *cyananthus -a -u* Blue-flower, κυαν-ανθος
cyanaster bluish; bluish-flowered, botanical Latin from κυανεος and *aster*
Cyanella feminine diminutive from *cyanus*
cyanescens turning blue, becoming blue, κυανεος
cyaneus -a -um, *cyano-* Prussian-blue, dark-blue, κυανεος
cyanocarpus -a -um with blue fruits, κυανεος-καρπος
cyanocentrus -a -um having a blue spur, κυανο-κεντρον
cyanocrocus -a -um saffron and blue, blue *Crocus*-like, κυανο-κροκος
cyanophyllus -a -um blue-leaved, κυανο-φυλλον

Cyanotis Blue-ear, κυανο-ωτος (for the petals)
cyanoxanthus -a -um blue to golden, κυανο-ξανθος (the variable colour of charcoal burner *Rusula*)
cyanus azure, blue, κυανεος (Meleager's Latin name for *Centaurea cyanus*)
Cyathea Little-cup, κυαθος (the basin-like indusium around the sorus) (***Cyatheaceae***)
cyatheoides resembling *Cyathea*, κυαθος-οειδης
cyathiflorus -a -um with wine-glass shaped flowers, *cyathus-florum*
cyathiformis -is -e shaped like a wine-glass, κυαθος, *cyathus*
cyathistipulus -a -um with cup-like (concave) stipules, *cyathus-(stipula, stipulae)*
Cyathodes Cup-shaped, κυαθος-ωδης (the five-toothed disc)
cyathophorus -a -um cup-bearing, κυαθ0-φορα
cyathulus -a -um with the shape of a small wine cup, diminutive of *cyathus*
-cybe -cover, -head, κυβη (the pileus or cap of a toadstool)
cybistax turned over, tumbled, κυβισταω
cybister tumbler-shaped, tumbling, deceptive, κυβισταω
Cybistetes Tumbler, κυβιστητηρ, to somersault (the wind-tumbled infructescence)
cycadifolius -a -um having leaves similar to those of *Cycas*
cycadinus -a -um like a small cycad, κοιξ, κοικας (κυκας)
Cycas Theophrastus' name, κοικας (wrongly transcribed as κυκας), for an unknown palm (sago palm)
cycl-, cyclo- circle-, circular-, disc-, wheel-, κυκλος, κυκλο-
Cyclamen Circle, κυκλος (Theophrastus' name, κυκλαμις, κυκλαμινος, for the coiled fruiting stalk); others relate it to the shape of the corms
cyclamineus -a -um, cyclaminus -a -um resembling *Cyclamen*
Cyclanthera Circled-anthers, κυκλος-ανθερα (their disposition)
Cyclanthus Flower-circles, κυκλος-ανθος (floral arrangement)
cyclius -a -um round, circular, κυκλιος
cyclobulbon having circular bulbs or pseudobulbs, botanical Latin from κυκλος and *bulbus*
cyclocarpus having circular fruits, fruiting on all sides, κυκλο-καρπος
Cyclocarya Circular-nut, κυκλο-καρυον (≡ *Pterocarya*)
Cyclocotyla Circular-cup, κυκλο-κοτυλη
cycloglossus -a -um round-tongued, κυκλο-γλωσσα
cyclophyllus -a -um with round leaves, κυκλο-φυλλον
cyclops round-eyed, gigantic, κυκλος-ωψ (the Cyclops were gigantic, one-eyed giants of Greek mythology)
cyclosectus -a -um cut around the edges, botanical Latin from κυκλο and *seco, secare, secui, sectum*
Cyclosorus, cyclosorus -a -um Circular sorus, κυκλο-σορος (have circular sori)
cycnocephalus -a -um turned over to resemble a swan's head, κυκνος-κεφαλη
Cydista Noblest, κυδιστος (the flower)
cydoni-, cydoniae- *Cydonia-*, quince-
Cydonia the Latin name for an 'apple' tree from Cydon (Khania), Crete (μελον κυδονιον, Cydonian apple, quince)
cygneus -a -um of swans, κυκνος; *cycnus, cycni; cygnus, cygni*
cygnoruus -a -um from the Swan River area of W Australia
cylindra-, cylindri-, cylindro- rolled, κυλινδεω; hollow, tubular, cylindric, *cylindrus, cylindri*
cylindraceus -a -um cylindrical, *cylindrus, cylindri* (flowers)
cylindricaulis -is -e having a hollow tubular stem, *cylindri-caulis*
cylindriceps having a hollow head, *cylindri-caput*
cylindricus -a -um, cylindro- long and round, cylindrical, *cylindrus, cylindri*
cylidrifolius -a -um having tubular or lengthwise-rolled leaves, *cylindri-folius*
cylindrobulbus -a -um having elongate cylindric bulbs or pseudobulbs, *cylindro-bulbus*
cylindrostachyus -a -um with cylindric spikes, botanical Latin from *cylindrus* and σταχυς

cylistus -a -um goblet- or chalice-shaped, κυλιξ, κυλικος
cylix goblet- or chalice-shaped, κυλιξ, κυλικος
cylleneus -a -um from Mount Killini, Korynthos, Greece (*Cyllene, Cyllenes*)
Cymbalaria Cymbal-like, *cymbalum, cymbali* (the peltate leaf shape)
cymbalarius -a -um cymbal-like, *cymbalum, cymbali* (the leaves of toad flax)
cymbi-, cymbidi- boat-shaped-, boat-, κυμβη, *cumba, cymba*
Cymbidium Boat-like, diminutive of κυμβη (the hollow recess in the lip)
cymbiferus -a -um bearing depressions, κυμβη-φερω
cymbifolius -a -um with boat-shaped leaves, *cymbi-folium*
cymbiformis -is -e boat-shaped, *cymbi-forma*
cymbispathus -a -um having a boat-shaped spathe, κυμβη-σπαθη
cymbispinus -a -um having hollowed, boat-shaped spines, *cymbi-spina* (myrmecophilous adaptation)
Cymbopogon Bearded-cup, κυμβη-πωγων
cyminum an old generic name, *Cuminum*, κυμινον from the Hebrew, kammon, for the aromatic seed (*Cuminum cyminum* exemplifies the use of Latin and Greek to make a permissible tautological name)
Cymodocea Waving, κυμα (undulating motion) (***Cymodoceaceae***)
Cymophyllus Undulate-leaved, κυμα-φυλλον
Cymopterus Undulate-winged, κυμα-πτερον (fruits)
cymosus -a -um having flowers borne in a cyme, κυμα, *cyma* (see Fig. 3a–d)
cymulosus -a -um having small cymes of flowers, diminutive of *cymosus*
cynanchicus -a -um of quinsy, κυναγκη (from its former medicinal use; literally, dog-throttling, κυν-αγχω)
cynanchoides resembling *Cynanchum*, κυναγκη-οειδης
Cynanchum, cynanchicus -a -um Dog-strangler, κυν-αγχω (some are poisonous, squinancy-wort, *Asperula cynanchica* was used for squinancy, tonsillitis; cognate with quinsy)
Cynapium Dog-parsley, *cyno-apium* (implying inferiority)
cynapius -a -um dangerous, inferior, κυνος (containing the alkaloid cynapine)
Cynara Dog, κυνος (the involucral spines of cardoon, *cardus*, or artichoke, Arabic, al-kharsuf)
cynaroides resembling *Cynara, Cynara-oides*
Cynastrum Blue-star, κυν-αστρον (the perianth)
cyno- dog-, κυων, κυν, κυνο-, *cynicus* (usually has derogatory undertone, implying inferiority)
cynobatifolius -a -um eglantine-leaved, dog-thorn-like, κυνο-βατος-φολιυμ
Cynocrambe, cynocrambe Dog-cabbage, κυνο-κραμβη (implying inferiority)
cynoctonus -a -um dog's-bane, κυνο-κτονος
Cynodon, cynodon Dog-tooth, κυν-οδων (the form of the spikelets)
Cynoglossum Hound's-tongue, κυνο-γλωσσα (Dioscorides' name, κυνογλωσσον, for the rough leaf texture)
Cynometra Dog-matrix, κυνο-μετρον (for the seed pods)
Cynomorium Dog-mulberry, κυνο-μοροεις (parasitic habit) (***Cynomoriaceae***)
cynophallophorus -a -um carrying a dog's penis(-like structure), κυνο-φαλλος-φορα
cynops the ancient Greek name, κυν-ωψ (for a plantain)
Cynorkis Blue-orchid, κυν-ορχις (flower colour)
cynosbati dog thornbush, κυνο-βατος
Cynosurus Dog-tail, κυνος-ουρα (for the paniculate form; also the appearance of the constellation of Ursa Minor, in which the Pole Star was the centre of attention (cynosure) for mariners)
cyparissias cypress(-like), κυπαρισσος (used in Pliny for a spurge)
Cypella Goblet, κυπελλον (for the form of the flowers) (≡ *Trimeza*)
cyperifolius -a -um sedge-leaved, *Cyperus-folius*
cyperinus -a -um of sedges, of *Cyperus*
cyperoides resembling *Cyperus, Cyperus-oides*
Cyperus, cyperus the Greek name, κυπειρος, κυπερος (for several species) (***Cyperaceae***)

cyph-, cypho- bent-, curved, stooping-, κυφος, κυφο-, κυφ-
Cyphanthera Curved-anthers, κυφος-ανθερα
cyphochilus -a -um having a humped lip, κυφο-χειλος
Cyphomandra Stooped-male, κυφος-ανηρ (tree tomato's humped anthers)
cypri of Venus, Kypris
cyprianus -a -um from Cyprus, Cypriot
Cypripedium Aphrodite's slipper, Κυπρις-πεδιλον (Kypris was also Aphrodite or Venus)
cyprius -a -um from Cyprus, Cypriot (*Cyprus, Cyprius*)
cyranostigmus -a -um with a curved stigma, for Savinien Cyrano de Bergerac (1619–55), as portrayed in Edmond Rostand's 1897 play, for his large, curved nose, and *stigma*
cyrenaicus -a -um from Cyrenaica, formerly a province of, now unified, Libya
Cyrilla for Dominica Cyrillo (1734–99), professor of medicine at Naples (***Cyrillaceae***)
cyrt- curved-, arched-, κυρτος, κυρτο-, κυρτ-
cyrtanthiflorus -a -um having flowers with curved anthers, *cyrt-anthera-florus*
Cyrtanthus Curved-flower, κυρτ-ανθος
cyrtobotryus -a -um curving bunched, κυρτο-βοτρυς (fruits)
Cyrtococcum Curved-fruit, κυρτο-κοκκος
cyrtodontus -a -um having curved teeth, κυρτ-οδοντος
Cyrtogonium Curved-knee, κυρτο-γονυ (the rhizome)
Cyrtogonone an anagram of *Crotonogyne* (a related genus)
Cyrtomiun Arched, κυρτοωμα (the fronds)
cyrtonema curved threads, κυρτο-νημα (filaments)
cyrtophyllus -a -um having curved leaves, κυρτο-φυλλον
Cyrtorchis Curved-orchid, κυρτ-ορχις (the spur)
Cyrtosperma Curved-seed, κυρτο-σπερμα (reniform seeded)
cyst-, cysti-, cysto- hollow-, pouched-, κυστις, κυστο-
cystolepidotus -a -um covered with cyst-like scales, κυστο-λεπις-ωτος
cystolepis -is -e having cyst-like (glandular) scales, κυστο-λεπις
Cystopteris Bladder-fern, κυστο-πτερυξ (from the inflated-looking indusia)
cystopteroides resembling *Cystopteris, Cystopteris-oides*
cystostegius -a -um having a bladder-like cover, κυστο-στεγος (the indusium)
cythereus -a -um from the Ionian island of Kithira (*Cythera*, once famed for the purple murex dye used by senatorial-class Romans)
cytisoides resembling *Cytisus*, κυτισος-οειδης
Cytisus the Greek name, κυτισος, for a clover-like plant (broom)

Daboecia (*Dabeocia*) for St Dabeoc, Welsh missionary to Ireland (name given by Edward Lhuyd (1660–1709), its discoverer)
dacicus -a -um from Bohemia (*Dacia*)
Dacrycarpus Weeping-fruit, δακρυ-καρπος (relation to *Dacrydium* and *Podocarpus*)
dacrydioides resembling *Dacrydium*, δακρυδιον-οειδης
Dacrydium Little-tear, δακρυδιον (its exudation of small resin droplets)
Dacryodes Tear-drop-like, δακρυ-ωδης
dactyl-, dactylo-, -dactylis finger-, δακτυλος
Dactyladenia With-finger-like-glands, δακτυλος-αδην
Dactylaena Finger-cloaked, δακτυλος-χλαινα
dactyliferus -a -um finger-bearing, *dactylis-fera*
dactylinus -a -um fingered, having finger-like lobing, δακτυλος
Dactyliophora Finger-carrying, δακτυλος-φορα
Dactylis Dactyl, δακτυλος, one stressed syllable followed by two unstressed syllables (the spikelets of cock's-foot grass are in a large terminal cluster with two lesser clusters below). Others suggest the interpretation as finger, or bunch of grapes

Dactyloctenium Digitate-*Ctenium*, δακτυλος-κτενος (the *Ctenium*-like spikes are aggregated into an umbellate head)
Dactyloglossum the composite name for hybrids between *Dactylorchis* and *Coeloglossum*
dactyloides finger-like-, δακτυλος-οειδης
dactylon finger, δακτυλος (the narrow spike-like branches of the inflorescence)
Dactylopsis Fingered-looking, δακτυλος-οψις (the succulent leaves)
Dactylorchis Finger orchid, δακτυλος-ορχις (the palmate arrangement of the root-tubers)
daedaleus -a -um skilful craft, for Daedalus the inventor and craftsman of mythology; curiously fashioned, δαιδαλεος, *daedalus* (the leaf apex division)
Daemonorops Devil-shrub, δαιμονος-ρωψ (palms armed with thorns), δαιμων, δαιμονος also translates as divine, death, guardian or fate
daemonus -a -um of genius, fate or superstition, δαιμων, δαιμονιος
daghestanicus -a -um from the Dagestan republic on the W shore of the Caspian Sea
daguensis -is -e from the area of the Dagua river, W Colombia
Dahlgrenodendron Dahlgren's-tree, for Rolf Martin Theodore Dahlgren (1932–87), Swedish systematic botanist, botanical Latin from Dahlgren and δενδρον
Dahlia for Andreas Dahl (1751–89), Swedish student under Linnaeus
dahliae of or upon *Dahlia* species (*Entyloma* smut fungus)
dahuricus -a -um, dauricus -a -um, davuricus -a -um from Dauria, NE Asia, near Chinese–Mongolian–Siberian borders
Dais Torch, δαις (the inflorescence of the pompon tree, and some suggest heat, for the caustic bark)
daisenensis -is -e from Dai Sen, Honshu, Japan
Dalbergaria for Karl Theodore Frieherr von Dalberg (1744–1817), statesman, cleric and Grand Duke of Frankfurt
Dalbergia for Nicholas (Nils) Dalberg (1736–1820), Swedish physician and botanist, and his brother Carl Gustav, who collected in the W Indies
Dalea for Dr Samuel Dale (1659–1739), English physician, botanist and writer, friend of John Ray
dalecarlicus -a -um from Dalarna province of Central Sweden (*Dalecarlia*)
Dalechampia, dalechampii for James Dalechamp (Jacques d'Alechamps) (1513–88), French physician and botanist, author of *Historia generalis plantarum* 1587
dalhousiae for Countess Dalhousie (1786–1839), Vicereine of India
Dalhousiea for James Andrew Broun Ramsay (1812–78), tenth Earl Dalhousie, Viceroy of British India
dalmaticus -a -um from Dalmatia, eastern Adriatic, Dalmatian
Dalzellia for Nicholas Alexander Dalzell (1817–78), Scottish botanist in India
Dalzielia for John McEwen Dalziel (1872–1948), of W African Medical Service and RBG Kew
damaranus -a -um from Damaraland, Namibia (or that part now occupied by the Bergdama people (Damara))
damascenus -a -um from Damascus, Syria; coloured like *Rosa damascena*
Damasonium a name, *damasonion*, in Pliny for *Alisma*
Dammara, dammara from an Indo-Malayan vernacular name, damar minyak, for the varnish-resin obtained from *Agathis loranthifolia* (*Dammara orientalis*) and several other genera
Damnacanthus Damaging-thorned-one, modern Latin *damnosus-acanthus*
Dampiera for Captain William Dampier (1651–1715), Royal Navy circumnavigator, author of *A New Voyage Around The World* (1697)
Danaë for Danaë, the daughter of Acrisius Persius, King of Argos, in Greek mythology
Danaea (Danaa) for J. P. M. Dana (1734–1801), Italian botanist
danfordiae for Mrs C. G. Danford, who collected *Crocus* etc in Asia Minor *c.* 1876–9
danicus -a -um from Denmark, Danish (*Dania*)

Daniellia, daniellii for Dr D. Daniell who collected in Sierra Leone and Senegal *c.* 1840–53, or William Freeman Daniell (1818–65), collector in W Indies and China
Danthonia, danthonii for Etienne Danthoine (fl. 1788), student of the grasses of Provence, France
Danthonidium Little-*Danthonia*, diminutive of *Danthonia*
Danthoniopsis *Danthonia*-like, botanical Latin from *Danthonia* and οψις
danubialis -is -e, danuviensis -is -e from the upper Danube (*Danuvius*)
Daphne old name for bay-laurel, from that of a Dryad nymph of chastity, Daphne, in Greek mythology (spurge laurel)
daphneolus -a -um like *Daphne*
daphnephylloides, daphniphylloides resembling *Daphniphyllum*
Daphnimorpha With-the form-of *Daphne*, δαφνε-μορφη
Daphniphyllum *Daphne*-leaved, δαφνε-φυλλον (***Daphniphyllaceae***)
daphnoides, daphnoideus -a -um resembling *Daphne, Daphne-oides*
darcyi for John d'Arcy, contemporary collector with James Compton and Martin Rix in Mexico
dareoides resembling *Darea* (≡ *Asplenium*)
darjeeligensis -is -e from Darjeeling, India
Darlingtonia for William Darlington (1782–1863), American physician, botanist and mycologist
Darmera for Karl Darmer, nineteenth-century German horticulturist (formerly *Peltiphyllum peltatum*)
darwasicus -a -um from the Darvaz range of the Pamir Mountains, Tajikistan
Darwinia for Dr Erasmus Darwin (1731–1802) author of *The Botanic Garden* and grandfather of Charles R. Darwin
darwinii for Charles Robert Darwin (1809–82) naturalist and evolutionist, author of *The Origin of Species by Means of Natural Selection*
Darwiniothamnus Darwin's-shrub, botanical Latin from Darwin and θαμνος, for Charles Robert Darwin (1809–82)
Dasispermum Thickly-haired-seed, δασυ-σπερμα
Dasistoma Woolly-mouthed, δασυ-στομα
dasy- thick-, thickly-hairy-, woolly-, δασυς, δασυ-
dasyacanthus -a -um having thick spines, δασυ-ακανθα
dasyanthus -a -um with very hairy flowers, δασυ-ανθος
dasycarpus -a -um with a thickly hairy ovary, δασυ-καρπος
dasychaetus -a -um having a thick mane, with dense hairiness, δασυ-χαιτη
dasyclados shaggy-twigged, δασυ-κλαδος
Dasylepis Thick-scales, δασυ-λεπις (the clustered scales on the stout pedicels)
Dasylirion Thick-lily, δασυ-λειριον (the thick stems)
Dasynotus Hairy-backed, δασυ-(νωτον, νωτος) (the throat of the corolla)
dasypetalus -a -um having hairy petals, δασυ-πεταλον
Dasyphyllum, dasyphyllus -a -um Shaggy-leaf, with thickly hairy leaves, δασυ-φυλλον
Dasypogon Shaggy-beard, δασυ-πωγων
Dasypyrum Rough-wheat, δασυ-πυρος
Dasystachys, dasystachys Dense-spiked-one, with shaggy spikes, δασυ-σταχυς
dasystemon with very hairy stamens, δασυ-στεμον
-dasys -hairy, δασυς, δασυ-
dasystylus -a -um having hairy styles, δασυ-στυλος
dasytrichus -a -um thickly haired, δασυ-τριχος
datil from the ancient volcanic Datil region of the Colorado plateau, New Mexico
Datisca derivation obscure, δατεομαι for the divided leaves? (***Datiscaceae***)
Datura from an Indian vernacular name, dhatura, Sanskrit, dhustura, (Arabic, tatorali) (thorn apple)
Daubenya for Dr Charles Daubeny (1795–1867), Professor of Chemistry, then Botany at Oxford
dauci- carrot-like, resembling *Daucus*

daucifolius -a -um having leaves resembling those of carrot, *Daucus-folium*
daucoides resembling *Daucus*, *Daucus-oides*
Daucosma Carrot-fragrant, δαυκον-οσμη
Daucus the Latin name, *daucus*, for a carrot, Greek δαυκον
dauricus -a -um from Dauria, NE Asia
Davallia, davallianus -a -um for Edmond Davall (1763–98), Swiss botanist (***Davalliaceae***)
davallioides, davalliodes resembling *Davallia*, like a hare's-foot fern, *Davallia-oides*
daveauanus -a -um for Jules Daveau (1852–1929), Director of the Botanic Garden at Lisbon
Davidia, davidii, davidianus -a -um for l'Abbé Armand David (1826–1900), missionary and collector of Chinese plants (dove tree) (***Davidiaceae***)
Davidsonia for J. E. Davidson, Australian sugar grower *c.* 1860 (***Davidsoniaceae***)
Daviesia for Reverend Hugh Davies (1739–1821), Welsh botanist and author of *Welsh Botanology* (1813)
davisianus -a -um, davisii for Peter Davis, collector for Veitch in Peru *c.* 1875
davuricus -a -um from Dauria, NE Asia
dayanus -a -um for John Day (1824–88), collector of orchids in India, Ceylon and Brazil
Dayaoshania from the Dayao Shan mountains in S China
de- downwards-, outwards-, from-, out of-, *de*
dealbatus -a -um with a white powdery covering, white-washed, whitened, adjective from *dealbo, dealbare*
debilis -is -e weak, feeble, frail, *debilis*
Debregeasia for Prosper Justin de Bregeas, French naval explorer of the Far East 1836–7
dec-, deca-, decem- ten-, tenfold-, δεκας, *decem*
Decagoniocarpus Ten-edged-fruit, δεκα-γωνια-καρπος
decagonus -a -um having ten angles, ridges or corners, δεκας-γονυ
Decaisnea, decaisneanus -a -um for Joseph Decaisne (1807–82), French botanist and plant illustrator
Decalepis Ten-scaled-one, δεκα-λεπις (perianth structure)
decalvans balding, becoming hairless, *de-(calva, calvae)*
decandrus -a -um ten-stamened, δεκα-(ανηρ, ανδρος)
decapetalus -a -um with ten petals, δεκα-πεταλον
decaphyllus -a -um having ten leaves, leaflets or perianth segments, δεκα-φυλλον
Decarydendron Decary's-tree, for Raymond Decary (1891–1973), French administrator in Madagascar, botanist and collector, botanical Latin from Decary and δενδρον
decemfidus -a -um splitting into ten sections, *decem-(findo, findere, fidi, fissum)*
deceptor, deceptrix deceiver, male and female gerundives of *decipio, decipere, decepi, deceptum*
deceptus -a -um beguiling, deceiving, passive participle of *decipio, decipere, decepi, deceptum*
deciduus -a -um not persisting, falling-off, deciduous, *decido, decidere, decidi*
decipiens misleading, deceiving, present participle of *decipio, decipere, decepi, deceptum*
Deckenia for Karl Klaus von der Decken (1833–65), German explorer and surveyor of Kilimanjaro, E Africa
declinatus -a -um turned aside, curved downwards, *declino, declinare, declinavi, declinatum*
declivis -is -e sloping downwards, growing at a steep downwards angle, *declivis*
decliviticolus -a -um living on steep inclines, *declivis-colo*
Decodon Ten-teeth, δεκας-οδων (from the horn-like processes in the calyx sinuses)
decolor discoloured, faded, *decolor, decoloris*
decolorans staining, discolouring, present participle of *decoloro, decolorare, decoloravi, decoloratum*

decompositus -a -um divided more than once (leaf structure), decompound, *de-(compositio, compositionis)*
decor, decorans decorating, present participle of *decoro, decorare, decoravi, decoratum*
decoratus -a -um handsome, elegant, decorous, *decoro, decorare, decoravi, decoratum*
decorticans with stripping bark, becoming barkless, present participle of *decorticatio, decorticare, decorticavi, decorticatum*
decorticatus -a -um without bark, without a husk or cortex, *de-(cortex, cortic)*
decorticus -a -um with shedding bark, *decorticis*
decorus -a -um handsome, elegant, decorous, *decoro, decorare, decoravi, decoratum (decus, decoris,* ornamented)
decumanus -a -um, (decimanus) very large, of largesse, *decumanus* (literally, the tenth legion of Roman soldiers)
Decumaria Ten-partite, *decuma* (the numbers of floral structures)
decumbens prostrate with tips turned up, decumbent, *decumbo, decumbere, decubui*
decurrens running down, decurrent, *decurro, decurrere, decucuri (decurri), decursum* (e.g. the bases of leaves down the stem)
decurrentialatus -a -um having stems decurrently winged, *decurro-alatus*
decursive-pinnata having decurrent blades of the pinnae, *decursus-(pinna, pinnae)*
decursivus -a -um running downwards, in a downwards series, *decurro, decurrere, (decucurri) decurri, decursum*
decurtatus -a -um mutilated, cut-short, *decurtatus*
decurvans curved downwards, present participle of *de-(curvo, curvare, curvavi, curvatum)*
decurvus -a -um bent down, curved downwards, *decurvus*
decussatus -a -um divided crosswise, at right-angles, decussate, *decusso, decussare* (as when the leaves are in two alternating ranks)
decussus -a -um decussate, with alternating pairs of opposed leaves, *de-(cusso, cussare)*
defectus -a -um eclipsed; failing, weak, past participle of *deficio, deficere, defeci, defectum*
defensus -a -um defended, protected, *defendo, defendere, defendi, defensum* (having thorns, stinging hairs, or other protective features)
deficiens becoming less, dwindling, weakening, present participle of *deficio, deficere, defeci, defectum*
deflectans turned aside, turned down, present participle of *deflecto, deflectere, deflexi, deflectum*
deflexicalyx with deflexed calyx lobes, *de-flexi-calyx*
deflexicaulis -is -e having downwards-bending stalks, *de-flexi-caulos*
deflexispinus -a -um having downwards-directed thorns, *de-flexi-spina*
deflexus -a -um bent sharply backwards, deflexed, *deflecto, deflectere, deflexi, deflexum*
defloratus -a -um without flowers, shedding its flowers, *de-florum*
defoliatus -a -um not leafy, not producing or producing small and transient leaves, *defolio, defoliare, defoliavi, defoliatum*
deformis -is -e misshapen, deformed, *deformis*
defossus -a -um hidden away, buried, *defodio, defodere, defodi, defossum*
Degeneria Degenerate, *degenero, degenerare, degenaravi, degenaratum* (not having the fully enclosed carpels that an *Angiosperm* should, by Angiosperm definition, have)
Degenia, degenianus -a -um, degenii for Dr Arpád von Degen (1866–1934), Director of the Seed Testing Station, Budapest
dehiscens splitting open, gaping, dehiscent, present participle of *dehisco, dehiscere*
dein-, deino- venerable, fearful, terrible, dangerous, extraordinary, mighty, strange, marvellous, δεινος
Deinanthe Extraordinary-flower, δεινος-ανθος (large-flowered)
deinorrhizus -a -um dangerous-root, δεινος-ριζα (poisonous properties)
dejectus -a -um debased, low-lying, *deicio, deicere, deieci, deiectum*

delavayanus -a -um, delavayi for l'Abbé Jean Marie Delavay (1834–95), French missionary and collector of plants in China
delectus -a -um choice, chosen, *deligio, deligere, delegi, delectum*
delegatensis -is -e from Delegate, New South Wales
delibutus -a -um smeared, *delibuo, delibuere, delibui, delibutum*
delicatissimus -a -um most charming, most delicate, superlative of *delicatus*
delicatus -a -um, delicatulus -a -um charming, voluptuous, soft, *delicatus*
deliciosus -a -um of pleasant flavour, delicious, delightful, *deliciae, deliciarum*
delicus -a -um soft, with fine (lines), botanical Latin
delileanus -a -um, delilei for Alire Raffeneau Delile (1778–1850), French botanist
delineatus -a -um seducing, soothing, *delenio, delenire, delenivi, delenitum*
deliquescens melting, turning to liquid, present participle of *deliquesco, deliquescere, delicui* (autolysing)
delitescens hiding away, skulking, growing under cover, present participle of *delitesco, delitescere, delitui*
delo-, delos- manifest-, visible-, evident-, clear-, plain-, δηλος
Delonix Distinct-claw, δηλος-ονυξ (on the petals)
Delosperma Evident-seed, δηλος-σπερμα
Delostoma Clear-mouthed-one, δηλος-στομα (wide-mouthed flowers)
delphicus -a -um from Delphi, Greece, Delphic, *Delphi, Delphorum, delphicus*
delphinanthus -a -um *Delphinium*-flowered dolphin-flowered, δελφινος-ανθος
delphinensis-is -e from the former Dauphiné province, SE France (the lands formerly held by the Dauphin, called the delphinate, *terrae delphinatus*)
Delphinium Dolphin, δελφις, δελφινος (the name, δελφινιον, used by Dioscorides, for the unopened flower's appearance)
deltanthus -a -um having triangular outlined flowers, δελτα-ανθος
deltodon having a triangular tooth, δελτα-οδοντος
deltoides, deltoideus -a -um triangular-shaped, deltoid, δελτα-οειδης (see Fig. 8b)
delus -a -um plain, clear, evident, visible, conspicuous, δηλος
-deme a definable grouping of individuals of a specified taxon, δημος
demersus -a -um underwater, submerged, *demergo, demergere, demersi, demersum*
demetrionis -is -e for Demeter, corn goddess and mother of Persephone
deminutus -a -um shrunken, small, *deminuo, deminuere, deminui, deminutm*
demissus -a -um hanging down, low, weak, dwarf, *demitto, demittere, demisi, demissum*
dendr-, dendri-, dendro-, -dendron, (-dendrum) tree-, tree-like-, on trees-, δενδρον, δενδρο-, δενδρος, δενδριτης
Dendranthema Tree-flower, δενδρο-ανθεμιον (woody *Chrysanthemum*)
dendricolus -a -um tree-dwelling, botanical Latin from δενδρον and *colo*
dendrobiopsis -is -e having the appearance of *Dendrobium*, δενδρο-βιος-οψις
Dendrobium Tree-dweller, δενδρο-βιος (epiphytic)
Dendrocalamus Tree-*Calamus*, δενδρο-καλαμος (taller than *Calamus*)
dendrocharis -is -e tree of beauty, δενδρον-χαρις
Dendrochilum Tree-lip, δενδρον and χειλος (tree dwelling and with a distinctive lip)
dendroides, dendroideus -a -um, dendromorphus -a -um tree-like, branched, δενδρον-οειδης
Dendromecon Tree-poppy, δενδρον-μηκων (shrubby habit)
dendromorphus -a -um having tree-like form, δενδρον-μορφη
Dendropanax Tree-*Panax*, δενδρο-πανακες
dendrophilus -a -um tree-loving, δενδρον-φιλος (arboreal habitat)
Dendroseris Tree-endive, δενδρον-σερις (the foliage)
dendrotragius -a -um goat-tree, δενδρον-τραγος (they climb it for its leaves)
Dennettia for Richard E. Dennett, an early twentieth-century forester in Nigeria
Dennstaedtia for August Wilhelm Denstaedt (*c.* 1818), German botanist (***Dennstaedtiaceae***)
densatus -a -um, densi- crowded, close, dense, *densus* (habit of stem growth)
dens-canis dog's tooth, *(dens, dentis)-canis*

densiflorus -a -um densely flowered, close-flowered, *densus-(floreo, florere, florui)*
densifolius -a -um with dense foliage, *densus-folium*
dens-leonis lion's tooth, (*dens, dentis*)-(*leo, leonis*)
densus -a -um compact, condensed, close, with short internodes, *densus*
Dentaria Toothwort, *dens, dentis* (the signature of the scales upon the roots)
dentatisepalus -a -um having toothed sepals, (*dens, dentis*)-*sepalus*
dentatus -a -um, dentosus -a -um having teeth, with outward-pointing teeth, dentate, *dentatus* (see Fig 4b)
denticulatus -a -um minutely toothed, diminutive from *dentatus*
dentifer -era -erum tooth-bearing, *dentis-fero*
denudans becoming naked, stripping, present participle of *denudo, denudare, denudavi, denudatum*
denudatus -a -um hairy or downy but becoming naked, denuded, naked (leafless at flowering) *denudo, denudare, denudavi, denudatum*
deodarus -a -um from the eponymous Indian state (Sanskrit, deva dara, divine tree)
deorsus -a -um downwards, hanging, *deorsum, deorsus*
deorum of the gods, *deus, dei* (feminine *dea, deae*)
depauperatus -a -um imperfectly formed, dwarfed, of poor appearance, impoverished, (*de-*)*paupero, pauperare*
depavitus -a -um beaten or trodden down, *de-*(*pavo, pavire, pavui, pavitum*)
dependens hanging down, pendent, derived from, present participle of *dependeo, dependere*
depraehensus -a -um held different from an earlier one, *de-prae-*(*hendo, hendere, hensi, hensum*)
depressinervius -a -um having depressed veins, botanical Latin from *depressus-nervus*
depressus -a -um flattened downwards, depressed, past participle of *deprimo, deprimere, depressi, depressum*
deregularis -is -e deviating or not conforming to the rule, *de-*(*regulo, regulare, regulavi, regulatum*)
derelictus -a -um abandoned, neglected, *derelinquo, derelinquere, deriliqui, derelictum*
deremensis -is -e from Derema, Tanzania
-dermis -is -a -skin, -outer-surface, δερμα
descendens flowering downwards, descendent flowering, *descendo, descendere, descendi, descensum*
Deschampsia for the French naturalist Louis Auguste Deschamps (1765–1842)
Descurainia (Descurania) for François Descourain (1658–1740), French physician
deserti-, desertorus -a -um, desertoris -is -e of deserts, from past participle of *desero, deserere, deserui, desertum*
deserticolus -a -um inhabiting deserts, *deserti-colo*
deserti-siriaci from the Syrian Desert, E Mediterranean (*desero, deserere, deserui, desertum*)
desertus -a -um of uninhabited places, of places left waste, of deserts, from past participle of *desero, deserere, deserui, desertum*
Desfontainia for René Louiche Desfontaines (1750–1833), French botanist (***Desfontainiaceae***)
desiccatus -a -um dried up, of dryness, *de-siccus*
desma-, desmo- bundle-, band-, thong-, fillet-, δεσμη, δεσμος
Desmabotrys Clustered-fruit, δεσμη-βοτρυς
desmacanthus -a -um with spines in bundles, δεσμη-ακανθα
Desmanthus, desmanthus -a -um Bundle-flower, δεσμη-ανθος (the appearance of the inflorescence)
Desmazeria (Demazeria) for Jean Baptiste Henri Joseph Desmazières (1786–1862), French botanist
Desmodium Band, δεσμος (the lobed fruits, or the united stamens)
desmoncoides resembling *Desmoncus, Desmoncus-oides*
Desmoncus Grapple-band, δεσμος-ογκος (the hooked tips of the leaf rachices)

Desmoschoenus Banded-*Schoenus*, botanical Latin from δεσμος and *Schoenus* (banded variegation)
desolatus -a -um abandoned, solitary, left alone, *desolo, deslare, desolavi, desolatum*
despectens despised, looked down upon, open to view, present participle of *despecto, despectare*
desquamatus -a -um lacking scales, peeled, *desquamo, desquamare*
destitutus -a -um deficient, wanting, forsaken, *destituo, destiuere, detstui, destitutum*
desuetus -a -um not-customary, unusual, unaccustomed, *de-(suesco, suescere, suevi, suetum)*
detergens breaking-off, clearing, cleaning, *detergo, detergere, detersi, detersum*
deterrinus -a -um off-putting, deterring, frightening off, *deterreo, deterrere, deterrui, deterritum* (the colour variations)
detersibilis -is -e broken off, clean, clear, *detergo, detergere, detersi, detersum*
detersus -a -um wiped clean, *detergo, detergere, detersi, detersum*
detonsus -a -um shorn, stripped, shaved, bald, *detondeo, detondere, detondi, detonsum*
detortus -a -um turned aside, distorted, *detorqueo, detorquere, detorsi, detortum*
detruncatus -a -um mutilated, beheaded, cut off, *detrunco, detruncare, detruncavi, detruncatum*
deumanus -a -um to beg for, to want or wish for, δεω, δεομαι
deustus -a -um scorched, burned, frosted, *deuro, deureri, deussi, deustum*
Deutzia for Johannes van der Deutz (1743–88), sheriff of Amsterdam and Karl Pehr Thunberg's patron
devastatrix laying waste, feminine of *devastor*, from *devasto, devastare*
devoniensis-is-e from Devon, England (*Devonia*)
devrieseanus -a -um for Willem Hendrik de Vriese (1806–62), Dutch botanist
devriesianus -a -um for Hugo de Vries (1848–1935) Professor of Botany at Amsterdam
dextro- right-, *dexter, dexteri, dextri*
dextrorsus -a -um twining to the right, *dextrorsum*, dextrorse (anticlockwise upwards as seen from above)
di-, dis- twice-, two-, between-, away from-, δυας, δυαδος, δις, δι-
dia- -through-, across-, δια-
diabolicus -a -um slanderous, two-horned, devilish, διαβολος
diacanthus -a -um double-thorned, δι-ακανθα
diacritus -a -um distinguished, separated, διακρισις
diadema, diadematus -a -um band or fillet, crown, crown-like, διαδεμα
Dialaeliocattleya the formulaic name for hybrids between *Diacrium*, *Laelia* and *Cattleya*
dialy- disbanded-, very deeply incised-, separated-, διαλυω
dialystemon separated stamens, free-stamened, διαλυω-στεμμα
diamensis -is -e from Tasmania (prior to 1855 called Van Diemen's Land)
Diandrolyra Two-stamened-rice, δι-ανηρ-ολυρα (affinity to *Olyra*)
diandrus -a -um two-stamened, δι-(ανηρ, ανδρος)
Dianella Diana, diminutive of *Diana* (for the goddess of hunting)
dianthiflorus -a -um *Dianthus*-flowered
dianthoidea, dianthoides like *Dianthus*, διος-ανθος-οειδης
Dianthus Zeus'-flower, διος-ανθος (a name, διοσανθος, used by Theophrastus)
Diapensia Twice-five, δια-πεντε (formerly an ancient Greek name for sanicle but re-applied by Linnaeus) (***Diapensiaceae***)
diapensioides resembling *Diapensia*, *Diapensia-oides*
Diaphananthe Transparent-flowered, διαφανης-ανθος (texture of the corolla of wax orchids)
diaphanoides resembling (*Hieracium*) *diaphanum* (in leaf form), διαφανης-οειδης
diaphanus -a -um transparent, διαφανης (leaves)
diaprepes distinguished, conspicuous, excellent, διαπρεπω
Diarrhena Two-males, δι-(αρρην-αρρενος)
Diascia Two-sacked, δι-ασκος (the two spurs) (some derive it from διασκαο, to adorn)

diastrophis -is -e two-banded, distorted, διαστροφος
dibotrys two-bunched, δι-βοτρυς (inflorescences)
Dicanthium Twin-flowered, διχα-ανθεμιον
Dicentra Twice-spurred, δι-κεντρον (the two-spurred flowers)
dicentrifolius -a -um *Dicentra*-leaved, *Dicentra-folium*
dicha-, dicho- double-, into two-, διχα, διχη, διχου, διχελος, διχο-, δι-
Dichaetanthera Two-spurred-stamens, δι-χαιτ-ανθερα (the two spurs below the anthers)
Dichapetalum Halved-petals, διχα-πεταλον (the petals are deeply bifid) (***Dichapetalaceae***, ≡ ***Chailletiaceae***)
Dichelostemma Divided-crown, (διχα, διχη, διχελος)-στεμμα (the bifid staminodes)
dichlamydeus -a -um two-cloaked, δι-χλαμυς (with calyx and corolla, or with two spathes)
Dichondra Two-lumped, δι-χονδρος (the two-lobed ovary)
dichondrifolius -a -um two-grained-leaved, *Dichondra-folium*
Dichorisandra Two-separated-men, δι-χοριζο-ανηρ (two of the stamens diverge from the remainder)
Dichostemma Twice-crowned, διχο-στεμμα (two bracts cover the flower heads)
Dichotomanthes Cut-in-two-flower, διχοτομεω-ανθος
dichotomus -a -um divided into two equal portions, equal-branched, split into two, dichotomous, διχοτομεω
dichrano- two-branched-, two-headed, δι-κρανιον
dichranotrichus -a -um with two-pointed hairs, δι-κρανο-τριχος
Dichroa, dichrous -a -um of two colours, δι-χροα, δι-χρωος (the flowers)
dichroanthus -a -um with two-coloured flowers, δι-χρω-ανθος
Dichrocephala Two-coloured-headed-one, διχρωος-κεφαλη
dichromatus -a -um, dichromus -a -um, dichrous -a -um of two colours, two-coloured, δι-χρωμα, δι-χρωος
Dicksonia, dicksonii for James Dickson (1738–1822), British nurseryman and botanist, founder member of the RHS and the Linnean Society of London
Dicliptera Two-fold-winged, δικλις-πτερον (the capsule)
Diclis Twice-folded (of a door), δικλις, δικλιδος
dicoccus -a -um having paired nuts, two-berried, δι-κοκκος
dicolor two-coloured, *di-*(*color, coloris*)
Dicranium Double-headed, δι-κρανιον (the peristome teeth are bifid)
dicrano- two-branched, δικρανος. δικρανο-
Dicranostigma Twice-branched-stigma, δικρανο-στιγμα
dicranotrichus -a -um having hairs divided to two tips, δικρανο-τριχος
Dictamnus, dictamnus Mount Dicte, in Crete (Virgil's name for dittany or fraxinella) (cognate with dittander, or from Mount Dikte)
dictyo-, dictyon netted-, -net, δικτυον
dictyocarpus -a -um netted-fruit, δικτυον-καρπος
Dictyonema Netted-thread, δικτυον-νημα
dictyophyllus -a -um with net-veined leaves, δικτυον-φυλλον
Dictyopteris Netted-fern, δικτυον-πτερυξ
Dictyosperma Netted-seeded-one, δικτυον-σπερμα (the net-like raphe)
Dictyota, dictyotus -a -um Net-like, δικτυον
Dicyrta Twice-curved, δι-κυρτος (folds in the throat of the corolla)
Didieria, didieri for E. Didier (1811–89), who studied the plants of Savoy (cactus-like ***Didieraceae***)
Didiscus Two-discs, δι-δισκος (the pistil of blue lac flower)
didistichus -a -um with a four-ranked panicle, δι-δι-στιχος
didymo-, didymus -a -um twin-, twinned-, double-, equally-divided, in pairs, διδυμος
didymobotryus -a -um with paired bunches, διδυμο-βοτρυς (of flowers)
Didymocarpus, didymocarpus -a -um Twin-fruit, διδυμο-καρπος
Didymochlaena Twin-cloak, διδυμο-(χ)λαινα (indusia attached at centre and base but free at sides and apex)

didymophyllus -a -um having paired leaflets or leaves, διδυμο-φυλλον
Didymosalpinx Two trumpeted-one, διδυμο-σαλπιγξ (the flower arrangement)
Didymosperma Two-seeded-one, διδυμο-σπερμα (fruits usually have two seeds)
didymus -a -um twofold, double, διδυμος (testicle) (as the dumb-bell-shaped fruit of *Coronopus didymus*)
Dieffenbachia, dieffenbachianus -a -um, dieffenbachii for Herr Dieffenbach, gardener of Schönbrunn, Austria (dumb cane)
dielsianus -a -um, dielsii for Frank L. E. Diels (1874–1945), of the Berlin Dahlem Botanic Garden
diemenicus -a -um from Tasmania (prior to 1855 called Van Diemen's Land)
Dierama Funnel, διεραμα, δι-εραμαι (the shape of the perianth)
Diervilla for M. Dièreville, French surgeon and traveller in Canada during 1699–1700
Dietes Two-years, διετης, διετια (wedding lily)
difficilis -is -e difficult, awkward, *difficilis*
difformis -is -e, diformis -is -e of unusual or abnormal form or shape, irregular, *deformo, deformare, deformavi, deformatum*
diffractus -a -um shattering, becoming tessellated with cracks, present participle of *diffringo, diffringere*
diffusiflorus -a -um with open inflorescences, (*diffundo, diffundere, diffudi, diffusum*)-*florum*
diffusus -a -um loosely spreading, diffuse, *diffundo, diffundere, diffudi, diffusum*
digamus -a -um having both sexes in the same inflorescence, δις-γαμος
digeneus -a -um produced sexually, containing both sexes, of two (dissimilar) species, δι-γενος (hybrid name)
digitaliflorus -a -um with flowers similar to *Digitalis, Digitalis-florum*
Digitalis, digitalis -is -e Finger-bonnet, Fingerstall (Fuchs' translation of the German Fingerhut) (foxglove)
digitaloides resembling *Digitalis*
Digitaria Fingered, *digitus, digiti* (the radiating spikes)
digitatus -a -um fingered, hand-like, lobed from one point, digitate, *digitus, digiti* (see Fig. 8c)
digitiformis -is -e fingered, finger-shaped, *digitus-forma*
dignabilis -is -e noteworthy, *digno, dignare, dignor, dignari*
Digraphis, digraphis -is -e Twice-inscribed, δι-γραφις (with lines of two colours, ≡ *Phalaris*)
digynus -a -um with an ovary having two carpels, δι-γυνη
Diheteropogon Doubled-*Heteropogon* (has paired racemes)
dijonensis -is -e from Dijon, France
dilatatopetiolaris -is -e with inflated petioles, *dilatatus-petiolus*
dilatatus -a -um, dilatus -a -um widened, spread out, dilated, *dilato, dilatare, dilatavi, dilatatum*
dilectus -a -um picking, selection; beloved, past participle of *diligo, diligere, dilexi, dilectum*
Dillenia for Johann Jacob Dillenius (1687–1747), Professor of Botany at Oxford, author of *Historia Muscorum* and *Hortus Elthamensis* (***Dilleniaceae***)
dilleniifolius -a -um, dillaniaefolius -a -um having leaves resembling those of *Dillenia*
diluculus -a -um of dawn, of the dawn, *diluculum, diluculi* (early flowering)
dilutus -a -um pale, washed-out, past participle of *diluo, diluere, dilui, dilutum*
dimidiatus -a -um with two equal parts, dimidiate, halved, *dimidiatus*
diminutus -a -um very small, *de*-(*minuo, minuere, minui, minutum*)
dimitrus -a -um having two covers or head-dresses, δι-μιτρα, *di*-(*mitra, mitrae*) (doubled corolla)
dimorph-, dimorpho-, dimorphus -a -um two-shaped, with two forms, δι-μορφη (of leaf or flower or fruit)
dimorphanthus -a -um having flowers of two forms, δι-μορφη-ανθος
dimorphoelytrus -a -um having two distinct coverings, δι-μορφη-ελυτρον

dimorphophyllus -a -um with leaves of two shapes, δι-μορφη-φυλλον
Dimorphotheca Two-kinds-of-container, δι-μορφη-θηκη (the disc florets of cape marigolds vary in structure)
dinaricus -a -um from the Dinaric Alps, Dalmatian coast between Slovenia and Albania
Dinklageanthus Dinklage's-flower, botanical Latin from Dinklage and ανθος
Dinklageella for Max Julius Dinklage (1864–1935), who collected in Liberia from 1898
Dinklageodoxa Dinklage's-glory, botanical Latin from Dinklage and δοξα
dino- whirling, δινος, or terrible-, δεινος (dinosaur)
Dinophora Whorl-bearer, δινος-φορα
Diodia Two-toothed, δι-οδους
diodon two-toothed, δι-οδων
dioicus -a -um, dioeca of two houses, δις-οικος (having separate staminate and pistillate plants)
Dionaea for *Dione*, the Titan mother of Aphrodite, synonymous with Venus (Venus fly trap)
Dioncophyllum Two-hooked-leaved-one, δι-ογκο-φυλλον (***Dioncophyllaceae***)
Dionysia for Dionysos, the Greek *Bacchus*, god of debauchery
Dioon Two-egged-one, δις-ωον (the paired ovules on each scale)
Dioscorea for Dioscorides Pedanios of Anazarbeus, Greek military physician (yams) (***Dioscoreaceae***)
dioscoreifolius -a -um with leaves resembling those of *Dioscorea*
Diosma Divine-fragrance, διος-οσμη
diosmeus -a -um having a 'heavenly' perfume, διος-οσμη
diosmifolius -a -um *Diosma*-leaved, *Diosma-folium*
Diosphaera Job's-orb, διος-σφαιρα
diospyroides having a similarity to Diospyros, διος-πυρος-οειδης
Diospyros Divine-food, διος-πυρος (Jove's-fruit, date plum's edible fruit)
Diotis Two-ears, δι-ωτος (the spurs of the corolla, ≡ *Otanthus*)
diotus -a -um two-eared or two-handled, δι-ωτωεις
Dipcadi from an oriental name for *Muscari*
Dipelta Twice-shielded, δι-πελτη (the capsules are included between persistent bracts)
dipetalus -a -um having two petals, δι-πεταλον
diphrocalyx with a flat and wide calyx shaped like a chariot board, διφρος, chair-shaped calyx, διφρος-καλυξ
Diphylleia Double-leaf, δις-φυλλον (the deeply divided umbrella leaf) (***Diphylleiaceae***)
diphyllus -a -um two-leaved, δι-φυλλον
dipl-, diplo- two fold-, double-, alternating, διπλοος, διπλους, διπλοω, διπλο-, διπλ-
Diplachne Double-chaff, διπλ-αχνη
Diplacrum Double-lobed, διπλ-ακρος (the glumes have two side lobes)
Diplacus Two-flat-bodies, δι-πλακος (the placenta) (≡ *Mimulus*)
Dipladenia Double-gland, διπλοω-αδην (the two glands on the ovary)
diplantherus -a -um double-stamened, *duplus-antherus*, double-flowering, διπλους-ανθηρος
Diplarche Two-commencements, διπλοω-αρχη (two series of stamens)
Diplarrhena Double-male, διπλοω-αρρην (two perfect stamens)
Diplazium Duplicate, διπλαζω (the double indusium)
diplocrater having a double bowl, διπλο-κρατηρ (perianth)
Diplocyclos Two-circled, διπλοω-κυκλος (the double border of the seed)
Diplomeris Two-partite, διπλο-μερις (the appearance of the widely separated halves of the anther)
diplomerus -a -um having two parts, divided into two, διπλο-μερος
Diplopappus Double-down, διπλο-παππος
diplospermus -a -um two-seeded, διπλο-σπερμα

diplostemonus -a -um having twice as many stamens as petals, διπλο-στεμων
diplostephioides like a double crown, διπλο-στεφαν-οειδης (the arrangement of the capitulum)
Diplotaxis Two-positions, διπλο-ταξις (the two-ranked seeds)
diplotrichus -a -um, diplothrix having two kinds of hairs, διπλοω-τριχος
Dipogon Two-bearded, δι-πωγων
dippelianus -a -um for Leopold Dippel (1827–1914) of Darmstadt, author of *Handbuch der Laubholzkunde* (broad-leaf timber trees)
dipsaceus -a -um teasel-like, resembling *Dipsacus*
dipsaci of teasel, living on *Dipsacus* (*Anguillulina*, nematode)
dipsacifolius -a -um having *Dipsacus*-like leaves
dipsacoides resembling a teasel, διψακος-οειδης
Dipsacus Dropsy, διψακος (Dioscorides' name in analogy of the water-collecting leaf-bases) (the mature heads of *Dipsacus fullonum* were the teasels, used to tease a nap on woven woollen cloth, known as fulling) (***Dipsacaceae***)
dipso- thirst-, διπσα, διπσο-
diptero-, dipterus -a -um two-winged, δι-πτερυξ
Dipterocarpus, dipterocarpus -a -um with two-winged seeds, δι-πτερο-καρπος (***Dipterocarpaceae***)
Dipteronia Twice-winged, δι-πτερυξ (the two-winged carpels of the fruits)
dipteryx with two wings, δι-πτερυξ
dipyrenus -a -um two-fruited, two-stoned, δι-πυρην
Dirca from Dirce, in Greek mythology, a fountain in Boeotia
dis- two-, different-, twice, δισσος, δις-
Disa from a S African vernacular name
Disanthus Two-flowers, δι-ανθος (the paired flowers)
discadenius -a -um having flat circular glands, δισκος-αδην
Discaria Discoid, δισκος (the prominent disc)
discerptus -a -um torn apart, dispersed, *discerpo, discerpere, discerpsi, discerptum*
dischianum twice cleft, δις-χιαζειν
Dischistocalyx Twice-split-calyx, δι-σχιστος-καλυξ
disci-, disco- disk-, δισκος, δισκο-, *discus, disci*
disciflorus -a -um flowers with a distinct disc, with rotate flowers, *discus-florum*
disciformis -is -e having radiate flowers, *discus-forma*
discigerus -a -um disc bearing, *discus-gero*
discipes with a disc-like stalk, *discus-pes*
Discoglypremna Engraved-disc-shrub, δισκο-γλυπτο-πρεμνον (the flowers have a deeply segmented disc)
discoides discoid, δισκ-οειδης
discolor of different colours, two-coloured, *dis-color*
discophorus -a -um disc carrying, δισκο-φορα
Discopodium Disc-footed, δισκο-ποδιον (the flat receptacle)
discotis -is -e disc-eared, δισκ-ωτος (lobed)
disermas with two glumes, with two defences, δις-ερμα
disjunctus -a -um distinct, not grown together, disjunct, *disiungo, disiungere, disiunxi, disiunctum*
dispar unequal, different, *dispar, disparis*
Disperis Two-pouched, δισ-πηρα (the two anther loculi)
dispermus -a -um two-seeded, δι-σπερμα
dispersus -a -um scattered, *dispergo, dispergere, disprsi, dispersum*
Disporopsis Resembling-*Disporum*, δι-σπορος-οψις
Disporum Two-seeded, δι-σπορος (fruits are usually two-seeded)
disporus -a -um two-spored, δι-σπορος
dissectus -a -um, (disectus) cut into many deep segments, *disseco, dissecare, dissecui, dissectum*
disseminatus -a -um broadcast, sown, *dissemino, disseminare*
dissimilis -is -e unlike, different, *dissimilis*

dissitiflorus -a -um with flowers not in compact heads, spaced at intervals, (*dissero, disserere,disserui, dissertum*)-(*flos-floris*)
dissolutus -a -um loose, lax, *dissolvo, dissolvere, dissolvi, dissolutum*
Dissomeria Two-fold-parts, δισσος-μερος (the petals are twice as many as the sepals)
Dissotis Two-kinds, δισσοι (the anthers are highly modified in two ways), two-eared, δις-ωτος (the lobes at the geniculate part of the filament)
distachyon, distachyus -a -um two-branched, δι-σταχυς, two-spiked, with two spikes
distans widely separated, distant, *disto, distare*
Disteganthus Double-covered-flower, δι-(στεγος, σρεγη)-ανθος (the position of the corolla)
Distegocarpus Double-coated-fruit, δι-στεγω-καρπος
distensifolius -a -um having thick, swollen-looking leaves, (*distendo*; *distenno-, destendere, distendi, distentum*)-*folium*
distentus -a -um full or swollen, distended, *distendo (distenno), destendere, distendi, distentum*
distichanthus -a -um having flowers in two alternating ranks, δι-στιχος-ανθος
distichius -a -um, distichus -a -um in two alternately opposed ranks, δι-στιχος (leaves or flowers)
distichophyllus -a -um distichous leaved, with two-ranked leaves, δι-στιχος-φυλλον
Distictis Double-spotted, δι-στικτος (the double rows of seeds)
distillatorius -a -um shedding drops, of the distillers, *de*-(*stillo, stillare, stillavi, stillatum*)
distinctus -a -um distinct, set apart, past participle of *distinguo, distinguere, distinxi, distinctum*
distortus -a -um malformed, grotesque, twisted, distorted, *distorqueo, distorquere, distorsi, distortum*
Distylium Two-styles, δι-στυλος (the conspicuous, separate styles)
distylus -a -um two-styled, δι-στυλος
ditopoda, ditopus enriched-stem, (*dito, ditare*)-(*pous, podus*) (the stipe develops a basal covering of cottony fibres)
diureticus -a -um diuretic, causing urination, διουρειν
Diuris Two-tailed-one, δι-ουρα (the sepals)
diurnus -a -um lasting for one day, day-flowering, of the day, daily, *diurnus*
diutinus -a -um, diuturnus -a -um long-lasting, *diutinus*
divaricatus -a -um wide-spreading, straggling, divaricate, *divarica, divaricare*
divensis -is -e from Chester (*Deva*)
divergens spreading out, wide-spreading, divergent, *di-*(*vergo, vergere*)
diversi-, diversus -a -um differing-, variable-, diversely-, *diverto, divertere, diverti, diversum*
diversicolor having various colours, *diversi-color*
diversifolius -a -um with different leaves, variable foliage, *diversi-folium*
diversilobus -a -um variably lobed, botanical Latin from *diversus* and λοβος (leaves)
dives, divus -a -um belonging to the gods, divine, *divus*
dividus -a -um divided, *divido, dividere, divisi, divisum*
divinorus -a -um of the divine, of the prophets, *divinus*
divionensis -is -e from Dijon, France (*Divio, Divionensis*)
divisiflorus -a -um having flowers with finely divided petals, *divisus-florum*
divisus -a -um divided, *divido, dividere, divisi, divisum*
divulgatus -a -um widespread, *divulgo, divulgare, divulgavi, divulgatum*
divulsus -a -um torn violently apart, estranged, *divello, divellere, divelli, divulsum*
dixanthus -a -um having two shades of yellow, δι-ξανθος
Dizygotheca Two-yoked-case, δι-ζυγο-θηκη (the four-lobed anthers)
Docynia an anagram of *Cydonia*
docynoides resembling itself, *Docynia-oides*, hence *Docynia docynoides*
dodartii for D. Dodart (1634–1707), Parisian botanist

dodec-, dodeca- twelve-, *duodecem*, δωδεκα
Dodecadenia Twelve-glands, δωδεκα-(αδην, αδενος)
dodecandrus -a -um having twelve stamens, δωδεκα-ανηρ
dodecapetalus -a -um with twelve-petalled flowers, δωδεκα-πεταλον
dodecaphyllus -a -um having (about) twelve leaflets or leaves, δωδεκα-φυλλον
Dodecatheon Twelve-gods, δωδεκα-θεος (an ancient name used in Pliny) (American cowslips)
dodentralis -is -e of one span, about nine inches (23 cm) across, *dodrans, dodrantis* three-quarters of a foot
Dodonaea, dodonaei for Rembert Dodoens, *Dodonaeus*, (1517–85), Dutch physician and botanical writer
dodrantalis -is -e of (about) nine inches (23 cm) in height or length, *dodrans* (the span from thumb tip to extended little finger tip)
dolabratus -a -um axed, axe-shaped, *dolabra, dolabrae*
dolabriformis -is -e hatchet-shaped, *dolabra-forma*
doliarius -a -um having the shape of a large wine jar or barrel, *dolium, dolii* (*doliaris* tubby)
dolich-, dolicho- long-, δολιχος, δολιχο-, δουλιχο-
dolichanthus -a -um with long flowers, δολιχος-ανθος
dolichocentrus -a -um having long spurs, long-spurred, δολιχο-κεντρον
dolichoceras long-horned, δολιχο-κερας
dolichomischon long-pedicelled, δολιχο-μισχος
dolichorrhizus -a -um having long roots or rhizomes, δολιχο-ριζα
Dolichos the ancient Greek name, δολιχος (for long-podded beans)
dolichostachyus -a -um long-spiked, δολιχο-σταχυς
dolichostemon with long stamens, δολιχο-στεμον
Dolichthrix Long-haired, δολιχο-θριξ
doliiformis -is -e tubby, shaped like a wine jar, *dolium*
dolobratus -a -um hatchet-shaped, see *dolabratus*
dolomiticus -a -um from the Italian Dolomites, of soils of dolomitic origin (dolomite is named for the French geologist Dolomieu (1750–1801)
-dolon -net, -snare, -trap, δολος
dolosus -a -um deceitful, δολος
Dombeya for Joseph Dombey, French botanist who accompanied Ruiz and Pavon in Chile and Peru
domesticus -a -um of the household, *domesticus*
dominans becoming dominant, prevailing, present participle of *dominor, dominari, dominatus*
domingensis -is -e from San Domingo, E Hispaniola, W Indies; from the Dominican Republic
dominicalis -is -e, dominicus -a -um of Dominica (named by Columbus when sighted on a Sunday (*dies dominica*, the Lord's day)
donarius -a -um of reeds, δοναξ (habitat)
donax an old Greek name, δοναξ, for a reed
Dondia an Adansonian name (≡ *Suaeda fruticosa, S. epipactis* ≡ *Hacquetia epipactis*)
donianus -a -um, donii for G. Don (1764–1814), Keeper of the Botanic Garden, Edinburgh, or either of his sons, David (1799–1841), the Linnean Society librarian, or George (1798–1856), who collected for the RHS in Brazil
Doodia for Samuel Doody (1656–1706), apothecary and Keeper of the Chelsea Physick Garden *c.* 1691
doratoxylon spear-wood, δορατιον-ξυλον (aboriginal use for weapons)
Dorema Gift, the name used by Dioscorides for another plant (gum ammoniac, *Dorema ammoniacum*)
Doritis Lance-like, δορυ (the long lip of the corolla)
dorius -a -um from Doria, the Peloponnese area once conquered by the Dorians; pole-like, δορυ (tall single-stemmed)
dormannianus -a -um for Charles Dorman, orchid grower of Sydenham *c.* 1880

dormiens seasonal, having a dormant period, present participle of *dormio, dormire, domivi, domitum*
doronicoides resembling *Doronicum*, *Doronicum-oides*
Doronicum from an Arabic name, doronigi (leopard's bane)
Dorotheanthus Dorothea-flower (for Dr Schwantes' mother, Dorothea)
dorsalis -is -e of mountain ridges, of the back, *dorsum*
dorsi-, -dorsus -a -um on the back-, -backed, outside, *dorsum, dorsi* (outer curve of a curved structure)
Dorstenia for Theodore Dorsten (1492–1552), German botanist
dortmanna, dortmanniana for Herr Dortmann (*c.* 1640), pharmacist of Gröningen
-dorus -a -um -bag-shaped, -bag, δορος
dory- pole-, spear-, δορυ, δορατος, δορυ-
Doryanthes Spear-flower, δορυ-ανθος (the long flowering-scape)
Dorycnium Dioscorides' name, δορυκνιον, was for a *Convolvulus*, reapplied by Miller
Doryopteris Spear-fern, δορυ-πτερυξ
doryphorus -a -um lance-bearing, δορυ-φορος (long tapering leaves or stems)
dosua from a vernacular name, dosi-swa, for *Indigofera dosua*
Douglasia, douglasii for David Douglas (1798–1834), of Perthshire, Scotland, plant collector in NW America for the RHS
dovrensis -is -e from the Dovre mountain plateau, Dovrefjell, of S Central Norway
-doxa -glory, δοξα
Draba, draba Acrid, a name, δραβη, used by Dioscorides for *Lepidium draba* (whitlow grass)
drabae-, drabi- *Draba*-like-
drabifolius -a -um *Draba*-leaved, *Draba-folium*
Dracaena Female-dragon, δρακαινα (dragon tree, a source of dragon's blood)
draco, draco- dragon-, δρακων (for dragon's blood sap or resin)
Dracocephalum, dracocephalus -a -um Dragon-head, δρακο-κεφαλη (the shape of the corolla)
dracocephalus -a -um dragon-headed, δρακο-κεφαλη
dracoglossus -a -um serpent-tongued, δρακο-γλωσσα (division of frond)
dracomontanus -a -um from the Drakensberg mountains, S Africa
draconsbergensis -is -e from Drakensberg, S Africa
Dracontioides, dracontoides *Dracontium* like, δρακοντιον-οειδης
Dracontium, dracontius -a -um Snake, δρακων, δρακοντος, an ancient Greek name, used by Pliny for plant with serpentine roots, *dracontia-radix*
Dracophyllum Dragon-leaf, δρακο-φυλλον
dracophyllus -a -um dragon-leaved, δρακο-φυλλον (the markings)
Dracula Sinister, from the name, dragwlya, given to fifteenth-century Vlad Tepes, Prince of Wallachia, renowned for his cruelty (this dark orchid has a jelly-like lip to entice pollinators)
Dracunculus, dracunculus Little-dragon, diminutive of *draco, draconis* (a name used by Pliny for a serpentine rooted plant)
Drapetes Breaking-up, δραπετης
Dregea, dregei, dregeanus -a -um for Johann Franz Drege (1794–1881), German plant collector in S Africa
drepano- arched, sickle-shaped, δρεπανον, δρεπαντο, δρεπανο-
Drepanocarpus Curved-fruit, δρεπανο-καρπος (leopard's claw)
drepanocentron curved spur, δρεπανο-κεντρον
Drepanocladus Curved-branch, δρεπανο-κλαδος (the arched lateral branches)
drepanolobius -a -um having small curved lobes, δρεπανο-λοβος (leaves or petals)
drepanophyllus -a -um having sickle-shaped leaves, δρεπανο-φυλλον
Drepanostachyum, drepanostachyus -a -um Curved-spike, δρεπανο-σταχυς
drepanus -a -um arched or curved like a sickle-blade, δρεπανον
Drimia Acrid, δριμυς (the pungent juice from the bulbs)

Drimioposis resembling *Drimia*, δριμυς-οψις
drimyphilus -a -um salt-loving, halophytic, δριμυς-φιλος
Drimys Acrid, δριμυς (the taste of the bark of winter-bark)
dros- dew, δροσος
Drosanthemum Dewey-flower, δροσος-ανθεμιον (glistens with epidermal hairs)
Drosella Little-dewy-one, feminine diminutive from δροσος
Drosera Dewy, δροσος, δροσερος (the glistening glandular hairs of the apothecaries' *ros solis*, the sundew) (***Droseraceae***)
droserifolius -a -um having leaves similar to those of *Drosera*, *Drosera-folius*
drosocalyx having a calyx with, or appearing to have, spots of dew, δροσος-καλυξ
Drosophyllum Dewy-leaved, δροσος-φυλλον (the droplet-tipped leaf-glands)
drucei for George Claridge Druce (1859–1932), British botanist
Drummondia for Thomas Drummond (*c.* 1790–1835), of Havana, who collected in N America and died collecting
drummondii for Dr James Larson Drummond (1783–1853) founder of the Belfast Botanic Garden, or James Drummond (1784–1863), Curator of Cork Botanic Garden, or James Ramsay Drummond (1851–1921), of the Indian Civil Service, or Thomas Drummond
drumonius -a -um of woodlands, δρυμος
drupaceus -a -um stone-fruited with a fleshy or leathery pericarp, drupe-like, δρυππα
drupiferus -a -um drupe-bearing, δρυππα-φερω, *drupa-fero*
drusorum of oak woods, δρυς, δρυος
dryadeus -a -um of oaks, δρυος (*Inonotus dryadeus* is a fungal parasite on oak)
dryadifolius -a -um having *Dryas* like foliage, *Dryas-folium*
dryadioides of shady habitats; resembling *Dryas*, δρυος-οειδης
Dryandra for Jonas Dryander (1746–1819), Swedish botanist
Dryas, dryas Oak-nymph, δρυας (the leaf shape) one of the mythological tree nymphs or Dryads (oak-like leaves of mountain avens); of woodland shade
drymeius -a -um of woods, woodland, δρυς, δρυμος
drymo- wood-, woodland-, δρυς, δρυμο-
Drymoglossum Wood-tongue, δρυμο-γλωσσα (the arboreal habitat)
Drynaria Woodland, δρυς (forest margin habitat)
drynarioides resembling *Drynaria*, *Drynaria-oides*
dryophilus -a -um woodland-, shade- or oak-loving, δρυς-φιλεω
dryophyllus -a -um oak-leaved, δρυς-φυλλον
Dryopteris, dryopteris Oak-fern, δρυς-πτερυξ, Dioscorides' name, δρυοπτερις, for a woodland fern (buckler ferns)
Drypetes Stone-fruits (the hard seeds), δρυπτω to lacerate (spiny)
Drypis Theophrastus' name, δρυπτω, for the scratching, spiny leaves
dualis -is -e two fold, δυας, *dualis*
dubayanus -a -um from Dubai, United Arab Emirates
dubitans doubting, wavering, present participle of *dubito, dubitare, dubitavi, dubitatum*
dubius -a -um uncertain, doubtful, *dubito, dubitare, dubitavi, dubitatum*
Duchesnea for Antoine Nicolas Duchesne (1747–1827), French botanist
duclouxianus -a -um, duclouxii for Mon. Father Ducloux (b. 1864), collector in Yunnan, China
duffii for Sir Mountstuart Elphinstone Grant Duff (1829–1906), botanist, Governor of Madras
dulcamara bitter-sweet, *dulcis-amara*
dulciferus -a -um bearing sweetness, *dulcis-fero*
dulcificus -a -um sweetening, making sweet, *dulcis-(fingo, fingere, finxi, fictum)*
dulcis -is -e sweet-tasted, mild, *dulcesco, dulcescere*
dumalis -is -e, dumosus -a -um thorny, compact, bushy, *dumosus*
Dumasia for Jean B. Dumas (1800–84), French pharmacist and chemist

dumetorum of bushy habitats, of thickets, *dumetum, dumeti*
dumetosus -a -um having a bushy habit, *dumetum, dumeti*
dumicolus -a -um inhabiting thickets, *dumetum-colo*
dumnoniensis -is -e from Devon, Devonian (*Dumnonia*)
Dumortiera, dumortieri for B. C. Dumortier (1797–1878), of Belgium
dumosus -a -um bushy, *dumosus*
dumulosus -a -um quite bushy, diminutive of *dumosus*
Dunalia, dunalianus -a -um for Michel Felix Dunal (1789–1856), Montpellier botanist
dunensis -is -e of sand-dunes, from Old English, dun, for a hill
duoformis -is -e having a double form, *duo-forma*
duplex growing in pairs, double, duplicate, *duplex, duplicis, duplici-*
duplicatus -a -um double, folded, bent, *duplico, duplicare, duplicavi, duplicatum*
duplicilobus -a -um twice-lobed or segmented, *duplici-lobus*
duploserratus -a -um twice-serrate, with toothed teeth, *duplici-serratus*
durabilis -is -e durable, tough, *duro, durare, duravi, duratum*
duracinus -a -um harsh-tasting, hard-berried, hard-fruited, *duresco, durescere, durui*
durandii for Elias Magloire Durand (1794–1873)
durangensis -is -e from either Durango state, N Mexico, or Durango, SW Colorado, USA
Duranta for Castor Durantes (d. 1590), physician and botanist from Rome
duratus -a -um lasting, enduring, hardy, *duro, durare, duravi, duratum*
durhamii for Frank Rogers Durham (1872–1947), Secretary of the RHS 1925–46
durifolius -a -um tough-leaved, *durus-folium*
Durio from the Malayan name, durian, for the fruit
durior, durius harder, comparative of *durus*
durispinus -a -um having hard spines, persistently spiny, *durus-spina*
durissimus -a -um most persistent or tough, superlative of *durus*
duriusculus -a -um rather hard or rough, diminutive comparative of *durus*
durmitoreus -a -um from the Durmitor mountains, former Yugoslavia
durobrivensis -is -e from Rochester (Kent or USA), or Dubrovnik, Croatia
durus -a -um hard, hardy, *durus*
Durvillaea for Jules Sébastien César Dumont d'Urville (1790–1842), French navigator
duthieanus -a -um, duthiei for John Firminger Duthie (1845–1922), Superintendent of the Botanic Garden, Saharanpur, Uttar Pradesh, India
Duvalia for Henri Auguste Duval (1777–1814), author of *Enumeratio plantarum succulentum in horto Alenconio*
dybowskii for Dybowski, French Inspector General of Colonial Agriculture *c.* 1908
Dyckia for Prince Salms Dyck (1773–1861), German writer on succulent plants (see *Salmia*)
dyerae for Lady Thistleton-Dyer (*née* Hooker) (1854–1945)
dyerianus -a -um for Sir William Thistleton-Dyer (1843–1928) of the Science Schools Building (forerunner of Imperial College), London
dykesii for William Rickatson Dykes (1877–1925), author of *Genus Iris* and Secretary of the RHS, 1920–25
Dypsis Dipping, δυπτω (δυψα thirsty) (slender-stemmed palms)
dys-, dyso- poor-, ill-, bad-, difficult-, unpleasant, δυσ-
Dyschoriste Poorly-divided, δυσ-χωρις (the stigma)
dyscritus -a -um difficult to assess, doubtful, δυσκριτος
dysentericus -a -um of dysentery, δυσεντερια (medicinal treatment for)
dysocarpus -a -um having foul-smelling fruit, δυσ(ο)-καρπος
Dysodea Evil-scented, δυσ-οδμος
dysodes unpleasant-smelling, δυσ-ωδης
dysophyllus -a -um having foul-smelling foliage, δυσ-φυλλον
dysosmius -a -um evil-smelling, δυσ-οσμη

e-, ef-, ex- without-, not-, from out of-, εξ-, εκ-, *e-*, *ex-* (privative)
Earinus, earinus -a -um Belonging-to-spring, εαρινος (flowering season)
eatonii for Amos Eaton (1776–1842)
ebenaceus -a -um ebony-like, εβενος
Ebenus, ebenus -a -um Ebony-black, *hebenus*, εβενος (Arabic, hebni, cognate with ebony) (*vide supra*, Family names re ***Ebenaceae***)
Ebermaiera for Karl Heinrich Ebermaier (1802–70), writer on medicinal plants (≡ *Chamaeranthemum*)
eboracensis -is -e from York (*Eboracum, Eburacum*)
eborinus -a -um ivory-white, like ivory, *ebur, eburis*
ebracteatus -a -um without bracts, privative *e-bractea*
ebracteolatus -a -um lacking bracteoles, *e-bracteolae*
ebudensis -is -e, ebudicus -a -um from the Hebridean Isles (*Ebudae Insulae*)
ebulbus -u -um without bulbs; not swollen, *e-bulbus*
Ebulus, ebulus a name, *ebulus, ebuli* in Pliny for danewort
eburneus -a -um, eburnus -a -um ivory-white with yellow tinge, *ebur, eburis*
ecae for Mrs E. C. Aitchison
ecalcaratus -a -um without a spur, spurless, privative *e-(calcar, calcaris)*
ecarinatus -a -um without a keel, un-ridged, *e-(carina, carinae)*
Ecballium (*Ecbalium*) Expeller, εκβαλλειν (at maturity, the squirting cucumber expels its seeds when touched, εκβαλλω to expel, εκβολη, expulsion, causing childbirth contractions, it also has cathartic properties)
ecbolius -a -um casting out, expelling, εκβολη, εκβολος (cathartic)
eccremo- pendent, hanging, εκ-κρεμαστος
Eccremocarpus Hanging-fruit, εκ-κρεμαστος-καρπος (Chilean glory flower)
Echeveria for Athanasio Echeverria y Godoy, one of the illustrators of *Flora Mexicana*
echidnis -is -e, echidnus -a -um serpentine, εχιδνα, *echidna*
Echinacea, echinaceus -a -um Spiny-one, εχινος (the spiny involucral bracts) (purple cone flower) (cf. *Erinacea*)
Echinaria Hedgehog-like, εχινος (prickly capitate inflorescence of hedgehog grass)
echinatoides resembling (*Rubus*) *echinatoides*, εχινος-οειδης
echinatus -a -um, echino- covered with prickles, hedgehog-like, εχινος
Echinella, echinellus -a -um Little hedgehog, slightly prickly achenes, *echinus* with diminutive suffix *-ellus*
echiniformis -is -e having hedgehog- or sea-urchin-like form, *echinus-forma*
echino- spiny like a hedgehog or sea urchin, εχινος, εχινο, *echinus, echini*
Echinocactus Hedgehog-cactus, εχινο-κακτος
echinocarpus -a -um having spine-covered fruits, εχινο-καρπος
echinocephalus -a -um having a prickly head, εχινος-κεφαλη (the warty surface of the pileus)
Echinocereus Hedgehog-*Cereus*, εχινο-*Cereus*
Echinochloa Hedgehog-grass, εχινο-χλοη (the awns of the scabrid spikelets)
Echinodorus Hedgehog-bag, εχινο-δορος (the fruiting heads of some species)
Echinofossulocactus Prickly-ditch-cactus, botanical Latin *echino-fossula-cactus* (wavy spine-clad ridges of the brain or barrel cactus)
echinoides sea-urchin- or hedgehog-like, εχινος-οειδης
Echinopanax Hedgehog-*Panax*, εχινο-*Panax*
Echinophora Prickle-carrier, εχινο-φορα (the spiny umbels)
echinophytus -a -um prickly or hedgehog-plant, εχινος-φυτον
Echinops Hedgehog-resembler, εχινος-ωψ (globe thistles)
echinosepalus -a -um with prickle-covered sepals, εχινο-σκεπη
echinosporus -a -um having spores covered in small prickles, εχινο-σπορος
echinus -a -um prickly like a hedgehog, εχινος
echioides resembling *Echium*, *Echium-oides*
Echites the name in Pliny for a twining or coiling plant, perhaps a *Clematis*, viper-like, εχις, εχιδνα

Echium Viper, εχις (a name, εχιον, used by Dioscorides for a plant to cure snakebite) (viper's bugloss)
ecirrhosus -a -um lacking tendrils, *e-cirrus*
ecklonianus -a -um, ecklonii for Christian Friedrich Ecklon (1785–1868), apothecary and student of the S African flora
eclectus -a -um selected, picked out, εκ-λεγω, εκλεγειν
Eclipta Deficient, εκλιπης (has few receptacular scales)
eco- habitat, dwelling place, οικος, οικο-
ecostatus -a -um without ribs, smooth, *e-costatus* (a comparative state)
ecristatus -a -um not crested, lacking a crest, *e-cristatus*
ect-, ecto- on the outside-, outwards-, εκτος, εκτο-, εκτ-
Ectadiopsis Appearing-far-distant, εκταδιος-οψις
ectophloeos living on the bark of another plant, εκτο-φλοιος (symbionts, parasites and saprophytes)
ectypus -a -um outside or not agreeing to the type, εκτ-τυπος, *e-typus*
edentatus -a -um, edentulus -a -um without teeth, toothless, *e-*(*dens, dentis*)
Edgeworthia, edgeworthii for M. P. Edgeworth (1821–81), botanist of the East India Company
edinensis -is -e, edinburgensis of Edinburgh, Scotland (*Edinburgum*)
editorum of the editors, productive, radiant; of heights, from *edo, ederi, edidi, editum*
edo, edoensis from Tokyo (formerly Edo)
edomensis -is -e from the Edom area of SW Jordan
Edraianthus Sedentary-flower, εδραιος-ανθος (the flower-clusters on the peduncles)
edulis -is -e of food, edible, *edo, esse, edi, esum*
efarinosus -a -um lacking farina, without a mealy indumentum, *e-farina*
effusus -a -um spread out, very loose-spreading, unrestrained, *effundo, effundere, effudi, effusum*
egalikensis -is -e from Igaliko, Greenland
Eglanteria, eglanterius -a -um from a French name, eglantois or eglantier, for *Rosa canina*
egregius -a -um outstanding, exciting, *egregius*
ehrenbergii for Karl August Ehrenberg (1801–49), collector in Port-au-Prince and Mexico
Ehretia for George Dionysius Ehret (1708–70), botanical artist
ehrhartii for J. F. Ehrhart (1742–95), of Switzerland
Eichhornia (Eichornia) for J. A. F. Eichhorn (1779–1856) of Prussia (water hyacinth)
eichleri, eichlerianus -a -um for Wilhelm Eichler of Baku, who, *c.* 1873, sent *Tulipa eichleri* to Regel
elachi-, elacho-, elachy- small-, ελαχυς, smallest ελαχιστος (followed by an organ or structure)
elachisanthus -a -um having small flowers, ελαχυς-ανθος
elachophyllus -a -um small-leaved, ελαχυς-φυλλον
elachycarpus -a -um small-fruited, ελαχυς-καρπος
elae-, elaeo- olive-, ελα, ελαια
elaeagnifolius -a -um, (elaeagrifolia) with *Elaeagnus*-like leaves, *Elaeagnus-folium*
elaeagnoides resembling *Elaeagnus*, ελαια-αγνος-οειδης
Elaeagnus, elaeagnos Olive-chaste-tree, ελαια-αγνος (oleaster) (***Elaeagnaceae***)
Elaeis (Elais) Olive, ελαιος (the fruit of the oil-palm, *Elaeis guineensis*, has assumed huge commercial importance, like the olive)
Elaeocarpus Olive-fruited, ελαια-καρπος (the fruit form and structure) (***Elaeocarpaceae***)
Elaeophorbia Olive-*Euphorbia*, ελαια-*Euphorbia* (composite name indicating the olive-like fruits)
elaeopyren olive-like kernel, having an oily-kernel, elaια- πυρην
elaeospermus -a -um having oil-rich seed, ελαια-σπερμα
elaidus -a -um oily, like the olive, ελαια
elaiophorus -a -um oil-bearing, ελαια-φορα (by analogy with the olive)

elaphinus -a -um tawny, fulvous, ελαφη, a fawn
elapho- stag's-, ελαφος, ελαφη, ελαφο-
Elaphoglossum Stag's-tongue, ελαφο-γλωσσα (shape and texture of the fronds)
Elaphomyces Stag-fungus, ελαφος-μυκες
elaphro- light-, easy-, nimble-, ελαφρος, ελαφρο-
elaphroxylon having light wood, ελαφρο-ξυλον
elasticus -a -um yielding an elastic substance, elastic, ελαυνειν, ελαστικος
elaterium Greek name, ελατηρος, for the squirting cucumber, ελατηρος driving away (squirting out seeds), (ελατηρ a driver, for the threads that aid cryptogamic spore dispersal)
Elatine, elatine Little-conifer, ελατινος, of pine wood (a name, ελατινη, used by Dioscorides) (***Elatinaceae***)
elatinoides resembling *Elatine*, ελατινος-οειδης
elatior, elatius taller, comparative of *elatus*
Elatostema Tall-crown, ελατη-στεμμα; High-renown, *elatus-stemma* (the inflorescence)
elatus -a -um exalted, tall, high, *effero, effere, extuli, elatum*
elbistanicus -a -um from the Elbistan area of eastern Anatolia, S Turkey
elbursius -a -um from the Elburz mountain range of N Iran
eldorado that of gold, golden one, of great abundance, Spanish fictional country of great plenty
electra-, electro- amber-, ηλεκτρον (mostly for the colour but also for minor electrostatic features, as shown by amber itself)
electrocarpus -a -um amber-fruited, ηλεκτρον-καρπος
electrus -a -um for Electra, daughter of Agamemnon and Clytemnestra in Greek mythology
electus -a -um select, *electo, electare*
elegans graceful, elegant, *elegans, elegantis*
elegantissimus -a -u most elegant, most graceful, superlative of *elegans*
elegantulus -a -um quite elegant or graceful, diminutive of *elegans*
eleo- marsh, ελος, ελεο-, cf. *heleo-*
Eleocharis (Heleocharis) Marsh-favour, ελεο-χαρις
Eleogiton (Heleogiton) Marsh-neighbour, ελεο-γειτων (in analogy with *Potamogeton*)
elephanticeps ivory or large headed, *elephantus-ceps*
elephantidens elephant's tooth, *elephantus-(dens, dentis)*
elephantinus -a -um large; having the appearance of ivory, ελεφαντινος
elephantipes like an elephant's foot, *elephantus-pes*, ελεφας-(πους, ποδος) (appearance of the stem or tuber)
Elephantopus Elephant's-foot, ελεφαντος-πους (achenes carried on feet and make some species troublesome weeds)
elephantotis -is -e elephant-eared, ελεφας-ωτος (large pendulous leaves)
elephantus -a -um of the elephant, ivory; large, ελεφας, ελεφαντος, ελεφαντο-, *elephantus, elephanti* (cognate with oliphant, a horn made of ivory)
elephas elephantine, ivory-like, ελεφας, ελεφαντος (also variously applied to pendulous structures)
Elettaria from a Malabar vernacular name
Eleusine from Eleusis, Greece
eleuther-, eleuthero- free-, ελευθερος
eleutherandrus -a -um having free stamens, ελευθερος-ανηρ
Eleutheranthera Free-stamened-one, ελευθερος-ανθηρος
eleutherantherus -a -um with stamens not united but free, ελευθερος-ανθηρος
Eleutherococcus Free-fruited, ελευθερος-κοκκος
eleutheropetalus -a -um having distinctly separate petals, ελευθερος-πεταλον
Eleutheropetalum Free-petalled, ελευθερος-πεταλον (polypetalous)
elevatus -a -um lifting, alleviating, *elevo, elevare*
eleyi for Charles Eley, who hybridized *Malus niedzwetskyana* with *M spectabilis* (*M.* × *eleyi*)

elgonicus -a -um from the volcanic Mount Elgon (Masai, elgonyi), Uganda
elimensis -is -e hardy, of the outdoors, *elimino, eliminare*
elinguis -is -e, elinguus -a -um lacking a tongue or labellum, *e-(lingua, linguae)*
Elionurus Sun-tailed, ηλιος-ουρα (the ciliate fringe of the lower glume)
elisabethae for Elisabeth of Wied, wife of King Karl Eitel Friedrich I of Romania (the poetess Carmen Sylva)
Elisena for Princess Elisa Buonaparte (b. 1777), sister to Napoleon
Elisma a variant of *Alisma*, some suggest for the ovule orientation, ελισσω, turned
ellacombianus -a -um for Henry Thomas Ellacombe (1790–1885) or his son H. N. Ellacombe (1822–1916), both Rectors of Bitton, Somerset
Elliottia for Stephen Elliott (1771–1830), American botanist, author of Flora of South Carolina
elliottii for either G. M. Scott-Elliott (1862–1934), boundary commissioner and botanist in Sierra Leone *c.* 1891, and Madagascar, or Captain Elliott, plant grower of Farnborough Park, Hants, or Clarence Elliott of the Six Hills Nursery, Stevenage
ellipsoidalis -is -e ellipsoidal, botanical Latin from ελλειψις, ελλειπειν (a solid having an oval profile)
ellipticus -a -um about twice as long as broad, oblong with rounded ends, elliptic, ελλειψις, ελλειπειν
elliptifolius -a -um having elliptic or oval leaves, *ellipticus-folium*
ellobo- into-lobes-, εν-λοβος
-ellus -ella -ellum -lesser, -ish (diminutive suffix)
Elodea Marsh, ελος-ωδης (Canadian pondweed grows in water)
elodeoides resembling *Elodea*, ελος-ωδης-οειδης
elodes of bogs and marshes, ελος-ωδης (cf. *helodes*)
elongatus -a -um lengthened out, elongated, *e-longus*, (*elongo, elongare*)
Elsholtzia for Johann Sigismund Elsholtz (1623–88), Prussian botanical writer
eludens warding off, teasing, deceiving, present participle of *eludo, eludere, elusi, elusum*
elwesianus -a -um, elwesii for Henry John Elwes (1846–1922), author of *The Genus Lilium* and co-author with Dr A. Henry of *The Trees of Great Britain and Ireland*
Elymandra Millet-flowered, ελυμος-ανηρ
Elymus Hippocrates' name, ελυμος, for a millet-like grass (Englished as lime grass)
Elytraria Covered, ελυτρον (the bracts of the inflorescence)
elytri-, elytro- covering-, ελυτρον, ελυτρο-
Elytrigia Husk, ελυτρον
Elytroa Covering-bearing, ελυτρο-φορα (spikelets resemble insects with wing-case-like glumes)
elytroglossus -a -um having a sheathing tongue or labellum, ελυτρο-γλωσσα
elytroides resembling *Elytroa*
Elytropus Covered-stem, ελυτρον-πους (the covering of numerous bracts)
em-, en- on-, for-, in-, into-, put into-, within-, made-, not-, εμ-
emarcescens not withering, retaining shape, *e-(marcesco, marcescere)*
emarcidus -a -um not flaccid or withered, *e-marcidus* (*marcesco, marcescere*)
emarginatus -a -um notched at the apex, *e-(margo, marginis)* (see Fig. 7h)
emasculus -a -um without-male, without functional stamens, *e-masculus*
Embelia from a Cingalese vernacular name
emblica an old Bengali name, amlaki, for the medicinal fruited *Emblic myrobalan* (≡ *Phyllanthus emblica*)
-embola -peg-like, εμβολος
Embothrium In-little-pits, εν-βοθριον (the anthers are inserted in slight depressions) (Chilean fire bush)
emeiensis -is -e from Mt Omei, Sitchuan
emeriflorus -a -um one day flowering, botanical Latin from ημερα and *florum*
emeritensis -is -e from Merida (*Emerita*)
emeritus -a -um out of desert, for service, *e-(mereor, mereri, merui, meritus)*

emeroides resembling *emerus*, vetch-like
emersus -a -um rising out, *emergo, emergere, emersi, emersum* (of the water)
emerus from an early Italian name for a vetch
emeticus -a -um causing vomiting, emetic, *emetica, emeticae*, εμετος
emetocatharticus -a -um of cleansing through vomiting, εμετος-καθαρτης
Emilia Etymology uncertain, some suggest it is commemorative for Emily
eminens noteworthy, outstanding, prominent, *eminens, eminentis*
eminii for Emin Schnitzer, 'Emin Pasha' (1840–92), physician of Egypt and the Congo
Eminium a name used by Dioscorides
emissus -a -um released, sent out, past participle of *e-*(*mitto, mittere, misi, missum*) (cognate with emissary and emission)
emmeno- lasting, enduring, εμμενης, εμμενο-
Emmenopteris Enduring-wing, εμμενο-πτερυξ (one lobe of the calyx is stalked and enlarged)
emodensis -is -e, emodi from the W Himalayas, 'Mount Emodus', N India
emodi from the Sanskrit, hima, for snow (Sanskrit, hima-alaya, identifies the Himalayas as the abode of snow)
empetrifolius -a -um with *Empetrum*-like leaves
empetrinus -a -um resembling *Empetrum*
Empetrum On-rocks, εμ-πετρος (Dioscorides' name refers to the habitat of *Frankenia pulverulenta*) (***Empetraceae***)
empusus -a -um purulent, εμπυος; suppurating, malicious-looking, *em-*(*pus, puris*)
Enantia Opposite, εναντιος (the one-seeded carpels contrasted to the usual state)
enantio- reverse, hostile, opposite-, εναντα, εναντι, εναντιος, εναντιο-
Enantiophylla Opposite-leaved, εναντιο-φυλλον
enantiophyllus -a -um having opposite leaves, opposed-leaved, εναντιο-φυλλον
Enantiosparton Opposed-ropes, εναντιο-σπαρτον
Enarthrocarpus Jointed-fruit, εναρθρο-καρπος
enatus -a -um escaping, *enato, enatare* (e.g. one organ from another, as the coronna of *Narcissus*)
Encelia Little-eel, εγχελιον (the form of the fruits)
Encephalartos In-a-head-bread, εν-κεφαλη-αρτος (the farinaceous centre of the stem apex yields sago, as in sago-palms)
encephalo- in a head-, εν-κεφαλη
encephalodes head- or knob-like; resembling the brain, εν-κεφαλη-ωδης
encephalus -a -um forming a head, εν-κεφαλη
Encheiridion Within-sleeves, εν-χειριδος
Enchytraeus Of-pots, εν-(χυτρα, χυτρις, χυτριδος, χυτρος) (Japanese pot-worm or white worm)
encleistocarpus -a -um with a closed fruit, εν-κλειστος-καρπος
encliandrus -a -um with enclosed-stamens, εν-χλειω-ανηρ
enculatus -a -um hooded, *en-culus* (between the buttocks)
end-, endo- internal-, inside-, within-, ενδο-
endecaphyllus -a -um having eleven leaves or leaflets, ενδεκα-φυλλον
endivia ancient Latin name, *endivia, indivia*, for chicory or endive (from Arabic, tybi, see *Intybus*)
endlicheri, endlicherianus -a -um for Stephan Ladislaus Endlicher (1804–49), Botanic Garden Director, Vienna
endo- within-, inner-, inside-, ενδοθι, ενδον, ενδο-
Endodesmia Inside-bundle, ενδο-δεσμος (the cup-like arrangement of the united stamens)
Endosiphon Inside-tube, ενδο-σιφον
Endostemon Inside-stamens, ενδο-στημων (included)
endotrachys having a rough inner surface, ενδο-τραχυς
Endymion Endymion was Selen's (Diana's) lover, of Greek mythology
enervis -is -e, enervius -a -um destitute of veins, apparently lacking nerves, *e-nervus* (veins)

Engelmannia, engelmannii for Georg Engelmann (1809–84), physician and author on American plants
Englera, Englerastrum, Englerella, engleri, englerianus -a -um for Heinrich Gustav Adolf Engler (1844–1930), systematist and director of Berlin (Dahlem) Botanic Garden
enki- swollen-, pregnant-, εγκυος
Enkianthus Pregnant-flower, εγκυος-ανθος (the coloured involucre full of flowers)
ennea- nine-, εννεα, εννε-
enneacanthus -a -um having thorns in groups of nine, εννε-ακανθος
enneagonus -a -um nine-angled, εννεα-γωνια
enneandrus -a -um having nine stamens, εννε-ανηρ
enneaphyllos, enneaphyllus -a -um nine-leaved, εννεα-φυλλον
Enneapogon Nine-bearded, εννεα-πωγων (the lemmas are divided into nine awns)
enneaspermus -a -um nine-seeded, εννεα-σπερμα
enodis -is -e without knots or nodes, smooth, *e-*(*nodus, nodi*)
enoplus -a -um armed, having spines, ενοπλιος, ενοπλος
enormis -is -e irregular, immense, *enormis*
enotatus -a -um worthy of note, *enoto, enotare*; unmarked, unbranded, *e-*(*noto, notare, notavi, notatum*)
ensatus -a -um, ensi- sword-shaped, *ensis* (leaves)
Ensete, ensete from the Abyssinian vernacular for *Musa ensete*
ensifolius -a -um with sword-shaped leaves, shaped like a sword, (*ensis, ensis*)*-folius*
ensiformis -is -e having sword-shaped leaves, *ensis-forma*
-ensis -is -e -belonging to, -from, -of (adjectival suffix after the name of a place)
Entada Adanson's use of a Malabar vernacular name
Entelea Complete, εντελης (all stamens are fertile)
entero- intestine-, gut-, entrails, εντερον, εντερο-
Enterolobium Entrail-pod, εντερο-λοβος (the spiral pods)
Enteropogon Intestine-bearded, εντερο-πωγων (the long curved spikes)
ento on the inside-, inwards-, within-, εντος, εντο-
Entolasia Woolly-within, εντο-λασιος (the villous upper lemma)
entomo-, entom- insecto-, εντομον (because they are εντομος, cut into segments)
entomanthus -a -um insect-flowered, εντομον-ανθος (floral structure)
entomophilus -a -um of insects, insect-loving, εντομον-φιλος
Enydra Water-dweller, ενυδρος (ενυδρις an otter)
enysii for J. D. Enys (1837–1912), who introduced *Carmichaelia enysii* to Britain
eo-, eos- of the dawn-, morning-, eastern-, early, for some time, εως, ηως, ηος (εως-φορος bringer of dawn, morning star)
eocarpus -a -um fruiting early, fruiting for some time, εως-καρπος
Eocene new dawn, εως-καινος
Eomecon Eastern-poppy, εως-μηκων
ep-, epi- upon-, on-, over-, towards-, somewhat-, επι-
epacridea like *Epacris*
epacrideus -a -um similar to *Epacris*
epacridoideum resembling *Epacris*
Epacris Upon-the-summit, επι-ακρα (some live on hilltops) (***Epacridaceae***)
Epeteium Annual, επετειος
epetiolatus -a -um lacking petioles, *e-petiolus*
Ephedra from an ancient Greek name, εφεδρος (επι-εδρα), used in Pliny for *Hippuris* (morphological similarity) (shrubby horsetails)
ephedroides resembling *Ephedra, Ephedra-oides*
ephemerus -a -um transient, ephemeral, εφημεριος
ephesius -a -um from Ephesus, site of the temple to Diana, Turkey
ephippius -a -um having a saddle-like depression, *ephippium* (concavity in a cylindrical structure)
epi- on, upon, at, by, near, with, in presence of, επι-
epibulbon produced on a bulb or pseudobulb, *epi-bulbus*

epidendroides resembling *Epidendron*, επι-δενδρον-οειδης
Epidendron (um), epidendron Upon-trees (living), επι-δενρον (the epiphytic habit)
Epifagus Upon-beech, *epi-Fagus* (root parasites, usually on oaks)
Epigaea, epigaeus -a -um Above-ground, επι-γαια, growing close to or on the ground's surface (American mayflower)
epigeios, epigejos of the earth, earthly, επι-γαια (distinct from aquatic or underground)
Epigeneium On-the-chin, επι-γενειον (the chin-like composite mentum at the base of the column)
epiglyptus -a -um appearing to have a carved upper surface, *epi-glyptus*
Epigynum Upon-the-ovary, επι-γυνη (the attachment of the stamens to the stigmatic head)
epigynus -a -um having a superior ovary, επι-γυνη
epihydrus -a -um of the water surface, επι-υδωρ
epilinum on flax, on *Linum*, επι-λινον (parasitic *Cuscuta*)
epilis -is -e, epilosus -a -um lacking hairs, *e-(pilus, pili)*
epilithicus -a -um growing on marble or rocks, επι-λιθος
epilobioides resembling *Epilobium*, επι-λοβος-οειδης
Epilobium Gesner's name, ιον επι λοβον, indicating the positioning of the corolla on top of the ovary, επι-λοβος (willowherbs)
Epimedium the name, επι-μηδιον, used by Dioscorides and then by Pliny
epipactidius -a -um somewhat like a helleborine, diminutive from *Epipactis*
Epipactis, epipactis -is -e a name, επι πακτις, used by Theophrastus for an ελλεβορος-like plant (Helleborine orchid)
Epiphyllum, epiphyllus -a -um Beside-the-leaf, επι-φυλλον, having flowers or other organs growing upon leaves or phyllodes
epiphyticus -a -um growing upon another plant, επι-φυτον
Epipogium (Epipogon) Bearded-above, επι-πωγων (the lip of the ghost-orchid is uppermost)
Epipremnum, epipremnus -a -um On-trees, επι-πρεμνον (epiphytic on tree stumps and stems)
epipsilus -a -um somewhat naked, επι-ψιλος (the sparse foliage of *Begonia epipsila*)
epipterus -a -um on a wing, επι-πτερον (fruits)
epipterygius -a -um upon feathers, with a feathered surface, επι-πτερυγος (processes covering the cheilocystidia)
epirensis -is -e, epirocticus -a -um, epirus -a um from the Epirus area of NW Greece and S Albania
epirotes living on dry land, ηπειρωτης
Episcia Of-the-shadows, επι-σκια (prefers shade)
episcopalis -e of bishops, of the overseer, convenient, worthy of attention; some interpret as resembling a bishop's mitre; worthy of attention, επισκοπεω, επισκοπος
episcopi for Bishop Hannington of Uganda
Epistephium Filled-to-the-crown, επι-στεφιον, (the enlarged top of the ovary)
epistomius -a -um snouted, closed at the mouth, επιστομιζω (flowers)
epiteius -a -um annual, covering, exciting, επιτεινω
Epithelantha Nipple-flowering, επι-θηλη-ανθος (the button cactus)
epithymoides dodder-like, επι-θυμος-οειδης
epithymum upon thyme, επι-θυμος (parasitic)
epuloticus -a -um festive, to be feasted upon, *epulor, epulare, epulatus*
equestris -is -e of horses or horsemen, equestrian, *equester, equestris*
equi-, equalis -is -e equal-, *aequus*
equinoctialis -is -e of the equinox, opening at a particular hour of the day
equinus of the horse, *equinus*
equisetaceus -a -um resembling *Equisetum*
equisetifolius -a -um with leaves like a horsetail, *equus-saeta-folium*
equisetiformis -is -e having a horsetail like habit, *equus-saeta-forma*
equisetinus -a -um somewhat horsetail-like, *equus-saeta* (stem morphology)

Equisetum Horse-bristle, *equus-saeta* (a name in Pliny for a horsetail)
equitans riding, as if astride, present participle of *equito, equitare* (leaf bases of some monocots)
equitantifolius -a -um having equitant leaves, with leaf-bases that are astride the stem or axis, *equitans-folium*
eradiatus -a -um lacking radiance, privative, *e-radiatus*
eradicatus -a -um having no roots, *e-*(*radix, radicis*); destructive, *eradico, eradicare*, to root out or destroy
Eragrostis, eragrostis Love-grass, ερασ-αγρωστις
Eranthemum Love-flower, ερασ-ανθεμιον
Eranthis Spring-flower, εαρ(ηρ)-ανθος (winter aconite's early flowering season)
erba-rotta red-herb (*Achillea*)
Ercilla for Don Alonso de Ercilla (1533–95)
erebius -a -um of the underworld, *erebeus* dark, gloomy(-coloured), ερεβεννος, ερεβος
erectiflorus -a -um having upright flowers, *erectus-florum*
erectus -a -um upright, erect, *erigo, erigere, erexi, erectum*
erem- solitary-, lonely-, helpless-, desert-, ερημια
Eremanthe Solitary-flower, ερημος-ανθος
Eremia Solitary, ερημια (single-seeded loculi)
eremicolus -a -um, eremocolus -a -um living in solitude or empty habitats (deserts etc), botanical Latin from ερημος and *colo*
eremo- lonely-, desert-, destitute-, solitary-, ερημος, ερημο-
Eremomastax Solitary-mouth, ερημο-μασταξ (the long-tubed corolla)
eremophilus -a -um desert-loving, ερημο-φιλος (living in desert conditions)
Eremopogon Solitary-bearded, ερημο-πωγων (the single racemes)
eremorus -a -um of deserts or solitary places, ερημος
Eremospatha Solitary-spathe, ερημο-σπαθη
Eremurus Solitary-tail, ερημος-ουρα (the long raceme)
eri-, erio- woolly-, εριον, εριο-
eriantherus -a -um with woolly stamens, ερι-ανθηρος
Erianthus, erianthus -a -um Woolly-flowers, ερι-ανθος
Erica Pliny's version of an ancient Greek name, ερεικη, used by Theophrastus (***Ericaceae***)
ericetinus -a -um, ericetorum of heathland, of *Erica* dominated vegetation
ericifolius -a -um *Erica*-leaved, *Erica-folium*
ericinus -a -um, ericoides, erici- of heaths, heath-like, resembling *Erica, Erica-oides*
ericssonii for Mr Ericsson, who collected for Sander in Malaya *c.* 1892
erigens rising-up, present participle of *erigo, erigere, erexi, erectum* (for horizontal branches which turn up at the end)
erigenus -a -um Irish-born, of Irish origin, Erin, archaic name for Ireland, *erin-genus*
Erigeron Early-old-man, εριο-γερων, Theophrastus' name (early-flowering fleabanes)
Erinacea, erinaceus -a -um Prickly-one, prickly, hedgehog-like, *er, eris* (hedgehog broom), resembling *Erinacea* (cf. *Echinacea*)
Erinus, erinus Of-spring, εαρινος (Dioscorides' name for an early-flowering basil-like plant) (Erinus was an avenging deity)
erio- woolly-, wool-, ειριον, ειριος, εριον, εριο-
erioblastus -a -um having woolly buds, εριο-βλαστος
Eriobotrya Woolly-cluster, εριο-βοτρυς (the indumentum almost hides the heads of small flowers of the loquat, Cantonese, lu kywit)
eriobotryoides resembling *Eriobotrya*, εριο-βοτρυς-οειδης
eriocalyx having a woolly-surfaced calyx, εριο-καλυξ
eriocarpus -a -um with woolly fruits, εριο-καρπος
Eriocaulon Woolly-stem, εριο-καυλος (***Eriocaulaceae***)
Eriocephalus, eriocephalus -a -um Woolly-headed-one, εριο-κεφαλη (with a woolly fruiting head)
Eriochloa Woolly-grass, εριο-χλοη

Eriochrysis Golden-fleeced, εριο-χρυσος (the ferrugineous to yellow hairs on the callus)
Eriogonum Woolly-joints, εριο-γονυ (the hairy jointed stems)
Eriophorum Wool-bearer, εριο-φορος (cotton grass)
eriophorus -a -um bearing wool, εριο-φορος
eriophyllus -a -um with woolly leaves, εριο-φυλλον
eriopodus -a -um woolly-stalked εριο-ποδος
Eriosema Woolly-standard, εριο-σημα
Eriospermum Woolly-seed, εριο-σπερμα
eriostachyus -a -um with woolly spikes, εριο-σταχυς
eriostemon with woolly stamens, εριο-στημωυ
Erismadelphus Brother-of-*Erisma*, Ερισμα-αδελφος (related to *Erisma*)
eristhales very luxuriant, *Eristhalis*-like
eritimus -a -um most highly prized, εριτιμος
Eritrichium Woolly-hair, εριο-(θριξ, τριχος) (the indumentum)
Erlangea from the University of Erlangen, Bavaria
ermanii for G. A. Erman (1806–77), of Berlin, traveller and collector
ermineus -a -um ermine-coloured, white broken with yellow, from medieval Latin *mus armenius*
erodioides *Erodium*-like, *Erodium-oides*
Erodium Heron, ερωδιος (the stork's-bill shape of the fruit)
Erophila Spring-lover, εαρ(ο)-φιλεω (εαρ, εαρος, ηρ, ηρος spring)
erophilus -a -um liking (growing or flowering) the spring, εαρ(ο)-φιλεω
erosilabius -a -um having a jagged-edged labellum or lip, *erosus-labium*
erosus -a -um jagged, as if nibbled irregularly, erose, *erodo, erodere*
erraticus -a -um differing, wandering, of no fixed habitat, *erro, errare, erravi, erratum*
-errimus -a -um -est, -very, -the most (superlative suffix)
erromenus -a -um vigorous, strong, robust, ερρομενος
erubescens shamed, blushing, turning red, *erubesco, erubescere, erubescui*
Eruca, eruca Belch, *eructo, eructare* (the ancient Latin name for colewort)
erucago *Eruca*-like, feminine suffix *-ago*
Erucastrum somewhat *Eruca*-like, *Eruca-astrum*
erucifolius -a -um with *Eruca*-like leaves, *Eruca-folium*
eruciformis -is -e looking like *Eruca, Eruca-forma*
erucaeformis -is -e caterpillar-shaped, *eruca*, a caterpillar (septate and blunt at each end)
erucoides resembling *Eruca, Eruca-oides*
erumpens bursting out, breaking through, *erumpo, erumpere, erumpi, eruptum* (vigorously suckering)
Ervum the Latin name for a vetch, *Vicia ervilia*, called οροβος by Theophrastus
eryngii of *Eryngium* (*Pleurotus eryngii* is a saprophyte on *Eryngium* remains and on other Umbelliferous species)
Eryngium Theophrastus' name, ηρυγγιον, for a spiny-leaved plant (sea holly)
eryo- woolly-, εριον, εριο- (*Eryngium giganteum* is known as Miss Wilmott's ghost)
eryogynus -a -um having a woolly ovary, εριο-γυνη
Erysimum a name, ερυσιμον, used by Theophrastus (perennial wallflowers)
Erythea for Erythea, one of the Hesperides, the daughter of night and the dragon Lado of mythology
erythra, erythri- red- (see *erythro-*), ερυθρος, ερυθρο-
Erythraea, erythraeus -a -um (*errythro-*) reddish, ερυθρος (≡ *Centaurium*)
Erythrina Red, ερυθρος (flower colour of some coral-tree species)
erithrinus -a -um red, ερυθρος
erythro- red, ερυθρος, ερυθρο-
erythrobalanus reddish acorns, ερυθρο-βαλανος
erythrocalyx having a red calyx, ερυθρο-καλυξ
erythrocarpus -a -um with red fruits, ερυθρο-καρπος
Erythrochiton Red-cloak, ερυθρο-χιτων

erythrochlamys cloaked in red, ερυθρο-χλαμυς
erythrococcus -a -um with red berries, ερυθρο-κοκκος
Erythronium Red, ερυθρος (flower colours) (dog's-tooth violet)
erythrophaeus -a -um dusky-red, ερυθρο-φαιος
erythrophyllus -a -um with red leaves, ερυθρο-φυλλον
erythropodus -a -um with red stalks, ερυθρο-ποδιον
erythropus red-stalked, ερυθρο-πους
erythrorhizus -a -um red-rooted, ερυθρο-ριζα
erythrosepalus -a -um with red sepals, ερυθρο-σκεπη
erythrosorus -a -um with red sori, ερυθρο-σωρος
erythrostachyus -a -um with red spikes, ερυθρο-σταχυς
erythrostictus -a -um with red dots, ερυθρο-στικτος
erythroxanthus -a -um orange, yellowish-red, ερυθρο-ξανθος
Erythroxylon (um) Red-wood, ερυθρο-ξυλον (***Erythroxylaceae***)
Escallonia for the Spaniard Antonio Escallon, eighteenth-century botanist, traveller and plant hunter in S America (***Escalloniaceae***)
Eschscholzia (Eschscholtzia), eschscholtzii for Johann Friedrich Gustav von Eschscholz (1793–1831), Estonian traveller and naturalist (Californian poppy)
-escens -ish, -part of, -becoming, -becoming more, -being, present participle of *edo, edere; esse, edi, esum* (*essentia*)
esculentus -a -um being fit to eat, edible by humans, full of food, *esca, escae* (*edo, edere, esse, edi, esum*)
esparto rope, the Spanish derivative of the name, *spartum*, in Pliny for the grass used for ropes, mats and wickerwork, σπαρτον
Espeletia for Don José de Espeleta, Viceroy of New Grenada
-esthes clothing, garment, layer of covering, εσθημα, εσθης, εσθητος, εσθησις (with a qualitative or quantitative prefix)
estriatus -a -um without stripes, *e-*(*striata, striatae*)
esula an old generic name, *esula*, in Rufinus for a spurge
etesiae yearly, ετησιος (applied to herbaceous growth from perennial rootstock)
Ethulia etymology uncertain
-etorus -a -um -community (indicating the dominant component of the habitat)
etruscus -a -um from Tuscany, the area of the Etruscans, between the Tiber and the Arno, (*Etruria*), Italy
ettae for Miss Etta Stainbank
eu- well-, good-, proper-, completely-, right-, ευ, ευ-
Euadenia Well-marked-glands, ευ-αδην (the five lobes at the base of the gynophore)
euanthemus -a -um showy, nicely-flowered, ευ-ανθεμον
eublepharus -a -um having nice eyelashes, well fringed, ευ-βλεφαρον
euboeus -a -um, euboicus -a -um from the Greek Aegean island of Evvoia (Euboea)
Eucalyptus Well-covered, ευ-καλυπτος (the operculum of the calyx conceals the floral parts at first) (gum trees)
Eucharis Graceful, ευ-χαρις
Euchlaena Beautiful-wool, ευ-χλαινα (the tasselled stigmas)
euchlorus -a -um of beautiful green, true green, ευ-χλωρος
euchlous -a -um of good appearance, with a good texture, ευ-χλοη
Euchresta Beneficial, ευχρηστος (used in Chinese medicine as a febrifuge)
euchrites fit for service or use, useful, ευχρηστος
euchromus -a -um, euchrous -a -um well-coloured, ευ-χρωμα
Euclea Good-fame, ευκλεια
Euclidium Great-beauty, ευ-χλιδη, or for the Greek mathematician Euclid (*c*. 330 BC), or well closed, ευ-κλειω (indehiscent)
Eucnide Good-nettle, ευ-κνιδη (stinging hairs)
Eucodonia Beautiful-trumpet, ευ-κωδων (the corolla tube)
Eucomis Beautiful-head, ευ-κομη (the head of leaves above the flowers)
Eucommia Good-gum, ευ-κομμι (some yield gutta-percha) (***Eucommiaceae***)
eucosmus -a -um well ordered, well decorated, ευ-κοσμος

Eucrosia Well-fringed, ευ-χροσσος (of the stamens)
eucrosioides resembling *Eucrosia, Eucrosia*-οειδης *(Amaryllidaceae)*
Eucryphia Well-covered, ευ-κρυφαιος (the leaves are clustered at the branch ends) (***Eucryphiaceae***)
eucyclius -a -um, eucyclus -a -um nicely circular, nicely rounded, ευ-κυκλιος
Eudesmia Beautiful-bundle, ευ-δεσμα (the groups of bundled stamens)
eudorus -a -um sweetly perfumed, fragrant, ευ-ωδες
eudoxus -a -um of good character, ευ-δοξος
Eufragia Well-separated, ευ-φραγμα (isolated growths of root parasite, ≡ *Bartsia*)
eugeneus -a -um of good birth, noble, generous, ευ-γενεια; well-bearded, ευ-γενειος
Eugenia for Prince Eugene of Savoy (1663–1736), patron of botany (clove tree)
eugenioides *Eugenia*-like, *Eugenia-oides*
euglaucus -a -um nicely shining or bluish, ευ-γλαυκος (no indumentum)
euglossus -a -um having a well formed tongue (lip), ευ-γλωσσα
euleucus -a -um of a true white colour, ευ-λευκος
Eulophia Beautiful-crest, ευ-λοφος (the crests of the lip)
Eulophidium *Eulophia*-like
Eulophiella feminine diminutive of *Eulophia*
eulophus -a -um beautifully crested, ευ-λοφος
Eumorphia Well-formed, ευ-μορφη
eumorphus -a -um well-shaped, ευ-μορφη
euneurus -a -um having nice veins, ευ-νευρα
eunuchus -a -um castrated, ευνουχος (flowers without stamens)
Euodia, (Evodia) euodes Well-fragranced, sweet-scented, ευ-ωδια, ευ-ωδης
euonymifolius -a -um having leaves similar to those of *Euonymus, Euonymus-folium*
euonymoides resembling *Euonymus, Euonymus-oides*
Euonymus (Evonymus) Famed, of-good-name, ευ-ωνυμος, Theophrastus' name (spindle trees)
eupalustris -is -e well-suited to marshy habitats, botanical Latin from ευ- and *palus, paludis*
Eupatorium for Mithridates Eupator, King of Pontus, reputedly immune to poisons through repeated experimentation with them upon himself to find their counters (mithridates)
eupatorius -a -um similar to *Eupatorium*
euphlebius -a -um well-veined, ευ-(φλεψ, φλεβος)
Euphorbia for Euphorbus, physician to the King of Mauritania, who used the latex of a spurge for medicinal purposes (***Euphorbiaceae***)
Euphrasia Healthy-mind, Gladdening, ευ-φραινο (signature of eyebright flowers as being of use in eye lotions)
euphrasioides resembling *Euphrasia*, ευ-φραινο-οειδης
euphraticus -a -um from the area of the river Euphrates, *Euphrates, Euphratis*
euphues well-grown, of good stature, ευ-φυη
euphyllus -a -um having good foliage, ευ-φυλλον
Euplotes Properly-floating, ευ-πλωτος (planktonic)
eupodus -a -um long-stalked, ευ-(πους, ποδος)
euprepes, eupristus -a -um comely, good-looking, ευ-πρεπης
Euptelea Handsome-elm, ευ-πτελεα (***Eupteleaceae***)
Eupteris Proper-*Pteris*, ευ-πτερυξ
eur-, euro-, eury- (*euri-*) wide-, broad-, good-, ευρυς, ευρυ-
eurisyphon broadly tubular, ευρυ-σιφον (flowers)
eurocarpus -a -um with broad fruits, ευρυ-καρπος
eurolepis -is -e with mouldy-looking scales, ευρως-λεπις
europaeus -a -um from Europe, European (*Europa*)
Eurotia Mouldy-one, ευρως (the pubescence)
eurotrophilus -a -um liking to feed on humus-rich soils, (ευρως, ευρωτος)-τροφη-φιλεω
Eurya etymology uncertain (ευρυς, ευρυ- broad- or wide-)

Euryale for one of the Gorgons of mythology, Euryale (had burning thorns in place of hair) (***Euryalaceae***)
eurycarpus -a -um with wide fruits, ευρυ-καρπος
Eurychone Good-cloud, κονια dust sand, ashes, κονιατος whitewashed, κονιορτος a cloud of dust (≡ *Angraecum*)
euryopoides resembling *Euryops*, ευρυ-ωψ-οειδης
Euryops Wide-eyed, ευρυ-ωψ
eurysiphon having a wide tubed (corolla), ευρυ-σιφον
-eus -ea -eum -resembling, -belonging to, -noted for
Euscaphis Good-vessel, ευ-σκαφη (the colour and shape of the dehiscent leathery pods)
Eusideroxylon New-*Sideroxylon*, ευ-σιδηρο-ζυλον (Borneo ironwood)
euspathus -a -um having a distinct spathe, ευ-σπαθη
eustachyus -a -um, eustachyon having long trusses of flowers, ευ-σταχυς
Eustephia Well-crowned, ευ-(στεφανη, στεφανος)
Eustoma Of-good-mouth, ευ-στομα (for the throat of the corolla, the Greek ευστομα meant speaking good words, or keeping silent)
Eustrephus Well-twisted, ευ-στρεφω (scandent habit)
Euterpe Attractive, Euterpe (the name of the muse for music and lyric poetry)
eutheles properly female, ευ-θηλυς (the nipple-like umbo of the pileus)
eutriphyllus -a -um three-leaved throughout, ευ-τρι-φυλλον
euxanthus -a -um of a pure yellow colour, ευ-ξανθος
euxinus -a -um from the Baltic (called the inhospitable, αξενος, sea, *Pontus Axeinus*, until settled and renamed *Pontus Euxinus*, the hospitable, ευξεινος, sea)
evalvis -is -e without valves, botanical Latin, *e-valvae*
evanescens quickly disappearing, vanishing, *evanesco, evanescere, evanescui*
evansianus -a -um, evansii for Thomas Evans of Stepney, London *c.* 1810
evectus -a -um lifted up, springing out, carried-forth, *e-(vecto, vectare)*
evenius -a -um, evenosus -a -um without evident veins, *e-(vena, venae)*
evernius -a -um resembling the lichen *Evernia* in colouration
evertus -a -um overturned, expelled, turned out, *e-(verso, vertere, verti, versum)*
Evodia (Euodia) Well-perfumed, ευ-ωδης
evolutus -a -um unfolding, rolling onward, *evolvo, evolvere, evolvi, evolutum*
evolvens becoming rolled back, *evolvo, evolvere, evolvi, evolutum*
Evolvulus Unentangled, *evolvo, evolvere, evolvi, evolutum* (not twining like *Convolvulus*)
ewersii for J. P. G. Ewers (1781–1830), German botanist who studied the Altai flora
ex- without-, outside-, over and above-, out of-, εκ-, εξ-, *ex-*, *e-*
exacoides resembling *Exacum*, *Exacum-oides*
exactus -a -um exact, thrusting out, demanding, *exigo, exigere, exegi, exactum*
Exaculum *Exacum*-like, diminutive of *Exacum*
Exacum a name in Pliny (may be derived from an earlier Gallic word, or refer, *ex-(ago, agere, egi, actum)*, to its expulsive property)
exalatus -a -um lacking wings, *ex-(ala, alae)*
exalbescens out of *albescens* (related to); turning pale, *exalbesco*
exaltatus -a -um, exaltus -a -um lofty, very tall, *ex-altus*
exappendiculatus -a -um lacking an appendage, *ex-appendicula* (the spadix)
exaratus -a -um ploughed, with embossed grooves, engraved, *exaro, exarare, exaravi, exaratum*
exaristatus -a -um lacking awns, *ex-(arista, aristae)*
exasperatus -a -um rough, roughened, *exaspero, exasperare, exasperavi, exasperatum* (surface texture)
excavatus -a -um hollowed out, excavated, *excavo, excavare*
excellens distinguished, excellent, present participle of *excello, excellere*
excelsior higher, taller, very tall, comparative of *excelsus*
excelsissimus -a -um the most lofty, superlative of *excelsus*
excelsus -a -um tall, eminent, illustrious, *excelsus*

exchlorophyllus -a -um lacking green pigmentation, εξ-χλωρος-φυλλον (does not refer solely to leaves but includes other albinoid structures)
excipuliformis -is -e basin-shaped, *excipula, excipulae* (the persistent *Calvatia* fruit body wall)
excisus -a -um cut away, cut out, *excido, excidere, excidi, excisum*
excoriatus -a -um with peeling bark or epidermis, *ex-(corium, corii)*
excorticatus -a -um without bark, stripped, *ex-(cortex, corticis)* (peeling bark)
excurrens with a vein extended into a marginal tooth, *excurro, excurrere, excucurri (excurri), excursum* (as on some leaves)
excurvus -a -um curving outwards, *ex-curvus*
exerens protruding, revealing, stretching out, present participle of *ex(s)ero, ex(s)erere, ex(s)erui, ex(s)ertum*
exhibens presenting, showing, holding out, present participle of *exhibeo, exhibere, exhibui, exhibitum*
exiguus -a -um very small, meagre, poor, petty, *exiguus*
exilicaulis -is -e straight- or slender-stemmed, *exilis-caulis*
exiliflorus -a -um having few or small flowers, *exilis-florum*
exilis -is -e, exili- meagre, small, few, slender, thin, *exilis, exili-*
eximius -a -um excellent in size or beauty, choice, distinguished, *eximius*
exitiosus -a -um fatal, deadly, pernicious, destructive, *exitiosus*
Exoascus Outside-ascus, εξω-ασκος (the asci are superficial, not in an infructescence)
Exochorda Outside-cord, εξω-χορδη (the vascular anatomy of the placental wall of the ovary)
exoletus -a -um fully grown, mature, *exoletus* (some interpret as empty or weak)
exoniensis -is -e from Exeter, Devon (*Isca* or *Exonia*)
exornatus -a -um embellished, adorned, *exorno, exornare, exornavi, exornatum*
exorrhizus -a -um having adventitious roots, εξω-ριζα
exosus -a -um odious, detestable, *exosus*
exotericus -a -um common, external, εξωτερικος
exoticus -a -um not native, foreign, exotic, εξωτικος
expallescens blanching, turning pale, having fading colour, present participle of *expallesco, expallescere, expallui*
expansus -a -um spread out, expanded, unfolding, *expando, expandere*
expatriatus -a -um not indigenous, without a country, squandering, *expatro, expatrare*
explanatus -a -um flattened out, spread out flat; distinct, adjective from *explano, explanare, explanavi, explanatum*
explodens dehiscing violently, exploding, adjective from *explodo, explodere, explosi, explosum*
exquisitus -a -um choice, larger than the norm, *exquisitus*
exscapus -a -um without a stem, *ex-scapus*
exsculptus -a -um with deep cavities, dug out, adjective from *exsculpo, exsculpere, exsculpsi, exsculptum*
exsectus -a -um cut out, castrate, *ex-(seco, secare, secui, sectum)*
exserens thrusting out, revealing, present participle from *exsero, exserere, exserui, exsertum*
exsertus -a -um projecting, protruding, held out, *exsero, exserere, exserui, exsertum*
exsiccatus -a -um uninteresting; dry, dried out, *exsiccatus* (especially as *flora exsiccata*, for preserved herbarium material)
exspersus -a -um scattered, diffuse, *exspergo, exspergere, exspersum*
exstans outstanding, present participle from *ex-(sto, stare, steti, statum)*
exstipulatus -a -um without stipules, *ex-(stipula, stipulae)*
exsul foreigner, exile, *exsul, exsuli, exul (exsulo, exsulare, exsulavi, exsulatum)*
exsulans, exulans being secluded or exiled, present participle of *exsulo, exsulare, exsulavi, exsulatum (exsul, exul*, an exile)
exsurgens lifting itself upwards, thrusting upwards, *exsurgo, exsurgere, exsurrexi, exsurrectum*

extensus -a -um reaching out, extended, past participle of *extendo, extendere, extendi, extensum*
extra- outside-, beyond-, over and above-, *extra*
extractus -a -um drawn out, removed from, *extraho, extrahere, extraxi, extractum*
extremiorientalis -is -e from the most eastern part of its range, *extremi-(oriens, orientis)* (classically, from the Indian subcontinent)
extrorsus -a -um beyond the start, directed outwards from the central axis, *extra-orsus* (outwards facing, extrorse, stamens)
extrusus -a -um thrust out, burgeoning, *extrudo, extrudere, extrusi, extrusum*
exudans producing a (sticky-)secretion, exuding, sweating, *ex-(sudo, sudare, sudavi, sudatum)*
exultatus -a -um leaping up, joyful, *ex-(salio, salire, salui, saltum)*
exuvialis -is -e, exuviatus -a -um moulting, with stripping or peeling (outer layer), *exuo, exuere, exui, exutum*

faba, fava the old Latin name for the broad bean, perhaps from φαγο, to eat (***Fabaceae***)
fabaceus -a -um, fabae- bean-like, resembling *Faba*
fabago Faba-like, with feminine suffix *-ago*
fabarius -a -um of beans, bean-like, *faba*
Fabiana for Archbishop Francisco Fabian y Fuero (1719–1801), Spanish patron of botanical studies (false heath)
fabri of artisans; skilfully produced, for building, *fabricor, fabricare, fabricatus* (timber of *Quercus*, *Abies* and *Acer fabri*)
Fabricia for Johann Christian Fabricius (1745–1808), Danish student of Linnaeus (NB: Hieronymus Fabricius at Aquapendente, Italy (1537–1619) is usually referred to simply as Fabricius, and was a pioneer anatomist and microscopist)
facetus -a -um humorous, elegant, fine, *facetus*
Fadyenia for James MacFadyen (1800–50), author of a Flora of Jamaica
faenum hay, fodder, *faenum* (*faenus* is profit or interest)
faenum-graecum Greek-hay, fenugreek, *faenum-Graecia*
faeroensis -is -e from the Faeroes group of islands, N Atlantic Ocean
fagi of beech, living on *Fagus* (*Phyllaphis*, homopteran gall insect)
fagi-, fagineus -a -um beech-like, *Fagus-*
fagifolius -a -um having leaves similar to those of *Fagus*, *Fagus-folium*
Fagonia for Monsignor Fagon (1638–1718), physician to Louis XIV of France
Fagopyrum (*Fagopyron*), *fagopyrum* Beech-kernel, φαγο-πυρην (buckwheat is from the Dutch boekweit)
Fagus the Latin name, *fagus*, for the beech tree, from φαγο, for the edible seed of beech (***Fagaceae***)
falacinus -a -um pillar-like, columnar, *fala, falae*
falc-, falci-, falco- curved like a scythe or sickle blade, *falx, falcis* (leaves, leaflets, petals or bracts) (cognate with falcon)
Falcaria Sickle, *falx, falcis* (the shape of the leaf-segments)
falcarius -a -um, falcatorius -a -um of the sickle maker, sickle shaped, *falx, falcis*
Falcatula Somewhat-sickle-shaped, feminine diminutive from *falcatus* (the pods)
falcatulus -a -um shaped like a small sickle, diminutive of *falcatus*
falcatus -a -um, falci- sickle-shaped, falcate, *falcatus*
falcifolius -a -um with sickle-shaped leaves, *falcis-folium*
falciformis -is -e sickle-like, *falcis-forma*
falcinellus -a -um like small scythes, diminutive of *falx, falcis* (the pinnae)
falcipetalus -a -um having curved petals, botanical Latin from *falcis* and πεταλον
falconeri for either Dr Hugh Falconer (1805–65), Superintendent of Botanic Garden at Saharanpur, Uttar Pradesh, India, or William Falconer (1850–1928), gardener at Harvard, USA
falcorostrus -a -um having a curved beak, *falcis-rostrum*

falklandicus -a -um from the Falkland Islands, S Atlantic
fallacinus somewhat deceptive, diminutive of *fallax, fallacis* (a *Rumex* hybrid epithet)
fallax deceitful, deceptive, false, bent, *fallax, fallacis*
Fallopia for Gabriello Fallopio (1523–62), Italian surgeon, anatomist and pharmacologist
fallowianus -a -um for George Fallow (1890–1915), of the Botanic Garden, Edinburgh
Fallugia for Virgilio Fallugi, seventeenth-century Italian botanical writer, Abbot of Vallombroso
falsotrifolium *Falcatula falsotrifolium* is a synonym for *Trifolium ornithopodioides* (showing an element of uncertainty)
famatimensis -is -e from the high-pampas Sierra de Famatina, Argentina
famelicus -a -um greedy; stunted, starved, hungry, *famelicus* (cognate with famished)
familiaris -ia -e domestic, common, *familia*
Faradaya for Michael Faraday (1791–1867), English scientist
farcatus -a -um solid, filled, *farcio, farcire, farsi, fartum* (not hollow)
farctus -a -um solid, not hollow, *farctus*
farfara with a mealy surface, a name in Pliny for butterbur (*far, faris* corn or meal)
Farfugium With-swiftly-passing-flour, *far-*(*fugio, fugere, fugi, fugitum*) (early loss of indumentum)
Fargesia, fargesii for Père Paul Guillaume Farges (1844–1912), plant collector in Szechwan, China
farinaceus -a -um of mealy texture, yielding farina (starch), farinaceous, *farina, farinae*
fariniferus -a -um bearing farina or flour, (*farina, farinae*)*-fero*
farinipes with a farinaceous or mealy stalk, (*farina, farinae*)-(*pes, pedis*)
farinolens smelling of meal, (*farina, farinae*)-(*oleo, olere, olui*)
farinosus -a -um with a mealy surface, mealy, powdery, *farina*
farleyensis -is -e from Farley Hill Gardens, Barbados, West Indies
farnesianus -a -um from the Farnese Palace gardens of Rome
farorna name for the hybrid *Gentiana farreri* × *ornata*
Farquharia, farquharianus -a -um for General William Farquhar (1770–1839), of Singapore
farreri for Reginald J. Farrer (1880–1920), English author and plant hunter
Farsetia for Philip (Filippo) Farseti, Venetian botanist
fasci- band-, burden-, bundle-, *fascis* (the *lictoris* who accompanied Roman consuls or magistrates carried a *fascis*, bundle of rods with an axe)
fascians fasciating, present participle of *fascio, fasciare, fasciavi, fasciatum* (bacterium or other agent causing stem deformity)
fasciarus -a -um elongate and with parallel edges, band-shaped, *fascia, fasciae*
fasciatus -a -um bound together, bundled, fasciated, *fascis*, as in the inflorescence of cockscomb (*Celosia argentea 'cristata'*)
Fascicularia Bundle-like, *fasciculus* (the habit)
fascicularis -is -e, fasciculatus -a -um clustered in bundles, fascicled, *fasciculatus*
fasciculiflorus -a -um with bundles of flowers, with clustered flowers, *fasciculatus-florum*
fasciculus -a -um having small groups or bundles, *fasciculus*
fascinator bewitcher, magician; very interesting, *fascino, fascinare*
fasciolus -a -um having branches in small tufts, diminutive of *fascis*
fassoglensis -is -e from Fazughli, the gold-bearing area of Sudan
fastigiatus -a -um with branches erect like the main stem, sloping, fastigiate, *fastigate*
fastuosus -a -um vain, proud, *fastus*
fatiflorus -a -um fate or misfortune-flower, *fatus, fati* (*fatifer, fatiferi*, deadly)
Fatshedera the composite name for hybrids between *Fatsia* and *Hedera*

Fatsia from a Japanese vernacular name, fa tsi
fatuosus -a -um silly, pompous, *fatuus*
fatuus -a -um not good, insipid, tasteless, simple, foolish, *fatuus*
Faucaria Throat, *fauces, faucium* (the leaves gape apart)
faucilalis -is -e, faucius -a -um wide-mouthed, throated, *fauces, faucium*
fauconnettii for Dr Charles Fauconnet (1811–75), of Geneva
faurei (*fauriei*) for either Abbé Faure, director of the Grenoble Seminary, or Abbé Urbain Faure (1847–1915), a missionary in Japan
faustus -a -um lucky, auspicious, *faustus*
favigerus -a -um bearing honey-glands, *favus-gero*
favoris-is -e favourable, agreeable; popular, supportive, *favor, favoris*
favosus -a -um cavitied, faveolate, honeycombed, *favus, favi*
febrifugus -a -um fever-dispelling (medicinal property) cognate through old English, feferfuge, with feverfew, (*febris, febris*)-(*fugo, fugare, fugavi, fugatum*)
fecetus -a -um synthesized, made, created, *facio, facere, feci, factum*
fecundator, fecundatrix fertilizer, *fecundo, fecundare* (botanically, a misnomer for *Andricus*, the oak-galling cynipid)
fecundus -a -um fruitful, fertile, fecund, *fecundus*
Fedia etymology uncertain
fedtschenkoianus -a -um, fedtschenkoi for either Olga Fedtschenko (1845–1921) or her son Boris Fedtschenko (1873–1947), Russian botanists
Feijoa for Don da Silva Feijoa, botanist of San Sebastian, Spain
fejeensis -is -e from the Fiji Islands, S Pacific
Felicia for a German official named Felix at Regensburg (d. 1846), but some interpret it as *felix, felicis* cheerful (blue marguerite)
felinus -a -um relating to or affecting cats, *feles, felis*; fruitful, favourable, *felix, felicis*
felis-linguus -a -um cat's-tongue, *felis-lingua*
Felix Fruitful, *felix, felicis*
felleus -a -um as bitter as gall, full of bile, *fel, fellis*
felosmus -a -um foul-smelling, φελ-οσμη
femina female, *femina, feminae*
fenas toxic, poisonous, φενω to murder (some interpret as hay-like, *faenum*, but *faenum habet in cornu*, he is dangerous)
Fendlera, fendleri for August Fendler (1813–83), German naturalist and explorer in New Mexico
Fendlerella diminutive from *Fendlera*
fenestralis -is -e, fenestratus -a -um with window-like holes or openings, *fenestra, fenestrae* (*Ouvirandra fenestralis*)
fenestrellatus -a -um latticed with small window-like holes, diminutive of *fenestra*
fennicus -a -um from Finland, Finnish (*Fennica*)
-fer, -ferus, -fera, -ferum -bearing, -carrying, φερω, *fero, ferre, tuli, latum*
ferax fruitful, *ferax, feracis*
ferdinandi-coburgii for King Ferdinand of Bulgaria (1861–1948), alpine plant grower
ferdinandi-regis as *ferdinandi-coburgii*
ferganensis -is -e, fergenicus -a -um from the Ferghana region of Uzbekistan
fergusonii for W. Ferguson (1820–87), collector in Ceylon
fernambucensis -is -e from Pernambuco state, Brazil
fero-, ferus -a -um wild, feral, *fera, ferae*
ferox very prickly, ferocious, *ferox, ferocis*
Ferraria for Giovani Battista Ferrari (1584–1655), Italian botanist
ferreus -a -um rusty-brown coloured, durable, iron-hard, of iron, *ferrum, ferri*
ferruginascens turning rusty-brown, *ferrugo, ferruginis*
ferrugineus -a -um rusty-brown in colour, *ferrugo, ferruginis*
ferruginiflorus -a -um having rusty-brown flowers, *ferrugineus-florum*
ferruginosus -a -um conspicuously rust-coloured, *ferrugo, ferruginis*
ferrum-equinum horse-shoe-like, *ferrum-equinus* (horse-shoe orchid)

ferrus -a -um sword-like, durable, iron-like, *ferrum, ferri* (was used for any iron object)
fertilis -is -e heavy-seeding, fruitful, fertile, *fertilis*
Ferula Staff, *ferula* (Pliny's classical Latin name) (giant fennel)
ferulaceus -a -um fennel-like, resembling *Ferula*, hollow-
Ferulago *Ferula*-like, *ferula* with feminine suffix
ferulifolius -a -um with *Ferula*-like leaves, *Ferula-folium*
-ferus, -fera, -ferum -carrying, *fero, ferre, tuli, latum*
ferus -a -um wild, untamed, feral, *ferus*
fervens, fervidus -a -um raging, blazing, passionate, agitated, *ferveo, fervere, ferbui; fervo, fervere, fervi; fervidus*
festalis -is -e, festivus -a -um agreeable, bright, pleasant, cheerful, festive, adjectival form of *festus*
festinus -a -um hasty, quick (-growing), *festinus*
Festuca Straw (a name used in Pliny, *festuca* also the rod used for manumitting Roman slaves to freedman), fescue
festucaceus -a -um similar to *Festuca*
festuciformis -is -e looking like *Festuca*
festucoides resembling *Festuca, Festuca-oides*
Festulolium the composite name for hybrids between *Festuca* and *Lolium*
festus -a -um sacred, used for festivals, *festus*
fetidus -a -um bad-smelling, stinking, foetid, *foetidus*
fibratus -a -um fibrous, *fibra, fibrae*
fibrillosus -a -um, fibrosus -a -um with copious fibres, fibrous, *fibra, fibrae*
fibuliformis -is -e shaped as a tapering cylinder, *fibula, fibulae*
fibulus -a -um broach, clamp, clasp, *fibula, fibulae* (mostly for the tapered shape)
Ficaria, ficarius -a -um small-fig, diminutive of *ficus*, an old generic name for the lesser celandine (the shape of the root tubers)
fici-, ficoides fig-like, resembling *Ficus, Ficus-oides*
ficifolius -a -um Fig-leaved, *Ficus-folium*
ficoideus -a -um similar to *Ficus*
ficto-, fictus -a -um false, *fictus*
fictolacteum false (*Rhododendron*) *lacteum*, *ficto-lacteus*
Ficus the ancient Latin name, *ficus, fici*, for the fig (and for haemorrhoids), from the Hebrew, fag
ficus-indica Indian fig, *ficus-*(*india, indiae*) (morphology of the *Opuntia* fruit)
-fid, -fidus -a -um -cleft,-divided, *findo, findere, fidi, fissum*
Fieldia, fieldii for Baron Field (1786–1846), Chief Justice of the Supreme Court, New South Wales
figo fixed, pierced, *figo, figere, fixi, fixum*
Filago Thread, *filum* with feminine suffix (the medieval name refers to the woolly indumentum)
filamentosus -a -um, filarius -a -um, fili- thread-like, with filaments or threads, *filum, fili*
filaris -is -e thread-like, *filum, fili*
fili- thread-like-, *filum, fili*
filicaulis -is -e having very slender stems, *fili-caulis*
filiceps having a narrow head (of flowers), *fili-ceps*
filicifolius -a -um with small fern-like leaves, *filix-folius*
filicinus -a -um, filici-, filicoides fern-like, *filix-oides*, living on ferns (gall midges)
filiculmis -is -e having thread-like stalks, *fili-culmus*
filiculoides like a small fern, *filicula-oides*
filiculus -a -um like a small fern, diminutive of *filix*
filiferus -a -um bearing threads or filaments, *fili-fero*
filifolius -a -um thread-leaved, *fili-folium*
filiformis -is -e thread-like, *fili-forma*
Filipendula, filipendula Thread-suspended, *fili-pendulus* (slender attachment of meadow-sweet tubers)

filipendulinus -a -um somewhat like *Filipendula*
filipendulus -a -um hanging by threads, *fili-pendulus*
filipes slender-stemmed, with thread-like stalks, *fili-pes*
Filix Latin for fern
filix-femina (*foemina*) female fern, *filix-*(*femina, feminae*)
filix-fragilis brittle-fern, *filix-fragilis*
filix-mas masculine (male) fern, *filix-*(*mas, maris*)
fimbri- fringe-, fringed-, *fimbriae, fimbriarum*
fimbriatulus -a -um finely fringed, diminutive of *fimbriatus*
fimbriatus -a -um with a fringe, fringed, *fimbriae, fimbriarum*
fimbripetalus -a -um having fringed petal-margins, *fimbri-petalus*
Fimbristylis Fringed-styled, *fimbri-stilus*
finisterrae from Finisterre range of mountains, Papua New Guinea (end of the land)
finitimus -a -um neighbouring, adjoining, akin, related, *finitimus* (linking related taxa)
finmarchicus -a -um from Finnmark county, N Norway
Firmiana for Karl Joseph von Firmian (1716–82), Governor of Lombardy
firmipes strong-stemmed, stout-stemmed, *firmus-pes*
firmulus -a -um quite firm or strong, diminutive of *firmus*
firmus -a -um strong, firm, lasting, *firmus*
fiscellarius -a -um resembling a wicker basket, intertwined, *fiscella, fiscellae*
fischeri, fischerianus -a -um for either Friedrich Ernest Ludwig von Fischer (1782–1854), Director of the Botanic Garden at St Petersburg, or Walter Fischer, who collected cacti *c.* 1914
Fischeria, fischeri for Dr Fischer of the Botanic Garden, St Petersburg
fissi-, fissilis -is -e cleft, divided, splitting, *fissum, fissi-*
Fissidens Split-teeth, *fissi-dens* (the 16 divided peristome teeth)
fissifolius -a -um with deeply split leaf blades, *fissi-folium*
fissipedus -a -um having a stalk divided near the base, *fissi-pes*
fissistipulus -a -um with split stipules, *fissi-*(*stipula, stipula*)
fissuratus -a -um having slits, *findo, findere, fidi, fissum*
fissus -a -um, -fissus cleft almost to the base, *findo, findere, fidi, fissum*
fistulosus -a -um, fistulus -a -um hollow, pipe-like, tubular, fistular, *fistula*
Fittonia for Elizabeth and Sarah Mary Fitton, nineteenth-century botanical writers
Fitzroya For Robert FitzRoy RN (1805–65), commander of the *Beagle* for the Survey Expedition to Patagonia
flabellaris -is -e, flabellatus -a -um fan-like, fan-shaped, flabellate, *flabellum, flabelli-*
flabellifer -era -erum fan-bearing, *flabellum-fero* (with flabellate leaves)
flabellifolius -a -um with fan-shaped leaves, *flabelli-folium*
flabelliformis -is -e pleated fanwise, *flabelli-forma*
flabellulatus -a -um resembling a small fan, diminutive of *flabellum*
flaccatus -a -um flaccid, flabby, *flacceo, flaccere*
flaccid- sagging, flagging, weakening, *flacceo, flaccere*
flaccidifolius -a -um having soft or flaccid leaves, *flaccidus-folium*
flaccidior more limp or feeble, comparative of *flaccus*
flaccidissimus -a -um most sagging or feeble, superlative of *flaccidus*
flaccidus -a -um limp, weak, feeble, soft, flabby, flaccid, *flaccidus*
flaccospermus -a -um with *flacca*-like seeds (*Carex*)
flaccus -a -um drooping, pendulous, flabby, *flacceo, flaccere* (flap-eared)
Flacourtia for Etienne de Flacourt (1607–61), French E India Company (***Flacourtiaceae***)
Flagellaria Tendrilled, *flagellum* (the leaves often have tendrillar apices)
flagellaris -is -e, flagellatus -a -um, flagelli- with long thin shoots, whip-like, stoloniferous, *flagellus* , *flagelli-*
flagelliferus -a -um bearing whips, *flagelli-fero* (elongate stems of New Zealand trip-me-up sedge)

flagelliflorus -a -um flowering on whip-like stems, *flagelli-florum*
flagellifolius -a -um having long whip-like leaves, *flagelli-folium*
flagelliformis -is -e long and slender, whip-like, flagelliform, *flagelli-forma*
flagellus -a -um whip-like, *flagellum, flagelli*
flammans flame-like, flaming, *flammeo, flammare, flammavi, flammatum*
flammeus -a -um flame-red, fiery-red, *flamma, flammae*
flammiferus -a -um flame-bearing, *flamma-fero* (having fiery flowers)
Flammula Little-flame, an old generic name for lesser spearwort, some suggest a reference to the burning taste
flammulus -a -um little flame, flame-coloured, diminutive of *flamma, flammae*
flandrius -a -um from lowland areas of Holland, Belgium and France, Flandrian (*Flandria*)
flav-, flavi-, flavo- yellowish, *flavus, flavi-, flavo-*
flavantherus -a -um having yellow flowers, botanical Latin from *flavus* and ανθηρος
flavens being yellow, *flavens, flaventis*
flaveolus -a -um somewhat yellow, *flaveus*
flaveoplenes fully yellow, *flavus-plenus*
flavescens pale-yellow, turning yellow, present participle of *flavesco, flavescere*
flavicans, flavidus -a -um somewhat yellow, comparatives of *flavus*
flavissimus -a -um the yellowest, superlative of *flavus*
flavo-albus -a -um yellow and white, *flavo-albus*
flavonutans yellow-drooping, *flavo-*(*nuto, nutare*)
flavovirens greenish yellow, *flavo-virens*
flavus -a -um bright almost pure yellow, *flavus*
flectens turning round or aside, *flecto, flectere, flexi, flexum*
fleischeri for M. Fleischer (1861–1930), of Mentone, France
fletcherianus -a -um for Reverend J. C. B. Fletcher, orchid grower of Mundam, Chichester
flexi-, flexilis -is -e pliant, flexible, *flecto, flectere, flexi, flexum*
flexicaulis -is -e with curved or bending stems, *flexi-caulis*
flexifolius -a -um pliant-leaved, *flexi-folius*
flexilis -is -e flexile, pliant, *flexilis*
flexipes pliant-stalked, *flexi-pes*
flexuosiformis -is -e zigzag-shaped, *flexuosus-forma* (stems or inflorescence axes)
flexuosus -a -um zigzag, sinuous, winding, much bent, tortuous, *flexuosus*
flexus -a -um, -flexus -a -um -turned, *flecto, flectere, flexi, flexum*
flocc-, flocci-, flocco- trivial; a bit of wool, *floccus, flocci*
floccifer -era -erum, flocciger -era -erum, floccosus -a -um bearing a woolly indumentum which falls away in tufts, floccose, (*floccus, flocci*)-*fero* or *gero*
flocciflorus -a -um having woolly flowers, *flocci-florum*
floccipes with floccose stalks, *flocci-pes*
floccopus floccose-stalked, *flocci-pes*
floccosus -a -um having woolly tufts, *floccosus*
flocculentus -a -um a little woolly, diminutive of *floccosus*
flocculosus -a -um woolly, wool-like, *floccus, flocci*
flora flowered, *flos, floris*; Flora was the Roman goddess of flowering plants
flore-albo white-flowered, *flore-albus*
florentinus -a -um from Florence, Florentine (*Florentia*)
flore-pleno double-flowered, full-flowered, *florum-plenus*
floribundus -a -um abounding in flowers, freely-flowering, *florum-abundus*
floridanus -a -um from Florida, USA
floridulus -a -um somewhat flowery, diminutive of *floridus*
floridus -a -um florid, ornate; free-flowering, flowery, *floridus*
floriferus -a -um flower-bearing, producing many flowers, *florum-fero*
florindae for Mrs Florinda N. Thompson
floripecten comb-flower, *florum-*(*pecten, pectinis*)
floripendulus -a -um having hanging flowers, *florum-pendulus*

florulentus -a -um abundantly flowery, comparative of *florum*
-florus -a -um -flowered, *flos, floris* (botanical Latin uses *florus -a -um* for flowered or flowering, in place of the original Latin meaning of beautiful)
-flos, floris -is -e flowered, *flos, floris*
flos-aeris air-flower, *florum-aerius* (epiphytic air-plants)
Floscopa Floribundant, *flos-(copia, copiae)*
flos-cuculi flower of cuckoo, *flos-cuculus* (flowering in the season of cuckoo song)
flosculosus -a -um with small flowers, very ornamental, *flosculus*
flos-jovis Jove's flower, *flos-Iovis*
flos-reginae flower of the queen, *flos-regina*
fluctuans inconstant, fluctuating, present participle of *fluctuo, fluctuare*
fluctuosus -a -um stormy; wavy, undulating, *fluctuo, fluctuare* (leaf margins)
fluellyn for St Llywelyn; Lyte (1578) used *llysiau fluellyn* for herbs flowering around 7 April
fluens flowing, *fluo, fluere, fluxi, fluxum*
fluitans floating on water, *fluito, fluitare* (flote-grass)
fluminensis -is -e of the river, *flumen, fluminis;* from Rio de Janeiro (*Flumen Ianuarius*)
fluminis -is -e flowing, of rivers, *flumen, fluminis*
fluvialis -is -e, fluviatilis -is -e growing in rivers and streams, of running water, *fluvius, fluvi; fluvius, fluvii*
foecundissimus -a -um the most fruitful, superlative of *fecundus*
foecundus -a -um fruitful, fecund, *fecundus*
foederatus -a -um forming a compact growth of individuals, federated, *foederatus*
foedus -a -um revolting, hideous, *foedus, foederis*
foemina, foeminius -a -um feminine, *femina, feminae*
foeneus -a -um hay-like, *faenum*
foeni- fennel-like-, *faeni-*
foeniculasius -a -um resembling *Foeniculum*
foeniculatus -a -um *Foeniculum*-like
Foeniculum, foeniculum Fodder, diminutive of *fenum, faenum*, the Latin name, *feniculum, faeniculum*, for fennel
foenisicii of mown hay, *faenum-seco*
foenum-graecum Greek-hay, *foenum-graecun*, fenugreek (the Romans used *Trigonella foenum-graecun* as fodder)
foenus -a -um fodder, hay, *faenum*
foerstermannii for J. F. Föstermann, who collected for Sanders in Assam *c.* 1885
foetans, foetens stinking, rank smelling, *foeteo, foetere*
foetidissimus -a -um most smelly, very stinking, superlative of *foetidus*
foetidolens malodorous, foul-smelling, *foetidus-(oleo, olere)*
foetidus -a -um (fetidus, foetidus) stinking, bad smelling, foetid, *foeteo, foetere*
Fokienia from Fujien (Fu Chien) province, SE China
foliaceus -a -um leaf-like, leafy, *folium*
foliatus -a -um, foliosus -a -um leafy, *folium*
-foliatus -a -um -leaflets, -leafleted, *folium* (usually preceded by a number)
foliiflorus -a -um flowering on the leaves, *folium-florum*
folio- leaflet-, diminutive of *folium*
-foliolatus -a -um -leafleted, *foliolus* (with a qualifying prefix)
foliosissimus -a -um having copious leaves, superlative of *foliosus*
foliosus -a -um leafy, well-leaved, *foliosus*
-folius -a -um -leaved, *-folium*
follicularis -is -e bearing follicles, *folliculus* (seed capsules as in hellebores, classically a small bag or egg-shell)
fomentarius -a -um of poultices, *fomentum, fomenti*
Fomes Kindling, touchwood, *fomes, fomitis*
fonsiflorus -a -um producing fountains of flowers, *(fons, fontis)-florum*

Fontanesia, fontanesii for Réné Louiche Desfontaines (1752–1833), French botanist, author of *Flora Atlantica*
fontanus -a -um, fontinalis -is -e of fountains springs or fast-running streams, *fons, fontis*
Fontinalis, fontinalis -is -e Spring-dweller (*fontanus* a spring)
foraminiferus -a -um bearing a much perforated surface, *foramen-fero*
foraminosus -a -um being pierced with small holes, *foramen, foraminis*
forbesianus -a -um, forbesii for Edward Forbes, Professor of Botany at Edinburgh (1815–54), or James Forbes (1773–1861), gardener and writer at Woburn Abbey, or John Forbes, who collected in Africa *c.* 1825, or H. O. Forbes, collector and writer in the far East *c.* 1886
forcipatus -a -um having a pincer-like shape, *forceps, forcipis, forcipi-*
fordii for either Charles Ford (1844–1927), superintendent of Hong Kong Botanic Garden, or Lyman Ford, of San Diego, California, USA
Forestiera for Charles Le-Forestier (*c.* 1800), French naturalist
forficatus -a -um scissor-like, *forfex, forficis, forfici-*
forficifolius -a -um having leaves arranged like the blades of shears, *forfici-folius*
forgetianus -a -um for Louis Forget (d. 1915), collector for Sanders in S and Central America
formanekianus -a -um for Dr Edward Formanek (d. 1900), Professor of Botany at Brünn (Brno), present-day Czech Republic
formicarius -a -um relating to ants, attracting ants, *formica* (sweet fluid exudates)
formicarus -a -um of ants, *formica, formicae* (morphological adaptations occupied by ants)
formiciferus -a -um bearing ants, *formica-fero* (commensal associations)
formidabilis -is -e capable of terrifying or inspiring respect, *formido, formidare, formidavi, formidatum*
-formis -is -e -sort, -kind, -resembling, -shaped, *forma, formae*
formosanus -a -um, formosensis -is -e from Taiwan (Formosa)
formosissimus -a -um the most handsome, the most beautiful, *formosus*
formosus -a -um handsome, beautiful, well-formed, *formosus*
fornicatus -a -um arched, arching, *fornicatus*
forniculatus -a -um slightly arched, diminutive of *fornicatus*
Forrestia for Peter Forrest, seventeenth-century botanist
forrestii for George Forrest (1873–1932), plant collector in China
forsteri, forsterianus -a -um for J. R. Forster (1729–98) or his son J. G. A. Forster (1754–94), of Halle
Forsythia for William Forsyth (1737–1804), superintendent of Kensington Royal Gardens and St James's Palace, founder member of the RHS a few months before his death
fortis -is -e durable, vigorous, strong, *fortis*
fortissimus -a -um the strongest, superlative of *fortis*
fortuitus -a -um casual, occasional, *fortuitus*
fortunatus -a -um rich, favourite, *fortuna, fortunae*
Fortunearia, fortuneanus -a -um, fortunei, fortuni for Robert Fortune (1812–80), Scottish plant collector for the RHS. in China
Fortunella diminutive from *Fortunearia* (≡ *Citrus*, the kumquats, Cantonese, kam kwat, little orange)
fossulatus -a -um having fine grooves, as if having been dug over, *fodio, fodere, fodi, fossum*
fosterianus -a -um, fosteri for Professor Sir Michael Foster FRS (1836–1907), physician and *Iris* grower of Cambridge
Fothergilla, fothergillii for Dr John Fothergill (1712–80), of Stratford, Essex, physician and introducer of American plants (American wych hazel)
fothergilloides resembling *Fothergilla, Fothergilla-oides*
foulaensis -is -e from the island of Foula, Shetland, Scotland

Fouquiera for Pierre Éloy Fouquier (1776–1850), French physician (***Fouquieriaceae***)
Fourcroya see *Furcraea*
fourcroydes similar to *Furcraea*, *Furcraea-oides*
fournieri for Eugene P. N. Fournier (1834–84), physician of Paris
foveatus -a -um having a pitted surface, *fovea*, *foveae*
foveolatus -a -um with small depressions or pits all over the surface, foveolate, *fovea*, *foveae*
foxii for Walter Fox (1858–1934), Singapore gardener
fracidus -a -um mellow-textured, slightly pulpy, *fracidus*
fractiflexus -a -um weakly twining, *fractus-flexus*
fragari-, *fragi-* strawberry-, *fraga*, *fragorum*
Fragaria Fragrance, *fragrans* (of the fruit)
fragarii of strawberries, living on *Fragaria* (*Aphelenchus*, nematode)
fragarifolius -a -um strawberry flowered, *fraga-folium*
fragarioides resembling *Fragaria*, *Fragaria-oides*
fragifer -era -erum strawberry-bearing, *fraga-fero*
fragiformis -is -e resembling strawberry, *Fragaria-form* (e.g. the warted red fruiting bodies of *Hypoxylon fragiforme* on beech)
fragilimus -a -um more fragile, comparative of *fragilis*
fragilis -is -e fleeting, brittle, fragile, *fragilis*
fragosus -a -um rough, breakable, *fragosus*
fragrans sweet-scented, odorous, fragrant, *fragrans*, *fragrantis*
fragrantissimus -a -um most fragrant, superlative of *fragrans*
frainetto from a Balkans vernacular name for an oak
franchetianus -a -um, *franchetii* for Adrien René Franchet (1834–1900), French botanist with particular interest in Chinese and Japanese plants
franciscanus -a -um, *fransiscanus -a -um* from San Francisco, USA
Francoa for Dr F. Franco of Valentia, sixteenth-century promoter of plant studies (bridal wreath) (***Francoaceae***)
francofurtanus -a -um from Frankfurt, Germany (*Francofurtum*)
Frangula, *frangula* Brittle, *frango* (medieval name refers to the brittle twigs of alder buckthorn)
franguloides resembling *Frangula*
frangulus -a -um breakable, fragile, *frango*, *frangere*, *fregi*, *fractum* (diminutive suffix)
Frankenia for John Frankenius (1590–1661), Swedish botanist (***Frankeniaceae***)
Franklinia for Benjamin Franklin (1706–90) inventor of the lightning conductor and President of the USA (≡ *Gordonia alatamaha*)
frankliniae, *franklinii* for Lady and Sir John Franklin (1786–1847), Arctic explorer and Governor of Tasmania
frankofurtanus -a -um from Frankfurt am Main, Germany
franzosinii for Signo Franzonsini, gardener at Intra, Lake Maggiore, Italy
Frasera, *fraserianus -a -um*, *fraseri* for John Fraser (1750–1811), nurseryman of Chelsea, England
fraternalis -is -e, *fraternus -a -um* closely related, brotherly, *frater*, *fratris*
fraudulosus -a -um full of deceit, *fraudo*, *fraudare*, *fraudavi*, *fraudatum*
fraxinellus -a -um like a small ash, diminutive of *Fraxinus*
fraxineus -ea -eum ash-like, *Fraxinus*
fraxini of ash, living on *Fraxinus* (symbionts, parasites and saprophytes)
fraxini-, *fraxineus -a -um* ash-like, resembling ash, *Fraxinus-*
fraxinifolius -a -um with leaves similar to *Fraxinus*
fraxinivorus ash-devouring, *fraxinus-*(*voro*, *vorare*, *voravi*, *voratum*) (inflorescences galled by *Eriophyes* gall mite)
fraxinoides ash-like, *Fraxinus-oides*
Fraxinus ancient Latin name, *fraxinus*, for ash, used by Virgil (ash tree)
Freesia for Friedrich Heinrich Theodor Freese (d. 1876), of Kiel, pupil of Ecklon

Fremontia, fremontii, Fremontodendron for Major General John Charles Fremont (1813–90) who explored W North America
frene- strap-, bridle-, curb-, *frenum, frena, freni*
fresnoensis -is -e from Fresno County, California
fretalis -is -e of Straits, *fretum, fretus* (*fretensis* Straits of Messina) (perhaps cognate with fret, a sea mist)
Freycinetia for Admiral Freycinet (1779–1842), French circumnavigator
Freylinia for L. Freylin, who compiled a catalogue of the plants of Buttigliera Marengo, N Italy *c.* 1810
freynianus -a -um for Joseph Freyn (1845–1903), Czech botanist of Prague
friburgensis -is -e from Friburgo, Brazil
friderici-augustii for Friederich August II of Bavaria (1797–1854)
frieseanus -a -um, friesii for Dr Elias Magnus Fries (1798–1874), Swedish cryptogamic botanist
friesianus -a -um for Thore Magnus Fries (1832–1913), son of Dr Elias Magnus Fries, explorer
friesicus -a -um from the N Sea coastal Friesland area of the Netherlands and Germany, including the Frisian Islands (*Frisia*)
frigescens cooling, becoming inactive, *frigesco, frigescere*
frigidus -a -um cold, of cold habitats, of cold localities, *frigidus*
frikartii for the Swiss nurseryman, Frikart (*Aster* × *frikartii*)
frisicus -a -um, frisius -a -um from Friesland, Friesian (*Frisia*)
frithii for Mathew Frith of the 1998 Kew Expedition to Cameroon
Fritillaria Dice-box, *fritillus* (the shape of fritillary flowers)
froebelii for Fröbel, nurserymen of Zurich *c.* 1874
Froelichia, froelichii for Joseph A. Froelich (1766–1841), German physician and botanist
frondeus -a -um having leafy frond-like branches, *frons, frondis*
frondiscentis -is -e having leaf-like petals, phyllodic, *frondis-essentia*
frondosus -a -um leafy, *frons, frondis*
fructifer -era -erum fruit-bearing, fruitful, *fructus-fero*
fructu- fruit-, *fructus, fructo-*
fructu-albo white-fruited, *fructus-albus* (berried)
fructu-coccineo red-fruited, *fructus-coccineus* (berried)
fructuosus -a -um fruitful, comparative of *fructus*
frumentaceus -a -um grain-producing, giving corn, *frumentum*
frumentarius -a -um pertaining to grain, *frumentum*
frustulentus -a -um appearing to be of many small pieces, full of crumbs, *frustulentus*
frutecorus -a -um, fruticorus -a -um of thickets, *fruticetum*
frutescens shrubby, becoming shrubby, *frutex-essentia*
frutetorus -a -um of scrubland or the bush, amongst shrubs, *frutex, fruticis*
frutex shrub, bush, *frutex, fruticis*
fruticans bushing, sprouting, present participle of *fruticor, fruticare*
fruticicolus -a -um living in bushy habitats, *fruticetum-colo*
fruticosus -a -um of shrub like habit, *frutex, fruticis*
fruticulosus -a -um dwarf-shrubby, diminutive of *fruticosus*
fucatus -a -um blushing, dyed, artificial-looking, *fuco, fucare, fucavi, fucatum*
fucescens turning red, blushing, *fuco, fucare, fucavi, fucatum*
Fuchsia, fuchsii for Leonhard Fuchs (1501–66), Professor at Tübingen and renaissance botanist
fuchsiaefolius -a -um with *Fuchsia*-like leaves, *Fuchsia-folium*
fuchsioides resembling *Fuchsia, Fuchsia-oides*
fuci- artificial, dyed (red), deceitful; drone, bee-glue, *fucus, fuci*
fucifer -era -erum drone-bearing, bee-glue bearing, rouge-bearing, *fucus-fero*
fuciflorus -a -um drone-flowered, *fucus-florum* (superficial resemblance of the flower to a drone bee)

fucifolius -a -um having red leaves, *fucus-folium*
fuciformis -is -e, fucoides seaweed-like, resembling *Fucus, Fucus-oides*
fucosus spurious, *fucosus*
Fucus, fucus Sea-weed, φυκος, *fucus* (Semitic origin for a red paint or cosmetic made from a rock lichen)
fugacissimus -a -um most transient or timorous, superlative of *fugax, fugacis*
fugax fleeting, rapidly withering, shy, fugacious, *fugax, fugacis*
fugiens dying, fleeting, present participle of *fugio, fugere, fugi, fugitum*
-fugius -a -um -vanishing, -escaping, -avoiding, *fugio, fugere, fugi, fugitum*
Fugosia see *Cienfuegosia*
-fugus -a -um -banishing, -putting-to-flight, *fugo, fugare, fugavi, fugatum*
Fuirena for G. Fuiren, Danish physician
fuji the Japanese vernacular name for *Wisteria*
fujianus -a -um from the environs of Mount Fuji (Fuji San or Fujiyama), Japan
fulcitus -a -um propped, supported, strengthened, *fulio, fulcire, fulsi, fultum*
fulcratus -a -um having supports, *fulcratus* (thorns, hooks or tendrils)
fulgens shining, glistening, *fulgeo, fulgere, fulsi* (often with red flowers)
fulgentis -is -e of brilliance or bright colours, *fulgeo, fulgere, fulsi*
fulgidus -a -um flashing, glowing, *fulgidus* (brightly coloured)
fuligineo-albus -a -um white and sooty-brown, (*fuligo, fuliginis*)-*albus* (colouration of stipe)
fulgineus -a -um, fuliginosus -a -um dirty-brown to blackish, sooty, *fuligo, fuliginis*
fullonum of cloth fullers, *fullo, fullonis*
fultus -a -um stiffened or supported, past participle of *fulcio, fulcire, fulsi, fultum*
fulvellus -a -um slightly tawny, diminutive of *fulvus*
fulvescens becoming tawny, *fulvus-essentia*
fulvi-, fulvo-, fulvus -a -um tawny, reddish-yellow, fulvous, *fulvus*
fulvicortex having a tawny surface, rind or bark, *fulvus-cortex*
fulvidus -a -um yellowish, tawny, *fulvus*
Fumana Smoke, *fumus*, (the colour of the foliage)
Fumaria Smoke, *fumus*, Dioscorides' name, καπνος, referred to the effect of the juice on the eyes being the same as that of smoke, but the medieval Latin name, *fumus terrae*, refers to the appearance of Fumitory plants as smoke arising from the ground) (***Fumariaceae***)
fumarioides resembling *Fumaria, Fumaria-oides*
fumatofoetens smelling of smoke, *fumus*-(*foeteo, foetere*)
fumeus -a -um steam-like, smoke-coloured, smoky, *fumus*
fumidus -a -um steamy, smoke-coloured, dull grey coloured, *fumidus*
fumosus -a -um smoky (colouration), smoked, *fumus*
funalis -is -e twisted together, rope-like, *funale, funalis*
funebris -is -e murderous, mournful, doleful, of graveyards, funereal, *funebris*
funereus -a -um from the Funeral Mountain group in the Amargosa range, E California, USA; fatal, *funereus*
funerius -a -um doleful, funereal; fatal, *funus, funeris*
fungosus -a -um spongy, fungus-like, pertaining to fungi, *fungus*
fungum-olens having a fungal smell, *fungus-olens*
funicaulis -is -e having thread-like stems, *funis-caulis*
funicularis -is -e, funiculatus -a -um like a thin cord, *funiculus, funiculi*
funiferus -a -um rope-bearing, cord-bearing, *funis-fero* (use as cordage)
funiformis -is -e like dreadlocks, ropes of hair, *funis-forma*
Funtumia from the W African vernacular names, funtum, o-funtum
furcans forking, dividing into two, *furca, furcae*
furcatus -a -um forked, furcate, *furca, furcae*
furcellatus -a -um having a small fork or notch (at the apex), diminutive of *furca, furcae*
Furcraea (Furcroyia) for Antoine François Fourcroy (1755–1809), French chemist
furfuraceus -a -um scurfy, mealy, scaly, *furfur, furfuris*

furfurascens becoming scurfy-surfaced, *furfur, furfuris*
furiens irritating, exciting to madness, present participle of *furio, furiare, furiavi, furiatum*
furiosus -a -um frantic, mad, *furiosus* (frenzied growth habit)
furtivus -a -um secretive, hiding away, *furtivus*
furtus -a -um intriguing, secretive, tricky, *furtum*
furvicolor dark-coloured, *furvus-color*
furvus -a -um black, very dark, *furvus*
Fusarium Spindle-like, *fusus* (the spore-bearing branches)
fusca-coreana Korean (*Clematis*) *fusca*
fuscans darkening, blackening, present participle of *fusco, fuscare*
fuscatus -a -um somewhat dusky-brown, *fuscus*
fuscescens turning swarthy, darkening, *fuscus-essentia*
fusci-, fusco-, fuscus -a -um bright-brown, swarthy, dark-coloured, *fuscus*
fuscinatus -a -um having trident-like form, *fiscina, fuscinae*
fuscoatrus -a -um literally swarthy-black, *fuscus-ater*
fuscomarginatus -a -um with brown margins, *fuscus-*(*margo, marginis*)
fusculus -a -um husky, blackish, diminutive of *fuscus*
fusiformis -is -e spindle-shaped, *fusus-forma*
fusipes with spindle-shaped stipes, *fusus-pes* (spindle-shanks' fruiting stipes)
futilis -is -e worthless, brittle, useless, *futilis, futtilis*
futurus -a -um of the future, coming, *futurus*

gabonensis -is -e, gabonicus -a-um from Gabon, equatorial W Africa
Gabunia from Gabon, equatorial W Africa
gaditanus -a -um from Cadiz (*Gades*), Spain
Gaertnera, gaertneri for J. Gärtner (1732–91), German physician
Gagea for Sir Thomas Gage (1781–1820), English botanist (an earlier Sir William Gage (1657–1727), of Bury St Edmunds, introduced the green-gage about 1725)
gagnepainii for François Gagnepain (1866–1952), botanist at the National Museum, Paris
Gahnia for H. Gahn (1747–1816), Swedish botanist
Gaillardia for Gaillard de Charentonneau (Marentonneau), patron of botany (blanket flowers)
-gala, galacto- milk, milky, milk-like γαλα, γαλακτος, γαλακτ-
galacifolius -a -um having leaves similar to those of *Galax*
galactanthus -a -um having milky or milk-white flowers, γαλακτ-ανθος
Galactia Milky, γαλακτος (the sap of the milk pea)
galactinus -a -um milky, γαλα, γαλακτος (flower or sap colour)
Galactites Milk-like, γαλα, γαλακτος, γαλακτιτης (for the white leaf venation)
galactodendron milk-tree, cow-tree, γαλακτος-δενδρον (the abundant sap's local use)
galanga an Asian vernacular name, galangal (from Arabic, kalanjan), for the ginger-like rhizome of *Kaempferia galanga*, which has culinary and medicinal uses
galantheus -a -um snowdrop-like, γαλα-ανθος
galanthi- *Galanthus-*, snowdrop-
Galanthus Milk-white flower, γαλα-ανθος (the colour of snowdrops)
galapageius -a -um from the Galapagos archipelago, Pacific Ocean
galaticus -a -um from Ankara, Turkey (*Galatia*)
Galax Milky, γαλακτος (the flower colour)
Galaxia Star-spangled, γαλαξια (simile, when flowering, with the milky way, γαλαξιας κυκλος)
galbanifluus -a -um with a yellowish exudate, *galbanus-*(*fluo, fluere, fluxi, fluxum*) (*Ferula galbaniflua* yields gum galbanum)
galbinus -a -um greenish-yellow, *galbinus*
gale from an old English vernacular name, gagel or gawl, used by J. Bauhin for bog-myrtle or sweet gale, *Myrica gale*

Galeandra Helmeted-stamen, (*galea, galeae*)-*andrus* (the cover of the another)
Galearia Helmeted, *galea* (the concave petals)
Galearis Hooded, *galea* (arrangement of the perianth)
galeatus -a -um, galericulatus -a -um helmet-shaped, like a skull-cap, *galea*
Galega Milk-bringing, γαλα-αγω (goat's rue is reputed to improve lactation)
galegi- resembling *Galega*
galegifolius -a -um having leaves resembling those of *Galega*
Galenia for Claudius Galenus of Pergamos (129–200), free-thinking mathematician, philosopher, physician, theist and experimentalist, author of πνευμα ψυχικον (*Spiritus animalis*) *Living Spirits*
Galeobdolon, galeobdolon Weasel-smell, γαλεν-βδολος (a name used in Pliny)
galeopsifolius -a -um having leaves similar to those of *Galeopsis*
Galeopsis Weasel-like, resembling *Galeobdolon* (γαλεν, γαλη)-οψις (an ancient Greek name used by Pliny), some derive from *galea*, a helmet
Galeottia, galeottianus -a -um for Henri Guillaume Galeotti (1811–58), Italian explorer and plant hunter in central America, director of Brussels Botanic Garden
galericulatus -a -um bonnet- or skull-cap-like, from diminutive of *galerum* (the calyx of skullcap, *Scutellaria galericulata*, the pileus of bonnet *Mycena*)
galeritus -a -um of rustic places, rustic, *galeritus*
gali- *Galium*-like-
galii of bedstraws, living on *Galium* (*Eriophyes*, acarine gall mite)
galiicola living on *Galium* (*Dasyneura*, dipteran gall midge)
galilaeus -a -um from Galilee, Israel
Galinsoga for Don Mariano Martinez de Galinsoga, eighteenth-century Spanish director of the botanic garden at Madrid (Englished as gallant soldier)
galioides bedstraw-like, resembling *Galium, Galium-oides*
Galipea the Amerindian vernacular name for the angostura tree, *Galipea officinalis*
Galium Milk, Dioscorides' name, γαλιον (the flowers of *G. verum* were used to curdle milk, γαλα, in cheese making)
Gallesia for Giorgio Gallesio (1772–1839), Italian botanist
gallicus -a -um from France, French, *Galli*; of the cock or rooster, *gallus, galli*
gallii for Nicolas Joseph Marie Le Gall de Kerinou (1787–1860), author of *Flore de Morbihan*, 1852
gallingar the medieval apothecaries' name for dried roots of *Alpinia officinarum*, from the Chinese, koa liang kiang, reapplied to the substitute, English galangal, or galingale, *Cyperus longus*
galopus -a -um with a milky stalk, γαλα-πους (exudate from damaged stipe)
Galphimia anagram of *Malpighia*
Galpinia, galpinii for Ernst Eduard Galpin of Barberton (1858–1941), collector in S Africa, author on the flora of Drakensberg, 1909
Galtonia for Sir Francis Galton (1822–1911) pioneer in eugenics, anthropology, fingerprinting and weather charting, writer on exploration (summer hyacinth)
Galvezia for Jose Galvez, a Spanish colonial administrator
gamandrea the medieval apothecaries' name for a medicinal plant (cognate with chamaedrys, χαμαιδρυς, and germander)
Gamanthera Fused-stamens, γαμο-ανθερα
Gamanthus Joined-flowers, γαμο-ανθος
Gambelia for William Gambel (1821–49), American biologist at the National Academy of Sciences
gambier the Malay vernacular name, gambir, for *Uncaria gambier* and the astringent catechu extract from it
gambogius -a -um from Cambodia, rich-yellow, gamboge (the resin obtained from *Garcinia gambogia* or *G. hanburyi*)
gamo- fused-, joined-, united-, married-, γαμος, γαμο-
Gamochaeta Fused-bristles, γαμο-χαιτη (the united pappus hairs)
Gamolepis United-scales, γαμο-λεπις (the involucral bracts)
gamopetalus -a -um having the petals united, γαμο-πεταλον

gamosepalus -a -um with a united calyx, γαμο-σκεπη
-gamus -a -um -union, -marriage, γαμος
gandavensis -is -e from either Ghent (*Gandavum*), Belgium, or Gandava, Pakistan
gangeticus -a -um from the Ganges region, *Ganges, Gangis*
gano- bright-, shining-, γαναω
Ganoderma Shining-skinned, γαναω-δερμα (the glossy surface)
Ganophyllum Shining-leaved-one, γαναω-φυλλον
Garaya for Leslie Andrew Garay (b. 1924), orchidologist at the Oakes Ames herbarium, Harvard, USA
Garcinia for Dr Laurent Garcin (1683–1751), a French naturalist with the East India company, travelled in the orient
Gardenia, gardenii for Dr Alexander Garden (1730–91), Anglo-American botanist, correspondent with Linnaeus
gardeniodorus -a -um having the fragrance of *Gardenia, Gardenia-odorus*
gardneri, gardnerianus -a -um for Hon. E. Gardner (b. 1784), political resident in Nepal, or George Gardner (1812–49), Superintendent of Botanic Garden at Peradeniya, Sri Lanka
garganicus -a -um from Monte Gargano, S Italy
gargaricus -a -um from the Gargara gorge of the Rif mountains, Morocco
gariepinus -a -um from the environs of the Gariep Dam, Orange River, S Africa
Garrya for Nicholas Garry, secretary of the Hudson's Bay Company, *c.* 1820–35 (silk tassel tree) (***Garryaceae***)
gasipaes from an Amazonian vernacular name for the peach-fruit of palm, *Guilielmia gasipaes*
gaspensis -is -e from the area of Gaspé Bay, Quebec, Canada (derived either from the eponymous discoverer Gaspar Corte-Real, or from an Indian vernacular, gespeg, the end of the world)
gaster-, gastro- belly-, bellied-, γαστηρ
Gasteranthus Bellied-flower, γαστηρ-ανθος (the expanded lower part of the perianth tube)
Gasteria Belly, γαστηρ (the swollen base on the corolla)
Gastonia for Gaston de Bourbon, son of Henry IV of France, patron of botany (≡ *Polyscias*)
Gastridium Little-paunch, diminutive of γαστηρ (the bulging of the glumes of nitgrass)
Gastrochilus Bellied-lip, γαστηρ-χειλος (the swollen lip)
Gastrodia Belly-like, γαστηρ-ωδες (the bell-shaped flowers)
Gastrolobium Bellied-pod, γαστηρ-λοβος (the inflated, segmented pods)
Gastronema Bellied-thread, γαστηρ-νημα (the base of the filaments)
gaudens pleasing, delighting, rejoicing, present participle of *gaudeo, gaudere, gavisus*
Gaudichaudia, gaudichaudianus -a -um, gaudichaudii for Charles Gaudichaud-Beaupré (1789–1854), naturalist on Freycinet's 1817–20 circumnavigation of the world
Gaudinia for Jean François Aimé Philipe Gaudin (1766–1833), Swiss botanist, author of *Flora Helvetica*, 1828–33
Gaulnettya the composite name for hybrids between *Gaultheria* and *Pernettya*
Gaultheria for Dr Jean François Gaulthier (1708–56), Swedish-Canadian botanist of Quebec (wintergreen)
Gaura Superb, γαυρος (stature and floral display of some)
gausapatus -a -um like woollen cloth, with a frieze, *gausapa, gausapis*
Gaussia for Carl Friedrich Gauss (1777–1855), German mathematician and cosmologist
gayanus -a -um for Jacques Etienne Gay (1786–1864), French botanist
gayi for either Jacques Etienne Gay or Claude Gay (1800–73), writers on S American plants
Gaylussacia for Joseph Louis Gay-Lussac (1778–1850), French philosopher and chemist (huckleberries)

Gazania for the Greek scholar Theodore of Gaza (1398–1478), who transcribed Theophrastus' works into Latin (1483); some interpret it as Riches, *gaza-ae* (treasure flowers)
Geastrum (*Geaster*) Earth-star, γη-(αστηρ, αστερος) (shape of the fungal fruiting body)
Geissanthera Overlapping-anthers, γεισσο-ανθερα
Geissanthus Tiled-flowers, γεισσο-ανθος
Geissaspis Tiled-with-shields, γεισσον-(ασπις, ασπιδος) (the large overlapping bracts of the flower-heads)
Geissois Tiled, γεισσον
Geissomeria Overlapping-parts, γεισσο-μερις (the overlapping bracts)
geisson-, *geisso-* overlapping-, tiled-, γεισσον
Geissorhiza Tiled-root, γεισσο-ριζα (appearance of the scaly-tunicated corns)
geito-, *geitono-*, *-geiton* neighbour-, γειτων, γειτονος, γειτονο-
Geitonoplesium Close-neighbour, γειτονο-πλησιος (scrambling habit)
Gelasine Smiling-dimple, γελασεινος (the flower structure)
gelatinosus -a -um with the appearance or consistency of jelly, *gelo, gelare; gelatus*
Geleznowia for Nokolai Ivanovich Zheleznow (1816–77), Russian agronomist at Moscow University
Gelidium Frost, *gelidus* (the alga produces gelose, from agar-agar, that is used in ice-cream making)
Gelidocalamus Frosted-*Calamus*, *gelidus-Calamus*
gelidus -a -um of icy regions, growing in icy places, *gelo, gelare; gelidus*
Gelsemium from the Italian, gelsomine, for true jasmine
gemelliflorus -a -um having flowers carried in pairs, *gemellus-florum*
gemellus -a -um in pairs, paired, twinned, double, *gemellus*
geminatus -a -um, *gemini-* united in pairs, twinned, *gemino, geminare, geminavi, geminatum*
geminiflorus -a -um with paired flowers, *geminus-florum*
gemma- bud-, jewel-, *gemma, gemmae*
gemmatus -a -um bejewelled, budding, having gemmae, *gemmo, gemmare; gemmatus*
gemmiferus -a -um, *gemmiparus -a -um* bearing gemmae or deciduous buds or propagules, *gemmae-fero, gemmae-pario* (*Brassica gemmifera* Brussels sprout)
gemmosus -a -um sprouting, with gemmae or bud-like propagules, *gemmo, gemmare*
genavensis -is -e from Geneva, Switzerland (*Genava*)
generalis -is -e normal, prevailing, usual, of the species, *generalis*
generosus -a -um well-stocked, noble, productive, *generosus*
-genes -descended-from, -born-of, -birth, γενεα, γενος, *genesis*
genevensis -is -e from Geneva, Switzerland (*Genava*)
geniculi-, *geniculatus -a -um* jointed, Pliny's name for a knee-like bend, *genu, geniculatus*
Genipa from the Tupi vernacular name, jenipapos, in Guiana (genipap fruit is used in making a preserve)
Genista a name in Virgil (*planta genista*, from which the Plantagenets took their name; some derive it from the Celtic, gen, for a small bush)
genistelloides resembling a small *Genista*, feminine diminutive *Genistella-oides*
genisti- broom-like, resembling *Genista*
genisticola living on *Genista*, *Genista-colo* (symbionts, parasites and saprophytes)
genistifolius -a -um having leaves similar to a *Genista*, *Genista-folium*
Gentiana a name, γεντιανη, in Pliny attributed to King Gentius of Illyria (180–167 BC), for his discovery of its medicinal properties (***Gentianaceae***)
Gentianella Gentian-like, feminine diminutive of *Gentiana*
gentianoides resembling *Gentiana*, *Gentiana-oides*
Gentianopsis Resembling *Gentiana*, γεντιανη-οψις, *Gentiana-opsis*
gentilianus -a -um, *gentlii*, *gentilis -is -e* for Louis Gentil (b. 1874), Curator of Brussels Botanic Garden, (*gentilis* hereditary, kinsman)

genuflexus -a -um kneeling, having bends at the nodes, *genu-(flecto, flectere, flexi, flexum)*
genuinus -a -um of the cheek or back teeth; natural, true, *genuinus*
-genus -a -um -borne, *genus, generis*; -begot, -produced, γενος, *gigno, gignere, genui, genitum*
Genyorchis Chinned-orchid, γενυς-ορχις (the foot to the column) (some interpret as Knee-orchid, *genu-orchis*)
geo- -on or under the earth-, γη, γηω
geocarpus -a -um earth-fruited, with fruits which ripen underground, γηω-καρπος
Geogenanthus Earth-borne-flower, γηω-γενος-ανθος
geogeneus -a -um earth-borne, γηω-γενος (fruiting stage of earth-petal fungus)
Geoglossum Earth-tongue, γηω-γλωσσα (the fruiting body)
geoides *Geum*-like, *Geum-oides*
geometrizans having very regular structure, symmetrical, γεω-μετριζανς
Geonoma (*Geonomos*) Skilled-in-agriculture, γεονομος (produces apical offset bulbs)
geonomiformis -is -e resembling *Geonoma* in shape
geophilus -a -um spreading horizontally, ground-loving, γηω- φιλος
geophyllus -a -um earth-leaf, γηω-φυλλον
georgei for George Forrest (1873–1932), collector in China
georgianus -a -um from Georgia, USA or any other Georgia
georgicus -a -um from Georgia, Caucasus
geotropus -a -um earth-turning, γηω-τροπη (ring- or group-forming growth of fruiting bodies)
-ger, -gerus, -gera, -gerum -carrying, wearing, producing, *gero, gerere, gessi, gestum*
geralensis -is -e from the Serra do Mar, or Geral mountains, Brazil
geranifolius -a -um, geraniifolius *Geranium*-leaved
geranioides resembling *Geranium, Geranium-oides*
Geranium Crane, γερανος (Dioscorides' name, γερανιον, refers to the shape of the fruit resembling the head of a crane, γερανον, cranesbill) (***Geraniaceae***)
Gerardia for John Gerard (1545–1612), gardener for Lord Burleigh (William Cecil), author of the *The Herball, or generall historie of plants* (1597)
gerardianus -a -um, gerardii for P. Gerard (1795–1835), who sent plants to Dr Nathaniel Wallich
gerardii for Jean Gerard, Madagascan guide who found *Voanioala gerardii*
gerascanthus -a -um having aged thorns, retaining thorns, γεραιος-ακανθος
Gerbera for Traugott Gerber, German naturalist and traveller (Barberton daisy)
germanicus -a -um from Germany, German (*Germania*)
germinans sprouting, budding, present participle of *germino, germinare*
geronti-, gero-, -geron old man, γερων, γεροντος, γεροντο-
gerontopogon carrying a beard, like an old man's beard, γεροντο-πωγων (a pappus)
-gerus -a -um -bearing, -carrying, -producing, *gero, gerere, gessi, gestum*
Gesneria (*Gesnera*), *gesnerianus -a -um* for Conrad Gesner (1516–65), Swiss botanist of Zurich (***Gesneriaceae***)
gesneriiflorus -a -um having *Gesneria*-like flowers, *Gesneria-florum*
Gethylis Small-leek, diminutive of γεθυον (some derive it from γηθεω, to rejoice)
-geton -neighbour, γειτων
Geum, geum a classical name, *geum*, in Pliny, (γευω stimulant, medicinal roots of some)
Gevuina from a Chilean vernacular name for the edible Chile hazel nut
ghaticus -a -um from the mountain ranges, Ghats, bordering the Deccan plateau, India
gibb-, gibbi-, gibbatus -a -um swollen on one side, gibbous, *gibbus, gibbi* (a gibbous moon is larger than a half-moon)
Gibbaeum Hump, *gibba* (the two unequal leaves)
gibberosus -a -um humped, hunchbacked, more convex on one side, *gibberosus*
gibbiflorus -a -um pouched-flowered, *gibbi-florum* (irregular form of perianth)

gibbosus -a -um somewhat swollen or enlarged on one side, *gibbus, gibbi*
gibbsii for Hon. Vicary Gibbs (1853–1933), of Aldenham, tree enthusiast
gibbus -a -um humped, with a hump, *gibbus, gibbi*
gibraltaricus -a -um from Gibraltar (*Calpe, Calpetanus*)
giennensis -is -e from Gien, France, or Jaén (Gienna), Spain
gigandrus -a -um having large stamens, γιγ-ανηρ
giganteus -a -um, giganticus -a -um unusually large or tall, gigantic, γιγαντειος (*giganteus*, of giants)
giganthes giant-flowered, γιγ-ανθος
gigantiflorus -a -um having large flowers, botanical Latin from γιγας and *florum*
gigas giant, γιγας (*gigantes, gigantum*, giant, giants)
Gigaspermum Giant-sperm, γιγας-σπερμα (the antherozooids)
gilbertensis -is -e from the environs of the Gilbert river, Cape York peninsula, N Australia
gileadensis -is -e from Gilead, N Jordan (or the area east of the River Jordan)
Gilia from a Hottentot name for a plant used to make a beverage, or for Felipe Salvadore Gil (*c.* 1790), Spanish writer on exotic plants
Gillenia for Arnoldus Gillenius, a seventeenth-century German botanist
Gilliesia, gilliesii for Dr J. Gillies (1792–1834) of Mendoza, Argentina
giluus -a -um, gilvo-, gilvus -a -um dull, pale yellow, *gilvus*
gilvescens turning dull-yellow, *gilvus-essentia*
gilviflorus -a -um having dull yellow flowers, *gilvus-flora*
gingidium gum, *gingiva, gingivae*, from an old name, γιγγιδιον, used by Dioscorides for a carrot-rooted plant (in medicine, *gingivitis* is inflammation of the gums)
Ginkgo derived from a Sino-Japanese name, gin-kyo (***Ginkgoaceae***)
ginnala a native name for *Acer ginnala*
ginseng from the Chinese name, ren-shen (man-herb) (the forked root)
giraffae of giraffes, from the Arabic, zarafa
giraldianus -a -um, giraldii for Giuseppe Girald (1848–1901), Italian missionary in Shensi, China
girondinus -a -um from the Gironde department of Aquitaine region, SW France
githago from generic name, *gith*, in Pliny (for *Nigella*) with feminine suffix (for resemblance of the seeds)
glabellus -a -um somewhat smooth, smoothish, diminutive of *glaber*
glaber -bra -brum, glabri-, glabro smooth, without hairs, glabrous, *glaber, glabri*
glaberrimus -a -um very smooth, smoothest, superlative of *glaber*
glabratus -a -um, glabrescens becoming smooth or glabrous, *glabri-essentia*
glabrius -a -um, glabrus -a -um smooth, bald, *glaber, glabri*
glabriusculus -a -um rather glabrous, a little glabrous, diminutive of *glaber*
glabrohirtus -a -um smooth and hairy, bald with a few hairs
glacialis -is -e of frozen habitats, of the ice, *glacies; glacialis*
gladiatus -a -um sword-like, *gladius, gladi*
gladiolatus -a -um like a small sword, diminutive of *gladius*
Gladiolus Small-sword, the name in Pliny, diminutive of *gladius* (cognate with gladdon, *Iris foetidissima*)
glandiformis -is -e shaped like an acorn or nut, (*glans, glandis*)*-forma*
Glandularia Glandular-one, *glandulae* (≡ *Verbena*, section *Glandularia*)
glandulicarpus -a -um having glandular or sticky fruits, botanical Latin from *glandulae* and καρπος
glandulifer -era -erum, glanduligerus -a -um gland-bearing, *glandulae-fero (gero)*
glanduliflorus -a -um having sticky, glandular flowers, *glandulae-florum*
glandulosissimus -a -um the most glandular, superlative of *glandulae*
glandulosus -a -um full of glands, glandular (from modern Latin, *glandulae*, for throat glands)
glareophilus -a -um liking screes or gravels, botanical Latin from *glarea* and φιλεω
glareosus -a -um gravelly, growing on gravel, *glarea, glareae*
glasti- *Isatis*-, woad-like- (from the Latin name, *glastum*, for woad)

glastifolius -a -um with woad-like leaves, *glastum-folius*
glaucescens, -glaucus -a -um developing a fine whitish bloom, bluish-green, sea-green, glaucous, *glaucus-essentia*
glauci-, glauco-, glaucus -a -um with a white or greyish bloom, glaucous (from Latin *glaucuma*, a cataract)
Glaucidium *Glaucium*-like
glaucifolius -a -um with grey-green leaves, *glaucus-folius*
glauciifolius -a -um with leaves resembling those of horned poppy, *Glaucium*
glaucinus -a -um a little clouded or bloomed, diminutive of *glaucus* (milky)
Glaucium Grey-green, γλαυκος (Dioscorides' name, γλαυκιον, for the colour of horned poppy latex)
glaucoalbus -a -um having a greyish bloom, *glaucus-albus*
glaucocarpus -a -um having a waxy bloom on the fruits, γλαυκο-καρπος
glaucochrous -a -um sea-green complexioned, γλαυκος-χρως
glaucoglossus -a -um with a glaucous tongue, γλαυκο-γλωσσα
glaucopeplus -a -um robed in sea-green, γλαυκο-πεπλος
glaucophylloides resembling (*Salix*) *glaucophylla*
glaucophyllus -a -um glaucous-leaved, γλαυκο-φυλλον
glaucopruinatus -a -um having a frosting of bluish green bloom, *glaucus-pruinosus*
glaucopsis -is -e glaucous-looking, γλαυκ-οψις
glaucopus -a -um glaucous-stalked, γλαυκο-πους
glaucosericeus -a -um having greyish-blue silky hairs, *glaucus-serricus*
glaucovirens greyish-green, *glaucus-virens*
glaucus -a -um clear, gleaming, sea-green, bluish: with a waxy bloom, γλαυκος, *glaucus*
Glaux a name, γλαυξ, used by Dioscorides
glaziouanus -a -um for A. M. F. Glaziou (1828–1906), French director of the Imperial Gardens, Rio de Janiero, Brazil
glebarius -a -um lump, clod, sod, forming tumps of tufts, *gleba, glaeba*
Glechoma, glechoma Dioscorides' name, βληχων, γληχων, for penny-royal
glechomae of ground ivy, living on *Glechoma* (*Dasyneuran*, dipteran gall midge)
Gleditsia (*Gleditschia*) for Johann Gottlieb Gleditsch (*Gleditsius*) (1714–86) of the Berlin Botanic Garden, locust trees
Gleichenia for F. W. Gleichen (1717–83), German director of Berlin Botanic Garden (***Gleicheniaceae***)
Gliricidia Dormouse-killer, (*glis, gliris*)-*cidium* (the poisonous seed and bark)
glischrus -a -um petty, slippery, glandular bristly, clammy, γλισχρος
globatus -a -um arranged or collected into a ball, *globus, globi*
Globba from an Amboina Island vernacular name, galoba
globiceps globe-headed, having crowded flower heads, *globus-ceps*
globiferus -a -um, globigerus -a -um carrying spheres, *globi-fero* (heads of flowers)
globiflorus -a -um globe-flowered, *globi-florum*
globispicus -a -um having short, globular spikes, *globi-spica*
globosus -a -um, globularis -is -e with small spherical parts, spherical, *globus, globi* (e.g. flowers)
Globularia Globe, *globulus* (the globose heads of flowers) (***Globulariaceae***)
globularifolius -a -um having leaves similar to those of *Globularia*
globularis -is -e globe shaped, *globulus, globuli*
globulifer -era -erum, globuligerus -a -um carrying small balls, *globuli-fero* (the sporocarps of pillwort)
globuligemmus -a -um having spherical buds, *globuli*-(*gemma, gemmae*)
Globulostylis Globular-styled-one, *globulus-stilus*
globulosus -a -um small round-headed, diminutive of *globosus*
globulus -a -um round or spherical, *globus, globi* (diminutive suffix)
glochi-, -glochin point-, -pointed, γλωχις
glochidiatus -a -um burred, with short barbed detachable bristles, botanical Latin *glochidiatus*

gloeo- sticky-, γλοια, γλοιος, γλοιο-
Gloeocapsa, gloeocapsus -a -um Sticky-box, γλοιο-καψα (the investing layers of mucilage)
Gloeophyllum Sticky-leaf, γλοιο-φυλλον
Gloeosporium Sticky-spored, γλοιο-σπορος (the spores of several parasitic imperfect fungi)
glomeratus -a -um collected into heads, aggregated, glomerate, *glomero, glomerare, glomeravi, glomeratum*
glomerulans clustering, accumulating, forming into balls, present participle of *glomero, glomerare, glomeravi, glomeratum*
glomerulatus -a -um with small clusters or heads, from the diminutive of *glomus*
glomeri- clustered-, crowded-, *glomero, glomerare, glomeravi, glomeratum*
glomeruliflorus -a -um with small flower heads, diminutive of *glomeri-florum*
gloriana glorious, a name for several *Saxifraga* hybrids, *glorior, gloriari, gloriatus*
Gloriosa, gloriosus -a -um Superb, full of glory, *gloriosus* (the flowers)
glossa-, glosso-, -glossus -a -um tongue-shaped-, -tongued, γλωσσα, γλωσσο-,
Glossocalyx Tongue-calyx, γλωσσο-καλυξ (the elongated calyx lobe)
Glossodia Tongue-like, γλωσσα-ειδω (the appendages on the labellum)
glossoides tongue-like, γλωσσα-οειδης
glossomystax having a moustached lip or labellum, γλωσσα-μυσταξ
Glossonema Tongue-thread, γλωσσο-νημα (the stamens)
Glossopetalon, Glossopetalum Tongue-petalled, γλωσσο-πεταλον (the narrow petals)
Glossorhyncha Tongue-snouted, γλωσσα-ρυγχος (the nectary-spur)
Glossostelma Tongue-crowned, γλωσσα-στελμα (the coronna)
Glossostemon Tongue-stamened, γλωσσα-στημων (the androecium)
Glossostigma Tongue-stigma, γλωσσα-στιγμα (the compressed stigma)
glotti-, glotto-, -glottis -little-tongued, γλωττα, γλωττη, γλωττο- (simile with the epiglottis, separating the oesophagus and trachea, below the tongue)
Glottidium Little-tongue, diminutive variant of γλωττα (for the mode of separation of the seeds)
Glottiphyllum Tongue-leaved, γλωττο-φυλλον (leaf-shape)
Gloxineria the composite name for hybrids between *Gloxinia* and *Gesneria*
Gloxinia for Peter Benjamin Gloxin of Comar, eighteenth-century physician and naturalist, author of *Observationes botanicae*, 1785 (florists' *Gloxinia* is *Sinningia*)
glumaceus -a -um with chaffy bracts, conspicuously glumed, with husks, *gluma*
glumaris -is -e husk or chaff-like, *gluma, glumae*
Glumicalyx Chaffy-calyx, botanical Latin from *gluma* and καλυξ
-glumis -is -e -glumed, *gluma*
Gluta Glue, *gluten, glutinis* (for their exudations of Lac)
glutiniferus -a -um having a sticky exudate, glue-bearing, *gluten-fero*
glutinosus -a -um sticky, viscous, glutinous, with glue, *gluten, glutinis*
Glyceria Sweet, γλυκυς, γλυκερος (the sweet grain of *Glyceria fluitans*)
glycicarpus -a -um having sweet fruit, γλυκυς-καρπος
Glycine Sweet, γλυκυς (the roots of some species) (*Glycine max* seeds are wild soya beans, sauces are prepared by fermentation from *G. soja*)
glycinoides resembling *Glycine, Glycine-oides*
glyciosmus -a -um sweet-smelling, γλυκο-οσμη
glyco-, glycy- sweet-tasting or -smelling, γλυκερος, γλυκυς, γλυκυ-, γλυκο-
Glycosmis, glycosmus -a -um Sweet-smelling, γλυκο-οδμη (fragrant flowers)
glycyphyllos sweet-leaved, γλυκο-φυλλον (the taste)
Glycyrrhiza (*Glycorrhiza*), *glycyrrhizus -a -um* Sweet-root, γλυκο-ριζα, γλυκυρριζα (the rhizomes are the source of the cognate, liquorice)
Glyphia Engraved, γλυφω (the elongate grooves on the fruit wall, γλυφις a notch)
glypho- marked, etched, engraved, γλυφω
Glyphochloa Engraved-grass, γλυφω-χλοη
glypto- cut-into-, carved-, γλυπτης, γλυπτο-
glyptocarya having deeply grooved kernels, γλυπτο-καρυον

glyptodontus -a -um having incised teeth, γλυπτο-οδοντος
Glyptopleura Carved-ribs, γλυπτο-πλευρα (on the walls of the fruits)
glyptospermus -a -um having sculptured seed coats, γλυπτο-σπερμα
glyptostroboides resembling-*Glyptostrobus, Glyptostrobus-oides*
Glyptostrobus Carved-cone, γλυπτο-στροβιλος (appearance of female cone scales)
Gmelina for Johann Georg Gmelin (1709–55), German naturalist and traveller in Siberia, professor of botany at Tübingen University
gmelinii for Johann Georg Gmelin (1709–55), and his nephew Samuel Gottlieb Gmelin (1743–74), German naturalist
Gnaphaliothamnus *Gnaphalium*-like shrub, γναφαλλον-θαμνος
Gnaphalium Soft-down, (from a Greek name, γναφαλλον, for a plant with felted leaves)
gnaphalobryum like a felt of moss, γναφαλλο-βρυον
gnaphalocarpus -a -um having fruits with a felty surface, γναφαλλο-καρπος
gnaphaloides felted like *Gnaphalium*, γναφαλλον-οειδης
gnemon from the Malayan vernacular name, genemo, for *Gnetum gnemon*
gnetaceus -a -um resembling *Gnetum*
Gnetum from the Malayan vernacular name, genemo, for *Gnetum gnemon* (***Gnetaceae***)
Gnidia, gnidium the Greek name for *Daphne*, from Gnidus, Crete
gnomus -a -um sundial-like, γνωμων (the long, conical pseudobulbs)
gobicus -a -um from the Gobi desert area, Mongolia
Godetia for Charles H. Godet (1797–1879), Swiss botanist
Godoya for E. Godoy (1764–1839), Spanish statesman and patron of Botany.
Goethea for Johann Wolfgang von Goethe (1749–1832), German philosopher, poet, anatomist and botanist
Goldbachia for Carl Ludwig Goldbach (1793–1824), writer on Russian medicinal plants
Gomesa for Bernardino Antonio Gomez (1769–1823), Portuguese naval surgeon who wrote on medicinal plants of Brazil
gomezoides like *Gomesa, Gomesa-oides*
Gomphandra Club-stamened-one, γομφος-ανηρ
Gomphichis Club-like, γομφος (the shape of the column)
gompho- peg-, nail-, bolt- or club-shaped, γομφος, γομφο- (with an enlarged distal part)
Gomphocarpus Club-fruited-one, γομφος-καρπος (the shape of the fruit)
gomphocephalus -a -um club-headed, γομφο-κεφαλη
Gomphogyne Clubbed-stigma, γομφο-γυνη
Gompholobium Club-podded-one, γομφος-λοβος (the shape of the inflated fruit)
Gomphostemma Club-headed-wreath, γομφο-στεμμα (the head of flowers)
Gomphostigma Clubbed-stigma, γομφο-στιγμα
Gomphrena the ancient Latin name used by Pliny, from γομφραενα
Gonatanthus Angled-flower, γονατ-ανθος (the bent spathe)
Gonatopus Jointed-stalk, γονατο-πους (the jointed petiole)
-gonatus -a -um -angled, -kneed, γονυ, γονατος, γονατο-, γονατ-
Gonatostylis Jointed-style, γονατο-στυλος
gongolanus -a -um from the environs of the Gongola river basin, NE Nigeria
Gongora for Don Antonio Caballero y Góngora (1740–1818), Viceroy of New Granada, Bishop of Cordoba, and patron of Jose Celestino Mutis
Gongronema Swollen-thread, γογγρος-νημα (the filaments)
Gongrostylus Swollen-style, γογγρος-στυλος
Gongylocarpus Knobbly-fruit, γογγυλη-καρπος
gongylodes roundish, knob-like, swollen, turnip-shaped, γογγυλη-ωδης
Gongylolepis Domed or rounded-scaled-one, γογγυλη-λεπις
Gongylosciadium Domed-shade, γογγυλη-σκια (the umbrella-shaped umbel)
Gongylosperma Knobbly-seed, γογγυλη-σπερμα
goni-, gono- offspring-, bud-, spore-, productive-, γονη, γονος, γονιμος

goniatus -a -um hook-shaped or angled, γωνια
gonio-, gono- angled-, prominently angled-, γωνια, γωνιωδης, γωνιο-, γωνο-
Gonioanthela Angled-head-of-small-flowers, γωνιο-ανθημιον
Goniocalyx, goniocalyx Angled-calyx, γωνιο-καλυξ
goniocarpus -a -um with angular fruits, γωνιο-καρπος
Goniocaulon Angled-stem, γωνιο-καυλος
Goniocheilus Angular-lipped, γωνιο-χειλος
Goniolimon Angular-fruited-citrus, botanical Latin from γωνια and *limon*
Gonioma Angular-swellings, γωνιο-ομα
Goniophlebium Angle-veined, γωνιο-φλεβος
Goniopteris Angled-wing, γωνιο-πτερυξ (frond morphology)
Goniothalamus Angular-fruiting-body, γωνιο-θαλαμος
Gonocalyx Angular-calyx, γωνο-καλυξ
Gonocarpus Angled-fruit, γωνο-καρπος (raspwort)
Gonocaryum Ridged-fruit, γωνο-καρυον
Gonolobus Angled-fruit, γωνο-λοβος (the fruits of some)
gonophorus -a -um having a gonophore (bearing the stamens and ovary), γωνο-φορεω
Gonospermum, gonospermus -a -um Angular-seed, with angular seeds, γωνο-σπερμα
-gonus -a -um angled, with blunt longitudinal ridges, γωνια (number- or feature-)
gony- knee, joint, γονυ, γονατος
Gonystylus Kneed-style, γονυ-στυλος
Goodenia for Dr Samuel Goodenough (1743–1827), Bishop of Carlisle, founder member of the Linnaean Society, Vice-President of the Royal Society, monographer of *Carex* (***Goodeniaceae***)
Goodia for Peter Good (d. 1803), plant collector in E Indies and with Robert Brown for Kew in Australia
Goodmania for George Jones Goodman (1904–99), American botanist
Goodyera for John Goodyer (1592–1664), English botanist who translated Dioscorides' *Materia medica* into English
Gordonia for James Gordon (*c.* 1708–80), English nurseryman of Mile End Nursery, London
gorganicus -a -um from Gurgan or Gorgan, N central Iran
gorgoneus -a -um fierce, terrible, γοργος; gorgon-like, resembling one of the snake-haired Gorgons of mythology (*Gorgo, Gorgonis*)
gorgonicus -a -um from Cape Verde islands (*Gorgades*)
Gorgonidium Somewhat-terrible, diminutive of γοργος
gorgonis -is -e wild, of wild appearance, γοργος
Gorteria for David de Gorter (1717–83), Dutch physician, botanist and collector
gortynius -a -um from Áyioi Dhéka, SW Crete (*Gortyn* or *Gortyna*)
goseloides resembling the S African genus *Gosela*
gossipiphorus -a -um cotton-bearing, *gossypium-fora* (*Saussurea gossipiphora* looks like a ball of cotton-wool)
Gossweilera for John Gossweiler (1873–1952), botanist and collector in Cabinda and Belgian Congo
Gossweilerodendron Gossweiler's tree, botanical Latin from Gossweiler and δενδρον
gossypi-, gossypinus -a -um cotton-plant-like, resembling *Gossypium*
Gossypioides, gossypiodes Similar-to-*Gossypium*, *Gossypium-oides*
gossypiphorus -a -um bearing cotton, botanical Latin from *gossypium* and φορα
Gossypium Soft (botanical Latin from an Arabic name, goz, for a soft substance) (cotton)
gothicus -a -um from Gothland, Sweden (Jordanes, sixth century, claimed that the (Visogoths) Goth tribe originated in southern Scandinavia)
gottingensis -is -e from the university city of Göttingen, Lower Saxony, Germany
Gouania for Antoine Gouan (1733–1821), French professor of botany at Montpellier, author of *Flora Monspeliaca*
Goupia from the Guyanese vernacular name for cupiuba (*Goupia glabra*)

gourianus -a -um from Gour, Bengal
Govenia, govenianus -a -um for James Robert Goven, *Rhododendron* specialist of Highclere, Secretary to the RHS 1845–50
Goyanzianthus Flower-of-Goias-state, Brazil, botanical Latin from Goias and ανθος
Grabowskia for Heinrich Emanuel Grabowski (1792–1842), German apothecary, botanist and collector
gracilentus -a -um slender and recurved, *gracilis-(lento, lentare)*
gracilescens slenderish, somewhat slender, *gracilis-essentia*
gracili-, gracilis -is -e slender, graceful, *gracilis, gracili-*
gracilifolius -a -um having slender leaves, *gracili-folium*
gracilior more slender, more graceful, comparative of *gracilis*
gracilistylus -a -um having a slender style, *gracili-stilus*
gracillimus -a -um very slender, most graceful, superlative of *gracilis*
Graderia an anagram of *Gerardia*, for John Gerard (1545–1612), see *Gerardia*
graebnerianus -a -um for Dr K. O. R. P. P. Graebner (1871–1933), of Berlin Botanic Garden
graecizans becoming widespread, *graecisso, graecissare* (aping the Greeks)
graecus -a -um Grecian, Greek, *Graecus*
grahamii for Robert Graham (1786–1845), Regius keeper at Edinburgh Botanic Garden
grallatorius -a -um stilt-rooted, having stilts, *grallator, grallatoris*
gramineus -a -um greensward, grassy, grass-like, *gramen, graminis, gramini-*
gramini-, graminis -is -e grass-like, *gramen, graminis*, of grasses (e.g. symbionts, parasites and saprophytes)
graminifolius -a -um having grass-like leaves, *gramini-folium*
graminoides grass like, *gramen-oides*
gramma-, -grammus -a -um outline-, lined-, γραμμη, -figured, -lettered, γραμμα, γραμματος
Grammangis Lined-receptacle, γραμμα-αγγιον (the perianth markings)
Grammatophyllum, grammatophyllus -a -um Figured-leaf, γραμματο-φυλλον (coloured markings or stripes of the perianth lobes, or leaves)
grammatus -a -um marked as with letters or raised lines or stripes, γραμματα, γραμματο-
grammicus -a -um from the Grámmos mountains of the Pindus range, on the Greek–Albanian border
Grammitis (Grammites) Short-line, diminutive of γραμμα (sori appear to join up like lines of writing at maturity) (***Grammitidaceae***)
grammopodius -a -um having linear marks on the stipe, γραμμα-ποδιον
Grammosciadium Lined-shade, γραμμη-σκιαδειον (the disposition of the umbels)
Grammosolen Figured-pipe, γραμμα-σωλην (the corolla tube)
Grammosperma Figured-seed, γραμμα-σπερμα
granadensis -is -e, granatensis -is -e either from Granada in Spain, or from Colombia, S America, formerly New Granada
Granadilla, granadilla Spanish diminutive of Granada, many seeds (pomegranate), for the fruit of *Passiflora edulis*
granatinus -a -um pale-scarlet, the colour of pomegranate, *Punica granatum*, flowers
granatus -a -um many-seeded (*Punica granatum*, apple with many seeds)
grandi- large-, great-, strong-, showy-, *grandis, grandi-*
grandiceps having a showy head (of flowers), large-headed, *grandis-caput*
grandidens with large teeth, *grandis-(dens, dentis)*
grandidentatus -a -um with large teeth, *grandis-(dens, dentis)*
grandiflora-alba with large white flowers, *grandis-florum-albus*
grandiflorus -a -um with large flowers, *grandis-florum*
grandifolius -a -um with large leaves, *grandis-folium*
grandilimosus -a -um of large muddy areas, swamps or marshes, *grandi-limosus*
grandis -is -e large, powerful, full-grown, showy, big, *grandis*

grandissimus -a -um most spectacular or imposing, superlative of *grandis*
Grangea an Adansonian name, possibly commemorative for Grange
Grangeopsis resembling *Grangea*, *Grangea-opsis*
Grangeria for N. Granger, an eighteenth-century traveller in Indian Ocean area
graniticolus -a -um of soils on granitic rock, *graniticus-colo*
graniticus -a -um of granitic rocks, grained, *graniticus* (modern Latin)
grantii for James Augustus Grant (1827–92), Scottish explorer of the Nile
granularis -is -e as if composed of granules, knots or tubercles, *granularis*
granulatus -a -um, *granulosus -a -um* as though covered with granules, tubercled, granulate, *granum*
granuliferus -a -um bearing granules of epidermal excretions, *granum-fero*
granum-paradisi grains of paradise (but more probably for the aromatic seeds of another species, *Afromonum melegueta*)
graph-, *-graphys* marked with lines, as though written on, γραφω, γραφις
Graphistylis Marked-style, γραφις-στυλος
Graphorkis Written-on-orchis, γραφις-ορχις (the marking on *G. scripta*)
grapto- lined-, marked-, γραπτος, γραπτο-
Graptopetalum Written-on-petalled-one, γραπτο-πεταλον (the bands across the petals of some)
Graptophyllum Written-on-leaf, γραπτο-φυλλον (the lines marking the leaves)
grat- pleasing, graceful, *gratus*
gratianopolitanus -a -um from Grenoble, France (*Gratianopolis*)
Gratiola Agreeableness, diminutive from *gratia* (medicinal effect)
gratiosus -a -um obliging; favoured, popular, *gratiosus*
gratissimus -a -um most pleasing or agreeable, superlative of *gratus*
gratus -a -um pleasing, agreeable, *gratus*
graveolens strong-smelling, *gravis-oleo*, rank-smelling, heavily scented, *graveolens, graveolentis*
gravesii for Robert Graves, surgeon and *Iris* grower
gravi- heavy-, strong-, offensive-, *gravis*
gravidus -a -um pregnant; laden, full, loaded, *gravidus*
Grayia, grayanus -a -um (Graya) for Asa Gray (1810–88), American Professor of Systematic Botany at Harvard, USA, author of *Manual of the Botany of the Northern United States* (1848)
Greenea, greenei for Dr David Greene (1793–1862), American botanist of Boston, USA
Greenovia for George Bellas Greenough (1778–1855), English geologist, founder of the Geological Society, London
Greenwayodendron Greenway's-tree, for P. J. Greenway (1897–1980), systematist with the E African Agricultural Research Station
gregarius -a -um growing together, common, *gregarius*
Greigia, greigii for General Samuel Alexjewitsch Greig (1827–87), President of the Russian Horticultural Society
grenadensis -is -e from the Caribbean Isle of Spice, Grenada, Lesser Antilles
Grevillea, grevilleanus -a -um for Charles F. Greville FRS, Earl of Warwick (1749–1809), Lord of the Admiralty, founder member of the RHS in 1804, Vice-President of the Royal Society
Grewia for Nehemiah Grew (1641–1712), British physician, plant anatomist and pioneer microscopist, author of *The Anatomy of Plants* (1682)
Greyia, greyi for Sir George Grey (1812–98), collector and patron of botany, Governor of S Australia, New Zealand and Cape Colony
Grias Edible, γραω, *grias, griadis* (cognate with graze) (fruit of anchovy pear)
Grielum Aged(-looking), γρηυς, γραυς (because of the grey indumentum)
Griffinia for William Griffin (d. 1827), English nurseryman, collector and patron
Griffithella, griffithianus -a -um, griffithii either for William Griffiths (1810–45), English botanist, Superintendent of Calcutta Botanic Garden, or for J. E. Griffith (1843–1933) of Bangor, Wales (*Potamogeton griffithii*)

grignonensis -is -e from Grignon, France
Grindelia for David Hieronymus Grindel (1766–1836), Latvian physician and botanist of Dorpat, Riga, professor at Tartu (Californian gum)
Grisebachia, grisebachianus -a -um, grisebachii for Heinrich Rudolf August Grisebach (1814–79), Professor of Botany at Göttingen, traveller in Balkans and S America
Griselinia for Francesco Griselini (1717–83), Italian botanist
griseoargenteus -a -um silvery-grey-coloured. *griseo-angentus*
griseofulvus -a -um silvery-reddish-yellow, *griseo-fulvus* (griseofulvin is used to treat fungal infections of the hair)
griseolilacinus -a -um silvery-lilac-coloured, *griseo-lilacinus*
griseolus -a -um somewhat greyish, diminutive of *griseus*
griseopallidus -a -um pearly-coloured, *griseo-pallidus*
griseus -a -um, (grizeus) bluish- or pearl-grey, *griseus, griseo-* (relates with grizzled)
Grobya for Lord Grey of Groby, English orchidologist and horticultural patron
Groenlandia for Johannes Groenland (1824–91), German Parisian gardener (*Potamogeton*-like aquatic)
groenlandicus -a -um from Grønland, Greenland (*Groenlandia*)
Gronovia for Johan Frederik Grovovius (1690–1760), botanist of Leiden, Holland, author of *Flora Virginica* (1739–43)
grosse-, grossi- very large, thick, coarse, from French gros, grosse, late Latin *grossus*
grossefibrosus -a -um being coarsely fibrous, *grossus-fibrosus*
grosseserratus -a -um having large teeth, *grossus-serratus* (leaf margin)
Grossularia, grossularia modern Latin from the French name, groseille, for the gooseberry (***Grossulariaceae***)
grossularifolius -a -um *Grossularia*-leaved, *Grossularia-folium*
grossularioides, grossuloides gooseberry-like, *Grossularia-oides*
grossus -a -um large, late Latin *grossus*
Grubbia for Mikael Grubb (1728–1808), Swedish mineralogist and botanist, collector in the Cape area of S Africa ***(Grubbiaceae)***
gruinus -a -um crane-like, *grus, gruis*
grumosus -a -um appearing as clusters of grains, tubercled, granular, *grumus*, a little hill
gryllus -a -um of the cricket, *gryllus* (haunt and scabrid texture)
grypoceras griffin-horned, γρυπος-κερας (γρυψ, γρυπος, *gryps, grypis* a griffin)
grypos griffin-like, *grypis* (the inflorescence of distant star-shaped spikes)
gryposepalus -a -um having hooked sepal-apices, *grypis-sepalus*
guacayanus -a -um from the area of the Guacaya river (named for the many gaucas, Colombian Indian graves, plundered for gold artefacts)
guadalupensis -is -e from Guadeloupe Island off lower California, USA, or from any of at least 10 other Guadeloupe states, counties, towns, rivers, mountains or islands
Guaiacum from the S American name, guayac, for the wood of life (*lignum vitae*) tree and its resin
guaianiensis -is -e from the Guainia department of E Colombia
guajava S American Spanish name, guayaba, for the guava, *Psidium guajava*
guamensis -is -e from Guam, Marianas island, S Pacific Ocean
Guamatela anagram of Guatemala
Guarea a S American vernacular name, guara, for one species
Guatteria for Giovanni Battista Guatteri, eighteenth-century Italian botanist
guaricanus -a -um, guaraniticus -a -um of the Guarani, S American Indians, of the Venezuelan Guarico area
guatamalensis -is -e from Guatemala, Central America
guavirobus -a -um from a Tupi-Guarani vernacular for the fruit of *Campomanesia guaviroba*
Guayania of the Guyana Highlands, Guyana region of the Guianas, northern S America

Guazuma the Mexican vernacular name for bastard cedar (*Guazuma ulmifolia*)
Gueldenstaedtia, gueldenstaedtianus -a -um for A. J. von Güldenstädt (1741–85), botanist in the Caucasus
Guettarda for Jean Etienne Guettard (1715–86), French natural historian, physician to Louis Duke of Orleans, involved with Lavoisier in the mapping of the geology of France
guianensis -is -e from Guiana, northern S America
Guibourtia for Nicholas Jean Baptiste Gaston Guibourt (1790–1861), French pharmacologist, author of *Histoire abregee des drogues simples* (1849–51)
guicciardii for Jacops Guicciard, who collected plants in Greece
Guichenotia for Antoine Guichenot, a French gardener and traveller on the Nicholas Baudin expedition to New Holland, NW Australia (1800–03)
Guilfoylia for William Robert Guilfoyle (1840–1912), Australian botanist, director of Melbourne Royal Botanic Garden
Guillenia for C. Guillen, seventeenth-century Mexican Jesuit missionary
guineensis -is -e from W Africa (Guinea Coast)
guizhouensis -is -e from Kweichow province, SW China (Guizhou)
Guizotia for François Pierre Guillaume Guizot (1787–1874), French historian and deposed from Premiership in 1849
gulestanicus -a -um from Guleston or Gulestan, E Uzbekistan
Gulubia New Guinea vernacular name for the gulubi palm
gummifer -era -erum producing gum, *gummi-fero*
gummi-gutta having drops of sticky exudate, *gummi-gutta*
gummosus a -um gummy, with a sticky exudate, comparative of *gummi*
Gundlachia for Johannes Christoph Gundlach (1810–96), German naturalist in Cuba
Gunnarella for Gunnar Seidenfadden, Danish botanist and orchidologist (*Siedenfaddenia*) (the feminine diminutive suffix emphasizes that it is a genus of small epiphytic orchids)
Gunnera for Johann Ernst Gunnerus of Trondheim (1718–73), Norwegian botanist, Bishop of Trondheim, author of *Flora Norvegiaca* (***Gunneraceae***)
Gunnia, gunnii for Ronald Campbell Gunn (1808–81), S African editor of the *Tasmanian Journal of Natural Sciences*, collector for William Jackson Hooker
Gurania anagram of *Anguira*
gussonii for Giovanni Gussone (1787–1866), Director of the Botanic Garden at Palermo
Gustavia for Linnaeus' patron, King Gustavus III of Sweden (1746–92)
Gutierrezia for P. Gutierrez, Spanish nobleman
gutta drop, *gutta, guttae* (*Dichopsis gutta* yields a latex, gutta percha, or chaoutchouc); some derive from Malay, getah
guttatus -a -um spotted, covered with small glandular dots, *gutta*
guttiferus -a -um gum- or resin-producing, *gutta-fero*
guttulatus -a -um slightly glandular, marked all over with small dots, diminutive of *gutta*
Guzmania for Anastasio Guzman (d. 1807), Spanish naturalist in S America
Gyminda anagram of *Myginda*
gymn-, gymno- exposed-, naked-, γυμνος, γυμνο-, γυμν-
Gymnacranthera Apical-exposed-stamen, γυμν-ακρα-ανθερος
Gymnadenia Naked-gland, γυμνος-αδην (exposed pollen viscidia)
gymnandrus -a -um with naked stamens, with exposed anthers, γυμν-ανδρος
Gymnanthera, gymnantherus -a -um Exposed stamens, γυμν-ανθηρος
Gymnanthes, gymnanthus -a -um Naked-flowered, γυμν-ανθος
Gymnarrhena Exposed androecium, γυμν-αρρην
Gymnema Naked-threads, γυμν-νημα (the exposed filaments)
Gymnemospis Looking-like-*Gymnema*, γυμν-νημα-οψις
Gymnocalycium Exposed-calyx, γυμνο-καλυξ (the protruding flower-buds)
Gymnocarpium Naked-fruit, γυμνος-καρπος (oak fern sori lack indusia)

gymnocarpus -a -um with naked carpels, with exposed ovary, γυμνο-καρπος
gymnocaulon with a clear stem, γυμνο-καυλος
Gymnocheilus naked-lip, γυμνο-χειλος (unmarked labellum)
Gymnocladus Bare-branch, γυμνο-κλαδος (foliage of Kentucky coffee-tree is mainly towards the ends of the branches)
Gymnocoronis Exposed-halo, γυμνο-κορωνις
Gymnodiscus Exposed-disc, γυμνο-δισκος (the receptacle)
gymnodontus -a -um with naked teeth, γυμνο-οδοντος
Gymnogramma (*Gymnogramme*) Naked-line, γυμνο-γραμμη (the sori lack a covering indusium)
Gymnogrammitis Exposed-lines, γυμνο-γραμμα (the naked sori)
Gymnomitrium Exposed-turban, γυμνο-μιτρα (the peristome)
Gymnopetalum Naked-petal, γυμνο-πεταλον
gymnophyllus -a -um with naked leaves, γυμνο-φυλλον
Gymnophyton Exposed-plant, γυμνο-φυτον (habitat)
Gymnopodium Naked-extremity, γυμνο-ποδεων
Gymnopogon Exposed-beard, γυμνο-πωγων (the excurrent leaf vein)
Gymnopteris Naked-fern, γυμνο-πτερυξ (the linear sori do not have an indusium)
gymnorrhizus -a -um having exposed roots, buttress-rooted, γυμνο-ριζα
Gymnoschoenus Exposed-*Schoenus*, γυμνο-σχοινος (the conspicuous inflorescences of button grass reed
Gymnosiphon Naked-tube, γυμνο-σιφον (colourless parasite)
gymnosorus -a -um having non-indusiate sori, naked sori, γυμνο-σορος
Gymnosperma, Gymnospermium Naked-seed, γυμνο-σπερμα
Gymnosporangium Naked-sporangium, γυμνο-σπορα-αγγειον (the emergent spathulate structures containing the telutospore stage)
Gymnostachys, Gymnostachium Naked-spike, Exposed-spike, γυμνο-σταχυς
Gymnostemon Exposed-stamens, γυμνο-στεμων
Gymnostephium Exposed-crown, γυμνο-στεφανος
Gymnostoma, Gymnostomium Naked-mouth, γυμνο-στομα (the mouth of the capsule of the beardless-moss lacks a fringe of teeth)
Gymnotheca Exposed-box, γυμνο-θηκη (ovaries of achlamydeous flowers)
gyn-, gyno-, -gynus -a -um relating to the ovary, female-, -pistillate, -carpelled, γυνη
Gynadropsis Female-and-male, γυνη-ανηρ-οψις (the disposition of the stamens and style)
Gynandriris Woman-man-*Iris*, γυνη-ανηρ-ιρις (the united stamens and pistil, ≡ *Iris*)
gynandrus -a -um having filaments and styles fused into a column, γυνη-ανηρ
Gynatrix Hairy-ovary, γυνη-(θριξ, τριχος)
Gynerium Woolly-ovary, γυν-εριον
Gynocardia Heart-of-the ovary, γυνη-καρδια (the seeds provide the one-time leprosy medicinal chaulmoogra oil, ≡ *Hydnocarpus*)
gynochlamydeus -a -um female cloak, γυνη-χλαμυς (enlarged calyx)
Gynochthodes Bank-like-ovary, γυν-οχθη-ωδης (the humped appearance)
Gynoxis Pointed-ovary, γυν-οξυς (the fruit shape)
Gynura Female-tail, γυν-ουρα (the elongated stigma)
gypo- vulture-, γυψ, γυπος a vulture (suggesting desolate habitats)
Gypothamnium Vulture-bush, (γυψ, γυπος)-θαμνος (provenance, the Atacama Desert, S Chile)
Gypsacanthus Vulture's-*Acanthus*, γυψ-ακανθος (provenance, the Mexican deserts)
gypsi- -chalk, γυψος, *gypsum, gypsi*
gypsicolus -a -um living on gypsum, living on calcium, *gypsum-colo*
Gypsophila Lover-of-chalk, γυψος-φιλος (the natural habitat)
gyrans revolving, moving in circles, γυρος, *gyrus*
Gyranthera Circled-stamens, γυρος-ανθερα (arrangement of the androecium)
Gyrinops, Gyrinopsis None-apparent-circle, γυρ-ιν-οψις (lacking or with a reduced corolla)
gyro-, -gyrus -a -um bent-, twisted-, -round, γυρος

Gyrocarpus Turning-fruit, γυρο-καρπος (the winged, dipterocarp-like fruit)
gyroflexus -a -um bent around, turned in a circle, γυρο-πλεξω
Gyromitra Curved-girdle, γυρο-μιτρα
Gyroporus Curved-pore, γυρο-πορος (opening of polypore tube)
Gyrostemon Turned-stamen, γυρο-στεμον (ring of divided stamens)
gyrosus -a -um bent backwards and forwards (cucurbit anthers)
Gyrotaenia Turned-filament, γυρο-ταινια (incurved tactic filaments)

haageanus -a -um for J. N. Haage (1826–78), seedsman of Erfurt, Germany
Haageocereus Haage's-*Cereus*, for Frederick Adolph Haage (1796–1866), botanist, collector and nurseryman of Erfurt
Haastia, haastii for Sir Johann Franz Julius von Haast (1824–87), government geologist in New Zealand, explored S Island
Habenaria Thong, *habena* a strap (the spur of some is long and flat)
Haberlea for Karl Konstantin Christian Haberle (1764–1832), Professor of Botany at Pest, now part of Budapest in Hungary
Hablitzia for Carl Ludwig van Hablitz (1752–1821), naturalist traveller in the Middle East
habr-, habro- soft-, graceful-, delicate-, luxuriant-, αβρος, αβρο-
Habracanthus Delicate-*Acanthus*, αβρο-ακανθος
Habranthus Elegant-flower, αβρο-ανθος
Habrochloa Graceful-grass, αβρο-χλοη
Habropetalum Delicate-petalled-one, αβρο-πεταλον
habrotrichus -a -um soft-haired, softly hairy, αβρο-τριχος
habyssinicus -a -um from Abyssinia (Ethiopia)
hachijoensis -is -e from Hachijo island, Izu archipelago, Japan
Hackelia, hackelii for either Joseph Hackel (1783–1869), an E European botanist, or P. Hackel, Professor of Agriculture at Leitmeritz, Bohemia (Czech Republic)
Hackelochloa Hackel's-grass, botanical Latin from Hackel and χλοη, for Eduard Hackel (1850–1926), Austrian grass taxonomist
Hacquetia for Balthasar Hacquet de la Molte (1740–1815), French-born naturalist, professor at Lemberg, author of *Plantae Alpinae Carniolica*
hadramawticus -a -um from Hadramaut (Hadramawt or Hadramout) Yemen
hadriaticus -a -um from the area around the Adriatic Sea (*Mare Hadriaticus*)
Haekeria for Gottfried Renatus Haeker (1789–1864), German pharmacist and botanist
haema-, haemo-, haemato- blood-red, the colour of blood, αιμα, αιμ-, αιματο-
haemaleus -a -um blood-red, αιμα
haemalus -a -um, haematodes blood-coloured, αιμα-ωδες
Haemanthus Blood-flower, αιμ-ανθος (for the red species called fireball lilies)
haemanthus -a -um with blood-red flowers, αιμ-ανθος
haemastomus -a -um with a blood-red mouth, αιμα-στομα
haematinus -a -um bluish-black coloured, αιμα, αιματο-
haematocalyx with a blood-red calyx, αιματο-καλυξ
Haematocarpus, haematocarpus -a -um Blood-red-berries, αιματο-καρπος
haematocephalus -a -um with blood-red heads, αιματο-κεφαλη (of flowers)
haematochilos with a blood-red lip or lips, αιματο-χειλος
haematochiton with a blood-red cloak, αιματο-χιτων (sheath of the inflorescence)
haematocodon having red bells, αιματο-κωδων (flowers)
Haematodendron Blood-tree, αιματο-δενδρον (the slash exudate)
haematodes blood-red, blood-like, αιματο-ωδες
haematopus -a -um appearing to have blood, with bleeding stalks, αιματο-πους (the red sappy exudate)
haematosiphon having a red (corolla) tube, αιματο-σιφον
haematospermus -a -um having blood-red seeds, αιματο-σπερμα
Haematostaphis Blood-grapes, αιματο-σταφυλη (fruit of blood-plum)

Haematostemon Blood-threads, αιματο-στεμον (inflorescence bracteoles)
Haematoxylon Blood-wood, αιματο-ζυλον (the heartwood, which is the source of the dye-stuff haematoxylin)
Haemodorum Blood-gift, αιματο-δωρον (Australian aboriginal use of blood-root-lily as food) (***Haemodoraceae***)
haemorrhoidalis -is -e blood-red (bleeding), αιμορροειδης (veins)
haemorrhoidarius -a -um having a blood(-like) flow, αιμορροεω (the cut flesh of *Agaricus haemorrhoidarius* turns red)
Hagenbachia for Karl Friedrich Hagenbach (1771–1849), Swiss professor of botany and anatomy at Basle
Hagenia for Karl Gittfrie Hagen (1749–1829), German pharmacist and botanist
hagenii for B. Hagen (1853–1919), physician and collector in the Pacific islands
hageniorus -a -um from Hagen Mountain of the central highlands of New Guinea
haichowensis -is -e from Lien-yün-kang, E China (salt area formerly Hai Chow)
Hainania, hainanensis -is -e from Hainan province of S China
Haitia, hatiensis -is -e from Haiti, W Indies
hajastanus -a -um from Armenia, the land of the Hauq people
Hakea for Baron Christian Ludwig von Hake (1745–1818), German horticulturalist and patron of botany
hakeifolius -a -um with leaves resembling those of *Hakea*
hakeoides resembling *Hakea, Hakea-oides*
hakkodensis -is -e from the Hakkodo mountains, N Honshu, Japan
Hakonechloa Hakone-grass, botanical Latin from Hakone and χλοη, from the hot spring area of Mount Hakone, Honshu, Japan
hakonecolus -a -um living in the Hakone hot-spring area, Honshu, Japan, botanical Latin from Hakone and *colo*
hakusanensis -is -e from Haku San, Honshu, Japan
hal-, halo- saline-, salt-, αλς
Halanthium Sea-flower, αλς-ανθος (strand habitat)
Halarchon Ruler-over-salt, αλς-αρχων (dominates saline habitats)
Halenia for Jonas Petrus Halenius, student of Linnaeus
halepensis -is -e, halepicus -a -um from Aleppo (Halab), northern Syria
Halesia for the Reverend Dr Stephen Hales (1677–1761), experimentalist and writer on plants (snowdrop trees)
halicacabus -a -um from an ancient Greek name, from *Halicarnassus*, Bodrum, Turkey
halicarnassius -a -um from Bodrum, SW Turkey (Caria or Halicarnassus, Gulf of Cerameicus)
halimi- orache-like, with silver-grey rounded leaves, of the sea, αλιμος
halimifolius -a -um *Atriplex-halimus*-leaved, *Halimium-folium*
Halimiocistus the name for hybrids between *Halimium* and *Cistus*
Halimione Daughter-of-the-sea, αλιμος-ωνη
Halimiphyllum *Halimium*-leaved, αλιμον-φυλλον
Halimium Dioscorides' name, αλιμον, for *Atriplex halimus* (≡ *Cistus*, with leaves resembling those of *Atriplex halimus*)
halimo- maritime-, saline-, αλς, αλος, αλιμος, αλιμο-
Halimocnemis Maritime-leggings, αλιμο-κνημις (habitat and internodes)
Halimodendron Maritime-tree, αλιμο-δενδρον (the habitat on saline soils)
halimus -a -um with silver-grey foliage, orache-like, αλιμον
halipedicolus -a -um living on saline plains, botanical Latin from αλιμος-πεδιον and *colo*
haliphloeos, haliphleos with salt-covered bark, αλιμος-φλοιος
Halleria, halleri for Albrecht von Haller (1708–77), Swiss polymath who wrote *Stirpes Helvetica*, professor at Göttingen and founder member of the Swedish Royal Academy of Science
Hallia for Birger Marten Hall (1741–1841), Swedish physician and botanist

hallianus -a -um for R. G. Hall, American physician in China
hallii for Elihu Hall (1822–82), American Rocky Mountains botanist, or Sir Daniel Hall (1864–1942), author of *The Genus Tulipa* (1940)
halo-, salt, saline, αλς, αλος, αλο- (the habitat), corn-, αλως
Halocarpus Corn-fruited, αλως-καρπος; some interpret as Sea-fruit
Halocharis Sea-beauty, αλο-χαρις
Halocnemum Sea-leggings, αλο-κνημις (jointed stems and habitat)
halodendron maritime tree, tree of saline soils, αλιμος-δενδρον
Halodule Salt-forest, αλος-υλη; or Enslaved-by-the-sea, αλο-δουλος (a sea grass or shoalweed)
Halogeton Sea-neighbour, αλο-γειτων
Halopegia Sea-fountain, αλο-πηγη (marantaceous continuity of emergent flowers); some interpret as Another-spring, αλλος-πηγη
Halopeplis Sea-spurge, αλο-πεπλις
Halophila, halophilus -a -um Salt-loving, αλο-φιλος (sea grass)
halophyticus -a -um salt-plant, halophyte, αλο-φυτον
Halophytum Salt-plant, αλο-φυτον
Halopyrum Sea-wheat, αλο-πυρος (dune grass)
Haloragis (Halorrhagis) Seaside-berry, αλο-(ρακ, ραγος) (the fruits of the coastal species) (***Haloragaceae***)
Halosicyos Sea-cucumber, αλο-σικυος (habitat and fruit)
Halothamnus Maritime-bush, αλο-θαμνος (habitat and habit)
Halotis Of-saline-habitats, αλς, αλος
Haloxylon Salt-wood, αλο-ζυλον (the habitat)
hama- together with-, αμα-
hamabo from Hama, N Syria
hamadae together, neighbours, αμα (the use of rattan to join things)
Hamadryas, hamadryas Wood-nymph (in mythology, dying at the same time as its associated tree, αμα-δρυς) (also used for a N African baboon of the stony deserts, hamada, and the king cobra)
Hamamelis Greek name, αμαμελις, for a tree with pear-shaped fruits, possibly a medlar, some interpret as At-the-same-time-tree-fruit, αμα-μηλον (***Hamamelidaceae***)
hamatocanthus -a -um with hooked spines, *hamatus-acanthus*
hamatopetalus -a -um having hook(-tipped) petals, botanical Latin from *hamatus* and πεταλον
hamatus -a -um, hamosus -a -um hooked at the tip, hooked, *hamatus*
hamiltonianus -a -um, hamiltonii for William Hamilton, American naturalist
Hammarbya Kuntze's name for Linnaeus, who had a house at Hammarby in Sweden
Hammatolobium Knotted-pod, (αμμα, αμματο)-λοβος (contorted legume)
hamosus -a -um hooked, *hamatus*
Hampea for Georg Ernst Ludwig Hampe (1795–1880), German botanist
hamulatus -a -um having a small hook, clawed, taloned, diminutive of *hamus*
hamulosus -a -um covered with little hooks, *hamus, hami*
Hanburia for Daniel Hanbury (1825–75), brother to Sir Thomas Hanbury (1832–1907), with whom he created La Mortola garden on the Italian Riviera, patrons of horticulture
Hancockia for William Hancock (1847–1914), Irish botanist and collector for Kew
Handelia for Heinrich Freiherr von Handel-Mazzetti (1882–1940), Austrian botanist, collector in China, keeper of Vienna Natural History Museum
hannibalensis -is -e from the environs of Hannibal, Missouri, USA
hapalacanthus -a -um with soft thorns, απαλο-ακανθα
Hapaline Delicate-one, απαλος (epiphytic aroid)
hapalus -a -um, hapal- soft, απαλος, απαλ-
hapl-, haplo- simple-, single-, plain-, απλοος, απλους, απλο-, απλ-
haplanthus -a -um having a single flower, απλ-ανθος

haplocalyx having the sepals united into a single structure, απλο-καλυξ
Haplocarpha Single-scaled-one, απλο-καρφη (the pappus)
Haplocoelum Single-chamber, απλο-κοιλος (the ovary)
Haploesthes Single-garment, απλο-εσθης (involucral bracts)
Haplolophium One-crested, απλο-λοφος (the seed wing)
Haplopappus Single-down, απλο-παππος (its one-whorled pappus)
Haplophragma Half-partitioned, απλο-φραγμα
Haplophyllum Simple-leaved, απλο-φυλλον (undivided leaves)
Haplophytum Simple-plant, απλο-φυτον
Haplosciadium Simple-umbel, απλο-σκιαδος
Haplostichanthus Single-rowed-flowers, απλο-στιχος-ανθος (inflorescence of island mint)
hapto- fastening-, attaching-, touching-, απτω, απτο-
Haptocarpum Attaching-fruit, απτο-καρπος (glandular)
Haraella Little-Hara, feminine diminutive for Yoshie Hara, Japanese botanist
Harbouria for J. Harbour, American naturalist and collector
Hardenbergia for Countess Franziska von Hardenberg von Huegel (1794–1870), sister of the traveller Count Hugel, and patron of science
Hardwickia for Thomas Hardwicke (1755–1835), General of Bengal artillery, botanist and collector for East India Company
Hariota for Thomas Harriot (1560–1621), English polymath, explorer, mentor of Sir Walter Raleigh
harlandii for Dr W. A. Harland (d. 1857), who collected in China
harmalus -a -um from an Arabic vernacular name, harmil, for its medicinal use
Harmsia, harmsii for Hermann August Theodore Harms (1870–1942), German professor at Prussian Academy of Sciences, editor of *Pflanzenreich*, Botanic Museum in Berlin
Harmsiodoxa Harms'-glory, for Hermann Harms (*supra*), botanical Latin from Harms and δοξα
harp-, harpago- snatching-, robbing-, αρπαξ, αρπαγος (covering of small hooks, burred)
Harpachne Snatching-chaff, (αρπαξ, αρπαγος)-αχνη (the glumes)
Harpalyce for the mythological amazon, daughter of Harpalycus of Thrace
Harpagophytum Grapple-plant, αρπαγος-φυτον, rapacious-plant, (the fruit is covered with barbed spines)
Harpanema Curved-threads, αρπα-νημα (staminal coronna)
harpe-, harpeodes sickle-, sickle-like, *harpe*, αρπη, αρπηωδης
Harpephyllum Sickle-leaved, αρπη-φυλλον
Harpochloa Sickle-grass, αρπη-χλοη
harpophyllus -a -um with sickle-shaped leaves, αρπη-φυλλον
Harpullia a Bengali vernacular name
Harrimanella feminine diminutive for Edward Henry Harriman (1848–1909), American founder of the Union Pacific Railroad, philanthropist
Harrisella, Harrisia for William Harris (1860–1920), Irish botanist, superintendent of Public Gardens and Plantations, Jamaica
harrisianus -a -um for T. Harris of Kingsbury, importer of Mexican plants *c.* 1840
Harrisonia, harrisonianus -a -um for Arnold Harrison and for his wife, of Aigburth, Liverpool
harrovianus -a -um for George Harrow (1858–1940), of Veitch's Coombe Wood nursery, Chelsea
harryanus -a -um for Sir Harry James Veitch (1840–1924), head of Veitch & Sons, nurserymen of Chelsea
Hartogia for J. Hartog (*c.* 1663–1722), early Dutch gardener in S Africa and Ceylon
Hartwegia, hartwegii for Carl Theodor Hartweg (1812–71), who collected for the RHS in Central America
hartwigii for Augustus Karl Julius Hartwig (1823–1913), German writer on horticulture

Harungana from the vernacular name, aronga, of the monotypic genus in Madagascar
Harveya for William Henry Harvey (1811–66), Irish systematic botanist, phycologist and artist, Colonial Secretary in S African Cape, curator at Trinity College Dublin, author on phycology etc
harveyanus -a -um for either J. C. Harvey, who collected in Mexico *c.* 1904, or W. H. Harvey (see above)
hascombensis -is -e from the garden at Hascombe Park, Godalming, of C. T. Musgrave, lawyer and RHS treasurer
haspan from a Ceylonese vernacular name
Hasseltia, Hasseltiopsis for Johan Conraad van Hasselt (1797–1823), Dutch surgeon and botanist
hastati-, hastatus -a -um formed like an arrow-head, spear-shaped, *hasta, hastae* (see Fig. 6a), hastate
hastatulus -a -um with small arrow-like thorns, diminutive of *hasta*
hastifer bearing a spear, *hasta-fero*
hastifolius -a -um having spear-shaped leaves, *hasta-folius*
hastiformis -is -e spear-shaped, *hasta-forma*
hastilabius -a -um having a hastate labellum or lip, *hasta-labellum*
hastilis -is -e like a long pole, shaft, javelin or vine prop, *hastile, hastilis*
hastulatus -a -um like a short spear, diminutive of *hasta, hastae*
Hatiora an anagram of *Hariota*
Hausknechtia for Heinrich Karl Haussknecht (1838–1906), German pharmacist, botanist and collector in the Middle East
havanensis -is -e from Havana, Cuba
hawkeri for Lieutenant Hawker, who collected on the South Sea Islands *c.* 1886
Haworthia, haworthii for Adrian Hardy Haworth (1765–1833), English naturalist, an authority on succulent plants, author of *Synopsis plantarum succulentarum* (1812)
hayachinensis -is -e from the environs of Hyachine-san, highest mountain of the Kitakami-Sammyaku range, NE Honshu, Japan
Hayatella feminine diminutive for Bunzo Hayata (1874–1934), Japanese botanist
Haylockia for Matthew Haylock, gardener to Dean Herbert at Spofforth
Haynaldia, haynaldii for Stephen Franz Ludwig Haynald (1816–91), botanist and Cardinal Bishop of Koloesa
Hazardia for Barclay Hazard (1852–1938), American botanist
Hebe for Hebe, Greek goddess of youth, daughter of Jupiter and wife of Hercules
hebe- pubescent-, sluggish-, soft-, ηβη
hebecarpus -a -um pubescent-fruited, ηβη-καρπος
hebecaulis -is -e slothful-stemmed, *hebes-caulis* (prostrate stems of *Rubus hebecaulis*)
Hebeclinium Downy-couch, ηβη-κλινη (the common receptacle)
hebegynus -a -um with a blunt or soft-styled ovary, with part of the ovary glandular-hairy (*Aconitum hebegynum*)
Hebeloma Pubescent-border, ηβη-λωμα
Hebenstreitia for Johan Ernst Hebenstreit (1703–57), Professor of Botany at Leipzig
hebepetalus -a -um with pubescent petals, with dull petals, ηβη-πεταλον
hebephyllus -a -um with pubescent leaves, with leaves resembling those of a *Hebe*
Heberdenia for William Heberden the Elder (1710–1801), 'learned physician' (Samuel Johnson)
Hebestigma Downy-stigma, ηβη-στιγμα
hebetatus -a -um dull, blunt, soft pointed, *hebes, hebetis*
hebraicus -a -um (hebriacus) Hebrew, *Hebrus, Hebri*
Hecastocleis Each-enclosed, εκαστος-κλειω (involucrate florets)
hecat-, hecato- one hundred, εκατοντασ, εκατομ-, εκατον-, εκατο-
hecatensis -is -e from the Hecate Strait, British Columbia, between Queen Charlotte Island and mainland Canada
Hecatonia Hundreds, εκατον (a Sectional name in *Ranunculus* for species with numerous carpels)

hecatophyllus -a -um many (one hundred)-leaved, εκατον-φυλλον
hecatopterus -a -um many (one hundred)-winged, εκατον-πτερον
Hecatostemon Hundred-stamens, (εκατοντας, εκατο-)-στημων
Hechtia for Julius Konrad Gottfried Hecht (1771–1837), Prussian counsellor and knight
hecisto- least-, smallest-, ηκιστος, ηκιστο-; ηκιστα
Hecistopteris Smallest-fern, ηκιστο-πτερυς (the minute moss fern)
Hectorella, hectori, hectorii, hectoris -is -e for Sir James Hector (1843–1907), Director of New Zealand Geological Survey (Hector, son of Priam of Troy, was killed by Achilles)
Hedeoma Sweet-fragrance, ηδυς-οδμη (a name, ηδυοσμον, in Pliny for a wild mint)
Hedera the Latin name, *hedera*, for ivy
hederaceus -a -um, hederi- ivy-like, resembling *Hedera* (usually in the leaf-shape)
hederae of ivy, *Hedera* (host to *Orobanche)*
hederi- ivy-like, *hedera* (mostly leaf shape but also for climbing)
hederifolius -a -um having leaves similar to those of *Hedera, Hedera-folius*
Hederorkis Climbing-orchid, *hedera-orchis*
hedraeanthus -a -um (hedraianthus) sedentary-flowered, with sessile flowers, εδραι-ανθος
hedwigii for Professor John Hedwig (1730–99), Leipzig bryologist
hedy-, hedys- sweet-, of pleasant taste or smell, ηδυς, ηδυνος, ηδυ-
hedyanthus -a -um having sweetly fragrant flowers. ηδυ-ανθος
hedycarpus -a -um with sweet fruits, ηδυ-καρπος
Hedycarya Sweet-nut, ηδυ-καρυον
Hedychium Sweet-snow, ηδυ-χιον (some have fragrant white flowers)
Hedyosmum, hedyosmus -a -um Sweet-smelling-one, ηδυ-οσμη (ηδυοσμον mint)
Hedyotis Sweetness, ηδυ-(ους, ωτος)
Hedypnois Sweet-breath, ηδυ-(πνοη, πνοιη) (πνεω, πνευμα)
hedysaroides resembling *Hedysarum, Hedysarum-oides*
Hedysarum Dioscorides' name, ηδυσαρον (meaning uncertain, ηδυς sweet)
Hedyscepe Sweet-cover, ηδυ-σκεπη (umbrella palms)
Hedythyrsus Sweet-thyrse, ηδυ-θυρσος
heeri for Oswald Heer (1809–83), Director of Zurich Botanic Garden
Heimia for Genheimerat (Privy Counsellor) Dr Ernst Ludwig Heim (1774–1834), German physician
Heinsia for *Heinsius*, philologist and translator of Theophrastus
Heisteria for Lorenz Heister (1683–1758), German anatomist
hekisto- least-, slowest-, slackest-, ηκιστος, ηκιστο-; ηκιστα
Hekistocarpa Smallest-fruit, ηκιστος-καρπος
Heldreichia, heldreichianus -a -um, heldreichii for Theodor Heinrich Hermann von Heldreich (1822–1902), Director of the Athens Botanic Garden
helena from Helenendorf, Transcaucasia
helenae for either Helen, the daughter of Sir W. Macgregor of Australia, or Helen, wife of Dr E. H. Wilson
helenioides resembling *Helenium, Helenium-oides*
Helenium, helenium for Helen of Troy (a name, ηλενιον, used by the Greeks for another plant and reapplied by Linneaus) (*Inula helenium*, elecampane, is supposed to have been taken to Pharos by Helen of Troy)
heleo- marsh-, ελος, ελεο-
Heleocharis, heleocharis Marsh-favour, ελεο-χαρις (*Eleocharis*)
heleogenus -a -um whose home is the marsh, ελεο-γενεα
Heleogiton Marsh-neighbour, ελος-γειτων
heleonastes confined to marshes, ελεο-ναστος (νασσειν)
heleophilus -a -um marsh-loving, ελεο-φιλεω
heli-, helio- sun-loving-, sun-, ηλιος, ηλιο, ηλι-
Heliamphora Sun-pitcher, ηλι-(αμφιφορευς, αμφορευς) (pitcher plant)
Helianthella feminine diminutive from *Helianthus*

helianthemoides resembling *Helianthemum*, *Helianthemum-oides*
Helianthemum Sun-flower, ηλι-ανθεμιον (rock rose)
helianthi resembling *Helianthus*
helanthoides resembling *Helianthus*, *Helianthus-oides*
Helianthus Sun-flower, ηλι-ανθος (the large golden heads of many species tend to follow the sun, *girare-sole*, cognate with Jerusalem [artichoke])
helianthus-aquatica humid or aquatic sunflower, *Helianthus-aquaticus*
helicanthus -a -um twisted-flower, ελικ-ανθος
helichrysoides resembling *Helichrysum*
Helichrysopsis Resembling *Helichrysum*, ηλι-χρυσους-οψις
Helichrysum Golden-sun, ηλι-χρυσους
helici- coiled like a snail-shell, wreathed, twisted, ελιξ, ελικος, ελικο-, ελικ-
Helicia Spiralled, ελικος (perianth lobes)
helicocephalus -a -um having a twisted (flowering) head, ελικο-κεφαλη
Helicodiceros Two-twisted-horns, ελικο-δις-κερας (the basal lobes of the leaves)
helicoides, helicoideus -a -um of a coiled or twisted appearance, ελιξ-οειδης
Heliconia for Mount Helicon, Boetea, Greece, sacred to the Muses of mythology (***Heliconiaceae***)
Helicostylis Coiled-style, ελικο-στυλος
Helicteres Ear-drop, ελικτηρ (the screw-shaped ear-ring-like carpels)
Helicteropsis resembling-*Helicteres*. ελικτηρ-οψις
helicto- twisted-, wreathed-, rolled-, ελικτος, ελικτο-
Helictonema Twisted-threads, ελικτο-νημα (filaments)
Helictotrichon (um) Twisted-hair, ελικτο-τριχος (the geniculate awns)
Helinus Tendrilled, ελιξ (climbing by spiral tendrils)
helio- sun-like, sun-, ηλιος, ηλιο-
Heliocarpus Sun-fruit, ηλιο-καρπος (the fringed fruits)
heliolepis -is -e sun-scaled, with golden scales, ηλιο-λεπις
Heliophila Sun-lover, ηλιο-φιλος (exposed habitats)
heliophyllus -a -um having leaves adapted to full sun, ηλιο-φυλλον
Heliopsis Sun-like, ηλιο-οψις (for the yellow flower-heads)
helioscopius -a -um sun-observing, sun-watching, Dioscorides' name, ηλιοσκοπιον (the flowers track the sun's course)
heliospermus -a -um having flat and round seeds, ηλιο-σπερμα
Heliotropium Turn-with-the-sun, ηλιο-(τροπη, τροπεω, τρεπειν) (turnsole)
Helipterum Sun-wing, ηλι-πτερον (the fruit's plumed pappus)
helix winding, ελιξ, ελικος ancient Greek name for twining plants
Helixanthera Twisted-anther, ελιξ-ανθερα
helleborifolius -a -um with *Helleborus*-like leaves, *Helleborus-folium*
helleborine hellebore-like, adjectival suffix *Helleborus-ina*
Helleborus Poison-food, ηλλειμ-(βορα, βοσις) (the ancient Greek name for the medicinal *H. orientalis*)
hellenicus -a -um from Greece, Grecian, Greek, Hellenic (*Hellas, Helladis*)
Helleriella for Alfonse Heller, botanist in Nicaragua (diminutive suffix for Heller's dotted orchid)
helmandicus -a -um from the environs of the Helmand river (*Erymandrus*), Afghanistan
Helmholtzia for Herman Ludwig Ferdinand von Helmholtz (1821–94), German physicist and physiologist
helminth-, helmintho- worm-, ελμινθος, ελμινθο-
Helminthia (Helmintia) Worm, ελμινς, ελμινθος (≡ *Picris*, for the elongate wrinkled fruits)
helminthorrhizus -a -um having worm-like roots, ελμινθο-ριζα
Helminthostachys Worm-like-spike, ελμινθο-σταχυς (the fertile spike of the 'flowering' fern)
Helminthotheca Worm-like-case, ελμινθο-θηκη (the beaked fruits)
helo-, helodes of bogs and marshes, ελωδης, ελοδες

Helobiae Marsh-life, ελος-βιοω
Helodea *vide Elodea*
helodes marshy, growing in marshes, ελωδης
helodoxus -a -um marsh-beauty, glory of the marsh, ελο-δοξα
Helonias Swamp-pasture, ελος-νομη
Heloniopsis resembling *Helonias*, ελος-οψις
helophorus -a -um thicket-forming, ελος-φορεω
Helosciadium Marsh-umbel, ελο-σκιαδειον
Helosis Marsh, ελος
helveticus -a -um from Switzerland, Swiss (*Helvetia*)
helvolus -a -um pale yellowish-brown, *helvus*
helvus -a -um dimly yellow, honey-coloured, dun-coloured, *helvus*
Helwingia for George Andreas Helwing (1666–1748), Prussian cleric, botanist and botanical writer
Helxine a name, ελξινη, used by Dioscorides formerly for pellitory
Hemarthria Half-joined, ημι-αρθρον (the pedicels are fused to the rachis)
Hemerocallis Day-beauty, ημερο-καλος, the Greek name, ημεροκαλλες (reflects that the flowers are short-lived)
hemi- half-, ημι, *hemi-* (used in the sense of looking-like, or half-way-to)
Hemiandra Halved-male, ημι-ανδρος (the dimidiate stamens)
Hemiboea Half-*Boea*, *hemi-boea*, (a related genus)
Hemibotrya Half-cluster, ημι-βοτρυς (staminate flowers have aborted ovules – and pistillate flowers have sterile anthers) (section of *Saxifraga*)
Hemichaena Half-agape, ημι-χαινα (the mouth of the corolla)
Hemichroa Half-coloured, ημι-χροα
hemicryptus -a -um half- or partially concealed, ημι-κρυπτος
hemidartus -a -um patchily covered with hair, half-flayed, ημι-δαρτος
Hemidesmus Half-banded, ημι-δεσμα
Hemidictyum Half-netted, ημι-δικτυον
Hemigenia Half-home, ημι-γενεα (the androecium lacks two stamens)
Hemigraphis Half-writing, ημι-γραφις (the filaments of the outer stamens bear 'brushes')
Hemimeris Half-portioned, ημι-μερις (the two apparent halves of each flower)
hemionitideus -a -um barren, like a mule, ημι-ονος, *hemicillus*
Hemionitis, *hemionitis* Mule, ημι-ονος, half an ass (barren-fern, non-flowering)
Hemiorchis Half-orchis(-like), ημι-ορχις (distinctive zingiberaceous flower structures)
hemiphloius -a -um partially covered in bark, ημι-φλοιος (stripping bark)
Hemiphragma Half-separated, ημι-φραγμα (the ovarian septum)
Hemipilia Half-felt-covered, ημι-πιλος (the partially covered pollinia)
hemipoa half-*Poa*, ημι-ποα
Hemipogon Half-bearded, ημι-πωγων (the anthers)
Hemiptelea Half-elm, ημι-πτελεα (the fruit has a crest-like wing only in the upper half)
Hemisphace name, ημι-σφακος, of a section of *Salvia*, perhaps *sedi incertis*
hemisphaericus -a -um hemispherical, ημι-σφαιρα (fruit or flower-head)
Hemistylus Half-styled, ημι-στυλος (the short style)
Hemitelia Half-complete, ημι-τελειος (indusium scale-like at lower side of the sorus and caducous)
hemitomus -a -um cut into two, ημι-τομη (deeply lobed along the length)
hemitrichotus -a -um half hairy, ημι-(θριξ, τριχος)
Hemizonia Half-embraced, ημι-ζωνη (the outer achenes)
Hemizygia Half-yoked, ημι-ζυγος
Hemsleya, *hemsleyanus -a -um*, *hemsleyi* for William Boting Hemsley (1843–1924), of the Royal Botanic Gardens, Kew
henchmanii for Francis Henchman (fl. 1824), nurseryman of Clapton, Middlesex
hendersonii for either A. Henderson of the RHS gardens at Chiswick (d. 1879), or for Louis Fourniquet Henderson (b. 1853), who collected in Oregon, USA

henryanus -a -um, henryi for Augustine Henry (1857–1930), Irish botanist who collected in China and co-authored *Trees of Great Britain and Ireland* with Elwes
Hepatica Of-the-liver, ηπαρ, ηπατος (medieval signature of leaf or thallus shape of *herba hepatica* as of use for liver complaints)
hepaticifolius -a -um with *Hepatica*-like leaves, *Hepatica-folium*
hepaticus -a -um dark purplish red, liver-coloured, ηπατος
Heppiella for Johann Philipp Hepp (1797–1867), German physician and botanist (with feminine diminutive suffix)
Heppimenes the composite name for hybrids between *Heppiella* and *Achimenes*
hepta-, hepto- seven-, επτα
Heptacodium Seven-headed-one, επτα-κωδεια (the inflorescence)
heptamerus -a -um having the parts arranged in sevens, επτα-μερος (μερις)
heptandrus -a -um having seven stamens. επτα-ανηρ
Heptanthus Seven-flowered, επτα-ανθος
heptapeta a misnomer by Buc'hoz, who described *Lassonia heptapeta* from a picture that showed only five erect tepals
heptaphyllus -a -um having seven leaves or leaflets, επτα-φυλλον
Heptapleurum Seven-ribs, επτα-πλευρα (≡ *Schefflera*)
Heptaptera Seven-winged , επτα-πτερον (the fruits)
heptemerus -a -um with parts arranged in sevens, επτα-μερος
heracleifolius -a -um with *Heracleum*-like leaves, *Heracleum-folium*
heracleoides resembling *Heracleum, Heracleum-oides*
heracleoticus -a -um from Iráklion (*Heracleum*), Crete
Heracleum Hercules' (Heracles')-all-healer (a name, πανακες ηρακλειον, used by Theophrastus)
heracleus -a -um resembling *Heracleum*
herae for the Greek goddess Hera (*Juno*)
herba weed, grass, young plant, herb, *herba, herbae*
herba-alba white-herb, *herba-albus* (*Artemisia*)
herba-barona fool's-herb, *herba-*(*baro, baronis*) (of the dunce or common man)
herbaceus -a -um not woody, low-growing, herbaceous, *herba, herbae*
herba-sardoa Sardinian poison, *herba-*(*Sardinia, Sardiniae*) (the island was famed for producing poisons)
herba-venti wind-herb, *herba-*(*ventus, venti*) (of the steppes)
Herbertia, herbertii for Reverend Dr William Herbert (1778–1847), botanist and Dean of Manchester, specialized in bulbous plants
herbiolus -a -um vegetable-herb, *herba-holus*
herbstii for Messrs Herbst & Rossiter of Rio de Janeiro, Brazil, *c.* 1859
hercegovinus -a -um from the Balkans around Mostar, Bosnia–Herzegovina
herco- fenced, a barrier, ερκιον, ερκος, ερκο-
hercoglossus -a -um with a coiled tongue, ερκο-γλωσσα
hercynicus -a -um from the Hertz mountains (*Hercynia silva*), central Germany
Heritiera, heritieri for Charles Louis l'Héritier de Brutelle (1746–1800), French botanist and patron of the flower illustrator Pierra Joseph Redouté
hermaeus -a -um from Mount Hermes, Greece
Hermannia, hermanniae for Paul Hermann (1646–95), Professor of Botany at Leiden, traveller in the Cape and Ceylon, and physician to East India Company in Djakarta
hermaphroditicus -a -um, hermaphriditus -a -um hermaphrodite, with flowers containing both stamens and ovary, ηρμαφροδιτος
Hermbstaedtia for Sigismund Friedrich Hermbstaedt (1760–1833), Prussian court physician and botanist
Herminium Buttress, ερμιν, ερμις (the pillar-like tubers, some liken them to bedposts)
Hermodactylus Hermes'-fingers, Ερμης-δακτυλος, the Greek name for several *Colchicums* with palmate (hand-like) tubers
hermoneus -a -um from Mount Hermon, Syria

Hernandia for Francisco Hernandez (1514–78), physician to Philip II of Spain and botanist, explorer and collector in S America
Herniaria Rupture-wort (*hernia*), Dodoens' name (1583) for its former medicinal use; some derive from Greek ερνος for a bud or shoot
herpe-, herp- creeping, ερπω, ερπειν, ερπυζω; reptile, worm, ερπετον
Herpetacanthus Creeping-*Acanthus*, ερπετον-ακανθος
herpeticus -a -um ringworm-like, (from ερπω I creep, ερπειν creeping); ερπετον a creeping animal
Herpolirion Creeping-lily, ερπω-λειριον (the rhizomes)
Herpysma Creeping, ερπυζω (the rhizomes)
Herrania for General Pedro Alcantara Herran (1800–45), President of the State of New Granada (Colombia) (the state existed from 1830 to 1858)
Herreanthus for Hans Herre (1895–1979), German botanist, author of *The Genera of Mesembryanthemaceae* (1971)
hersii for Joseph Hers, who collected for the Arnold Arboretum in China *c.* 1920
hesper- evening-; western-, εσπεριος, *hesperius*
Hesperaloe Western-aloe, εσπερ-αλοη (*Aloe*-like with Texan–Mexican provenance, false *Yucca*)
Hesperantha Evening-flower, εσπερ-ανθος (time of anthesis)
Hesperevax Western-*Evax*, *hesperius-Evax* (American genus resembling former genus *Evax* (≡ *Filago*)
hesperides of the far west (for the classical world being the Iberian peninsula), εσπεριος, *hesperides*
hesperidius -a -um of the evening, *hesperus* (cognate with vespers)
hesperidus -a -um from the far west, of the nymph guardians of the golden apple tree; *Citrus*-fruited (Iberia and N Africa), εσπεριδος, *hesperides*
Hesperis Evening, εσπερος, *hesperus* (Theophrastus' name, also the name for Venus or Hesperis the evening star, of the west, becoming Lucifer, the morning star, of the east)
hespero-, hesperius -a -um western-, evening-, εσπερα, εσπερος, εσπεριος (Hesperia was the land of the west)
Hesperocallis Western-beauty, εσπερο-καλος (American desert lily)
Hesperochiron Western-*Chironia*, *hespero-Chironia*
Hesperocnide Western-nettle, εσπερο-κνιδη
Hesperodoria Gift-of-the-west, εσπερο-δωρον
Hesperolinon Western-flax, εσπερο-λινον
Hesperomecon Western-poppy, εσπερο-μηκων
Hesperopeuce Western-fir, εσπερο-πευκη
Hesperoxiphion Western-sword (Western-*Iris*), εσπερο-ξιφος
hessei for Paul Hesse, botanist and traveller
Hetaeria Companionship or Brotherhood, εταιρεια, εταιρια
heter-, hetero- varying-, differing-, diversely-, other-, ετερος, ετερο-
heteracanthus -a -um having variously shaped spines, ετερο-ακανθος
Heterachne Differing-chaff, ετερο-αχνη (the differing fertile and sterile lemmas)
heteradenus -a -um with varying glands, ετερο-αδην
heterandrus -a -um having variable stamens, ετερο-ανηρ
Heteranthemis Different-oxeye, ετερο-ανθεμις
Heteranthera, heteranthus -a -um Differing-anthers, ετερο-ανθερα (mud plantain has one only in cleistogamous flowers but three in normal flowers)
Heteranthoecia Varying-spikelets, ετερο-ανθ-οικος (the unequal glumes and lemmas)
heterocarpus -a -um having variably shaped fruits, ετερο-καρπος (the two halves of the binary fruits of *Turgenia heterocarpa*)
Heterocentron Variable-spurred-one, ετερο-κεντρον (the two processes on the larger anthers)
heterochlamydius -a -um with different cloaks, with both calyx and corolla, ετερο-χλαμυς

heterochromus -a -um of varying colour, ετερο-χρωμα
heterochrous -a -um having a variable surface covering, ετερο-χρως
heterocladus -a -um with diverse branching, ετερο-κλαδος
heteroclinus -a -um different-beds, having staminate and pistillate organs on separate receptacles, ετερο-κλινη
heteroclitus -a -um different-declensions; anomalous in formation, ετεροκλιτος
Heterocodon Differing-bells, ετερο-κωδων (enantiostyly of the dimorphic corollas)
heterocyclus -a -um with varying circles, ετερο-κυκλος (the zigzagging scars around the stems of tortoise shell bamboo)
Heterodelphia Much-related, ετερο-αδελφος
heterodontus -a -um with varying teeth, ετερο-(οδους, οδοντος)
heterodoxus -a -um of changing glory, ετερο-δοξα
Heterodraba Differing-*Draba*, ετερο-δραβη
heterogamus -a -um having variation in flower sexuality in the same inflorescence; having abnormal arrangement of the stamens and ovary; heterogamic, ετερο-γαμος
Heterogonium Varying-angles, ετερο-γωνια
Heterolepis, heterolepis -is -e Differing scales, ετερο-λεπις (of strobili or spikelets)
heterolobus -a -um with variable lobing, ετερο-λοβος
heteromallus -a -um with variable woolly indumentum, ετερο-μαλλος
Heteromeles Different-*Meles*, ετερο-μηλον
heteromerus -a -um having variable numbers of floral parts; having different numbers of parts (petals and sepals); heteromerous, ετερο-μερος
Heteromma Different-appearances, ετερο-ομμα
Heteromorpha Differing-forms, ετερο-μορφη
heteronemus -a -um diverse-stemmed, ετερο-νημα
Heteropanax Different-*Panax*, ετερο-παναξ
Heteropappus Variable-pappus, ετερο-παππος (the two forms of pappus)
heteropetalus -a -um having unequal or variable petals, ετερο-πεταλον
heterophyllus -a -um diversely-leaved, ετερο-φυλλον
heteropodus -a -um with variable stalks, ετερο-(πους, ποδος)
Heteropogon Varying-beard, ετερο-πωγων (the twisting awns of the female spikelets)
Heteroporus Variable-pored-one, ετερο-πορος (the pores of the tubules)
Heteropterys Variable-winged, ετερο-πτερον
Heteropyxis Different-box, ετερο-πυξις (the capsules)
Heterorachis Variable-rachis, ετερο-ραχις
Heterosmilax Different-*Smilax*, ετερο-σμιλαξ
Heterospathe Variable-spathe, ετερο-σπαθη (the length varies)
Heterosperma Variable-seed, ετερο-σπερμα
Heterostemma Variable-crown, ετερο-στεμμα (the coronna)
Heterostemon Differing-stamens, ετερο-στημων (different lengths)
Heterothalamus Varying-receptacles, ετερο-θαλαμος
Heterotheca Variable-cases, ετερο-θηκη (the honeycombed receptacle)
Heterotoma Variously-cut, ετερο-τομη (the corolla lobes)
Heterotrichum Varied-hairiness, ετερο-τριχος (the leaf indumentum)
Heterozostera Different-*Zostera*, ετερο-ζωστηρ (Tasmanian eel grass)
Heuchera for Johann Heinrich von Heucher (1677–1747), German professor of medicine at Wittenberg (coral flowers)
Heucherella the composite name for hybrids between *Heuchera* and *Tiarella*
heucherifolius -a -um with leaves resembling those of *Heuchera*
heucheriformis -is -e looking like *Heuchera*
heuffelianus -a -um, heuffelii for Johann Heuffel (1800–57), Hungarian physician and botanist
Hevea from the Brazilian vernacular name, heve, for the Para-rubber tree (called rubber from its use to make pencil erasers)
Hewittia for Mr Hewitt, editor of the *Madras Journal of Science* (1837)

hex-, hexa-, hexae- six-, εξ-, εξα-
hexacanthus -a -um having spines in groups of six, εξ-ακανθος
Hexachlamys Six covers, εξα-χλαμυς (the ovary)
hexaedropus -a -um having a stalk with six flattened sides, εξα-εδρος-πους
hexafarreri name of a hybrid *Gentiana farreri*
Hexaglottis Six-tongued, εξα-γλωττα (the stigma)
hexagonopterus -a -um six-angled-wing (frond), εξα-γωνια- πτερον
hexagonus -a -um six-angled, εξα-γωνια
Hexalectris Six-cock-combed, εξ-(αλεκρυων, αλεκτρονος; αλεκτωρ, αλεκτορος) (the six-crested labellum)
Hexalobus Six-lobed, εξα-λοβος (the six equal petals)
hexamerus -a -um six-partite, εξα-μερος
hexandrus -a -um six-stamened, εξ-ανηρ
hexapetalus -a -um with six petals, εζα-πεταλον
hexaphyllus -a -um six leaved, εξα-φυλλον
Hexatheca Six-thecae, εξα-θηκη
Hexisia Look-alike, εξις-ιδεα (morphology of lip and sepals)
Heyderia, heyderi for Herr Heyder (1808–84), Berlin cactus specialist
Heynea, heyneanus -a -um for Dr Benjamin Heyne (1770–1819), German botanist, missionary and collector near Travancore, Kerala, SW India
hians gaping, present participle of *hio, hiare* (*hiatus* or gaping corolla)
Hibbertia for George Hibbert (Hibbard) (1715–1838), English merchant with the East India Company, patron of botany, Member of Parliament (1806)
hibernalis -is -e of winter, *hiberna, hibernorum* (flowering or leafing)
hibernicus -a -um from Ireland, Irish (*Hibernia*)
hiberniflorus -a -um winter-flowering, *hiberna-floris*
hibernus -a -um of winter, flowering or green in winter, *hibernus*
Hibiscus an old Greek name, ιβισκος, from Virgil, *hibiscum* for marsh mallow, *Althaea officinalis*
Hicoria from the Algonquin vernacular, pawcohiccora, for Hickory (≡ *Carya*)
hidakanus -a -um from Hidaka region, Hokkaido, Japan
hidalgensis -is -e from Hidalgo state, Mexico, or Hidalgo county New Mexico
Hidalgoa for Miguel Gregorio Ignacio Hidalgo, nineteenth-century Mexican naturalist, or from the eponymous Hidalgo, Mexico (climbing *Dahlia*) (hidalgo is Spanish for 'a man of substance')
hiemalis -is -e of winter, *hiems, hiemis* (persisting or flowering)
hieraci- Hieracium-, hawkweed-like-
hieracifolius -a -um, hieraciifolius -a -um with leaves similar to those of *Hieracium*, *Hieracium-folium*
hieracii of hawkweeds, living on *Hieracium* (*Aulacidea*, hymenopteran gall wasp)
hieracioides resembling *Hieracium, Hieracium-oides*
Hieracium Hawkweed (Dioscorides' name, ιερακιον, which Pliny claimed was used by hawks, ιεραξ, ιερακος, to give them acute sight)
Hierochloe Holy-grass, ιερο-χλοη (religious use and association)
hierochunticus -a -um from the classical name, *Hierochuntia*, for Jericho (*Anastatica hierochuntica* is the rose of Jericho)
hieroglyphicus -a -um with sacred-carvings; marked as if with signs, ιερο-γλυφη
hierosolymitanus -a -um from Jerusalem (the Roman name, *Hierosolyma*)
hifacensis -is -e from Haifa, Israel (*Hefa*)
highdownensis -is -e connected with Sir Frederick Stern's garden at Highdown, Worthing, from Highdown nursery, Goring by Sea, W Sussex
higoensis -is -e from Higo province, Kyushu, Japan
Hilaria for Auguste François de Saint-Hilaire (1779–1853), French botanist and explorer in S America
hilaris -is -e cheerful, merry, *hilaris*
Hildebrandtia, hildebrandtii for Johann Maria Hildebrandt (1847–81), collector on the E African coast and Madagascar *c.* 1872–81

Hildegardia for Saint Hildegard (1098–1179), German Abbess Hildegard von Bingen, polymath and writer
Hillebrandia, hillebrandii for Wilhelm Hillebrand (1821–86), author of a Flora of Hawaii
Hillia for Sir John Hill (1716–75), English apothecary, naturalist and botanical writer, editor of *The British Magazine* (1746–50)
hillieri for Sir Harold Hillier (1905–85), of Jermyn House, or the Hillier arboretum and nursery
himalaicus -a -um, himalayae from the Himalayan mountains
Himalayacalamus botanical Latin from Himalaya and *Calamus*
himalayanus -a -um, himalayensis -is -e from the Himalayan mountains, Sanskrit, hima-alaya, land of snow
himalensis -is -e from the Himalayan mountains
himanto- leather thong, strap-, ιμας, ιμαντος
himantodes strap-like, ιμαντο-ωδης
Hinantoglossum Strap-tongue, ιμαντος-γλωσσα, (lizard orchid's narrow labellum)
Himatanthus Cloak-flower, ιματιον-ανθος (the floral bracts)
Hintonella for George Hinton (1882–1943), metallurgist and botanist in Mexico (feminine diminutive suffix)
Hindsia, hindsii for Richard Brinsley Hinds, surgeon naturalist on the HMS *Sulphur* expedition (1836–42) under Sir Edward Belcher
hindustanicus -a -um from N India, the land of the Hindus, Hindustan
hinnuleus -a -um tawny-cinnamon-coloured (like a fawn)
hinnuleus, hinnulei Hippeasprekelia the composite name for hybrids between *Hippeastrum* and *Sprekelia*
Hippeastrum Knight-star, (ιππειος, ιππικος)-αστερος (the equitant leaves suggest being astride a horse)
Hippeophyllum Equitant-leaved-one, ιππειος-φυλλον
Hippia etymology uncertain; some suggest for Hippias of Elis, Greek contemporary of Socrates
hippo- horse-, ιππειος, ιππικος, ιππο- (usually infers coarseness or inferiority)
Hippobromus Horse-stench, ιππο-βρωμα (the odour of bruised horsewood)
Hippocastanum Horse-chestnut, ιππο-καστανον (Matthiolus attributed the name to the Turk's use of the fruits to treat breathing problems in horses, see *Aesculus*) (***Hippocastanaceae***)
hippocastanus -a -um resembling horse chestnut, chestnut-brown, ιππο-καστανον
Hippocratea for Hippocrates (460–377 BC), Greek physician who first divorced medicine from myth and suspicion and is regarded as the father of medicine
Hippocrepis Horse-shoe, ιππο-κρηπις (the shape of the fruit)
Hippolytia, hippolytii for Hippolytus (in mythology, Hippolytus was son of Theseus and the Amazon Hippolyte)
Hippomane, hippomanes Horse-madness, ιππο-μανια, Theophrastus' name, ιππο-μανης, for a spurge causing horses to become frenzied
hippomanicus -a -um eagerly eaten by horses, ιππο-μανιας
hippomarathrum horse-fennel, ιππο-μαραθον, Dioscorides' name for an Arcadian plant which caused madness in horses
Hippophae Horse-killer, ιππο-φενω (name, ιπποφαες, used by Theophrastus for a spiny plant); some interpret as ιππο-φαους, horse deliverance?
hippophaeoides resembling *Hippophae*
hippophaifolius -a -um with *Hippophae*-like leaves
Hippuris Horse's-tail, ιππο-ουρα (for the tail of a horse, ιππουρις, also for *Equisetum*) (***Hippuridaceae***)
Hiptage Flying-one, ιπταμαι (the three-winged samaras)
Hiraea for Jean Nicolas de la Hire, French physician and botanist
hircanicus -a -um from the environs of Gorgan, SE of the Caspian, northern Iran (*Hyrcania* or Varkana)
hircinicornis -is -e goat-horned, *hircus-cornu*

hircinus -a -um of goats, smelling of a male goat, *hircus, hirci*
hirculus from a plant name, diminutive of *hircus*, in Pliny (a small goat)
Hirschfeldia for Christian Cayus Lorenz Hirschfeld, eighteenth-century Austrian botanist and horticulturalist of Holstein
hirsutellus -a -um somewhat hairy, with very short hairs, diminutive of *hirsutus*
hirsutiformis -is -e with a hairy appearance, *hirsuti-forma*
hirsutipetalus -a -um with hairy petals, *hirsuti-petalus*
hirsutissimus -a -um very hairy, hairiest, superlative of *hirsutus*
hirsutulus -a -um, hirtellus -a -um, hirtulus -a -um somewhat hairy (diminutive suffix)
hirsutus -a -um rough-haired, covered in long hairs, *hirsutus*
Hirtella Hairy, feminine diminutive of *hirtus*
hirtellus -a -um somewhat hairy, diminutive of *hirtus*
hirti- having shaggy hair, hairy-, *hirtus*
hirticaulis -is -e hairy-stemmed, *hirtus-caulis*
hirtifolius -a -um having a hairy leaf, *hirtus-folium*
hirtipes hairy-stemmed, *hirtus-pes*
hirtovaginatus -a -um hairy-sheathed, *hirtus-vagina*
hirtulus -a -um weakly hairy, diminutive of *hirtus*
hirtus -a -um hairy, shaggy-hairy, *hirtus*
hirundinaceus -a -um, hirundinarius -a -um pertaining to swallows, *hirundo, hirundinis*
hirundinis -is -e, hirundo swallow-like, *hirundo* (curved)
hispalensis -is -e from Seville, southern Spain (*Hispalis*)
hispanicus -a -um from Spain, Spanish, Hispanic (*Hispania*)
hispaniolicus -a -um from Haiti and the Dominican Republic, Hispaniola (Española), Greater Antilles, W Indies
hisperides *vide hesperides*, of the far west (to the Romans, Spain)
hispi-, hispid- bristly-, *hispidus*
Hispidella Bristly-one, feminine diminutive of *hispidus*
hispidissimus -a -um the most hispid (of related species), superlative of *hispidus*
hispidovillosus -a -um having long, rough hairs, *hispidus-villosus*
hispidulus -a -um slightly bristly, diminutive of *hispidus*
hispidus -a -um bristly, with stiff hairs, *hispidus*
histio- sail-, ιστιον, ιστιο- (some derive as mast or web, ιστος)
Histiopteris Sail-fern, ιστιο-πτερυξ (the frond of bat's-wing fern)
histrio- of varied colouring, theatrical, *histrio, histrionis*, actor
histrioides resembling (*Iris*) *histrio, Iris histrio-oides*
histrionicus -a -um showing off, of actors, of the stage, *histrionia, histrioniae*
histrix showy, theatrical, *histricus* of the stage, *histrio* an actor (see also *hystrix*)
Hitchcockella for Albert Spear Hitchcock (1865–1935), American botanist of Missouri botanic garden and USDA, Washington
hittiticus -a -um from the Turkish peninsula; of the Hittites, from the land of the Hittites, Anatolian (Land of Hatti)
Hodgkinsonia for Clement Hodgkinson, English naturalist of the Crown Lands Survey, Australia (1861–74)
Hodgsonia, hodgsonii for Brian Houghton Hodgson (1800–94), English orientalist, East India Company resident in Nepal
Hoehnea, Hoehneela, Hoehnophytum for Frederico Carlos Hoehne (1882–1959), Brazilian botanist, professor at Rio de Janeiro
Hoffmannia, hoffmannianus -a -um for Georg Franz Hoffmann (1761–1826), Professor of Botany at Göttingen
hoffmannii for Herr Hoffmann, Austrian discoverer of *Symphyandra hoffmanni* in Bosnia *c.* 1880
Hoffmannseggia for Count Johann Centurius Graf von Hoffmannsegg (1766–1849), German naturalist, founder of the Berlin Zoological Museum
Hofmeisterella for William Friedrich Benedict Hofmeister (1824–77), German botanist and reproductive cytologist, professor at Heidelberg

Hohenackeria for Rudolf Friedrich Hohenacker (1798–1874), Swiss physician and plant collector
Hohenbergia for the Hohenberg branch of the Hohenzollern dynasty of Imperial Germany, originating in Baden-Württemberg
Hoheria from a Maori name, houhere or hoihere, for lacebark
Holarrhena Complete-male, ολος-αρρην (the anthers)
Holboellia for Frederik Ludvig Holboell (1765–1829), Danish botanist, superintendent of Copenhagen Botanic Garden
Holcoglossum Strap-tongued-one, ολκος-γλωσσα (the labellum)
Holcolemma Strap-shaped-lemma, ολκος-λεμμα
Holcus Millet, ολκος (the name in Pliny for a grain)
holdtii for Fredrich von Holdt, who raised × *Robinia holdtii* in Colorado
holitoris -is -e of the market gardener, *holitor, holitoris*
hollandicus -a -um from either NE New Guinea (*hollandia*) or Holland
Hollisteria for William Welles Hollister (1818–86), California rancher
Holmskioldia for Theodore Holmskjold (1732–1794), Danish physician and botanist, professor at Sorø Academy, Denmark
holo- completely-, entirely-, entire-, whole-, ολος, ολο-
Holocalyx Entire-calyx, ολο-καλυξ
Holocarpha Entire-stalk, ολο-καρφη (the solid receptacle)
holocarpus -a -um entire-fruited, ολο-καρπος
Holocheilus Entire-lipped-one, ολο-χειλος
holocheilus -a -um, holochilus -a -um having an entire lip, ολο-χειλος
holochrysus -a -um completely gilded, all golden, ολο-χρυσος
Holodiscus Entire-disc, ολο-δισκος (refers to the entire floral disc)
holodontus -a -um covered all over with teeth, spiny throughout, ολος-(οδους, οδοντος)
holophyllus -a -um with entire leaves; fully leaved, ολο-φυλλον
Holoptelea Entire-elm, ολο-πτελεα (the fruit is winged all round)
holopus -a -um fully stalked, ολο-πους
Holoschoenus a name, ολοσχοινος, used by Theophrastus
holosericeus -a -um, holosericus -a -um silky-haired throughout, completely wrapped in silk, ολο-σηρικος
holosteoides resembling *Holosteum, Holosteum-oides*
Holosteum, holosteus -a -um Whole-bone, ολος-οστεον (Dioscorides' name, ολοστεον, for a chickweed-like plant)
Holostylis Whole-style, ολο-στυλος
Holothrix Covered-with-hair, ολο-θριξ (pubescent)
holotrichus -a -um with hairy all over, ολο-τριχος (both surfaces of the leaves)
holstii for C. H. E. W. Holst (1865–94), German traveller in E Africa
homal-, homalo- smooth-, flat-, equal-, ομαλης, ομαλος, ομαλο-, ομαλ-
Homalanthus Like-a-flower, ομαλ-ανθος (the inflorescence due to the colouration of older leaves)
Homalium Equal, ομαλος (the petals are equal in number to the sepals – see *Dissomeria*)
Homalocalyx Whole-calyx, ομαλο-καλυξ (falls entire as a lid)
Homalocenchrus Smooth-*Cenchrus*, ομαλο-κεντρον (lacks the involucre of prickly sterile flowers found in *Cenchrus*, ≡ *Leersia*)
Homalocephala Flat-head, ομαλο-κεφαλη (the apices of the plants)
Homalocheilos Smooth-lipped, ομαλο-χειλος
Homalocladium Flat-branched, ομαλο-κλαδος (≡ *Muelenbeckia platyclados*)
Homalopetalum Even-petals, ομαλο-πεταλον (the uniform perianth segments)
homalophyllus -a -um with smooth leaves, ομαλο-φυλλον
Homalosciadium Flat-umbel, ομαλο-σκιας
Homalostigma Flat-stigma. ομαλο-στιγμα
Homalotheca Smooth-cased, ομαλο-θηκη
Homeria for the Greek epic poet Homer (*c.* 850 BC); some derive as Meeting-together, ομηρεω, for the united filament bases around the style

homo- one and the same-, ομος, ομο-; not varying-, agreeing with-, uniformly-, together, ομως ομω-
homocarpus -a -um having uniform fruits, ομω-καρπος
Homogyne Not-differing-female, ομω-γυνη (the styles of neuter and female florets are not different)
homoio-, homolo- similar-, almost identical, ομοιος, ομοιο-
Homoioceltis Resembling-*Celtis*, ομοιο-κηλτις
homoiolepis -is -e having uniform scales, ομοιο-λεπις
homoiophyllus -a -um with uniform leaves, ομοιο-φυλλον
Homolepis, homolepis -is -e Uniformly-scaled, ομο-λεπις (the glumes); uniformly covered with scales
Homonoia Uniform-meaning, ομο-(νοος, νους) (the united stamens)
homonymus -a -um of the same name, ομοιος-ονομα
Homopholis Uniformly-scaled, ομο-φολιδος
homophyllus -a -um having regular or uniform leaves, ομο-φυλλον
honanensis -is -e from Honan (Henan) province, central N China
hondensis -is -e from Honda, on the Magdalena river, Colombia
hondoensis -is -e from Hondo island (Honshu, Japan's largest island)
hongkongensis -is -e from Hong Kong
Honkenya for Gerhard August Honkeny (1724–1805), German botanist and author of *Synopsis plantarum Germaniae*
Hoodia, hoodii for Mr Hood, a London surgeon who cultivated succulents *c.* 1830 (important appetite suppressants used by the San, Kalahari desert bushmen)
Hoodiopsis Hoodia-resembling, *Hoodia-opsis*
hoogianus -a -um for Johannes (John) Hoog (1865–1950), head of Messrs van Tubergen, Dutch bulb growers, and nephew of the founder
hookerae for Lady Maria Hooker (1797–1872), wife of Sir William Jackson Hooker
Hookeria, hookeri, hookerianus -a -um for either Sir William Jackson Hooker (1785–1865) or his son Sir Joseph Dalton Hooker (1817–1911), both directors of the Royal Botanic Gardens, Kew
hoopesii for Thomas Hoopes, collector in N America *c.* 1859
Hopea for Dr John Hope (1725–86), Scottish botanist of the Edinburgh Botanic Garden
hopeanus -a -um for Thomas Hope (1770–1831) or his wife Louise, of Deepdene, Dorking
Hoplestigma Cloven-stigma, οπλη-στιγμα
Hoplophyllum Armed-leaf, οπλον-φυλλον
hoppeanus -a -um, hoppei for David Heinrich Hoppe (1760–1846), apothecary and Professor at Regensburg
hoppenstedtii for Señor Hoppenstedt, landowner in Mexico
Horaninovia for Paul Fedorowitsch Horaninow (1796–1865), Russian botanist
horarius -a -um lasting for one hour, *hora, horae* (the expanded petals of *Cistus*)
hordeaceus -a -um barley-like, *hordeum*
Hordelymus Barley-lime-grass, botanical Latin *Hordeum-Elymus*
hordestichos with barley-like ranks, botanical Latin from *hordeum* and στιχος (the inflorescence)
Hordeum Latin name, *hordeum*, for barley
horizontalis -is -e flat on the ground, spreading towards the horizon, οριζων
Horkelia, Horkeliella for Johann Horkel (1769–1846), German plant physiologist
horminoides clary-like, *Horminium-oides*
Horminum, horminium Exciter, ορμαινω (the Greek name, ορμαινον, for sage used as an aphrodisiac, ≡ *Salvia horminum*)
hormo- chain-, necklace-, ορμος
hormophorus -a -um bearing a chain or necklace, ορμος-φορεω
Horneophyton Horne's-plant, botanical Latin from Horne and φυτον

Hornschuchia, hornschuchianus -a -um, hornschuchii for Christian Friedrich Hornschuch (1793–1850), German bryologist and naturalist, Director of Greifswald University Botanic Garden
Hornungia for Ernst Gottfried Hornung (1795–1862), German scientific writer
hornus -a -um this year's, of the current year, *hornus*
horologicus -a -um with flowers that open and close at set times of day, *horologium* (literally, a clock; once a design favourite for floral clocks)
horrens bristling, shaggy, present participle of *horreo, horrere, horrui*
horribarbis -is -e having a bristly beard, *horridus-(barba, barbae)*
horridulus -a -um uncouth, protruding a little, somewhat thorny or prickly, *horridus*
horridus -a -um very thorny, rough, horridly armed, *horridus*
horripilus -a -um having erect bristles or spiny hairs, *horridus-(pilus, pili)*
horsfalliae, horsfalii for Mrs Horsfall (fl. 1830s) of Liverpool
Horsfieldia, horsfieldii for Dr Thomas Horsfield (1773–1859), American physician and naturalist who collected in Java and Sumatra
Horsfordia for F. H. Horsford, a New England naturalist
Hort used as an authority, for Arthur Hort (1864–1935) (*Lychnis flos-jovis* and *Globularia meridionalis* cultivars bear his name), or Fenton J. A. Hort (1828–1892), cleric and botanist
hort. used to indicate 'in the sense of gardeners' for plant names not agreeing with the same name attributed to an authority
Hortensia, Hortensis, hortensia A synonym for *Hydrangea*, for Hortense van Nassau
hortensis -is -e, hortorum cultivated, of the garden, *hortus*
hortulanus -a -um, hortulanorum of the gardener, of food-producers, *hortulanus* (the ornamental gardener was a *topiarius*)
Horvatia for Adolf Oliver Horvat (b. 1907), Hungarian botanist
hosmariensis -is -e from the mountainous neighbourhood of Beni Hosmar, Morocco
Hosta, hosteanus -a -um for Nicolaus Tomas Host (1761–1834), Austrian physician and botanist (***Hostaceae***)
hostilis -is -e of the enemy, foreign, hostile, *hostilis*
hottentotorus -a -um of the Khoikhoi people of southern Africa (formerly called Hottentots)
Hottonia for Pieter Hotton (1648–1709), Dutch botanist and professor at Leyden
Houstonia, houstonianus -a -um for Dr William Houston (1695–1733), writer on American plants, bluets
Houttea, houtteanus -a -um (houttianus) for Louis van Houtte (1810–76), Belgian nurseryman
Houttuynia for Martin (Maarten) Houttuyn (1720–94), Dutch naturalist and writer
Hovea for Anton Pantaleon Hove (Hoveau) (fl. 1780s–1820s), Polish botanist and collector for Kew
Hovenia for David ten Hoven (1724–87), a Dutch senator
Howea (Howeia) from the Lord Howe Islands, East of Australia, or for Admiral Lord Richard Howe (1726–99)
Howellia, Howelliella, howellii for Thomas Jefferson Howell (1842–1912), and Joseph Howell (1830–1912), who collected in California *c.* 1884–97
Hoya, Hoyella for Thomas Hoy (*c.* 1750–1822), gardener at Sion House for the Duke of Northumberland, wax flower
huachucanus -a -um from the Huachuca mountains of Arizona, USA
Hubbardia, Hubbardochloa for Charles Edward Hubbard (1900–80), English botanist, herbarium keeper of the Royal Botanic Gardens at Kew
Huberia for François Huber (1750–1831), Swiss apiarist who, despite being blind, studied the life history of the honey bee and published *Nouvelles observations sur les abeilles* (1792)
Hudsonia for William Hudson (1730–93), English apothecary, botanist, Keeper of the Chelsea Physic Garden and author of *Flora Anglica*

Huegelia, huegelii, hugelii for Baron Karl von Hügel (1794–1870), who travelled in the Philippines
Huernia, Huerniopsis for Justus Heurnius (1587–1652), Dutch missionary and first collector in the Cape, S Africa. (Robert Brown's error in spelling)
hugonis for Father Hugh Scallon, collector in West China *c.* 1899
Hugueninia for Auguste Huguenin (1780–1860), French botanist
Hulsea for Gilbert W. Hulse (1807–83), US Army surgeon and plant collector
Hulteniella for Eric Gunnar Hulten (1894–1981), Swedish botanist and collector in Alaska
Humata Earth, *humus, humi* (creeping rhizome)
Humbertia, Humbertianthus, Humbertiella, Humbertiodendron, Humbertochloa for Jean Henri Humbert (1887–1967), French botanist and collector in Madagascar, author of *Flore de Madagascar et des Comores*
Humboldtia, humboldtianus -a -um, humboldtii for Friedrich Wilhelm Heinrich Alexander Baron von Humboldt (1769–1859), who explored central America and wrote on natural history and meteorology; from Humboldt Bay, New Guinea
hume-, humi- wet, damp, dank, moist, *umidus, umi-*
Humea for Lady Amelia Hume (1751–1809) of Wormleybury, Hertfordshire
humeanus -a -um for David Hume of Edinburgh Botanic Gardens, or W. Burnley Hume, another gardener
humicolus -a -um of damp habitats, living in damp, *umi-cola*
humidicolus -a -um of damp habitats, living in damp, *umidus-cola*
humidorus -a -um of confined humid places, *umidus* (etymological analogy with cuspidor)
humiflorus -a -um flowering at ground level, *humi-florum*
humifructus -a -um fruiting at ground level (or in the soil surface), *humi-fructus*
humifusus -a -um trailing, sprawling, spreading over the ground, *humi-fusus*
humilis -is -e low-growing, close to the ground, of the ground, *humilis*
humilior smaller than most of its kind, comparative of *humus, humi*
humillimus -a -um very small or short, superlative of *humilis*
Humiria, Humiriastrum from a S American vernacular name, umiri
humistratus -a -um forming a blanket over the ground, *humi-stratus*
humulifolius -a -um hop-leaved, with leaves resembling those of *Humulus*
Humulus from the Slavic-German, chmeli, Latinized from Germanic, humela
Hunaniopanax Hunan-*Panax*
hungaricus -a -um from Hungary, Hungarian
Hunnemannia for John Hunneman (d. 1839), English botanist and plant introducer
hunnewellianus -a -um for the New England horticultural family Hunnewell
Hunteria for William Hunter (1755–1812), Scottish naturalist and surgeon with the East India Company
hupehanus -a -um, hupehensis -is -e from Hupeh (Hupei, Hubei) province, China
Huperzia for Johann Peter Huperz (1771–1816), German physician and botanist
Hura from a S American vernacular name for the sand-box tree, *Hura crepitans*
huronensis -is -e from Huron, S Dakota, USA, or Lake Huron, USA/Canada
Hutchinsia, hutchinsiae, hutchinsii for Miss Ellen Hutchins (1785– 1815), cryptogamic botanist of Bantry and Ballylickey, Co. Cork
Hutchinsonia for John Hutchinson (1884–1972), systematic botanist at the RBG Kew, author of several standard works
Huthamnus for Hu Hsen-Hsu (1894–1968), Chinese botanist
Huttonaea for Mrs Henry Hutton, orchid collector in S Africa
huttonii for J. Hutton, who collected in S Africa *c.* 1860, or Henry Hutton, who collected for the James Veitch company.
Huxleya for Thomas Henry Huxley (1825–95), English polymath, traveller, author of numerous works and staunch supporter of Darwinian theory
Hyacinthella Little-hyacinth, feminine diminutive of *Hyacinthus*
hyacinthiflorus -a -um with dark blue-purple flowers, *Hyacinthus-florum*
hyacinthifolius -a -um with dark purple foliage, *Hyacinthus-folium*

Hyacinthoides, hyacinthoides Hyacinth-like, υακινθος-οειδης
Hyacinthus Homer's name for the flower(s) which sprang from the blood of Υακινθος, or from an earlier Thraco-Pelasgian word, υακινθος, for the blue colour of water, cognate with jacinth (***Hyacinthaceae***)
hyacinthus -a -um, hyacinthinus -a -um dark purplish-blue, resembling *Hyacinthus*
Hyaenanche Hyena-strangler, υαινα-αγχονη
hyalinolepis -is -e having transparent or papery scales, υαλος-λεπις
hyalinus -a -um nearly transparent, crystal, hyaline, υαλος, of glass, υαλινος
Hyalis Crystal, υαλος
Hyalochlamys Glassy-covering, υαλος-χλαμυς
Hyalolaena Glassy-mantle, υαλος-χλαινα
Hybanthus Curved-flower, υβος-ανθος (the arched corolla tube)
hybernalis -is -e of winter, *hibernus*
hybernicus -a -um from Ireland, *Hibernia*
hybernus -a -um either from Ireland, *Hibernia*, or of winter, *hiberna, hibernorum*
Hybochilus Arched-lip, υβος-χειλος
hybridinus -a -um hybrid, adjective from *hybrida*
hybridogagnepainii the name for a hybrid (*Berberis*) *gagnepainii*
hybridus -a -um mongrel, half-breed, hybrid, *hibrida, hibridae; hybrida, hybridae*
hydaticus -a -um wet, watery, υδατος, υδατικος
Hydnocarpus Truffle-fruit, υδνος-καρπος (for the oil-producing fruit and seeds) (≡ *Taractogenos*)
Hydnophytum Truffle-plant, υδνος-φυτον (their epiphytic modification)
Hydnora Truffle, υδνος (for the somatic structure of this root parasite) (*Hydnoraceae*)
Hydra Water-serpent, υδρα (compares the several arms around the stomum with the mythological many-headed serpent)
Hydrangea (Hortensia) Water-vessel, υδωρ-(αγγος, αγγειον) (for the shape of the capsules) (***Hydrangeaceae***)
hydrangeiformis -is -e looking like a *Hydrangea*
hydrangeoides resembling *Hydrangea*
Hydrastis (Hydrastes) etymology uncertain
Hydriastele Water-pot-post, υδρα-στηλη (can yield a drink)
Hydrilla Water-serpent, diminutive form of υδρα
hydro- water-, of water-, υδωρ, υδατος, υδρος, υδρο-
Hydrobryum Water-moss, υδρο-βρυον (*Podostemaceae*)
Hydrocera Water-horn, υδρο-κερας (the nectar filled spur)
Hydrocharis Water-beauty, υδρο-χαρις (***Hydrocharitaceae***)
Hydrochloa Water-grass, υδρο-χλοη
Hydrocleys Water-key, υδρ-κλεις
Hydrocotyle Water-cup, υδρο-κοτυλη
hydrocotylifolius -a -um with *Hydrocotyle*-like leaves, *Hydrocotyle-folium*
Hydrodictyon Water-net, υδρο-δικτυον (the reticulate structure of water net algae)
Hydrodysodia Aquatic-*Dyssodia*, υδρο-δυσ-οδμη
hydrogrammus -a -um watermarked, υδρο-γραμμα (the soaked-appearance of the [hygrophanous] pilea)
Hydrolapathum, hydrolapathus -a -um a name in Pliny, υδρο-λαπαθον, for a water dock
Hydrolea Water-oil, υδωρ-ελαια (greasy texture of the leaves)
Hydropectis Aquatic-comb (*Pectis*), botanical Latin from υδωρ and *pecten*
Hydrophilus, hydrophilus -a -um Liking water, υδρο-φιλος
Hydrophylax Water-sentinel, υδρο-φυλαξ (*Hydrophylax maritima* is a sand-dune colonizer)
Hydrophyllum Water-leaf, υδρο-φυλλον (leaf texture) (***Hydrophyllaceae***)
hydropiper water pepper, botanical Latin from υδωρ and *piper*
Hydrostachys Water-spike, υδρο-σταχυς (submerged aquatics)
Hydrothauma Aquatic-marvel, υδρο-θαυμα (aquatic grass)
hyemalis -is -e pertaining to winter, of stormy weather, of winter, *hiemalis* (flowering season)

Hyeronima for Georg Hans Wolfgang Hieronymus (1846–1921), German botanist and collector in S America
hygro- languid, pliant, supple, moist, υγροτης, υγρος, υγρο-
Hygrocheilus Moist-lip, υγρο-χειλος (the nectary)
Hygrochloa Flowing-grass, υγρο-χλοη (aquatic)
hygrometricus -a -um responding to moisture level, υγρο-μετρεω (earth-star fruiting bodies open in the wet and close on drying)
Hygrophila Moisture-loving, υγρο-φιλεω (spiny plant of arid habitats, flowers in response to moisture)
hygrophilus -a -um loving water, υγρο-φιλεω
hygroscopicus -a -um heeding or sensitive to water, υγρο-σκοπεω (absorptive)
hylaeus -a -um of woods, of forests, υληεις
Hylandia for Bernard Patrick Matthew Hyland (b. 1937), Australian botanist
hylematicus -a -um of woods or forests, υληματος
hylo- forest, woodland, υλη-
Hylocereus Forest-cactus, botanical Latin from υλη and *cereus* (climbing cactus)
hylocharis -is -e joy of woodlands, υλη-χαρις
hylocolus -a -um forest-dwelling, botanical Latin from υλη and *cola*
Hylodendron Forest-tree, υλη-δενδρον
hylogeiton woodland neighbour, υλη-γειτων
Hylomecon Wood-poppy, modern Latin from υλη-μηκων
hylonomus -a -um woodland-dwelling, υλη-νομος
Hylophila, hylophilus -a -um Forest-loving, living in forests, wood-loving, υλη-φιλος
hylothreptus -a -um woodland-feeding, υλη-θρεπτρα (xylophagous)
hymen-, hymeno- membrane-, membranous-, υμην, υμενος (*Hymen, Hymenis* was the god of marriage, υμεναιος)
Hymenachne Membranous-chaff, υμην-αχνη (the thin glumes of marsh grass)
Hymenaea, hymenaeus for *Hymen, Hymenis*, the god of marriage, *hymenaeus*, of weddings (the leaflets are joined)
Hymenandra Membraned-anther, υμην-ανηρ (the connective)
Hymenanthera Membranous-stamen, υμην-ανθερα (the membranous appendages of the anthers)
hymenanthus -a -um membranous flowered, υμην-ανθος
hymenelytrus -a -um having a membranous covering, υμην-ελυτρον
Hymenocallis Membraned-beauty, υμενος-καλος (the filament-cup of spider lily)
Hymenocardia Membraned-heart, υμενος-καρδια (the winged, heart-shaped capsule)
Hymenocarpos (*us*) Membranous-fruit, υμενος-καρπος (the thin wall)
Hymenoclea Membranous-enclosure, υμενος-κλειω
hymenodes, hymenoides membranous textured, υμενος-ωδες; membranous, υμην-οειδης
Hymenodictyon Membranous-net, υμενος-δικτυον (the reticulate membrane around the seeds)
Hymenogyne Membranous-ovary, υμενος-γυνη
Hymenoleana Membranous-cover, υμενος-χλαινα
Hymenolepis Membranous-scaled-one, υμενος-λεπις
Hymenolobus Membranous-pod, υμενος-λοβος
Hymenolophus Membranous-crest, υμενος-λοφος (on the seed)
Hymenomycetes Membranous-fungi, υμενος-μυκες (have exposed gills)
hymenophorus -a -um membrane-bearing, υμενος-φορα
Hymenophyllopsis Resembling-*Hymenophyllum*, υμενος-φυλλον-οψις (***Hymenophyllopsidaceae***)
Hymenophyllum, hymenophyllus -a -um Membranous-leaf, υμενος-φυλλον (delicate frond of the filmy fern) (***Hymenophyllaceae***)
Hymenorchis Membranous-orchid, υμενος-ορχις (the perianth)
hymenorrhizus -a -um having membranous roots, υμενος-ριζα
hymenospathus -a -um having a membranous spathe, υμενος-σπαθη
Hymenosporum Membranous-spored-one, υμενος-σπορος (the winged seed)
Hymenostegia Membranous-cover, υμενος-στεγη

Hymenoxys Pointed-membrane, υμενος-οξυς
hyo- pig-, hog-, swine-, υς, υος, υο-
Hyobanche Pig-strangler, υο-αγχω (total parasite)
Hyophorbe Pig-fodder, υο-φορβη (pigs are fed on the fruit)
Hyoscyamus Hog-bean, υς-κυαμος (a derogatory name, υοςκυαμος, used by Dioscorides; Pliny refers to henbane's poisonous nature)
Hyoseris Pig-salad, υο-σερις (swine's succory)
hypanicus -a -um from the region of the Hypanis river, Ukraine
hyparcticus -a -um beneath the arctic, sub-arctic, botanical Latin from υπ-αρκτος (hypo-arctic distribution)
hypargeius -a -um shining white below, υπο-αργης (lower leaf surfaces)
hypargenteus -a -um having silvery undersides, botanical Latin from υπο and *argentum*
hypargyreus -a -um silvery beneath, υπο-αργυρος
Hyparrhenia Male-beneath, υπ-αρρην (the arrangement of the spikelets)
Hypecoum Rattle, υπεχεο (Dioscorides' name, υπεκοων, for the loose seeds in the flat curved pods) (***Hypecoaceae***)
Hypelate a name in Pliny for a holly, re-used by P. Brown for inkwood
hyper- above-, over-, υπερ-
Hyperacanthus Thorned-above, υπερ-ακανθος
hyperacrion beyond the heights, υπερ-ακρα
hyperaizoon above, or better than, (*Sedum*) *aizoon*
hyperanthus -a -um with bearded flowers, υπερ-ανθος
hyperboreus -a -um beyond the north wind, of the far north, northern, υπερ-βορεας
hyperici- *Hypericum*-like-
hypericifolius -a -um having leaves similar to those of *Hypericum*
hypericoides resembling *Hypericum*
Hypericophyllum *Hypericum*-leaved, υπερ-εικων-φυλλον
Hypericopsis Resembling-*Hypericum*, υπερ-εικων-οψις
Hypericum Above-pictures, υπερ-εικων (Dioscorides' name, υπερεικον, for its use over shrines to repel evil spirits); some derive it from υπ-ερεικε, from heath-like habitats (***Hypericaceae***)
Hyperthelia Female-below, υπερ-θηλυς (spikelet arrangement)
hypertrophicus -a -um hypertrophied, misshapen, abnormally enlarged, υπερ-τροφις
hyperythrus -a -um having reddish undersurfaces, υπ-ερυθρος
hyphaematicus -a -um composed of interwoven threads, υπερ-ματικη
Hyphaene Entwined-one, υφαινο (the entwining fibres in the fruit wall)
Hypholoma thread-fringe, υφη-λωμα
hypnicolus -a -um living on and with mosses, botanical Latin from υπνος and *colo*
hypnoides moss-like, *Hypnum-oides*
hypnophilus -a -um liking mossy habitats, υπνος-φιλεω
Hypnum Sleep, υπνος
hypo- under-, by-, through-, beneath-, υπο-
Hypobathrum Below-a-step, υπο-βαθρον (one seed is superposed on the other)
Hypocalymma Beneath-a-veil, υπο-καλυμμα (the deciduous calyx)
Hypocalyptus Enveloped-in-a-veil, υπο-καλυπτος
hypochaeonius -a -um growing under snow, υπο-(χιων, χιονος)
hypochaeridis -is -e of cat's ear, living on *Hypochaeris* (*Phanacis*, gall wasp)
Hypochaeris (*Hypochoeris*) a name, υποχοιρις, used by Theophrastus; some suggest derivation as υπω-χουρος, comparing the pig's belly bristles to those on the abaxial surface of some species
hypochaeroides (*hypochoeroides*) resembling *Hypochaeris*, *Hypochaeris-oides*
hypochlorus -a -um green beneath, υπο-χλωρος
hypochondriacus -a -um sombre, melancholy, υποκονδριακος (colour) (υπο-χονδρος is the soft area below the sternal cartilage; melancholy was supposed to be located in the liver)

hypochrysus -a -um golden underside, golden beneath, υπο-χρυσος
hypocrateriformis -is -e of tubular flowers surmounted with lobes forming a shallow cup, υπο-κρατηρ
Hypodaphnis Inferior-laurel, υπο-Δαφνε (the inferior ovary is unusual in the *Lauraceae*)
Hypoderris Under-a-skin, υπο-δερρις (the hair vein fern indusia)
Hypoestes Under-cover, υπο-εστες (the enveloping bracts)
hypogaeus -a -um underground, subterranean, υπο-γαια (as in groundnut fruiting)
hypoglaucus -a -um glaucous beneath, υπο-γλαυκος
hypoglossus -a -um (hypoglottis) beneath-a-tongue, sheathed-below, υπο-γλωσσα (*Ruscus hypoglossus* cladodes have a large scale-like fract subtending the flowers)
hypogynus -a -um having a superior ovary, υπο-γυνη
Hypolepis Under-scale, υπο-λεπις (the sori are additionally protected by the deflexed margin of the pinna)
hypoleucus -a -um whitish, pale below, υπο-λευκος
hypoliarus -a -um soft or tender beneath, υπο-λιαρος
Hypolytrum Gore-beneath, υπο-λυθρον (colouration of leaf sheaths); some interpret as under a cover, υπο-ελυτρον
hypomelas with black undersides, υπο-(μελας, μελαινα)
hypophegeus -a -um from beneath oak trees, υπο-φηγος (φηγος was a kind of oak and *Monotropa hypophegia* is a parasitic on *Quercus*); some translate it as beneath beech trees, *fagus* being cognate with φηγος
hypophyllus -a -um produced or growing on the underside of a leaf, υπο-φυλλον
hypopilinus -a -um having softly felted undersurfaces, υπο-πιλος
Hypopitys, hypopithys, hypopitys growing under pine trees, υπο-πιτυς
hypopsilus -a -um having bald undersurfaces, υπο-ψιλος
hypostomus -a -um having stomata only on the lower leaf surfaces, υπο-στομα
hypotheius -a -um brimstone(-coloured) beneath, υπο-θειον, (orange-red bruising)
Hypoxis Sharp-below, υπο-οξις (the seed pod shape) (***Hypoxidaceae***)
hypoxylon from under wood, υπο-ζυλον (rot fungus)
Hypsela Lofty, υψηλος (montane provenance)
Hypselochloa Stately-grass, υψηλος-χλοη
Hypselodelphys Stately-sister, υψηλος- αδελφη (the grouped fruits); some say High-womb
Hypseocharis Beauties-of-the-heights, υψι-χαριεις (montane provenance)
Hypseochloa Mountain-grass, υψι-χλοη
Hyptidendron Tree-*Hyptis*, υπτιος-δενδρον (arborescent habit)
Hyptis Turned-back, υπτιος (the resupinate corolla limb)
hyrcanicus -a -um, hyrcanus -a -um from the Caspian Sea area, Hyrcanian
hysginus -a -um dark reddish pink, υσγυνον
hyssopi- hyssop-like, resembling *Hyssopus*
hyssopifolius -a -um with *Hyssopus*-like leaves, *Hyssopus-folium*
Hyssopus Dioscorides' name, υσσωπος, for another plant, *Origanum vulgare*, earlier from a Semitic word, ezob
hyster- inferior-, later-, υστερος, υστερο-; or womb, υστερα
hysteranthus -a -um with inferior flowers, υστερο-ανθος
hystri-, hystrix spiny, like a porcupine, υστριξ (the spiny corm of *Isoetes hystrix*)
hystricinus -a -um spiny, υστριξ, υστριχος
Hystricophora Hedgehog-bearing, υστριχος-φορα (the spiny involucral bracts)
hystriculus -a -um somewhat prickly, diminutive from υστριξ, υστριχος
Hystrix, hystrix Hedgehog (Porcupine), υστριξ, υστριχος (the inflorescence is covered in spines)

iacinthus -a -um reddish-orange coloured (ιακυντος, relates to *Hyacinthus*)
iandinus -a -m the colour (green) of jade, from Spanish, piedra de ijada, stone of colic (supposed curative)

ianthinus -a -um, -ianthus -a -um bluish-purple, violet-coloured, ιανθινος
-ianus -a -um -pertaining to (possessive of a person or place)
iaponicus -a -um see *japonicus*
-ias -much resembling, -like
ibaguensis -is -e from Ibangué, Tolima department, central Colombia
ibericus -a -um, ibiricus -a -um either from Spain and Portugal (*Hiberes*) or from the Georgian Caucasus
iberideus -a -um from the Iberian peninsula, *Hiberes*
iberidi- *Iberis*-like
iberidifolius -a -um *Iberis*-leaved, *Iberis-folium*
Iberis Spanish, *hiberes* (Dioscorides' name, ιβηρις, for an Iberian cress-like plant)
Ibicella Goat-like, feminine diminutive of *ibex* (the two curved processes on the unicorn plant fruit)
-ibilis -is -e -ible, -capable of (suffix turning nouns into adjectives)
ibiricus -a -um from the Iberian peninsula, *Hibericus*
ibota the Japanese name for *Ligustrum ibota*
ibukiensis -is -e from Ibuki-yama area, Honshu, Japan
Icacina Icaco-like, resembling *Chrysobalanus icaco* (branching resembles that of coco-plum) (***Icacinaceae***)
icaco the W Indian vernacular name for *Chrysobalanus icaco*
-icans -becoming, -resembling (present participle of verbs)
icaricus -a -um from Ikaria island, Greece
ichanganus -a -um, ichangensis -is -e from I-ch'ang (Yichang), W Hupeh, China
Ichnanthus Vestige-flower, ιχν-ανθος (the winged callus of the upper florets) (common on forest tracks)
ichneumoneus -a -um resembling a wasp, ichneumon-like, ιχνευτης
Ichthyothere Of-fish-hunting, ιχθυς-θηρα (used as a fish poison)
ichthyotoxicus -a -um fish-poisoning, ιχθυς-τοξικος
-icolus -a -um -of, -dwelling in, *colo, colere, colui, cultum*
icos-, icosa- twenty-, εικοσι
icosagonus -a -um having twenty (or thereabouts) angles, εικοσι-γωνια
icosandrus -a -um twenty-stamened, εικοσι-ανηρ
ictalurus -a -um having yellowish barbel-like structures, catfish-like, *Ictalurus*
icteranus -a -um of yellowing, *ictericus* jaundiced
ictericus -a -um, icterinus -a -um yellowed, jaundiced, *ictericus*
-icus -a -um (location)-from
idaeus -a -um from Mount Ida in Crete, or Mount Ida in NW Turkey
Idahoa, idahoensis -is -e from Idaho, USA
ida-maia, ida-maya for Ida May Burke, daughter of the discoverer of *Brodiaea ida-maia* (*Dichelostemma ida-maia*), (*c.* 1867)
ideobatus thorn bush from Mount Ida, *idaeus-batus*
-ides -resembling, -similar to, -like, ειδος, ειδω, ειδον, ειδης
Idesia for Eberhard Ysbrants Ides, seventeenth-century Dutch explorer in China
idio- peculiar-, different-, ιδιος, ιδιο-
-idion a Greek diminutive suffix, -ιδιον
Idiospermum Distinctive-seed, ιδιο-σπερμα (they have up to six cotyledons) (***Idiospermaceae***)
-idius -a -um -resembling
idomenaeus for Idomeneus, in Homer's *Odyssey*, King of Crete
idoneus -a -um worthy, apt, proper, suitable, sufficient, *idoneus*
Ifloga an anagram of *Filago*
ignatii for Saint Ignatius Loyola (1491–1556)
ignavus -a -um lazy, listless, relaxing, slothful, *ignavus*
igneiflorus -a -um having fiery-coloured flowers, *igneus-florum*
ignescens, igneus -a -um kindling, fiery-red-and-yellow, glowing, *ignesco, ignescere*
igneus -a -um fiery, *igneus*
ignevenosus -a -um having fiery-coloured veins, *igneus-vena*

igniarius -a -um burning, *igneus* (*Phellinus igniarius* is black, cracked and reddish edged)
ignivolvatus -a -um having a flame-coloured volva, *ignis* (the bright orange velar remnants)
ignotus -a -um unknown, obscured, overlooked, inferior, *ignotus*
iguaneus -a -um chameleon-like, of variable colour, botanical Latin from Spanish iguana (Arawak, iwana)
Iguanura Lizard-tail, botanical Latin from Spanish iguana (Arawak, iwana) and ουρα (for the inflorescence)
ikariae from Ikaria island, Samos department, Greece
il-, im-, in- in-, into-, for-, contrary-, contrariwise-
Ilex, ilex Holly, the Latin name, *ilex* (*ilignus*, for the holm-oak, *Quercus ilex*)
ilici-, ilicinus -a -um holly-, *Ilex-*, *ilex, ilicis*
ilicifolius -a -um holly-leaved, *Ilex-folium*
ilicis of holly, living on *Ilex* (symbionts, parasites and saprophytes)
iliensis -is -e from the area of the Ili river, China / Kazakhstan
-ilis -is -e -able, -having, -like, -resembling (suffix turning a noun into an adjective)
illaquens ensnaring, entangling, present participle of *illaqueo, illaqueare*
illecebrosus -a -um alluring, enticing, charming, seductive, *illecebrosus*
Illecebrum Attraction, *illicio, illicere, illexi, illectum* (a name, *illecebra*, in Pliny) (***Illecebraceae***)
Illicium Seductive, *illicio, illicere, illexi, illectum* (the fragrance) (***Illiciaceae***)
-illimus -a -um -est, -the best, -the most (superlative),
illinatus -a -um, illinitus -a -um smeared, smudged, *illino, illinere, illevi, illitum*
illinoiensis -is -e from Illinois, USA
-illius -a -um -lesser (a diminutive suffix)
illustratus -a -um pictured, painted, as if painted upon, *illustro, illustrare, illustravi, illustratum*
illustris -is -e distinguished, distinct, bright, clear, *illustris*
illyricus -a -um from western former Yugoslavia (*Illyria*)
ilvensis -is -e from the Isle of Elba, Italy, or the river Elbe, Germany (*Ilva*)
Ilysanthes Mud-flower, ιλυς-ανθος (the habitat of most)
imanto- leathery, thong, ιμαντινος, ιμαντο-
Imantophyllum Thong-leaf, ιμαντο-φυλλον
imbecillis -is -e, imbecillus -a -um feeble, weak, frail, *imbecillus*
imberbis -is -e without hair, unbearded, *in-barba*
imbricans, imbricatus -a -um overlapping like tiles, *imbrex, imbricis* (leaves, corolla, bracts, scales), imbricate
imeretinus -a -um (imeritina) from Imeretie, W Caucasus (? without merit, *i-meritus*)
Imerinaea from Imerina county of Madagascar
imitans imitating, copying, present participle of *imitor, imitare, imitatus*
Imitaria Look-alike, *imitor, imitare, imitatus* (the cactus-like form)
immaculatus -a -um unblemished, immaculate, without spots, *im-*(*macula, maculae*)
immanis -is -e enormous; monstrous, savage, *immanis*
immarginatus -a -um without a rim or border, *im-*(*margo, marginis*)
immersus -a -um growing under water, *immergo, immergere, immersi, immersum*
immodestus -a -um extravagant, without restraint or modesty, *immodestus*
impar, impari- unpaired-, unequal-, *impar, imparis*
Impatiens, impatiens Impatient, *impatiens, impatientis* (touch-sensitive fruits)
impeditus -a -um tangled, impeding, obstructing, *impedio, impedire, impedivi(-ii), impeditum*
Imperata, imperati for Ferrante Imperato (1550–1625), Italian botanist of Naples, author of *Del l'historia naturale* (1599)
Imperatoria, imperator, imperatoria Imperial, emperor, ruler, master, *imperator, imperatoris* (masterwort)
imperatricis for Napoleon I's Empress, Marie Josephine Rose Tascher de la Pagerie (1763–1814), feminine form of *imperator*

imperfectus -a -um unfinished, lacking perfection, *imperfectus* (anatomical deficiencies)
imperforatus -a -um without perforations or apparent perforations, *im-(perforo, imperforare, imperforavi, imperforatum)*
imperialis -is -e very noble, imperial, of nobility, *imperialis*
imperiorus -a -um of the empire, *imperium*
implexus -a -um tangled, interlaced, *im-(pleco, plecere, plexi, plectum)*
impolitus -a -um unpolished, not ornamental, inelegant, *impolitus*
imponens deceptive, cheating, present participle of *impono, imponere, imposui, impositum*
impressus -a -um marked with slight depressions, sunken, impressed, *imprimo, imprimere, impressi, impressum* (e.g. leaf-veins)
impudicus -a -um lewd, shameless, impudent, immodest, *impudicus*
imschootianus -a -um for A. van Imschoot of Ghent, Belgium *c.* 1895
in- not-, un-, en-, -em
inaccessus -a -um unapproachable, *inaccessus*
inaequaliflorus -a -um unequally flowered, *inaequalis-florum*
inaequalis -is -e unequal-sided, unequal-sized, *inaequalis* (veins or other feature)
inaequidens with unequal teeth, unevenly toothed, *in-aeque-dens*
inaequilateralis -is -e unequal-sided, *in-aeque-(latus, lateris)* (leaves)
inaequilobus -a -um irregularly lobed, *in-aeque-lobus*
inaequisepalus -a -um having unequal sepals, *in-aeque-sepalus*
inaguensis -is -e from the Inagua islands, Bahamas
inalatus -a -um lacking a wing, *in-alatus*
inamoenus -a -um lacking charm, *in-amoenitas*
inanis -is -e empty; poor; worthless, *inanis*
inapertus -a -um without an opening, unexposed, not opened, *in-(aperio, aperire, aperui, apertum)*
inarticulatus -a -um not jointed, indistinct, *in-(articulo, articulare, articulavi, articulatum)* (nodes)
inatophyllus -a -um thong-leaved, ενατο-φυλλον
inauritus -a -um lacking a cortinal ring, un-eared, *in-auritus* (the ring-less stipe)
incaeduus -a -um uncut, *in-(caedo, caedere, cecidi, caesum)*
incandescens turning white, present participle of *incandesco, incandescere, incandui*
incanescens turning grey, becoming hoary, present participle of *incanesco, incanescere, incanui*
incanus -a -um quite grey, hoary-white, grey, *incanus*
incarnatus -a -um flesh-coloured, *carneus, in-caro, in-carnis; incarnare*
Incarvillea for Pierre Nicholas de Cheron d'Incarville (1706–57), French missionary and correspondent of Bernard de Jussieu from China, and writer of *Mémoire sur le vernis de la Chine.*
incasicus -a -um of the S American Inca people
incertus -a -um doubtful, uncertain, *incertus* (*sedi incertis*, of uncertain placing)
incisi-, incisis -is -e, incisus -a -um sharply and deeply cut into, incised, *incisus*
incisifolius -a -um having incised leaves, *incisus-folius*
incisodentatus -a -um sharply toothed, with deeply sharp-toothed margins, *incisus-dentatus*
inclaudens not closing, present participle from *in-(claudo, claudere, clausi, clausum)*
inclinatus -a -um not upright, leaning, inclined, *inclino, inclinare, inclinavi, inclinatum* (growing from the sides of oak stumps)
inclinis -is -e bent, turned back, *inclino, inclinare, inclinavi, inclinatum*
includens encompassed, enclosed, *includo, includere, inclusi, inclusum*
inclusus -a -um not protruding, included, *inclusus* (e.g. corolla longer than the style)
incognitus -a -um untried; unrecognized, *incognitus*
-incolus -a -um -resident, -inhabitant, *incola, incolae*
incommodus -a -um troublesome or inconvenient, *incommodus*

incomparabilis -is -e beyond compare, incomparable, *in-(comparo, comparare, comparavi, comparatum)*
incompletus -a -um lacking parts, *in-(compleo, complere, complevi, completum)* (of the flowers)
incomptus -a -um unadorned, rough, inelegant, *incomptus*
inconcessus -a -um forbidden, *inconcessus*
inconcinnus -a -um awkward; inelegant, *inconcinnus*
inconspicuus -a -um small, *in-(conspicio, conspicere, conspexi, conspectum)*
inconstans not constant, fickle, varying, *in-(consto, constare, constiti, constatum)*
inconstrictus -a -um not constricted, *in-(constringo, constringere, constrinxi, constrictum)* (corolla tube)
incrassatus -a -um very thick, made stout, *incrassus* (e.g. *Sempervivum* leaves)
incredibilis -is -e extraordinary, *incredibilis*
incrustans encrusting, present participle of *incrusto, incrustare* (encrusting growth habit)
incrustatus -a -um encrusted or packed together, *incrusto, incrustare*
incubaceus -a -um lying in or on, sitting (upon the ground), *incubo, incubare, incubui, incubitum*
incubus -a -um lying upon, incubous (when a lower distichous leaf overlaps the next upper on the dorsal side); a male demon, Latin for a nightmare (see *succubus*)
incumbens leaning, reclining upon, present participle of *incumbo, incumbere, incubui, incubitum*
incurvus -a -um, incurvatus -a -um inflexed, incurved, *incurvo, incurvare*
indecorus -a -um unbecoming, *indecorus*
indefensus -a -um undefended, without thorns, *indefensus*
indehiscens not dehiscing, not splitting at maturity, present participle of *in-(dehisco, dehiscere)*
indicus -a -um from India, Indian, was used loosely for the Orient, *india, indiae*
Indigofera Indigo-bearer, ινδικος-φερω, *indicum-fero* (ινδικον φαρμακον, Indian dye)
indivisus -a -um whole, undivided, *in-(divido, dividere, divisi, divisum)*
indo- prefix to indicate an Indian characteristic genus resembling one limited to another geographical area
Indocalamus Indian-*Calamus*
indochinesnsis -is -e from Indo-China (Vietnam, Laos, Cambodia)
Indofevillea Indian-*Fevillea* (an American genus)
Indoneesiella Indian-*Neesiella*
Indopiptadenia Indian-*Piptadenia*
Indopoa Indian-*Poa*
Indorouchera Indian-*Rouchera*
Indosasa Indian-*Sasa*
induratus -a -um hard, indurate, *induro, indurare* (usually of an outer surface)
indurescens hardening, present participle of *induresco, indurescere, indurui*
indusiatus -a -um having a protective cover, annulus or ring of hairs, *indusium*
indutus -a -um entangled; dressed, *induo, induere, indui, indutum*
induvialis -is -e, induviatus -a -um clothed, *induvia, induviarum* (with dead remnants of leaves or other structure)
inebrians able to intoxicate, inebriating, *inebrio, inebriare*
inermis -is -e defenceless, without spines or thorns, unarmed, *inermis, inermus*
-ineus -a -um -ish, -like
inexpectans not expected, *in-(exspecto, exspectare, exspectavi, exspectatum)* (found where not expected)
infarctus -a -um stuffed into, turgid, *in-(farcio, farcire, rarsi, fartum)*
infaustus -a -um inauspicious, unlucky, unfortunate, *infaustus*
infectorius -a -um, infectoris dyed, used for dying, of the dyers, *infector, infectoris*
infectus -a -um discoloured, stained, *inficio, inficere, infeti, infectum*
infecundus -a -um unfruitful, *infecundus*
infernalis -is -e infernal, as of the lower world, *infernus*

infestans attacking, infesting, present participle of *infesto, infestare*
infestus -a -um troublesome, hostile, dangerous, invasive weed, *infesto, infestare*
infirmus -a -um weak, feeble, trivial, *infirmo, infirmare*
inflatus -a -um swollen, puffed-up, inflated, *inflo, inflare, inflavi, inflatum*
inflexus -a -um bent or curved abruptly inwards, inflexed, *inflectus, inflectere, inflexi, inflexum*
infortunatus -a -um unfortunate, *infortunatus* (poisonous)
infossus -a -um deeply sunken, buried, *infodio, infodere, infodi, infossum*
infra- below-, *infra*
infractus -a -um broken, bent, curved inwards, *infractus*
infrapurpureus -a -um being purple beneath, *infra-purpureus* (leaves)
infundibularis -is -e, infundibulus -a -um funnel shaped, *infundibulum*
infundibulifolius -a -um having funnel-shaped leaves, *infundibulum-folium*
infundibuliformis -is -e trumpet-shaped, funnel-shaped, *infundibulum-forma*
infuscatus -a -um spoilt, tarnished, darkened, *infusco, infuscare*
infusus -a -um spreading, *infundo, infundere, infudi, infusum*
Inga from the W Indian vernacular name
ingens huge, enormous, *ingens, ingentis*
ingratus -a -um disagreeable, unwelcome, *ingratus*
injucundus -a -um not pleasing, *in-iucundus*
injunctus -a -um unattached, not conjoined, *iniungo, iniungere, iniunxi, iniunctum*
innatus -a -um natural, inborn, innate, *innascor, innasci, innatus*
innocuus -a -um not harmful, *in-nocuus*
innominatus -a -um not named, unnamed, *in-nominatus*
innovans adapting, renewing, *innovo, innovare*
innovatus -a -um adapted, renewed, *innovo, innovare*
ino- sinew, nerve, fibre-, fibrous-, ις, ινος, ινο-
Inocarpus Fibrous-fruit, ινο-καρπος
inocephalus -a -um fibrous-headed, ινο-κεφαλη
Inocybe Fibrous-head, ινο-κυβη (the rough surfaced pileus of most)
inocybeoides resembling *Inocybe*, ινος-κυβη-οειδης
inodorus -a -um without smell, scentless, *in-(odor(odos), odoris)*
inominatus -a -um unlucky, inauspicious, *inominatus*
inopertus -a -um unconcealed, *in-(operio, operire, operui, opertum)*
inophyllus -a -um fibrous-leaved, with fine thread-like veins, ινο-φυλλον
inopinatus -a -um, inopinus -a -um unimagined, unexpected, *in-(opinor, opinare, opinatus)*
inops destitute, deficient, poor, weak, *inops, inopsis*
inordinatus -a -um disordered, irregular, *inordinatus*
inornatus -a -um without ornament, unadorned, *inornatus*
inoxius -a -um without prickles, harmless, *i-noxius*
inquilinus -a -um, inquillinus -a -um introduced, inhabitant, tenant, *inquillanus*
inquinans becoming defiled, turning brown, staining, discolouring, present participle of *inquino, inquinare, inquinavi, inquinatum*
insanus -a -um outrageous, extravagant, frantic, *insanus*
inscalptus -a -um carved, engraved, *in-(scalpo, scalpere, scalpsi, scalptum)*
inscriptus -a -um as though written upon, inscribed, *in-(scibo, scribere, scripsi, scriptum)*
insculptus -a -um carved, engraved, *in-(sculpto, sculoere, sculpsi, sculptum)*
insect- modern Latin from *insectus*, segmented or notched
insectifer -era -erum bearing insects, *insectum-fero* (mimetic fly orchid)
insectivorus -a -um insect-eating, *insectum-(voro, vorare, voravi, voratum)*
insertus -a -um inserted, *insero, inserere, inserui, insertum* (the scattered inflorescences)
insidiosus -a -um artful, treacherous, deceitful, *insidiosus*
insignis -is -e remarkable, decorative, striking, conspicuous, distinguished, *insignio, insignire*

insiticius -a -um, insititius -a -um, insitivus -a -um grafted, *insero, inserere, insevi, insitum*
insolitantherus -a -um having unusual or distinctive anthers, *insolitus-anthera*
insolitus -a -um unusual, *insolitus*
inspersus -a -um appearing to have been sprinkled upon, *inspergo, inspergere, inspersi, inspersum*
instar worthy, as good as; resemblance, form, *instar*
insubricus -a -um from the Lapontine Alps (*Insubria*) between Lake Maggiore and Lake Lucerne
insulanus -a -um, insularis -is -e growing on islands, insular, *insula, insulae*
insularimontanus -a -um from insular or isolated mountains, *insula-montanus*
insulus -a -um growing in scattered blocks, *insula, insulae*
intactus -a -um unopened, untouched, undefiled, chaste, *intactus* (the flowers, especially when self-pollinated)
intaminatus -a -um chaste, unsullied, *intaminatus*
integer -era -erum, integra, integrum, integri- undivided, entire, intact, whole, *integer, integri*
integerrimus -a -um whole, not divided or lacking parts, *integerrimus*
integrifolius -a -um with entire leaves, *integri-folium*
inter- between-, *inter*
interamericanas between North and South America, *inter-americana*
intercedens between the parents, intermediate, coming between, *intercedeo*
intercursus -a -um intervening, crossing-over, *intercurso, intercursare* (to attack between the lines)
interiorubrus -a -um having red on the inside, *interior-ruber*
interjacens lying between, *interiaceo, interiacere* (intermediate)
interjectus -a -um intermediate in form, interposed, *intericio, intericere, interieci, interiectum* (between two other species)
intermedius -a -um between extremes, intermediate, *inter-medium*
intermis -is -e, intermissus -a -um lacking continuity, neglected, interrupted, *intermitto, intermittere, intermisi, intermissum*
intermixtus -a -um intermingled, mixed together, *intermixtus*
interpositus -a -um introduced, placed between, *interpono, interponere, interposui, interpositum*
interruptus -a -um with scattered leaves or flowers, with gaps in the infructescence, *interrumpo, interrumpere, interrupi, interruptum*
intersitus -a -um, interstes interposed, *intersero, interserere, interserui, intersertum*
interstitius -a -um having evident air spaces (smaller than lacunae), *interstitium*
intertextus -a -um interwoven, *intertextus*
intonsus -a -um bearded, unshaven, long-haired, unshorn, *intonsus*
intortus -a -um curled, twisted, *intorqueo, intorquere, intorsi, intortum*
intra-, intro- within-, inside-, *intra*
intramarginalis -is -e distinctly within the margin, *intra-*(*margo, marginis*) (a conspicuous vein, sori etc.)
intranervatus -a -um sparsely veined, *intra-nerva*
intricatissimus -a -um completely entangled, superlative of *intricatus*
intricatus -a -um entangled, *intrico, intricare*
introflexus -a -um turned or bent backwards, *intro-*(*flecto, flectere, flexi, flexum*)
introlobus -a -um having lobes inside (the corolla), *intro-lobus*
introrsus -a -um facing inwards, turned towards the axis, introrse, *introrsum, introrsus*
intrusus -a -um projecting forwards, *intrusus*
Intsia etymology doubtful (an Indian vernacular name?)
intumescens becoming swollen, swelling, present participle of *intumesco, intumescere, intumui*
intybus from a name, *indivia*, in Virgil for wild chicory or endive (εντυβιον, εντυβον, from Arabic, tybi, for its harvest time January)

Inula a name, *enula campana*, in Pliny for *Inula helenium*, elecampane; some derive it as cognate with ελενιον
Inulanthera *Inula*-bloomed, botanical Latin from *Inula* and ανθερος
inuncans covered with hooked hairs or glochidia, *in-uncus*
inunctus -a -um having an oily surface, anointed, *inunguo, inunguere, inunxi, inunctum*
inundatus -a -um of marshes or places which flood periodically, flooded, *inundo, inundare, inundavi, inundatum*
-inus -a -um -ine, -ish, -like, -resembling, -from, (adjectival suffix to a noun)
inutilis -is -e harmful, useless, *inutilis*
invaginatus -a -um enclosed in a sheath, *in-vagina*
invasorius -a -um invasive, *invado, invadere, invasi, invasum*
invenustus -a -um lacking charm, unattractive, *invenustus*
inversus -a -um turned over, inverted, *inverto, invertere, inverti, inversum*
invisus -a -um detested, hostile; not obvious, not visible, *invisus* (creeping below other vegetation)
involucratus -a -um surrounded with bracts, involucrate, with an involucre, *involucrum* (the flowers)
involutus -a -um obscured, rolled inwards, involute, *involvo, involvere, involvi, involutum*
involvens entangling, enveloping, wrapping up, present participle of *involvo, involvere, involvi, involutum*
iocastus -a -um for Jocasta, mother and wife of Oedipus
Iocenes an anagram of the related genus, *Senecio*
Iochroma Violet-colour, ιο-χρωμα (flower colour)
Iodanthus, iodanthus -a -um Violet-flowered, ιοδο-ανθος
Iodes, iodes Violet-like, violet-coloured, ιωδης, resembling *Viola*, ιον-(ειδες, ωδης) (late Latin from the colour of iodine vapour, iod-ine)
Iodina, iodinus -a -um violet-coloured, ιοδινος, adjectival suffix -ινος
Iodocephalus Iodine-coloured-head, ιοδο-κεφαλη (violet-brown)
ioensis -is -e from Iowa, USA
ioessus -a -um violet-coloured, ιοεις
ion-, iono- violet-, ιον- (formerly used for various plants with fragrant flowers, e.g. stock, or wallflower)
-ion -occurring
Ionacanthus Violet-*Acanthus*, ιον-ακανθα
Ionactis Violet-rayed, ιον-(ακτις, ακτινος)
ionandrus -a -um having violet stamens, ιον-ανηρ
ionantherus -a -um, ionanthes violet-flowered, ιον-ανθηρος
ionanthus -a -um with violet-coloured flowers, ιον-ανθος
ionenis -is -e of the Iones, from the Ionian islands or sea, W Greece
ionicus -a -um from the Ionian islands, W Greece, *Ionia*
ionidiflorus -a -um having violet-like flowers, botanical Latin from ιον and *florum*
Ionidium Violet-like, ιον-οειδης
ionochlorus -a -um violet-green, ιον-χλωρος (variable ochreous bluish green colouration)
ionophthalus -a -um violet-eyed, ιον-οφθαλμος
Ionopsidium Appearing-like a-small-violet, ιον-οψις (diminutive suffix)
Ionopsis Violet-looking, ιον-οψις (flower colour of violet cress)
ionosmus -a -um violet-scented, ιον-οσμη
Iostephane Violet-crown, ιο-(στεφνος, στεφανη)
ipecacuanha a Tupi vernacular name, ipekaaguebe, for the drug used against dysentery from the rhizomes of *Cephaelis ipecacuanha*
Ipheion a name, ιφυον, used by Theophrastus (ιφιος strong or fat)
Iphigenia Valiant-occurrence, ιφι-γενεα (for the deep crimson flowers); in mythology, Iphigenia was the brave daughter of Agamemnon and Clytemnestra
Ipomoea Worm-resembling, ιψ-ομοιος (the sinuous twining stems)

Ipomopsis Resembling-*Ipomoea*, ιψ-ομοιος-οψις; some interpret as Conspicuous, υπομ-οψις
iquiquensis -is -e from Iquique, N Chile
Iranecio Iranian-plant-resembling-*Senecio*
iranicus -a -um from Iran, Iranian
ircutianus -a -um from Irkutsk province, W and N of Lake Baikal, Russia
irenaeus -a -um peaceful, ειρηναιος (Irene was the goddess of peace)
Iresine Woolly, ειρος (the indumentum on the flowers)
Iriartea for Don Bernardo de Yriarte, eighteenth-century Spanish botanist and patron of science
iricolor of *Iris* colours, *iris-color*
iricus -a -um from Ireland, Irish (Eire)
iridescens iridescent, *iris-essentia* (having many colours when seen from different angles)
iridi- rainbow-coloured, *Iris*-like, ιρις, ιριδος, *iris, iridis*
iridiflorus -a -um *Iris*-flowered, *iridi-florum*
iridioides similar to *Iris*, ιριδος-οειδης, *Iris-oides*
Iridodictyum *Iris*-net, ιριδος-δικτυον (≡ *Iris* of the section *Reticulata*)
Iridosma Iris-scented. (ιρις, ιριδος)-(οδμη, οσμη)
irio the name in Pliny for a cruciferous plant (σισυμβριον of the Greeks)
Iris the name, Iris, of the mythological messenger of the gods of the rainbow, cognate with orris (***Iridaceae***)
irisanus -a -um from Irisan, Luzon, Philippines
irradians irradiating, shining out, present participle of *ir-(radio, radiare)*
irrasus -a -um unshaven, rough, stubbly, *irrasus*
irregularis not of the rules, with irregularly sized parts, *ir-(regula, regulae)* (floral organs)
irrigatus -a -um of wet places, watered, flooded, *irrigo, irrigare, irrigavi, irrigatum*
irriguus -a -um watered, watery, *irriguus* (has clammy hairs)
irritabilis -is -e sensitive, excitable, *irritabilis*
irritans causing irritation, present participle of *irrito, irritare, irritavi, irritatum*
irroratus -a -um bedewed, dewy, *irroro, irrorare* (to bedew)
Irvingia, irvingii for Dr Edward George Irving (1816–55), Scottish surgeon and collector in S Nigeria (***Irvingiaceae***)
Isabella for Isabel Countess d'Eu, Brazilian patroness of science
isabellae for Isabel Forrest, daughter of the plant collector, George Forrest
isabellinus -a -um drab-yellowish, tawny-grey, uncomplimentarily for Isabella (1451–1504), Queen of Spain
Isachne Equal-scales, ισο-αχνη (the lemmas in some are identical)
isandrus -a -um equal-stamened, with equal stamens, ισο-ανηρ
isatidea like *Isatis*
Isatis Hippocrates' and Dioscorides' name, ισατις, for woad (the Latin name was *Glastum*)
isauricus -a -um from Isauria, Anatolia, the birthplace of the Byzantine emperor Zeno (Isaurian Ascendancy 474–491)
Ischaemum, ischaemum Blood-stopper, ισχανω-αιμα (a name in Pliny for its styptic property)
Ischnoderma Thin-skinned-one, ισχνος-δερμα (resin exudes)
Ischnogyne Slender-ovary, ισχνος-γυνη (the column)
Ischnolepis Slender-scaled-one, ισχνος-λεπις (coronnal scales)
ischnophyllus -a -um having thin, weak or dry leaves, ισχνος-φυλλον
ischnopodus -a -um slender-stalked, ισχνος-(πους, ποδος)
ischnopus -a -um thin stalked, with slender stems, ισχνος-πους
Ischnosiphon Slender-tubed-one, ισχνος-σιφον (the hollow stems)
Ischnostemma Slender-crown, ισχνος-στεμμα
Ischyrolepis Strongly-scaled, ισχυρος-λεπις
-iscus -a -um -lesser (diminutive suffix)

iseanus -a -um, isensis -is -e from Ise-shima national park, Honshu, Japan
Iseilema Equal-covers, ισ-ειλυμα
Isertia for Paul Erdmann Isert (1756–89), Danish doctor in W Africa and Guiana
Isidorea for Isidorus Hispalensis (560–636), Bishop of Seville, author of the encyclopaedic *Etymologiarum*
Isidrogalvia for Isidro Galvez
islandicus -a -um from Iceland, Icelandic
islayensis -is -e from Islay region, S Peru
Ismene for Ismene, the daughter of Oedipus and Jocasta (≡ *Hymenocalis*)
Isnardia for A. T. Danty d'Isnard (1663–1743), professor of botany at Paris
iso- equal-, ισος, ισο-
isobasis -is -e equal-footed; equal-founded, with regular bases, ισο-βασις
Isoberlinia Equal-to-*Berlinia*, botanical Latin from ισο and *Berlinia* (related genus)
Isocheilus Equal-lip, ισο-χειλος (the laterals equal the labellum)
Isodendron Equal-to-a-tree, ισο-δενδρον (arborescent *Violaceae*)
Isodictyophorus Bearing -regular-net, ισο-δικτυον-φερω (*Isodictyophorus reticulatus*)
Isoëtes Equal-to-a-year, ισο-ετος (Pliny's name, *isoetes*, implies green throughout the year) (***Isoetaceae***)
isoetifolius -a -um having leaves similar to those of *Isoetes, isoetes-folium*
Isoetopsis Resembling-*Isoetes*, ισο-ετος-οψις (physical similarity)
Isoglossa Equal-tongued-one, ισο-γλωσσα
Isolepis Equal-scales, ισο-λεπις (the upper and lower glumes, ≡ *Scirpus*)
Isoloma Equal-border, ισο-λωμα (the equal lobes of the perianth)
Isolona Equal-petals, ισο-λωνα (the equal petals); some interpret as Equal-to-*Annona* (related genus)
Isomeris, isomeris -is -e Equal-parts, ισο-μερος (floral parts not equal; malodorous throughout)
Isometrum Equal-dimension, ισο-μετρον (very regular symmetry)
Isonandra Equal-stamens, ισο-ανηρ (equal numbers of fertile and sterile stamens)
Isonema Equal-threads, ισο-νημα (the exerted stamens)
isopetalus -a -um having uniformly shaped petals, ισο-πεταλον
isophyllus -a -um equal-leaved, uniformly leaved, ισο-φυλλον
Isophysis Equal-parts, ισο-φυσις
Isoplexis Equal-folds, ισο-πλεκω (upper corolla lobe and the lip)
Isopogon Equal-beard, ισο-πωγων (fringed flowers)
isopyroides resembling *Isopyrum, Isopyrum-oides*
Isopyrum Equalling-wheat, ισο-πυρος (fruits similar to wheat grains)
Isotoma Equal-division, ισο-τομη (the equal corolla segments, ≡ *Lobelia*)
Isotria Three-equal, ισο-(τρεις, τρια) or Equal-triad, ισο-τριας (the sepals)
Isotropis Equal-keeled, ισο-τροπις (the carina)
israeliticus -a -um of the Israelites
issicus -a -um from Issus, Cilicia, Turkey
-issimus -a -um -est, -the best, -the most (superlative suffix)
-ister -ra -rum see *-aster* (this suffix is added to genera based on a vowel stem, e.g. *Sinapistrum, sinapi-istrum*)
Isthmia Girdle, ισθμος (the diatom's necklace-like band)
isthmius -a -um necklaced, ισθμιον
istria, istriacus -a -um from Istria, Croatia
itabiritensis -is -e from Itabira, the Iron Mountain area, Minas Gerais, Brazil
italicus -a -um from Italy, Italian (*Italia*)
itatiaiae from the Pico de Itatiaia, Itatiaia national park, Brazil
Itea Greek, ιτεα, for a willow (***Iteaceae***)
iteaphyllus -a -um, iteophyllus willow-leaved, ιτεο-φυλλον
iteratus -a -um repeated, *iteratus* (growth cycle)
-ites,-itis -closely resembling, -having, -related to, -ιτης, -ιτις
ithy- straight-, erect-, ιθυς, ευθυς
Ithycaulon straight-stemmed, ιθυς-καυλος

ithypetalus -a -um having erect petals, ιθυς-πεταλον
Itoa for Tokutaro Ito (1868–1941), Japanese botanist
ituriensis -is -e from Zaire's Ituri forest
-ium -lesser (diminutive suffix)
Iva, iva an old name used by Rufinus, applied to various fragrant plants and re-applied by Linneaus as a fragrant genus of Composites)
Ivesia for Eli Ives (1778–1861), American physician and botanist, professor at Yale
Ivodea an anagram of *Evodia* (≡ *Euodia* q.v.)
ivorensis -is -e from the Ivory Coast, W Africa
Ixerba an anagram of *Brexia* (related genus)
Ixeridium Resembling-*Ixeris*
Ixeris etymology uncertain
Ixia Bird-lime, ιξος (Theophrastus' name refers to a thistle with clammy sap)
Ixianthes Sticky-flower, ιξος-ανθος
Ixiochlamys Sticky-cover, ιξος-χλαμυς
ixioides resembling *Ixia*, ιξος-οειδης
Ixiolaena Sticky-mantle, ιξος-χλαινα
Ixiolirion *Ixia*-lily, ιξια-λειριον (the superficial resemblance)
Ixocactus Sticky-cactus, ιξος-κακτος (*Loranthaceae*)
ixocarpus -a -um sticky-fruited, ιξος-καρπος
Ixodia Resembling-mistletoe, ιξοδης (the sap and similar leaf-shape)
Ixonanthes Sticky-flower, ιξος-ανθος (***Ixonathaceae***, ≡ ***Irvingiaceae***)
Ixophorus Bird-lime-carrying, ιξος-φορα
Ixora the name of a Malabar deity, Iswara (Sanskrit, icvara)
izuensis -is -e from the islands of Izu peninsula, Japan

jaborandi, jaburan from the Tupi-Guarani, jaburandiba, for 'he who spits' (salivation is caused by chewing its leaves)
Jaborosa from the Arabic name for a *Mandragora*
Jacaranda from the Tupi Guarani S American name, jakara'nda, for *Jacaranda cuspidifolia*
Jacea, jacea medieval name with Spanish roots, for knapweed
jacens hanging loose; lying flat, present participle of *iaceo, iacere, iacui*
jacinthinus -a -um reddish-orange coloured (*iacuntus*, relates to *Hyacinthus*)
Jackia, jackii for John George Jack (1861–1949), Canadian dendrologist at Arnold Arboretum
Jackiopsis Resembling-*Jackia*, botanical Latin from *Jackia* and οψις
jackmanii for G. Jackman, plant breeder of Woking *c.* 1865
Jacksonia, jacksonii for George Jackson (1790–1811), Scottish botanist and illustrator
Jacmaia an anagram of Jamaica, its provenance
jacobaeae of ragwort, living on *Senecio jacobaea* (*Contarinia*, dipteran gall midge)
jacobaeus -a -um either for Saint James (*herba sancti Jacobi*, flowering about 25 July) or from Iago Island, Cape Verde
Jacobinia from Jacobina, Brazil
jacobinianus -a -um for G. A. von Jacobi (1805–74) of Berlin
Jacobsenia for Hermann Johannes Heinrich Jacobsen (1898–1978), German botanist and Curator at Kiel botanic garden
Jacquemontia, jacquemontianus -a -um, jacquemontii for Victor J. Jacquemont (1801–32), French traveller in the E Indies and naturalist
Jacquinia, Jaquiniella for Nicholas Joseph Franz de Jacquin (Baron von Jacquin) (1727–1817), Professor of Botany at Leiden
jactus -a -um spreading out, scattered, *iacto, iactare, iactavi, iactatum*
jaegeri for Paul Jaeger (b. 1905), French botanist in W Africa
Jagera for Herbert de Jager, Dutch botanist and collector in Indonesia
jalapa from Jalapa, Veracruz (*Mirabilis jalapa* false jalap); true purgative jalap is derived from *Ipomoea purga* (*Exogonium purga*).

Jaliscoa, jaliscanus -a -um from Jalisco, Mexico
Jaltomata a Mexican vernacular name for false holly
jamaicensis -is -e from Jamaica
jambolana from a Hindu name, jambosa, for *Eugenia jambolana*
Jambosa, jambos from a Malaysian name, shamba, for rose-apple (*Eugenia jambos*)
jambosellus -a -um like a small rose-apple, diminutive from *Jambosa*
Jamesbrittenia for James Britten (1846–1924), English botanist at Kew and British Museum (NH)
Jamesia, jamesii for Dr Edwin P. James (1797–1861), American botanist on Major Long's Rocky Mountains expedition of 1820
jamesianus -a -um, jamesii for George Forrest's brother, James Forrest
Jamesonia, jamesonii for Dr William Jameson (1796–1873), Scottish surgeon botanist on S American expeditions, Professor of Botany at Quito, Ecuador
Jancaea for Victor von Janka (1837–1900), Austrian botanist who studied the plants of the Danube
Jankaea, jankae, jankiae for Victor von Janka
Jansonia for Joseph Janson (1789–1846), English botanist
janthinus -a -um bluish-purple, violet-coloured, ιανθινος
januensis -is -e from Genoa, N Italy, Genoan (*Genua*)
japonese, japonicus -a -um (iaponicus -a -um) from Japan, Japanese
japonica-verschafeltii Verschaffelt's *Zelkova japonica*, ≡ *Z.* ×*verschaffeltii*
Japonolirion Japanese-lily, botanical Latin from Japan and λειριον (genus of *Melianthaceae*)
japurus -a -um from the environs of the Japura river, Amazonia
Jaquemontia for Victor Jaquemont (1801–32), French collector for the Paris Jardin des Plantes
Jasarum for Julian Alfred Steyermark (1909–88), American systematist (botanical Latin from his initials JAS and *Arum*)
Jasione Healing-one, ιασις (from Theophrastus' name, ιασιωνη, for *Convolvulus*)
jasmineus -a -um resembling *Jasminum*
jasminiflorus -a -um with jasmine-like flowers, *Jasminum-florum*
Jasminocereus Jasmine-cactus, *Jasminum-cereus* (floral fragrance of candelabrum cactus)
jasminodorus -a -um jasmine-scented, *Jasminum-odoris*
jasminoides jasmine-like, *Jasminum-oides*
Jasminum Latinized from the Persian name, yasemin, Arabic, yasamin, for perfumed plants
Jasonia for Jason, son of Iolcos and leader of the Argonauts in the search for the Golden Fleece
jaspidius -a -um, iaspidius -a -um jasper-like, striped or finely spotted in many colours
jatamansi a Sanskrit vernacular name for *Nardostachys* (the rhizomes of which are used to prepare the ointment called Spikenard)
Jateorhiza Physician's-root, (ιατηρ, ιατηρος, ιατρος)-ριζα (*Radix colomba* tonic)
Jatropha Physician's-food, ιατρος-τροφη (medicinal use) (the signature of the swollen stem base of *Jatropha podagrica* confers its name of gout plant)
Jaubertia for Comte Hyppolyte François de Jaubert (1798–1874), French botanist, founder of the French Société Botanique
Jaumea for Jean Henri Jaume Saint Hilaire (1772–1845), French botanist and illustrator, author of the illustrated *Plantes de France* (1808–22)
javalensis -is -e, javanicus -a -um from Java, Javanese
javariensis -is -e from the Yavari region, Amazonia
Jeffersonia for Thomas Jefferson (1743–1826), American naturalist, historian, philanthropist, President who strove to end slavery
jeffreyi for J. Jeffrey, Scottish gardener and collector in Oregon 1850–54
jejunifolius -a -um insignificant-leaved, *ieiunus-folium*
jejunus -a -um barren, poor, meagre, small, *ieiunus*

jemtlandicus -a -um from Jemtland, W Sweden
Jenmaniella, jenmannii for George Samuel Jenman (1845–1902), English Curator of the Botanic Garden in Jamaica and writer on ferns
Jepsonia for Willis Linn Jepson (1867–1946), American surgeon and naturalist, professor at University of California
jessoensis -is -e, jezoensis -is -e from Jezo (Yezo), Hokkaido, Japan
Joannesia for Joannes (João V, 1706–50), King of Portugal
jocundus -a -um see *jucundus*
Johannesteijsmannia for Johannes Elias Teijsmann (1808–82), Dutch botanist and Curator of Bogor, Buitenzorg Gardens, Java
johannis -is -e from Port St John, S Africa (*joannis*)
johimbe from a vernacular name, yohimbine, for the stimulant derivative
johnsonii for J. E. Johnson (1817–82), American botanist
johnstonii for either Mr Johnston of Oporto, *c.* 1886, or Sir Henry Hamilton Johnston (1858–1927), Governor of the Uganda Protectorate
Joinvillea for Prince François Ferdinand Philippe Louis Marie d'Orleans de Joinville, naval officer, for a time exiled, author of *Essais sur la marine française* (1852), son of Louis-Philippe, Duc d'Orleans
jolonensis -is -e from Jolo Island, Philippines
jonesii for Morgan Jones, orchid grower
jonquilleus -a -um the bright yellow of *Narcissus odorus*
jonquillus -a -um from the Spanish vernacular name, jonquillo (little rush) for jonquil, *Narcissus jonquilla*
jorullensis -is -e from area around El Jorullo, recent volcanic area to the east of the Sierra Madre, Mexico
Josephinia for the Empress Marie Josephine Rose Tascher de la Pagerie (1763–1814), wife of Napoleon Bonaparte
josikaea for the Hungarian Baroness Rosa von Josika, *c.* 1831
Jovellana for Don Caspari Melchior de Jovellanos (Jove Llanos) (1744–1811), Spanish statesman and patron of botany, student of Peruvian plant life
Jovibarba Jupiter's-beard, *Iovis-barba* (the fringed petals)
jovis-tonantis for the Roman state god, Jupiter (*Jovis pater*), in his guise as Jove the Thunderer (*Jovis tonantis*)
juanensis -is -e from Genoa, N Italy, or from San Juan, Argentina
Juania from the islands west of Valparaiso, found by and named for the navigator Juan Fernandez (1536–1604)
Juanulloa for George Juan and Antonio Ulloa, Spanish explorers of Peru
Jubaea for King Juba of Numidia (Algeria), who wrote on Arabian natural history
Jubaeopsis With-the-appearance-of-Jubaea, botanical Latin from *Jubaea*, and οψις
jubatus -a -um maned, with a crest, *juba, jubae* (crested with long awns)
jucundus -a -um pleasing, delightful, *iucundus*
judaicus -a -um of Judaea, Jewish, from Palestine
judenbergensis -is -e from the Judenburg mountains, Austria
jugalis -is -e, jugatus -a -um joined together, yoked, *iugalis*
juglandi- *Juglans*-like-
juglandifolius -a -um having leaves resembling those of walnut, *Juglans-folium*
Juglans Jupiter's-nut, *Iuglans* (*glans Jovis* in Pliny) (brought to England by the Romans, in old English it was walh-hnut or foreign-nut, walnut) (***Juglandaceae***)
jugosus -a -um hilly, ridged, *iugo, iugare, iugavi, iugatum*
-jugus -a -um -yoked, -paired, *iugum, iugi*
jujuba from an Arabic name, jujube, for *Zizyphus jujuba* (both Latin words are cognates of the Greek, ζιζυφον)
jujuyensis -is -e from Jujuy province, N Argentina
juliae for Julia Ludovicowna Mlokosjewitsch, who, *c.* 1900, discovered *Primula juliae*
julibrissin silken, from the Persian name for *Acacia julibrissin*
julibrissius -a -um silken, see *julibrissin*

juliflorus -a -um silken-flowered, *julibrissin-florum*
juliformis -is -e downy, *julibrissin-forma*
julii for Julius Derenberg of Hamburg, succulent grower
Jumellea, Jumelleanthus for Henri Lucien Jumelle (1866–1935), French botanist
Juncago Slender-rush, feminine suffix on *juncus* (Tournefort's name for *Triglochin*) (***Juncaginaceae***)
junceiformis -is -e resembling (*Agropyron*) *junceum*, or resembling *Juncus*
juncellus -a -um like a small rush, diminutive of *Juncus*
junceus -a -um, juncei-, junci- rush-like, resembling *Juncus*
juncifolius -a -um rush leaved, *Juncus-folium*
juncorus -a -um of rushes, living on *Juncus* (*Livia*, homopteran, psyllid gall insect)
Juncus Binder, *iungo, iungere, iunxi, iunctum* (classical Latin name refers to use of rushes for weaving and basketry) (***Juncaceae***)
jungens linking, joining together, present participle of *iungo, iungere, iunxi, iunctum*
Jungia for Joachim Jung (1587–1657), German polymath who made early landmarks in plant terminology and nomenclature
juninensis -is -e from Junin department, central Peru
junipericolus -a -um living on *Juniperus, Juniperus-colo*
juniperifolius -a -um with leaves resembling those of *Juniperus*
juniperinus -a -um bluish-brown, juniper-like, resembling *Juniperus* or its berry colour, living on *Juniperus* (*Oligotrophus*, dipteran gall midge)
juniperoides resembling *Juniperus*
Juniperus the ancient Latin name, *iuniperus* (with cognates such as *Juncus*, for binding, Geneva and gin)
junonia, junonis -is -e, junos for the Roman goddess Juno, wife of Jupiter
juranus -a -um from the Jura mountains, France–Switzerland
jurassicus -a -um from the Jura mountains, France–Switzerland
Jurinea for Louis Jurine (1751–1819), Professor of Medicine
juruanensis -is -e from the environs of the Juruá river that joins the Solemöes (Amazon) river at Tamaniquá
Jussieua (Jussiaea), jussieui for Bernard de Jussieu (1699–1777) who made a major contribution to establishing the concept of the taxonomic species and of natural classification
Justicia for James Justice (1698–1763), Scottish legal clerk and horticulturalist
Juttadinteria for Jutta Dinter, wife of German botanist Moritz Kurt Dinter
juvenalis -is -e youthful, *iuvenilis*; for Juvenal the Roman satirist (descriptive of the juvenile phase of plants that go on to adopt a mature phase with distinct morphological and biological features, *Hedera, Retinopspora, Chamaecyparis*)

Kabulia from Kabul, Afghanistan
Kadsura from the Japanese vernacular name for *Kadsura japonica*
Kaempferia, kaempferi for Engelbert (Englebrecht) Kaempfer (1651–1715), German physician and botanist, of the Swedish Embassy to Persia, author of *Amoenitatum exoticarum* (1712)
kahiricus -a -um from Cairo, Egypt (El Qahirah)
kaido a Japanese name for *Malus* × *micromalus* (*spectabilis* × *ringo*)
kaki from the Japanese name, kaki-no-ki, for persimmon (*Diospyros kaki*)
Kalaharia from the Kalahari desert of Namibia, SW Africa
Kalanchoe from a Chinese vernacular name
kalbreyeri for M. W. Kalbreyer (1847–1912), collector for Veitch in W Africa and Brazil
kali, kali- either from the Persian for a carpet, or a reference to the ashes of saltworts being alkaline (al-kali); cognate with *kalium* (Potassium)
Kalimeris Beautiful-parts, καλος-μερις
Kaliphora Bearer-of-beauty, καλος-φορα

Kalmia, kalmianus -a -um for Peter Kalm (1716–79), a highly reputed Finnish student of Linnaeus
kalmiiflorus -a -um with flowers resembling those of *Kalmia, Kalmia-florum*
Kalmiodendron a composite name for hybrids between *Kalmia* and *Rhododendron*
Kalmiopsis resembling *Kalmia*, botanical Latin from *Kalmia* and οψις
Kalmiothamnus the composite name for hybrids between *Kalmia* and *Rhodothamnus*
kalo- beautiful-, καλος, καλο-
kaloides of beautiful appearance, καλ-οειδης
Kalopanax Beautiful-*Panax*, καλο-παν-ακεσις
kamaonensis -is -e from Kumaon, Nepal
kamerunicus -a -um from Cameroon (Cameroun), W Africa
kamtschaticus -a -um from the Kamchatka peninsula, E Russia
kanabensis -is -e from Kanab, Utah, and Kanab Creek, Arizona, USA
Kanahia from an E African vernacular name
Kandaharia from Kandahar, Afghanistan
Kandelia from the Malabari vernacular name for *Kandelia candel*
kansuensis -is -e from Kansu (Gansu) province, China
kapelus -a -um from Kapela mountains, Croatia
karadaghensis -is -e from Karadagh region, NE of Tabriz, Iran (Persian)
karakoramicus -a -um from the Karakoram mountain range, Kashmir
karasbergensis -is -e from Karasberg, Namibia, SW Africa
Karatavicus -a -um, karataviensis -is -e from Kara Tau mountains of Kazakhstan
Karelinia, karelinii for Gregor Karelin (1801–72), Russian botanist explorer, who found *Fritillaria karelinii*
karibaensis -is -e from the environs of Lake Kariba, Zambia/Zimbabwe, Africa
karooensis -is -, karooicus -a -ume from the S African Karoo (*vide infra*)
karpathensis -is -e from Karpathos island, Aegean, Greece
karroo from the name of the S African semi-desert plateau, Karoo
karsensis -is -e, karsianus -a -um from Kars river area, N Turkey
karsticolus -a -um inhabiting the Karst mountains, Dalmatian Adriatic
Karwinskia, karwinskii, karwinskianus -a -um for Wilhelm Friederich Karwinsky von Karwin (1780–1855), German plant collector in Brazil and Mexico
kashgaricus -a -um from K'a Shih (Kashgar), Sinkiang region, China
Kashmiria from Kashmir
kashmirianus -a -um from Kashmir
katangensis -is -e from Katanga (Shaba), Democratic Republic of the Congo
katherinae for Mrs Katherine Saunders (1824–1901), who collected plants in Natal
kattegatensis -is -e from the Swedish 'cat's throat' strait (Kattegat) connecting the Baltic to the North Sea
kauensis -is -e from the volcanic Kau littoral desert, SE Hawaii
Kaulfussia, kaulfussii for Dr G. F. Kaulfuss (d. 1830), Professor of Botany at Halle, Germany
kawakamii for Takija Kawakami (1871–1915), Formosan botanist
Kearnemalvastrum Kearney's *Malvastrum*, botanical Latin for Thomas Henry Kearney (1874–1956) and *Malvastrum*
Keckia, Keckiella for David Daniels Keck (1903–95), American experimental taxonomist, who did pioneering work with Jens Clausen and William M. Heisey
Kedrostis a name in Pliny for a white vine
Keetia for Johan Diederik Mohr Keet (1882–1967), S African botanist
keetmanshoopensis -is -e from Keetmanshoop, SE Namibia
Kefersteinia for Adolf Keferstein, German orchidologist and lepidopterist
Kegeliella for Hermann Kegel (1819–56), head gardener at Halle botanic garden
keleticus -a -um charming, κηλητικος (κηλητηριον a charm)
Kelleria for Engelhardt Keller, nineteenth-century German writer
Kelloggia for Albert Kellogg (1813–87), American physician and botanist
Kelseya for Harlan P. Kelsey (1872–1958), nurseryman of Massachusetts, USA
keniensis -is -e, kensiensis -is -e from Kenya, E Africa

Kennedya, kennedyi for John Kennedy (1759–1842), of the Lee & Kennedy nursery in Hammersmith, London
kent-, kentro-, -kentron goad-, spur-, -spurred, κεντρον, κεντρο-, κεντρ-
Kentia, Kentiopsis for Colonel Kent, English soldier and botanist
Kentranthus Spurred-flower, κεντρ-ανθος (the spur at the base of the corolla) (*vide Centranthus*)
kentrophyllus -a -um having spurred leaves, κεντρο-φυλλον
kentrophytus -a -um painful or prickly plant, κεντρο-φυτον (petiolar thorns of *Astragalus kentrophyta*)
kentuckiensis -is -e from Kentucky, USA
keriensis -is -e from Kerry, Ireland
Kermadecia, kermadecensis -is -e from the volcanic Kermadec islands, NE of Auckland, New Zealand
kermesinus -a -um carmine-coloured, carmine (the kermes oak, *Quercus coccifera*, is host to the insect, *Kermes ilicis*, from which is obtained the red dye, Arabic, kirmiz or qirmiz) cognate with crimson
Kernera, kernerianus -a -um, kerneri for Johann Simon von Kerner (1755–1830), Professor of Botany, Stuttgart
kero- bees-wax, κηρος
Kerria, Kerriochloa, Kerriodoxa for William Kerr (d. 1814), superintendent of the Botanic Garden Ceylon and collector of Chinese plants at Kew
kerrii for Arthur Francis George Kerr (1877–1942), collector in Siam
kesiyus -a -um from the Khasi Hills, Assam, N India
Keteleeria for Jean Baptiste Keteleer (1813–1903), French nurseryman
kevachensis -is -e from Kevachi volcano area, Solomon Islands
kewensis -is -e of Kew Gardens
keyensis -is -e from the environs of the Florida Keys, USA
keysii for Mrs Keys, a friend of Thomas Nuttall (q.v.)
khasianus -a -um from the Khasi Hills, Assam, N India
Khaya from W African Wolof and Fulani vernacular names, kaye and khaye, for *Khaya senegalensis*
khuzestanicus -a -um from the Khusistan region of Iran
Kibatalia from the Sudanese vernacular name, ki batali
kibbiensis -is -e from the Kibbi hills, Ghana, W Africa
Kickxia for Jean Jaques Kickx (1775–1831), Belgian cryptogamic botanist, author of *Flora Bruxellensis* (1812)
Kigelia from the native Mozambique vernacular name, kigeli keia, for the sausage tree
kilimandscharicus -a -um from Mount Kilimanjaro, NE Tanzania
Killipia, Killipiodendron for Ellsworth Paine Killip (1890–1968), American botanist at the Smithsonian National Herbarium
Kimjongillia used in 2003 for a plant to be grown competitively in N Korea, for Kim Jong Il (1942–), the country's ruler
kinabaluensis -is -e from Mount Kinabalu, N Borneo, E Malaysia
Kingdonia, kingdonii for Captain Francis Kingdon-Ward (1885–1958), botanist, extensive traveller in the East, writer and plant introducer
Kingiella, Kingiodendron, kingianus -a -um, kingii for Sir George King (1840–1909), English physician and botanist, Director of the Botanic Garden, Calcutta
kiotensis -is -e from Kyoto, Honshu, Japan
Kirengeshoma from the Japanese, ki- (yellow) -renge-shoma
kirilowii for Ivan Kirilow (1821–42), Russian botanist
Kirkia for Sir John Kirk (1832–1922), Scottish physician and naturalist, Consul at Zanzibar and botanist
kirkii for either Sir John Kirk (1832–1922), or Thomas Kirk, writer on New Zealand plants
kirro- tawny, citron-coloured-, κιρρος, κιρρο-
kirroanthus -a -um citron-flowered, κιρρο-ανθος

kishtvariensis-is -e from Kishtwar, Kashmir
kisso- ivy-, ivy-like-, κισσος
Kissodendron Ivy-tree, κισσος-δενδρον (Queensland, Australia)
kitadakensis -is -e from Kita Dake, Japan
Kitaibela (Kitaibelia), kitaibaleianus -a -um, kitaibelii for Professor Pal Kitaibel (1757–1817), botanist at Pécs, Hungary
kiushianus -a -um, kiusianus -a -um from Kyushu, one of the major islands forming S Japan
kiyosumiensis -is -e from Kiyosumi, Kominato, E Japan
Klattia, klattianus -a -um, klatii for Friedrich Wilhelm G. Klatt (1825–97), a contributor to *Flora Brasiliensis*
Kleinhovia for Dr Christiaan Kleinhoff (d. 1777), Director of the Botanic Garden, Batavia
Kleinia for Jacob Theodore Klein (1685–1759), German botanist and ornithologist
Klotzschia for Johann Friedrich Klotsch (1805–60)
Knappia, knappii for Joseph Armin Knapp (1843–99), Viennese writer on the flora of E Europe
Knautia for Christian (Christoph) Knaut (1654–1716), German botanist and author of *Methodus plantarum genuina*
Knema Internode, κνημη (distinctly nodal)
Knightia, knightii for Thomas Andrew Knight (1758–1838), President of the Horticultural Society of London, who raised the monarch pear, dedicated to William IV, in 1830
Kniphofia for Johannes Hieronymus Kniphof (1704–1763), Professor of Medicine at Erfurt and botanist
Knowltonia for Thomas Knowleton (1692–1782), curator of Eltham botanic garden
Kobresia (Cobresia) for (Paul) Carl von Cobres (1747–1823), Austrian botanist
kobus from a Japanese name, kobushi, for some *Magnolia* species
Kochia, kochianus -a -um for Wilhelm Daniel Joseph Koch (1771–1849), Professor of Botany at Erlangen, or Dr Heinrich Koch (1805–87), a botanist, of Bremen
Koeberlinia for Christoph Ludwig Koeberlin, nineteenth-century German cleric and botanist
koehneanus -a -um, koehnei for Bernhard Adelbert Emil Koehne (1848–1918), Professor of Botany at Berlin
Koeleria for Georg Ludwig Koeler (1765–1807), German physician, botanist and writer on grasses
Koellikeria for Albrecht Kölliker (Koelliker) (1817–1905), Swiss anatomist, professor at Wurzburg
Koelreuteria (Köelreuteria) for Joseph Gottlieb Kölreuter (Koelreuter) (1733–1806), Professor of Natural History, Karlsruhe, pioneer hybridizer
Koenigia (Koenigia, Koeniga) for Johann Gerhard König (Koenig) (1728–85), Latvian student of Linnaeus, missionary and botanist in India
Kohleria for Michael Kohler, nineteenth-century Swiss natural historian
Kohlrauschia for F. H. Kohlrausch, assiduous German lady botanist of Berlin
koilo- hollow-, κοιλος
koilolepis -is -e having hollow scales, κοιλο-λεπις
Kokia from a Hawaiian vernacular name for tree cotton
Kokoona a vernacular name from Sri Lanka
kola as *Cola*, a Mende vernacular name, ngola
Kolkwitzia for Richard Kolkwitz (1873–1956), Professor of Botany, Berlin
Kolobopetalum Shortened-leaf, κολοβοω-πεταλον
kolomicta a vernacular name from Amur, eastern Russia, for *Actinidia kolomicta* (for its varying leaf colour?)
kolpakowskianus -a -um for Gerasim Alexzewitsch Kolpakewsky (d. 1980), Governor of Siebenstrombezirks, Turkestan
komarovi, komarowii for Vladimir Leontiewitch Komarow (1869–1945), Russian botanist

kongboensis -is -e from Kongbo, Tibet
kongosanensis -is -e from Kyongsan province, S Korea
konjac the vernacular name, konjaku, for *Amorphophallus konjac*
Koompassia from a Malayan vernacular name
Koordersiodendron for Sijfert Hendrik Koorders (1863–1919), Dutch botanist at the Bogor Buitenzorg gardens
kopatdaghensis -is -e, kopetdaghensis -is -e from the Kopet-Dag range of mountains, Turkmenistan
kophophyllus -a -um having dull, weak or blunt leaves, κωφος-φυλλον
Kopsia, Kopsiopsis for Jan Kops (1765–1849), Dutch agronomist, professor at Utrecht
koreanus -a -um, koraiensis -is -e from Korea, Korean
korsakoviensis -is -e from Korsakov, Sakhalin island, Russia
Korthalsia for Pieter Willem Korthals (1807–92), Dutch botanist and plant collector in Java, Sumatra and Borneo
kosanini for Nedeljko Košanin (1874–1934) of Belgrade, Serbia
Kosteletzkya for Vincenz Franz Kosteletzky (1801–87), Bohemian physician and botanist
Kostermansia, Kostermanthus for André Joseph Guillaume Henri Kostermans (1907–94), Dutch botanist in Java, Indonesia
kotschianus -a -um, kotschyanus -a -um, kotschyi for Theodore Kotschy (1813–66), Austrian botanist
kousa a Japanese name for *Cornus kousa*
kouytchensis -is -e from Kweichow (Guizou) province, SW China
Krameria for Johann Georg Heinrich Kramer (1684–1744) and his son William Henry Kramer (d. 1765), Austrian botanists (***Krameriaceae***)
Krascheninnikovia for Stephan Petrovich Krascheninnikov (1713–55), Russian botanist and explorer of Siberia
Kraussia, krausianus -a -um, krausii for Christian Ferdinand Friedrich von Krauss (1812–90), German zoologist of Stuttgart, naturalist and collector in S Africa
Kreodanthus Fleshy-flowered-one, (κρεας, κρεως)-ανθος
Krigia for David Krieg (d. 1713), German physician and botanist, collector in N America
Krugiodendron Krug's-tree, for Carl Wilhelm Leopold Krug (1833–98), German botanist
Kuhlhasseltia for Heinrich Kuhl (1796–1821) and Johan Conraad van Hasselt (1796–1823), Dutch botanists
kumaonensis -is -e from Kumaon, N India
Kumlienia for Thure Ludwig Theodore Kumlien (1819–88), Swedish naturalist in N America
Kunstleria for H. H. Kunstler (1837–87), plant collector in Malaysia
Kuntheria, kunthianus -a -um, kunthii for Carl Sigismund Kunth (1788–1850), German systematist and botanical author on plants of the New World
Kunzea for Gustave Kunze (1793–1851), German physician and botanist, director of Leipzig botanic garden
kurdicus -a -um of the Kurds, from Kurdistan (parts of Iraq, Iran, Armenia and Turkey)
kurdistanicus -a -um from Kurdistan
kurilensis -is -e from the Kuril archipelago between the Russian Kamchatka Peninsula and the Japanese island of Hokkaido
kurroo from the Gahrwain vernacular name for *Gentiana kurroo*
kuschkensis -is -e from Kasak Turkestan
kwangsiensis -is -e from Chuang autonomous region of Kwangsi, S China
kwantungensis -is -e from Kwangtung (Guangdong), China
kweitschoviensis -is -e from Kwechow (Guizhou), SW China
Kydia for Col. Robert Kyd (1746–93), soldier in Bengal who founded the Calcutta Botanic Garden

Kyllingia, Kyllingiella for Peter Kylling, seventeenth-century Danish apothecary and botanist
kweo the Congolese vernacular name, mkweo, for *Beilschmiedia kweo*

labdanus -a -um see *ladanum*
-labellus -a -um -lipped, -with a small lip, *labellum, labelli*
labialis -is -e, labiatus -a -um lip-shaped, lipped, labiate, *labia, labiae*
Labichea for M. Labiche, nineteenth-century naval officer
labilis -is -e slippery; unstable, labile, *labilis*
labillardieri for Jaques Julien Houtou de la Billardière (see *Billardiera*)
labiosus -a -um conspicuously lipped, comparative of *labia, labiae*
labius -a -um lip, *labia, labii*
lablab the Turkish name for hyacinth bean, *Dolichos lablab*, from Arabic, lubia; others attribute it to a Hindu plant name
labradoricus -a -um from Labrador, Newfoundland
labrosus -a -um with a pronounced lip, *labrum, labri*
labruscus -a -um a wild vine, *labrusca, labruscae*
labukensis -is -e from Labuan island, Brunei Bay, Sabah, Borneo
laburnifolius -a -um having *Laburnum*-like leaves, *Laburnum-folium*
Laburnocytisus the composite name for the chimaera involving *Laburnum* and *Chamaecytisus*
Laburnum, laburnum an ancient Latin name used by Pliny
labyrinthicus -a -um having a complex pattern of lines, intricately lined, λαβυριν-θος, *labyrinthus*
lac-, lacto- milky-, *lac, lactis*
-lacca -resin, varnished-, from the Italian, lacca
laccatus -a -um with a varnished appearance, *lacca*
lacciferus -a -um producing a milky juice, *lacca-fero*
lacco- pit-, cistern-, pond-, λακκος, λακκο-
Laccodiscus Pit-like-disc, λακκος-δισκος (the concave floral disc)
Laccospadix Cistern-like-spadix, λακκος-σπαδιξ
lacer, lacerus -era -erum, laceratus -a -um torn into a fringe, as if finely cut into, *lacero, lacerare, laceravi, laceratum; laceri-, lacerti-*
lacerifolius -a -um with torn edged or fringed leaves, *laceri-folium*
lacertiferus -a -um having appendages resembling lizard tails, *lacerti-fero*
lacertinus -a -um lizard-tailed (the common garden lizard is *Lacerta vivipara*)
lacerus -a -um torn, having narrow segments as if torn, *lacero, lacerare, laceravi, laceratum*
lach-, lachno- downy-, woolly-, λαχνη, λαχν-, λαχ-
lachanthus -a -um with downy flowers, λαχ-ανθος
Lachenalia, lachenalii for Werner de La Chenal (de Lachenal) (1763–1800), Professor of Botany at Basle, Switzerland, Cape cowslips
lachenensis -is -e from Lachen (Lachung), N Sikkim, India
Lachnellula Little-shaggy-haired-one, feminine diminutive from λαχνη (the small fruiting body has a covering of shaggy hair)
lachnoglossus -a -um woolly-mouthed, λαχνη-γλωσσα
lachnogynus -a -um having a downy ovary, λαχνη-γυνη
Lachnopilis Woolly-felted, λαχνη-πιλος
lachnopus -a -um woolly-stemmed, downy-stalked, λαχνη-πους
Lachnostoma Woolly-mouth, λαχνη-στομα (the throat of the corolla is bearded)
lachungensis -is -e from Lachung (Lachen), N Sikkim, India
lacidus -a -um torn, *lacero, lacerare, laceravi, laceratum*
lacinatus -a -um lobed, *lacina, lacinae*
laciniatus -a -um, laciniosus -a -um jagged, fringed, slashed, with many flaps, *lacinia, laciniae* (see Fig. 4f)
lacinulatus -a -um as if finely cut at the margin, diminutive of *laciniatus*

lacistophyllus -a -um having torn leaves, λακιστος-φυλλον
laconicus -a -um from Lakonía (Laconía), Peloponnese, S Greece
lacrimans (*lachrymans*) weeping habit, causing tears, *lacrimo, lacrimare, lacrimavi, lacrimatum*
lacryma-jobi Job's-tears, *lacrimae-jobi* (the shape and colour of fruit)
lacrymus -a -um drops, tears, *lacrimo, lacrimare, lacrimavi, lacrimatum* (gum-drop of sap or resin)
Lactarius, lactarius -a -um Milky, *lactans* (the sap from damaged gills of milk-cap fungi)
lactescens having *lac*, having sap that turns milky, *lac-essentia*
lacteus -a -um, lact-, lacti- milk-coloured, milky-white, *lacteus*
lacticolor milky or creamy coloured, *lactis-color*
lactifer(us) -era -erum producing a milky juice, *lac, lactis*
lactiflorus -a -um with milk-white flowers, *lacteus-florum*
lactifluus -a -um flowing with milky sap, *lactis-(fluo, fluere, fluxi, fluxum)*
Lactuca the Latin name, *lactuca, lactucae* (it has a milky sap, *lac, lactis*, giving the cognate lettuce)
lactucellus -a -um somewhat *Lactuca* like, diminutive of *Lactuca*
lacunarius -a -um growing in dykes and ditches, *lacuna, lacunae*
lacunosus -a -um with gaps, furrows, pits or deep holes, *lacuna, lacunae*
lacuster, lacustris -is -e of lakes or ponds (*lacus* lake)
ladanifer -era -erum bearing *ladanum*, ληδανον (the resin called myrrh)
Ladanum *Lada*, the Latin name, *ladanum*, for the resin of *Cistus creticus*
ladanus -a -um gummy, sticky (glandular calyx of *Galeopsis ladanum*)
ladysmithensis -is -e from the environs of Ladysmith, S Africa
Laelia for Laelia, one of the Vestal Virgins
laetans rejoicing, gladdening, present participle of *laetor, laetari, laetatus*
laetevirens bright-green, *laete-virens*
laeti-, laetis -is -e, laetus -a -um pleasing, vivid, bright, *laetus*
laetificus -a -um joyful, *laetificus*
laetiviolaceus -a -um bright-violet, *laete-violaceus*
laevi-, laevis -is -e smooth, not rough, beardless, delicate, *laevis* (more correctly *levis*)
laevicaulis -is -e smooth-stemmed, *levis-caulis*
laevigatus -a -um polished, not rough, smooth, *levis, laevo* (see *levigatus*)
laevipes smooth-stemmed, *levis-pes*
laevirostris -is -e with a polished beak, *levis-rostrum* (the fruit)
laevispermus -a -um smooth-seeded, botanical Latin from *levis* and σπερμα
lag- hare's-, λαγος, λαγως, λαγ-, λαγο-
lagaro-, lagaros- lanky-, long-, thin-, narrow-, λαγαρος
lagarocladus -a -um thin-branched, λαγαρο-κλαδος
Lagarosiphon Narrow-tube, λαγαρο-σιφον (the corolla of the fruiting flowers)
Lagarostrobus Narrow-coned-one, λαγαρο-στροβιλος
Lagascea (Lagasca), lagascae for M. Lagasca (1776–1839), Professor of Botany at Madrid
lagen-, lagenae-, lageni- bottle-, *lagena*
lagenaeflorus -a -um with flask-shaped flowers, *lagena-florum*
Lagenaria Flask, *lagenaria* (the bottle-gourd fruit of *Lagenaria siceraria*)
lagenarius -a -um with the appearance of a bottle or flask, *lagena*
lagenicaulis -is -e with flask- or bottle-shaped stems, *lagena-caulis*
lageniformis -is -e bottle-shaped, *lagena-forma*
Lagenophora Flask-bearing, *lagena-fero* (Gourd trees)
Lagerstroemia for Magnus von Lagerström of Göteborg (1696–1759), friend of Linnaeus
Lagetta, lagetto Jamaican vernacular name for the tree and the lace-bark that it provides
lago- hare-, λαγως, λαγωος
lagodechianus -a -um from Lagodechi (Lagodekhi), Caucasus

lagopides resembling (*Coprinus*) *lagopus*, λαγωπους-ειδης
lagopinus -a -um resembling a small *Lagopus*
lagopodus -a -um hare's foot, λαγωπους, λαγωποδος
lagopoides resembling *Lagopus*
Lagopus, lagopus Hare's-foot, λαγωπους (Dioscorides' name was used for a trefoil), hare-like (colour and surface texture) (*Ochroma lagopus* seed fibre is called pattes-de-lièvre or rabbit's paws)
Lagotis Hare's-eared, λαγως-ωτος
lagunae-blancae of the white lagoons, botanical Latin from Spanish, laguna, and French, blanc
Lagunaria *Lagunaea*-like, for Andrés de Laguna (1494–1560), Spanish botanist (≡ *Hibiscus*-like)
Laguncularia Small-bottle, *lagoena, lagoenae* (the fruit)
lagunensis -is -e from Laguna district of N Mexico, or other Lagunas
Lagurus Hare's-tail, λαγως-ουρα (the inflorescence)
lahue from the Lahu region of SW China
lahulensis -is -e from the Lahul district of Himachal Pradesh, India
-lainus -a -um cloaked, see *chlainus*
lakka from a vernacular name for the palm *Cyrtostachys lakka*
Lamarckia (Lamarkia) for Jean Baptiste Antoine Pierre Monnet Chevalier de la Marck (Lamarck) (1744–1829), evolutionist
Lambertia, lambertii for Aylmer Bourke Lambert (1761–1842), who wrote the *Genus Pinus* (1803–24)
lamellatus -a -um layered, lamellate (diminutive of *lamina* sheet)
lamelliferus -a -um bearing plates or scales, (*lamina, laminae*)*-fero* (e.g. bark form)
lamii- deadnettle-like, resembling *Lamium*
laminatus -a -um layered, laminated, plated, *laminatus*
Lamiopsis, lamiopsis -is -e Looking-like-*Lamium, Lamium-opsis*
Lamium Gullet, λαιμος (the name in Pliny refers to the gaping mouth of the corolla) (***Lamiaceae***, ≡ ***Labiatae***)
lampas torch, lamp-like, bright, *lampas*, λαμπω
lampongus -a -um from Lampung province of S Sumatra, Indonesia
lampr-, lampro- shining-, glossy-, λαμπρος
Lampranthus, lampranthus -a -um Shining-flower, having glossy flowers, λαμπρος-ανθος
lamprocarpus -a -um shining-fruit, λαμπρος-καρπος
lamprochlorus -a -um bright-green, λαμπρος-χλωρος
lamprophyllus -a -um having glossy leaves, λαμπρος-φυλλον
lamprospermus -a -um with shining seed, λαμπρος-σπερμα
Lamprothamnium Shining-bush-like, λαμπρο-θαμνος
Lamprothamnus Brilliant-shrub, λαμπρο-θαμνος
lamprotyrius -a -um brilliant purple, botanical Latin from λαμπρο and *Tyrus* (for the mollusc-derived Imperial Tyrian purple dye, from Tyre)
lanatoides resembling (*Rhododendron*) *lanatus*
lanatus -a -um woolly, *lana, lanae, laneus*
lancastriensis -is -e from Lancashire, Lancastrian (*Lancastria*)
Lancea for John Henry Lance (1793–1878), orchidologist of Dorking
lanceolatus -a -um, lanci- narrowed and tapered at both ends, lanceolate, *lancea, lanceae*
lanceolus -a -um resembling a short lance, diminutive of *lancea, lanceae*
lancerottensis -is -e from Lanzarote (Lancerotte), Canary Isles
lanceus -a -um spear-shaped, *lancea, lanceae*
lanciferus -a -um lance-bearing, (*lancea, lanceae*)*-fero* (processes at the base of the column in *Catasetum*)
lancifolius -a -um with sharply pointed leaves, *lanceae-folium*
Landolphia for M. Landolph (1765–1825), who commanded the W African expedition *c.* 1786

landra from the Latin and Italian name, *landra*, for a radish
lanearius -a -um woollen, velvety, *laneus*
langeanus -a -um, langei for Johann Martin Christian Lange (1818–98), Professor of Botany at Copenhagen
langleyensis -is -e from Veitch's Langley Nursery, England
langsdorffii for G. H. von Langsdorff (1774–1852), physician of Freiburg
languidus -a -um dull, weak, drooping, *langueo, languere*
lani- woolly-, *lana*
lanicaulis -is -e having a woolly stalk, (*lana, lanae*)-*caulis*
laniceps woolly-headed, *lana-ceps*
laniger -era -erum, lanigerus -a -um, lanigeri softly hairy, woolly or cottony, (*lana, lanae*)-(*gero, gerere, gessi, gestum*)
lanipes woolly-stalked, *lana-pes*
Lankesteria for Dr Edwin Lankester (1814–74) English botanist
lannesianus -a -um for Lannes de Montebello, who sent *Prunus lannesiana* to Paris from Japan in 1870
lanosus -a -um softly hairy, *lana*
Lantana an old Latin name for *Viburnum*
lantanoides resembling *Lantana, Lantana-oides*
lanthanum inconspicuous, λανθανω (to escape notice)
lanuginosus -a -um with a woolly or cottony covering, like wool, *lana, lanae*
lanugo downy, woolly, *lanugo, lanuginis*
lanyensis -is -e from Lan Yu island, east of the southernmost tip of Taiwan
Lapageria for Marie Joséphine Rose Tascher de la Pagerie (1763–1814), Napoleon's Empress Joséphine, avid collector of plants at Malmaison
lapathi- sorrel-like-, dock-like-, λαπαθον
lapathifolius -a -um with sorrel-like leaves, *Lapathum-folium*
Lapathum Adanson's use of the Latin name for sorrel, from Greek λαπαθον (by derivation through the French, lapatience, we have the English, patience, for *Rumex patientia*)
Lapeirousia (Lapeyrousia), lapeyrousei for J. F. G. de la Peyrouse (1741–88), French circumnavigator
lapideus -a -um stone-, stone-like, *lapideus*
lapidicolus -a -um living in stony places, *lapidi-colo*
lapidiformis -is -e stone-shaped, *lapidi-forma*
lapidius -a -um hard, stony, *lapis, lapidis*
lapidosus -a -um of stony places, stony, *lapidosus*
lapithicus -a -um from Lapithos, Cyprus
Laplacea for Pierre-Simon de Laplace (1749–1827) French mathematician
Laportea for M. Laporte, nineteenth-century entomologist
lappa Latin name, *lappa, lappae*, for goose-grass (e.g. burrs of goose-grass and burdock, *Arctium lappa*)
lappaceus -a -um bearing buds, bud-like, burdock-like, burr-like, *lappa*
Lappago Burr, *lappa-ago* (upper glume has a recurved flattened tail with hooked hairs)
lapponicus -a -um, lapponus -a -um from Lapland, of the Lapps (*Lappones*)
lappulus -a -um with small burs, diminutive of *lappa* (the nutlets)
Lapsana (*Lampsana*) Purge, Dioscorides' name, λαψανη, λαμψανη, for a salad plant
laramiensis -is -e from Laramie mountains, SE Wyoming, USA
Lardizabala for Señor M. Lardizalay of Uribe, Spanish naturalist (***Lardizabalaceae***)
largus -a -um ample, plentiful, liberal, *largus*
larici-, laricinus -a -um larch-like, resembling *Larix*
laricifolius -a -um, laricinifolius -a -um larch-leaved, *Larix-folium*
laricinus -a -um larch-like, *Larix*
laricio the Italian name for several pines
Larix Dioscorides' name, *larix*, for a larch (cognate, via Turner's reference to the German name, larchem baum, with larch)

larpentae, larpentiae for Lady Larpent of Roehampton, *c.* 1846
lascivius -a -um running wild, impudent, lustful, *lascivo, lascivire*
Laser a Latin name for several umbellifers (*laserpicium*, silphium)
laserpitifolius -a -um with *Laserpitium*-like leaves, *Laserpitium-folium*
Laserpitium an ancient Latin name, *laserpicium, laserpicii* for silphium
lasi-, lasio- shaggy-, rough-, woolly-, λασιος, λασιο-
lasiacanthus -a -um having hairy spines, λασι-ακανθος
lasiagrostis shaggy *Agrostis*, λασις-αγρωστις
Lasiandra, lasiandrus -a -um with shaggy haired stamens, λασιος-ανηρ (≡ *Tibouchina*)
Lasianthus, lasianthos, lasianthus -a -um Shaggy-flowered, λασι-ανθος
lasiocarpus -a -um having woolly fruits, λασιο-καρπος
Lasiocaryum Woolly-nut, λασιο-καρυον
lasiocladus -a -um with shaggy branches, λασιο-κλαδος
lasiogynis -is -e, lasiogynus -a -um having a woolly ovary, λασιο-γυνη
lasiolaenus -a -um shaggy-cloaked, woolly-coated, λασιο-(χ)λαινος
Lasiopetalum Woolly-petals, λασιο-πεταλον (the sepals are downy and petaloid)
lasiophyllus -a -um with woolly leaves, λασιο-φυλλον
lasiostipes having a woolly stem, botanical Latin from λασιο and *stipes*
lasiostylus -a -um with woolly styles, λασιο-στυλος
lasiosus -a -um very shaggy-haired, λασιος
Lasiurus Woolly-tailed, λασι-ουρα (the densely villous racemes)
Lastrea for Charles Jean Louis de Lastre (1792–1859), French botanical writer
lataevirens see *laetevirens*
latakiensis -is -e from Al Ladhiqiyah (Latakia), NW Syria
Latania Mauritian vernacular name for Mascarene palm
latebrosus -a -um (lataebrosus) porous, full of hiding places, *latebra, latebrae*
lateralis -is -e, lateri- on the side, laterally-, *latus, lateris; lateralis*
latericius -a -um brick-red, *later, lateris* bricks
lateriflorus -a -um with a one-sided inflorescence, *latus-florum*
laterifolius -a -um growing to the side of a leaf, *latus-folium*
laterinervius -a -um straight-veined, *latus-nervus*
lateritius -a -um brick-red, *later, lateris* bricks
Lathraea Clandestine, λαθραιος (*Lathraea clandestina* is a root parasite, inconspicuous until flowering time)
lathyris the ancient name for a kind of spurge (*Euphorbia lathyris*)
lathyroides resembling *Lathyrus, Lathyrus-oides*
Lathyrus the ancient name, λαθυρος, for the chickling pea (*Lathyrus sativus*) used by Theophrastus, λαω-θουριος grasping enthusiastically
lati-, latisi- broad-, wide-, *latus*
laticeps having a wide head, *latus-ceps* (inflorescence)
latici- latex-, juice-, *latex, laticis*
laticostatus -a -um having broad veins, *latus*-(*costa, costae*)
latidens broad-toothed, *latus-dens*
latiflorus -a -um having wide flowers, *latus-florum*
latifolius -a -um with broad leaves, *latus-folium*
latifrons with broad fronds, *latus-frons*
latiglumis -is -e having broad glumes, *latus-gluma*
latilobus -a -um broad lobed, *latus-lobus*
latinus -a -um, latius -a -um of *Latium*, the ancient Italian district that included Rome (Latinus was the king of the Latins, in mythology)
latipes broad-stalked, thick-stemmed, *latus-pes*
latisectus -a -um with broad divisions or cuts, past participle of *seco, secare, secui, sectum*
latissimus -a -um very broad, superlative of *latus*
latiusculus -a -um somewhat broad, *latus* with diminutive suffix
latobrigorum from the area of the *Latobrigi*, Rhinelands people or Belgic Gauls

latopinna with broad pinnae, *latus-pinna*
latus -a -um wide, broad, *latus*
lauchianus -a -um for Wilhelm Georg Lauche (1827–83), German gardener
laudatus -a -um praised, lauded, excellent, *laudo, laudare, laudavi, laudatum*
Laurelia *Laurus*-like, resembling the bay-tree
Laurentia for M. A. Laurenti, seventeenth-century Italian botanist
laureolus -a -um Italian name, diminutive of *laurea*, for *Daphne laureola*, from its resemblance and its use in garlands
lauri- laurel-, *Laurus*-like-
lauricatus -a -um wreathed, resembling laurel or bays, *laurus*
laurifolius -a -um laurel-leaved, *Laurus-folium*
laurinus -a -um of laurel, laurel-like, *laurus*
laurisylvaticus -a -um woodland or wild laurel, *laurus-*(*silva, silvae*)
laurocerasi of cherry-laurels, *Prunus laurocerasus* (ground layer habitat under broad-leaves)
laurocerasus laurel-cherry, *laurus-cerasus* (cherry-laurel)
Laurus the Latin name, *laurus*, for laurel or bay (Celtic, laur, green) (***Lauraceae***)
Laurustinus Laurel-like-*Tinus*
lautus -a -um washed, elegant, neat, fine, *lavo, lavare, lavi, lautum*
Lavandula To-wash, a diminutive from *lavo, lavare, lavi, lautum* (its use in the cleansing process)
lavandulaceus -a -um resembling *Lavandula*
lavandulae- lavender-, *Lavandula-*
lavandulifolius -a -um with leaves resembling *Lavandula*
Lavatera for the brothers Lavater, eighteenth-century Swiss naturalists
lavateroides *Lavatera*-like, *Lavatera-oides*
lavatus -a -um washed, *lavo, lavare, lavi, lautum*
lawrenceanus -a -um, lawrenci for Sir Trevor Lawrence (1831–1913), orchid grower and President of the Royal Horticultural Society 1885–1913
Lawsonia for Dr Isaac Lawson, eighteenth-century Scottish botanical traveller (henna plant, *Lawsonia inermis*)
lawsonianus -a -um for P. Lawson (d. 1820), Edinburgh nurseryman
laxi- open, loose, not crowded, spreading, distant, lax, *laxus, laxi-*
laxicaulis -is -e loose-stemmed, not having rigid stems, *laxus-caulis*
laxiflorus -a -um loosely flowered; wide-flowered, *laxus-florum*
laxifolius -a -um loosely leaved, with open foliage, *laxus-folium*
laxissimus -a -um the most loose or spreading, superlative of *laxus*
laxiusculus -a -um somewhat weak , drooping or loose, diminutive suffix on *laxus*
laxus -a -um open, loose, not crowded, spreading, distant, lax, from *laxo, laxare, laxavi, laxatum*, to loosen
Layia for George Tradescant Lay (1799–1845), naturalist with Beechey (tidy tips, the ligulate florets)
lazicus -a -um from the Black Sea area of NE Turkey (Lazistan)
lazulinus -a -um ultramarine, diminutive of *lazulum* (blackish-blue to violet-blue but not so intense as lapis-lazuli)
lebbek an Arabian vernacular name for *Albizzia lebbek*
lebomboensis -is -e from the Lebombo Mountains (Big-nose mountains), Swaziland/Mozambique, SE Africa
lecano- basin-, dish-, ληκανη, ληκανο-
Lecanodiscus Basin-disc, ληκανο-δισκος (the concave floral disc)
lecontianua -a -um, lecontei for Dr John Lawrence le Conte (1825–83), who found *Ferocactus leconei* in Arizona
Lecythis Oil-jar, ληκυθος (the shape of the leathery fruit from which the lid falls when mature) (***Lecythidaceae***)
ledanon, ledanus -a -um, ledo-, ledi- gum; gummy, sticky, λεδανον
Ledebouria for Carl Friedrich von Ledebour (1785–1851), student of the Russian flora

ledifolius -a -um with *Ledum*-like leaves, *Ledum-folium*
Ledodendron the composite name for hybrids between *Ledum* and *Rhododendron*
ledon gummy, ληδανον
ledophyllus -a -um having sticky leaves, ληδανον-φυλλον (as *Cistus* spp.)
Ledum an ancient Greek name, ληδον, for the ladanum-resin producing *Cistus ladaniferus*
Leea, leeanus -a -um for James Lee (1715–95), Hammersmith nurseryman (***Leeaceae***)
leeoides resembling *Leea, Leea-oides*
Leersia for Johann Daniel Leers (1727–74), German botanist
legionensis -is -e from León, NW Spain (corruption of *legio*, since the town was established by the Roman seventh Gemina Legion)
Legousia uncertain commemorative etymology
lei-, leio- smooth-, λειος, λειο-
leiantherus -a -um with smooth stamens, λειος-ανθηρος
Leianthus, leianthus -a -um Hairless-flower, λειος-ανθος
leichtlinii for Max Leichtlin (1831–1910), plant introducer of Baden-Baden
leiobotrys smooth bunches, λειο-βοτρυς
leiocarpus -a -um with smooth spores or fruits, λειο-καρπος
leiodermus -a -um having a smooth epidermis, λειο-δερμα
leiomerus -a -um smooth, with smooth parts, λειο-μερος
leiopetalus -a -um having smooth or glossy petals, λειο-πεταλον
Leiophyllum, leiophyllus -a -um Smooth-leaf, λειο-φυλλον
leiopodius -a -um having a smooth stalk, λειο-ποδιον
leiospermus -a -um with smooth seeds, λειο-σπερμα
Leitneria for Edward F. Leitner, nineteenth-century German naturalist (***Leitneraceae***)
Lemna Theophrastus' name, λεμνα, for a water-plant (***Lemnaceae***)
lemniscatoides resembling (*Bulbophyllum*) *lemniscatum*
lemniscatus -a -um beribboned (the Roman victor had ribbons, *lemnisci*, from his crown)
lemoinei for Victor Lemoine (1823–1911), shrub nurseryman of Nancy, France
Lemonia, lemonianus -a -um, lemonii for Sir Charles Lemon MP (1784–1868), British horticulturalist
lemosii for Dr Lemos of Para, who found *Catasetum lemosii*
lendiger -era -erum nit-carrier, (*lens, lendis*)-*gero* (the appearance of the spikelets)
lendyanus -a -um for A. F. Lendy (1826–89), orchid grower of Sunbury on Thames
lenneus -a -um for P. J. Lenné (1789–1866), director of Prussia's Royal Gardens
Lennoa derivation uncertain (***Lennoaceae***)
Lens the classical name for the lentil, *lens, lentis*
lentago pliant, lasting, *lentus* with feminine suffix
lenti- spotted-, freckled-, lenticelled-, diminutive from *lens, lentis*
Lentibularia etymology of Gesner's name is uncertain, usually regarded as referring to the lentil-shaped bladders (***Lentibulariaceae***)
lenticularis -is -e lens-shaped, bi-convex, *lens, lentis*
lenticulatus -a -um with conspicuous lenticels on the bark, lenticulate, *lens, lentis*
lentiformis -is -e lens-shaped, bi-convex, *lentis-forma*
lentiginosus -a -um mottled, freckled, *lentigo, lentiginis*
lentiscifolius -a -um having leaves resembling those of *Pistacia lentiscus*
lentiscus Latin name for the mastic tree, *Pistacia lentiscus*
lentus -a -um sticky, tough, pliant, lasting, *lentus*
leo-, leon- lion-, λεων, λεοντος, λεοντο-, *leo, leonis*
Leochilus Lion-lip, λεο-χειλος (obscure name unless the nectary on the lip is related to Aesop's story of honey from the lion's ear; 'out of the strong comes sweetness')
leodensis -is -e from Liege, Belgium (*Leodium*)
leonensis -is -e from Sierra Leone, W Africa
leoninus -a -um tawny-coloured like the lion, *leo, leonis*

leonis -is -e toothed or coloured like a lion, *leo, leonis*
Leonotis Lion's-ear, λεων-ωτος
Leontice Lion's (-footprint) (the shape of the leaf) (***Leonticaceae***)
leontinus -a -um from Leontini, Sicily
leonto- lion's-, λεοντος, λεοντο-, *leo, leonis*
Leontodon, leontodon Lion's-tooth, λεοντο-οδων
Leontopodium, leontopodium Lion's-foot, λεοντο-ποδιον
Leonurus, leonurus Lion's-tail, λεων-ουρα (*Leonurus cardiaca* is a motherwort)
leopardinus -a -um conspicuously spotted, leopard-like, λεοπαρδος, *leopardus*
Lepidagathis Scaly-*Agathis*, λεπις-αγαθις
lepidanthus -a -um having scaly flowers, λεπις-ανθος
Lepidium Little-scale, diminutive of λεπις (Dioscorides' name, λεπιδιον, for a cress, refers to the fruit)
lepido-, lepiro- flaky-, scaly-, λεπις, λεπιδος, λεπιδο-, λεπιδ- (the scales may be minute as on butterflies' and moths' wings)
lepidobalanus scaly acorn, λεπιδο-βαλανος (the scaly cupules)
Lepidobotrys Scale-cluster, λεπιδο-βοτρυς (the flowers emerge from strobilus-like groups of subtending bracts)
lepidocarpos, lepidocarpus -a -um having scaly fruits, λεπιδο-καρπος
lepidocaulon with a scaly stem, λεπιδο-καυλος
lepidophyllus -a -um with scaly leaf surfaces, λεπιδο-φυλλον
lepidopteris -is -a scale-winged, λεπιδο-πτερυξ
lepidopus -a -um scaly-stalked, λεπιδο-πους
lepidostylus -a -um having scales on the style, λεπιδο-στυλος
Lepidothamnus Scaly-shrub, λεπιδο-θαμνος
Lepidotis Scaly, (λεπις, λεπιδος)-ωτος
lepidotus -a -um scurfy, scaly, lepidote, λεπιδωτος
Lepidozamia Scaled-*Zamia*
lepidus -a -um neat, elegant, graceful, *lepidus, lepide*
Lepigonum Scaly-nodes, λεπι-γονυ, or Scale-seed, λεπι-γονος (≡ *Spergularia*)
Lepiota Scaly-ear, λεπι-(ους, ωτος)
-lepis -scaly, -scaled, -λεπις
Lepistemon Stamen-scale, λεπις-στημων (the scale on the corolla, below the insertion of each stamen)
Lepiurus Scale-tail, λεπι-ουρα (the inflorescence of sea hard grass, cf. *Pholiurus*)
leporinus -a -um hare-like, *lepus, leporis* (*Carex leporinus* spikes suggest hare's paws)
leprosus -a -um scurfy, leprosied, λεπρα, λεπρος
lept-, lepta-, lepto- husk-free-, slender-, fine-, small-, delicate-, λεπτος, λεπτο-, λεπτ-
leptacanthus -a -um having slender or weak spines, λεπτος-ακανθος
Leptactinia Slender-rayed, λεπτος-ακτινος (the circlet of fine corolla lobes)
Leptadenia Slender-glanded, λεπτος-αδην (on the staminal column)
leptandrus -a -um with slender stamens, λεπτ-ανηρ
leptanthus -a -um with delicate or slender flowers, λεπτ-ανθος
Leptarrhenia Small-male, λεπτ-αρρην
Leptaspis Small-shield, λεπτ-ασπις (the short glumes of male florets)
Leptinella Small-slender-one, feminine diminutive from λεπτος
leptocarpus -a -um having small fruits, λεπτο-καρπος
leptocaulis -is -e having a slender stalk, λεπτο-καυλος
leptocephalus -a -um small-headed, λεπτο-κεφαλη (the small pileus)
leptoceras having a slender horn, λεπτο-κερας (nectary)
leptochilus -a -um with a slender lip, λεπτο-χειλος
Leptochloa Delicate-grass, λεπτο-χλοη
leptoclados with slender shoots, λεπτο-κλαδος
Leptodactylon Slender-fingered, λεπτο-δακτυλος (the digitate leaf segments)
Leptodermis Thin-skin, λεπτο-δερμα (the inner fruit-wall)
Leptodontium Fine-toothed, λεπτο-οδοντος
Leptoglossa Thin-tongue, λεπτο-γλωσσα

Leptogramma Slender-lined, λεπτο-γραμμα (the sori)
leptolepis -is -e with slender scales, λεπτο-λεπις
Leptomeria Slender-parts, λεπτο-μερις (the stems)
Leptonychia Slender-clawed, λεπτ-ονυχος (the staminodes)
leptophis -is -e slender, snake-like, λεπτ-οφις
leptophyllus -a -um fine- or slender-leaved, λεπτο-φυλλον
leptopodius -a -um, leptopus with slender stalks, λεπτο-(πους, ποδος)
leptorrhizus -a -um with slender roots, λεπτο-ριζα
leptosepalus -a -um having slender sepals, λεπτο-σκεπη
leptospadix with a narrow spadix, λεπτο-σπαδιξ
Leptospermum, leptospermus -a -um Narrow-seed, λεπτο-σπερμα (slender seeded)
leptostelis -is -e having a slender trunk, λεπτο-στελεκος
leptostemon with slender stamens, λεπτο-στεμμα
leptotes with delicate or slender ears, λεπτ-ωτος (leaves or florets)
Leptothrium, leptothrium Cleaned of husks, λεπτος, (λεπτ-οθριξ, λεπτο-θριξ delicate-haired, the spikelets are burrs, carried off with their peduncles by passing animals)
Lepturus, lepturus Hare's-tail, λεπτ-ουρα (the inflorescence)
lerchenfeldianus -a -um for Josef Radnitzky von Lerchenfeld (1753–1812)
lesbis -is -e from Lesvos (Lesbos), Greece (one-time home of Sappho (610–580 BC), poetess whose writing gave rise to the concept of the lesbian)
Leschenaultia, leschenaultii for L. T. Leschenault de la Tour (1773–1826), French botanist
Lespedeza for V. M. de Lespedez, Spanish Governor of Florida
Lesquerella for Leo Lesquereux (1806–89), American paleobotanist
Lessertia for Baron Benjamin de Lessert (1773–1847), author of *Icones plantarum*
lesuticus -a -um from the kingdom of Lesotho (Basutoland), S Africa
lettonicus from Latvia (Lettland)
leuc-, leuco- bright-, brilliant-, white-, pale-, λευκος, λευκο-, λευκ-
Leucadendron White-tree, λευκος-δενδρον
Leucaena Bright, λευκος (for the flowers) (≡ *Acacia*)
Leucanthemella Little-white-flower, feminine diminutive from *Leucanthemum* (≡ *Chrysanthemum)*
leucanthemifolius -a -um with leaves resembling those of *Leucanthemum, Leucanthemum-folium*
Leucanthemopsis Resembling-*Leucanthemum*, λευκος-ανθεμιον-οψις
Leucanthemum White-flower (Dioscorides' name, λευκανθεμον, for a plant also called ανθημις and χαμαιμηλον) (≡ *Chrysanthemum)*
leucanthus -a -um white-flowered, λευκ-ανθος
Leucas White, λευκη (the flowers)
leucaspis -is -e white shield, λευκ-ασπις (the flat corolla face of *Rhododendron leucaspis*)
leuce a name, λευκη, for the white poplar
leucensis -is -e from Lecco, Lake Como, Italy (*Leucum*)
Leuceria White, λευκηρης (some have an indumentum similar to that of white poplar, λευκη)
leuco- brilliant-, grey-, pale-, white-, λευκος, λευκο-
Leucobryum White-*Bryum*, λευκο-βρυον (the greyish-white appearance)
leucocaulos with white stem, λευκο-καυλος
leucocephalus -a -um white-headed, λευκο-κεφαλη (inflorescence)
leucochilus -a -um having a white lip or labellum, λευκο-χειλος
leucochroa white-coloured, pale, λευκο-χροα
Leucochrysum White-and-gold, λευκο-(χρυσεος, χρυσους)
Leucocoryne White-clubbed-one, λευκο-κορυνη (the stamens of glory of the sun)
Leucocrinum White-*Crinum*
leucodendron white tree, whited branched, λευκο-δενδρον (bark colour)
leucodermis -is -e white-skinned, λευκο-δερμα (bark colour)

leucogalus -a -um having white milk, λευκο-γαλα (sap)
Leucogenes White-noble, λευκο-ευγενες (its morphological parallelism with edelweiss, lady-white)
Leucojum White-violet, λευκο-ιον (Hippocrates' name, λευκοιον, for a snowflake)
leucolaenus -a -um clothed in white, λευκο-(χ)λαινα
leucomelas pale and gloomy, λευκο-μελας (the outer and inner colours of *Paxina leucolemas*)
leuconeurus -a -um having white nerves, λευκο-νευρα (veins)
leucophaeus -a -um ashen, pale brown, λευκο-φαιος
leucophalus -a -um white-ridged, λευκο-φαλος (the low ridges have abundant silky white hairs)
leukopharynx having a white throat, λευκο-(φαρυγξ, φαρυγγος)
leucophyllus -a -um with white leaves, λευκο-φυλλον (very pale green)
Leucophyta White-plant, λευκο-φυτον (its woolly indumentum)
Leucopogon White-beard, λευκο-πωγων (the corolla lobes of some)
Leucoraoulia the composite name for hybrids between *Leucogenes* and *Raoulia*
Leucorchis White-orchid, λευκ-ορχις
leucorrhizus -a -um having white roots, λευκο-ριζα
Leucosceptrum White-staff, λευκο-σκηπτρον (the hairy inflorescence)
leucosiphon with a white (corolla) tube, λευκο-σιφον
Leucospermum White-seed, λευκο-σπερμα (the smooth, glossy seeds)
leucostachyus -a -um white-panicled, λευκο-σταχυς
leucostelis -is -e white-pillared, λευκο-στηλη (the columnar stems)
leucothites whitish, close to white, λευκος
Leucothoe an ancient Greek name, Leucothoe was daughter of King Orchanus of Babylon, and loved by Apollo
leucotrichophorus -a -um bearing white hairs, λευκο-(θριξ, τριχος)-φορος
leucotrichus -a -um white-haired, λευκο-(θριξ, τριχος)
leucoxylem with white wood, λευκο-ξυλον
Leuzea, leutzeanus -a -um for J. P. F. de Leuze (1753–1826), French botanist traveller and friend of de Candolle
levantinus -a -um from the Levant (Mediterranean coastal regions from Greece to Egypt)
leveillei for A. A. Hector Léveillé (1863–1918), of Le Mans, France
leviculus -a -um rather vain, *leviculus*
levigatus -a -um smooth, polished, *levigo, levigare, levigati, levigatum*
levipes having smooth stalks, *levis-pes*
levis -is -e smooth, not rough, *levis* (see *laevis*)
Levisticum Reliever, *levo* (the Latin equivalent of the Dioscorides' Greek name λιγυστιχος)
Lewisia, lewisianus -a -um, lewisii for Captain Meriwether Lewis (1774–1809), of the trans-American expedition, bitter roots
leyanus -a -um for Reverend Augustin Ley (1842–1911), Gloucestershire cleric and botanist
Leycesteria for William Leycester (1775–1831), judge and horticulturalist in Bengal
Leymus an anagram of *Elymus*
leysianus -a -um for Dr P. Leys of Labuan *c.* 1879
lhasicus -a -um from Lhasa, Tibet
lheritieranus -a -um for Charles Louis L'Heritier de Brutelle (1746–1800)
Lhotzkya (Lotskya) for Dr John Lotzky (1739–1843), Austrian botanist and traveller
liaotungensis -is -e from the Liaotung peninsula, S China
Liatris derivation uncertain, λεια booty, λειος, smooth or bald
libanensis -is -e, libanoticus -a -um from Mount Lebanon, Syria
libani from the lands of the *Libani*, from Lebanon, Lebanese
libanotis -is -e from Mount Lebanon or of incense, strongly scented, λιβανωτος
liber unrestricted, undisturbed, *liber, liberi*
libericus -a -um, liberiensis -is -e from Liberia, W Africa

libero- book-; bark-, *liber, libri* (a characteristic)
liberoruber with red bark, *liber-ruber*
Libertia for Marie A. Libert (1782–1865), Belgian writer on hepatics
libo- frankincense-, λιβανος
Libocedrus Frankincense-cedar, λιβανος-κεδρος (the resin exuding incense cedar)
libonianus -a -um for Joseph Libon (1821–61), collector for the de Jonghes in Brazil
libratus -a -um powerful; balanced, well-proportioned, *libro, librare, libravi, libratum*
liburnicus -a -um from Croatia (*Liburnia*) on the Adriatic
libycus -a -um from Libya, Libyan
licens free, bold, unrestricted, vigorous, *licens, licentis*
lichenastrus -a -um lichen-like, botanical Latin from λειχην and *ad-instar* (growth habit)
lichiangensis -is -e, licjiangensis -is -e from Lijiang, Yunnan province, China
Licuala from a Moluccan vernacular name for *Licuala rumphii*
lidjiangensis -is -e from Lijiang, Yunnan province, China
Liebigia for Justus Freiherr von Liebig (1803–73), German bio-chemist
ligatus -a -um united, bandaged, bound, *ligo, ligare, ligavi, ligatum*
ligericus -a -um from the Loire river valley area (*Liger, Ligeris*)
Lightfootia for Reverend J. Lightfoot (1735–88), Scottish botanist, author of a Flora of Scotland
lignatilis -is -e of wood or trees, *lignum, ligni* (*Pleurotus* saprophyte on dead (hollow) trees
lignescens turning woody, becoming woody, *lignum-essentia*
ligni- woody-, wood-, of woods-, *lignum, ligni-*
lignosus -a -um woody, *lignum, ligni*
lignum-vitae wood-of-life, *lignum-vita*, (the remarkably durable timber of *Guaiacum officinale*)
ligtu from a Chilean name for St Martin's flower (*Alstoemeria ligtu* cv. pulchra)
Ligularia Strap, *ligula* (the shape of the ray florets)
ligularis -is -e strap-shaped, ligule-like, *ligula, ligulae*
ligulatus -a -um with a ligule, with a membranous projection, ligulate, *ligula, ligulae*
ligulistylis -is -e with a strap-like style, botanical Latin *ligula-stilus*
Ligusticum, ligusticus -a -um Dioscorides' name, λιγυοστικος, for a plant from Liguria, NE Italy
ligustifolius -a -um privet-leaved, *Ligustrum-folium*
Ligustrina, ligustrinus -a -um Privet-like, resembling *Ligustrum*
Ligustrum Binder, *ligula* (a name used in Pliny and Virgil)
likiangensis -is -e from Lijiang, Yunnan province, China
likipiensis -is -e from the Laikipia plateau, Tanzania/Kenya
lilaciflorus -a -um with lilac-like flowers, *lilac-florum*
lilacinus -a -um lilac-coloured, lilac-like, *lilacinus*
Lilaea Of-desire, λιλαιος
Lilaeopsis *Lilaea*-like, λιλαιος-οψις
lili-, lilii- lily-, *lilium, lili, lilii-*
liliaceus -a -um lily-like, resembling *Lilium*
liliago silvery, lily-like, *Lilium* with feminine suffix
liliflorus -a -um, liliiflorus -a -um with lily-like flowers, *lilii-florum*
lilioasphodelus -a -um having flowers suggesting both *Lilium* and *Asphodelus* (*Hemerocallis lilioasphodelus*)
liliohyacinthus -a -um having flowers suggesting both *Lilium* and *Hyacinthus* (*Scilla liliohyacinthus*)
Lilium the name, *lilium*, in Virgil (Celtic, li, white) (***Liliaceae***)
lilliputianus -a -um of very small growth, Lilliputian, botanical Latin from Jonathan Swift's imaginary country, Lilliput
limaci- slug-, *limax, limacis*
limaeus -a -um of stagnant waters, λιμναιος

limbatus -a -um bordered, with a margin or fringe, *limbus, limbi*
limbo- border-, margin-, *limbus, limbi*
limbospermus -a -um with fringed spores or seeds, producing spores around the margins of the pinnae, botanical Latin from *limbus* and σπερμα
-limbus -a -um -bordered, -fringed, *limbus, limbi*
limensis -is -e from Lima, Peru (*Phaseolus limensis*, Lima bean)
limettoides resembling (*Citrus*) *limetta*
limettus -a -um little lemon, *limon* (the smaller fruited sweet lemon)
limicolus -a -um living in mud, *limicola*
limitaris -is -e at the boundary, *limitaris*
limitatus -a -um restricted, *limitatus*
limn-, limno- lake-, pool-, pond-, λιμνη, marsh-, λιμνωδης
Limnanthemum Pond-flower, λιμνη-ανθεμιον (spreads over surface)
Limnanthes Pond-flower, λιμνη-ανθος (***Limnanthaceae***)
limneticus -a -um growing in or around lakes, ponds, pools, swamps or the sea λιμνη
Limnobotrys Pond-grape, λιμνη-βοτρυς
Limnocharis Marsh-beauty, λιμνη-χαρις (the habitat) (***Limnocharitaceae***)
Limnophila, limnophilus -a -um Marsh-loving, λιμνη-φιλος (the habitat)
Limnophyton Marsh-plant, λιμνη-φυτον
limon, limonius -a -um from the Persian name, limoun, Arabic, limun, for the lemon and other *Citrus* fruits
limonifolius -a -um with leaves resembling those of *Limonium, Limonium-folium*
Limonium Meadow-plant, λειμων (Dioscorides' name, λειμωνιον, for a meadow plant)
Limosella Muddy, feminine diminutive of *limus*
limosellifolius -a -um having mudwort-like leaves, *Limosella*-leaved
limosus -a -um muddy, slimy, living on mud λνμα, *limus*
limus -a -um mud, slime, dirt, *limus, limi*
lin-, linarii-, lini- thread-, flax-, λινον
linaceus -a -um flax-like, resembling *Linum*
Linanthus Flax-flowered, λινον-ανθος
Linaria Flax-like, λινον (the leaf similarity of some species)
linarifolius -a -um, linariifolius -a -um with *Linaria* like leaves, *Linaria-folium*
linarioides resembling *Linaria, Linaria-oides*
lindavicus -a -um from Lindau, Germany (*Lindavia*)
Lindelofia for Friedrich von Lindelof of Darmstadt, patron of botany
Lindenia, lindenianus -a -um, lindenii for J. J. Linden (1817–98), Belgian horticulturalist
Lindera for Johann Linder (1678–1723), Swedish botanist
linderifolius -a -um having leaves resembling those of *Lindera, Lindera-folium*
lindheimeri for Ferdinand Lindheimer (1801–79), who found *Linheimera texana*
Lindleyella, lindleyanus -a -um for Dr John Lindley (1799–1865), saviour of the Royal Botanic Gardens, Kew
Lindsaea, lindsayi for Dr John Lindsay (1785–1803), of Jamaica (***Lindsayaceae***)
linearicarpus -a -um with elongate fruits, botanical Latin from *linearis* and καρπος
linearifolius -a -um having narrow and parallel-sided leaves, *linearis-folium*
linearilobus -a -um with narrow parallel-sided lobes, botanical Latin from *linearis* and λοβος
linearis -is -e narrow and parallel-sided (usually the leaves), linear, *linearis*
lineatus -a -um marked with lines (usually parallel and coloured), striped, *lineatus*
linggensis -is -e from Lingga archipelago, Sumatra, Indonesia
lingua, linguae-, lingui- tongue-shaped-, tongue-, *lingua* (some aspect or part)
-linguatus -a -um, -lingus -a -um -tongued, *lingua, linguae*
linguellus -a -um resembling a small tongue, diminutive of *lingua*
linguiferus -a -um bearing a tongue or tongues, *lingua-fero*
linguiformis -is -e tongue-shaped, *lingua-forma*

lingularis -is -e, lingulatus -a -um, linguus -a -um tongue-shaped, lingulate, *lingua, linguae* (*Linguus* was a name in Pliny)
linicolus -a -um of flax-fields, living in flax fields, *linum-colo*
liniflorus -a -um flax-flowered, *Linum-florum*
linifolius -a -um flax-leaved, with leaves resembling *Linum, Linum-folium*
linitus -a -um smeared, *lino, linere, levi, litum*
Linnaea by Gronovius, at request of Carolus Linnaeus, for its lowly, insignificant and transient nature
linnaeanus -a -um, linnaei for Carolus Linnaeus (1707–78)
linnaeoides resembling *Linnaea, Linnaea-oides*
linoides flax-like, resembling *Linum*, λινον-οειδης
Linosyris, linosyris Osyris'-flax, an old generic name by l'Obel, λινον-οσυρις (≡ *Chrysocoma*)
lintearius -a -um resembling weaving, *linteum, lintei* (gauze or lace-bark)
Linum the ancient name for flax, λινον, *linum* (***Linaceae***)
lio- smooth-, λειος, λειο-
liolaenus -a -um smooth-cloaked, glabrous, λειο-(χ)λαινα
Liparia Shining, λιπαρος (for the shining leaves), λιπαρος, shining, oily, fat, greasy
Liparis Shining, λιπαρος (the shining leaf-texture)
liparocarpus -a -um smooth- or oily-fruited, λιπος-καρπος
lipo- grease-, oil-, fat-, λιπος, λιπο-
Lipocarpha Greasy-stem, λιπος-καρφη
Lippia for Augustin Lippi (1678–1701), French/Italian naturalist
lipsicus -a -um from Leipzig, Germany (*Lipsia*)
lipsiensis -is -e from Lipsoi, Greece
Liquidambar Liquid-amber, *liquidus ambar* (the fragrant resin, *balsamum liquidambrae*, from the bark of sweet gum, *Liquidambar styraciflua*)
liratus -a -um ridged, having a ridge, *lira*
lirelli- with a central furrow-, *lira*
lirio- lily-white-, λιριον, λειριο-
Liriodendron Lily-tree, λειριο-δενδρον (the showy flowers of the tulip tree)
Liriope for one of the Nymphs of Greek mythology, (*liriopipium* is the tail-end of an academic hood)
Lisianthus (Lisyanthus) Divorce, λυσις-ανθος (the intense bitterness of the flowers, ≡ *Eustoma*)
liss-, lisso- smooth-, λις, λισσος, λισσο-
Lissochilus Smooth-lip, λισσο-χειλος (of the corolla)
lissopleurus -a -um having smooth nerves or ridges, λισσο-πλευρα
lissospermus -a -um having smooth seeds, λισσο-σπερμα
listadus -a -um praying, in prayer, λισσομαι, to pray or beg (the bracts are held together)
Listera for Dr Martin Lister (1638–1712), physician to Queen Anne and pioneer palaeontologist
listeri for J. L. Lister of the Bhoton Cinchona Association *c.* 1898
listrophorus -a -um bearing spade-like structures, λιστρον-φορεω
Listrostachys Spade-ear, λιστρον-σταχυς (the quadrate lip)
litangensis -is -e from Litang, W China
Litchi from the Mandarin vernacular name, li chih
literatus -a -um with the appearance of being written upon, *littera, litterae, litteratus*
lithicolus -a -um living on or amongst stones, botanical Latin from λιθος and *colo*
litho- stone-, λιθος, λιθο-, λιθι-
Lithocarpus Stone-fruit, λιθο-καρπος (the hard shell of *Lithocarpus javensis*)
Lithodora Stone-skinned, λιθο-δορα (the fruits, ≡ *Lithocarpus*)
lithophilus -a -um living in stony places, stone-loving, λιθο-φιλος
Lithophragma Stone-wall, λιθο-φραγμα (the habitat of some)
lithophytus -a -um stone-plant, λιθο-φυτον (mimetic form and habitat)

Lithops, lithops, lithopius -a -um Stone-like, λιθο-(ωψ, ωπος) (the mimetic appearance of stone-cacti)
lithospermoides resembling *Lithospermum, Lithospermum-oides*
Lithospermum Stone-seed, λιθος-σπερμα (Dioscorides' name, λιθοσπερμον, for the glistening, whitish nutlets)
lithuanicus -a -um from Lithuania, Lithuanian
litigiosus -a -um disputed, contentious, *litigo, litigare,* to quarrel or go to law
litoralis -is -e, littoralis -is -e, littoreus -a -um growing by the sea-shore, *litus, litoris*
Litsea from a Japanese vernacular name, li tse
litticolus -a -um inhabiting the beach, *littoralis-colo*
Littonia for Dr Samuel Litton (1781–1847), Professor of Botany at Dublin
Littorella Shore, diminutive of *litus* (the habitat)
lituiflorus -a -um trumpet-flower, *lituus-florum*
lituiformis -is -e shaped like an augur's staff or trumpet, *lituus-forma*
lituus -a -um forked and with the ends turned outwards, like an augur's staff, *lituus, litui*
liukiuensis -is -e from the Ryukyu-shoto archipelago, S Japan
livens becoming bluish or black and blue, present participle of *liveo, livere*
lividus -a -um lead-coloured, bluish-grey, leaden, black and blue, *lividus*
Livistonia for Patrick Murray of Livingston, whose garden formed the nucleus of the Edinburgh Royal Botanic Garden, 1670
Lizei for the Lizé Frères, nurserymen of Nantes, France, *c.* 1912
llano- of treeless savanna-, through Spanish, llanos, from *planum*
Llavea, llavea for M. de Llave, who found the fern *Llavea cordifolia*
Lloydia for Edward Lloyd (1660–1709), Keeper of the Ashmolean Museum, Oxford
lloydii for either Curtis G. Lloyd (1859–1926), American botanist, or James Lloyd (1810–96), of London and Nantes, or Francis Ernest Lloyd (1868–1947) of Tucson, Arizona Desert Laboratory
Loasa from a S American vernacular name (**Loasaceae**)
lobatus -a -um, lobus -a -um with lobes, lobed, λοβος, lobus (see Fig. 4e)
lobbianus -a -um, lobbii for the brothers William Lobb (1809–63) and Thomas Lobb (1820–94)
Lobelia, lobelii for Matthias de l'Obel (1538–1616), Flemish renaissance pioneer of botany and herbalist to James I of England, author of *Plantarum seu stirpium historia* (***Lobeliaceae***)
lobi- lobes-, λοβος
lobiferus -a -um having lobes, *lobi-fero*
-lobium, -lobion -pod-fruited, -podded, λοβος (literally the lobe of an ear or the liver)
Lobivia an anagram of Bolivia, provenance of the genus
lobo-, -lobus -a -um lobed-, -lobed, λοβος, λοβο-, *lobus*
lobocarpus -a -um having lobed fruits, λοβος-καρπος
lobophyllus -a -um having lobed leaves, λοβος-φυλλον
Lobostemon Lobe-stamened, λοβος-στεμον (the stamens are opposite the corolla lobes)
Lobularia Small-pod, feminine diminutive of λοβος, *lobus*
lobularis -is -e, lobulatus -a -um with small lobes, diminutive of *lobus*
lobuliferus -a -um carrying lobed structures, λοβος-φερω, *lobus-fero*
localis -is -e local, of restricted distribution, *loco, locare, locavi, locatum*
lochabrensis -is -e from Lochaber, Scotland
lochmius -a -um coppice-dweller, of thickets, λοχμη a lair or thicket
-locularis -is -e -celled, *loculus, locularis* (usually the ovary)
locuples reliable; rich, opulent, *locupleto, locupletare; locuples, locupletis*
locusta in botanical Latin, spikeleted (an old generic name for *Valerianella locusta* (classical Latin, crayfish or locust)
Loddigesia, loddigesii for Conrad Loddiges (1743–1826), nurseryman in Hackney

loderi for Gerald Loder (Lord Wakehurst) (1861–1936), plantsman and owner of Wakehurst Place from 1903 to 1936, or Simon Loder (1932–94), of Clapton Court, Somerset

Lodoicea for Louis XV of France (1710–74) (Lodewijk, to his Polish father-in-law, King Stanislav); the extravagant signature of the coco-de-mer was behind a rapid spread of venereal diseases from the Seychelles

Loeselia, loeselii for Johann Loesel (1607–57), author of *Flora Prussica*

loeseneri for Dr L. E. T. Loesene (1865–1941), German botanist

Logania, loganii for James Logan (1674–1751), Irish Governor of Pennsylvania and author of *Experimenta de plantarum generatione* (***Loganiaceae***)

loganobaccus -a -um Logan's berry, after the developer Judge James Harvey Logan (1841–1928), of Santa Cruz, California

Loiseleuria for Jean Louis August Loiseleur-Deslongchamps (1774–1849), French botanist and physician

loliaceus -a -um resembling *Lolium*

Lolium a name in Virgil for a weed grass (Italian, loglio)

loma-, -loma -fringe, -border, hem, edge- λωμα (the name Loma applies also to Peruvian grass steppe, Argentinian 'slopes', a W African tribe, and a Dominican Republic peak)

Lomandra Edged-anthers, λωμα-(ανηρ, ανδρος)

Lomaria Bordered, λωμα (the marginal sori)

lomariifolius -a -um having leaves resembling fronds of *Lomaria, Lomaria-folium*

Lomariopsis *Lomaria*-like, *Lomaria-opsis* (***Lomariopsidaceae***)

Lomatia Fringed, λωμα (the seeds are bordered with a wing)

Lomatium Fringed, λωμα (the winged seeds)

lomensis -is -e from Lome, Togo, W Africa

lomentiferus -a -um bearing constricted pods that break up into one seeded portions, *lomentum-fero* (literally, bearing bean meal)

Lonchitis, lonchitis -is -e lance-shaped, shaped like a spear, λογχη (a name, λογχιτις, used by Dioscorides for an orchid)

lonchitoides resembling *Lonchitis, Lonchitis-oides*

loncho- lance-, λογχη

Lonchocarpus Lance-fruit, λογχη-καρπος, (the flat, indehiscent pods)

lonchophyllus -a -um with spear-like leaves, λογχη-φυλλον

londinensis -os -e from London (*Londinium*)

longaevus -a -um long-maturing, aged, *longaevus* (monocarpic after several tens of years)

longan, longanus -a -um from an Indian vernacular name, linkeng or longyen, for the fruit of *Euphoria longana*

longe-, longi- elongated-, long-, *longe*

longebracteatus -a -um with long bracts, *longe-bracteatus*

longeracemosus -a -um with long racemes, *longe-racemosus*

longesquamatus -a -um with long scales, *longe-squamata*

longiauritus -a -um with long ears, *longe-auritus* (basal lobes)

longibarabatus -a -um long-bearded, *longe-barbatus*

longibracteatus -a -um with long bracts, *longe-bracteatus*

longibulbus -a -um having elongate bulbs, *longe-bulbus*

longicalcaratus -a -um having long spurs, *longe-*(*calcar, calcaris*)

longicalyx with an elongate calyx, *longe-calyx*

longicauda long-tailed, with a long appendage, *longe-caudatus*

longicaulis -is -e long-stemmed, *longe-caulis* (usually clear-stemmed)

longicollis -is -e having a long neck, *longe-collum*

longicornu, cornutus -a -um with long horns, *longe-cornu*

longicuspis -is -e with long cusps, *longe-cuspis* (i.e. a small but pronounced apical point)

longicystis -is -e with a long cystidioles, *longe-*(κυστις) *cystis* (sterile cells amongst spore-producing basidia)

longifimbriatus -a -um with a long fringe, *longe-fimbriatus*
longiflorus -a -um with long flowers, *longe-florum*
longifolius -a -um with long leaves, *longus-folium*
longiformis -is -e elongate, *longus-forma*
longigemmis -is -e with long buds, *longus-gemma*, from *gemmo, gemmare*, to sprout
longilobus -a -um with long lobes, *longus-lobus* (leaf or other structure)
longipedicellatus -a -um having flowers borne on long pedicels, *longe-pedicellus*
longipedunculatus -a -um having long stalks, *longe-pedunculus*
longipes long-stalked, *longus-pes*
longipetalus -a -um with long petals, *longe-petalum*
longipetiolatus -a -um with long petioles, *longe-petiolaris*
longiracemosus -a -um with long racemes, *longe-racemosus*
longiradiatus -a -um with long ray, *longe-radiatus* (florets or pedicels etc.)
longiscapus -a -um with long flowering scapes or peduncles, *longe-scapus*
longisiliquus -a -um with long fruiting pods or siliqua, *longe-siliquus*
longispathus -a -um with long spathes, *longe-spatha*
longispicatus -a -um with long flowering spikes, *longe-spicatus*
longissimus -a -um the longest, superlative of *longe* (of the species' character)
longistylis -is -e, longistylus -a -um long-styled, *longe-stylus*
longitubus -a -um with long tubular flowers, *longe-tubus*
longulus -a -um somewhat lengthened, diminutive of *longe*
longus -a -um long, elongated, *longe*
Lonicera, lonicera for Adam Lonitzer (1528–86), German physician and botanist
lonicerifolius -a -um with leaves similar to honeysuckle, *Lonicera-folium*
Lophanthus, lophanthus -a -um Crested-flower, with tufts of flowers, λοφος-ανθος
Lophhira Crested, diminutive of λοφος (one of the sepals enlarges to a wing which aids fruit dispersal)
lopho- tufted-, crest-, crested-, λοφος, λοφο-, λοφ-, bristled-, maned-, λοφια
Lophochloa Crested-grass, λοφο-χλοη
lophogonus -a -um crested-angular, with crested angles, λοφο-γωνια (as on a stem or fruit)
Lophomyrtus Crested-myrtle, λοφο-μυρτον
lophophilus -a -um living on hills, hill-loving, λοφο-φιλος
Lophophora Crest-bearer, λοφο-φορα (has tufts of glochidiate hairs, *L. williamsii* is the hallucinogenic peyote button cactus)
lophophorus -a -um maned, having a crest, crest-bearing, λοφο-φορα
Lophospermum, lophospermus -a -um Crested-seed, λοφο-σπερμα
lora-, loratus -a -um, lori-, loro- strap-, shaped like a strap, λορος, λορο-, *lorum, lori*
loranthiflorus -a -um with *Loranthus*-like flowers, *Loranthus-florum*
Loranthus Strap-flower, λορος-ανθος (the shape of the 'petals') (***Loranthaceae***)
Lordhowea vide Howea (*Lordhowea insularis* is endemic on Lord Howe Island)
lorentzianus -a -um for H. A. Lorentz (b. 1869), who explored in New Guinea
loricatus -a -um clothed in mail, with a hard protective outer layer, *lorica, loricae*
loriceus -a -um armoured, with a breast-plate, *lorica, loricae*
lorifolius -a -um with long narrow leaves, strap-leaved, *lori-folium*
Loroglossum Strap-tongue, λορο-γλωσσα (the elongate lip, ≡ *Himantoglossum*)
Loropetalum Strap-petalled, λορο-πεταλον
loti on trefoil, living on *Lotus* (*Continaria*, dipteran gall midge)
loti- trefoil-like, *Lotus-oides*
lotiflorus -a -um *Lotus*-flowered, *Lotus-florum*
lotifolius -a -um with *Lotus*-like leaves, *Lotus-folium*
lotoides resembling *Lotus, Lotus-oides*
Lotononis shared features, *Lotus-Ononis*
Lotus the ancient Greek name, λωτος, for various leguminous plants, used by Theophrastus for *Zizyphus lotus*
Loudetiopsis *Loudetia*-like, *Loudetia-opsis*

loudonii for John Claudius Loudon (1783–1843), Scottish gardener, architect and author of several major gardening works
louisianicus -a -um, louisianus -a -um from Louisiana, USA
loureirii for João de Loureiro (1715–96), Portuguese missionary in Cochin China (Vietnam)
Lowia for Sir Hugh Low (1824–93), collector in Borneo (***Lowiaceae***)
loxo- slanting, oblique-, λοξος, λοξο-
Loxococcus Oblique-fruit, λοξο-κοκκος
Loxodera Oblique-callus, λοξο-δερω
Loxogramma Oblique-lined, λοξο-γραμμη (the sori)
loxophlebus -a -um having oblique veins, λοξο-(φλεψ, φλεβος)
Loxoscaphe Oblique-bowl, λοξο-σκαφη (the shape of the indusium)
Loxosoma Oblique-body, λοξο-σωμα (the sporangial annulus is incomplete)
Loxostylis Oblique-style, λοξο-στυλος
lubbersianus -a -um for C. Lubbers (1832–1905), Professor of Botany at Brussels
lubricatus -a -um, lubricus -a -um smooth, slippery, hazardous, *lubrico, lubricare*
lucalensis -is -e from the environs of the Lucala river, Angola
lucanius -a -um from Luca, Malta
lucens shining, present participle of *luceo, lucere, luxi*
lucernensis -is -e from Lucerne, Switzerland
lucernus -a -um from Lucerne, *Luceria*, Switzerland, *lucerna*, a lamp
lucescens shining, glittering, present participle of *lucesco, lucescere*
luchuensis -is -e from the Ryukyu-shoto archipelago, S Japan
luciae for Madame Lucie Savatier
lucianus -a -um from St Lucia, W Indies
lucidrys the epithet for the hybrid *Teucrium chamaedrys* × *lucidum*
lucidus -a -um bright, clear, shining, *luceo, lucere, luxi*
luciferus -a -um of the morning star (Venus or *Lucifer*); shining, flowering in the morning
luciliae for Lucile Boissier (1822–49), wife of Edouard Boissier
lucis -is -e of exposed habitats; bright, light, *lux, lucis*
lucombianus -a -um for William Lucombe (*c.* 1696–1794) of Exeter, nurseryman
luconianus -a -um from Luzon, Philippines
lucorum of woodland or woods, *lucus, luci*
luctuosus -a -um lamentable; sad, sorrowful, *luctuosus*
luculentus -a -um bright, excellent, full of brightness, *lux, lucis*
Luculia from a Nepalese vernacular name, luculi swa
ludens of games, sportive, *ludo, luderi, lusi, lusum*
ludibundus -a -um safe, easy, playful, *ludibundus*
ludicrus -a -um sporty, showy, theatrical, *ludicer, ludicri*
ludificans ridiculing, thwarting, *ludifico, ludificare; ludificor, ludificari, ludificatus*
ludovicianus -a -um from Louisiana, USA (Ludovicia), or for Louis XIV
Ludwigia for Christian Gottlieb Ludwig (1709–73), German botany professor at Leipzig
lueddemannianus -a -um for Gustave Adolphe Lüddermann (d. 1884), Paris nurseryman
Luetkea for F. Lütke (1797–1882), commander of the fourth Russian voyage around the world
Luffa Loofah, from the Arabic name, louff, for *Luffa cylindrica*
lugdunensis -is -e from Lyons, *Lugdunum*, France
lugens mourning, downcast, present participle of *lugeo, lugere, luxi*
lugubris -is -e mournful, *lugubris*
lujiangensis -is -e from Liu Chiang (Liujiang), Kwangsi, China
Luma, luma from a Chilean vernacular name for *Myrtus luma*
lumbricalis -is -e worm-like, *lumbricus, lumbrici*
Lumbricus, lumbricus Earthworm, *lumbricus, lumbrici* (some algae are worm-shaped, lumbricoid)

luminiferus -a -um lamp or ornament bearing, (*lumen, luminis*)-*fero* (possibly for the gland-covered lower leaf surface)
luminosus -a -um brilliant, *luminosus*
lumutensis-is -e from the environs of the port of Lumut, W Malaysia
luna of the moon, crescent-shaped, of a month, *luna, lunae*
Lunaria, lunaria Moon, *luna* (a name used by Fuchs and Mattioli for the shape and colour of the septum (or replum) of the fruit of honesty) (the fern *Botrychium lunaria* is moonwort because of the shape of its pinnae)
lunariifolius -a -um with leaves resembling those of *Lunaria*
lunarioides resembling *Lunaria, Lunaria-oides*
lunatus -a -um half-moon-shaped, lunate, *luna, lunae*
lunulatus -a -um crescent-moon-shaped, diminutive of *luna*
lupicidius -a -um wolf's-bane, *lupus-caedere*
lupinellus -a -um like a small *Lupinus*
lupinoides resembling *Lupinus*
Lupinus the ancient Latin name, *lupinus, lupini*, for the white lupin, diminutive of *lupus*
lupuli-, lupulinus -a -um hop-like, with the rampant habit of hops, *Humulus lupulus*
Lupulus, lupulus Brunfels' name, in reference to its straggling habit on other plants (the ancient Latin name for hop was *Lupus salictarius* – willow wolf)
luquensis -is -e from Luque, S Paraguay
luquillensis -is -e from the Sierre de Luquillo, Puerto Rica
luridus -a -um sallow, dingy yellow or brown, wan, lurid, ghastly, *luridus*
luristanicus -a -um from Lorestan (Luristan) province, W Iran
Luronium Rafineque's name for a water plantain, *Alisma*
lusitanicus -a -um from Portugal (*Lusitania*), Portuguese
lutarius -a -um of muddy places, living on mud, *lutum*
luteiflorus -a -um yellow-flowered, *luteus-florum*
lutensis -is -e from the Lut desert, E Iran
luteo- yellow-, *luteus*
luteoalbus -a -um yellow with white, *luteus-albus*
luteocarpus -a -um yellow fruited, *luteus*-(καρπος) *carpus*
luteolus -a -um yellowish, diminutive of *luteus*
luteorubra yellow with red, *luteus-rubrum*
luteotactus -a -um touched with yellow, *luteus*-(*tango, tangere, tetigi, tactum*)
luteovenosus -a -um yellow-veined, *luteus*-(*vena, venae*)
luteovirens, luteovirescens becoming yellowish-white, *lutum-essentia*
luteoviridis -is -e yellowish green, yellow with green, *luteus-viridis*
luteovitellinus -a -um orange-yellow, *luteus-vitellus*
lutescens turning yellow, yellowing, yellowish, *luteus-essentia*
lutetianus -a -um from Paris (*Lutetia*), Parisian
luteus -a -um yellow; vile, of clay, muddy, *lutum, luti*
lutosus -a -um dyer's weed, yellow, of mires or clay, *lutum, luti*
lutrus -a -um of the otter, *lutra*
luxatus -a -um excessive, debauched, pompous, *luxus*
Luxembergia for the Duke of Luxembourg, sponsor of Auguste Saint Hilaire's expedition to Brazil
luxurians rank, exuberant, luxuriant, of rapid growth, *luxurio, luxuriare, luxurior, luxuriari*
luxuriosus -a -um luxurious, voluptuous, *luxuria, luxuriae*
luzonicus -a -um from Luzon island, Philippines
Luzula an ancient name of obscure meaning
luzuli- *Luzula*-like
luzuloides resembling *Luzula*
Luzuriaga for Don Ignatio M. R. de Luzuriaga, Spanish botanist
lyallii for David Lyall (1817–95), who collected in New Zealand from HMS *Terror*

lycaonicus -a -um from the ancient region, Lycaonia, of Anatolia, Turkey
Lycaste for Lycaste, daughter of King Priam of Troy
Lycene the composite name for hybrids between *Lychnis* and *Silene*
Lychnis Lamp, λυχνος (Theophrastus' name, λυχνις, for the hairy leaves were used as wicks for oil lamps)
lychnitis from a name in Pliny meaning of lamps, λυχνος
lychno-, lychnoides *Lychnis*-like, λυχνος-οειδης
Lychnothamnus Light-bush, λυχνος-θαμνος
lycioides box-thorn-like, *Lycium-oides*
Lycium the ancient Greek name, λυκιον, for a thorn tree (*Rhamnus*) from Lycia, re-applied by Linnaeus
lycius -a -um from Lycia, λυκιον, SW Turkey
lyco- wolf-, λυκος, λυκο- (usually implying inferior wild, or rampant)
Lycocarpus, lycocarpus -a -um Wolf-fruit, λυκο-καρπος (clawed at the upper end)
lycoctonus -a -um wolf-murder, λυκο-κτονος (poisonous wolf's-bane, *Aconitum lycoctonum*)
lycoperdoides puff-ball-like, resembling *Lycoperdon*, λυκο-περδειν-οειδης
Lycoperdon Wolf's-fart, λυκο-περδειν (for the emission of clouds of spores)
Lycopersicum (on) Wolf-peach, λυκο-περσικον, Galen's name, λυκοπερσιον (for an Egyptian plant) (*Lycopersicum esculentum* is the tomato, from Nahuatl, tomatl)
lycopodioides resembling *Lycopodium*, λυκο-ποδιον-οειδης
Lycopodium Wolf's-foot, λυκο-ποδιον (Tabernaemontana's translation of the German, Wolfsklauen, for a clubmoss) (***Lycopodiaceae***)
Lycopsis Wolf-like, λυκος-οψις (Dioscorides' derogatory name, λυκοψις)
lycopsoides resembling *Lycopsis*, λυκοψις-οειδης
Lycopus Wolf's-foot, λυκος-πους
Lycoris for Lycoris the actress, and Marc Antony's mistress
lydenburgensis -is -e from Lydenberg, Transvaal, S Africa
lydius -a -um from the ancient region of Lydia, SW Turkey
Lygeum Pliant, λυγιζω to bend, λυγος a willow twig
Lygodium Twining-one, λυγοδης (the climbing fern's stems)
lynceus -a -um lynx-like, of the lynx (Lynkeos was a keen-sighted Argonaut)
lyonii for John Lyon (*c.* 1765–1814), introducer of American plants
Lyonothamnus Lyon's-shrub, for W. S. Lyon (1851–1916), its discoverer
lyratus -a -um lyre-shaped, λυρα, *lyratus* (rounded above with small lobes below – usually of leaves)
lyrifolius -a -um having lyre-shaped leaves, *lyratus-folium* (with enlarged upper lobe)
lyroglossus -a -um having a lyrate tongue, λυρα-γλωσσα
lyrophyllus -a -um having lyre-shaped leaves, λυρα-φυλλον (with enlarged upper lobe)
lysi-, lysio-, lyso- loose-, loosening-, λυσις, λυσι-
lysicephalus -a -um having loose (flower-)heads, λυσις-κεφαλη
Lysichiton (um) Loose-cloak, λυσι-χιτων, (the open, deciduous spathe)
Lysimachia Ending-strife, λυσι-μαχη, Dioscorides' name λυσιμαχειος, λυσιμαχια (Pliny relates that the Thracian king Lysimachos discovered it)
lysimachoides resembling *Lysimachia*
Lysionotus Rear-opening, λυσις-νωτος (capsules open elastically along dorsal suture)
-lysis loosening, dissolution, decay, metamorphosis, λυσις
lysistemon with loose stamens, λυσις-στεμον (stamens not conjoined or having rigid filaments)
lysolepis -is -e with loose scales, λυσις-λεπις
Lythrum Gore, λυθρον (Dioscorides' name, λυτρον, may refer to the flower colour of some species) (***Lythraceae***)
Lytocaryum Releasing-nut, λυτο-καρυον (λυτεριος, releasing, λυτικος able to release, λυτρον, ransom)

Maakia for Richard Maack (1825–86), Russian naturalist
Maba from a Tongan vernacular name
Macadamia for Dr John Macadam (1827–65), Secretary to the Philosophical Institute of Victoria (Queensland nut)
Macaranga from the Malayan vernacular name, umbrella tree (the large leaves)
macaronesicus -a -um from the Macronesian islands, E Atlantic
macarthurii for either Captain J. Macarthur, or Sir William Macarthur (1800–82), who collected in Australia
macdougallii for Dr Daniel Trembly MacDougall (1865–1958), Director of the Desert Laboratory, Tucson, Arizona, USA
macedonicus -a -um from Macedonia, Macedonian
macellarius -a -um, macellus -a -um of the market, *macellum, macelli*
macer -era -erum meagre, poor, *macer, macri*
macerispicus -a -um having thin or poor spikes, *macer-(spica, spicae)*
Macfadyena for James Macfadyen (1800–50), author of *Flora of Jamaica*
Machaeranthera Dagger-anthered-one, μαχαιρα-ανθερα
macilentus -a -um thin, lean, *macies*
Mackaya, mackaianus -a -um, mackaii for James Townend Mackay (1775–1862), Curator of the Botanic Garden, Trinity College Dublin
mackenii for Mark Johnston M'Ken (1823–72), curator of the Durban Botanic Garden
Macleaya, macleayanus -a -um for either Alexander Macleay (1767–1848), Secretary to the Linnean Society of London, or Sir George Macleay (1809–91)
Maclura for William Maclure (1763–1840), American geologist
macowanii for Peter MacOwan (1830–1909) of Huddersfield, Director of Cape Colony Botanic Garden
macpalxochitlquahuitl a Mexican vernacular name for *Cheirostemon platanoides*
macr-, macro- big-, large-, long-, tall-, deep-, far-, μακρος, μακρο, μακρ-
macracanthus -a -um with large thorns, μακρο-ακανθος
Macradenia Long-gland, μακρ-αδην (the pollinial attachments)
macradenius -a -um having large glands, μακρ-αδην
macrandrus -a -um having large anthers, μακρ-ανηρ
macranthus -a -um large-flowered, μακρ-ανθος
macroblastus -a -um with large shoots, μακρο-βλαστος
macrobotrys having large clusters, μακρο-βοτρυς
macrobulbon having large bulbs, μακρο-βολβος, *macro-bulbus*
macrocalyx with a large calyx, μακρο-καλυξ
macrocapnos large *Fumitory* like, μακρο-καπνος (big smoke)
macrocarpus large-fruited, μακρο-καρπος
macrocentrus -a -um large-spurred, μακρο-κεντρον
macrocephalus -a -um with large heads, μακρο-κεφαλη (of flowers)
macrochaetae large-awned, μακρο-χαιτη
macrococcus -a -um having large berries, μακρο-κοκκος
Macrocystis, macrocystis Large-bladder, μακρο-κυστις
macrodon, macrodontus -a -um having large teeth, μακρ-(οδους, οδοντος)
macrodus -a -um large-toothed, μακρ-οδους
macroglossus -a -um large-lipped, with a large tongue, μακρο-γλωσσα
macrogonus -a -um very angular, with large nodes, μακρο-γονυ (stout ribbed stems)
macrolepis -is -e with large scales, μακρο-λεπις
macromeris -is -e with large parts, μακρο-μερις
macronemus -a -um having large or long stamens, μακρο-νημα
macronychius -a -um, macronyx with large claws, μακρ-(ονυξ, ονυχος) (the petals)
macropetalus -a -um having large petals, μακρο-πεταλον
macrophyllus -a -um with large leaves, μακρο-φυλλον
Macropiper Large-pepper, botanical Latin from μακρο and *Piper* (Maori pepper-tree, kawa-kawa)
macropodianus -a -um from Kangaroo Island, SE Australia, botanical Latin, μακρο-(πους, ποδος), to emphasize the abundance of the island's kangaroo population

macropodus -a -um, macropus with a large stalk, μακρο-(πους, ποδος)
macropterus -a -um having large wings, μακρο-πτερον
macropunctatus -a -um with large spots, *macro-(pungo, pungere, pupugi, punctum)*
macrorhizus -a -um, macrorrhizus -a -um large-rooted, μακρο-ρυζα
macrorrhabdos large-stemmed, heavily branched, μακρο-ραβδος
macrosiphon large-tubular, long-tubed, μακρο-σιφον
macrosmithii large (*Rhododendron*) *smithii*
macrosolen with a long tube, μακρο-σωλην (the corolla)
macrospermus -a -um large seeded, μακρο-σπερμα
Macrosphyra Large-globed, μακρο-σφαιρα
macrosporus -a -um large-spored, μακρο-σπορος (spores about 10 m × 6 m)
macrostachus -a -um, macrostachyus -a -um with large spikes, μακρο-σταχυς
macrostegius -a -um with a large cover, μακρο-στεγη (bract or spathe)
macrostemon with large stamens, μακρο-στεμον
macrostephanus -a -um with a large crown, μακρο-στεφανος (coronna)
macrostigmus -a -um with large stigmas, μακρο-στιγμα
macrostylus -a -um with long styles, μακρο-στυλος
macrosyphon with a long tubed corolla, μακρο-σιφον
macrothyrsus -a -um with large thyrsoid inflorescences, μακρο-θυρσος
macrotomius -a -um long segments, μακρο-τεμνειν (the 'cut' of the calyx)
Macrozamia Large-*Zamia*
macrurus -a -um, macrourus -a -um long-tailed, μακρ-ουρα
macrus -a -um long-lasting, tall, high, large, long, μακρος, μακρο-, μακρ-
maculatus -a -um, maculosus -a -um, maculifer -era -erum spotted, blotched, bearing spots, *macula, maculae*
maculi- spot-like-, *macula, maculae*
maculiferus -a -um, maculigerus -a -um bearing or carrying spots or blemishes, *macula-fero, macula-gero*
maculiflorus -a -um having flowers with spotted petals, *maculae-florum*
madagascariensis -is -e from Madagascar, Madagascan
Maddenia for Major E. Madden, a writer on Indian botany
madeirensis -is -e from Madeira, Macronesian
maderaspatanus -a -um, maderaspatensis -is -e from the Madras region of India
maderensis -is -e from Madeira, Madeiran (*Madera*)
Madia from a Chilean vernacular name for *Madia sativa*
madidus -a -um drunk; soaked, wet, sodden, *madidus*
madrensis -is -e from the Sierra Madre, N Mexico
madritensis -is -e from Madrid, Spain (*Matritum*)
madurensis -is -e from Madura Island, Indonesia, or Madura, S India
Maerua from an Arabic vernacular name, meru
Maesa from the Arabic vernacular name, maas
maesiacus -a -um from the Bulgarian/Serbian region once called *Maesia*
Maesobotrya *Maesa*-like-fruited (similarity of the fruiting clusters)
maestus -a -um mournful, sorrowful, *maestus*
magdalenicus -a -um from the valley of the Magdalena river, N central Colombia
magdalensis -is -e from the Madeleine Islands, Quebec, Canada
magdeburgensis -is -e from Magdeburg, E central Germany
magellanicus -a -um from the Straits of Magellan, S America (named for Ferdinand Magellan (1480–1521), Portuguese explorer)
magellensis -is -e from Monte Majella, Italy
magentius -a -um from Magenta, N Italy (name commemorates a bloody battle fought there)
Magistrantia medieval Latin name for *Peucedanum ostruthium*, masterwort
magistratus -a -um demanding attention, *magister, magistri*
magnatus -a -um the greatest, the most prized, comparative of *magnus* (the white truffle)
magni-, magno-, magnus -a -um large, *magnus, maior, maximus*

magnicalcaratus -a -um having notable spurs, *magnus-(calcar, calcaris)*
magnificus -a -um great, eminent, distinguished, magnificent, *magnifico, magnificare*
magniflorus -a -um large-flowered, *magnus-florum*
magnifolius -a -um large-leaved, *magnus-folium*
Magnolia for Pierre Magnol (1638–1715), Professor of Botany and Director of Montpellier Botanic Garden (***Magnoliaceae***)
magnus -a -um great, large, high, noble, *magnus*
mahagoni mahogany, from a S American vernacular name for *Swietenia mahagoni*
mahaleb an Arabic vernacular name for *Prunus mahaleb*
Mahernia an anagram of *Hermannia*, a related genus
Mahoberberis the composite name for hybrids between *Mahonia* and *Berberis*
Mahonia for Bernard McMahon (1775–1816), American horticulturalist
mahonii for John Mahon (1870–1906), of the Uganda Botanical Garden
mai-, maj- May-, *maius*
Maianthemum May-flower, μαι-ανθεμιον (μαιμακτηριων, the Attic fifth month was our November)
mairei for Professor Edouard le Maire of Ghent, who collected in S Africa
maius -a -um of the month of May, *maius, mai*
majalis -is -e (magalis) of the month of May, *maius* (flowering time)
majesticus -a -um majestic, *maiestas, maiestatis*
major -or -us larger, greater, bigger, *maior, maioris*, comparative of *magnus*
Majorana medieval Latin, *maiorane*, for sweet marjoram, *Majorana hortensis* (wild marjoram is *Origanum vulgare*)
majoricus -a -um of Majorca (*Majorica*) Balearics
majusculus -a -um somewhat larger, *maior* (diminutive suffix)
makinoi for Tomitaro Makino (1863–1957), Japanese botanist
makoyanus -a -um for Jacob Mackoy (1790–1873), nurseryman of Liège
malabaricus -a -um from the Malabar coast, S India
Malacantha Very-thorny, μαλα-ακανθος
malaccensis -is -e from Maleka (Malacca), Malaysia
Malachium Tenderness, μαλακια, μαλακος
malachius -a -um soft, delicate, luxurious, μαλακια, μαλακος
malaco-, malako-, mollusc-, μαλακια soft, tender, weak, mucilaginous, mallow-like, μαλακος, μαλακο-
Malacocarpus Soft-fruit, μαλακος-καρπος (the fleshy fruit)
Malacodendron, malacodendron Soft-tree, μαλακος-δενδρον (pubescence of branches, leaves and calyx)
malacoides *Malva*-like, soft to touch, μαλακος-οειδης
malacophilus -a -um pollinated by snails, snail-loving, μαλακοια-φιλος
malacophyllus -a -um with soft or fleshy leaves, μαλακος-φυλλον
malagasius -a -um from Madagascar, Madagascan (Malagasy Republic)
malaianus -a -um from Malaya, Malaysian
malaitensis -is -e from the volcanic Mala island (Malaita), Solomons
malawiensis -is -e from Malawi, central Africa
Malaxis Tenderness, μαλαξις (hard to cultivate because adapted to *Sphagnum* bog conditions, ≡ *Hammarbya*); some interpret as Soft, for the soft foliage
Malcomia (Malcolmia) for William Malcolm, eighteenth-century English nurseryman
maleolens of bad fragrance, very much stinking, *male-olens*
malesicus -a -um from the Malesian Islands (Indo/Polynesia)
malevolus -a -um malicious, ill-disposed, *malevolus* (with spines)
maliflorus -a -um apple-blossomed, *Malus-florum*
malifolius -a -um having leaves resembling those of *Malus, Malus-folium*
maliformis -is -e apple-shaped, *Malus-forma* (fruits)
mallei-, malleo- hammer-like, *malleolus, malleoli*
malleiferus -a -um hammer-carrying, *malleolus-fero* (usually a floral structure)
malleolabrus -a -um having a hammer-shaped lip, *malleolus-labrum*

mallo-, -mallus -a -um fleecy, woolly, with woolly hair, μαλλος, μαλλο-, μαλλοτος
mallococcus -a -um downy-fruited, μαλλος-κοκκος
mallophorus -a -um wool-bearing, μαλλος-φορα
mallophyllus -a -um having woolly leaves, μαλλος-φυλλον
Mallotus, mallotus -a -um Woolly, μαλλοτος (the fruits of some species)
Malope a name for mallow in Pliny, μαλλος-ωπος woolly-looking?
Malpighia for Marcello Malpighi (1628–94), Italian professor at Pisa (***Malpighiaceae***)
Malus, malus the ancient Latin name, *malus, mali*, for an apple tree
Malva Soft, *malva* (the name in Pliny), cognate via old English, mealwe, with mallow (***Malvaceae***)
malvacearus -a -um of *Malva*, living on *Malva* (symbionts, parasites and saprophytes)
malvaceus -a -um mallow-like, resembling *Malva*
malvastroides resembling *Malvastrum, Malvastrum-oides*
Malvastrum *Malva*-like, *Malva-astrum*
Malvaviscus, malvaviscus -a -um Mallow-glue, *malva-viscus* (the wax mallow)
malverniensis -is -e from the Malvern hills, Herefordshire / Worcestershire
malvicolor coloured mauve, like *Malva, Malva-color*
malviflorus -a -um with *Malva*-like flowers, *Malva-florum*
malvifolius -a -um having leaves resembling those of *Malva*
malvinus -a -um deep mauve, mallow-like, *Malva*
malyi for M. Maly of Vienna Botanic Garden *c.* 1870
mammaeformis -is -e, mammiformis -is -e shaped like a nipple, *mamma, mammae*
Mammea from a W Indian vernacular name for mammee apple
Mammillaria (Mamillaria) Nippled, *mammilla, mammillae* (conspicuous tubercles)
mammillaris -is -e, mamillarius -a -um, mammillatus -a -um having nipple-like structures, mammillate, *mammilla, mammillae*
mammilliferus -a -um bearing nipples, *mammilla-fero* (epidermal tubercles)
mammosus -a -um full-breasted, *mamma, mammae* (covered with nipple-like outgrowths)
manan a vernacular name, manao or rotan manan, in SE Asia for the rattan palm, *Calamus manan*
manchuricus -a -um from Manchuria, E Asia
mancinellus -a -um from Spanish, manzana, manzanilla, for the bitter apple-like fruit of *Hippomane mancinella*
mancus -a -um crippled, deficient, inferior, *mancus*
mandarinorus -a -um of the cultured classes or counsellors, from Portuguese, mandarim
mandchuricus -a -um, mandshuricus -a -um from Manchuria, Manchurian
Mandevilla for Henry John Mandeville (1773–1861), diplomat in Buenos Aires who introduced Chilean jasmine
mandibularis -is -e jaw-like, having jaws, late Latin *mandibula*, from *mando, mandere, mandi, mansum*
Mandragora, mandragor us -a -um Man-dragon, a Greek name, μανδραγορας, derived from Syrian, namta ira, for mandrake
manescaui for M. Manescau of Pau, France, *c.* 1875
Manettia, manettii for Xaviero Manetti (1723–85), Prefect of the Florence Botanic Garden
manghas a vernacular name for the mango-like fruit of *Cerbera manghas*
Mangifera Mango-bearer, from the Hindu name, mangu, or Tamil, man kai, for the mango fruit, and *fero*
Manglesia, mangle, manglesii for Captain James Mangles (1786–1867), and Robert Mangles (d. 1860) of Sunningdale, or Harry Mangles, nineteenth-century *Rhododendron* breeder of Valewood, Surrey
mangostana from the Malayan vernacular name, mang gistan, for *Garcinia mangostana*, mangosteen

Manicaria Glove, *manica* (the spathe of the inflorescence)
manicatus -a -um with a felty covering which can be stripped off, manicate, *manicae, manicarum* (literally, with long sleeves)
manifestus -a -um obvious, evident, *manifestus*
Manihot from the Tupi-Guarani name, manioca, for cassava (the flour prepared from the root is tapioca, from Tupi-Guarani, typyoca)
manillanus -a -um from Manilla, Philippines
manipuliflorus -a -um grouped, with few-flowered clusters, *manipulatum*
manipuranus -a -um, manipurensis -is -e from Manipur, Assam, India
manna having a sweet exudate, Arabic, mann, μαννα
mannensis -is -e from the environs of the Mann river, Australia
Mannia, mannii for Gustav Mann (1835–1916), collector for Kew in W Africa 1859–63, Inspector of Assam Forests
Manniella diminutive of *Mannia*
manniferus -a -um manna-bearing, μαννα-φερω (Aramaic, manna, for the exudate from *Tamarix mannifera*)
mano- scanty-, μανος
manopeplus -a -um with a thin cloak, scantily covered, μανος-πεπλος
manriqueorum for Manrique de Lara, of the Manriques
manshuricus -a -um, manshuriensis -is -e from Manchuria, Manchurian
mantegazzianus -a -um for Paulo Mantegazzi (1831–1910), Italian traveller and anthropologist
manticus -a -um of seers or soothsayers, μαντις, μαντικος
Mantissia Prophet, μαντις (the flowers resemble a praying mantis)
mantoniae for Professor Irene Manton (1904–88), fern cytologist of Leeds University and first woman president of the Linnean Society of London
manuanus -a -um from the Manua Islands, SW Pacific
manubriatus -a -um having a handle-like structure, *manubrium, manubri*
manzanita Spanish vernacular for a small apple
Mapania from a W African vernacular name
mappa napkin(-textured), cloth(-textured), *mappa, mappae*
maracandicus -a -um from Samarkand (Maracanda), Uzbekistan
Maranta for Bartolomea Maranti, sixteenth-century Venetian botanist (***Marantaceae***)
marantifolius -a -um having foliage resembling *Maranta, Maranta-folium*
Marantochloa Maranti's-grass, botanical Latin from Maranti and χλοη
Marasmius Withering, μαρασμος (the scorched turf phase of the fairy ring, or the senescent fruiting body's leathery texture)
marathon fennel, μαραθον
Marattia for Giovanni Francesco Maratti (1723–77), Italian botanist, author of *De floribus filicum* (***Marattiaceae***)
marcescens not putrefying, persisting, retaining dead leaves and/or flowers, *marcesco, marcescere*
Marcgravia for Georg Markgraf (Marcgraf) (1610–1644), German engineer and geographer in Brazil (***Marcgraviaceae***)
marckii for Jean Baptiste Antoine Pierre Monnet de la Marck (1744–1829); French pre-Darwinian evolutionist (Lamarck)
margaretae for Margaret Mee (1909–88), botanical illustrator of Brazilian plants
margaritaceus -a -um, margaritus -a -um pearly, of pearls, μαργαριτης, *margarita, margaritae*
margaritae for Mlle Marguerite Closon
margaritiferus -a -um bearing pearl, μαργαριτης-φερω, *margarita-fero*
magaritisporus -a -um having pearl-like spores, μαργαριτης-σπορος
marginalis -is -e of the margins, margined, *margo, marginis*
marginatus -a -um having a distinct margin, *margo, marginis* (the leaves)
Margyricarpus Pearl-fruit, μαργαριτης-καρπος (the white berry-like achenes)
mariae for either Mrs Mary Burbidge or Miss Mary Anderson

marianus -a -um for the Virgin Mary, *Maria*; or from Maryland, USA; or from the Sierra Morena, Spain (*Montes Mariani*)
Marica Flagging, μαραινω (the flowers die away early) (≡ *Neomarica*)
marientalensis -is -e from the environs of Mariental, bordering the Kalahari, Namibia
mariesii for Charles Maries (1850–1902), English plant collector in Japan for Veitch *c.* 1880
marifolius -a -um having leaves similar to those of *Teucrium marum*
marilandicus -a -um, marylandicus -a -um from the Maryland region, USA
marinus -a -um marine, growing by or in the sea, *mare*
mariorika a hybrid epithet for *Picea mariana* × *ormorika*
maris see *mas*
marisculus -a -um like a small rush, diminutive of *Mariscus*
Mariscus, mariscus -a -um the name for a rush-like plant in Pliny
maritimus -a -um growing by the sea, maritime, of the sea, *mare*
marjoletti for Joseph Marie Marjolett (1823–94), who found *Tulipa marjoletti*
marjoranus -a -um derived from the Latin name, *maiorana* (sweet marjoram)
Markhamia for Sir Clements Robert Markham (1830–1916), explorer and writer
marmelos a Portuguese vernacular name, marmelo, for marmalade
marmorarius -a -um marbled, with coloured veins, *marmor, marmoris* (in the corolla)
marmoratus -a -um, marmoreus -a -um with veins of colour, sparkling, marbled, μαρμαρος
maroccanus -a -um, marocanus -a -um from Morocco, NW Africa, Moroccan
marrubialis -is -e *Marrubium*-like
Marrubium the name in Pliny, either from the Hebrew, marrob, for the bitter-juice, or from the town of *Marrubium* in *Latium*
Marsdenia for Willam Marsden (1754–1836), author of a history of Sumatra
Marshallia for Humphrey Marshall (1722–1801), who compiled the first list of American trees in 1785
marshallianus -a -um, marshallii for Marschall von Bieberstein (see *Biebersteinia*)
marsicus -a -um from the central Apennines, Italy (the land of the *Marsi*)
Marsilea for Luigi Fernando Marsigli (1656–1730), Italian patron of botany (***Marsileaceae***)
marsupialis -is -e pouched, having pouches, μαρσυπιον
marsupiflorus -a -um, marsupiiflorus -a -um with pouch-like flowers, botanical Latin from μαρσιπος and *florum*
martagon either from *herba martis, herba martina*, herb of Mars (German, Goldwürtz) used in alchemy (Pierandrea Mattioli, 1501–77), or resembling a kind of Turkish turban (Turk's cap)
martellianus -a -um for Conte Ugolino Martelli (1860–1934), author of floras for Italy and Eritraea
Martia, Martiusia, martianus -a -um for K. F. P. von Martius (1794–1868), German botanist in Brazil
martinicensis -is -e, martinicus -a -um from Martinique
martinii for Claude Martin (1731–1800), correspondent of Roxburgh (q.v.)
Martynia for John Martyn FRS (1699–1768), Professor of Botany at Cambridge (his son Thomas Martyn was also Professor of Botany at Cambridge for 63 years between 1762 and 1825) (***Martyniaceae***)
maru a vernacular name for mastic
maruta the Italian vernacular name for *Anthemis cotula*
marylandicus -a -um from Maryland, USA
mas, maris bold, with stamens, male, man, *mas, maris*
mascaratus -a -um masked, darkened, Arabic, maskara
Mascarenhasia from the Mascarene islands
Maschalocephalus Overpowered-head, μασχαλιζω (inflorescences almost concealed by bracts amongst leaf-bases)

masculinus -a -um, masculus -a -um male, staminate, vigorous, with testicle-like tubers
Masdevallia (Masdevillia) for Dr Jose Masdevall (d. 1801), Spanish physician and botanist
masius -a -um from the volcanic Mount Karaca (*Masius*) on the Arabian platform, SE Turkey
masonianus -a -um, masonorus -a -um, masoniorus -a -um for Canon G. E. and Miss M. H. Mason, plant collectors at Umtata, S Africa
massiliensis -is -e from Marseilles, France (Massilia)
Massonia for Francis Masson (1741–1805), who collected plants in S Africa
mastacanthus -a -um mouth-flower, morsel-flower, μασταξ-ανθος (μαστακος-ανθος)
mastersianus -a -um for Dr Maxwell Tylden Masters FRS (1833–1907), editor of *Gardeners' Chronicle*
mastichinus -a -um similar to mastic, μαστιχη, the exudate from *Pistacia lentiscus* (Dioscorides' μαστιχη was chewed, μασαομαι, masticated, to sweeten the breath)
mastigophorus -a -um (producing gum, gum-bearing, μαστιχη-φορα) whip-bearing, (μαστιξ, μαστιγος)-φορα (μαστιγο-φορος a constable)
mastoideus -a -um breast-shaped, μαστοειδες
Matonia for Dr William George Maton (1774–1835), Vice-President of the Linnean Society
Matricaria Of-the-womb, *matrix, matricis* (former medicinal use in treatment of uterine infections)
matricarioides resembling *Matricaria, Matricaria-oides*
matritensis -is -e from Madrid (*Matritum*), Spain
matronalis -is -e of the married woman, *matrona, matronae* (the Roman matronal festival was held on 1 March) (*Hesperis matronalis* or *viola flos matronalis*, dame's violet)
matsudana for Sadahisa Matsudo (1857–1921), Japanese botanist
Matteuccia, Matteucia for C. Matteucci (1800–68), Italian physicist
matthewsii for Mr Matthews, Curator of the Botanic Garden, Glasgow
Matthiola (Mathiola) for Pierandrea Mattioli (1501–77), Italian physician and botanist and author of *Commentarii in sex libros Pedanii Dioscoridis*
Matucana from the village of Matucana, Peruvian provenance of *Matucana haynei*
matutinalis -is -e, matutinus -a -um morning, of the morning, early, *matutinus*
maulei for William Maule of Bristol, who introduced *Chaenomeles maulei c.* 1874
Maurandella *Maurandia*-ish, feminine diminutive from *Maurandia*
Maurandia (Maurandya) for Catharina Pancratia Maurandi, a botany student at Carthagena *c.* 1797
mauritanicus -a -um from Morocco or N Africa generally (*Mauretania*)
mauritianus -a -um from the island of Mauritius, Indian Ocean
maurorum of the Moors (μαυρος), Moorish, of *Mauretania*
maurus -a -um from Morocco, *Mauretania*, Moorish (colour)
maweanus -a -um, mawii for George Maw (1832–1912), author of a monograph on *Crocus*
max the most, biggest or best, modern Latin from *magnus, maximus*
Maxillaria, maxillaris -is -e Jaws, *maxilla, maxillae* (column and lip resemble an insect's jaws)
Maximiliana for Prince Maximilian von Wied-Neuwied (1782–1867), German botanist in Brazil and N America
maximilianii for King Maximilian Joseph of Bavaria (1811–64)
Maximowiczia, maximowiczianus -a -um, maximowiczii for K. J. Maximowicz (1827–91), Conservator of the Botanic Garden at St Petersburg and traveller in the East
maximus -a -um largest, greatest, superlative of *magnus*
maya supernatural, illusory, Sanskrit, ma, maya
Mayaca Moss-like, μυιακος (moss-like freshwater aquatic herbs) (***Mayacaceae***)
maydis of or upon maize (*Urocystis* smut fungus)

mays from the Taino/Mexican name, mahiz, for Indian corn
Maytenus from the Chilean vernacular name, maiten
mazatlanensis -is -e from the Mazatlán peninsula, W central Mexico
Mazus Nipple, μαζος (the shape of the corolla)
meandriformis -is -e of winding form, much convoluted, *maeander-forma* (Μαιανδρος was a river)
meanthus -a -um small-flowered, μει-ανθος
-mecon -poppy, μηκων
Meconopsis Poppy-like, μηκων-οψις
Medeola for the sorceress Medea, daughter of Aëdes and Colchis, who aided Jason and the Argonauts
medeus -a -um remedial, healing, curing, of cures, *medeor, mederi*
medi-, medio- middle-sized, between-, intermediate-, from the centre, *medius*
Medicago Median-grass, Dioscorides' name, μηδικη, from a Persian name for lucerne, or medick (*medica* with feminine suffix *-ago*)
medicinalis -is -e having medicinal properties, medicinal, *medicina, medicinae*
medicus -a -um from Media (Iran), curative, medicinal (doctors in Rome were frequently from the East)
Medinilla for J. de Medinilla y Pineda, Governor of the Marianne Islands (Mauritius) in 1820
mediocris -is -e ordinary, average, *mediocris*
mediolanensis -is -e from Milan (Mediolanum), Italy
medioloides florist's *Smilax* (from Mediolanum)
medioluteus -a -um mid-yellow, with a yellow centre, *medius-luteus*
mediopictus -a -um with a coloured stripe down the centre-line, *medius-pictus* (of a leaf)
medioradiatus -a -um with a radiant centre, radiating from the centre, *medius-radiatus*
mediosorus -a -um having centrally arranged sori, botanical Latin from *medius* and σωρος
mediterranea, mediterraneus -a -um from the Mediterranean region; from well inland, *medi-terra*
Medium a plant name, μηδιον, in Dioscorides
medius -a -um between, intermediate, mid-sized, centre, *medius*
medocinus -a -um from the Médoc district, France
medullaris -is -e, medullus -a -um pithy, soft-wooded
medullarius -a -um, medullosus -a -um with a large pith
medusae like Medusa's head (the long threadlike sepals of the heads of flowers likened to the serpents hair of Medusa)
medwediewii for H. Medwediew, who collected *Betula medwediewii* in Trans-Caucasus *c.* 1888
meeana for Margaret Mee (1909–88), illustrator of Brazilian plants
Meehania, meehanii for Thomas Meehan (1826–1901), London nurseryman and writer on plants
Meeusella, meeuse, meeusei for Adriaan Dirk Jacob Meeuse (1914–), Dutch systematist
mega-, megali-, megas- biggest-, bigger-, big-, μεγα, μεγαλη, μεγας (comparatives)
megacalyx having a very large calyx, μεγα-καλυξ
megacarpus -a -um large fruited, μεγα-καρπος
megacephalus -a -um large-headed, μεγα-κεφαλη (of composite inflorescences)
Megaclinium Large-bed, μεγα-(κλινη, κλιναριον) (the many-flowered rachis)
Megacodon Massive-bell, μεγα-κωδον (massive flowered)
megalanthus -a -um large-flowered, μεγαλη-ανθος
megalobotrys having large berries or bunches, μεγαλη-βοτρυς
megalocarpus -a -um large-fruited, μεγαλη-καρπος
megalophyllus -a -um with very large leaves, μεγαλη-φυλλον
megalorrhizus -a -um large-rooted, μεγαλη-ριζα

megalurus -a -um long-tailed, μεγαλη-ουρα
megalus -a -um rapid, quick, fast, μεγαλως
Megaphrynium Large-*Phrynium* (*Phrynium* is a tropical Asiatic genus)
megaphyllus -a -um large-leaved, μεγα-φυλλον
megapotamicus -a -um of the big river, μεγα-ποταμος, from the Rio Grande or Amazon
megarhizus -a -um, megarrhizus -a -um large-rooted, μεγα-ριζα
megaseiflorus -a -um *Megasea*-flowered (≡ *Bergenia*-flowered), *Megasea-florum*
megaseifolius -a -um *Megasea*-leaved (≡ *Bergenia*-leaved), *Megasea-folium*
Megastachya Large-eared, μεγα-σταχυς (spikelets up to 20-flowered)
megastigma with a large stigma, μεγα-στιγμα
megathurus -a -um large entrance, μεγα-θυρα (the open throat of the corolla)
megeratus -a -um very beautiful, *megeratus*
megisto-, megistus -a -um the largest or biggest, μεγιστος, superlative of μεγας
megistocarpus -a -um the largest-fruited, μεγιστο-καρπος
megistophyllus -a -um the largest-leaved, μεγιστο-φυλλον
megistostictus -a -um most prominently spotted, μεγιστο-στικτος
mei- less-, μειων, comparative of μικρος
meiacanthus -a -um having fewer thorns or spines, μει-ακανθος
meiandrus -a -um with few stamens, μει-ανηρ
meifolius -a -um with fewer leaves, botanical Latin from μειων and *folium*; *Meum*-leaved, *Meum-folium*
meio- (meon-) fewer-, less than-, μειων, μειοω, μειον, μειο- (prefixed to an organ of reference), meiosis, μειωσις, is the reduction division during spore formation
meiophyllus -a -um with smaller or fewer leaves, μειο-φυλλον (in each successive whorl)
meiostemonus -a -um with fewer stamens, μειο-στεμμα
mekongensis -is -e from the environs of the Mekong river, SE Asia
mela-, melan-, melano- black-, μελας, μελανος, μελανο-
Melaleuca Black-and-white, μελας-λευκος (the colours of the bark on trunk and branches)
melaleucus -a -um black and white coloured, μελας-λευκος (outer surfaces)
Melampyrum (*Melampyron*) Black-wheat, μελας-πυρος (a name, μελαμπυρον, used by Theophrastus for a weed of wheat crops)
melan-, melano- dark-, black-, μελας, μελαινα, μελαν, μελανο-
melanacmis -is -e dark-tipped, μελαν-ακμη
melanandrus -a -um having dark or black stamens, μελαινα-ανηρ
melananthus -a -um with black flowers, μελαινα-ανθος
melancholicus -a -um sad-looking, drooping, melancholy, μελανχολια (μελας-χολη, black-bile)
melanciclus -a -um with dark circular markings, μελανο-κυκλος
melandriifolius -a -um with leaves resembling those of *Melandrium, Melandrium-folium*
Melandrium the name, *malandrum*, used in Pliny of uncertain meaning (the dark anthers of some?) (≡ *Silene*)
melanion black violet, μελαν-ιον
melano- black-, μελας, μελανος
melanocalyx with a very dark or black calyx, μελανο-καλυξ
melanocarpus -a -um with very dark or black fruits, μελανο-καρπος
melanocentron, melanocentrus -a -um having a black spur, μελανο-κεντρον
melanocerasus -a -um black-cherry-like, *melano-Cerasus* (garden huckleberry)
melanochaetes with black bristles, μελανο-χαιτη
melanochlamys clothed in black, μελανο-χλαμυς
melanochrysus -a -um black gold, μελανο-χρυσος (dark leaves have a varnished surface that glows golden in sunlight)
Melanodiscus -a -um Black-disc, μελανο-δισκος (floral feature)
melanolasius -a -um with black shaggy hair, μελανο-λασιος

melanophloeus -a -um black-barked, μελανο-φλοιος
melanops black-eyed, μελαν-ωψ
melanopsis -is -e very dark looking, μελαν-οψις
melanorhodus -a -um very dark red, black and red, μελανο-ροδον
Melanoselinum Black-parsley, μελανο-σελινον
melanosporus -a -um with black spores, μελανο-σπορος
melanostachys having black spikes (catkins), μελανο-σταχυς
melanostictus -a -um with black spotting, μελανο-στικτος
melanotrichus -a -um having very dark or black hair, μελανο-τριχος
melanoxylon black-wooded, μελανο-ξυλον
Melanthera Black-stamened-one, μελας-ανθηρα
Melanthium Black-flower, μελας-ανθεμιον (the dark senescent tepals)
melanus -a -um blackened, μελανω, μελαινω
Melasphaerula Little-black-globe, μελας-σφαιρα, Latin diminutive from σφαιρα
Melastoma Black-mouth, μελας-στομα (the fruits stain the lips black) (***Melastomataceae***)
Melchrus Honey-coloured, μελι-χρωμα (the floral glands)
Meleagris, meleagris -is -e Greek name for Meleager of Calydon, chequered as is a guinea fowl (*Numidia meleagris*) and snake's head fritillary (*Fritillaria meleagris*)
melegueta probably from a Portuguese vernacular name for peppers, including *Afromomum melegueta*, grains of paradise
meles badger, *meles*
meli- honey-, μελι, μελιτος, *mel, mellis*
Melia from the Greek name, μελιη, for ash tree (the resemblance of the leaves) (***Meliaceae***)
Melianthus Honey-flower, μελι-ανθος (***Melianthaceae***)
Melica Honey-grass, μελι, μελιτος (Cesalpino's name for a sorghum)
meliciferus -era -erum music-bearing, (μελος, μελικος)-φερω
Melicocca Honey-berry, μελι-κοκκος (the genip tree's very sweet fruit)
Melicope Honey-parts, μελι-κοπη (the four nectaries)
Melicytus etymology uncertain (Honey-shield, μελι-κυτος, for the scale on the extended connective?)
melifluus -a -um with copious nectar, flowing with honey, (*mel, mellis*)-*fluo, fluere, fluxi, fluxum*
Melilotus Honey-clover, μελι-λωτος (Theophrastus' name, μελιλωτος, refers to melilot's attractiveness to honeybees)
melinanthus -a -um with quince-like flowers, μελινος-ανθος, honeyed flowers
Melinis Ashen-looking, μελινος (leaves are sticky haired)
melinocarpus -a -um ashen-fruited, μελινος-καρπος
melinus -a -um quince-like, quince-coloured; ash-like, μελινος
Meliosma, meliosmus -a -um Honey-perfumed, μελι-οσμη (the fragrance of the flowers)
meliosmifolius -a -um with leaves resembling those of *Meliosma, Meliosma-folium*
Melissa Honeybee, μελισσα, μελιττα (named for the nymph, Μελισσα, who, in mythology kept bees, and for the plant's use in apiculture)
melissaefolius -a -um, melissifolius -a -um with *Melissa*-like leaves, *Melissa-folium*
melissophyllus -a -um, (*mellisifolius*) a name in Pliny, balm-leaved, with *Melissa*-like leaves
melitensis -is -e from Malta, Maltese (*Melita*)
Melittis Honey-rich, μελιτοεις (bastard balm attracts bees, μελισσα, μελιττα)
melius, melior better, improved, comparative of *bonus*
melleus -a -um of honey, honeyed, *mel, mellis* (smelling or coloured)
mellifer -era -erum honey-bearing, honey-making, *mellifer, melliferi*
mellifluus -a -um with copious nectar, flowing with honey, (*mel, mellis*)-*fluo, fluere, fluxi, fluxum*
mellinus -a -um the colour of honey, sweet, *mellitus*
melliodorus -a -um, honey-smelling, *mellis*-(*odor, odoris*)

melliolens honey-fragrant, *mellis-(olens, olentis)*
mellitus -a -um honeyed, sweet, *mellitus*
melo- apple-, μηλον (from μηλοπεπον, apple-gourd)
Melocactus Melon-cactus, μηλον-κακτος (the shape)
meloctonus -a -um badger slaughtering, botanical Latin from *Meles, Meles* and κτονος (*Aconitum meloctonum*, badger's bane)
melongena apple-bearer, μηλον-γενος (producing a tree-fruit, the egg plant)
meloniformis -is -e, (meloformis -is -e) like a ribbed sphere, melon-shaped, botanical Latin from μηλον and *forma*
Melothria the Greek name, μηλοθρον, for bryony
membraneus -a -um, membranaceus -a -um thin in texture, parchment-like, membranous, μεμβρανα, *membrana, membranae*
membranifolius -a -um having very thin leaves, *membrana-folium*
Memecylon Imitation, μιμημα (from the Greek name for the fruits of *Arbutus*, which are similar)
memnonius -a -um dark brown, brownish-black, changeable, *memnon*
memoralis -is -e remembered, famed, *memoro, memorare, memoravi, memoratum*
mendelii for Mr Mendel of Manley Hall, Manchester, orchid grower
mendocicus -a -um from the Mendoza river, Argentina
mendocinensis -is -e from Mendocino, California
mene-, meni- crescent-, moon-, μην, μηνη, μηνος
-mene membrane, μενινξ, μενινγ-
meniscatus -a -um curved-cylindrical, μηνισκος
meniscoides meniscus-shaped, concavo-convex, μηνισκος-οειδης
Menisorus Lunate-sorus, μηνη-σορος (shape of the sori)
menispermifolius -a -um having leaves resembling those of *Menispermum, Menispermum-folium*
menispermoides resembling *Menispermum, Menispermum-oides*
Menispermum Moon-seed, μηνη-σπερμα (the compressed, curved stone of the fruit) (***Menispermaceae***)
mensalis -is -e of plateaux or table-lands, *mensa, mensae*
Mentha the name, *menta*, in Pliny, μινθη
menthifolius -a -um mint-leaved, with leaves resembling *Mentha, Mentha-folium*
menthoides mint-like, *Mentha-oides*
mentiens deceptive, false, present participle of *mentior, mentire, mentitus*
mentorensis -is -e from Mentor, Ohio, USA
mentosus -a -um, -mentum having a chin, botanically a lip or labellum, *mentum, menti*
Mentzelia for Christian Mentzel (1622–1701), early plant name lexicographer
Menyanthes Moon-flower, μηνη-ανθος (various derivations have been proposed. Theophrastus' name, μηνηανθος, for *Nymphoides peltata*) (***Menyanthaceae***)
Menziesia, menziesii for Archibald Menzies (1754–1842), Scottish naturalist on the *Discovery* (1790–5) with Vancouver
meonanthus -a -um few or small-flowered, μειον-ανθος
mephiticus -a -um emitting a foul odour, *mephitis; mephiticus*
meracus -a -um individual, alone, pure, *meracus*
Meratia for François Victor Mérat de Vaumartoise (1780–1851), French physician and botanist
mercadensis -is -e from the Cerro del Mercado area, Mexico
Mercurialis Of-the-god-Mercury, *herba mercurialis*, named by Cato for Mercury, messenger of the gods
merdarius -a -um of dung, coprophilous, *merda, merdae*
Merendera from the Spanish vernacular name, quita meriendas
merguensis -is -e from the Mergui archipelago, SE Myanmar (Burma)
meri-, meros- partly-, part-, μερις, μεριδος, μερι-
-meria, -meris -is -e -parts, -μερις, μεριζω
meridensis -is -e from either the Mérida cordillera in W Venezuela, or Mérida in SW Mexico, or Mérida in W Spain.

meridianus -a -um, meridionalis -is -e southern, midday, of noon, *merides, merideie* (flowering at midday)
Merinthosorus Divided-sorus, (μεριζω) μερινθος-σορος
merismoides divided-looking, μερισμος-οειδης (the surface of the fruiting body has numerous radiating ridges)
meritus -a -um deserved, to be acquired, past participle of *mereo, merere, merui; mereor, mereri, meritus*
Mertensia for Franz Karl Mertens (1764–1831), Professor of Botany at Bremen
mertensianus -a -um for Karl Heinrich Mertens (1795–1830), who collected in Alaska
mertonensis -is -e from Merton College, Oxford
-merus -a -um -partite, -divided into, -merous, -parts, -μερος
merus undiluted, pure, bare, mere, *merus, mera*
Meryta Rolled, μερυς (the appearance of the male flowers)
mes-, mesi-, meso- somewhat-, between-, middle-, μεσος, μεσο-, μεσ-
mesacanthus -a -um with moderately large spines, μεσ-ακανθος
mesadenius -a -um with central glands (petiolar), moderately glandular, μεσ-αδην
mesaeus -a -um intermediate, neutral, μεσευω
Mesanthemum Middle-flowered, μεσο-ανθεμον (the flowers are surrounded by involucral bracts)
mesargyreus -a -um with silver towards the middle, μεσ-αργυρος (leaf colouration)
mesariticus -a -um from the Mesará plain of S Crete
Mesembryanthemum (Mesembrianthemum) originally, Midday-flower, μεσ-ημβρια-ανθεμον, μεσ-ημβρινος-ανθεμον (flowers of some open in full midday sun) but current name recognizes night-flowering components and derives as μεσος-εμβρυον-ανθεμον, flower with a central embryo
mesentericus -a -um mesentery-like, μεσεντεριον (the texture and colouration of the tripe-fungus, *Auricularia mesenterica*)
meso- intermediate, middle, μεσος
mesochoreus -a -um from the middle region, country or land, from the midlands, μεσο-χωρα
mesogaeus -a -um of middle earth, μεσο-(γη, γαια)
mesoleucus -a -um with white centres, μεσο-λευκος
mesophaeus -a -um with a dark centre, μεσο-φαιος (colouring)
mesopolius -a -um with a grey centre, μεσο-πολιος
mesoponticus -a -um from the middle sea, μεσο-ποντιος (lakes of central Africa)
mesopotamicus -a -um from between the rivers, μεσο-ποταμος
mesozygius -a -um yoked or united in the middle, μεσο-ζυγος
mespiliformis -is -e resembling a medlar, *Mespilus-forma*
Mespilus Half-felted, μεσο-πιλος (Theophrastus' name, μεσπιλη σατανειος, for the medlar)
messanensis -is -e, messanius -a -um from Messina area, Italy (*Messana*)
messeniacus -a -um, messeniensis -is -e from Messenia, Morea, Greece
-mestris -is -e -months, *mensis, mensis* (the period of growth or flowering) (*semester* six-months)
met-, meta- amongst-, next to-, after-, behind-, later-, with-, μετα
metallicus -a -um lustrous, metallic in appearance, μεταλλικος
metalliferus -a -um bearing (metal or) a metallic lustre, μεταλλικος-φερω
Metasequoia Close-to-*Sequoia*, botanical Latin from μετα and *Sequoia* (resemblance of the dawn redwood)
metel an Arabic vernacular name for *Datura metel*
meteloides resembling (*Datura*) *metel*
meteoris -is -e dependent upon the weather, μετεωρος (flowering)
methipticos intoxicating, μεθυσκω (to make drunk)
methistico-, methysticus -a -um intoxicating, μεθυσκω (*Piper methysticum* is intoxicating pepper, or [Tongan] kava)
metrius -a -um estimable, measured, passed over, μετρεω

metro- mother-, centre-, heart-, μητρα; size, standard, μετρον
Metrosideros Heart-of-iron, μητρα-σιδηρος (the hard timber)
Metroxylon Heart-wood, μητρα-ζυλον (the large medulla)
Metternichia for Prince Metternich of Winneburg, Austria (1772–1859)
metulifer, metuliferus -a -um carrying apprehension, (*metuo, metuere, metui, metutum*)-*fero* (the fruits)
Meum (*Meon*) an old Greek name, μηον, in Dioscorides (*meu* of the apothecaries)
mexicanus -a -um from Mexico, Mexican
meyenianus -a -um, meyenii for Franz Julius Ferdinand Meyen (1804–40), plant illustrator
meyeri for either Johann Karl Friedrich Meyer (1765–1805), apothecary of Stettin, or G. F. W. Meyer (1782–1856), who wrote the Flora of Hanover, or Adolph Bernard Meyer (1840–1911), explorer, or Abraham Julien Meyer (1770–1843), of Batavia, or E. H. F. Meyer (1791–1858), botanical author, or Reverend G. Meyer (fl. 1929), who collected S African succulents
Mezereum (*mezereon*) a name used by Avicenna (Ibn Sina) (980–1037), from the Persian, mazarjun
Mibora an Adansonian name of uncertain meaning
micaceus -a -um from micaceous soils, mica-like, *mica, micae* (colour appearance)
micans shining, sparkling, glistening, *mico, micare, micui*
Michauxia, michauxii for André Michaux (1746–1803), French botanist who collected in N America to supply the Rambouillet forest near Paris
Michelia, michelianus -a -um for either Pietro Antonio Micheli (1679–1737), Florentine botanist, or Marc Micheli (1844–1902) of Geneva
michiganensis -is -e from Michigan, USA
micholitzianus -a -um, micholitzii for Wilhelm Micholitz (1854–1932), who collected for Sander in the Philippines, New Guinea, etc.
michuacanus -a -um from Michoacan state, W central Mexico
micr-, micra-, micro- small-, μικρος, μικρο-, μικρ-
micrandus -a -um with small stamens, μικρ-ανηρ
Micranthes Small-flower, μικρος-ανθος
micranthidifolius -a -um with *Micranthus*-like foliage
Micranthus, micranthus -a -um Small-flowered, having small flowers, μικρ-ανθος
Microbiota Small-*Thuja* (*Biota* was an earlier synonym for *Thuja*)
microbotryus -a -um having small bunches, μικρο-βοτρυς (of fruits or flowers)
Microcachrys Little-*Cachrys*
Microcala Little-beauty, μικρο-καλος
Microcalamus Small-*Calamus*-like
microcarpus -a -um small-fruited, μικρο-καρπος
microcephalus -a -um small-headed, μικρο-κεφαλη (inflorescences)
Microchloa Small-grass, μικρο-χλοη
Microcoelia Small-belly, μικρο-κοιλος (obscure lobes of lip)
Microcycas Small-*Cycas*
microdasys small and hairy, with short shaggy hair, μικρο-δασυς
Microdesmis Small-clusters, μικρο-δεσμη (refers to the clustered flowers)
microdon small-toothed, μικρο-οδων
Microdracroides Small-*Dracaena*, μικρο-δρακαινα-οειδης (mature plants form false 'trunks' and resemble small *Dracaena* plants)
microglochin small-point, μικρο-γλωχις (the extended tip of the flowering axis)
Microglossa Small-tongue, μικρο-γλωσσα (the short ligulate florets)
Microgramma Small-lines, μικρο-γραμμη (the sori)
Microlepis Small-scale, μικρο-λεπις (thin outward-facing indusium is attached at the base and sides)
microlepis -is -e with small scales, scurfy, μικρο-λεπις
microleucum small (*Rhododendron*) *leucum*
Microloma Small-fringe, μικρο-λωμα (the hair groups in the corolla tube)
micromalus small apple, μικρο-μηλεα, *micro-Malus* (fruiting body)

Micromeles Small-apple, μικρο-μελον (the fruit's size)
Micromeria Small-parts, μικρο-μερις (the diminutive flowers)
micromerus -a -um with small parts or divisions, μικρο-μερις
Micromonas Small-unit, μικρο-μονας (the smallest monocaryotic alga)
micropetalus -a -um small-petalled, μικρο-πεταλον
microphyllus -a -um small-leaved, μικρο-φυλλον
microphyton small plant, μικρο-φυτον (as distinct from other species)
microps of small appearance, tiny, μικρ-ωψ
microptilon small wing, μικρος-πτιλον (phyllaries)
microrhizus -a -um with a small root or rooting base, μικρο-ριζα
microsepalus -a -um having small sepals, μικρο-σκεπη
Microseris Small-*Seris*, μικρο-σηρις
Microsisymbrium Little-*Sisymbrium*, from μικρο and *Sisymbrium*
Microsorium Small-sori, μικρο-σορος (restricted to junctions of three veins)
microspermus -a -um small-seeded, with little seed, μικρο-σπερμα
microstachyus -a -um with small spikes, μικρο-σταχυς (aments)
Microstegium, microsteius -a -um Small-covers, μικρο-στεγη (the minute lemmas or bracts)
microstipulus -a -um having small stipules, botanical Latin from μικρο and *stipula*
Microstrobos Small-cone, μικρο-στροβιλος (comparison with *Pinus strobus*)
microthelis -is -e having small (nipples or) tubercles, μικρο-θηλη
microthyrsus -a -um having small thyrses, μικρο-θυρσος (see Fig. 3d)
microtrichus -a -um with very short hairs, μικρο-(θριξ, τριχος)
microxiphion with small swords, μικρο-ξιφος (the leaves)
micrugosa small (*Rosa*) *rugosa, micro-rugosa*
-mict- -mixed-, -mixture-, μικτος
middendorfianus -a -um, middendorfii for Alexander Theodor von Middendorf (1815–94), Russian collector in N India and Siberia
Mikania for Joseph G. Mikan (1743–1814), Professor of Botany at Prague
mikanioides resembling climbing hemp-weed, *Mikania scandens, Mikania-oides*
Mikaniopsis Resembling-*Mikania*, botanical Latin from Mikan and οψις
Mila an anagram of Lima, the genus' Peruvian provenance
miliaceus -a -um millet-like, pertaining to millet, *Milium*
miliaris -is -e thousands, minutely glandular-spotted, *milia, milium*
militaris -is -e upright, resembling part of a uniform, soldierly, *miles, militis*
Milium the Latin name, *milium, mili*, for a millet grass
Milla for Juliani Milla, eighteenth-century gardener at the Madrid court
mille- a thousand- (usually means 'very many'), *mille, millia*
milleflorus -a -um having (thousands or) many flowers or florets, *mille-florum*
millefolii of millfoil (*Rhopalomyia* gall midge in axillary buds)
millefolius -a -um, millefoliatus -a -um thousand-leaved (much-divided leaves of milfoil), milfoil-like, *mille-folium*
millegranus -a -um having (thousands or) very large numbers of seeds, *mille-granum*
milleri for Philip Miller (1691–1771), Curator of Chelsea Physic Garden, author of *The Gardener's Dictionary* (1731)
Milletia for J. A. Millet, eighteenth-century French botanist
Miltonia for Earl Charles Fitzwilliam (1786–1857), Viscount Milton
mimetes mimicking, μιμησις
Mimosa Mimic, μιμος (the sensitivity of the leaves, an imitator or mime) (***Mimosaceae***)
mimosifolius -a -um with leaves resembling those of *Mimosa, Mimosa-folium*
mimosoides resembling *Mimosa, Mimosa-oides*
mimuloides resembling *Mimulus, Mimulus-oides*
Mimulopsis Resembling-*Mimulus*, botanical Latin from *Mimulus* and οψις
Mimulus Mask-flower, μιμεομαι (diminutive of *mimus*, the flowers somewhat mimic a face)

mimus -a -um theatrical, farcical, mimetic, μιμος, *mimus, mimi*
Mimusops Monkey-face, μιμος-οψις (imaginary resemblance of the corolla)
minax extending, projecting, threatening, *minax, minacis*
miniatus -a -um cinnabar-red, the colour of red lead, *minium*
minim, minimus -a -um least, smallest, superlative of *parvus*
minimiflorus -a -um having the smallest flowers, *minimus-florum*
minisculus -a -um somewhat smaller, comparative of *parvus*
minor -or -us smaller, comparative of *parvus*
minorcensis -is -e from Menorca, Balearic Islands
minous -a -um for King Minos of Crete, son of Zeus and Europa
Minuartia for Juan Minuart (1693–1768), botanical writer of Barcelona
minus -a -um small, less, *minus*, comparative of *parvus*
minusculus -a -um smallish, *minusculus*
minutalis -is -e very small, *minutal, minutalis* (literally, like mince)
minutiflorus -a -um with very small flowers, *minuti-florum*
minutifolius -a -um with very small leaves, *minuti-folium*
minutissimus -a -um extremely small, smallest, superlative of *parvus*
minutulus -a -um somewhat small, diminutive of *minutus*
minutus -a -um very small, minute, inconspicuous, *minuta, minutus*
mio- see *meio-*
miquelianus -a -um for Dr Friedrich Anton Wilhelm Miquel (1811–71), Director of Botanic Garden at Utrecht, Holland
Mirabilis, mirabilis -is -e Wonderful, extraordinary, astonishing (*Mirabilis jalapa*, marvel of Peru)
mirabundus -a -um astonishing, *mirabundus*
miraguanus -a -um from the Miraguâne peninsula, Haiti
mirandanus -a -um from Miranda state, Venezuela
mirandus -a -um extraordinary, wonderful, *mirandus*
mirificus -a -um most wonderful, *mirifice*
miris -is -e, mirus -a -um wonderful, strange, *mire, mirus*
mirissimus -a -um most wonderful or extraordinary, superlative of *mire*
Miscanthus Pedicelled-flowered, μισχος-ανθος (the conspicuous inflorescence)
miscellus -a -um variable, mixed, *misceo, miscere, miscui, mixtum*
misellus -a -um little, poor, *misellus*
miser -era -erum, miserus -a -um wretched, inferior, poor, pitiful, *miser, miseri*
miserrimus -a -um more insignificant, comparative of *miser*
mishmiensis -is -e, mishimiensis -is -e from the Mishmi hills, outliers of the Himalayas, NE India
Misopates Reluctant-to-open, botanical Latin from μισος and *pateo, patere, patui*
mississippiensis -is -e from the Mississippi, USA
missuricus -a -um, missouriensis -is -e from Missouri, USA
mistassininus -a -um from the area around Lake Mistassini, Quebec
mistiensis -is -e from the Andean Misti or Arequipa volcano, S Peru
Mitchellia (*Mitchella*), *mitchellianus -a -um, mitchellii* for Dr John Mitchell (1711–68), botanist in Virginia, USA
Mitella Little-mitre, diminutive of μιτρα (the shape of the fruit)
mithridatus -a -um for Mithridates Eupator, king of Pontus (mithridates give protection against poisons, cf. antidotes)
mitifolius -a -um having soft leaves, *mitis-folium*
mitior, mitius -a -um softer, comparative of *mitis*
mitis -is -e gentle, mild, bland, not acid, without spines, *mitis*
mitissimus -a -um most gentle or mild, superlative of *mitis*
Mitracarpus Capped-ovary, μιτρα-γυνη (the circumscissile fruit)
Mitragyna Mitred-ovary, μιτρα-γυνη (the cap-like stigma)
Mitraria Capped, μιτρα (the bracteate inflorescence)
mitratus -a -um turbaned, mitred, μιτρα (head-dress)
mitreolus -a -um with a small cap, diminutive of *mitra, mitrae*

mitriformis -is -e, mitraeformis -is -e mitre-shaped, turban-shaped, *mitra-forma*
Mitriostigma Mitred-stigma, μιτρα-στιγμα (the cap-like stigma)
mixo- mixing-, mingling-, μιξις, μιξο-
mixomycetes half-fungus, μιξο-μυκητος (slime fungi)
mixtus -a -um mixed, mingled, *mixtura, mixturae*
miyabeanus -a -um, miyabei for Professor Kingo Miyabe (1860–1951), Director of Sapporo Botanic Garden, Japan
mlokosewitschii for Ludwik Franciszek Mlokosewitsch (1831–1909), who found his *Paeonia* in the central Caucasus
-mnemon -memorable, -unforgettable, μνημων
mnio- moss-, *Mnium-*
mniophilus -a -um moss-loving, living amongst mosses, μνιον-φιλος
Mniopsis Moss-like, μνιον-οψις (genus of the aquatic *Podostemaceae*)
Mnium Moss, μνιον
moabiticus -a -um, (moabaticus) from the biblical land of Moab, Jordan
mobilis -is -e pliant; excitable, fickle; rapidly, *mobilis*
mocambicanus -a -um from Mozambique (Moçambique)
modestissimus -a -um most restrained or unassuming, superlative of *modestus*
modestus -a -um modest, unpretentious, restrained, *modestus*
modicus -a -um mean, small, *modicus*
Moehringia (Möhringia) for Paul Heinrich Gerhard Möhring (1710–92), naturalist and physician of Oldenberg
Moenchia for Conrad Moench (1744–1805), German botanist
moesiacus -a -um from the Balkans (*Moesia*)
mohavensis -is -e from the Mojave desert, California, USA
Mohria, mohrii for Daniel Mohr (1780–1808), German botanist
Mohrodendron Mohr's tree (see *Mohria*) (≡ *Halesia carolina*)
molaris -is -e of the millstone, ground, granular, *mola, molae*
moldavicus -a -um from Moldova, from the Danube area (Romania and Ukraine)
molendinaceus -a -um, molendinaris -is -e shaped like a mill-sail, with a wing-like expansion, *mola, molae; molaris* (literally, related to millstones)
molestus -a -um annoying, troublesome, *molestus*
Molinia, molinae for Juan Ignacio (Giovanni Ignazio) Molina (1740–1829), writer on Chilean plants
Molium Magic-garlic, from μωλη (after *Allium moly*)
molle from Peruvian name, mulli, for *Schinus molle*
mollearis -is -e resembling *Schinus molle*
molli-, mollis -is -e softly hairy, soft, *mollis* (cognate with mullein)
molliaris -is -e supple, graceful, pleasant, *mollio, mollire, mollivi, mollitum*
mollicaulis -is -e soft-stemmed, *mollis-caulis*
mollicellus -a -um somewhat soft or tender, diminutive of *mollis*
molliceps soft-headed, *mollis-caput*
mollicomatus -a -um, mollicomus -a -um having long, soft hair, *mollis-comatus* (indumentum)
molliferus -a -um bearing a soft hairy covering, *mollis-fero*
molliformis -is -e resembling (*Bromus*) *mollis*
mollissimus -a -um the softest, superlative of *mollis*
molliusculus -a -um quite pubescent, soft or tender, diminutive of *mollis*
Mollugo, mollugo Tender, *mollis* with feminine suffix *-ugo* (a name in Pliny) (***Molluginaceae***)
Molopospermum Striped-seed, μωλωψ-σπερμα (the yellow fruit has brown vittae)
Moltkia, moltkei for Count Joachim Gadake Moltke (1746–1818), of Denmark
moluccanus -a -um from Maluku, Indonesia (the Moluccas)
Moluccella derivation obscure (Bells of Ireland, from Molucca?)
moluccellifolius -a -um having leaves resembling those of *Moluccella*
moly the Greek name, μολυ, of a magic herb used against Circe by Odysseus (Homer); Lyte associated it with *Allium*

molybdeus -a -um, molybdos sad, neutral-grey, lead-coloured, μολυβδος, μολιβος
mombin a W Indian vernacular name for hog plum, *Spondias mombin*
Momordica Bitten, *mordeo, mordere, momordi, morsum* (the jagged seeds of balsam pear appear to have been nibbled)
mon-, mona- one-, single-, alone-, μονας, μονα-, μον-
monacanthus -a -um with single spines, μον-ακανθος
monacensis -is -e from Monaco, S Europe, or Munich, Germany
monachorus -a -um of monks, *monachus, monachi*, (the apothecaries' *Rhabarbarum monachorum* (≡ *Rumex patientia*) is monk's rhubarb)
monadelphus -a -um having a single brotherhood, μον-αδελφος (conjoined stamens)
Monadenium Single-glanded-one, μονος-αδην (the allied *Euphorbia* has two glands)
monadenus -a -um having a single gland, μον-αδην (the nectary)
monancistrus -a -um with single barbs, μον-αγκιστριον
monandrus -a -um one-stamened, with a single stamen, μονο-ανηρ
monanthus -a -um, monanthos having a single flower, or flowers borne singly, μονο-ανθος
Monarda for Nicholas Monardes (1493–1588) of Seville, first herbal writer to include newly discovered American plants
Monardella diminutive of *Monarda*
monarensis -is -e from the Monaro plateau, New South Wales, Australia
mondo from a Japanese vernacular name
monensis -is -e from Anglesey or the Isle of Man, both formerly known as *Mona*
Monerma Single-pendant, μονος-ερμα, an old generic name referring to the single glume, ≡ *Psilurus*
Moneses One-desire, μονος-(ιημι, εσις) (for the solitary flower)
mongholicus -a -um, mongolicus -a -um from Mongolia, Mongolian
moniliferus -a -um necklaced, carrying beads, *monile-fero*
moniliformis -is -e necklace-like, like a string of beads, *monile, monilis*
monilis -is -e necklace-like, *monile*
Monimia Mascarene islands vernacular name (***Monimiaceae***)
mono- one-, single-, μονος, μονο-
monocarpus -a -um monocarpic, fruiting is followed by death, μονο-καρπος
monocephalus -a -um with a single head, μονο-κεφαλη
monochlamys having a single cover, with a single perianth whorl, μονο-χλαμυς
Monochoria Single-membrane, μονο-χωριον (the persistent perianth contains the fruit)
monoclinus -a -um hermaphrodite, with stamens and ovary in one flower, μονο-κλινη (literally, with a single bed)
monoclonos single-branched, μονο-κλων
monococcus -a -um one-fruited or -berried, μονο-κοκκος
monocolor self-coloured, of a single colour, *mono-color*
Monocymbium Single-keel, *mono-cymba-forma* (the spatheole subtending the raceme)
Monodora Single-gift, μονο-δωρον (the solitary flowers; *M. myristica*, orchid nutmeg)
monogynus -a -um with a single ovary, μονος-γυνη (a compound ovary)
monoicus -a -um separate staminate and pistillate flowers on the same plant, of a single house, monoecious, μονο-οικος
monopetalus -a -um one-petalled; having the corolla united, gamopetalous, μονο-πεταλον
monophyllus -a -um, monophyllos having a single leaf, μονο-φυλλον
Monopsis Single-featured, μονο-οψις (regular, whereas *Lobeliaceae* are mostly bi-labiate)
monopyrenus -a -um with single nutlets or pyrenes, separated drupes, μονο-πυρην
monorchis -is -e one-testicle, μονο-ορχις (*Herminium* has a single tuber at anthesis)
monorhizus -a -um with a single root, μονο-ριζα
monosematus -a -um with a single mark, μονο-σεμειον (the blotch at the base of the corolla)

monospermus -a -um single-seeded, μονο-σπερμα
monostahcyus -a -um with single spikes, μονο-σταχυς
Monotes Taken-apart, μονοωτης (the first, and only genus of dipterocarps in Africa when erected; later, taken apart, μονοω)
Monotoca Single-offspring, μονος-τοκος (one-seeded fruits)
Monotropa, monotropus -a -um Turned-to-one-side, μονοτροπος, (the flowering habit), or One-turn, μονο-τροπη (the secund flowers) (***Monotropaceae***)
Monsonia for Lady Ann Monson (*c.* 1714–76), who corresponded with Linnaeus
monspeliacus -a -um, monspeliensis -is -e, monspeliensius -a -um from Montpellier, S France
monspessulanus -a -um from Montpellier, S France
Monstera Monstrous, *monstrum*, of huge size or monstrous foliage, but the derivation is uncertain
monstrosus -a -um, monstrus -a -um marvellous, monstrous, wonderful, horrible, *monstrum, monstri*
montanensis -is -e from Montana, USA
montanus -a -um of mountains, *mons, montis*
Montbretia (Montbrettia) for Antoine François Ernest Conquebert de Montbret (1781–1801), died in Cairo when botanist to the French expedition to Egypt (≡ *Crocosmia*)
monteiroi, monteiroae for Joachim John Monteiro (1833–78) and his wife, Portuguese naturalists
montenigrinus -a -um from Montenegro
montereyensis -is -e from Monterey county, California, USA
montevidensis -is -e from Montevideo, Uruguay
Montezuma, montezunae for Montezuma, fifteenth-century King of Mexico
Montia for Guiseppe L. Monti (1712–97), Italian Professor of Botany at Bologna
monticolus -a -um mountain-living, mountain-dweller, *monti-colo*
montigenus -a -um borne of mountains, *monti-(gigno, gignere, genui, genitum)* (montane habitat)
montis-draconis -is -e from Drakensberg, S Africa
montis-duidus -a -um from Mount Duida, Venezuela
montis-lous -a -um from the volcanic Mauna Loa, Hawaii
montivagus -a -um wandering on mountains, *montis-vagus*
montregalensis -is -e from Mount Royal, Montreal, Canada
montuosus -a -um of mountainous habitats, *montuosus*
moorei for Thomas Moore (1821–87), writer on ferns, or Dr David Moore (1807–79), Curator of Glasnevin Botanic Garden, or Sir Frederick Moore (1857–1949), also Curator at Glasnevin, or Charles Moore (1820–1905), Director of Sydney Botanic Garden, Australia
Moraea, moraea for R. Moore, English botanist (in 1739 Linnaeus married Sara Moraea)
moranensis -is -e from Real du Moran, Mexico
moratus -a -um of steady nature or manner, *moratus*
morbilosus -a -um diseased(-looking), pustuled, *morbillus, morbilli* (*Drynaria* fronds)
Morchella Little-Moor (*maurus*, suggested by the morel's small yellow cap) (French moré, German morchel)
mordenensis -is -e from Morden, Manitoba
morettianus -a -um for Professor Giuseppe Moretti (1782–1853) of Pavia Botanic Garden, Italy
Moricandia for Moïse Etienne (Stefano) Moricand (1779–1854), Swiss botanist
morifolius -a -um mulberry-leaved, with *Morus*-like leaves, *Morus-folium*
Morina for Louis Morin (1636–1715), French botanist (***Morinaceae*** ≡ ***Dipsacaceae***)
Morinda, morinda Indian mulberry, *morus indica* (the leaves of *Morinda cikrifolia*, horse-radish tree or noni, are chewed for the psychoactive effect of their amphetamine content)

morindoides resembling *Morinda*, *Morinda-oides*
Moringa from a Malabar vernacular name, moringo, for the horse-radish-tree (***Moringaceae***)
morinifolius -a -um with leaves resembling those of *Morina*, *Morina-folium*
morio the name, μωριον, of a plant causing madness (μωρια, folly)
Morisia for Giuseppe Giacinto (Josephi Hyacinthi) Moris (1796–1869), professor of botany at Turin, Italy
Morisonia for Robert Morison (1620–83), Director of Royal Garden at Blois, Physician to Charles II, Professor of Botany at Oxford, author of *Praeludia botanica* (1669), influenced Linnaeus on classification of plants
moritzianus -a -um for Johann Wilhelm Karl Moritz (1796–1866), who travelled in Venezuela and the West Indies
-morius -a -um -divisions, -parts, -merous, μοριον (of the flower)
Mormodes Goblin-like, μορμο-ωδης (suggested by the flower shape)
mormomicus -a -um annoying, spectral, μορμως
mormos bugbear, spectre, goblin, μορμω, μορμων, μορμος, μορμους (*Botrychium mormos*, goblin fern)
morpho-, *-morphus -a -um* appearance-, -shaped, -formed, μορφη
morrenianus -a -um, *morrenii* for Professor Charles François Antoine Morren (1807–58) of Liège, Belgium
morrisonensis -is -e, *morrisonicolus -a -um* from Morrison, Illinois, USA
morrisonmontanus -a -um from the environs of the Chung Yang (Morrison) range, Taiwan
morsus-ranae bite of the frog, *morsus-(rana, ranae)* (frog-bit)
mortefontanensis -is -e from the Motrefontaine nursery of Chantrier brothers, France
mortuiflumis -is -e of dead water, growing in stagnant water, *mortuus-(flumen, fluminis)*
Morus the ancient Latin name, *morus*, for the mulberry (***Moraceae***)
mosaicus -a -um tessellated, parti-coloured, coloured like a mosaic, modern Latin, *musaicus*
moschatellinus -a -um a little bit musky, an old generic name for *Adoxa*, moschatel, feminine double diminutive of *moschatus* (has a musky fragrance when wet)
moschatus -a -um musk-like, musky-scented, μοσχος (sprout, descendant, young bull)
moscheutos a vernacular name for swamp rose-mallow, *Hibiscus moscheutos*
moschiferus -a -um bearing a musky fragrance, μοσχος-φορος
Moschosma Musk-fragrant, μοσχος-οσμη
moschus -a -um fragranced, musk-like, μοσχος
mosera, *moseri*, *moserianus -a -um* for Moser, the French nurserymen
mossiae, *mossii* for Mrs Moss of Otterspool, Liverpool *c.* 1838
motorius -a -um agitated, kept in motion, *moto*, *motare*
moulmainensis -is -e from Moulmein, Mayanmar (Burma)
moupinensis -is -e from Mupin, W China
moxa a vernacular name for the woolly leaves of *Artemisia moxa*
mucidus -a -um snivelling, mouldy, *mucidus*
Mucor Mould (*mucidus* mouldy)
mucosus -a -um slimy, *mucosus*
mucro-, *mucroni-* pointed-, sharp-pointed-, *mucro*, *mucronis*
mucronatus -a -um mucronate, with a hard sharp-pointed tip, *mucro*, *mucronis* (see Fig. 7b)
mucroniferus -a -um bearing short straight points, *mucronis-fero* (at leaf apex)
mucronifolius -a -um with mucronate leaves, *mucronis-folium*
mucronulatus -a -um with a hard, very short, pointed tip, diminutive of *mucronatus*
Mucuna the Brazilian vernacular name for cow-itch, *Mucuna pruriens*
Muehlenbeckia for Dr H. Gustave Muehlenbeck (1798–1845), Alsatian physician
muelleri for Otto Frederik Müller (1730–84), author of *Flora Danica*, or Ferdinand von Müller (1825–96)

mughus, mugo an old Italian vernacular name for the Tyrolean or dwarf pine, *Pinus mugo*
mugodscharicus -a -um from the Mughalzar (Mugodzhar) hills, NW Kazakhstan
Muhlenbergia (*Muehlenbergia*) for Henri Ludwig Mühlenberg (1756–1817), of Pennsylvania, USA
Mulgedium Milker, *mulgeo, mulgere, mulsi* (Cassini's name refers to the possession of latex, as in *Lactuca*)
muliensis -is -e from the lands of the Muli people, E India (muli is an Indian vernacular name for *Melocanna bambusoides*)
mult-, multi-, multus -a -um many, *multi*
multiangularis -is -e having many angles, corners or ridges, *multi-angulus*
multibracteatus -a -um with many bracts, *multi-bractea* (*brattea*)
multibulbosus -a -um producing many bulbs, *multi-bulbus*
multicaulis -is -e many-stemmed, *multi-caulis*
multicavus -a -um with many hollows, many-cavitied, *multi-cavus*
multiceps many-headed, *multi-ceps*
multiclavus -a -um with many club-like knotty branches, many-branched, *multi-clava*
multiculmis with many culms, *multi-*(*culmus, culmi*)
multifidus -a -um much divided, deeply incised, with many deep divisions, *multi-*(*findo, findere, fidi, fissum*)
multiflorus -a -um many-flowered, floriferous, *multi-florum*
multifoliolatus -a -um having many leaflets, *multi-foliola*
multifolius -a -um many-leaved, foliaceous, *multi-folium*
multiformis -is -a many-shaped, variable, *multi-forma*
multijugus -a -um pinnate, with many pairs of leaflets, *multi-iugis*
multinerius -a -um, multinervis -is -e many-nerved, *multi-nerva*
multinominatus -a -um having many names, *multi-*(*nomen, nominis*)
multipartitus -a -um much divided, *multi-*(*pars, partis*)
multipedatus -a -um with many feet, *multi-pedatus* (stalks)
multiplex with very many parts, very-double, manifold, *multiplex, multiplicis*
multiradiatus -a -um with many rays, *multi-*(*radio, radiare*)
multiramosus -a -um many-branched, *multi-*(*ramus, rami*)
multiscapoideus -a -um having many flowering scapes, *multi-*(*scapus, scapi*)
multisectus -a -um much divided, cut into many segments, *multi-*(*seco, secare, secui, sectum*)
multisiliquosus -a -um many (siliqua) fruited, *multi-*(*siliqua, siliquae*)
multisulcatus -a -um having a much grooved or furrowed surface, *multi-*(*sulcus, sulci*)
multizonatus -a -um marked with many zones, *multi-*(*zona, zonae*) (colouration)
multratus -a -um having many qualities, very settled, *multi-*(*traho, trahere, traxi, tractum*)
mume from the Japanese name, ume
mundulus -a -um quite neat, neatish (diminutive of *mundus*)
mundus -a -um, mundi neat, elegant, of the world, heavenly, *mundus, mundi*
munitus -a -um fortified, armed, *munitio, munitionis*
mupinensis-is -e from Mupin, China
muralis -is -e growing on walls, of the walls, *murus, muri*
muralius -a -um covering walls, *murus, muri*
Murbeckiella, murbeckii for Svante Murbeck (1859–1946), botany professor at Lund, Sweden
murex jagged rock; purple (*Murex* is a genus of spiny molluscs yielding a purple dye), *murex, muricis*
muricatus -a -um rough with short superficial tubercles, muricate (the tip of the shell of *murex*)
murice from a vernacular name for the bark of *Byrsophyllum* species
muricidus -a -um mouse-killer, *muri-*(*caedes, caedis*) (the poisonous seeds)

muriculatus -a -um somewhat rough-surfaced, diminutive of *muricatus*
murinus -a -um mouse-grey, of mice, *mus, muris*
murorum of walls, *murus, muri*
murra myrrh, *murra, murrae*
Murraya, murrayanus -a -um either for Johan Andreas Murray (1740–91), Swedish pupil of Linnaeus and, as Professor of Botany at Göttingen, editor of his works, or for Stewart Murray (?1789–1858) of Glasgow Botanic Garden
Musa for Antonio Musa (63–14 BC), physician to Emperor Augustus; from Egyptian, mauz or mouz (Sanscrit, moka); Musa was a Roman inspirational goddess (***Musaceae***)
musaicus -a -um mottled like a mosaic; resembling *Musa*
musalae from Mount Musala, Bulgaria
Musanga from a Congo vernacular name, given by Christian Smith (*Musanga smithii*)
Muscadinia, muscadinus -a -um Muscadine, from French, muscade, for nutmeg (musky-flavoured grapes of SE USA, ≡ *Vitis rotundifolia*)
muscaetoxicus -a -um, muscitoxicus -a -um fly-poisoning, *musca-toxicum* (*Zigadenus muscaetoxicus* has been used to prepare a fly poison)
Muscari Musk-like (from the Turkish, moscos, fragrance)
muscari- fly-, like *Muscari* inflorescence-
muscariformis -is -e resembling Muscari, *Muscari-forma*; shaped like a fly trap, *muscarium-forma* (fly-brush-like)
Muscarimia, muscarimi *Muscari*-like
muscarioides resembling *Muscari, Muscari-oides*
muscarius -a -um of flies, *musca, muscae* (use of fly agaric in milk to attract and stupefy flies); forming a loose irregular corymb, *muscarium*
musci- fly-, *musca, muscae*; moss-, *muscus, musci*; mouse-, *mus, muris*
muscicolus -a -um living amongst mosses, *musci-colo*
muscifer -era -erum fly-bearing, *musci-fero* (floral resemblance)
musciformis -is -e moss-like, *musci-forma*
muscipulus -a -um fly-catching, *musca-capio* (*muscipula* was a mousetrap, *musculus* a mouse; *Dionaea muscipula*, Venus' flytrap)
muscitoxicus -a -um fly-inebriating, *musci-toxicus*
muscivorus -a -um fly-eating, *musci-(voro, vorare, voravi, voratum)*
muscoides moss-like, *musci-oides* (moss saxifrage)
muscorus -a -um of mosses, of mossy habitats, *muscus, musci*
muscosus -a -um musky; moss-like, mossy, *muscus*
musi- banana-, *Musa-*
musicolus -a -um growing on or with bananas, *Musa-colo*
musifolius -a -um banana-leaved, with leaves resembling *Musa, Musa-folium*
musimomum 10000 to 1, μυσιοι-ομος
muskingumensis -is -e from the Muskingum river in E central Ohio, USA
Mussaenda from a Singhalese vernacular name for *M. frondosa*
Musschia for J. H. Mussche (1765–1834), director of Ghent Botanic Garden and author of its catalogue in 1810
mussini for Count Grafen Apollos Apollosowitsch Mussin-Puschkin (d. 1805), phytochemist from the Caucasus (*Nepeta mussini*)
mustangensis -is -e from Mustang island or creek, USA
mustellinus -a -um brown like a weasel, *mustela, mustelae*
mutabilis -is -e changeable (in colour), mutable, *muto, mutare, mutavi, mutatum*
mutans changing, variable, mutant, *muto, mutare, mutavi, mutatum*
mutatus -a -um changed, altered, *muto, mutare, mutavi, mutatum*
muticus -a -um, mutilatus -a -um cut off, without a point, not pointed, blunt, *mutilo, mutilare, mutilavi, mutilatum*
mutilus -a -um rudimentary, maimed, *mutilus*
Mutisia (Mutisa) for José Celestino Bruno Mutis y Bosio (1732–1808), Spanish writer on the flora of Colombia and discoverer of *Cinchona*

myagroides resembling *Myagrum, Myagrum-oides*
Myagrum, myagrum Fly-hunt, μυια-αγρον (Dioscorides' name, μυαγρα, μυαγρον)
Myanthus Fly-flower, μυια-ανθος (appearance of the drying flowers, ≡ *Catasetum*)
Mycelis de l'Obel's name has no clear meaning (μυκης, μυκελος, fungal hyphae)
mycenopsis -is -e having fruiting bodies resembling the bonnet fungus, *Mycena galericulata*
-myces, myco- -fungi, fungus-, mushroom-, μυκης, μυκο-
-mycetes -fungus, μυκης, μυκητυς
Mycoacia Spiny-fungus, μυκης-ακη (yellow spine-covered fruiting stage)
myconis -is -e fungus-like, of fungi, μυκης
-myia -fly, μυια (suffix for plant pests such as *Poamyia*, galling *Poa nemoralis, Taxomyia*, causing artichoke gall on yew, and *Rhabdomyia*, galling various plants)
myiagrus -a -um fly-hunt, μυια-αγρον (glutinous)
myo-, my- mouse-, closed-, μυς, μυος, μυ- (also muscle-, as in myocardial)
myoctonus -a -um mouse-death, μυος-κτονος (poisonous to mice)
myodes fly-like, μυαι-ωδης
Myoporum Closed-pore, μυο-πορος (the window-like leaf spots) (***Myoporaceae***)
myosodes coloured or smelling of mice, μυος-ωδης
myosorensis -is -e from Mysore (now Karnataka state), S India
myosotidiflorus -a -um having flowers similar to those of *Myosotidium, Myosotidium-florum*
Myosotidium *Myosotis*-like, diminutive of μυος-ωτος
Myosotis Mouse-ear, μυς-(ους, ωτος) (Dioscorides' name μυοσωτα, μυοσωτις)
myosotidiflorus -a -um with *Myosotis*-like flowers
Myosoton Mouse-ear, μυος-ωτος (Dioscorides' name synonymous with μυοσωτις)
myosuroides mousetail like, *Myosurus-oides* (inflorescence)
Myosurus Mouse-tail, μυος-ουρα (the fruiting receptacle)
myr-, myro- myrrh-, *Myrrhis-*
Myrcia a name from mythology equating to Venus
myri-, myrio- numerous-, myriad-, countless, flowing, μυριος, μυριο-
myriacanthus -a -um with very many thorns, μυριο-ακανθος
Myriandra Myriad-stamens, μυριος-ανδηρος (≡ *Hypericum*)
myrianthus -a -um with a large number of flowers, μυρι-ανθος
Myrica Fragrance, μυρικη (the ancient Greek, Homeric name, μυρικη, for *Tamarix*) (***Myricaceae***)
Myricaria *Myrica*-like, a Homeric name, μυρικη, for a tamarisk
myricoides resembling *Myrica, Myrica-oides*
myriocarpus -a -um prolific fruiting, μυριο-καρπος
Myriophyllum, myriophyllus -a -um Numerous-leaves, μυριο-φυλλον (Dioscorides' name for the much divided foliage)
Myriostoma Many-outlets, μυριο-στομα (the pores of the spore-sac)
Myristica Myrrh-fragrant, μυριστικος (true nutmeg, *Myristica fragrans*, the dried outer covering of which provides mace, μακιρ) (***Myristicaceae***)
myristicaeformis -is -e, myristiciformis -is -e somewhat like *Myristica*
myristicus -a -um myrrh-like, μυριστικος (calabash-nutmeg, *Monodora myristca*)
myrmeco- ant-, μυρμηξ, μυρμηκος, μυρμηκο- (many ant/plant symbioses involve plant structure modifications)
myrmecophilus -a -um ant-loving, μυρμηκο-φιλος (plants with special ant accommodations and associations)
myro- fragrant-, unguent-, balsam-, μυρον, μυρο-
myrobalanus -a -um perfumed nut, fragrant acorn, μυρο-βαλανος (the vernacular name for the fruit of *Terminalia myrobalanus*, cognate with mirabelle)
Myrodia Balsam-fragrant, μυρρ-οδμη
Myrosma, myrosmus -a -um Balsam-fragrant, μυρρ-οσμη
myrothamnus -a -um fragrant-shrub, μυρο-θαμνος
myrrhifolius -a -um having leaves resembling those of *Myrrhis*
Myrrhis Dioscorides' ancient name, μυρρα, for true myrrh, *Myrrhis odorata*

myrrhus -a -um myrrh, *Myrrhis* (*Commiphora myrrha*)
Myrsine Dioscorides' ancient name, μυρσινη, for the myrtle (***Myrsinaceae***)
myrsinites myrtle-like, μυρσινη-ιτης
myrsinoides *Myrsine*-like, *Myrsine-oides*
myrti- myrtle-, *Myrtus-*
myrticolus -a -um growing on or amongst myrtles, *Myrtus-colo*
myrtifolius -a -um, myrsinaefolius -a -um myrtle-leaved, *Myrtus-folium*
myrtilloides resembling a small *Myrtus, myrtillus-oides*
myrtillus little myrtle, diminutive of *Myrtus*
myrtinervius -a -um myrtle-veined, *Myrtus-nerva*
Myrtus the Greek name, μυρτον, for myrtle (the apothecaries' name for the berries was *myrtilli*) (***Myrtaceae***)
mysorensis -is -e from Mysore, India
Mystacidium Moustache-like, μυσταξ, μυστακιδος (coronal fringe)
mystacinus -a -um moustached, whiskered, μυσταξ, μυστακινος
mystropetalus -a -um having spoon-shaped petals, μυστρο-πεταλον
mystrophyllus -a -um having spoon-shaped leaves, μυστρο-φυλλον
Mystroxylon Spoon-wood, μυστρο-ξυλον (use of timber)
mysurensis -is -e from Mysore, India
myuros, myurus -a -um mouse-tailed, μυ-ουρα
Myurus Mouse-tail, μυ-ουρα (the panicle)
myx-, myxo- amoeboid-, mucus-, mucilage-, slime-, μυξα
Myxomphalia Slimy-navel, μυξα-ομφαλος (depression of the cap)
Myxomycetes Slime-fungus, μυξα-μυκητος
Mzimbanus -a -um from the Mzimba plain, NW Malawi

nacreus -a -um mother-of-pearl-like, nacre, of uncertain French etymology
Naegelia (Nagelia) for Karl von Nägeli (1817–91), professor of botany at Zurich and Munich
naevatus -a -um, naevius -a -um, naevosus -a -um freckled, with mole-like blotches, *naevus, naevi*
Naias, Najas Naias, one of the three mythological freshwater nymphs, or Naiads (see also *Nymphaea* and *Nyssa*) (***Najadaceae***)
naio- dwelling-, inhabiting-, ναιω, ναω
Naiocrene Fountain-dweller, ναιω-κρηνη
nairobensis -is -e from Nairobi, Kenya
Nama, nama-, namato- brook-, stream-, fountain-, ναμα
namaensis -is -e from the area of the Nama people of Namibia
namaquanus -a -um, namaquensis -is -e from Namaqualand, western S Africa
namatophilus -a -um brook-loving, ναματο-φιλος
namibensis -is -e from the Namib coastal desert area of SW Africa
namulensis -is -e from the environs of Mount Namúli, N Mozambique, E Africa
nan, nana-, nanae-, nani-, nano-, nanoe-, nanno- very small, dwarf, νανος, ναννος
Nandina from the Japanese name, nandin, for the sacred bamboo (***Nandinaceae***)
nanellus -a -um very dwarf, diminutive of *nanus*
nangkinensis -is -e, nankinensis -is -e from Nanking (Nanjing), China
nanifolius -a -um having very small leaves, *nanus-folium*
nannopetalus -a -um having very small petals, ναννο-πεταλον
nannophyllus -a -um very small-leaved, ναννο-φυλλον
Nannorhops Dwarf-bush, ναννο-ρωψ (≡ *Chamaerhops*)
nanodes of dwarf appearance, ναν-ωδης
nanothamnus -a -um dwarf-thorn-bush, νανο-θαμνος
nanshanicus -a -um from the Nan Shan (Qilian Shan) mountains, Tsinghai, central China
nanus -a -um dwarf, *nanus* (modern Latin from νανος)
napaeifolius -a -um, napeaefolius -a -um mallow-leaved, *Napaea-folium*

napaeus -a -um of woodland glades, glen or dell, ναπη, ναπαιος
napalensis -is -e, napaulensis -is -e from Nepal, Nepalese
napellus -a -um swollen, turnip-rooted, like a small turnip, diminutive of *napus*
napi- turnip-, *napus, napi*
napifolius -a -um turnip-leaved, *Napus-folium*
napiformis -is -e having a turnip-like root, *Napus-forma*
napipes with a nap of hairs on the stipe, botanical Latin from Old English, noppe and *pes*
napobrassica sectional name for *Brassica* whose components produce rutabaga or swedes (*Napus-Brassica*, turnip-cabbage)
Napoleona (Napoleonaea) for Napoleone Buonaparte (1769–1821), Emperor Napoleon I of France (1804–14)
napolitanus -a -um from Naples, Napoli, Italy
napuliferus -a -um, napuligerus -a -um turnip-bearing, *Napus-fero* or *-gero* (fleshy cylindrical rootstock)
Napus the name, *napus,* in Pliny for a turnip
narbonensis -is -e from Narbonne (*Narbona*), Languedoc-Rousillon, S France
narcissiflorus -a -um with *Narcissus*-like flowers
narcissifolius -a -um with leaves resembling those of *Narcissus, Narcissus-folium*
Narcissus the name, Narcissus, of a youth in Greek mythology who spurned the nymph, Echo, and fell in love with his own reflection (Pliny prefers derivation from ναρκη, torpor, for the narcotic effect if eaten)
narcoticus -a -um narcotic, numbing, ναρκαω, ναρκοτικος
nardiformis -is -e *Nardus*-like, mat-forming, *Nardus-forma*
Nardophyllum Fragrant-leaf, ναρδος-φυλλον
Nardosmia Spikenard-scented, ναρδος-οσμη (≡ *Petasites*)
Nardostachys, nardostachys Fragrant-bush, nard spike, ναρδοσταχυς (the fragrant ointment is made from the fusiform roots)
Nardurus *Nardus*-tail, ναρδος-ουρα (the narrow inflorescence)
Nardus, nardus Spikenard-like, ναρδος (the lower parts of *Nardus stricta* are a little like the biblical spikenard *Nardostachys jatamansi*)
narinosus -a -um wide-tubed, broad-nosed, with wide nostrils, superlative of *naris*
Narthecium Little-reed, diminutive of ναρθηξ (the cane-like stem; also an anagram of *Anthericum*)
Narthex, narthex Cane, ναρθηξ (an old Greek name for *Ferula narthex*)
narynensis -is -e from the environs of the Naryb river, Kyrgyzstan–Uzbekistan
naso a name used by Ovid
Nasonia Nose, *nasus* (the shape of the anther and column)
nasturtiifolius -a -um with leaves resembling those of *Nasturtium, Nasturtium-folium*
Nasturtium Nose-twist (from Pliny's *nasturcium, quod nasum torqueat,* and *nomen accipit a narium tormento,* for the mustard-oil smell)
nasturtium-aquaticum water *Nasturtium, Nasturtium-aquaticus*
nasutus -a -um satirical; large-nosed, *nasus, nasi; nasutus*
natalensis -is -e from Natal, S Africa
natalitius -a -um of births or birthdays, *natalis, natalicius*
natans floating on water, swimming, *nato, natare*
nathaliae for Queen Natholia, wife of a former king of Milan
nativo created, native, natural, *nativus*
natrix grass or water snake, *natrix, natricis* (common habitat)
Nauclea Hull-enclosed, ναυς-κλειω (shape of the two valves of the fruit)
Naumbergia for S. J. Naumberg (1768–99), Professor of Botany at Erfurt
nauseosus -a -um nauseating, disgusting, *nauseo, nauseare* (the odour)
nauticus -a -um, nautiformis -is -e shaped like a boat-, ναυς, *nauticus, nauticus-forma*
navajous -a -um for the Navaho (Navajo) tribe of native Americans
navicularis -is -e, naviculatus -a -um boat-shaped, *navicula, naviculae*
naviculifolius -a -um having leaves with a pronounced keel, *navicula-folium*

navus -a -um energetic, vigorous, *gnavus, navus*
nayaritensis -is -e from Nayarit state, W central Mexico
Neanthe derivation unclear (≡ *Chamaedorea elegans*)
neapolitanus -a -um from Naples, Italy (*Neapolis*)
nebrodensis -is -e from Mount Nebrodi, Sicily
nebularis -is -e clouded, *nebula, nebulae* (the colour and blooming of clouded agaric)
nebulicolus -a -um growing at altitude, cloud-dwelling, *nebula-colo*
nebulosus -a -um cloud-like, clouded, vaporous, nebulous, *nebula, nebulae*
necopinus -a -um surprising, unexpected, *necopinus*
necro- dead-, decayed-, νεκρος, νεκρο
necrophagus -a -um saprophytic, eating dead matter, νεκρο-φαγω
nectar- nectar-, honey-, νεκταρ, νεκταρος
nectariferus -a -um bearing nectar, νεκταρος-φερω
Nectaroscordum Nectar-garlic, νεκταρος-σκοροδον
Nectria Honey, νεκταρ (the mass of yellow, drop-like fruiting bodies)
neerlandicus -a -um from the Netherlands
negevensis -is -e from the Negev (Ha-Negev, Hebrew, ngb) desert area, Israel
neglectus -a -um (formerly) overlooked, disregarded, neglected, *neglegio, neglegere, neglexi, neglectum*
negrosensis -is -e from Negros Island, central Philippines
Negundo, negundo from a Sanskrit name, nirgundi, for a tree with leaves like box-elder
neilgherrensis -is -e from the Nilgiri hills, Tamil Nadu, S India
Neillia for Patrick Neill (1776–1851), Edinburgh botanist
nelsonii for either Reverend John Nelson (1818–82), who grew narcissi, or William Nelson (1852–1922) of Natal (*Albuca nelsonii*)
nelumbifolius -a -um with *Nelumbo*-like leaves, *Nelumbo-folium*
Nelumbo (Nelumbium), nelumbo from a Singhalese vernacular name, nelumbi, for water-bean (***Nelumbonaceae***)
-nema, nema-, nemato- -thread, thread-, thread-like-, νημα, νηματο
Nemastylys Thread-styled-one, νημα-στυλος
Nematanthus Thread-flower, νηματο-ανθος (thread-like pedicels of *N. longipes*)
nematocaulon with a slender stalk. νηματο-καυλος
Nemesia a name, νεμεσιον, used by Dioscorides for another plant
nemo- of clearings-, of glades-, glade-, νεμος, νεμο-, κνεμος
Nemocharis Joy-of-the-glades, νεμο-(χαρις, χαριτος)
Nemopanthus (*Nemopanthes*) Thread-flower, νημα-ανθος (the slender pedicels)
Nemophila, nemophilus -a -um Glade-loving, νεμο-φιλος (woodland habitat)
nemoralis -is -e, nemorosus -a -um, nemorum of woods, sylvan, *nemus, nemoris*
nemorensis -is -e from woodlands, *nemus, nemoris*
nemus -a -um of glades, *nemus, nemoris*
nemusculus medieval Latin for undergrowth or scrub, *nemus*
neo- new-, νεος, νεο- (as a generic prefix denoting systematic relationship)
neoalaskana new (*Betula*) *alaskana*
Neobaumannia New-*Baumannia*
Neobenthamia New-*Benthamia*, for George Bentham (1800–84)
Neobesseya New-*Besseya*, for Charles Bessey
neo-britanniae from New Britain, Papua New Guinea, botanical Latin from νεος and *Britannia*
neo-caledoniae from New Caledonia, French territory, SW Pacific Ocean, botanical Latin from νεος and *Caledonia*
neocorymbosus -a -um new (*Hieracium*) *corymbosum*
Neodypsis New-*Dypsis*, νεο-δυψις
neoelegans new (*Aster*) *elegans*, botanical Latin from νεος and *elegans*
neogaeus -a -um from the New World, νεο-(γη, γαια)
neo-hibernicus -a -um from New Ireland, Papua New Guinea, botanical Latin from νεος and *Hibernia*

Neohyptis New-*Hyptis*
Neolitsea New-*Litsea*
neolobatum new (*Polyscias*) *lobatum*, botanical Latin from νεος and *lobatus*
Neomarica New-*Marica*
neomexicanus -a -um from New Mexico, USA, botanical Latin from νεος and Mexico
neomontanus -a -um from Neuberg, Germany, of Neuberg
Neopaxia New-*Paxia*
neopolitanus -a -um from Naples, Neapolitan, νεος-πολις, (*Neapolis*, new town, was founded 600 BC to accommodate the earlier, Greek-populated *Palaepolis*)
Neoregelia New-*Regelia*, for Constantin von Regel (1890–1970), Russian botanist
Neoschumannia New-*Schumannia*
neoscoticus -a -um from Nova Scotia, Canada, botanical Latin from νεος and *Scotia*
Neostenanthera New-*Stenanthera*, νεος-στεν-ανθερα
Neotinea (*Neotinnea*) New-*Tinea*
Neotostema Youthful-stamens, νεοτης-στεμων (literally, stamens like a body of young men)
neotropicus -a -um from the New Tropics (tropics of the New World), botanical Latin from νεος and Middle English, tropic (τροπη)
Neottia Nest-of-fledglings, νεοσσια, νεοττια (the appearance of the roots of *Neottia nidus-avis*, or 'bird's nest bird's nest')
nepalensis -is -e, nepaulensis -is -e from Nepal, Nepalese
Nepenthes Grief-assuaging, νε-πενθης (its reputed drug property of removing anxiety) (***Nepenthaceae***)
nepenthoides resembling *Nepenthes*, νε-πενθης-οειδης
Nepeta, nepeta the name in Pliny, for a plant from Nepi, Etruria, Italy
nepetellus -a -um little mint, diminutive of *nepeta*
nepetifolius -a -um having leaves resembling those of *Nepeta*, *Nepeta-folium*
nepetoides resembling *Nepeta*, *Nepeta-oides*
nephelophilus -a -um cloud-loving, νεθελη-φιλος
nephr-, nephro- kidney-shaped-, kidney-, νεφρος, νεφρο-
Nephrangis Kidney-vessel, νεφρος-αγγειον (the two lateral lobes of the lip)
nephrocarpus -a -um having kidney-shaped fruits, νεφρο-καρπος
Nephrodium Kidneys, νεφρος, νεφρωδης (the shape of the indusia of the sori)
nephroideus -a -um reniform, kidney-shaped, νεφρος-οειδης
Nephrolepis, nephrolepis -is -e Kidney-scale, νεφρο-λεπις (the shape of the indusia of the sori)
nephrophyllus -a -um having kidney-shaped leaves, νεφρος-φυλλον
Nephthytis for Nephthys of mythology, who bore Typhon's son Anubis
neptunicus -a -um of the seas, for *Neptunus*, the god of the sea
nericius -a -um from the province of Närke, S Sweden
nerii- oleander-like-, *Nerium-*
neriiflorus -a -um having oleander-like flowers, *Nerium-florum*
neriifolius -a -um (nereifolius -a -um) oleander-leaved, *Nerium-folium*
Nerine Nerine, a sea nymph, daughter of Nereus
neriniflorus -a -um having *Nerine*-like flowers, *Nerine-florum*
Nerium Dioscorides' ancient Greek name for oleander, *Nerium oleander*
nerineoides resembling *Nerine*, *Nerine-oides*
Nertera Lowly, νερτερος (for its small stature, not because it is infernal)
nerterioides resembling bead plants, *Nertera-oides*
nervalis -is -e loculicidal on the mid-rib, with a tendril-like prolongation of the mid-nerve, *nerva, nervae*
nervatus -a -um, nervis -is -e nerved or veined, *nerva, nervae*
Nervilia Veined, *nerva, nervae* (the prominent leaf veins of some)
nervosus -a -um with prominent nerves or veins, *nerva, nervae*
nervulosus -a -um with delicate or fine veins, diminutive of *nervus*
nervus -a -um nerve or vein, *nervus, nervi* (literally a sinew)

Nesaea Neseia, the name of a sea nymph
nesioticus -a -um of islands, islander, insular, νησος, νησιωτης, νησιωτις
Neslia for the eighteenth-to nineteenth-centruy French botanist, Nesles
neso- island-, νησος, νησο-
Nesogordonia Island-*Gordonia* (it was originally thought to be confined to Madagascar)
nesophilus -a -um island-loving, νησο-φιλος
nessensis -is -e from Loch Ness, Scotland, or Ness Botanic Garden, Cheshire
neuro-, -neurus -a -um ribbed-, -nerved, -veined, νευρα, νευρη, νευρο-
neurolobus -a -um with veined lobes, νευρο-λοβος
Neuropeltis Veined-shielded-one, νευρο-πελτη (the veined and enlarged bract at fruiting)
neuropetalus -a -um having conspicuously marked veins in the petals, νευρο-πεταλον
neurosus -a -um having pronounced or prominent nerves or veins, νευρα
Neurotheca Ribbed-container, νευρα-θηκη (the ridged calyx)
neutrus -a -um nondescript, neutral, *neuter, neutri*
nevadensis -is -e from Nevada or the Sierra Nevada, USA, or from the Sierra Nevada, Spain
Neviusia (Neviusa), nevii for its finder, R. D. Nevius
newellii for Mr Newell of Downham Market, Norfolk *c.* 1880
newryensis -is -e from Newry, County Down, Northern Ireland
Newtonia for Sir Isaac Newton (1642–1727), proponent of the laws of motion
nexus -a -um enslaved; connected, entwined, *necto, nectere, nexi (nexui), nexum*
niamniamensis -is -e, niamjamensis -is -e from Nia Nia, Congo
nicaeensis -is -e either from Nice (*Nicaea*), SE France or from Nicaea, Bithynia, NW Turkey
Nicandra for Nicander of Calophon (100 BC), writer on plants and antidotes
nicaraguae from Nicaragua, Central America, Nicaraguan
nichollsii for Mr Nicholls, of New Zealand (*Leptospermum nichollsii*)
nicobaricus -a -um from the Nicobar Islands, Bay of Bengal, India
nicolai either for King Nicolas of Montenegro, or for Tsar Nicholas of Russia
Nicotiana for Jean Nicot (1530–1600), who introduced tobacco to France in the late sixteenth century
nicotianifolius -a -um having leaves similar to those of *Nicotiana, Nicotiana-folium*
nicoyanus -a -um from the Nicoya peninsula, W Costa Rica
nictitans nodding, moving, blinking, winking, present participle of *nicto, nictare*
nidi-, nidus nest, nest-like, *nidus, nidi*
nidificus -a -um nest-like, *nidus, nidi* (fertile whorls of charaphyte, *Tolypella*)
nidiformis -is -e shaped like a nest, *nidi-forma*
nidorosus -a -um gnawed nest, *nidus-(rodo, rodere, rosi, rosum)* (concave pileus)
nidulans crouching like a bird in its nest, nidulant, present participle of *nidifico, nidificare* (lying in a slight hollow)
Nidularia, Nidularium, nidularius -a -um Little-nest, *nidulus, niduli* (organs lying in a nest-like structure or arrangement)
nidus-aves bird's-nest, *nidus-avis* (resemblance)
niedzwetzkyanus -a -um for the Russian Judge Niedzwetzky
Nierembergia for Juan Eusebia Nieremberg (1594–1658), Spanish Jesuit naturalist
Nigella Blackish, diminutive of *niger* (the seed coats)
nigellastrum medieval Latin name for corn-cockle, *Nigella-ad-instar*
niger -gra -grum black, *niger, nigeri*
nigercors dark-centred, black-hearted, *niger-cor*
nigericus -a -um from Nigeria, W Africa
nigramargus -a -um with black edges, dark-edged, *nigra-(margo, marginis)*
nigrans dusky, darkening, present participle of *nigro, nigrare*
nigratus -a -um blackened, *nigresco, nigrescere*
nigrescens blackish, darkening, turning black, present participle of *nigresco, nigrescere*

nigri-, nigro- black-, dark-, *niger, nigri, nigro-*
nigricans almost black, blackish with age, present participle of *nigro, nigrare*
nigripes black-stalked, *niger, nigri-pes*
nigripetalus -a -um black-petalled, botanical Latin from *niger, nigri* and πεταλον
nigristernus -a -um with black breast or heart, botanical Latin from *niger, nigri* and στερνον
nigritanus -a -um from Nigeria, W Africa
nigroglandulosus -a -um having black glands, *niger, nigri-(glans, glandis)*
nigropaleaceus -a -um bearing black scales, *niger, nigri-(palea, paleae)*
nigropunctatus -a -um marked with black dots, *niger, nigri-(pungo, pungere, pupugi, punctum)*
nigropurpureus -a -um dark-purple, *nigro-purpureus*
nigrus -a -um black, *niger, nigra, nigrum*
nikkoensis -is -e, nikoensis -is -e from Nike or Niko National Park, Honshu, Japan
nil the Arabic name for *Pharbitis hederacea*
niliacus -a -um from the River Nile (*Nilus, Nili*)
niloticus -a -um from the Nile valley (*niliacus*)
nimbicolus -a -um dwelling with clouds or rain-storms, *nimbus-colo*
nimborus -a -um of clouds or rain-storms, *nimbus, nimbi*
ningpoensis -is -e from Ningpo (Ningbo), China,
nintooa the Japanese name for *Lonicera japonica*
nipalensis -is -e, nipaulensis -is -e *vide nepalensis*
Niphaea Snowy, νιφας, νιφαδος (the white flowers)
Niphimenes the composite name for hybrids between *Niphaea* and *Achimenes*
nipho- snow-, νιφας, νιφο-
niphophilus -a -um snow-loving, νιφο-φιλος
Nipponanthemum Flower-of-Japan (*N. japonicum*)
nipponicus -a -um, (niponicus -a -um) from Japan (Nippon), Japanese
nissanus -a -um from Nish, SE Serbia
Nissolia, nissolia, nissolicus -a -um for Guillaume Nissole (1647–1735), botanist of Montpellier
Nitella Little-shining-one, diminutive of *nitidus*
Nitellopsis Resembling-*Nitella*
nitens, nitidi-, nitidus -a -um with a polished surface, neat, shining, *niteo, nitere*
nitescens becoming glossy, present participle of *nitesco, nitescere, nitui*
nitidellus -a -um having somewhat smooth or polished surfaces, diminutive of *nitidus*
nitidifolius -a -um with glossy leaves, *nitidus-folium*
nitidissimus -a -um with the most glossy surfaces, superlative of *nitidus*
nitidulus -a -um quite smooth-surfaced, diminutive of *nitidus*
nitidus -a -um bright, shining, clear, lustrous, *nitidus*
Nitraria Soda-producer, *nitrum, nitri* (grows in saline deserts, burnt yields nitre, νιτρον)
nitrariaceus -a -um of alkaline soils, *nitrum, nitri*
nitratus -a -um nitrous, *nitrum, nitri* (smell)
nitrophilus -a -um alkali-loving, νιτρο-φιλος (growing on soda- or potash-rich soils)
nivalis -is -e snow-white, growing near snow, *nix, nivis*
niveus -a -um, nivosus -a -um purest white, snow-white, *nix, nivis*
nivicolus -a -um living (flowering) in snow, *nivi-colo*
nobilior more grand, more noble, comparative of *nobilis*
nobilis -is -e famous, grand, noble, notable, *nobilis*
nobilissimus -a -um the most notable, the grandest, superlative of *nobilis*
nocteolens night-smelling or -stinking, present participle from *noctis-(oleo, olere)*
nocti- night-, *nox, noctis*
noctiflorus -a -um night-flowering, *(nox, noctis)-florum*
Noctiluca Night-light *(nox, noctis)-(lux, lucis)* (phosphorescent marine organism; literally, moon-light)

nocturnalis -is -e, nocturnus -a -um at night, for one night, *nocturnus* (flowering)
nodiferus -a -um carrying knobs or girdles, *nodus-fero* (at the nodes)
nodiflorus -a -um flowering at the nodes, (*nodus, nodi*)-*florum*
nodosus -a -um many-jointed, conspicuously jointed, knotty, *nodosus*
nodulosus -a -um with swellings (on the roots), noduled, diminutive of *nodosus*
noeanus -a -um for either Wilhelm Noe, or Frank Vicomte de Noë
Nolana Small-bell, diminutive of *nola* (***Nolanaceae***)
Nolina for P. C. Nolin, French writer on agriculture *c.* 1755
noli-tangeri do not touch, (*noli, nolite*)-(*tango, tangere, tetigi, tactum*) (on being touched, the ripe fruit ruptures, expelling seed)
noma-, nomo- meadow-, dwelling-, pasture-, νομη
nomados wandering, roaming, νομας, νομαδος
Nomalxochia the Mexican vernacular name
Nomimium, nominius -a -um Customary-violet, νομιμος-ιον
Nomocharis Meadow-grace, νομο-χαρις
non- un-, no-, not-, *non*
Nonnea (*Nonea*) for J. P. Nonne (1729–72), botanical writer from Erfurt, Germany
nonpictus -a -um of plain colour, not painted, *non-pictus*
nonscriptus -a -um, (*nondescriptus -a -um*) unmarked, not written upon, *non-scriptus*
nootkatensis -is -e, nutkatensis -is -e from Nootka (Nutka) Island or Nootka Sound, British Columbia, area of the Nootka Indians, Vancouver Island
norbitonensis -is -e from Norbiton
Nordmannia, nordmannianus -a -um for either Alexander von Nordmann (1843–1866), zoologist of Odessa and Helsingfors, or M. Nordmann, German botanist
normalis -is -e representative of the genus, usual, around the norm, *norma*
northiae, northianus -a -um for Miss Marianne North (1830–96), botanical artist at Kew
northlandicus -a -um from the North Island, New Zealand
norvegicus -a -um from Norway (*Norvegia*), Norwegian (of high altitude in Scotland)
nossibensis -is -e from the volcanic Nosy Be (Nossi-Bé) Island, NW Madagascar
Nostoc an alchemical name used by Paracelsus, who assumed that the slime came from shooting stars (falling star)
notabilis -is -e notorious, remarkable, *notabilis*
notatus -a -um distinguished, spotted, lined, noted, marked, *nota, notae*
Notelaea Southern-olive, νοτος-ελαια
notero- moist-, southern-, νοτερος, νοτερο-
notho-, nothos-, nothus -a -um false-, bastard-, νοθος, νοθο-, νοθ-, *nothus, notho-*
Nothofagus False-beech, *notho-fagus*
Notholaena Spurious-cloak, νοθος-χλαινα, *notho-laena*
Notholirion False-lily, νοθος-λιριον
Nothoscordum Bastard-garlic, νοθο-σκοροδον
nothoxys falsely pointed, pungent, passionate or bold, νοθ-οξυς
noti, notio, noto- of the southwest wind, southern-, νοτος. νοτο-, νοτ-
noto- southern-, νοτος, νοτο-; the back-, νωτον, νωτος, νωτο-
Notospartium Southern-*Spartium* (New Zealand)
Notothlaspi Southern-*Thlaspi* (New Zealand)
-notus -a -um -at the back, νωτον, νωτος
novae-angliae, nova-anglica from New England, *novus-Anglia, novus-Anglicus*
novae-belgii (novi-belgae) from New Belgium, *novus-Belgae* (New Netherlands or New York)
novae-britanniae from New Britain Island, Papua New Guinea (named by William Dampier 1699)
novae-caesareae (novi-caesareae) from New Jersey, *novus-Caesaria*, USA
novae-guineae from New Guinea, botanical (including Papua New Guinea, Irian Jaya and Indonesia)

novae-hiberniae from New Ireland, *novus-Hibernia*, Papua New Guinea
novae-hollandiae from Australia (named New Holland by Abel Tasman, 1644), *novus-Hollandia*
novae-zelandiae from New Zealand, botanical of New Zealand
noveboracensis -is -e from New York, *novus-Eburacum*, USA
novem- nine-, *novem*
novemfolius -a -um nine-leaved (or leafleted), *novem-folium* (not exact)
novi-belgii from New York, (formerly called *Novum Belgium*), USA
novi-caesareae from New Jersey, USA
novo-granatensis -is -e from the state of New Granada, Colombia
novo-mexicanus -a -um from New Mexico, USA
novus -a -um, nov-, novae-, novi- new-, *novus*
noxius -a -um hurtful, harmful, *noxa, noxae*
nubi- cloud-, gloom-, *nubes, nubis*
nubicolus -a -um of cloudy places, *nubes-colo*
nubicus -a -um from the Sudan (Nubia), NE Africa (land of the *Nubae*)
nubigenus -a -um (nubiginus -a -um) cloud-formed, cloud-born, *nubes-genus*
nubilorum from high peaks, of clouds, *nubes, nubis*
nubilus -a -um gloomy, sad, dusky, greyish-blue, *nubilus*
nucifer -era -erum nut-bearing, (*nux, nucis*)*-fero*
nuciformis -is -e shaped like nuts, *nucis-forma*
nuculosus -a -um containing hard, nut-like seeds, diminutive of *nucula*
nuculus -a -um with small nuts, *nucula*, diminutive of *nux, nucis*
nudatus -a -um stripped, bared, exposed, *nudo, nudare, nudavi, nudatum*
nudi-, nudus -a -um bare, naked, thornless, *nudo, nudare, nudavi, nudatum*
nudicarpos, nudicarpus -a -um having naked fruits, botanical Latin from *nudus* and καρπος
nudicaulis -is -e naked-stemmed, leafless, *nudus-caulis*
nudiflorus -a -um with fully exposed or naked flowers, *nudus-florum*
nudifolius -a -um with simple, exposed leaves, *nudus-folium* (contrasting with revolute clusters)
nudipes clear-stemmed, *nudus-pes*
nudiusculus -a -um somewhat bared or denuded, diminutive of *nudus*
nuevo-mexicanus -a -um from New Mexico, USA
numerosus -a -um populous, *numerosus*
numidicus -a -um from Algeria, Algerian (Numidia the land of the *Nomas, Nomadis*)
numinus -a -um divine, powerful; nodding, *numen, numinis*
numismatus -a -um coin-like, νομισμα, νομισμτικος
nummatus -a -um moneyed; coin-like, *nummus, nummulus*
nummularifolius -a -um with circular, coin-like leaves, *nummus-folium*
nummularis -is -e circular, coin-like, *nummus, nummulus* (the leaves)
nummularius -a -um money-wort-like, having leaves like small coins, *nummus nummi*
nuperus -a -um of recent times, fresh, new, *nuper*
Nuphar the Persian name, ninufar, for a water lily (ancient Latin *nenufar, ninufer*) or from Mosul (Nineveh)
nuristanicus -a -um from Nurestan (Nuristan), E Afghanistan
nurricus -a -um from Nurri (Nurria), Sardinia
nutabilis -is -e sad-looking, drooping, nodding, *nuto, nutare*
nutaniflorus -a -um having drooping flowers, *nutans-florum*
nutans drooping, nodding, present participle of *nuto, nutare* (the flowers)
nutkanus -a -um see *nootkatensis*
Nuttallia, nuttallianus -a -um, nuttallii for Thomas Nuttall (1786– 1859), of Long Preston, Yorkshire, grower of American plants at Rainhill, Lancashire
nux- nut-, *nux, nucis*
nux-muscata musk-fragrant nut, medieval Latin name for the nutmeg (*Myristica fragrans*)

nux-vomica nut-of-abscesses, *nux-*(*vomica, vomicae*), with nuts causing vomiting, *vomo, vomere, vomui, vomitum* (*Strychnos nux-vomica* contains the alkaloid strychnine)
nyassicus -a -um, nyassanus -a -um from Malawi (formerly Nyassaland)
nyct-, nycto- night-, *nox, noctis;* νυξ, νυκτως, νυκτο-
nyctagineus -a -um night-flowering, born of the night, νυκτο-γιγνομαι
Nyctaginia Nocturnal, νυκτο-γιγνομαι (***Nyctaginaceae***)
Nyctanthes, Nyctanthus Night-flower, νυκτο-ανθος
nyctanthus -a -um nocturnal-flowering, νυκτο-ανθος
nycticalus -a -um beautiful at night, νυκτο-καλος
nyctitropus -a -um having night movements, having nyctinastic sleep movements, νυκτο-(τροπη, τροπαω)
Nyctocalos Night-beauty, νυκτο-καλος
nymansensis -is -e from Nymans Gardens, near Haywards Heath, W Sussex (cultivarietal name 'Nymansay')
nymphae- waterlily-like-, *Nymphaea*-like-
Nymphaea Nymphe, a mythological freshwater Naiad (***Nymphaeaceae***)
nymphalis -is -e *Nymphe*-like (one of the water nymphs)
Nymphe the name, Nymphe, used by Theophrastus
Nymphoides, nymphoides resembling *Nymphaea, Nymphaea-oides*
Nypa a Japanese vernacular name
Nyssa Nyssa, a mythological fresh-water Naiad (tupelo, cotton-gum and other swamp trees), νυσσα was the turning post or the winning post on a race track (***Nyssaceae***)
nyssanus -a -um from Nis (*Naissus*), Serbia

oaxacanus -a -um from Oaxacan, Mexico
ob-, oc-, of-, op- completely-, against-, contrary-, opposite-, inverted-, inversely-
obassia the Japanese vernacular name for *Styrax obassia*
obconellus -a -um like a small inverted cone, diminutive of *ob-*(*conus, coni*)
obconicus -a -um like an inverted cone, *ob-*(*conus, coni*)
obcordatus -a -um inversely cordate, *ob-cordatus* (stalked at narrowed end of a heart-shaped leaf), obcordate
obductus -a -um spreading, covering, *obduco, obducere, obduxi, obdutum*
obesifolius -a -um having thick or fleshy leaves, *obesus-folium*
obesus -a -um succulent, fat, coarse, *obesus*
obfuscatus -a -um clouded over, confused, *ob-fusco*
Obione Daughter-of-the-Obi (a Siberian river), botanical Latin from Obi and ωνη
obispoensis -is -e from San Luis Obispo, California
oblanceolatus -a -um narrow and tapering towards the base, *ob-lanceolatus*
oblancifolius -a -um having leaves tapering to the base, oblanceolate, *ob-lancea-folium*
oblatus -a -um somewhat flattened at the ends, oval, oblate, modern Latin, *ob-latus*
obliquatus -a -um turned aside, *obliquo, obliquare* (flower position)
obliqui- slanting-, asymmetrical-, *obliquus. obliqui-*
obliquinervis -is -e with oblique veins, *obliquus-*(*nervus, nervi*)
obliquistigmus -a -um having an obliquely held stigma, botanical Latin from *obliquus* and στιγμα
obliquus -a -um slanting, unequal-sided, oblique, *obliquus*
oblongatus -a -um, oblongi-, oblongus -a -um elliptic with blunt ends, oblong
oblongifolius -a -um oblong-leaved, *oblongus-folium*
obovati-, obovalis -is -e, obovatus -a -um egg-shaped in outline and with the narrow end lowermost, obovate, *ob-ovatus*
obpyramidatus -a -um like an inverted pyramid, narrow at the base, *ob-*(*pyramis, pyramidis*)
obrienianus -a -um, obrienii for James O'Brien (1842–1930), orchid grower of Harrow

obscissus -a -um   with a squared-off end, cut off, *ob-scissus*
obscuratus -a -um, obscurus -a -um   dark, indistinct, obscure, of uncertain affinity, *obscuro, obscurare, obscuravi, obscuratum*
obsoletus -a -um   rudimentary, decayed, worn out, *obsolesco, obsolescere, obsolevi, obsoletum*
obstructus -a -um   hindered, blocked, with the throat of the corolla restricted by hairs or appendages, *obstuo, obstruere, obstruxi, obstructum*
obtectus -a -um   covered over, *obtego, obtegere, obtexi, obtectum*
obturbinatus -a -um   reverse top-shaped, wide at the base and tapered to the apex, *ob-turbinatus*
obtusangulus -a -um   having blunt ridges or angles, obtuse-angled, *obtusus-angulus* (branching)
obtusatus -a -um, obtusi-, obtusus -a -um   blunt, rounded, obtuse, *obtusus*
obtusifolius -a -um   obtuse-leaved, *obtusus-folium*
obtusilobus -a -um   with obtuse lobing, botanical Latin from *obtusus* and λοβος
obtusior   more obtuse, comparative of *obtusus* (than the type)
obtusipetalus -a -um   having obtuse petal apices, botanical Latin from *obtusus* and πεταλον
obtusiusculus -a -um   somewhat obtuse, diminutive of *obtusus*
obtusulus -a -um   rounded, somewhat obtuse, diminutive of *obtusus*
obvallaris -is -e, obvallatus -a -um (obvalearis)   walled around, enclosed, fortified, *obvallatus* (with a corona)
obvolutus -a -um   half-amplexicaule, with one leaf margin overlapping that of its neighbour, half-equitant, *obvolutus*
occidentalis -is -e   western, occidental, of the West, *occidens, occidentis*
occitanicus -a -um   from the Languedoc area of France (*Occitania*)
occlusus -a -um   closed up, shut, *occludo, occludere, occlusi, occlusum*
occultus -a -um   hidden, secretive, concealed, *occulto, occultare, occultavi, occultatum*
oceanicus -a -um   growing near the sea, *oceanus, oceani*
ocellatus -a -um (ocelatus -a -um)   gem; like a small eye, with a colour-spot bordered with another colour, *ocellus, ocelli*
Ochagavia   for the nineteenth-century Chilean statesman, Silvestri Ochagavia
Ochna   an ancient Greek name, οχνη, used by Homer for a wild pear (***Ochnaceae***)
ochnaceus -a -um   resembling *Ochna*
ochr-, ochro-   ochre-, pale-yellow-, ωχρος, ωχρο-
ochraceus -a -um   ochre-coloured, yellowish, ωχρος
ochratus -a -um   pale-yellowish, ωχρος
ochreatus -a -um   greaved, with stipules clasping the stem, with an ochrea, *ocrea, ocreae* (the greave-like stipular structure in *Polygonaceae*)
ochrocarpus -a -um   with pale yellowish fruits, ωχρος-καρπος
ochrochlorus -a -um   pale yellowish-green, ωχρος-χλωρος
ochroleucus -a -um (*ochroleucon*)   buff-coloured, yellowish-white, ωχρο-λευκος
Ochroma   Pale-yellow(-flower), ωχρος (the flower colour of balsa-wood)
Ochromonas   Pale-yellow-unit, ωχρο-μονας (Chrysophyceaen)
Ochrosia   Pale-yellow(-flower), ωχρος
ochrus -a -um   pale yellowish, ωχρος *ochrus -a -um*
ochth-, ochtho-   slope-, dyke-, bank-, οχθη, οχθο-
Ochthocosmus   Hill-decoration, οχθο-κοσμος (distinctive leaves, persistent flowers and montane habitat)
ochthophilus -a -um   living on banks, bank-loving, οχθο-φιλος
ocimoides, ocymoides   resembling sweet basil, *Ocimum*-like
Ocimum (Ocymum)   Theophrastus' name, οκιμον, for an aromatic plant, οζω smell
oct-, octa-, octo-   eight-, οκτω, οκτα-, *octo-*
Octadesmia   Eight-bundles, οκτα-δεσμη (there are eight pollinial masses)
octandrus -a -um   eight-stamened
-octanus -a -um   slaying, killing, murdering, -κτεινω, κτονος

Octoknema Eight-legs, οκτω-κνημη (the 3–5 style arms are bifid) (***Octoknemataceae***)
Octolepis Eight-scales, οκτω-λεπις (the paired scale-like petals)
Octolobus Eight-lobed, οκτω-λοβος (the calyx)
-octonus -a -um -slaughtering, -killing, κτονος
octopetalus -a -um with eight petals, οκτω-πεταλον
octophyllus -a -um having (about) eight leaves or leaflets, οκτω-φυλλον
octopodes with eight stalks, οκτω-ποδος
ocularia an apothecaries' name for *Euphrasia officinalis*, eyebright
oculatus -a -um eyed, with eyes, *oculatus*
oculus-christi eye of Christ, *oculus-*(*Christus, Christi*) (*Inula oculus-christi*)
oculus-draconis dragon's eye, *oculus-*(*draco, draconis*)
oculus-solis sun's-eye-, *oculus-*(*sol, solis*)
ocymastrus -a -um somewhat resembling *Ocimum, Ocimum-astrum*
ocymifolius -a -um with leaves resembling *Ocimum, Ocimum-folium*
ocymoides *Ocimum*-like, οκιμον-οειδης
odaesanensis -is -e from Mount Odae, NW South Korea
-odes, -oides -allied, -resembling, -shaped, -similar to, ωδης, -οειδης
odessanus -a -um from Odessa, Black Sea area of Ukraine
odont-, odonto- tooth-, οδους, οδων, οδοντος, οδοντο-
odontadenus -a -um with glandular teeth, οδοντο-αδην
Odontites, odontites Tooth-related, οδοντο-ιτης (the name in Pliny refers to its use for treating toothache)
odontocarpus -a -um with toothed fruits, οδοντο-καρπος
odontochilus -a -um with a toothed lip, οδοντο-χειλος
Odontoglossum Toothed-tongue, οδοντο-γλωσσα (the toothed lip)
odontoides tooth-like, dentate, οδοντο-οειδης
odontolepis -is -e having toothed scales, οδοντο-λεπις
odontolomus -a -um with a toothed fringe, οδοντο-λωμα
Odontonema Tooth-thread, οδοντο-νημα (remnant of fifth stamen)
odontopetalus -a -um having toothed or indented petals, οδοντο-πεταλον
odontophyllus -a -um having toothed leaves, οδοντο-φυλλον
odoratissimus -a -um the most fragrant, superlative of *odor (odos), odoris*
odoratus -a -um, odorifer -era -erum fragrant, sweet-scented, bearing perfume, *odor(odos), odoris* (*Lathyrus odoratus*, the sweet-smelling pea, sweet pea)
odorus -a -um fragrant, *odor(odos), odoris*
-odus -a -um -joined
Oecoclades Living-on-branches, οικεω-κλαδος (epiphytic, ≡ *Trichoglottis*)
oeconomicus -a -um of the household, economical, οικονομικος
Oedera, oederi for George Christian Oeder (1728–91), Professor of Botany at Copenhagen, author of *Flora Danica*
oedo- swelling-, becoming swollen-, οιδημα, οιδμα, οιδαω, οιδεω
oedocarpus -a -um swelling fruit, οιδαω-καρπος
oedogonatus -a -um with swollen nodes, οιδαω-γονυ (comparison with the reproductive state of the filamentous alga, *Oedogonium*)
Oedogonium Swollen-ovary, οιδαω-γυνη (the enlarged gynoecial cells)
oelandicus -a -um from Öland, Sweden
Oenanthe Wine-fragrant-flower, οινος-ανθος
oenanthemus -a -um having wine-red flowers, οινος-ανθεμιον
oenipontanus -a -um from Innsbruck (*Oenipons*)
oeno- wine-, οινος, οινο-
Oenocarpus Wine-fruit, οινος-καρπος (fruits used to make palm-wine)
Oenothera, oenothera Ass-catcher, ονο-θηρας, or Wine-seeking, οινο-θηρα (Greek name for another plant but the etymology is uncertain) (***Oenotheraceae***)
oerstedii, oerstedtii for Anders Sandoe Oersted (1816–72), Danish collector in Costa Rica and Colombia
oestriferus -a -um causing frenzy; (with appearance of) bearing gad-flies, *oestrus-fero*

officinalis -is -e, officinarum of the apothecaries, officinal medicines, sold in shops, *officina*
Oftia a name by Adanson with no clear meaning
ogeche an American vernacular name for Ogeechee lime *(Nyssa ogeche)*
ohiensis -is -e from Ohio, USA
oianthus -a -um with egg-shaped flowers, ωον-ανθος (the ovoid-bell-shaped tube)
-oides, -oideus -a -um -allied, -like, -resembling, -shaped, ειδος, ειδω, οειδης, οιδες; botanical Latin *-oides*
oido- a swelling, οιδημα, (οιδαω, οιδεω, to swell)
oidocarpus -a -um having swollen fruits, οιδαω-καρπος
oistophyllus -a -um sagittate-leaved, with arrow-shaped leaves, οιστος-φυλλον
okanoganensis -is -e from the area of Okanogan Lakes, British Columbia, Canada, or the river in Washington, USA
oklahomensis -is -e from Oklahoma, USA
olacogonus -a -um having furrowed joints (internodes), ολαξ-γονυ
Olax Furrow, ολαξ (the appearance given by of the two-ranked leaves) (***Olacaceae***)
olbanus -a -um from the area of Olbia, N Sardinia, Italy
olbia, olbios rich, ολβιος, or from Hyères (*Olbia*), France, or several former Greek places named Olbia
Oldenlandia for Henrik Bernard Oldenland (1663–97), Danish collector in S Africa
Oldfieldia for Dr Oldfield, who was on the 1832–34 Niger Expedition
oldhamianus -a -um, oldhamii for Richard Oldham (1837–64), who collected in China and Formosa for Kew
Olea Oily-one, ελαα, ελαια (the ancient name for the olive) (***Oleaceae***)
oleaceus -a -um olive-like, resembling *Olea*
oleaefolius -a -um with olive-like leaves, *olea-folium*
oleagineus -a -um, oleaginosus -a -um fleshy, rich in oil, *oleagineus*
oleander a medieval Latin name, *oleander, oliandrum, lauriendrum* (Italian, oleandra, for the olive-like foliage)
Oleandra Oleander-like, ολεανδρη (the fronds somewhat resemble oleander leaves)
Olearia for Adam Oelenschlager (*Olearius*) (b. 1600), German botanist; some derive it as Olive-like, for the similarity of the leaves of some species
olearius -a -um of the olive, *Olea* (this phosphorescent fungus grows also on oak and chestnut in Britain)
Oleaster, oleaster Wild-olive, *Olea-astrum* (≡ *Eleagnus*, Theophrastus used the name for a willow)
olei- olive-, *Olea-*
oleifer -era -erum oil-bearing, (*oleum, olei*)*-fero*
oleifolius -a -um with olive-like leaves, *Olea-folium*
-olens, -olentis fragrant, musty, stinking, smelling, present participle of *oleo, olere, olui*
-olentus -a -um -fullness of, -abundance, -ulent
oleoides olive- or oil-like, ελαια-οειδης, *Olea-oides*
oleospermus -a -um oil-seeded, ελαια-σπερμα
oleosus -a -um greasy, oily, *oleum, olei*
oleraceus -a -um of cultivation, aromatic, esculent, vegetable, (*h*)*olus*, (*h*)*oleris*
olgae for Olga Fedtschenko (1845–1921)
olibanum from the Arabic, al luban, for the resinous secretion, λιβανος or frankincense, of *Boswellia*
olidus -a -um stinking, smelling, rank, *olidus*
olig-, oligo- feeble-, small-, ολιγος, few-, ολιγοι, ολιγο-, ολιγ-
oligandrus -a -um having few stamens, ολιγ-ανδρος
oliganthus -a -um with small or few flowers, ολιγ-ανθος
oligocarpus -a -um with few fruits, ολιγο-καρπος
oligococcus -a -um with few berries, ολιγο-κοκκος
Oligocodon Small-bell, ολιγο-κωδων

oligodon few-toothed, feebly toothed, ολιγ-οδων
oligomerus -a -um having few parts, ολιγο-μερος
oligophlebius -a -um with few veins, indistinctly veined, ολιγο-φλεψ
oligophyllus -a -um with few leaves, ολιγο-φυλλον
oligorrhizus -a -um having few or underdeveloped roots, ολιγο-ριζα
oligospermus -a -um with few seeds, ολιγο-σπερμα
oligostachyus -a -um with few stems, ολιγο-σταχυς
oligostromus -a -um with few swellings, ολιγο-στρωμα
olisiponensis -is -e from Lisbon (*Olisipo*), Portugal
olitorius -a -um of gardens or the gardener, (*h*)*olitorius*, salad vegetable, culinary
olivaceomarginatus -a -um with dull olive or greenish-brown margins, *olivaceus*-(*margo, marginis*)
olivaceus -a -um olive-coloured, greenish-brown, *oliva, olivaceus*
oliveus -a -um having the colour of a ripe olive, *oliva*
olivieri, oliverianaus -a -um for Antoine Olivier (1756–1814), French collector on Mount Elwend
oliviformis -is -e olive-shaped, *oliva-forma*
ollarius -a -um resembling a pot or jar, *olla, ollae*
olorinus -a -um swan's, of swans, *olorinus*
oltensis -is -e from the Oltenia region of SW Romania
olusatrum Pliny's name for a black-seeded pot-herb, (*h*)*olus-ater*
olympicus -a -um from Mount Olympus, Greece, Olympian
Olyra Spelt, ολυρα (an old Greek name)
omalo-, omal- smooth, ομαλος, ομαλο- (see *homalo*)
omalosanthus -a -um uniformly or evenly flowered, ομαλος-ανθος
ombriosus -a -um of wet or rainy locations, ομβριος
ombro- rain-, storm-, moisture, ομβρος; shade-, *umbra, umbrae*
omeiensis -is -e either from Mount Omei, Omei Shan, China (Szechwan), or from Ome, Honshu, Japan
omeiocalamus *Calamus* of Mount Omei, China (Sichuan)
omiophyllus -a -um lacking reduced (submerged) leaves, ο-μειο-φυλλον
omissus -a -um overlooked, *omitto, omittere, omisi, omissum*
omni- totally-, entirely-, all-, *omnis, omni-*
omniglabrus -a -um entirely smooth, *omni-glaber*
omnivorus -a -um devouring all, *omni-*(*voro, vorare, voravi, voratum*)
-omoeus -a -um resembling, like, equal to, a match for, ομοιος
omorika from the Serbian name for *Picea omorika*
-omorius -a -um -resembling, -similar to, -bordering upon, ομορος, ομοριος
Omphalina (*Omphalia*) Little-navel, ομφαλος (the depressed centre of the mature cap)
omphalo- navel-, ομφαλος, ομφαλο-
Omphalobium Navel-pod, ομφαλος-λοβος (the pod of zebra-wood)
Omphalocarpum Navel-fruited-one, ομφαλος-καρπος
Omphalodes, omphalodes Navel-like, ομφαλος-ωδης (the fruit shape of navelwort)
Omphalogramma Navel-lines, ομφαλος-γραμμα (the seed testa)
omphalosporus -a -um with navelled seeds, ομφαλος-σπορα
Omphalotus Navel-looking, ομφαλοτικος (depressed cap)
-on -clan, -family
Onagra a former generic name, οναγρος a wild ass (ονος-αγριος), ≡ *Oenothera* (***Onagraceae***)
onc-, onco- tumour-, hook-, ογκος, ογκο-
oncidioides resembling *Oncidium, Oncidium-oides*
Oncidium Tumour, ογκος (the warted crest of the lip)
Oncinotis Hook-eared, ογκος-ωτος (scales alternating with the corolla lobes)
onciocarpus -a -um grapple-fruited, ογκος-καρπος box-fruited, ογχιον-καρπος
Oncoba from the Arabic vernacular name, onkob, for *Oncoba spinosa*
Oncocalamus Hooked-*Calamus*, ογκο-καλαμος

oncogynis -is -e with a warted ovary, ογκο-γυνη
oncophyllus -a -um having leaves with a hooked apex, ογκο-φυλλον
-one, -onis -daughter of, ωνη or -son of, ωνος
onegensis -is -e from Onega, Russia
onites of asses, ονος-ιτης (used by Dioscorides of an ass or donkey)
Onobrychis Ass-bray, ονο-βρυχαομα (a name used by Dioscorides and Galen – in Pliny, *palmes-asini* – for a legume eaten greedily by asses, ονος-βρυκω)
onobrychoides resembling *Onobrychis*, ονο-βρυχω-οειδης
Onoclea Enclosed-cup, ονος-κλειω, Dioscorides' name, ονοκλειον, for the sori being concealed by the rolled frond margins (cognate with αγχουσα, anchusa)
onomatologia the rules to be followed in forming names, ονοματο-λογος
Ononis the classical name, ονωνις, ανωνις, used by Dioscorides
Onopordum (-on) Ass-fart, ονος-πορδον (its flatulent effect on donkeys)
onopteris ass-fern, ονο-πτερυξ from a name used by Tabernaemontanus
Onosma Ass-smell, ονος-οσμη (said to attract asses)
onoticus -a -um ass-eared, ονος-ωτος (fruiting stage of hare's ear fungus)
ontariensis-is -e from Ontario, Canada
onustus -a -um burdened, full (of flower or fruit), *onustus*
onychimus -a -um onyx-like, with layers of colour, *onix, onichis*
Onychium Claw, ονυξ, ονυχος (the shape of the frond pinnules)
-onychius -a -um -clawed, ονυξ, ονυχος
oo- egg-shaped-, ωον, ωο-
oocarpus -a -um having egg-shaped fruit, ωον-καρπος
oophorus -a -um bearing egg(-shaped structures), ωον-φερω
oophyllus -a -um having egg-shaped leaves, ωον-φυλλον
oothelis -is -e having an ovoid ovary; having nippled ovoid tubercles, ωον θηλις
opacus -a -um darkened, dull, shady, not glossy or transparent, *opacus*
opalinus -a -um translucent, *opalus* (Sanskrit, upala, for a milky gemstone)
opalus from the old Latin name, *opulus*, for maple
Opercularia Lidded-one, *operculum* (the calyx)
operculatus -a -um lidded, with a lid, *operculum*
opertus -a -um hidden, concealed, *operio, operire, operui, opertum*
Ophelia Useful, οφελος (medicinal uses, ≡ *Swertia*)
ophianthus -a -um long-flowered, with serpentine flowers, οφις-ανθος
ophio- snake-like, snake-, οφις, οφιο-
Ophiobotrys Serpentine-raceme, οφιο-βοτρυς (the slender branches of the inflorescence)
ophiocarpus -a -um with an elongate fruit, snake-like-fruited, οφιο-καρπος
ophiocephalus -a -um snake-headed, having a head of elongate structures, οφιο-κεφαλη
ophioglossifolius -a -um snake's-tongue-leaved, *Ophioglossum-folium*
ophioglossoides resembling *Ophioglossum*
Ophioglossum Snake-tongue, οφιο-γλωσσα (appearance of fertile part of frond of adder's tongue fern) (***Ophioglossaceae***)
ophioides, ophiodes serpentine, οφιο-οειδης
ophiolithicus -a -um of serpentine rock habitats, οφιο-λιθος
Ophiopogon Snake-beard, οφιο-πωγων
Ophioscordon Snake-garlic, οφις-σκοροδον (the irregular stem)
ophites marbled; snake-like, serpentine, οφις (applies to serpentine rocks because of their structural patterning)
ophiuroides snake's-tails-like, resembling *Ophiurus, Ophiurus-oides*
Ophiurus Snake's-tails, οφις-ουρα (centipede grass)
ophrydeus -a -um, ophrydis -is -e similar to *Ophrys*
Ophrys Eyebrow, οφρυς (a name in Pliny)
ophthalmica *vide ocularia*
-ophthalmus -a -um -eyed, -eye-like, οφθαλμος
opiparus -a -um sumptuous, rich, *opiparus*

-opis -looking, -οπις
opistho- back-, behind-, οπισθε, οπισθιος, οπισθο-
Opithandra Backward-stamens, οπισθε-ανηρ
Oplismenus Armoured, οπλισμενος (armoured with awns)
oplitis -is -e heavily armed, οπλιτης
oplocarpus -a -um with armoured fruit, οπλο-καρπος
Oplopanax Armed-*Panax*, οπλο-παν-αχος (≡ *Echinopanax*)
opo- juice-, sap-, οπος (feeding, of parasites) (cognate with οπιον, diminutive of οπος opium)
opobalsamum balsamic-juiced, resin-balsam, οπος-βαλσαμον (Balm of Gilead)
Opopanax Panacea, sap-that-is-all-healing-, οπος-παναξ; Chiron told Hercules of the virtues of *Opopanax chironium* (Dioscorides listed a wide range of conditions to be treated with it)
Oporanthus Autumn-flower, οπορα-ανθος
oporinus -a -um autumnal, of late summer, οπωρινος
oppositi-, oppositus -a -um opposite-, opposed-, *oppono, opponere, opposui, oppositum*
oppositiflorus -a -um having paired or opposite flowers, *oppositi-florum*
oppositifolius -a -um opposite-leaved, *oppositi-folium*
oppositipinnus -a -um having opposed pinnae, *oppositi-(pinna, pinnae)*
-ops, opseo- -eyed, ωψ, ωπος
-opsis -like, -looking like, -appearance of, οψις, countenance, ωψ, ωπος
optatus -a -um desired, longed for, *opto, optare, optavi, optatum*
opticus -a -um *vide ocularia*
optimus -a -um the best, superlative of *bonus*
optivus -a -um chosen, *optivus*
opulentus -a -um sumptuous, splendid, enriched, *opulentus*
opuli- guelder-rose-like, *Opulus*
opulifolius -a -um *Opulus*-leaved, with leaves resembling the guelder rose, *Opulus-folium*
opuloides resembling the guelder rose, *Opulus-oides*
Opulus, opulus -a -um an old generic name for the guelder rose, *Viburnum*, initially thought to have been for some kind of *Acer*
Opuntia Tournefort's name for succulent plants from Opous, Boeotia
opuntiiflorus -a -um (opuntiaeflorus -a -um) *Opuntia*-flowered
opuntioides resembling *Opuntia, Opuntia-oides*
-opus -foot, πους, ποδος, eyed, looking, ωπος
orarius -a -um of the shoreline, *ora*
orbi- circular-, disc-like, *orbis, orbis, orbi-*
orbicularis -is -e, orbiculatus -a -um disc-shaped, circular in outline, orbicular, *orbis, orbis*
orbifolius -a -um with orbicular leaves, *orbi-folium*
orbus -a -um orphaned, childless, destitute, *orbus*; circular *orbis*
orcadensis -is -e from the Orkney Isles, Scotland, Orcadian (*Orcades*)
orchioides resembling *Orchis, Orchis-oides*
orchidastrus -a -um somewhat like an orchid, *Orchis-astrum*
orchidiflorus -a -um orchid-flowered, *Orchis-florum*
orchidiformis -is -e looking like an orchid, *Orchis-forma*
orchidis -is -e *Orchis*-like
orchioides resembling *Orchis*, ορχις-οειδης
Orchis Testicle, ορχις (the shape of the root-tubers) (***Orchidaceae***)
orculae- small barrel-, small cask-, diminutive of *orca, orcae*
orculiflorus -a -um having flowers shaped like small barrels, *orcula-florum*
ordensis -is -e from Ordenes, France
ordinatus -a -um neat, orderly, *ordino, ordinare, ordinavi, ordinatum*
Oreacanthus Mountain-*Acanthus*, ορειος-ακανθος
oreades, oreadis -is -e, oreadus -a -um montane, of the sun, heliophytic, ορειας (the Oreads were mythical mountain nymphs, ορειας-αδος)

oreadoides resembling (*Marismus*) *oreades* (*Marismus oreadoides* ≡ *Collybia oreadoides*)
orectopus -a -um with an elongated or stretched-out stalk, ορεκτος-πους
oreganus -a -um, oregonensis -is -e, oregonus -a -um from Oregon, USA
orellanus -a -um from a pre-Linnaean name for annatto, the red dye from *Bixa*
oreo-, ores-, ori- mountain-, ορος, ορεος, ορεο-
Oreobatus Mountain-thorn-bush, ορεο-βατος, or Mountain-ranging, ορειβατης (≡ *Rubus deliciosus*)
oreocharis -is -e mountain-joy, beauty of the mountain, ορεο-χαρις
oreocreticus -a -um from the mountains of Crete, ορεο-κρετικος
Oreodoxa, oreodoxus -a -um Mountain-glory, ορεο-δοξα
oreogenes, oreogenus -a -um born of the mountains, ορεο-γενος
oreogeton neighbour of mountains, ορεο-γειτων
oreonastes occupying, pressed-, or clinging-to the mountain, ορεο-νασσειν
Oreopanax Mountain-*Panax*, ορεο-παν-ακεσις
oreopedionis -is -e of mountain plateaux, ορεο-(πεδιας, πεδιαδος)
oreophilus -a -um mountain loving, ορεο-φιλος (habitat)
Oreopteris mountain-fern, ορεος-πτερις
Oreorchis Mountain-orchid, ορεο-ορχις
oreothaumus -a -um marvel of the mountains, ορεο-θαυμα
oreotrephes (oreostrephes) nurtured, or living, on mountains, ορεο-τρεφω
oresbius -a -um living on mountains, ορεινος, ορεο-βιος
oresigenus -a -um born of the mountains, ορεσι-γενος
oreus -a -um of mountains, montain, ορεος
organensis -is -e from any of the Organ mountains in New Mexico, USA, or Brazil
orgyalis -is -e a fathom in length, about 6 feet tall, οργυια (the distance from finger-tip to finger-tip with arms outstretched)
orientaletibeticus -a -um from E Tibet, botanical Latin from *oriens* and Tibet
orientalis -is -e eastern, oriental, of the East, *oriens, orientis*
origanifolius -a -um with leaves resembling *Origanum, Origanum-folium*
Origanum Joy-of-the-mountain, ορος-γανυμαι (Theophrastus' name, οριγανον, for an aromatic herb)
oritrephus -a -um nurtured on mountains, or nurtured in the east, ορε-τρεφω (from N China)
-orius -a -um -able, -capable of, -functioning
Orixa from the Japanese name for *Orixa japonica*
orizabensis -is -e from the volcanic Citlaltépeti (Pico de Orizaba), E central Mexico
ormenis -is -e etymology uncertain; sprouting, ορμενος ορμος, a necklace
ormo- necklace-like-, necklace-, ορμος
Ormosia Necklace, ορμος
ornans resembling manna-ash, *ornus*, embellishing, present participle of *orno, ornare, ornavi, ornatum* (with a manna-like exudate)
ornatipes having a decorative stalk, *ornatus-pes*
ornatissimus -a -um most ornate, showiest, superlative of *ornatus*
ornatulus -a -um somewhat showy, diminutive of *ornatus*
ornatus -a -um adorned, showy, *ornatus*
ornithanthus -a -um bird-flowered; flower of birds, ορνιθ-ανθος (appearance or food)
ornitheurus -a -um resembling a bird's tail, ορνις-, ουρα
ornitho-, ornith bird-like-, bird-, ορνις, ορνιθος, ορνιθο, ορνιθ-
ornithobromus -a -um bird-food, ορνιθο-βρωμα
ornithocephalus -a -um bird-headed, ορνιθο-κεφαλν (appearance of flowers)
Ornithogalum Bird-milk, ορνιθο-γαλα (Dioscorides' name, ορνιθογαλον, for a plant yielding bird-lime)
ornithommus -a -um bird-like, bird's eye, ορνιθο-ομμα
ornithopodioides, ornithopodus -a -um bird-footed, like a bird's foot, ορνιθοπους-οειδης (the arrangement of the fruits or inflorescence)

Ornithopus Bird-foot, ορνιθο-πους (the disposition of the fruits)
ornithorhynchus -a -um like a bird's beak, ορνιθο-ρυγχος
Ornus, ornus from the ancient Latin, *ornus*, for manna-ash, *Fraxinus ornus*
Orobanche Legume-strangler, οροβος αγχω (one species parasitizes legumes – see also *rapum-genistae*) (***Orobanchaceae***)
oroboides similar to *Orobus, Orobus-oides*
Orobus, orobus an old generic name, οροβος, for *Vicia ervilia*, ? ορα-βους
Orontium from the Orontes river, Syria, an old generic name, οροντιον, in Galen for an aquatic plant, golden club
Oropetium Mountain-retiring, ορο-πτηνος, ορο-πετεηνος (in small soil-pockets on rocks and ironstone outcrops)
orophilus -a -um mountain-loving, montain, ορο-φιλος
orospendanus -a -um hanging from mountains, *oro-*(*pendeo, pendere, pependi pensum*)
Orostachys Mountain-*Stachys*, ορο-σταχυς
oroyensis -is -e from Oroya (La Oroya), central Peru
Orphanidesia for Theodoros Geogios Orphanides (1817–86), Professor of Botany at Athens
orphanidis -is -e from the Balkan area once the land of the Orpheans (*Tulipa or Campanula orphanidea*)
orphanidius -a -um fatherless, unrelated, ορφανος
orphanis -is -e destitute, bereft, orphan, ορφανος
Orphium for Orpheus, the Greek poet–musician
orphnophilus -a -um liking shade, ορφναιος-φιλος
ortgeisii for Eduard Ortgeis, (1829–1916), of Zurich Botanic Garden (*Oxalis ortgeisii* tree oxalis)
orth-, ortho- correct-, straight-, erect-, upright-, ορθος, ορθο-
Orthilia Straight, ορθος (the style, but the etymology is uncertain)
Orthocarpus, orthocarpus -a -um Upright-fruit, ορθος-καρπος
orthocladus -a -um with straight branches, ορθος-κλαδος
Orthodontium Straight-toothed, ορθος-οδοντος
orthoglossus -a -um with a straight lip, ορθος-γλωσσα
ortholobus -a -um with straight lobes, ορθος-λοβος (cotyledons)
orthoplectron with straight plectrum (spur), ορθος-πλεχτρον (literally a punting pole)
Orthosanthus Erect-flower, ορθος-ανθος
orthosepalus -a -um with straight sepals, ορθος-σκεπη
Orthosiphon Straight-tubed-one, ορθος-σιφον
orthostates correctly placed, fixed, standing erect, ορθος-στατος
orthotrichus -a -um having standing or erect hairs, ορθος-τριχος
Orthrosanthus Daybreak-flower, ορθος-ανθος (time of anthesis)
ortubae from the region of Lake Maggiore, Italy
orubicus -a -um from Oruba Island, Caribbean
orvala origin obscure, possibly from Greek for a sage, ορμιν-like plant
Oryza from the Arabic name, eruz
oryzicolus -a -um living on or around rice, *Oryza-colo*
oryzifolius -a -um having leaves resembling those of *Oryza, Oryza-folium*
oryzoides resembling *Oryza*, rice like, *Oryza-oides*
Oryzopsis *Oryza*-resembler, *Oryza-opsis*
Osbeckia for Reverend Peter Osbeck (1723–1805), Swedish naturalist
osbeckiifolius -a -um having leaves resembling those of *Osbeckia*
Oscillatoria Oscillator, to swing, *oscillo, oscillare* (the slow motion exhibited by the alga)
oscillatorius -a -um able to move about a central attachment, versatile, *oscillo, oscillare*
oscitans drowsy, listless; yawning, gaping, present participle of *oscito, oscitare; oscitor, oscitari*
Oscularia, oscularis -is -e Kissing, *osculo, osculari, osculatus* (the touching leaf-margins)

oshimensis -is -e from Oshima island, Japan
-osis denotes a condition, e.g. *gummosis*, producing much gum
osmanicus -a -um for the Osman dynasty, who founded the Ottoman Empire (Osman Gazi (1258–1324), Genç Osman (1603–1622), to Osman Nuri Pasa (1832–1900))
Osmanthus, osmanthus -a -um Fragrant-flower, οσμη-ανθος (for the perfumed *Osmanthus fragrans*)
Osmarea the composite name for hybrids between *Osmanthus* and *Phillyrea*
Osmaronia Fragrant-*Aronia*, (the derivation is doubtful)
-osmius -a -um, -osmus -a -um -scented, fragrant-, οδμη, οσμη
osmo- thrust-, pressure-, ωσμος, ωσμο-
osmophloeus -a -um having fragrant bark, οσμη-φλοιος
osmophorus -a -um scent carrying, οσμη-φορα (the strong fragrance)
Osmorhiza Fragrant-root, οσμη-ριζα
Osmunda an old English name, in Lyte, either for Osmund the waterman (because of its boggy habitat), or for the Anglo-Saxon god of thunder, Osmund, equivalent of the Norse Thor (***Osmundaceae***)
osmundioides resembling *Osmunda, Osmunda-oides*
osproleon an ancient name, οσπρολεων, for a legume-damaging plant, οσπριον-λεων bean lion
Ossaea, ossea for Don José Antonio de la Ossa (d. 1831), Director of Havana Botanic Garden
osseticus -a -um from Ossetia (Osetiya), SW Russia and Georgia
osseus -a -um of very hard texture, bony, *os, ossis*
ossifragus -a -um of broken bones, *ossis-(frango, frangere, fregi, fractum)* (said to cause fractures in cattle when abundant in lime-free pastures)
osteo- bone-like-, bone-, οστεον
Osteomeles Bone-apple, οστεο-μελεα (the hard fruit)
Osteospermum, osteospermus -a -um Bone-seed, οστεο-σπερμα (the hard-coated fruits)
ostiolatus -a -um having a small opening, *ostiolus*, diminutive of *ostium, ostii*
ostraco- hard-shelled-, οστρακον (a potsherd)
ostreatus -a -um resembling an oyster, *ostrea* (the shape and blue-grey colour of the oyster mushroom)
Ostrowskia, ostrowskianus -a -um for Michael Nicolajewitsch von Ostrowsky, Minister of the Russian Imperial Domains and patron of botany *c*. 1880
ostruthiopsis ostrich-like, στρυθιον-οψις, (the foliage of young plants)
Ostruthium, ostruthius -a -um etymology uncertain; purplish, *ostrum*, ostrich-like, στρυθιον
Ostrya Hard-scale, οστρυς (a name in Pliny for a tree with hard wood)
Ostryopsis similar to *Ostrya*, οστρυς-οψις
osumiensis -is -e from the Osumi peninsula or archipelago, Kyushu, Japan
-osus -a -um -abundant, -large, -very much, -being-conspicuous
-osyne -notably
Osyris Dioscorides' name, Οσυρις (οζος much branched)
ot-, oto- ear-like-, ear-, ους, ωτος, ωτο-
Otacanthus Thorny-ears, (ους, οτος)-ακανθος
Otanthus Ear-flower, ωτ-ανθος (the two-spurred shape of the corolla)
otaviensis -is -e from Otava, Czech Republic, or the Otavi mountains of Namibia
-otes -appearing, -looking, ωτης
Othonna Linen, οθονη (its covering of downy hairs)
Othonnopsis resembling *Othonna*, οθονη-οψις
-otis -is -e -eared, ους, ωτιον, ωτος, ωτ-
otites, -otites relating to ears, ωτ-ιτης, Rupius' generic name referring to the shape of the lower leaves of *Silene otites*
oto- ear-, ους, ωτος, ωτο-
otocarpus -a -um with ear-like or shell-like fruits, ωτο-καρπος

otolepis -is -e with ear-like scales, with shell-like scales, ωτο-λεπις
otophorus -a -um bearing ears, ωτο-φορος (-shaped structures)
otrubae for Josef Otruba (1889–1953) of Moravia, Czech Republic
ottawaensis -is -e, ottawensis -is -e from Ottawa, Canada
Ottelia reputedly from the Malabar vernacular, ottel
ottophyllus -a -um having ear-shaped leaves, ωτο-φυλλον
-otus -a -um -looking-like, -resembling, -having
Oubanguia from the name of the river Oubangui, Nigeria
ouletrichus -a -um, oulotrichus -a -um with curly hair, ουλο-τριχος
Ouratea from the S American vernacular name
Ourisia for Governor Ouris of the Falkland Islands (some derive it as ουρις, a breeze, for its montane habitat)
ova-avis bird's-egg-like, (*ovum, ovi*)-*avis*
ovali-, ovalis -is -e egg-shaped (in outline), oval, *ovalis*
ovalifoliolatus -a -um having oval leaflets, diminutive of *ovalifolium*
ovalifolius -a -um with oval leaves, *ovali-folium*
ovati-, ovatus -a -um egg-shaped, *ovatus* (in the solid or in outline) with the broad end lowermost
ovatifolius -a -um with ovate leaf-blades, *ovatus-folium*
ovifer -era -erum, oviger -era -erum bearing egg-like structures, *ovi-fero*
oviformis -s -e egg-shaped, *ovi-forma* (in the solid), ovoid
ovinus -a -um of sheep, *ovis* (*Festuca ovina* is sheep's fescue)
ovoideus -a -um egg-shaped, ovoid, *ovoideus*
ovularis -is -e having little eggs (buds), *ovulum*, diminutive of *ovum*
Oxalis Acid-salt, οξυς-αλς (the name, οξαλις, in Nicander refers to the taste of sorrel) (***Oxalidaceae***)
oxicus -a -um sharp, οξυς (leaf apex or marginal teeth)
oxodus -a -um of sour or acid humic soils, οξωδης
oxonianus -a -um, oxoniensis -is -e from Oxford, England (*Oxonia*)
oxy-, -oxys acid-, -pungent, sharp-, -pointed, -fiery, bold-, οξυς, οξυ-
Oxyacantha Sharp-thorn, οξυ-ακανθος (Theophrastus' name)
oxyacanthoides resembling *oxyacanthus -a -um* of the same genus, οξυ-ακανθα-οειδης
oxyacanthus -a -um having sharp thorns or prickles, οξυ-ακανθα
Oxyanthus Sharp-flower, οξυ-ανθος (the acute calyx lobes)
Oxybaphus Acid-dye, οξυ-βαφη (≡ *Mirabilis viscosa*)
oxycarpus -a -um having a sharp-pointed fruit, οξυ-καρπος
oxycedri of pungent juniper, οξυ-κεδρος (semi-parasitic on *Juniperus*)
Oxycedrus, oxycedrus Pungent-juniper, οξυ-κεδρος
oxyceras sharp-horned, οξυ-κερας
Oxycoccus, oxycoccus -a -um Acid-berry, οξυ-κοκκος, having sharp, acid or bitter berries, οξυ-κοκκος
oxydabilis -is -e oxidizable, modern Latin from *oxygene* (for the ochraceous colour)
Oxydendrum (on) Sour-tree, οξυ-δενδρον (the acid taste of sourwood leaves)
oxygonus -a -um with sharp angles, sharp-angled, οξυ-γωνια
Oxygyne Pointed-style, οξυ-γυνη
Oxylobium, oxylobus -a -um with sharp-pointed pods or lobes, οξυ-λοβος
oxylophilus -a -um of acid or humus-rich soils, humus-loving, οξυ-φιλος
oxyodon, oxyodontus -a -um sharp-toothed, with sharp teeth, οξυ-οδους, οξυ-οδοντος
Oxypetalum, oxypetalus -a -um Sharp-petalled, οξυ-πεταλον
oxyphilus -a -um of acidic soils, acid soil-loving, οξυ-φιλος
oxyphyllus -a -um with sharp pointed leaves, οξυ-φυλλον
oxypterus -a -um with sharp wings, οξυς-πτερον
Oxyrachis Sharp-rachis, οξυ-ραχις (the pointed internodes of the disarticulated rachis)
Oxyramphis Sharp-beak, οξυ-ραμφος (the fruit)
Oxyria Acidic, οξυς (the taste)
oxysepalus -a -um with sharp sepals, οξυ-σκεπη

Oxyspora Sharp-seed, οξυ-σπορα (the seeds being awned at both ends)
Oxystelma Sharp-crown, οξυ-στελμα (the acute corolla lobes)
Oxytenanthera Sharp-narrow-flower, οξυ-τεν-ανθερα (the spikelet shape)
Oxytropis Sharp-keel, οξυ-τροπις (the pointed keel petal)
ozarkensis -is -e from the Ozarka Mountains of S central USA
Ozothamnus Fragrant-shrub, οζω-θαμνος

pabularis -is -e, pabularius -a -um of forage or pastures, *pabulum, pabuli*
Pachira (Pachyra) from the Guyanese vernacular name
pachy- stout-, thick-, παχυς, παχυ-
Pachycarpus, pachycarpus -a -um Thick-fruited, παχυ-καρπος (follicle shape, some being inflated)
pachycaulon thick-stemmed, παχυ-καυλος
pachycladus -a -um, pachyclados thick-branched, παχυ-κλαδος
pachygaster large-bellied, παχυ-(γαστηρ, γαστρος)
Pachylaena Thick-cloak, παχυ-(χ)λαινα (the indumentum)
pachypes having a thick stalk, botanical Latin from παχυς and *pes*
pachyphloeus -a -um thick-barked, παχυ-φλοιος
Pachyphragma Stout-partition, παχυ-φραγμα (the ribbed septum of the fruit)
pachyphyllus -a -um having thick leaves, παχυ-φυλλον
Pachyphytum Sturdy-plant, παχυς-φυτον (thick stems and leaves)
Pachypodanthium Thick-footed-flowers, παχυς-ποδ-ανθεμιον (the crowded stalkless carpels)
Pachypodium, pachypodius -a -um Stout-foot, παχυς-ποδεων (the fleshy roots)
pachypus thick-stem, παχυ-πους
pachyrhizus -a -um, pachyrrhizus -a -um having thick roots, παχυ-ριζα
Pachysandra Thick-stamens, παχυς-ανηρ (the filaments)
pachysanthus -a -um having thick flowers, παχυς-ανθος (somewhat fleshy corolla)
pachyscapus -a -um with a thick scape or peduncle, botanical Latin from παχυ and *scapus*
pachyspermus -a -um with thick (almost spherical) spores, παχυς-σπερμα
Pachystachys, pachystachis -is -e with thick spikes, παχυ-σταχυς (inflorescences)
Pachystela Thick-style, παχυς-στηλη
Pachystema, Pachistima Thick-crown, παχυς-στεμμα
Pachystigma Thick-stigma, παχυς-στιγμα see *Paxistima*
pachytrichus -a -um thickly haired (moss like indumentum of *Rhododendron pachytrichum*)
pachyurus -a -um having a thick tail, παχυ-ουρα
pacificus -a -um of the W American coast, *pacificus* (literally peacemaking)
padi- *Prunus-padus*-like
padifolius -a -um, padophyllus -a -um with leaves resembling those of *Prunus padus*
padus Theophrastus' name, παδος, for St Lucie Cherry or from the River Po (*Padus*), Italy
Paederia Malodorous, *paedor* (the crushed flowers)
Paeonia named παιονια by Theophrastus for Paeon, the physician to the gods who, in mythology, was changed into a flower by Pluto (Pliny gave *Peony* the same attribution) (***Paeoniaceae***)
Paepalanthus Rock-flowered, παιπαλ-ανθος (the hard calyx at fruiting time)
Paesia for Fernando Dias Paes Leme, Portuguese administrator of Minas Gerais *c.* 1660
paganus -a -um from the wild, of country areas, *pagus*
pageanus -a -um for Miss Mary Page (1867–1925), botanical illustrator, Bolus Herbarium, Cape Town 1914–24
pago- foothill-, παγος
pagoda with the habit of a pagoda, from Persian, butkada
pagodaefolius -a -um with leaves having a pagoda-shaped outline, *pagoda-folium*

pagophilus -a -um hill-loving, παγο-φιλος
pahudii for Charles F. Pahud (1803–73), Governor General of Dutch E Indies
palachilus -a -um with a spade-like lip or edge, παλα-χειλος (with edges adherent to supports)
palaeo- (paleo-) ancient-, παλαιος (Palaeolithic, ancient stones, παλαιος-λιθος)
palaestinus -a -um from Palestine, Palestinian
Palaquium from the Philippino vernacular name, palak-palak, for the gutta-percha tree, *Palaquium gutta*
paleaceus -a -um covered with chaffy scales, chaffy, *palea, paleae*
palibinianus -a -um for Ivan Vladimirovich Palibin (1872–1949), Director of Leningrad Botanic Museum
palilabris -is -e with a spade-shaped lip, *pala-labrum*
palinuri from Palinuro, Italy
Palisota for Ambroise Marie François Joseph Palisot de Beauvois (1752–1820), French botanist and plant collector
Paliurus, paliurus Dioscorides' ancient Greek name for Christ-thorn
pallasianus -a -um, pallasii for Peter Simon Pallas (1741–1811), German naturalist and explorer
pallens pale, greenish, present participle of *palleo, pallere, pallui*
pallescens (palescens) becoming pale, fading, present participle of *pallesco, pallescere, pallescui*
palliatus -a -um, palliolatus -a -um hooded, *palliatus*, as if wearing a Greek cloak, *pallium, pallii*
pallidiflorus -a -um with pale green flowers, *pallidus-florum*
pallidior more pale, comparative of *pallidus*
pallidissimus -a -um the palest, superlative of *pallidus*
pallidus -a -um greenish, somewhat pale, *pallidus*
palma-christi a medieval Latin name for the shape of the leaf of castor oil plant, *Ricinus communis*
palmaris -is -e of a hand's breadth, about three inches wide, *palma, palmae*; excellent, *palmaris* (*palmarius*, prize-winning)
palmati-, palmatus -a -um with five or more veins arising from one point (usually on divided leaves), hand-shaped, palmate, *palma, palmae* (see Fig. 5a)
palmatifidus -a -um with hand-like division, *palmati-(findo, findere, fidi, fissum)*
palmatilobus -a -um palmately lobed, *palmati-lobus* (*vide* Fig. 5a)
palmensis -is -e from Las Palmas, Canary Isles
palmeri for Dr Edward Palmer (1831–1911), who explored for plants in Mexico
palmetto W Indian vernacular name for *Sabal palmetto*
palmi- date-palm-, palm-of-the-hand-, *palma, palmae*
palmicolus -a -um living on or with palms, *palma-colo*
palmifolius -a -um with palm-like leaves, *palma-folium*
palmifrons having large, much-divided leaves or leafy branches, *palma-(frons, frondis)*
palmunculus small palm, diminutive of *palma*
palpebrae eyelashed, with fringe of hairs, *palpebra* an eyelid
paludaffinis -is -e related to swamps, *palus-affinis*
paludicolus -a -um dwelling in swamps, *paludis-colo*
paludis -is -e of swamps, *palus, paludis*
paludosus -a -um growing in boggy or marshy ground, wetter than *palus, paludis*
palumbinus -a -um lead-coloured (the colour of woodpigeons)
paluster -tris -tre of swampy ground, *palus, paludis* (*palustris* is often used as a masculine ending in botanical names)
pam- entirely, quite, παμ, παν
pamiricus -a -um of the Pamir mountain range, Tajikistan and surrounding area
pamirolaicus -a -um from the N Alay range of the Pamir mountains, central Asia
pampanus -a -um from the pampas, grass plains of S America
pamphylicus -a -um from Murtana (*Pamphylia*), Turkey

pampini- tendrillar-, tendril-, vine-shoot-like-, *pampinus, pampini*
pampinosus -a -um leafy, with many tendrils, vine-leaved, *pampinus, pampini*
pamplonensis -is -e from Pamplona, Colombia
pan-, panto- all-, πας, πασα, παν, παντως, παντο-, παντ-
panaci- *Panax-*
panaciformis -is -e resembling *Panax, Panax-forma*
Panaeolina diminutive from *Panaeolus*
Panaeolus Variegated, παν-αιολος (mottled appearance of irregularly ripening spores)
panamensis -is-e from Panama, Central America
panamintensis -is -e from the Panamint mountain range, E California, USA
Panax Total-remedy, Theophrastus' name, πανακης, παν-(ακεσις, ακεως) (the ancient virtues of ginseng)
panayensis -is -e from Panay island, Philippines
pancicii for Joseph Pančic (1814–88), Croatian botanist
Pancratium All-potent, παν-κρατος (a name used by Dioscorides)
pandani- similar to *Pandanus*
pandanifolius -a -um with leaves arranged like those of *Pandanus*
Pandanus Malayan name, pandang, for screw-pines (***Pandanaceae***)
pandorana Pandora's (surprising, objects of desire, the changing form of the Wonga Wonga vine, *Pandorea pandorana*)
Pandorea Pandora, *vide supra* (Wonga Wonga vine)
Pandorina Resembling *Pandorea*
pandoensis -is -e from Pando, the northernmost department of Bolivia
panduratus -a -um fiddle-shaped, pandurate, panduriform, *panduratus*
panduriformis -is -e fiddle-shaped, *pandura-forma* (leaves)
pandurilabius -a -um having a fiddle-shaped labellum or lip, *pandura-labium*
panguicensis -is -e from Panguich, Utah, USA
paniceus -a -um like millet grain, *panicum*
panicoides similar to *Panicum, Panicum-oides*
paniculatus -a -um, paniculosus -a -um with a branched-racemose or cymose inflorescence, tufted, paniculate, *panicula* (see Fig. 2c)
paniculiferus -a -um bearing panicles, *panicula-fero*
Panicum the ancient Latin name, *panicum*, for the grass *Setaria italica*
panneformis -is -e with a surface texture like felt or cloth, *pannus*
panneus -a -um felted or cloth-like, *pannus*
pannifolius -a -um cloth-leaved, *pannus-folium*
pannonicus -a -um from SW Hungary (*Pannonia*)
pannosus -a -um woolly, tattered, coarse, ragged, *pannus*
panormitanus -a -um from Palermo, *Panormus* (παν-ορμος, always open harbour), Sicily
panteumorphus -a -um well formed all round, παντ-ευ-μορφος
pantherinus -a -um panther-like, *panthera, pantherae* (the white patches of veil remnant on the ochreous pileus)
pantothrix hairy all round, παντο-θριξ
panuoides rag-like, *pannus-oides* (grey felted surface texture)
Papaver the Latin name, *papaver, papaveris*, for poppies, including the opium poppy (***Papaveraceae***)
papaveris -is -e of poppies, living on *Papaver* (symbionts, parasites and saprophytes)
papaviferus -a -um poppy-bearing, *Papaver-fero* (flower or fruit)
papaya from a Carib vernacular name for pawpaw, *Carica papaya*
Paphinia, paphinius -a -um Paphos, Venus' (Paphos, in Cyprus, was sacred to Venus)
paphio- Venus'-, *Paphos* (*vide supra*)
Paphiopedilum Venus'-slipper (see *Cypripedium*) (Venus' temple was at Paphos, Cyprus)

papil-, papilio-, papilio butterfly-, butterfly-like, *papilio, papilionis* (flowers)
papilionaceus -a -um resembling a pea flower, *Papilionaceae* (resemblance of corolla of most)
papillatus -a -um, papillosus -a -um covered with nipples or minute lobes, papillate, *papilla, papillae*
papillifer -era -erum, papilliger -era -erum producing or bearing papillae, (*papilla, papillae*)*-fero*
pappi-, -pappus pappus-, downy-, down-; -pappus, *pappus*, a woolly seed (botanically used for the variously modified calyx surmounting the pseudo-nuts of composites)
pappophoroides down-carrying-like, παππο-φορα-οειδης (*Schmidtia* is pubescent all over)
papposus -a -um downy, *pappus*
papuanus -a -um from Papua New Guinea
papulentus -a -um with small pimples, diminutive of *papula*
papuli- pimple-, pimpled-, *papula, papulae*
papulosus -a -um pimpled with small soft tubercles, *papula, papulae*
papyraceus -a -um with the texture of paper, papery, *papyrus*
papyrifer -era -erum paper-bearing, *papyrus-fero*
Papyrus Paper, παπυρος, via Syrian, babeer, Greek, βιβλιον, βιβλος, for the paper made from the Egyptian bulrush, for book and letter (is cognate with bible)
para- near-, beside-, wrong, irregular-, παρα
parabolicus -a -um ovate-elliptic, parabolic in outline, late Latin from παραβολικος, placed side-by-side or application
paradisae from the Paradise Nursery
Paradisea for Count Giovani Paradisi (1760–1826) of Modena
paradisi, paradisiacus -a -um of parks, of gardens, of paradise, ecclesiastical Latin from παραδεισος, an enclosed royal park (grapefruits, *Citrus paradisi*, originated in Barbados and have their fruits in bunches)
Paradisianthus Heavenly-flower, παραδεισος-ανθος
paradoxus -a -um strange, unusual, unexpected, παρα-δοξος
paraensis -is -e from the Pará state or river of Brazil
paraguariensis -is -e, paraguayensis -is -e from Paraguay (*Ilex paraguayensis* provides maté, a beverage drunk from a small gourd which, in Quechua, is called a mati)
Parahebe Near-*Hebe*
Parahyparrhenia Near-*Hyparrhenia*
paraibicus -a -um from the environs of the Paraiba river, Brazil
Parajubaea Near-*Jubaea*
paraleucus -a -um almost white, παρα-λευχος
paralias seaside, by the beach, παρ-αλος (ancient Greek name, παραλιος, for a maritime plant)
parallelogrammus -a -um having parallel lines or markings, παρα-αλληλος-γραμμα
parallelus -a -um being equidistant along the length, side by side, παρα-αλλελος
paramutabilis next to (*Hibiscus*) *mutabilis*
Parapentas Near-*Pentas*, παρα-πενταχα (relationship)
Parapholis Irregular-scales, παρα-φολις (the position of the glumes)
paraplesius -a -um about equal to or resembling, παρα-πλησιος (*Salix pentandra*)
Paraquilegia Near-*Aquilegia*
Paraserianthes Near-*Serianthes*
parasiticus -a -um living at another's expense, parasitic, παρα-σιτεω (formerly applied to epiphytes)
Parastranthus Upside-down-flower, παρα-στρεφο-ανθος (floral presentation)
Paratheria Near-wild-beasts, παρα-θηριος (growing in water at drinking sites)
parazureus -a -um almost blue (varying through violet, violet-grey, wine and olive)
parci- with few-, scanty, sparing, frugal, *parcus, parce*
parcibarbatus -a -um having a small beard or few bristles, *parci-*(*barba, barbae*)
parciflorus -a -um few-flowered, *parci-florum*

parcifrondiferus -a -um bearing few or small leafy shoots, with few-leaved fronds, *parci-frondis-fero*
parciovulatus -a -um having few ovules, *parcus-ovulum* (diminutive of *ovum*)
pardalianches leopard-strangling, παρδαλις-αγχω (a name, παρδαλιαγχες, in Aristotle for plants poisonous to wild animals; an undeserved name for *Doronicum pardalianches*, leopard's-bane)
pardalianthes spotted-flowered, παρδαλι-ανθος
pardalinus -a -um, pardinus -a -um spotted or marked like a leopard, παρδαλις, *pardus*
Pardancanda the composite name for hybrids between *Pardanthus* and *Belamcanda*
pardanthinus -a -um resembling *Belamcanda* (*Pardanthus*)
Pardanthopsis resembling *Pardanthus*, παρδος-ανθος-οψις
Pardanthus Leopard-flower, παρδος-ανθος (spotting of the corolla)
parellinus -a -um, parellus -a -um litmus-violet, modern Latin from the French, parelle, for the dye-lichen, *Lecanora parella*)
Parentucellia for Thommaso Parentucelli (1397–1455), Pope Nicholas V, who founded the Vatican Apostolic Library and Botanic Garden
pari-, parilis -is -e uniform-, paired-, equal-, *parilis*
paricymus -a -um having equal or uniform cymes, *parilis-cyma*
parietalis -is -e, parietarius -a -um, parietinus -a -um of walls, parietal (also, the placentas on the wall within the ovary)
Parietaria Wall-dweller (a name, *herba parietaria*, in Pliny used for a plant growing on walls, *paries, parietis*); cognate via old French, peletre, is pellitory
parilicus -a -um of the Roman *Parilia* festival for the goddess *Pales*, equal, *parilis*
Parinarium (*Parinaria*) from a Brazilian vernacular name, parinari
paripinnatus -a -um with an equal number of leaflets and no odd terminal one
Paris Equality, *par, paris* (the regularity of its leaves and floral parts); in mythology, Paris was the son of Priam and declared Venus the most beautiful goddess
parishii for Reverend Charles S. Pollock Parish (1822–97), specialist on Burmese orchids
parisiensis -is -e French, *parisiensis* (continental)
Parkia, parkii for Mungo Park (1771–1806), Scottish explorer whose 1795 Niger expedition failed and who died on his second Niger expedition of 1805
Parkinsonia for either John Parkinson (1567–1629), author of *Paradisi in sole*, or Sydney Parkinson (1745–71), illustrator on Joseph Banks' *Endeavour* trip of 1768–71
parkinsonianus -a -um for John P. Parkinson FLS (*c.* 1772–1847), Consul General in Mexico, orchid collector
Parlatoria, parlatorei, parlatoris -is -e for Filippo Parlatore (1816–77), Professor of Botany at Florence and author of *Flora Italiana*
Parmentiera for Antoine-Augustin Parmentier (1737–1813), French writer on edible plants (*P. cerifera* is the candle-tree)
parmularius -a -um like a small round shield, *parmula*
parmulatus -a -um with a small round shield, *parmula, parmulae*
parnassi, parnassiacus -a -um from Mount Parnassus, Greece
Parnassia l'Obel's name for *Gramen Parnassium* – grass of Parnassus (Dioscorides' name, αγρωστις εν παρνασσο, gave the Latin name, *gramen parnasium*) (***Parnassiaceae***)
parnassifolius -a -um having leaves similar to those of *Parnassia, Parnassia-folium*
Parochetus Brookside, παρ-οχετος
Parodia for Lorenzo Raimundo Parodi (1895–1966) of Buenos Aires, botanist and writer on grasses
Paronychia Beside-nail, παρ-ονυξ (Dioscorides' name, παρονυχος, for its former use to treat whitlows)
paronychioides resembling *Paronychia, Paronychia-oides*
Paropsis Dish-of-food, παρ-οψις (a small dish)
parqui from the Chilean name for *Cestrum parqui*

Parrotia for F. W. Parrot (1792–1841), German naturalist and traveller (Persian ironwood tree)
Parrotiopsis resembling *Parrotia*, *Parrotia-opsis*
Parrya for Captain Sir William Edward Parry (1790–1855), Arctic navigator
parryi for Charles Christopher Parry (1823–90), English-born American botanist (*Lilium parryi*)
Parsonsia for Dr John Parsons (1705–70), Scottish physician and writer on natural history
Parthenium, *parthenium* Virginal, παρθενιον (Theophrastus' name, παρθενος, for composites with white ray florets)
Parthenocissus Virgin-ivy, παρθενος-κισσος (French name Virginia creeper)
parthenus -a -um virgin, of the virgin, virginal, παρθενος
-partitus -a -um -deeply divided, -partite, -parted, *pars*, *partis*
-parus -a -um -bearing, -producing, *pario*, *parere*, *peperi*, *petum*
parvi- small-, *parvus*, *parvi-*
parviflorus -a -um small-flowered, *parvus-florum*
parvifolius -a -um with small leaves, *parvus-folium*
parvissimus -a -um the smallest, superlative of *parvus*
parvulus -a -um very small, least, comparative of *parvus*
parvus -a -um small, *parvus*
pascuus -a -um of pastures, *pascuus*
pashia the Nepalese vernacular name for *Pyrus pashia*
Pasithea another name, Pasithea, for the Grace Aglaia, of mythology
Paspalidium *Paspalum*-resembling, πασπαλος-ειδιον
paspalodes looking like *Paspalum*, πασπαλος-ωδες
Paspalum a Greek name, πασπαλος, for millet grass
Passerina Sparrow, *passer*, *passeris* (the beaked seed)
passerinianus -a -um resembling *Passerina*; of sparrows, sparrow-like
passerinoides resembling *Passerina*, *Passerina-oides*
Passiflora Passion-flower, (*patior*, *pati*, *passus*)-*florum* (the signature of the numbers of parts in the flower related to the events of the Passion) (***Passifloraceae***)
passionis -is -e of Passion-tide, late Latin *passio* (in the sense of suffering)
Pastinaca Earth-food, from a trench in the ground (formerly for carrot and parsnip, *pastinare*, to dig)
pastoralis -is -e, *pastoris -is -e* growing in pastures, of shepherds, *pastor*, *pastoris* (*acus pastoris* was the herbalist's name for shepherd's needle, *Scandix pecten-veneris)*
patagonicus -a -um from Patagonia, Argentina/Chile, S America
patagua the Chilean vernacular name for *Crinodendron patagua*
patavinus -a -um from Padua (Patavina), Italy
patchouli the Tamil vernacular name, pacculi, for the aromatic oil from *Microtoena patchouli*
patellaris -is -e, *patelliformis -is -e* knee-cap-shaped, small dish-shaped, *patella*, *patellae*
patens, *patenti-* spreading out from the stem, patent, *pateo*, *patere*, *patui*
patentiflorus -a -um having flowers spreading out from the rachis, *patenti-florum*
patentifolius -a -um having leaves spreading away from the stem, *patenti-folium*
pateri- saucer-, *patera*, *paterae*
Patersonia, *patersonii* for Colonel William Paterson (1755–1810), Scottish traveller in S Africa
patientia enduring, *patiens*, *patientis* (French, lapatience, Italian, lapazio; cognate with *Lapathum*)
patinatus -a -um convex like a dish, *patina*, *patinae*
Patrinia for Eugène Louis Melchior Patrin (1742–1815), French traveller in Siberia
patulus -a -um spreading, opened up, broad, *patulus*
paucandrus -a -um with few stamens, botanical Latin from *paucus* and ανηρ
pauci- few-, small-, little-, *paucus*
paucicapitatus -a -um few-headed, *paucus-caput*

paucicostatus -a -um with few nerves or ribs, *paucus-*(*costa, costae*)
pauciflorus -a -um few-flowered, *paucus-florum*
paucifoliatus -a -um sparsely-leaved, *paucus-foliatus*
paucifolius -a -um with few leaves, *paucus-folium*
paucinervis -is -e few-nerved, *paucus-nerva*
pauciramosus -a -um with few branches, with little branching, *paucus-ramosus*
paucistamineus -a -um having few stamens, *paucus-*(*stamen, stamenis*)
paucivolutus -a -um slightly turned or rolled, *parcus-*(*volvo, volvere, volvi, volutum*)
paucus -a -um little-, few, *paucus*
Paullinia for Simon Paulli (1603–80), or Charles Frederick Paulli (1643–1742), Danish botanist
paulopolitanus -a -um, paulensis -is -e from São Paulo, Brazil
Paulownia for Princess Anna Paulovna (Paulowna) (1795–1865), consort of King William II of the Netherlands, and daughter of Czar Paul I of Russia
paulus -a -um small, *paulus*
pauper-, pauperi- poor-, *pauper, pauperis*
pauperculus -a -um of poor appearance, diminutive of *pauper, pauperis*
pauperiflorus -a -um having meagre or poor flowers, *pauperis-florum*
Pauridiantha Small-flowered-one, παυρος-ανθος
Paurotis Small-ear, παυρος-ωτος
pausiacus -a -um olive-green, *pausicus* (*pausia*, a kind of olive)
Pavetta from the Malabari vernacular name for *Pavetta indica*
Pavia, pavius -a -um from Pavia, Italy (≡ *Aesculus*)
pavimentatus -a -um pavement, paving, floor, *pavimentum* (mode of growth)
Pavonia, pavonianus -a -um, pavonii for Don José Antonio Pavón y Jiménez (1790–1844), Spanish botanist in Peru, author with H. Ruiz Lopez of *Flora Peruviana et Chilensis prodromus*
pavonicus -a -um, pavoninus -a -um peacock-blue, showy, *pavo, pavonis*
pavonius -a -um peacock-blue, resembling *Pavonia*
paxianus -a -um, paxii for Ferdinand Pax (1858–1942), Director of Breslau Botanic Garden
Paxistima Thick-stigma, παχυ-στιγμα (the short style of the immersed ovary)
Paxtonia, paxtonii for Sir Joseph Paxton (1801–65), gardener at Chatsworth to the Duke of Devonshire and designer of the 'Crystal Palace'
pecan from a N American Algonquin vernacular name, paccan
pechei for George Peché of Moulmein
Pecteilis Comb-like, *pecten-ilis* (the outward pointing teeth)
pecten-aboriginus -a -um native's comb, modern Latin from *pecten* and *ab-origine*, from the beginning
pecten-veneris Venus' comb, *pecten, pectinis* (a name used in Pliny)
pectinatus -a -um comb-like, (scalloped) pectinate, *pectino*
pectinellus -a -um like a small comb, *pecten, pectinis* (the thorny midribs)
pectinifer -era -erum with a finely divided crest, like a comb, *pecten, pectinis*
Pectis Comb, *pecten*
pectoralis -is -e of the chest, *pectus, pectoris* (used to treat coughs)
peculiaris -is -e one's own, special, *peculiaris*
ped- stalk-, foot-
pedalis -is -e, pedali- about a foot in length or stature, *pes, pedis; pedalis*
Pedalium A-foot, *pes, pedis* (about 12 inches in stature; some disperse fruits with hooks attaching to animal's feet) (***Pedaliaceae***)
pedati-, pedatus -a -um palmate but with the lower lateral lobes divided, pedate, *pes, pedis* (see Fig. 5b)
pedatifidus -a -um divided nearly to the base in a pedate manner, *pedatus-fidus* (see Fig. 5b)
pedatisectus -a -um pedately cut almost to the veins, *pedatus-*(*seco, secare, secui, sectum*)
pedatoradiatus -a -um spreading out like a (bird's) foot, *pedatus-radiatus*

pedemontanus -a -um from Piedmont, N Italy (foot of the hills)
pedialis -is -e with a long flower-stalk, *pes, pedis*
pedicellatus -a -um, pedicellaris -is -e (pediculatus) each flower clearly borne on its own individual stalk in the inflorescence, pedicellate, modern Latin *pedicellus*
Pedicularis Louse-wort, *pedis, pedis; pediculus*
pedicularis -is -e of lice, *pedis, pedis* (a name for a plant in Columella thought to be associated with lice)
pedifidus -a -um shaped like a (bird's) foot, *pedis-(findo, findere, findi, fissum)*
pedil-, pedilo- shoe-, slipper-, πεδιλον
Pedilanthus Shoe-flower, πεδιλ-ανθος (involucre of bird cactus)
pediophilus -a -um growing in upland areas, πεδιο-φιλος
peduncularis -is -e, pedunculatus -a -um with the inflorescence supported on a distinct stalk, pedunculate, diminutive from *pedatus*
pedunculosus -a -um with many or conspicuous peduncles, *pedunculus*
Peganum Theophrastus' name, πηγανον, for rue
pekinensis -is -e from Peking (Beijing), N China; or for Pekin, Illinois, USA
pel- through-
pelargoniifolius -a -um with leaves resembling those of *Pelargonium, Pelargonium-folium*
Pelargonium Stork, πελαργος (Greek name compares the fruit shape of florists' geranium with a stork's head)
pelegrina from a vernacular name for *Alstroemeria pelegrina*
pelewensis -is -e from the volcanic islands of Pelew (Palau), E Pacific Ocean
pelianthinus -a -um clay-flowery, πηλινος-ανθινος
pelicanos pelican-like, πελεκαν
peliorrhincus -a -um, peliorrhynchus -a -um like a stork's beak, πελαργος-ρυγχος
pelios- black-, livid-, πελιος
Peliosanthes Livid-flowered-one, πελιος-ανθος
pelisserianus -a -um for Guillaume Pelisser, sixteenth-century Bishop of Montpellier, mentioned by Tournefort as discoverer of *Teucrium scordium* and *Linaria pelisseriana*
Pellaea Dusky, πελλος (the fronds of most)
pellitus -a -um skinned, covered with a skin-like film, *pellis*
pellucidus -a -um, perlucidus -a -um through which light passes, transparent, clear, pellucid, *perluceo, perlucere, perluxi; perlucidus*
pellucipes transparent-stalked, *perlucidus(pellucidus)-pes* (*perluceo, perlucere, perluxi*)
pellus -a -um dusky, πελλος
pelo-, pelonus -a -um clay-, muddy-, πηλος
pelocarpus -a -um mud-fruit, fruit of the mud, πηλος-καρπος
pelochtho- mud-bank-, river's edge, πηλος-οχθος
peloponnesiacus -a -um from the Peloponnese (Peloponnisos), Greece
peloritanus -a -um from the Peloritani mountains, Messina, Sicily, Italy
pelorius -a -um monstrous, peloric, πελωριος (e.g. radial forms of normally bilateral flowers)
pelta-, pelti-, pelto- shield-, πελτη, *pelta, peltae*
peltafidus -a -um with peltate leaves that are cut into segments, *pelta-(findo, findere, findi, fissum)*
Peltandra Shield-stamen, πελτη-ανηρ
Peltaria Small-shield, diminutive from πελτη
peltastes lightly armed, πελταστης
peltatus -a -um stalked from the surface (not the edge), peltate, πελτη (see Fig. 5d)
peltifolius -a -um having peltate leaves, *pelta-folium*
Peltiphyllum Shield-leaf, πελτη-φυλλον (the large leaves that follow the flowers); see *Darmera*
Peltophorum, peltophorum Shield-bearer, πελτη-φορος (the shape of the stigma)
peltophorus -a -um with flat scales, shield-bearing, πελτη-φορα

pelviformis -is -e shallowly cupped, shaped like a shallow bowl, *pelvis-forma*
pemakoensis -is -e from the Tsangpo gorge, Pemako province, Tibet
pembanus -a -um from Pemba island, Zanzibar Protectorate
pen-, pent-, penta- five-, πεντε
Penaea for Pierre Pena, sixteenth-century French botanist (***Penaeaceae***)
penangianus -a -um, penanianus -a -um from Penang island, Malaysia
pendens, penduli-, pendulinus -a -um, pendulus -a -um drooping, hanging down, *pendeo, pendere, pependi*
pendulicaulis -is -e having lax or hanging stems, *pendulus-caulis*
penduliflorus -a -um with pendulous flowers, *pendulus-florum*
pendulifolius -a -um with hanging leaves, *pendulus-folium*
penetrans piercing, penetrating, present participle of *penetro, penetrare, penetravi, penetratum*
penicillaris -is -e, penicillatus -a -um, penicillius -a -um (penicellatus) covered with tufts of hair, brush-like, *penicillus, penicilli*
Penicillium Paint-brush, *penicillus, penicilli* (the sporulating state)
peninsularis -is -e living on a peninsula, *paene-insula* (almost an island)
penna-, penni- feather-, feathered-, winged-, *penna, pinna*
penna-marina sea-feather, *penna-marinus*
pennatifidus -a -um pinnately divided, *pennati-(findo, findere, fidi, fissum)*
pennatifolius -a -um having feathery foliage, *penna-folium*
pennatus -a -um, penniger -era -erum arranged like the barbs of a feather, feathered, *penna; pina, pinnae*
pennigerus -a -um bearing feathery leaves, *penna-gero*
penninervis -is -e, penninervius -a -um pinnately nerved, *penna-nerva*
Pennisetum Feathery-bristle, *penna-seta* (the feathery bristles of the inflorescence)
pennivenius -a -um pinnately veined, *penna-(vena, venae)*
pennivesiculatus -a -um with vesicles arranged pinnately, *penna-vesicula* (on the leaves)
pennsylvanicus -a -um, pensylvanicus -a -um from Pennsylvania, USA
pennulus -a -um feather-like, *penna; pina, pinnae*
pensilis -is -e hanging down, pensile, *pensilis*
Penstemon (Pentstemon) Five-stamens, πεντε-στεμον (five are present but the fifth is sterile)
penstemonoides resembling *Penstemon, Penstemon-oides*
pent-, penta- five-, πεντε
Pentactina Five-rayed, πεντ-ακτις (the linear petals)
pentadactylon five-fingered, πεντε-δακτυλος (leaves)
pentadelphus -a -um with the stamens coupled in five bundles, πεντε-αδελφος
Pentadesma Five-bundles, πεντε-δεσμη (the grouping of the many stamens)
Pentaglottis Five-tongues, πεντε-γλωττα (the scales in the throat of the corolla)
Pentagonanthus Pentagonal-flower, πεντε-γωνια-ανθος
Pentagonia, pentagonus -a -um Five-angled, πεντε-γωνια (the corolla or leaves)
pentagynus -a -um five-styled, with a five partite ovary, πεντε-γυνη
pentalobus -a -um five-lobed, πεντε-λοβος
pentamerus -a -um having the (floral) parts in fives, πεντε-μερος
pentandrus -a -um with five stamens in the flower, πεντε-ανηρ
pentapetalus -a -um with five petals, πεντε-πεταλον
pentaphyllus -a -um five-leaved, with five-partite leaves, πεντε-φυλλον (Cinquefoil)
pentapotamicus -a -um from the environs of Punjab, or the five rivers, πεντε-ποταμος (Jhelum, Chenab, Ravi, Beas, Sutlej – joining the Indus)
pentapterus -a -um with five wings, πεντε-πτερυξ (e.g. on the fruit)
Pentapterygium Five-small-winged, πεντε-πτερυγιον
Pentas Fivefold, πενταχα
Pentaschistis Five-partite, πεντε-σχιστος (the awn and lemma bristles)
pentaschistus -a -um having five clefts or splits, πεντε-σχιστος
pentaspermus -a -um five-seeded, πεντε-σπερμα

Penthorum Five-columns, πεντε-ορος (the beaks on the fruit)
pentops having an eye-like mark on each petal, five-eyed, πεντε-ωψ
Pentparea Five-pursed, πεντα-πηρα (the five-locular ovary)
penzanceanus -a -um from Penzance, Cornwall
Peperomia Pepper-like, *Piper*-ομοιος (some resemble *Piper*)
peperomiodes resembling *Peperomia, Peperomia-odes*
peplis Dioscorides' name, πεπλις, for a Mediterranean coastal spurge
peploides spurge-like, πεπλις-οειδης, *Peplus-oides*
peplum robed, *peplum, pepli* (the state robe of Athena)
peplus Dioscorides' name, πεπλος, for a northern equivalent of *peplis*
Pepo Sun-cooked (name for a pumpkin, οικνος-πεπων, ripening to become edible, i.e. fully ripened), or πεπων, a gourd
per- around-, through-, more than-, extra-, very-, περι-, περ-, *per-*
pera- over-, much-, περα
peracutus -a -um very acutely pointed, *per-acutus*
perado from a Canary Isles vernacular name for *Ilex perado*
peramabilis -is -e very loveable or lovely, *per-amabilis*
peramoenus -a -um very beautiful, very pleasing, very delightful, *per-amoenus*
Peraphyllum Over-leafy, περα-φυλλον (the crowded foliage)
perarmatus -a -um very thorny, heavily armed, *per-armatus*
perbellus -a -um very pretty, *per-bellus*
percarneus -a -um deep-red, *per-carneus*
percinctus -a -um through the surround, *cingo, cingere, cinxi, cinctum* (the ring becomes loose and slips down the stipe)
percurrens running through, along the whole length, *per-currens*
percursus -a -um running about through (the soil), *percursatio*
percussus -a -um actually or appearing to be perforated, striking, *percussio*
perdulcis -is -e very sweet, pleasant throughout, *per-dulcis*
peregrinans spreading, wandering abroad, present participle of *peregrinor, peregrinare, peregrinatus*
peregrinus -a -um strange, foreign, exotic, *peregrinus*
perennans, perennis -is -e continuing, perennial, through the year, *per-(annus, anni)*
perennitas continuing, of the perennial state, *perennis*
Perenospora Durable-spored, περανος (περηνος)-σπορος (with spores lasting a whole year)
Pereskia for Nicholas Claude Fabry de Pieresc (1580–1637)
pereskiifolius -a -um with leaves similar to those of *Pereskia*
Pereskiopsis resembling *Pereskia, Pereskia-opsis*
Perezia for Lorenzo Pérez of Toledo, apothecary and writer of a history of drugs in 1575
perfectus -a -um complete, not lacking part (of the essential organs), *perficio, perficere, perfaci, perfectum*
perfoliatus -a -um, perfossus -a -um the stem appearing to pass through the completely embracing leaves, *per-folium* (Turner's description of this in *Bupleurum rotundifolium* was as 'waxeth thorow', giving the common name thorow-wax)
perfoliosus -a -um having good foliage, *per-foliosus*
perforatus -a -um pierced or apparently pierced with small round holes, *perforo, perforare, perforavi, perforatum*
perfossus -a -um excavated, pierced through, perfoliate, *perfodio, perfodere, perfodi, perfossum*
perfusus -a -um dyed; sprinkled, drenched, *perfundo, perfundere, perfudi, perfusum*
pergamenus -a -um with a texture like that of parchment, *pergamena* (from Pergamon, a town in Mysia famed for its libraries; now Bergama, Turkey)
pergratus -a -um very pleasing, *per-gratus*
Pergularia Arbour, *pergula* trellis or arbour (the twining growth)
peri- around-, about-, peri-, περι-

periacanthus -a -um with rings of or surrounded with thorns, περι-ακανθος (*Daemonorops*)
perianthomegus -a -um having an enlarged perianth, περι-ανθος-μεγας
Pericallis All-round-beauty, περι-καλλος
periclymenus -a -um, periclymenoides from Dioscorides' name, περικλυμενον, for a twining plant
periculosus -a -um hazardous, dangerous, *periculosus*
periens twining or wrapping around, present participle of περιειλεω
Perilla thought to be from a Hindu vernacular name
Periploca Twine-around, περι-πλοκος (the silk vine's twining habit)
periplocifolius -a -um having leaves similar to *Periploca, Periploca-folium*
Peristrophe Girdled-around, περι-στροφος (the involucre)
perlarius -a -um, perlatus -a -um with a pearly lustre, having pearl-like appendages, *perlarius, perlatus*
perlatus -a -um carried through, enduring, past participle of *perfero, perferre, pertuli, perlatum*
permeabilis -is -e penetrable, *permeo, permeare*
permixtus -a -um confusing, promiscuous, disordered, *permisceo, permiscere, permiscui, permixtum*
permollis -is -e very tender, soft or pliant, *per-mollis*
permutatus -a -um completely changed, *permuto, permutare, permutavi, permutatum*
Pernettya for Dom Antoine Joseph Pernetty (1716–1801), who accompanied Bougainville and wrote *A Voyage to the Falkland Islands*
perniciosus -a -um ruinous, destructive, irritant, *perniciosus*
peronatus -a -um with a woolly-mealy covering, booted, *pero, peronis* (on fungal fruiting bodies),
Perotis Through-the-ear, περ-ωτος (the auricled leaf-bases)
Perovskia (Perowskia), perovskianus -a -um for V. A. Perovski (*c.* 1840), provincial governor of Orenburg, Russia
perpelis -is -e living on rocks which turn to clay, περι-πελος
perplexans intricate, causing confusion, obscuring, present participle of περ-πλεκω; *perplexor, perplexare*
perplexissimus -a -um most obscure or intricate, superlative from περ-πλεκω
perpropinquus -a -um very closely related, *per-propinquus*
perpusillus -a -um exceptionally small, very small, weak, *per-pussilus*
perralderianus -a -um for Henri René le Tourneaux de la Perraudière (1831–61)
Perrottetia for G. S. Perrottet (1793–1870), Director of Agriculture, Senegal
perscandens wide-spreading, present participle from *per-(scando, scandere)*
Persea Theophrastus' Greek name for an oriental tree; Perseus was the hero of Greek legend
persepolitanus -a -um from Iran (formerly Persia; Persepolis was the capital of Persia from *c.* 522 BC until *c.* AD 300)
persetosus -a -um very prickly, *per-setosus*
persi-, persici-, persicoides peach-, περσικον-οειδης
Persica Persian (≡ *Prunus*)
Persicaria Peach-like, περσικον (Rufinus' name refers to the leaves)
persicariae of bistort, living on *Polygonum persicaria* (symbionts, parasites and saprophytes)
persicarius -a -um resembling peach (the leaves), an old name for *Polygonum hydropiper*
persicifolius -a -um peach-leaved, with leaves like *persica, Persica-folium*
persicus -a -um from Persia (Iran), Persian, *persicus*
persimilis -is -a very like, *persimilis*
persistens persistent, present participle of *persisto, persistere, perstiti*
persistentifolius -a -um retaining foliage, with persistent leaves, *persistens-folium*
persolutus -a -um loose, lazy, released, *persolutus* (literally paid up or explained)
personatus -a -um with a two-lipped mouth, masked, *personatus* (bilabiate flower)

Persoonia, persoonii for Christiaan Hendrik Persoon (1755–1837), S African botanist and author of *Synopsis plantarum*
perspectus -a -um well-known, past participle of *perspecto, perspectare*
perspicuus -a -um transparent, clear, bright, *perspicio, perspicere, perspexi, perspectum*
pertusus -a -um leaky, perforate, with holes, pierced through, *pertundo, pertundere, pertudi, pertusum*
Pertya for Josef Anton Maximilian Perty (1804–84), professor of natural history, Berne, Switzerland
perulatus -a -um wallet-like, with conspicuous scales, *perulatus* (e.g. on buds)
perulus -a -um having small pouches or wallets, *perula, perulae*
perutilis -is -e always useful or ready, *per-utilis*
peruvianus -a -um from Peru, Peruvian
pervetus -a -um ancient, very old, *pervetus, peveteris*
perviridis -is -e deep-green, *per-viridis*
pes-, -pes -stalk, -foot, *pes, pedis*
pes-caprae (*pes-capriae*) nanny-goat's foot, *pes-*(*capra, caprae*) (leaf shape of *Oxalis pes-caprae*)
Pescatorea for M. Pescatore, orchidologist
pes-corvi crow-foot, *pes-*(*corvus, corvi*)
pestaloziae for Johann Heinrich Pestalozzi (1746–1827), Swiss educational reformer
pestifer -era -erum pestilential, destructive, baleful, *pestifer, pestiferi*
pes-tigridis tiger's foot, *pes-*(*tigris, tigridis*)
petalocalyx having a petaloid calyx, πεταλον-καλυξ
petalodes looking like petals, πεταλον-ωδες
petaloideus -a -um petal-like, πεταλον-οειδης (the early caducous large sepals)
-petalus -a -um -petalled, πεταλον
Petasites Wide-brimmed-hat, πετασος-ιτης (Dioscorides' name refers to the large leaves)
petaso- wide-brimmed, parachute-like-, πετασος
petecticalis -is -e blemished with spots, *petechia*
petiolaris -is -e, petiolatus -a -um having a petiole, not sessile, distinctly petiolate, *petiolus*, from French, pétiole
petioli living in petioles, *petiolus* (gall insect on aspen)
petiolosus -a -um with conspicuous petioles, *petiolus*
Petiveria for James Petiver (1665–1718), London apothecary and botanist
petr-, petra-, petro- rock-like-, rock-, πετρα, πετρος, πετρο-, *petra*
petraeus -a -um rocky, of rocky places, πετραιος
petranus -a -um from Petra, Jordan
Petrea (Petraea) for Robert James, Lord Petre (1713–43), patron of botany
petricolus -a -um dwelling amongst rocks, *petra-colo*
Petrocallis, petrocallis -is -e Rock-beauty, πετρο-καλος
Petrocoptis Rock-breaker, πετρο-κοπτω
Petrocosmea Rock-ornament, πετρο-κοσμος (the habitat)
petrodo- of rock-strewn-areas-, πετρωδες, πετραιος
Petrophila, petrophilus -a -um Rock-lover, πετρο-φιλος (habitat preference)
Petrophytum Rock-plant, πετρο-φυτον (the habitat)
Petrorhagia Rock-bursting, πετρο-ραγας (stem of ρηγνυμι)
Petroselinum Dioscorides' name, πετρο-σελινον (rock-parsley) for parsley
petroselinus -a -um parsley-like, *Petroselinum*
Petteria for Franz Petter (1798–1853), Austrian author of a botanical journey through Dalmatia, schoolteacher at Split, Croatia
Petunia from the Brazilian Tupi-Guarani name, petun, for tobacco
peuce an ancient Greek name for *Pinus peuce*
Peucedanum a name, πευκεδανον, used by Theophrastus for hog fennel. Gilbert-Carter suggests derivation from πευκη, for the pine-like resin produced (πευκεδανος bitter, destructive)

Peumus from a Chilean vernacular name for the fruit and tree of *P. boldus*
pexatus -a -um, pexus -a -um having a surface with an apparent nap, combed, *pecto, pectere, pexi, pexum*
Peyrousea see *Lapeirousia*
Peziza, pezizus -a -um Cup, πεζιζα (the matured fruiting body)
pezizoideus -a -um cup-shaped, πεζιζα-οειδης, orange-coloured (as the fungus *Peziza aurantia*)
Phaca Dioscorides' name, φακη, for a legume (≡ *Astragalus*)
-phace, -phacos lentil-like, φακη, φακος, φακο-
Phacelia, phacelius -a -um Bundle, φακελος (the clustered flowers)
phae-, phaeo- swarthy, brown, φαιος, φαιο-, φαι-
phaeacanthus - a -um with dark thorns, φαιος-ακανθος
phaeantherus -a -um with dark flowers, φαιος-ανθηρος
phaedr-, phaedro- gay-, φαιδρος, φαιδρο-, φαιδρ-
Phaedranassa Gay-queen, φαιδρος-ανασσα
Phaedranthus Gay-flower, φαιδρ-ανθος (the colourful flowers of the climber *P. buccinatorius*)
phaedropus -a -um of gay appearance, jolly-looking, φαιδρ-ωπος
phaen-, phaeno- shining-, apparent-, obvious-, revealed, φαεινω, φαεινος, φαινο (anglicised to phan-)
phaenocaulis -is -e with dark stems, φαεινω-καυλος
Phaenocoma Shining-hair, φαεινω-κομη (the large red flower-heads with spreading purple bracts)
phaenopyrum with shining grains, φαεινω-πυρος
Phaenospermum Shining-seeded-one, φαεινω-σπερμα
phaeo- dark-, dusky-brown, swarthy, φαιος
phaeocarpus -a -um dark-fruited, φαιο-καρπος
phaeochrysus -a -um dull-yellow, φαιο-χρυσος
phaeodon, phaeodontus -a -um having dark teeth, θαι-(οδους, οδοντος)
Phaeomeria Dark(-purple)-parts, φαιο-μερος
Phaeonychium Dark-*Onychium*, Dark-claw, φαιο-(ονυξ, ονυχος)
phaeostachys with a dark spike, φαιο-σταχυς
phaeus -a -um dark, dun, dusky, φαιος
phaidro- beaming, gay, φαιδρος
Phaiophleps Shining-veined-one, φαιο-(φληψ, φλεβος)
Phaius (*Phajus*) shining, dun, φαινω
phalacrocarpus -a -um having hairless fruits, bald-fruited, φαλακρος-καρπος
phalaenophorus -a -um bearing moths, φαλαινα-φορα (the floral appearance)
Phalaenopsis Moth-like, φαλαινα-οψις (flower form of the Moth orchid)
phalangiferus -a -um bearing spiders, φαλαγγιον-φορα (the floral appearance)
Phalaris Helmet-ridge, φαλος, φαληρος (Dioscorides' name, φαλαρις, for a plume-like grass); some derive it as Shining (for the seeds)
phalaroides resembling *Phalaris*, φαλαρις-οειδης
phaleratus -a -um (phalleratus) shining-white, ornamental, decorated, φαληρος, *phaleratus* (wearing medals)
phalliferus -a -um bearing a phallus, φαλλος-φερω
phalloides phallus-like, φαλλος-οειδης; resembling stinkhorn (*Phallus impudicus*)
Phallus *Membrum-virile*, φαλλος (the suggestive shape of stinkhorn fruiting body)
Phalocallis Beautiful-cone, φαλος-καλλος (the crested limbs)
phanero-, phanerus -a -um conspicuous-, manifest-, visible-, φανερος, φανερο-
Phanerophlebia, phanerophlebius -a -um Prominent-veined-one, φανερο-(φλεψ, φλεβος)
Pharbitis derivation uncertain, φαρβη, coloured-flowers?
pharmaco- drug-, poisonous-, φαρμακον, φαρμακο-
Pharnaceum for King Pharnaces II of Pontus, son of Mythradartes VI, beaten in battle by Caesar, who famously summed up the action as '*veni, vidi, vici.*'
-pharyngeus -a -um -throated, φαρυγξ, φαρυγγος, φαρυγος
phasoloides similar to *Phaseolus*, *Phaseolus-oides*

Phaseolus Dioscorides' name for a kind of bean, Latin *phaselus, phaseli*
pheb- myrtle-, φιβαλη
Phebalium Myrtle-like, φιβαλη
Phegopteris Oak-fern, φηγο-πτερυξ (a name created by Linnaeus from φηγος, an oak)
Phellandrium a name in Pliny, *phellandrion*, for an ivy-leaved plant
phello-, phellos corky-, cork, φελλος
phellocarpus -a -um having seed with a corky testa, φελλος-καρπος
Phellodendron Cork-tree, φελλο-δενδρον (the thick bark of the type species)
phellomanus -a -um with thin, loose or scanty bark, φελλος-μανος
Phelypaea (Phelipaea) for Louis Phelipeaux, Count of Ponchartrain, Tourneforte's patron
phen- see *phaen-*
phil-, philo-, -philus -a -um loving-, liking-, -fond of, φιλεω, φιλη, φιλος, φιλο-, φιλ-
philadelphicus -a -um from Philadelphia, USA
Philadelphus Brotherly-love, φιλ-αδελφος (some suggest that Athenaeus' name could be for Ptolemy II Philadelphus (308–246 BC), King of Egypt) (***Philadelphaceae***)
philaeus -a -um loveable, liking to increase, φιλ-αυξω
Philesia Loved-one, φιλεω (***Philesiaceae***)
Philibertia for J. C. Philibert, French writer on botany
philippensis -is -e, philippicus -a -um, philippinus -a -um from the Philippines
Phillyrea Leafy, φυλλον (from an ancient Greek name)
phillyreaefolius -a -um, phillyreifolius *Phillyrea*-leaved, *Phillyrea-folium*
phillyreoides *Phillyrea*-like, *Phillyrea-oides*
philocremmus -a -um loving cliffs or craggs, φιλος-κρεμνος
Philodendron Tree-loving, φιλεο-δενδρον (habit of epiphytic aroid)
philonotis -is -e moisture-loving, φιλο-νοτις
-philus -a -um -loving, -friend, φιλη, φιλος
Philydrum Water-loving, φιλεο-υδορ (***Philydraceae***)
phleb- vein-, φλεψ, φλεβος, φλεβο-, φλεβ-
phlebanthus -a -um with veined flowers, φλεβ-ανθος
Phlebodium Veined, φλεψ (pronounced frond venation)
phlebophyllus -a -um having (nicely) veined leaves, φλεβο-φυλλον
phlebotrichus -a -um with hairy veins, φλεβο-τριχος
phleioides rush-like, resembling the grass *Phleum, Phleum*-οειδης
Phleum Copious, φλεων (Greek name for a kind of dense-headed rush)
-phloebius -a -um -veined, φλεψ, φλεβος
-phloem with veined flowers, φλοιος-εμα
-phloeus -a -um -barked, -bark, φλοιος
phlog-, phlogi- *Phlox*-like, flame-, φλοξ, φλογος, φλογο-, φλογι-
Phlogacanthus Flame-*Acanthus*, φλογ-ακανθος (some have red flowers)
phlogifolius -a -um red-leaved, from φλογος and *folium*
phlogoflorus -a -um flame-red-flowered, botanical Latin from φλογος and *florum*
phlogopappus -a -um bearing red down, φλογο-παππος
Phlomis Flame, φλοξ, φλογος (the hairy leaves were used as lamp wicks)
phlomoides resembling *Phlomis*, φλομις-οειδης
Phlox Flame, φλοξ (Theophrastus' name for a plant with flame coloured flowers)
phocaena seal or porpoise, φοκε, φωκη
phocaicus -a -um from the Fokís (Phocis) district, central Greece
Phoenicaulis Scarlet-stem, φοινιξ-καυλος
phoeniceus -a -um scarlet, red with a little yellow, φοινιξ, φοινικος
phoenicius -a -um from Tyre and Sidon (*Phoenicia*, φοινικη, now Lebanon), purple-crimson, φοινικεος
phoenicodus -a -um with a purple entry, φοινικος-οδος (to the corolla-tube)
phoenicoides resembling *Phoenix*
phoenicolasius -a -um red-purple-haired, φοινικο-λασιος

Phoenix Phoenician, φοινιξ (who introduced the date palm to the Greeks)
phoenix date-palm, date, purple-red, purple dye, fabulous bird, lyre, φοινιξ
pholideus -a -um scaly, φολιδος
Pholidocarpus Scaly-fruit, φολιδος-καρπος
Pholidota Ear-scaled, φολιδος-οτις (*Pholidota* (*Pholidotis*), Scale-ear, φολιδος-ωτος (the shape of the bracts of rattle-snake orchid)
pholidotus -a -um with scaly ears, scaly, φολιδος-οτος
-pholis -scaled, φολιδος, φολιδο-, φολι-
Pholiurus Scale-tail, φολι-ουρα (the elongate spikes with scale-like glumes)
-phonis -is -e -murder, φονος, φονη
Phoradendron Tree-burden, φορα-δενδρον (the parasitic habit)
Phormium Little-basket, θορμιον, diminutive of φορμος (the leaf-fibres were used for weaving) (***Phormiaceae***)
-phorus -a -um -bearing, -carrying, -φορος, φορα, φερω
phosphoreus -a -um bright, light bringing, φως-φορος (seventeenth-century chemical Latin)
photeinocarpus -a -um brightly-fruited, φωτεινος-καρπος
Photinia Shining-one, φωτεινος (from the brilliant young foliage)
photiniphyllus -a -um having leaves resembling those of *Photinia*, φωτεινος-φυλλον
phoxinus minnow, with a pointed apex, φοξος
phragma-, -phragma fence-, enclosure-, φραγμα
Phragmites, phragmites Reed-of-hedges, Adanson's name, καλαμος-φραγμα-ιτης (*Arundo donax* is used for hedging in S Europe)
phryganais -is -e like a thorny bush-woodland, φρυγανα (literally firewood)
phrygius -a -um from Phrygia, Asia Minor (land of the Phryges, now Turkey)
Phrynium Toad, φρυνος (onomatopoeic for croaking, liking for moist places)
phthitis of death or decay, φθιτον
phthora of corruption, φθορα
phu foul-smelling, φοω
Phuopsis Valerian-like, resembling *Valeriana phu*, φοω-οψις
-phyceae, phyco- seaweed-, φυκος, φυκο-
Phycella Seaweed, diminutive of φυκος (≡ *Hippeastrum*)
phycofolius -a -um seaweed-leaved, botanical Latin from φυκος and *folium*
phyctidocalyx with a split or deciduous calyx, φυκτος-καλυξ
Phygelius Fugitive, φυγα-ηλιος (dislike of direct sunlight)
Phyla Tribe, φυλη, φυλον (derivation uncertain)
Phylica Leafy, φυλλικος (a name, φυλικη, in Theophrastus for a buckthorn-like plant with copious foliage)
phylicifolius -a -um with leaves like those of *Phylica, Phylica-folium*
phylicoides resembling *Phylica*, φυλλικος-οειδης
phyll-, phylla-, phyllo- leaf-, φυλλον, φυλλο-, φυλλ-
phyllanthes leaf-flowering, φυλλ-ανθος (the green perianth of *Epipactis phyllanthes*)
phyllanthoides resembling *Phyllanthus*, φυλλ-ανθος-οειδης
Phyllanthus Leaf-flower, φυλλ-ανθος (some flower from edges of leaf-like phyllodes)
Phylliopsis the compound name for hybrids between *Phyllodoce* and *Kalmiopsis*
phyllitidis *Phyllitis*-like
Phyllitis Dioscorides' name, φυλλιτις,refers to the simple leaf-like frond
phyllobolus -a -um leaf-shedding, throwing off leaves, φυλλο-βολις
phyllochlamys with a leafy cloak, φυλλο-χλαμυς (large floral bracts)
Phyllocladus (os) Leaf-branch, φυλλο-κλαδος (the flattened leaf-like cladodes)
Phyllodoce the name of one of Cyrene's attendant sea nymphs
Phyllodolon Leaf-trap, φυλλον-δολος (the leaf sheaths)
phyllomaniacus -a -um excessively leafy, a riot of foliage, φυλλο-μανικος
phyllophilus -a -um leaf-loving, φυλλον-φιλος (grows on leaf-litter)
phyllorhizus -a -um having photosynthesizing roots, φυλλον-ριζα
Phyllospora Leafy-spored, φυλλον-σπορος

Phyllostachys Leafy-spike, φυλλο-σταχυς (the leafy inflorescence)
phyllostachyus -a -um with leafy spikes, φυλλον-σταχυς
Phyllothamnus the composite name for hybrids between *Phyllodoce* and *Rhodothamnus*
Phylloxera Dry-leaf, φυλλον-ξερος (vine-damaging aphid)
-phyllus -a -um -leaved, φυλλον
phyma- swollen or of ulcerated appearance, φυμα, φυματο
phymatocarpus -a -um having swollen fruits, φυματο-καρπος
phymatochilus -a -um with a swollen or swelling on the labellum, φυματο-χειλος
Phymatodes Ulcerated-looking, φυματο-ωδες (the sori are in depressions)
phymatodeus -a -um warted, verrucose, φυματο-ωδες
physa-, physo- bladder-, swelling-, inflated-, bellows-, φυσαω, φυσα, φυσ-
Physacanthus Inflated-flower, φυσα-ανθος (the calyx, but *Physacanthus nematosiphon* is never inflated)
Physalis Bellows, φυσα (the inflated fruiting calyx resembles a bellows or bladder, φυσαλλις)
physalodes, physaloides resembling *Physalis*, φυσα(λ)-ωδες, φυσα(λ)-οειδης
Physaria Bladder-like, φυσα
Physianthus Inflated-flower, φυσα-ανθος (≡ *Araujia*)
Physocarpus, physocarpus -a -um Bladder-fruit, φυσα-καρπος
Physochlaina Inflated-cloak, φυσα-χλαινα (the calyx)
physodes puffed out, inflated-looking, φυσα-ωδες
physophyllus -a -um with inflated leaves, φυσα-φυλλον
Physoplexis Inflated-entanglement, φυσα-πλεξις (crowded inflorescence)
Physospermum Inflated-seed, φυσα-σπερμα (fruit of bladder seed)
Physostegia Inflated-cover, φυσα-στεγη (the inflated calyx)
phyt-, -phyta, phyto- plant-, φυτον
Phytelephas Vegetable-ivory, φυτον-ελεφας (the large hemicellulose seed of ivory nut, at one time turned for billiard balls)
Phyteuma, phyteuma That-which-is-planted, φυτευω-μα (name, φυτευμα, used by Dioscorides)
phyto- plant-, φυτον, φυτο-
Phytolacca Plant-dye, φυτον-λακ (the sap of the fruit) (**Phytolaccaceae**)
Phytophthora Plant-destruction, φυτον-φθορα (pathogenic fungi)
piassabus -a -um the Tupi vernacular name, piaçába, for piassava palm-fibre
piauhyensis -is -e from Piauí state, NE Brazil
pica ornate (*pica, picae* magpie; *picus pici* woodpecker)
picaceus -a -um magpie-like, *pica* (black and white colouration of the magpie fungus)
Picea, picea Pitch (the ancient Latin name, *pix*, refers to the resinous product)
piceopaleaceus -a -um having blackened scales or paleas, *piceus-(palea, paleae)*
piceus -a -um blackening, pitch-black, *piceus*
pichinchensis -is -e from Pichincha province, N central Ecuador
picolanus -a -um resinous, *pix, picis*
Picrasma Bitterness, πικραζειν (the bitter-tasting bark)
picridis -is -e ox-tongue-like, *Picris*-like
Picridium *Picris*-like, πικρια-ειδιον
Picris Bitter, πικρις (Theophrastus' name for a bitter, πικρος, potherb)
picro-, -picros -os -on bitter-, -bitter, πικρια, πικρος, πικρο-
picroides resembling *Picris*
picrorhizus -a -um having a bitter-tasting root, πικρο-ριζα
picrus -a -um bitter to the taste, πικρος
pictifolius -a -um with decorated leaves, (*pingo, pingere, pinxi, pictum*)*-folium*
picturatus -a -um embroidered, variegated, picture-like, *picturatus*
pictus -a -um, -pictus -a -um (*pichtus*) brightly marked, ornamental, painted, *pingo, pingere, pinxi, pictum*
Pieris from a name, Pierides (collective name for the muses of Greek mythology)

pigrus -a -um sluggish or slow-growing, *pigro, pigrare; pigror, pigrari*
pilanthus -a -um having flowers with a felted texture, πιλος-ανθος
pilaris -is -e pilose, πιλος, πιλο-, *pilus, pili*
Pilea Felt-cap, *pileus*
pileatus -a -um capped, having a cap, *pileus*
pileo- cap-, πιλος, *pileus* (literally, the felt cap presented when a slave was manumitted)
Pileostegia Felt-cap-covered, πιλος-στεγη
piliferus, pilifer -era -erum bearing hairs, with short soft hairs, ending in a long fine hair, (*pilus, pili*)-*fero*
pilo- felted with long soft hairs, πιλος, πιλο-, *pilus, pili*
Pilocarpus, pilocarous -a -um Felted-fruit, πιλος-καρπος
Pilosella, pilosella Soft-haired, feminine diminutive of *pilosus* (Rufinus' name for *Hieracium pilosella*)
pilosellae of hawkweed, living on *Hieracium pilosella (Cystiphora*, dipteran gall midge)
piloselloides hawkweed-like, *Pilosella-oides*
pilosellus -a -um tomentose, finely felted with soft hairs, diminutive of *pilosus*
pilosissimus -a -um very pilose, superlative of *pilosus*
pilosiusculus -a -um hairy-ish, with sparse very fine hairs, somewhat pilose, diminutive of *pilosus*
pilosulus -a -um loosely pilose, somewhat hairy, diminutive of *pilosus*
pilosus -a -um covered with soft distinct hairs, pilose, *pilosus*
piluiformis -is -e globular, *pilula-forma*
Pilularia Small-balls, diminutive of *pila* (the shape of the sporocarps)
pilularis -is -e, pilulifer -era -erum having glands or globular structures, bearing small balls, *pilula-fero*
Pimelea Fat, πιμελη (the oily seeds of rice flower)
pimeleoides resembling *Pimelea*
pimelus -a -um oily or fatty, πιμελη
Pimenta from the Spanish name, pimienta, for allspice, the dried fruit of *Pimenta officinalis*, pimento (Latin *pigmenta*, spices)
pimentoides allspice-like, *Pimenta-oides*
Pimpinella, pimpinella a medieval name of uncertain meaning, first used by Matthaeus Sylvaticus (cognates include pimpernel and, probably, *piper* and *prunella*)
pimpinellae on burnet saxifrage, living on *Pimpinella* (*Kiefferia*, dipteran gall midge)
pimpinellifolius -a -um Pimpernel-leaved, *Pimpinella-folium*
pimpinelloides resembling *Pimpinella*
pinaster Wild-pine, *pinus-aster*, Pliny's name for *Pinus sylvestris*
pindicola, pindicus -a -um living in, or from, the Pindus mountain range, Greece, botanical Latin from Pindus and *colo*
pindrow the W Himalayan vernacular name for *Abies pindrow*
Pinellia for Giovanni Vincenzo Pinelli (1535–1601), of the Naples botanic garden
pineolens smelling of pine, present participle from *pinus*-(*oleo, olere, olui*)
pineticolus -a -um dwelling on or amongst pines, *pinus-colo*
pinetorum associated with pines, of pine woods, genitive plural of *pinus*
pineus -a -um cone-producing, of pines, resembling a pine, *pinus*
pingui- fat-, *pinguis* (*pinguior* fatter)
Pinguicula Grease, feminine diminutive of *pinguis* (the fatty appearance of the leaves of butterwort)
pinguifolius -a -um waxy-leaved, thick-leaved, *pinguis-folium*
pini-, pini pine-like, pine-, living on *Pinus* (*Eriophyes*, acarine gall mite)
pinicolus -a -um living amongst pines, *pini-colo*
pinifolius -a -um pine-leaved, with needle shaped leaves, *pini-folium* (see Fig. 5c)
pinnati-, pinnatus -a -um set in two opposite ranks, winged, feathered, pinnate, *pinnatus* (see Fig. 5c)

pinnatifidus -a -um pinnately divided almost to the midrib, *pinnatus-(findo, findere, fidi, fissum)*
pinnatinervius -a -um with pinnate veins, *pinnatus-nerva*
pinnatisectus -a -um cut into pinnate segments, *pinnatus-(seco, secare, secui, sectum)*
pinnatistipulus -a -um with pinnately divided stipules, *pinnatus-(stipula, stipulae)*
pinnato-ramosa with pinnate branching, *pinnatus-(ramus, rami)*
pinnatus -a -um with pinnate leaves or branches, *pinnatus*
pinsapo from the Spanish name, pinapares, for *Abies pinsapo*
Pinus the ancient Latin name, *pinus*, for a cone-bearing tree, pine (cognate with pineus, piñon and pine[-apple]) (***Pinaceae***)
Piper from the Sanskrit pippali, Greek πεπερι, for pepper (***Piperaceae***)
piperascens pepper-like, resembling *Piper*
piperatus -a -um, piperitus -a -um with a hot biting taste, peppered, pepper-like, πεπεριζω (the taste)
piperinus -a -um peppery, *piper* (scented)
Piptadenia Falling-glands, πιπτω-αδην (those of the stamens)
Piptanthus Falling-flower, πιπτω-ανθος (with quickly deciduous floral parts)
pipto- falling or being thrown down, πιπτω
piptolepis -is -e having deciduous scales, πιπτω-λεπις
piptopetalus -a -um having early-caducous petals, πιπτω-πεταλον
Piptostigma Falling-stigma, πιπτω-στυγμα (the stigma falls off after flowering)
piri- pear-, *Pyrus*
piriformis -is -e pear-shaped, *pyrus-forma*
pirinensis -is -e, pirinicus -a -um from the Pirin Planina, Bulgaria
Pirola Small-pear, diminutive of *Pyrus* (similarity of foliage)
pirus the Latin name, *pirus, piri*, for a pear tree
pisanus -a -um from Pisa, Italy
pisacensis, pisacomensis -is -e from Pisac, near Cuzco, Peru
piscatorus -a -um of fishermen, *piscator, piscatoris* (fish poison, stupefies fish)
Piscidea Fish-poison, *piscis-(caedo, caedere, cecidi, caesum)*
piscidermis -is -e having a scaly epidermis, botanical Latin from *piscis* and δερμα
piscinalis -is -e of ponds or pools, *piscina, piscinae*
pisi-, piso- pea-like-, pea-, πισος, *pisum, pisi*
pisifer -era -erum bearing peas, *pisi-fero*
pisinus -a -um pea-green, *pisum, pisi*
pisocarpus -a -um pea-fruited, πισος-καρπος
Pisolithus Pea-stone, πισος-λιθος (the stony pea-shaped peridioles within the dung-like fruiting bodies)
pissardii (pissardi, pissarti) for M. Pissard, who introduced *Prunus cerasifera* 'Pissardii' to France from Iran in 1880
pissatorius -a -um of (preferred by) pitch makers, πισσα
Pistacia the Greek name, πιστακε, used by Nicander in 200 BC, Arabic, foustag (πιστακιον was the Greek name for the pistachio nut)
Pistia Watery, πιστος (habitat of the water lettuce)
pistillaris -is -e pestle-shaped, *pistilla, pistillae* (the club-shaped fruiting body)
Pisum the Latin name, *pisum*, for the pea
pitanga a S American Indian name for *Eugenia pitanga*
Pitcairnia for Dr William Pitcairn (1711–91), London physician
pitcairniifolius -a -um with leaves resembling those of *Pitcairnia*
pitcheri for Zina Pitcher (1799–1872), American army physician
pithece-, pitheco- ape-, monkey-, πιθηκος
Pithecolobium (Pithecellobium) Monkey-ears, πιθηκος-λοβος (the name is the Latinization of the vernacular name for the rain tree, alluding to the shape of the fruit)
pithyusus -a -um from Prinkipo (Pityoussa), one of the Nine Islands in the Sea of Marmara, SE of Istanbul, Turkey
pittonii for Joseph Claudius Pittoni (1797–1878), Knight of Dannenfeldt

Pittosporum Tar-seed, πιττα-σπορος, πισσα-σπορος (the resinous coating of the seed) (***Pittosporaceae***)
pitui- mucus-, phlegm, *pituita*
pityoides, -pitys pine-like, πιτυς-οειδης
pityophyllus -a -um with pine-like foliage, πυτιο-φυλλον
pityro- husk-, scurf-, πιτυρα, πιτυρον
Pityrogramma Scurf-lined, πιτυρα-γραμμα (lower surface of fronds becomes obscured by rod-like scaly secretions)
pityrophyllus -a -um having scurfy leaf epidermis, πιτυρον-φυλλον
-pitys, pityoides pine-like, πιτυς-οειδης
pityusa resembling a small pine, πιτυς
Pixidanthera Box-anthers, *pyxis-anthera* (they dehisce with a lid, see *pyxidatus*)
placatus -a -um quiet, calm, gentle, *placatus*
placenti- flat-cake-; placenta-, *placenta, placentae* (biologically, the tissue that supports generative bodies, regardless of shape)
placentiflorus -a -um with flat or disciform perianths, *placenta-florum*
placitus -a -um pleasing, past participle of *placeo, placere, placui, placitum*
placo- flat-body-, flat-, πλαξ, πλακος, πλακο-
Placodiscus Flat-disc, πλακος-δισκος (the floral disc)
placomyces flat mushroom, πλακος-μυκης
plagi-, plagio- sideways-, slanting-, oblique-, πλαγιος, πλαγιο-, πλαγι-; side-, flank-, παλγιον, πλαγιο-
Plagianthus Flank-flowered, πλαγιος-ανθος (axillary flowering)
plagiocarpus -a -um with oblique fruits, πλαγιο-καρπος
Plagiomnium Oblique-*Mnium*, πλαγιο-μνιον
Plagiospermum Oblique-seed, πλαγιο-σπερμα (the compressed ovoid seed)
Plagiostyles Oblique-styled, πλαγιο-στυλος (the short, fat stigma is to one side of the ovary)
-planatus -a -um -sided, -level, -flat, *planus*
planctonicus -a -um, planktonicus -a -um roaming, wandering, πλαγκτος (carried by water movements)
Planera for J. J. Planer (1743–89), Professor of Medicine at Erfurt, Germany
planeta, planetes not stationary, planet-like, wandering, πλανη, πλανημα; πλανης, πλανητης
planetus -a -um not fixed, wandering, πλανητος
plani- flat-, even-, *planus*
planibulbus -a -um having flattened bulbs, *planus-bulbus*
planiceps flat-headed, *planus-ceps* (the inflorescence)
planiflorus -a -um flat-flowered, *planus-florum*
planifolius -a -um having flat leaves, *planus-folium*
planipes having flat stalks, *planus-pes*
planipetalus -a -um flat-petalled, botanical Latin from *planus* and πεταλον (πλανος-πεταλον, variable perianth)
planiscapus -a -um flat stemmed, *planus-scapus* (*Ophiopogon planiscapus*)
planiusculus -a -um somewhat flat (diminutive of *planus*)
plantagineus -a -um (plentigineus) rib-wort-like, plantain-like, *Plantago*
plantaginifolius -a -um having rib-wort like leaves *Plantago-folium*
Plantago Foot-sole-like, feminine termination of *planta* (ancient Latin, *plantaginem*, for the way the leaves of some lie flat on the ground), cognate with the French derivative, plantain (***Plantaginaceae***)
plantago-aquatica water-plantain, *Plantago-aquaticus*
planus -a -um flat-, smooth, *planus*
plasmo-, plasmodio- moulded-, that formed-, πλασμα, πλαστος (biologically, the cytoplasm or protoplast)
plat-, platy- broad-, wide-, flat-, πλατυς
platanifolius -a -um having leaves resembling those of *Platanus, Platanus-folium*
platanoides plane-tree-like, *Platanus-oides*

Platanthera Flat-anthers, πλατυς-ανθερα (divergent thecae)
Platanus, platanthus -a -um Broad-crown, πλατυς (the Greek name, πλατανος,for *Platanus orientalis*; cognate with plane, in the sense of flat) (***Platanaceae***)
Platostoma Wide-mouthed, πλατυς-στωμα (of the corolla)
platy- flat-, level-; wide-, broad-, πλατυς, πλατυ-
platyacanthus -a -um with flattened spines, πλατυ-ακανθος
platyanthus -a -um having flat flowers, with a radiate corolla, πλατυ-ανθος
platybasis -is -e broad-based, πλατυς-βασις (of the stem)
platycalyx having a radiate calyx, πλατυ-καλυξ
platycarpus -a -um, platycarpos with flattened (but not discoid) fruits, πλατυ-καρπος
Platycarya Broad-nut, πλατυς-καρυον (the compressed nutlet)
platycaulis -is -e thick-stemmed, πλατυ-καυλος
platycentrus -a -um wide-eyed, broad-centred, broad-spur, πλατυς-κεντρον
platyceras with flattened horns, πλατυ-κερας
Platycerium Broad-horned, πλατυς-κερας (the stag's-horn-like, dichotomous lobing of the fertile fronds)
Platycladus, platycladus -a -um Flat-branched, with flattened branches, πλατυ-κλαδος
Platycodon Wide-bell, πλατυ-κωδον (the flower form)
Platycoryne Wide-club, πλατυ-κορυνη (the stigmas)
Platycrater Wide-bowl, πλατυ-κρατηρ (the broad calyx of the sterile flowers)
platyglossus -a -um with a broad lip, πλατυς-γλωσσα, (the labellum)
Platylepis, platylepis -is -e Broad-scaled, πλατυ-λεπις (the inflorescence bracts)
platylobus -a -um with flat lobes, πλατυ-λοβος
platyneuron with flat veins, πλατυ-νευρα
platypetalus -a -um with flat petals, πλατυ-πεταλον
platyphyllus -a -um, platyphyllos broad-leaved, πλατυ-φυλλον
platypodus -a -um with a flat foot, πλατυ-ποδος (stem)
platysepalus -a -um having flat sepals, πλατυ-σκεπη
platystegius -a -um with a flat covering, πλατυ-στεγη
platystigma with a flattened stigmatic surface, πλατυ-στιγμα
plausus -a -um pleasing, praiseworthy, laudable, *plaudo, plauderi, plausi, plausum*
plebeius -a -um (plebejum) common, *plebeius*
plebio-, -plebius -a -um veined, φλεψ, φλεβος
pleco- plaited-, πλεκω
plecolepis -is -e having joined scales, πλεκω-λεπις (capitular bracts of composites)
plecto-, plectus -a -um woven-, twisted-, pleated-, πλεκτος, πλεκτο-
plectocarpus -a -um with twisted fruits, πλεκτο-καρπος
Plectocephalus Twisted-head, πλεκτο-κεφαλη
Plectocolea Plaited-sheath, πλεκτο-κολεος
plectolobus -a -um with twisted lobes, πλεκτο-λοβος
Plectostachys Twisted-spike, πλεκτο-σταχυς
Plectranthus Spurred-flower, πληκτρον-ανθος
Plectrelminthus Worm-spurred, πληκτρον-ελμινθος (the 25-cm-long twisted spur)
plectro-, plectrus -a -um spur-, spurred, πληκτρον, πληκτρο-
Pleea for Auguste Plée (1787–1825), author of a Flora of Paris
pleio-, pleo- many-, several-, full-, large-, thick-, more-, πλειος, πλεως-, πλειο-
pleianthus -a -um having clusters of flowers, full of flowers, πλειο-ανθος
pleio- well-provided-, greater than, more-, πλειον, πλειος, πλειο-, πλεως, πληρης
pleioblastoides resembling *Pleioblastus* (≡ *Arundinaria*)
Pleioblastus (*Plioblastus*) Many-budded, Greater than its forebears, πλειο-βλαστος (also used to describe lichen spores that germinate at several points)
Pleiocarpa Free-fruiting, πλειο-καρπος
Pleioceras, pleiocerasus like a large (*Prunus*) *cerasus*, botanical Latin from πλειο and *cerasus*
pleiochromus -a -um richly coloured, πλειο-χρωμα
pleiocladus -a -um much-branched, πλειο-κλαδος

pleiogonus -a -um very angular, πλειο-γονυ (pattern of branching)
Pleiomele Full-of-honey, πλεως-μελι (≡ *Dracaena*)
Pleione mother of the Pleiades in Greek mythology
pleiospermus -a -um numerous-seeded, πλειο-σπερμα
Pleiospilos Many-spotted, πλειο-σπιλος (the punctate marking of the leaves)
pleisto- most, πλειστος, πλειστ-
pleistranthus -a -um most floriferous, heavily flowered, πλειστ-ανθος
pleni-, plenus -a -um double, full, *plenus*
pleniflorus -a -um double-flowered, *plenus-florum*
pleniradiatus -a -um very radiant, having numerous rays, *plenus-radii*
plenissimus -a -um very full or double-flowered, superlative of *plenus*
pleo- full-, complete-, well provided-, πληρης, πλεος, πλεως, πλεο-
Pleodorina Complete-purse, πλεο-δορος (diminutive, for the chloroplast)
Pleomele Honey-full, πλεο-μελε
Pleopeltis Full-of-scales, πλεο-πελτη
pler-, pleri-, -pleris full-, many-, πληρης
Pleroma Fullness or Filling-up, πληρωμα (many-ovuled loculi)
plesio-, -plesius -a -um near to-, close by-, -neighbouring, πλησιος, πλησιο-
pletho- many-, crowded-, πληθος
pleura-, pleuri-, pleuro- ribs-, edge-, side-, of the veins-, πλευρα, πλευρον, πλευρο-
Pleurandra Sideways-stamens, πλευρ-ανηρ (the laterally placed stamens)
Pleurochaete Ribbed-hairs, πλευρο-χαιτη (the peristome)
Pleurococcus ribbed coccus, πλευρο-κοκκος
Pleurogyne Lateral-stigma, πλευρο-γυνη
Pleuropetalum Veined-petals, πλευρο-πεταλον
pleuropterus -a -um with winged nerves, πλευρο-πτερον
Pleurospermum Ribbed-seeded-one, πλευρο-σπερμα
plexi-, -plexus -a -um knitted-, -braided, -network, πλεξι- (πλεξις knitting)
Plexipus Twining-stalk, πλεξι-πους
plicati-, plicatus -a -um folded-together-, -doubled, -folded, *plico, plicare, plicavi (plicui), plicatum*
plicatifolius -a -um with folded leaves, *plicatus-folium* (along the midrib)
plicatilis -is -e much folded together, *plicatilis*
plicatilobus -a -um fan-shaped, with folded lobes, *plicatus-lobus*
plicatulus -a -um slightly folded, diminutive of *plicatus*
plicatus -a -um pleated, folded, *plicatus*
plici- pleated, folded lengthwise, plicate, *plico, plicare, plicavi (plicui), plicatum*
plinianus -a -um for Gaius Plinius Secundus, Pliny the Elder, (24–79), Roman soldier and scholar, author of *Naturalis historia*
Plioblastus Well-provided-with-buds, πλειο-βλαστος
plocao-, ploco- folded-, chapletted-, πλοκος, πλοκο-
plococarpus with whorled fruits, with a chaplet of follicles, πλοκος-καρπος
Plocoglottis Folded-tongue, πλοκος-γλωττα
pluma soft feather, *pluma, plumae* (frond texture)
plumarius -a -um, plumatus -a -um plumed, plumose, feathery, *pluma*
plumbaginoides resembling *Plumbago, Plumbago-oides*
Plumbago Leaden, feminine suffix on *plumbum* (Pliny's name refers to a plant also called μολυβδαινα, for the flower colour) (***Plumbaginaceae***)
plumbaguneus -a -um resembling *Plumbago*
plumbeitinctus -a -um leaden coloured, *plumbum-tinctus*
plumbeus -a -um lead-coloured, the colour of lead, *plumbum, plumbi*
Plumeria, plumieri for Charles Plumier (1646–1704), French botanist and writer on tropical American plants (the common name, frangipane, is a French commemoration of the Italian Marchese Muzio Frangipani *c.* 1588)
plumerioides resembling *Plumeria, Plumeria-oides*
plumeus -a -um feathered, plumed, *plumeus*
plumosus -a -um feathery, *plumeus*

plur-, pluri- many-, several-, the most, *plurimus, pluri-*, superlative of *multus*
pluricapitatus -a -um several- or many-headed, *pluri-capitatus*
pluricaulis -is -e many-stemmed, *pluri-caulis*
pluridens many-toothed, *pluri-dens*
pluriflorus -a -um many-flowered, *pluri-florum*
pluriformis -is -e not uniform but assuming several forms, *pluri-forma*
plurijugus -a -um having many yokes (pairs) or ridges, *pluri-iugum* (leaflets or vittae)
plurinervis -is -e many-veined, *pluri-nerva*
plurisectus -a -um divided several times, *pluri-(seco, secare, secui, sectum)*
plus- more-, greater-, *plus, plus-*, comparative of *multus*
Pluteolus Little-shelter-like, *pluteus, plutei*
Pluteus Shelter, *pluteus, plutei*
pluvialis -is -e announcing rain, of the rains, *pluvia, pluviae*
pluviatilis -is -e growing in rainy places, *pluvia, pluviae*
pluvisilvaticus -a -um of rainforests, *pluvia-silvestris*
pneuma-, pneumato- air-, respiratory-, πνευμα, πνευματος, πνευμων, πηευμο-
pneumatophorus -a -um having breathing (roots) or pneumatophores, πνευματοφορος
Pneumonanthe, pneumonanthe (-*us*) Lung-flower, πνευμων-ανθος (floral signature and the former use of marsh gentian, *Gentiana pneumonanthe* for respiratory disorders)
Poa Pasturage, ποα (the Greek name for a fodder grass) (***Poaceae*** ≡ ***Gramineae***)
poae of *Poa*, living on *Poa* (*Poamyia*, gall midge)
Pocilla Small-cup, diminutive of *poculum* (the seeds)
pocophorus -a -um woolly, fleece-bearing, ποκο-φορα
poculatus -a -um cup-like, *poculum*
poculiformis -is -e goblet-shaped, *poculum-forma* (with upright limbs of the corolla)
pod-, podo-, -podius -a -um foot-, stalk, -foot, πους, ποδιον, ποδος, ποδο-; extremity, ποδεων
podagrarius -a -um, podagricus -a -um snare, of gout, *podagra* (the apothecaries' *herba podagraria*, or goutweed, was used to treat gout)
Podalyria for Podalyrius, son of Aesculapius
podalyriaefolius -a -um, podalyriifolius -a -um with leaves resembling those of *Podalyria, Podalyria-folium*
Podangis Stalked-vessel, ποδ-αγγειον (the tip of the spur)
Podanthus Stalked-flowered, ποδος-ανθος (peduncled heads)
podeti- stalk-, ποδος
-podioides -foot-like, ποδος-οειδης
-podion -little foot, ποδιον, diminutive of πους, ποδος; ποδεων, end or extremity
-podius -a -um, podo-, -podus -a -um foot, stalk, ποδεω, πους, ποδιον, ποδο-
Podocarpus, podocarpus -a -um Stalked-fruit, ποδο-καρπος (the characteristic shape of the fleshy fruit-stalks of some) (***Podocarpaceae***)
Podococcus Foot-fruit, ποδο-κοκκος (shape of the baccate fruit)
podogynus -a -um with a gynophore or stalk to the ovary, ποδο-γυνη
Podolepis Scaly-stalked-one, ποδος-λεπις (the peduncles)
podolicus -a -um from Podolia, W Ukraine
Podophyllum, podophyllus -a -um Foot-leaf, ποδο-φυλλον (the leaf suggests a webbed foot) (***Podophyllaceae***)
podospileus -a -um with a stained stalk, ποδος-σπιλοω
Podostemon Foot-stamened, ποδο-στεμον (the much reduced 'flower' arises from a thallose stem on rocks in fast-running water) (***Podostemaceae***)
Podranea an anagram of *Pandorea*
Poecilandra Variably-stamened, ποικιλος-ανηρ
poecilo- variable-, variegated-, variously-, ποικιλος, ποικιλο- (see *poikilo-*)
poecilobotrys with variously coloured clusters (of fruit), ποικιλο-βοτρυς
Poecilochroma Of-varying-colour, ποικιλο-χρωμα

poeppigianus -a -um, poeppigii for Eduard Friedrich Poeppig, student of the Chilean flora
poetarum, poeticus -a -um of poets, *poeta, poetae* (Greek gardens included games areas and theatres)
Poga from a vernacular name, mpoga, from Gabon, for the fruit of *Poga oleosa*
Pogogyne Bearded-ovary, πωγων-γυνη
pogon-, -pogon bearded-, -haired, -bearded, πωγων, πωγωνος
Pogonantherum Bearded-anther, πωγων-ανθερος
pogonanthus -a -um with bearded flowers, πωγων-ανθος
Pogonarthria Bearded-joints, πωγων-αρθρον (the nodes)
pogonoides beard-like, πωγων-οειδης
pogonopetalus -a -um with bearded petals, πωγωνος-πεταλον
Pogonostemon Bearded-stamen, πωγων(ο)-στεμον (the hairs mid-way up the filaments)
pogonstylus -a -um with bearded styles, πωγων-στυλος
poiformis -is -e grass-like, *poa-forma*
poikilo- variable-, variegated-, spotted-,ποικιλος
poikilophyllus -a -um with variegated leaves, ποικιλο-φυλλον
Poinciana for M. de Poinci, Governor of the Antilles and patron of botany
Poinsettia for Joel Roberts Poinsett (1779–1851), American statesman, in Mexico, who found *Euphorbia pulcherrima c.* 1828 (≡ *Euphorbia*)
Polanisia Great-variety, πολυ-ανισος (unequal length of the numerous stamens)
poissonii for M. Poisson (1833–1919), French botanist
polaris -is -e polar; of the North Star, *polus*, modern Latin from πολος, a pivot
Polemonium for King Polemon of Pontus (the name used by Pliny), πολεμος war (***Polemoniaceae***)
poli-, polio- grey-, πολιος, πολιο-, πολι-
Polianthes Grey-flowered, πολι-ανθος
polifolius -a -um, poliofolius -a -um *Teucrium*-leaved, grey-leaved, πολιος-φυλλον
poliochrous -a -um grey-skinned or complexioned, πολιος-χρως
Poliomintha Grey-mint, πολιο-μιντη
Poliothyrsis Greyish-panicle, πολιο-θυρσος (the colour of the inflorescence)
politus -a -um elegant, polished, *polio, polire, polivi, politum*
polius -a -um greyish-white, πολιος (*Teucrium polium* foliage)
pollacanthus -a -um flowering repeatedly, flowering often, πολλαχ-ανθος
Pollia Large, πολυς (stature and sometimes colonies)
pollicaris -is -e as long as the end joint of the thumb (*pollex*), about one inch
pollinosus -a -um as though dusted with fine flour or pollen, *pollen, pollinis*
polonicus -a -um from Poland, Polish (*Polonia*, land of the Polanei)
poluniniana for Oleg Polunin (1914–85)
poly- separate-, many-, πολυς, πολλη, πολυ
polyacanthus -a -um many-spined, πολυ-ακανθος
polyactinus -a -um many-rayed, πολυ-ακτις (cactus glochidia)
polyadenius -a -um with many glands, πολυ-αδην
Polyalthia Many-healing, πολυ-αλθομαι (the supposed properties of the flowers)
polyandrus -a -um with many stamens, πολυ-ανηρ
polyanthemos, polyanthus -a -um many-flowered, πολυ-ανθος
polyastrus -a -um with many stars, πολυ-αστηρ (flowers)
polyblepharus -a -um with many eyelashes, πολυ-βλεφαρον, (the leaf divisions)
polybotrya many-bunched, πολυ-βοτρυς (flowers)
Polycarpon, polycarpus -a -um Many-fruited, πολυ-καρπος (a name, πολυκαρπον, used by Hippocrates); fruiting repeatedly, polycarpic, πολυ-καρπος
Polyceratocarpus Many-horned-fruits, πολυ-κερατο-καρπος
polyceratus -a -um (polyceratius) many-horned, πολυ-κερας
polychromus -a -um many-coloured, πολυ-χρωμα
Polycodium Many-fleeced, πολυ-(κωας, κωδιον)
polycladus -a -um much branched, πολυ-κλαδος

polyedrus -a -um many-sided, πολυ-εδρος
polyepsis -is -e long-flowering, of many dawns, πολυ-εψος
Polygala Much-milk, πολυ-γαλα (Dioscorides' name, πολυγαλον, refers to the improved lactation in cattle fed on milkworts) (***Polygalaceae***)
polygalifolius -a -um with leaves resembling those of *Polygala*
polygaloides resembling *Polygala*, πολυ-γαλα-οειδης
polygamus -a -um the flowers having various combinations of the reproductive structures, πολυ-γαμος (of male, hermaphrodite and/or female)
Polygonatum Many-knees, πολυ-γονατον (the structure of the rhizome)
Polygonella Small-dock-like, feminine diminutive from *Polygonum*
polygonifolius -a -um with leaves like *Polygonum*, *Polygonum-folium*
polygonoides resembling *Polygonum*, *Polygonum-oides*
Polygonum Many-joints, πολυ-γοννα (Dioscorides' name may have referred to the fecundity, γονον, of docks, but others suggest a reference to the swollen nodes or knees, γονυ) (***Polygonaceae***)
polygrammus -a -um having many lines, πολυ-γραμμα (the striate stipe)
polygyrus -a -um twining, of many turns, πολυ-γυρος
polylepis -is -e very scaly, πολυ-λεπις
polylophus -a -um with many crests, πολυ-λοφος (areolar hairs)
polymorphus -a -um variable, of many forms, πολυ-μορφη
polymyces many-mushroomed, πολυ-μυκες (the large clusters of fruiting bodies)
polynesicus -a -um from the Polynesian islands
polyneurus -a -um many-veined, πολυ-νευρα
polynodus -a -um many-noded, πολυ-νοδα
polyodon many-toothed, πολυ-οδων (πολυ-οδους)
polypetalus -a -um having many petals, having separate petals, πολυ-πεταλον
polyphyllus -a -um with many leaf-segments, many-leaved, πολυ-φυλλον
polypodioides resembling polypody, *Polypodium-oides*
Polypodium Many-feet, πολυ-ποδιον, Dioscorides' reference to the rhizome growth pattern, polypody (***Polypodiaceae***)
Polypogon Many-bearded, πολυ-πωγων (the much-awned inflorescence)
Polyporus Many-pored, πολυ-πορος
polyrhizus -a -um with many roots, πολυ-ριζα
Polyscias Many-shades, πολυ-σκιας (the foliage)
Polyspatha Many-spathed, πολυ-σπαθη (each inflorescence has several recurved spathes)
polyspermus -a -um many-seeded, πολυ-σπερμα
polysphaerus -a -um with many globular heads, πολψ σφαιρα
Polystachya, polystachyus -a -um Many-spiked, πολυ-σταχυς (many spike-like panicles)
polystichoides resembling *Polystichum*, *Polystichum-oides*
polystichon with many rows, πολυ-στιχος (of spikelets)
Polystichum Many-rows, πολυ-στιχος (the arrangement of the sori on the fronds)
polystomus -a -um with many suckers or haustoria, πολυ-στομα
polytomus -a -um much incised or cut, πολυ-τομη
polytrichoides resembling *Polytrichum*, *Polytrichum-oides*
Polytrichum, polytrichus -a -um Many-hairs, πολυ-τριχος (the surface covering of the calyptra)
Polyxena For Polyxena, daughter of Priam and loved by Achilles
pomaceus -a -um pome-bearing, apple-green, apple-like, *pomum*
pomacochanus -a -um from the environs of the Pomacocha reservoir, central Peru
Pomaderris Lid-of-skin, πομα-δερρυς (the membrane covering the capsule)
pomanensis -is -e from Poman, NW Argentina
Pomax Operculum, πομα
pomedosus -a -um from Pomerania, po-morze, area around the S coast of the Baltic sea
pomeranicus -a -um from Pomerania, po-morze, Polish Baltic coastal area between the rivers Oder and Vistula

pomeridians, pomeridianus -a -um of the afternoon, pm, *post-meridianus* (afternoon flowering)
pomi- apple-like-, *pomum*
pomifer -era -erum pome-bearing, bearing apple-like fruits, *pomum-fero*
pomiformis -is -e apple-shaped, *pomum-forma*
pomponius -a -um having a top-knot or pompon, of great splendour, pompous, *pompa*
ponapensis -is -e from Ponapé, Caroline Islands, W Pacific
Poncirus from the French name, poncire, for Japanese bitter orange
ponderosus -a -um heavy, large, ponderous, *ponderosus*
pondoensis -is -e from Pondoland, eastern province of S Africa
poneanthus -a -um behind-flowered, with flowers not conspicuous, *pone-anthus*
Pontederia, pontederae for Guillo Pontedera (1688–1757), Professor of Botany at Padua (***Pontederiaceae***)
ponticus -a -um of the Black Sea's southern area, *Pontus* or *Pontica*
pontophilus -a -um living in the deep sea, ποντο-φιλος
poocolus -a -um inhabiting meadows, botanical Latin from ποα and *colo*
poophilus -a -um meadow-loving, ποα-φιλος
popayanus -a -um from Popayán, beneath the Puracé volcano, Colombia
popinalis of restaurants, *popina, popinonis* (*Rhodocybe popinalis* is of uncertain edibility)
populago women-like, crowding, *populus* with feminine suffix
populeus -a -um blackish-green (colour of leaves of *Populus nigra*)
populifolius -a -um poplar-leaved, *populeus-folium*
populinus -a -um of the poplar, *Populus* (bracket fungus also on other deciduous trees)
populneus -a -um (populnaeus) poplar-like, related to *Populus* (inquiline beetles)
populnifolius -a -um poplar-leaved, *populus-folium*
Populus the ancient name for poplar, old French pouplier, *arbor populi*, tree of the people
por- passage-, pore-, πορος
Porania a far-eastern vernacular name for snow-creeper, *Porania racemosa*
porcatus -a -um ridged, *porca, porcae*
porcinus -a -um of pigs, *porcus, porci*
poronaicus -a -um from the environs of Poronaysk and the Poronai river, S Sakhalin, Russia
Poronia Tuffa-like, πωρος (the appearance of the polyporous fruiting body with the dark dots of emerging perithecia)
porophilus -a -um loving soft stony ground, πορο-φιλος
porophyllus -a -um having (or appearing to have) holes in the leaves, πορο-φυλλον
porosporus -a -um with spores having pore(s) (the truncate-pored spores of *Boletus porosporus)*
Porphyra Purple, πορφυρα, the Greek name for the mollusc from which the dye was made (*Porphyra umbilicalis* is *Rhodophycean* laver of Welsh cuisine; porphyry is a reddish rock containing crystals, and porphyria is an hereditary disease involving light sensitivity and breakdown of haemoglobin)
Porphyrella the feminine diminutive from *Porphyra*
porphyreus -a -um, porphyrio n, porphyrius -a -um purple, bloody, warm-coloured, reddish, πορφυρεος (Porphyrion, one of 24 sons of Ge and Uranus, was a giant in Roman mythology)
porphyricolus -a -um living on porphyry rock formations, *porphyry-colo*
porphyrocephalus -a -um with a reddish head, πορφυρα-κεφαλη (the pileus)
porphyroglossus -a -um with a dark reddish-purple tongue, πορφυρο-γλωσσα
porphyroneurus -a -um with purple veins, πορφυρο-νευρα
porphyrophaeus -a -um bright reddish-purple, πορφυρο-φαιος
porphyrostelis -is -e having a purple stalk, πορφυρεος-στηλη
porosus -a -um with holes or pores, πορος, πορο-

Porpax Handle, πορπαξ (flower-shape suggests a shield-handle); some prefer Button (for the shape of the pseudobulbs)
porra-, porri- leek-like-, leek-, *porrum*-like-
porrectus -a -um spreading, long, protracted, *porrigo, porrigere, porrexi, porrectum*
porrifolius -a -um leek-leaved, *porrum-folium*
porrigens spreading, present participle of *porrigo, porrigere, porrexi, porrectum*
porrigentiformis -is -e porrigens-like, *porrigens-forma* (the leaf-margin teeth point outwards and forwards)
porriginosus -a -um with a very scurfy surface, with dandruff, *porrigo, porriginis*
porrum a Latin name used for various *Allium* species
Portenschlagia, portenschlagianus -a -um for Franz Elder von Portenschlag-Ledermeyer (1772–1822), Austrian botanist
portensis -is -e from Oporto (*Porto*), Portugal
portentosus -a -um unnatural, *potrentosus*
Portlandia for Margaret Cavendish Bentinck (1715–85), the Duchess of Portland, who corresponded with Rousseau
portlandicus -a -um from Portland Bill, England, or any other Portland (e.g. Jamaica or Oregon)
portoricensis -is -e from Puerto Rico, W Indies
portosanctanus -a -um from Porto Sano island, Madeira
portucasadianus -a -um from the environs of Puerto Casado, Chaco, Paraguay
portuguesanus -a -um from the Portuguesa state of NW Venezuela
portula abbreviated form of *Portulaca*
Portulaca from a name, *porcilacca*, in Pliny (cognate with porcelain and purslane) (***Portulacaceae***)
portulaceus -a -um *Portulaca*-like
portulacifolius -a -um having *Portulaca*-like leaves, *Portulaca-folium*
portulacoides resembling *Portulaca, Portulaca-oides*
porulus -a -um somewhat porous, πορος, diminutive of *porus*
poscharskyanus -a -um for Gustav Adolf Poscharsky (1832–1914), one-time garden inspector in Laubegast, Dresden
Posidonia, Poseidonia for Ποσειδον, god of the sea (*Neptune*) (*Posidonia oceanica*, used for packing glassware, ≡ *Potamogeton*) (***Posidoniaceae***)
post- behind-, after-, later-, *post-*
postianus -a -um, postii for Reverend George Edward Post (1838–1909), author of the *Flora of Syria, Palestine and Sinai*
posticus -a -um back, behind, turned outwards from the axis, extrorse, *posticus*
postmeridianus -a -um of the afternoon, *post-meridianus*
potam-, potamo- watercourse-, of watercourses-, river-, ποταμος, ποταμο-
Potamogeton Watercourse-neighbour, ποταμος-γειτων (the habitat) (***Potamogetonaceae***)
potamophilus -a -um watercourse- or river-loving, ποταμο-φιλος
potaninii for Grigori Nicholaevich Potanin (1835–1920), Russian explorer
potatorum of drinkers, *potor, potoris* (*Agave potatorum* is used for fermentation)
Potentilla Quite-powerful, diminutive of *potens* (as a medicinal herb)
potentillae living on *Potentilla* (symbionts, parasites or saprophytes)
potentillinus -a -um somewhat like *Potentilla*
potentilloides resembling *Potentilla*
poteriifolius -a -um with leaves resembling those of *Poterium*
Poterium Drinking-cup, ποτηριον (Dioscorides' name, ποτιρριον, for another plant)
pothinus -a -um longed-for, desired, ποθος
Pothos from a Sinhalese (Cingalese) vernacular name
potosiensis -is -e from Potosí mountain area, S Bolivia
pottsii for either John Potts (d. 1822), collector in China and Bengal, or George Honington Potts (1830–1907) of Lasswade, Midlothian, who introduced *Crocosmia pottsii*
poukhanensis -is -e from Pouk Han, Korea

-pous -foot, -stalk, -stalked, πους, ποδος
powellii for either Thomas Powell (1809–87), missionary in Samoa, or C. Baden Powell, of Tunbridge Wells (*Crinum powellii c.* 1885), or James Thomas Powell (1833–1904)
pradhanii for D. S. Pradhan of Chandra nursery, Sikkim *c.* 1930
prae-, pre- before-, compared-with-, in front-, *prae-*
praealtus -a -um very tall or high or deep, outstanding, *prae-altus*
praeandinus -a -um from the Andean foothills, botanical Latin from *prae-* and Andes
praecalvus -a -um with a transient indumentum, prematurely bald, *prae-calvus*
praecipuus -a -um special, outstanding, *praecipuus*
praeclarus -a -um beautiful, distinguished, very bright, *praeclarus*
praecocissimus -a -um the most quick or early to mature, superlative from *praecox*
praecox, praecocius -a -um premature, early-ripening, *prae-coxi* (*coquo, coqueri, coxi, coctum*)
praecultus -a -um of early cultivation, *praecolo, praecolere, praecolui, praecultum*
praecurrens spreading, running forwards, present participle of *prae-*(*curro, currere, cucurri, cursum*)
praeflorens early-flowering, *prae-florens;* tarnishing, *praefloro, praeflorare*
praegeri for Robert Lloyd Praeger (1865–1953), Dublin librarian and writer on *Sedum* and *Sempervivum* etc.
praegerianus -a -um for Dr Robert Lloyd Praeger (*vide supra*)
praegnacanthus -a -um having swollen thorns, *prae-*(*g*)*nasci-acanthus*
praegnans full, swollen, pregnant(-looking), *praegnans, praegnantis*
praelongus -a -um very long, *praelongus*
praemorsus -a -um as if nibbled at the tip, *prae-*(*mordeo, mordere, momordi, morsum*)
praenitens shining forth, seeming more attractive, *praeniteo, praeniter, praenitui*
praepinguis -is -e very rich, *prae-pinguis*
praeproperus -a -um rash, over-hasty, *praeproperus*
praerosus -a -um appearing to have been gnawed off, past participle of *praerodo, praerodere, praerosum*
praeruptorum of rough places, *praerupta* (living on screes)
praesignis -is -e conspicuous, *praesignis*
praestans pre-eminent, outstanding, *praestans*
praeteritus -a -um escaped, neglected, surpassed, excluded, past participle of *praetereo, praeterirr, praeterii, praeteritum*
praetermissus -a -um overlooked, omitted, *praetermitto, praetermittere, praetermisi, praetermissum*
praeteruptorus -a -um beyond violation, *praeter-*(*ruptor, ruptoris*)
praetervisus -a -um beyond the faculty of seeing, *praeter-visus* (the complexity of balistospore discharge in *Peziza*)
praetextus -a -um bordered, fringed, disguised, *praetextum*
praeustus -a -um appearing to have been scorched or frost-bitten, hardened at the tip, *praeustus*
praevernalis -is -e, praevernus -a -um before spring, early, pre-vernal, *prae-vernus*
pragensis -is -e from Prague, Czech Republic (*Praga, Pragensis*)
praireus -a -um of American open grasslands or prairies, via French, praerie, from Latin *pratum*
prasinatus -a -um, prasinus -a -um, prasus -a -um leek-green, leek-like, *prasinus* (for various *Allium* species)
prasophyllus -a -um with leaves similar to those of a leek, *prasinus*
pratensis -is -e of the meadows, *pratum, prati*
pratericolus -a -um of meadows, inhabitant of grassy places, *pratum-colo*
Pratia for M. C. L. Prat-Bernon (d. 1817), French naval officer who died on Freycinet's expedition
praticolus -a -um of meadows, inhabitant of grassy places, *pratum-colo*
prattii for Antwerp E. Pratt (fl. 1880–1915), British explorer in China, Tibet and New Guinea

pravissimus -a -um perversest, worst, most crooked, superlative of *pravus*
pravus -a -um crooked, deformed, *pravus*
precatorius -a -um relating to prayer, of petitions, *precor(ar), precatus* (*Abrus precatorius* seeds are used as rosary beads)
prehensilis -is -e grasping, *prehendo, prehendere, prehendi, prehensum* (flowers pollinated by insects that grasp the style or stamens)
preissii (*preisii*) for Dr Lugwig Preiss (1811–83), who collected in W Australia
Premna Stump-like, πρεμνον (the low habit)
premnoides resembling *Premna, Premna-oides*
prenans bent forwards, drooping, πρανης, πρηνης
Prenanthes, prenanthus -a -um Drooping-flower, πρηνη-ανθος (the nodding flowers)
prenanthoides resembling *Prenanthes, Prenanthes-oides*
preptus -a -um eminent, conspicuous, πρεπω
Prestonia for Charles Preston (1660–1711), Professor of Botany at Edinburgh
pretiosus -a -um valuable, precious, extravagant, *pretiosus*
Priestleya for Dr Joseph Priestley (1733–1804), who demonstrated that respiratory products (CO_2) of mice improve the growth of plants
primavernus -a -um earliest of spring, *prima-vernus*
primeria not a valid name (*primarius* first-rate?)
primitivus -a -um typical, first of its kind, *primitivus* (in contrast to hybrids and varieties)
primiveris -is -e first of spring, *primus-(ver, veris)*
Primula Little-firstling, feminine diminutive of *primus* (spring flowering) (***Primulaceae***)
primulaceus -a -um *Primula*-like
primulaize resembling *Primula*, a name for a *Saxifruga* hybrid, botanical Latin from *primula-izare* or *primula* with ιδειν
primuliflorus -a -um with primrose-like flowers, *Primula-florum*
primulifolius -a -um with leaves resembling those of *Primula-folium*
primulinus -a -um primrose-coloured, *Primula*-like
primuloides resembling *Primula, Primula-oides*
princeps, principis -is -e most eminent or distinguished, first-head, *princeps, principis*
pringlei for Cyrus Guernsey Pringle (1838–1911) of Vermont, USA, who collected in Mexico
prinoides resembling *Prinos, Prinos-oides* (oak-like)
prinophyllus -a -um having *Prinos*-like leaves, πριν-φυλλον
Prinos, prinos (*Prinus*) Earliest, πριν (ancient, preceding, before) (≡ *Quercus* section)
Prinsepia for James Prinsep (1778–1840), meteorologist of the Asiatic Society of Bengal
prio-, priono- serrated-, saw-toothed-, πριων, πριω-, πριονος, πριονο-
priochilus -a -um saw-lipped, πριο-χειλος
Prionitis Saw-like, πριωνιτις (an ancient name, πριονιτις, for *Stachys alopecurus*, for the similarly sharply serrated leaf margins)
Prionium Small-saw, πριωνιον (the serrated margins of the terminally clustered leaves)
prionochilus -a -um having a serrate lip or labellum, πριονο-χειλος
prionopetalus -a -um with serrated petals, πριονο-πεταλον
prionotes like a small saw, with a few serrations, πριονων, πριονος
priscus -a -um ancient, old-fashioned, *priscus*
prismati-, prismaticus -a -um prism-, prism-like-, angular, as if sawn, πριζειν
pristisepalus -a -um having cut or sawn petals, πριστος-σκεπη
pro- forwards-, for-, instead of-, before-, pro-, προ
proboscidius -a -um snout-like, προ-βοσκις (the spadix of the mouse plant *Arisarum proboscidium*)
Proboscoidea Snout-like, προ-βοσκις-οειδης (προ-βοσκειν, for obtaining food)

proboscoides, proboscoideus -a -um snout-like, trunk-like, προβοσκις-οειδης
probus -a -um upright, excellent, *probus*
procerus -a -um very tall or long, *procerus*
proclivis -is -e drooping, sloping downwards, *pro-clivus*
procumbens lying flat on the ground, creeping forwards, procumbent, *procumbo, procumbere, procumbi, procumbitum*
procurrens spreading below ground, running forwards, *procurro, procurrere, procurri, procursum*
prodigiosus -a -um unnatural, marvellous, prodigious, *prodigiosus*
productus-a -um stretched out, extended, produced, *produco, producere, produxi, productum*
profugus -a -um exiled, nomadic, *profugus*
profundeincisus -a -um deeply incised, *profundus-incisus* (*Acaena* leaves)
profundus -a -um large, very tall, *profundus*
profusoflorus -a -um very free-flowering, *profusus-florum*
profusus -a -um very abundant, profuse, *pro-*(*fundo, fundere, fusi, fusum*)
prolatus -a -um widened, extended, past participle of *profero, proferre, protuli, prolatum*
prolepticus -a -um developing early, precocious, προλεψις
prolifer -era -erum proliferous, producing bunched growth or offsets or young plantlets, *proles-fero*
prolificans proliferating, present participle from *proles-(fico, ficare)*
prolificus -a -um very fruitful, modern Latin *prolificus*
prolixus -a -um long, full, wide, spreading, *prolicio, prolicere, prolixi*
prominens outstanding, extending, projecting, present participle of *promineo, prominere, prominui*
pronatus -a -um, pronus -a -um lying flat, with a forward tilt, *pronus*
propaguliferus -a -um prolific, multiplying by vegetative propagules, *propago-fero*
propendens, propensus -a -um hanging down, present participle of *propendeo, propendere, propendi, propensum*
propinqua-grandiflora close to (*Campanula*) *grandiflora, propinquus-*(*Campanula*) *grandiflora*
propinquus -a -um closely allied, of near relationship, related, *propinquo, propinquare*
proponticus -a -um from around the Sea of Marmara (*propontis*), between the Bosporus and the Dardanelles, Turkey
prorepens creeping out, creeping forwards, *pro-repens*
pros- near-, in addition-, also-, against-, towards-, προς, προ-, *pro*
Proserpinaca an ancient name used in Pliny, προσερπτω to creep along (Proserpine, or Persephone, was the wife of Pluto)
proserpinacoides resembling *Proserinaca*, προσ-ερπω-οειδης
proserpinensis -is -e from Proserpine, E Queensland, Australia
proso-, prostho- towards-, to the front-, before-, on-the-side-of-, προς
Prosopis the name used by Dioscorides for butterbur
Prostanthera Standing-before-the-anther, προιστημι-ανθηρα (appendages)
prostigiatus -a -um having a forward exposed stigma, previously branded, *pro-*(*stigma, stigmatis*) a name for a *Rhododendron* hybrid,
prostratus -a -um lying flat but not rooting, prostrate, *prostratus*
Protea for Proteus (the sea god's versatility in changing form) (***Proteaceae***)
proteanus -a -um able to assume many shapes, pretending, προτεινω (proteism)
proteiflorus -a -um with *Protea*-like flowers, *Protea-florum*
proteioides resembling *Protea, Protea-oides*
proteino- stretched out-, exposed, feigning-, προ-τεινω (protein derives via French, protéine, from προτος, προτειος, primary)
protentus -a -um stretching out, spreading, *protendo, protendere, protendi, protentum*
proter-, protero-, proto- first-, προτερος, προτερο-
proterandrus -a -um male-first, with anthers maturing before the stigma, προτερος-ανηρ

proterogynus -a -um female-first, with the stigma becoming receptive before self pollen is released, προτερο-γυνη
protervus -a -um violent, precocious, bold, *protervus*
protistus -a -um earliest, number-one, the first of the first, superlative of προτος
Protomegabaria Former-*Megabarya*, botanical Latin from προτερος and *Megabarya* (relationship to the genus *Megabarya*)
protopunicus -a -um ancestral pomegranate, botanical Latin from πτροτερεω and *Punica*
protractus -a -um revealing, drawn-out, *protraho, protrahere, protraxi, protractum*
protrusus -a -um thrusting forwards, pushing out, *protrudo, protrudere, protusi, protrusum*
protuberans bulging outwards, *pro-*(*tuber, tuberis*)
provincialis -is -e from Provence, France (*Provincia*)
proximus -a -um next, nearest, *proximus*
pruhonicianus -a -um, pruhonicus -a -um from Pruhonice, Czech Republic
pruinatus -a -um, pruinosus -a -um powdered, with a hoary bloom as though frosted-over, *pruina, pruinatus*
pruinocarpus -a -um having waxy-bloomed fruit, botanical Latin from *pruinus* and καρπος
Prumnopitys Hindmost-pine, πρυμνος-πιτυς
Prunella (Brunella), prunella from the German name, die Braüne, for quinsy, for which it was used as a cure
prunelloides resembling *Prunella, Prunella-oides*
pruni- plum-like, plum-, προυμνον, *Prunus, Pruni*
pruni living on *Prunus* (symbionts, parasites and saprophytes)
pruniferus -a -um bearing plum-like fruits, *Prunus-fero*
pruniflorus -a -um having plum-like flowers, *Prunus-florum*
prunifolius -a -um plum-leaved, *Prunus-folium*
pruniformis -is -e plum-shaped, *Prunus-forma* (*Sarcococca* fruits)
prunophorus -a -um bearing plums, *Prunus-fero*
prunulus -a -um somewhat-prune-like, comparative of *prunus* (the ridged spores)
Prunus the ancient Latin name, *Prunus, Pruni*, for a plum tree
pruriens irritant, stinging, itch-causing, present participle of *prurio, prurire* (hairs on the fruits of *Mucuna pruriens*)
prussicus -a -um from Prussia (the areas ruled by the Hohenzollern dynasty)
pruthenicus -a -um from the environs of the Pruth river dividing Romania and Moldova
przewalskii for Nicholas M. Przewalski (1839–88), Russian explorer and collector in China (for whom, Przewalski's horse)
psamma-, psammo- sand-, ψαμμος
Psamma Strand-dweller, ψαμμος (the old generic name for marram grass refers to its habitat)
Psammisia for Psammis, Egyptian ruler *c.* 376 BC
psammophilus -a -um liking sandy habitats, ψαμμος-φιλος
psammopus -a -um sandy-stalked, downy-stalked, ψαμμος-πουσ
psaridis -is -e having surface markings like a starling, ψαρος
pseud-, pseudo- sham-, false-, ψευδης, ψευδος, ψευδο-, ψευδ-,
pseudacacia false or pseudo *Acacia* (the similar appearance of *Robinia pseudacacia*)
pseudacantholimon false or pseudo *Acantholimon*
pseudachillea false or pseudo *Achillea*
pseudacorus false *Acorus*, ψευδος-ακορον
Pseudagrostistachys False-grass-like-spike, ψευδος-αγρωστις-σταχυς (refers to the short axillary racemes)
pseudarmenia false or pseudo *Armeria*
Pseuderanthemum false or pseudo *Eranthemum*
Pseudoaegle False-*Aegle*, ψευδος-αεγλε
pseudoalpina false or pseudo (*Clematis*) *alpina*

pseudoambiguus false or pseudo (*Cotoneaster*) *ambiguus*
pseaudoaxillaris -is -e false or pseudo (*Carex*) *axillaris*
Pseudobombax false or pseudo *Bombax*
pseudocapsicastrum false or pseudo (*Solanum*) *capsicastrum*
pseudocapsicum false or pseudo (*Solanum*) *capsicum*
pseudocerasus false or pseudo (*Prunus*) *cerasus*
pseudochina false or pseudo Chinese (*Smilax*)
pseudochrysanthum false or pseudo (*Rhododendron*) *chrysanthemum*
pseudococcinius -a -um false or pseudo *coccineus*
Pseudocydonia false or pseudo *Cydonia* (*Chaenomeles*)
pseudocyperus false or pseudo *Cyperus* (*Carex*)
pseudocystopteris false or pseudo *Cystopteris* (*Ariostegia*)
Pseudocytisus false or pseudo *Cytisus* (*Vella*)
pseudodictamnus false or pseudo *Dictamnus* (*Ballota*)
pseudodistans false (*Pucinellia*) *distans*
Pseudoechinolaena false or pseudo *Echinolaena*
pseudofarinaceus -a -um with a false farina, *pseudo-farina*
pseudofennica false or pseudo (*Sorbus*) *fennica*
pseudofumarioides false or pseudo (*Pelargonium*) *fumarioides*
pseudoginseng false or pseudo (*Panax*) *ginseng*
pseudoglutinosum false or pseudo (*Pelargonium*) *glutinosum*
Pseudognaphalium false or pseudo *Gnaphalium*
pseudograndidentata false or pseudo (*Populus*) *grandidentata*
pseudohelvola false or pseudo *Helvola*
pseudohenryi false or pseudo (*Lysimachia*) *henryi*
Pseudohydnum false or pseudo *Hydnum*
pseudoibericum false or pseudo (*Cyclamen*) *ibericum*
pseudointegrifolia false or pseudo (*Meconopsis*) *integrifolia*
pseudointegrus -a -um incomplete, false completeness, *pseudo-integritas* (partial network of lines)
pseudokotschyi false or pseudo (*Saxifraga*) *kotschyi*
pseudolanuginosus false or pseudo (*Thymus*) *lanuginosus*
Pseudolarix false or pseudo *Larix*
pseudolaxiflora false or pseudo (*Linaria*) *laxiflora*
pseudomas, pseudo-mas false or pseudo (*Dryopteris* or *Lastraea*) *mas*
pseudomezereum false or pseudo (*Daphne*) *mezereum*
Pseudomonas False-monad (broad name for certain ciliate rod-bacteria, some being pathogenic)
pseudomuscari false or pseudo (*Muscari*) *muscari*
pseudomyrsinites false or pseudo (*Salix*) *myrsinites*
pseudonarcissus false or pseudo *Narcissus poeticus*
pseudonatronatus false or pseudo (*Rumex*) *natronatus*
Pseudopanax false or pseudo *Panax*
pseudoparadoxa false or pseudo (*Carex*) *paradoxa*
pseudopetiolatum false or pseudo (*Hypericum*) *petiolatum*
Pseudophegopteris false or pseudo *Phegopteris*
Pseudophoenix false or pseudo *Phoenix* (resemblance to date palm)
pseudopilosella false or pseudo *Pilosella*, with a false felting, false (*Conocybe*) *pilosella*
pseudoplatanus false or pseudo (*Acer*) *platanus*
pseudopraevignus false step-son, *praevignus, praevigni* (meaning unclear)
pseudopulcher false or pseudo (*Rumex*) *pulcher*
pseudopumilum false or pseudo (*Iris*) *pumilum*
pseudopura false or pseudo (*Mycena*) *pura*
Pseudorchis false or pseudo *Orchis*
pseudoreticulata false or pseudo (*Vitis*) *reticulata*
Pseudorlaya false or pseudo *Orlaya* (≡ false or pseudo *Daucus*)
Pseudosassa false or pseudo *Sassa*

pseudoscaber false or pseudo (*Asparagus*) *scaber*, false or pseudo (*Porphyrellus*) *scaber*
pseudosecalinus -a -um false or pseudo (*Bromus*) *secalinus*
pseudosibiricum false or pseudo (*Geranium*) *sibiricum*
pseudosieboldianum false or pseudo (*Acer*) *sieboldianum*
pseudosikkimensis -is -e false or pseudo (*Primula*) *sikkimensis*
pseudospathulata false or pseudo (*Alstoemeria*) *spathulata*
Pseudostellaria false or pseudo *Stellaria*
pseudostrobus false or pseudo (*Pinus*) *strobus*
pseudosuber false or pseudo (*Quercus*) *suber*
pseudotinctoria false or pseudo (*Inidigofera*) *tinctoria*
Pseudotsuga false or pseudo *Tsuga*
pseudoturneri false or pseudo (*Quercus*) *turneri*
pseudoversicolor false or pseudo (*Pedicularis*) *versicolor*
pseudoviola false or pseudo (*Impatiens*) *viola*
pseudovirgata false or pseudo (*Euphorbia*) *virgata*
Pseudowintera false or pseudo *Wintera*
Psidium a Greek name, ψιδιον (formerly for the pomegranate, for the similarity of the fruits)
psilanthus -a -um slender-, smooth- or naked-flowered, ψιλο-ανθος
psilicolus -a -um prairie-dwelling, living in empty places, botanical Latin from ψιλος and *colo*
psilo- slender-, smooth-, bare-, ψιλος, ψιλο-
psilocaulis -is -e slender or bare stemmed, ψιλος-καυλος
Psilocybe Bald-head, ψιλος-κυβη (the pileus of the liberty cap toadstool resembles the Phrygean bonnet given to freed Roman slaves)
psiloglottis -is -e having a smooth tongue, lip or labellum, ψιλος-γλωττα
psilostachyus -a -um having slender spikes, ψιλος-σταχυς
Psilostemon, psilostemon Slender- or Naked-stamens, ψιλος-στεμον
Psilostrophe Naked-carrier, ψιλος-(τροφος, τροφευς) (the naked receptacle as the 'nurse')
Psilotum Naked, ψιλος (hairless) (***Psilotaceae***)
psittacinus -a -um parrot-like, ψιττακος (contrasted colouration of parrot toadstool)
psittacoides resembling a parrot, ψιττακος-οειδης (brightly coloured flower-heads of *Cirrhopetalum*)
psittacorus -a -um of parrots, ψιττακος, *psittacus*
Psittacula Parrot-like, feminine diminutive from ψιττακος, *psittacus* (the troublesome ring-necked parakeet)
Psoralea Warted-one, ψοραλεος (the dot-marked vegetative parts)
Psyche Love, for Psyche (one of the Dryad nymphs married to Cupid, often personified in female form as the soul of man or as a butterfly)
psychodes, psycodes butterfly-like, ψυχη-ωδες (like Psyche the Dryad nymph)
Psychotria Refreshment, ψυχη (for the reputed medicinal properties of some)
psychrophilus -a -um cold-loving, ψυχρος-πιλος
Psylliostachys Bare-spike, ψιλο-σταχυς
Psyllium, psyllium of fleas, ψυλλα (from a Greek name, ψυλλιον, refers to the resemblance of *Plantago psyllium* seed to a fleas, ψυλλα) (gall-forming jumping plant lice, *Psylla*, are homopterans, not fleas)
psyllocephalus -a -um having a head of seeds looking like fleas, ψυλλα-κεφαλη
Ptarmica, ptarmica causing sneezes, πταρμικος (Dioscorides' onomatopoeic generic name, πταρμικη) (Gerard translated the apothecaries' *herba sternutatoria* as sneezewort, *Achillea ptarmica*)
ptarmiciflorus -a -um with sneezewort-like flower heads, *ptarmica-florum*
ptarmicoides resembling *Achillea ptarmica*, *ptarmica-oides*
Ptelea the ancient Greek name, πτελεα (for elm, transferred for the similarity of the fruit)
Pteleopsis *Ptelea*-like, πτελεα-οψις (resembling the hop-tree)

pteno- deciduous-, πτηνος
ptera-, ptero-, -pteris -is -e, ptery- with a wing-, winged-, πτερον, πτερο-, πτερυξ
Pteracanthus Winged-*Acanthus*, πτερ-ακανθος
Pteranthus, pteranthus -a -um with winged flowers, πτερ-ανθος
pteridifolius -a -um having fern-like leaves, *pteris-folium*
pteridioides resembling *Pteridium, Pteridium-oides*
pteridis -is -e of ferns, living on *Pteris* (*Eriophyes*, acarine gall mite)
Pteridium Small-fern, πτεριδιον (diminutive of πτερις)
Pteridophyllum Fern-leaf, πτερις-φυλλον (the pinnatisect leaf-blades)
pterifolius -a -um fern-leaved, *Pteris-folium*
pteriphilus -a -um fern-loving, πτερυξ-φιλος (epiphytic orchid)
Pteris Feathery, πτερυξ (the Greek name for a fern)
-pteris -fern, -wing-like, -winged, πτερυξ
Pterocarpus, pterocarpus -a -um Winged-fruit, πτερο-καρπος
Pterocarya Winged-nut, πτερο-καρπος (the winged fruits of most)
pterocaulis -is -e, pterocaulon with winged stems, πτερο-καυλος
Pteroceltis Winged-*Celtis*, πτερο-κηλτις (has winged seeds)
Pterocephalus, pterocephalus -a -um Winged-head, πτερο-κεφαλη (the appearance of the senescent flower-heads)
pterocladon having winged branches, πτερο-κλαδος
Pteroglossaspis Winged-tongue-shield, πτερο-γλωσσα-ασπις
pteroneurus -a -um with feathered venation or winged veins, πτερο-νευρα
pterospermus -a -um with winged seeds, πτερο-σπερμα
pterosporus -a -um with winged spore, πτερο-σπορα
pterostoechas winged (*Lavandula*) *stoechas*
Pterostylis Winged-style, πτερο-στυλος (the column is winged)
Pterostyrax Winged-*Styrax*, πτερο-στυραξ (one species has winged fruits)
-pterus -a -um -winged, πτερυς, πτερυγος (mostly meaning frond, or -fronded)
-pterygius -a -um -small-winged, πτερυγιον
Pterygota Winged-ear, πτερυγιον-ωτος (the *Acer*-fruit-like seed)
ptilanthus -a -um having downy (feathery) flowers, πτιλον-ανθος
ptilo- feathery-, πτιλον, πτιλο-
ptiloglossus -a -um with a feathery-textured surface of the lip, πτιλο-γλωσσα
Ptilostemon Feathery-stamened-one, πτιλον-στεμων
Ptilotus (Ptilotum) Feathered, πτιλον (≡ *Trichinium*)
ptolemaicus -a -um for the Greek astronomer and geographer Claudius Ptolemy (127–145 BC), author of μεγαλη συνταξις της αστρονομαις, which, as Almagest, was a standard astronomical treatise until Copernicus and Kepler
-ptosis -is -e -flapped, πτοσυς
ptycho- layered-, folded-, of dells or clefts-, πτυξη, πτυξ, πτυχος, πτυχο-
Ptychopyxis Folded-capsule, πτυχος-πυξις
Ptychosperma Folded- or Cleft-seed, πτυχος-σπερμα
pubens full-grown, juicy, *pubens*
pubera grown-up; downy, *pubes (puber), puberis*
puberulus -a -um somewhat downy, diminutive of *puber*
pubescens maturing, attaining maturity; becoming downy-hairy, present participle of *pubesco, pubescere, pubescui*
pubi-, pubigenus -a -um, pubigerus -a -um hairy, *pubes-gero*
pubibundus -a -um with much downy hair, *pubes-abunde*
pubicalyx with a downy calyx, botanical Latin from *pubes* and καλυξ
pubiflorus -a -um with downy flower surfaces, *pubes-florum*
pubigerus -a -um hairy, bearing hairs, *pubes-gero*
Pubilaria Hairy, *pubes* (the clothing of fibrous leaf remains on the rhizome)
pubinervis -is -e having hairs along the veins, *pubes-nerva*
pubirameus -a -um having pubescent branches, *pubes-ramus*
Puccinellia for Benedetto Puccinelli (1808–50), Professor of Botany at Lucca
puddum from a Hindi name for a cherry

pudens drooping, modest, present participle of *pudet, pudere, puduit*
pudibundus -a -um modest, *pudibundus*
pudicus -a -um retiring, chaste, modest, bashful, *pudicus* (*Mimosa pudica*, the humble plant, 'hides' itself when touched, by folding together all its leaflets)
pudorosus -a -um very bashful, *pudet, pudere, puduit* (*puditum*) (shy flowers)
Puelia, puelii for Timothée Puel (1812–90), French botanist in Syria
puellaris -is -e girlish, young wife; youthful, *puella, puellae*
Pueraria for Marc Nicolas Puerari (1765–1845), Swiss Professor of Botany at Copenhagen
puertoricensis -is -e from Puerto Rica, W Indies
pugionacanthus -a -um having dagger-shaped thorns, botanical Latin from *pugionis* and ακανθος
pugioniformis -is -e (us) dagger-shaped, (*pugio, pugionis*)-*forma*
pugnax obstinate, aggressive, *pugnax, pugnacis*
pugniformis -is -e with the shape of a fist, botanical Latin from (πυξ, πυγμη) and *forma*
pulchellus -a -um beautiful, pretty, diminutive of *pulcher*
pulcher -chra -chrum beautiful, handsome, fair, *pulcher, pulchri*
pulcherrimus -a -um most beautiful, most handsome, superlative of *pulcher*
pulchriflorus -a -um having beautiful flowers, *pulchri-florum*
pulegioides resembling *Pulegium, Pulegium-oides*
Pulegium, pulegius -a -um Flea-dispeller, *pulex, pulicis* (Pliny's name, *pulegium*, for a plant whose burning leaves kill fleas)
Pulicaria, pulicarius -a -um Fleabane, Latin name for a plant which wards off fleas, *pulex, pulicis*
pulicaris -is -e of fleas, *pulex, pulicis* (e.g. the shape of the fruits)
pullatus -a -um clothed in black, sad-looking, mournful, *pullus*
pulloides resembling (*Campanula*) *pulla*
pullus -a -um raven-black, almost dead-black, *pullus*
Pulmonaria Lung-wort, *pulmones* (the signature of the spotted leaves, *herba pulmonariae maculosae*, as indicative of efficacy in the treatment of respiratory disorders)
pulmonarioides resembling *Pulmonaria, Pulmonaria-oides*
pulmonarius -a -um like the lungs, *pulmo, pulmonis* (appearance or texture)
pulposus -a -um fleshy, pulpy, *pulpa, pulpae*
Pulsatilla, pulsatillus -a -um Quiverer, *pulsata* (Brunfels' name for the movement of the flowers in the wind, *'pulsatione floris vento'* in Linnaeus)
pulveratus -a -um, pulverulus -a -um, pulverulentus -a -um covered with powder, powdery, full of dust, *pulvis, pulveris*
pulviger -era -erum dusted, powdered, bearing dust, *pulvis-gero*
pulvinaris -is -e sacred couch, cushioned, pulvinate, *pulvinus, pulvini* (the swollen appearance of these organs of movement)
pulvinatus -a -um cushion-like, cushion-shaped, with pulvini, *pulvinus, pulvini pulvinus, pulvini*
pulviniformis -is -e having swellings resembling pulvini, *pulvinus-forma*
pumilio, pumilus -a -um very small, low, small, dwarf, *pumilio, pumilionis*
pumilionus -a -um dwarf, *pumilio, pumilionis* (habit)
punctati-, puncti-, punctatus -a -um punctate, with a pock-marked surface, spotted, *pungo, pungere, pupugi, punctum*
punctatifolius -a -um having leaves bearing dots or small indentations, *punctus-folium*
punctatissimus -a -um the most puncate or spotted, superlative of *punctatus*
puncticulatus -a -um, puncticulosus -a -um covered in small spots, *punctum, puncti*
punctilobulus -a -um dotted-lobed, botanical Latin from *punctatus* and λοβος
punctorius -a -um dotted, spotted, punctate, *punctum, puncti*
punctulatus -a -um covered in small dots, minutely punctate, *punctum, puncti*
pungens ending in a sharp point, pricking, *pungo, pungere, pupungi, punctum*

Punica from a name, *malum punicum*, Carthaginian apple; in Pliny, *malum granatum*, many-seeded apple, gave the old French, pume grenate, and our pomegranate (***Punicaceae***)
puniceus -a -um crimson, carmine-red, pomegranate-coloured, *puniceus*
punici- pomegranate-like, *Punica-*
punicus -a -um from Tunisia, Phoenician, (φοενικες); pomegranate, reddish-purple
punjabensis -is -e from the Punjab area of NW India and E Pakistan
puralbus -a -um plain- or pure-white, *purus-albus*
purdomii for William Purdom (1880–1921), collected for Veitch in China with Farrer
purdyi for Carl Purdy (1861–1945), student of the California flora
purgans purging, cleansing, present participle of *purgo, purgare, purgavi, purgatum*
purgus -a -um purgative (the officinal root, purga de Jalapa, of *Ipomoea purga*)
purpurascens, purpurescens becoming purple, *purpureus-essentia*
purpuratus -a -um empurpled, purplish, *purpureus*
purpureifolius -a -um purple-leaved, *purpureus-folium*
purpurellus -a -um somewhat empurpled, diminutive of *purpureus*
purpureoauratus -a -um purple and golden coloured, *purpureus-aureus*
purpureo-badius -a -um purple and reddish-brown, *purpureus-badius*
purpureobracteatus -a -um with purple bracts, *purpureus-brattea*
purpureocaeruleus -a -um purple and blue coloured, *purpureus-caeruleus*
purpureo-maculatus -a -um having purple spots, *purpureus-macula*
purpureopilosus -a -um having a purple indumentum, *purpureus-pilosus*
purpureospathus -a -um with purple spathes, *purpureus-spatha*
purpureosplendes glowing purple, present participle from *purpureus-(splendeo, splendere)*
purpureus -a -um reddish-purple, *purpureus*
purpurinus -a -um somewhat purplish, diminutive of *purpureus*
purpusii for either of the brothers J. A. and C. A. Purpus of Darmstadt
Purshia, purshianus -a -um for Frederick Traugott Pursh (Pursch) (1774–1820), author of *Flora Americae septentrionalis (Rhamnus purshiana* yields cascara sagrada*)*
purus -a -um clear, spotless, pure, chaste, unadorned, *purus*
-pus -foot, πους, ποδος
Puschkinia for Count Graffen Apollos Apollosovitsch Mussin-Puschkin (d. 1805), Russian phytochemist and plant collector in the Caucasus
pusilliflorus -a -um having very small flowers, *pussilus-florum*
pusillus -a -um weak, insignificant, minute, very small, slender, *pusillus*
pustulatus -a -um as though covered with blisters or pimples, *pustula, pustulae*
pustulosus -a -um pustuled, pimpled, *pustula, pustulae*
putatus -a -um, puteolatus -a -um pitted, *puteus, putei*
puteanus -a -um of rot, *puter, putris* (*Coniophora* is a wet-rot fungus and its fruiting body is rough with pits, *puteus, putei*)
putens foetid, stinking, present participle of *puteo, putere*
puteorum of the pits, *puteus, putei*
Putoria Stinker, *puteo, putere* to stink
putumayensis -is -e, putumayo from Putumayo department, S Colombia
putus -a -um entirely pure, *putus*
Puya from the Chilean vernacular name
pycn-, pycno- close-, densely-, compact-, dense-, πυκνος, πυκνο-, πυκν-
pycnacanthus -a -um having dense spines or thorns, πυκνος-ακανθος
Pycnanthemum Densely-flowered, πυκν-ανθεμος
Pycnanthus, pycnanthus -a -um Densely-flowered, πυκν-ανθος
Pycnobotrya Dense-bunched, πυκνος-βοτρυς
pycnocephalus -a -um with a dense head, πυκνο-κεφαλη (of flowers)
pycnophyllus -a -um with densely arranged leaves, πυκνο-φυλλον
Pycnoplinthopsis Resembling-compact-*Plinthus*, πυκνο-πλινθος-οψις (*Plinthus* is a Cape coast *Aizoaceaen* genus)

pycnosorus -a -um with a dense covering of sori, πυκνος-σωρος
Pycnostachys, pycnostachyus -a -um close-spiked, πυκνο-σταχυς
pycnotrichus -a -um with dense hairs, πυκνο-τριχος
pygmaeus -a -um, pygmeus -a -um dwarf, πυξ, πυγμαιος (the size of a fist, some say the length from elbow to fist)
pylzowianus -a -um for Mikhai Alexandrovich Pylzov, who collected in China *c.* 1870
pyr- fire-, πυρ, πυρος, πυρρος
Pyracantha Fire-thorn, πυρ-ακανθα, could equally be for flower colour or the lasting effects of pricks by its thorns (Dioscorides' name, πυρακανθα)
pyracanthus -a -um fire-thorned, πυρ-ακανθα (persistent irritation caused by the thorns)
Pyracomeles name of hybrids between *Pyrus* and *Chaenomeles*
pyrainus a name used by Rafinesque-Schmaltz for a *Pyrus* synonymous with *P. amygdaliformis* (πυρα, a funeral place or pyre, a fire place)
pyrami for Pyramus, Thisbe's lover
pyramidalis -is -e, pyramidatus -a -um conical, pyramidal, πυραμις
pyraster an old, derogatory generic name, *Pyrus-aster*
-pyren, pyreno- kernel-, stone-, πυρην
pyrenaeus -a -um, pyrenaicus -a -um from the Pyrenees mountain range (*Pyrene, Pyrenes*)
pyrenomyces fire-fungi, πυρην-μυκες (appear on burnt earth)
pyrethrifolius -a -um having leaves resembling those of *Pyrethrum, Pyrethrum-folium*
Pyrethrum, pyrethrum Fire, πυρ-εθρον (medicinal use in treating fevers)
pyri-, pyri pear-, *pirus, pyrus*, living on *Pyrus* (*Lastodiplosis*, dipteran gall midge)
pyriferus -a -um bearing fruits resembling pears, *pyrus-fero*
pyrifolius -a -um having leaves similar to those of pear, *Pyrus-folium*
pyriformis -is -e pear-shaped, *Pyrus-forma*
pyriodorus -a -um pear-scented, *Pyrus-odor*
pyro-, pyrro-, pyrrho- fire-, πυρ, πυρος, πυρο-, πυρρος, πυρρο-
Pyrocrataegus the name formula for hybrids between *Pyrus* and *Crataegus*
Pyrocydonia the name formula for graft hybrids, chimaeras, between *Pyrus* and *Cydonia*
pyrogalus -a -um fiery or burning milk, (πυρ, πυρος)-γαλα (the taste of the sap)
Pyrola Pear-like, *Pyrus* (compares the leaves) (Turner Englished the German name Wintergrün, as wintergreen) (***Pyrolaceae***)
pyrolifolius -a -um, pyrolaefolius -a -um with *Pyrola*-like leaves, *Pyrola-folium*
pyroliflorus -a -um, pyrolaeflorus -a -um having *Pyrola*-like flowers, *Pyrola-florum*
Pyrolirion Fire-lily, πυρο-λειριον (flower-colour)
pyroloides resembling *Pyrola, Pyrola-oides*
Pyronia the composite name for hybrids between *Pyrus* and *Cydonia*
pyropaeus -a -um fiery-eyed; deep-red, bronzed, πυροπος, *pyropus*
pyrophilus -a -um fire-loving, growing on burnt earth, πυρο-φιλος
pyrotechnicus -a -um fiery art, flamboyant, (πυρ, πυρος)-(τεχνη, τεχνικος)
pyrotrichus -a -um with flame-coloured hairs, πυρο-τριχος
Pyrostegia Fiery-roof, πυρο-στεγη (the ruddy colour of the upper corolla lobes)
Pyrrheima Red-clothed, πυρρος-ειμα (the covering of red hair, ≡ *Tradescantia*)
Pyrrocoma Flame-leaved, πυρρος-κομη
Pyrrosia Fire-coloured, πυρρος
Pyrrhula Little-flame, feminine diminutive from πυρρος (the troublesome bullfinch)
Pyrularia, pyrularius -a -um Little-pear, diminutive of *Pyrus* (in allusion to shape of the fruit)
-pyrum -wheat, πυρος
Pyrus from the ancient Latin name, *pirus*, for a pear tree
Pythium Rot-causing, πυθω (saprophytic fungi)

Pyxidanthera Lidded-box-anthers, πυξιδ-ανθερα
pyxidarius -a -um like a small lidded box, πυξις
pyxidiferus -a -um carrying a small box-like structure, πυξιδιον-φερω
pyxidatus -a -um small-box-like, πυξιδιον, diminutive of πυξις (e.g. some stamens)

quad-, quadri- four-, *quattuor; quad-, quadra-, quadri-, quadro-*
quadrangularis -is -e, quadrangulatus -a -um with four angles, quadrangular, *quadra-angularis*
quadratus -a -um four-sided, square-stemmed, made square, *quadro, quadrare*
quadriauritus -a -um four-lobed, four-eared, *quadri-auritus*
quadribracteatus -a -um having four bracts, *quadri-bracteatus*
quadricolor with four colours, *quadri-color* (in the flowers)
quadrifarius -a -um four-partite, four-ranked, *quadri-fariam*
quadrifidus -a -um divided into four, cut into four, *quadri-fidus*
quadrifoliolatus -a -um having four leaflets, *quadri-foliolatus*
quadrifolius -a -um four-leaved, *quadri-folium*
quadriglandulosus -a -um with four glands, *quadri-glandulosus* (on *Passiflora* petioles)
quadrijugatus -a -um with four pairs of leaflets. *quadri-jugatus*
quadrilocularis -is -e having a four-chambered ovary, *quadri-loculus*
quadrinatus -a -um having four digitate leaflets, *quadrinus*
quadripartitus -a -um four-partite, *quadri-(partio, partire, partivi, partitum)*
quadripetalus -a -um having four-petalled flowers, *quadri-petalum*
quadripinnatus -a -um four-times pinnate, *quadri-pinnatus*
quadriquetrus -a -um square-sided, four-sided, *quadri-quetrus*
quadriradiatus -a -um with four rays, *quadri-radiatus* (florets)
quadrispermus -a -um four-seeded, botanical Latin from *quadri* and σπερμα
quadrivalvis -is -e having a four-valved fruiting body, *quadri-(valvae, valvarum)*
quaesitus -a -um sought after, *quaero, quaerere, quaesivi (quaesii), quaesitum*
Qualea from the Guyanese vernacular name
quamash from the N American Indian name for *Camassia* bulbs, used as food
Quamoclit from the Mexican vernacular name for *Ipomaea quamoclit*, Indian pink; some interpret as Dwarf-kidney, κυαμος-κλιτος
Quaqua from a Khoikhoi vernacular name
quaquaversus -a -um growing in all directions, *quaqua-versus*
quartinianus -a -um from Rub'al Khali (the Empty Quarter), Saudi Arabia (the world's largest sand desert)
quartzitorus -a -um of soils derived from quartzite rock, modern Latin from German, Quartz
quasidivaricatus -a -um as if spreading, *quasi-divaricatus*
Quassia Linnaeus' name, for the Surinamese slave, Graman Quassi, who discovered the medicinal properties of *Quassia amara*, in 1730
quassioides resembling *Quassia, Quassia*-οειδης
quater- fourfold-, four times-, *quater*
quaternarius -a -um, quaternatus -a -um structures arranged in fours, *quaterni, quaternorum*
quaternellus -a -um with four divisions, four-partite, tetramerous, *quaternatus* with feminine diminutive suffix
quebracho Argentinian vernacular meaning axe-breaker, for *Schinopsis* hardwoods of the Gran Chaco, that are exploited for tannin
quelimanensis -is -e from Quelimane, Mozambique
quelpartensis -is -e from Jeju (Quelpart), Korea
querceticolus -a -um living in oak woodland, *Quercus-colo*
quercetorus -a -um of communities dominated by *Quercus*
querci-, quercinus -a -um oak-, oak-like, resembling *Quercus*
Quercifilix Oak-fern, *Quercus-filix*

quercifolius -a -um having leaves resembling those of oak, *Quercus-folium*
quercinus -a -um of oaks (saprophytes typically on *Quercus* remains)
quercitorum of communities dominated by *Quercus*
Quercus the old Latin name, *quercus, quercus*, for an oak (cognate with Arabic, al-qurq, and cork)
quercus-baccarum of oak berries (*Neuroterus*, cynipid oak-currant gall insect)
quercus-folii of oak leaves (*Cynips*, gall wasp)
quercus-radicis of oak roots (*Andricus*, gall wasp)
queretaroensis -is -e from Qerétaro state, central Mexico
Quesnelia for M. Quesnel, French Consul at Cayenne
-quetrus -a -um -angled, -acutely-angled, sided-, from *quadra* a square
quezaltecus -a -um from Quetzaltenango (Quezaltenango), SW Guatemala
quichiotis chimaeral, quixotic, modern Latin from the literary character, Don Quixote
quietus -a -um peaceful, calm, *quiesco, quiescere, quievi, quietum*
quilius -a -um tube-like, from Middle English for a hollow shaft or tube
Quillaja from the Chilean vernacular name, culay, for *Quillaja saponaria*, the soap-bark-tree
quin-, quini-, quinque- five-, *quinque; quini, quinorum; quini-*
Quinaria, quinarius -a -um Five-partite, *quinatus* (the leaflets, ≡ *Parthenocissus*)
quinatus -a -um five-partite, divided into five, *quinatus* (lobes)
quincuncialis -is -e five-twelfths; arranged like the spots on the five-side of a dice (*quincunx, quincuncis*) or aestivated with two members internal, two members external and the fifth half external and half internal, in five ranks
quindiuensis -is -e from Quindio, Colombia
quinghainicus -a -um from Tsing-hai (Ching-ahi or Qinghai) province, China
quinoa the Andean vernacular name, kanua, for the food staple, *Chenopodium quinoa*
quinquangularis -is -e five-cornered, five-angled, *quinque-angularis*
quinquefarius -a -um five-branched, *quinque-farius*
quinqueflorus -a -um five-flowered, *quinque-florum*
quinquefoliolatus -a -um with five leaflets, *quinque-foliolatum*
quinquefolius -a -um five-leaved, *quinque-folium*
quinquelobus -a -um, quinquelobatus -a -um with five complete or partial lobes, *quinque-lobus*
quinquelocularis -is -e five-celled, five-locular, *quinque-loculus* (the ovary)
quinquenervis -is -e having five veins, *quinque-nervus*
quinquepeta a misnomer by Buc'hoz, who described *Lassoia quinquepeta* from a picture that showed seven erect tepals
quinquevulnerus -a -um with five wounds, *quinque-vulneris* (e.g. red marks on the corolla)
quintuplex in multiples of five, fivefold, *quintuplex*
quintupli- five-, *quintuplex*
quintuplinervius -a -um with five veins, *quintuplex-nervus*
quintus -a -um fifth, *quintus*
Quiongzhuea from Kwangsi Chuang autonomous region, China
quiriguanus -a -um from Quirigua, E Guatemala
Quisqualis Who? What-kind? (from a Malay name, udani, which Rumphius transliterated as Dutch hoedanig, for how? what? to reflect the variable habit and colouring)
quitensis -is -e from Quito, Ecuador

rabdo- see *rhabdo-*, ραβδος, ραβδο-
Rabdosia Rod-like, ραβδος
racem-, racemi- with flowers arranged in a raceme, *racemus, racemi* (see Fig. 2b)
racemiflorus -a -um having racemose inflorescences, raceme-flowered, *racemus-florum*

racemosus -a -um having racemose inflorescences, *racemus* (see Fig. 2b)
racemus -a -um racemose, having pedicelled flowers arranged singly along the rachis, *racemus* (literally a bunch of grapes)
rache-, rachi-, -rachis rachis-, -rachis, backbone, ραχις (used botanically for the axis of compound structures such as leaves)
rachimorphus -a -um back bone-like, with a zigzag central axis, ραχις-μορφη (as in *Rottboellia*)
raco- ragged-, tattered-, patched-, ρακος, ρακο-
Racopilum Tattered-felt-hat, ρακος-πιλος (the calyptra)
Racosperma Patched-seed, ρακος-σπερμα (the testa)
raddeanus -a -um for Gustav Ferdinand Richard Radde (1831–1903), Director of the Tiflis Caucasian Museum
radens rasping, scraping, present participle from *rado, radere, rasi, rasum* (the rough surface)
radialis -is -e of long olives; actinomorphic, radial, *radius*
radians shining, radiating, present participle of *radio, radiare*
radiatiformis -is -e with the ligulate florets increasing in length toward the outside of the capitulum, *radius-forma*
radiatus -a -um radiating outwards, radiant, *radiatus*
radicalis -is -e arising from a root or a crown, *radix, radicis*
radicans with rooting stems or leaves, present participle from *radico, radicare, radicavi, radicatum*
radicantissimus -a -um having the most adventitious roots, superlative of *radicans*
radicatus -a -um with roots or root-like structures, past participle of *radico*, to strike root
radicicola living in roots, *radix-colo* (*Rhizobium* nodule bacterium)
radiciflorus -a -um flowering from roots or rootstocks, *radix-florum*
radicosus -a -um with a large, conspicuous or numerous roots, comparative of *radico*
radiiflorus -a -um with radiating flowers or perianth segments, *radius-florum*
radinus -a -um ray-like, slender, ραδινος, *radius, radii*
radioferens light bearing, glittering, present participle of *radio-(fero, ferre, tuli, latum)*
Radiola Radiating, diminutive of *radius* (the branches)
radiosus -a -um having many rays, *radius, radii*
radula scraping, rough, rasping, like a rasp, modern Latin from *rado, radere, rasi, rasum*
radulifolius -a -um having leaves similar to (*Pelargonium*) *radula, radula-folium*
raffia, roffia see *Raphia*
Rafflesia, rafflesianus -a -um for Sir Thomas Stamford Raffles (1781–1826), diplomat, orientalist, naturalist and a founder of London Zoo (***Rafflesiaceae***)
ragas fissured, ραγας
ragusinus -a -um from Dubrovnik (Ragusa), Croatia
rajah ruler, from the Hindi, raja, prince (Sanskrit, rajan, king)
rakaiensis -is -e from the Rakai Valley, Canterbury, New Zealand
rakiurus -a -um ragged-tailed, ραγος-ουρα; from Stewart Island, New Zealand
ramalanus -a -um from Mount Ramala, W China; having twigs, *ramalia*
Ramaria Twiggy, *ramalia, ramalium* (the numerous branches of the fruiting body)
ramealis -is -e of twigs or branches, *rameus*
ramellosus -a -um like brushwood, twiggy, diminutive of *ramus*
ramentaceus -a -um covered with scales, *ramentum, ramenti* (ramenta)
-rameus -a -um -branched, *ramus, rami*
rami- branches-, of branches-, branching-, *ramus, rami*
ramiferus -a -um bearing branches, branched, *ramus-fero*
ramiflorus -a -um with flowers on the branches, *ramus-florum*
ramipressus -a -um having very closely arranged branches, *ramus-(premo, premere, pressi, pressum)*

Ramischia for F. X. Ramisch (1798–1859), botanist of Prague, Czech Republic
Ramonda for Louis François Elisabeth Ramond de Carbonnières (1755–1827), French botanist and explorer in the Pyrenees
ramosior more branched, comparative of *ramosus*
ramosissimus -a -um greatly branched, superlative of *ramosus*
ramosus -a -um much branched, branching, *ramus, rami*
ramuensis -is -e from the environs of the Ramu river, Papua New Guinea
ramulosus -a -um very twiggy, *ramulus*, diminutive of *ramus*
Ranalisma Frog-*Alisma* (resemblance to an aquatic *Ranunculus*)
rancidus -a -um rank, rancid, disgusting, *rancidus*
Randia for Isaac Rand (1674–1743), Praefectus of Chelsea Physic Garden
Ranevea (*Ravenea*) for Paul Ranevé, Berlin horticulturalist
rangiferinus -a -um of reindeer or their territory (*Rangifer tarandus* is the reindeer (*Cladonia rangiferina*))
raniferus -a -um bearing frogs, supporting frogs, (*rana, ranae*)-*fero* (in the water contained in the leaf bases of many epiphytic bromeliads)
ranunculifolius -a -um *Ranunculus*-leaved, *Ranunculus-folium*
ranunculinus somewhat *Ranunculus*-like, diminutive of *Ranunculus*
ranunculoides *Ranunculus*-like, *Ranunculus-oides*
ranunculophyllus -a -um *Ranunculus*-leaved, botanical Latin from *Ranunculus* and φυλλον
Ranunculus Little-frog, diminutive of *rana*, (the amphibious habit of many) (***Ranunculaceae***)
Raoulia, raoulii for Edouard F. A. Raoul (1815–52), French surgeon and author of *Choix de plantes de la Nouvelle Zélande*
rapa, rapum an old Latin name, *rapum*, for a turnip, or rape
rapaceus -a -um of turnips, *Rapa*-like, *rapa, rapum*
raphani- radish-, radish-like-, ραπφανος, *raphanus, raphani*
raphanifolius -a -um with leaves resembling those of *Raphanus, Raphanus-folium*
raphanistrum like a wild *Raphanus, Raphanus-istrum*
raphanorhizus -a -um radish-like-rooted, ραπφανος-ριζα
Raphanus the Latin name, *raphanus*, for a radish, from ραπφανος, for cabbage or radish
raphe- seam-, ραφη
raphi-, raphio- needle-, ραφις, ραφιο-, ραφιδος, ραφιδο-
Raphia from the Malagasy name, raffia or roffia, for the fibres from *Raphia pedunculata*, or needle (the sharply pointed fruit)
raphidacanthus -a -um having needle-like thorns, ραφις-ακανθος
raphifolius -a -um having needle-like leaves, botanical Latin from ραφις and *folium*
Raphiolepis Needle-scale, ραφιο-λεπις (the subulate bracts)
Rapistrum Wild-turnip-like, *rapum-istrum* (implies inferiority of wild mustard)
rapum-genistae broom-turnip, (*rapio, rapere, rapui, raptum*)-*Genista* (the cormose base of *Orobanche* on roots of *Sarothamnus*)
rapunculoides resembling rampion, *Rapunculus-oides*
Rapunculus, rapunculus Little-turnip, diminutive of *rapum* (Bock's reference to *rapunculum, quasi parvum rapum*, referring to the swollen roots, gives the cognate rampion)
rari- thin-, scattered-, loose-, *rarus*
rariflorus -a -um having scattered flowers, *rarus-florum*
rarus -a -um scanty, porous, scattered, uncommon, rare, *rarus*
ratisbonensis -is -e from Regensburg (*Ratisbon*), Bavaria, Germany
Rauvolfia, Rauwolfia for Leonard Rauwolf (1535–96), Augsburg physician and traveller in Palestine etc.
Ravenala from the Madagascan name for the travellers' tree
Ravenea for Paul Ranevé, Berlin horticulturalist
ravidus -a -um greyish or tawny, *ravidus*
ravus -a -um tawny- or grey-coloured, *ravus*

re- back-, again-, against-, repeated-
Reaumuria for René A. Ferchault de Reaumur (1683–1757), French entomologist
recedens retiring, receding, present participle of *recedo, recidere, reccidi, recasum*
recens fresh, young, recent, new, *recens, recentis*
recisus -a -um cut off or cut back, *recido, recidere, recidi, recisum*
reclinatus -a -um drooping to the ground, deflexed, bent back, reclined,*reclino, reclinare, reclinavi, reclinatum*
reclusus -a -um see *inclusus*
recognitus -a -um authentic, the true one, examined, *recognosco, recognoscere, recognovi, recognitum*
reconditus -a -um hidden, not conspicuous, concealed, secluded, *recondo, recondere, reconditi, reconditum*
rectangularis -is -e rectangular, *rectus-angulus*
recti- straight-, upright-, erect-, *rectus*
recticaulis -is -e straight-stemmed, *rectus-caulis*
rectiflorus -a -um virtuous, straight or upright-flowered, *rectus-florum*
rectifolius -a -um erect-leaved, *rectus-folium*
rectinervis -is -e, rectinervius -a -um straight-veined, *rectus-nervis*
rectiramus -a -um having ascending or straight branches, *rectus-ramus*
rectus -a -um straight, upright, erect, *rego, regere, rexi, rectum*
recurvans arching, bending back, present participle of *recurvo, recurvare*
recurvatus -a -um, recurvi- curved backwards, recurved, *recurvo, recurvare*
recurvifolius -a -um having recurved leaves, *recurvus-folium*
recurvus -a -um bent or curved backwards, *recurvo, recurvare*
recutitus -a -um skinned, circumcised, *re-cutis* (the appearance caused by the reflexed ray florets of the flower head)
redactus -a -um reduced, rendered fruitless, *redigo, redigere, redegi, redactum*
redimitus -a -um bound, wreathed, crowned, *redimio, redimire, redimii, redimitum*
redivivus -a -um coming back to life, renewed, *redivivus* (perennial habit or reviving after drought)
redolens promising, smelling of, perfuming, scenting, present participle of *redoleo, redolere, redolui*
reductus -a -um drawn back, reduced, *reduco, reducere, reduxi, reductum*
reduncus -a -um curved back, *reduncus*
redundans abounding, overflowing, present participle of *redundo, redundare, redundavi, redundatum*
Reevesia for John Reeves (1774–1856), botanist in Canton, China
reficiens restoring, reviving, refreshing, present participle of *reficio, reficere, refeci, refectum*
reflexipetalus -a -um having petals sharply bent backwards upon themselves, botanical Latin from *reflexus* and πεταλον
reflexus -a -um bent back upon itself, reflexed, *reflecto, reflctere, reflexi, reflexum*
refractus -a -um abruptly bent backwards, broken(-looking), *refringo, refringere, refregi, refractum*
refulgens reflecting, flashing, present participle of *refulgeo, refulgere, refulsi*
regalis -is -e outstanding, kingly, royal, regal, *rex, regis*
Regelia, regelii for Dr Eduard Albert von Regel (1815–92), superintendent of St Petersburg Imperial Botanic Gardens
regeneratus -a -um regenerating (regrowing after cutting), *re-(genero, generare, generavi, gerenatum)*
regerminans re-budding, re-shooting, present participle of *re-(germino, germinare)*
regerminatus -a -um freely re-sprouting, *re-germinans*
regina, reginae queen, of the queen (has been used for several queens)
reginae-amelaiae for Caroline Amelia Elizabeth von Braunschweig-Lüneberg (1768–1821) briefly wife of King George IV
reginae-olgae for Queen Olga (890–969), first Russian ruler of Kiev
regis-jubae King-Juba, who was a king of Numidia (Algeria)

registanicus -a -um from Rigistan (Registan or the land of sand), SW Afghanistan
regius -a -um splendid, royal, kingly, *rex, regis*
regma- breaking-, fracture-, ρηγμα
regmacarpius -a -um with a schizocarp breaking into cocci, with a dehiscent fruit, ρηγμα-καρπος
regnans lording it, prevailing, becoming supreme, present participle of *regno, regnare, regnavi, regnatum*
regularis -is -e uniform, actinomorphic, standard, *regula, regulae*
regulus goldcrest (*Regulus regulus*); petty ruler, prince, *regulus*
rehderi, rehderianus -a -um for either Jacob Heinrich Rehder (1790–1852), Parks Inspector of Moscow, or Professor Alfred Rehder (1863–1949), see below
Rehderodendron Rehder's-tree, for Professor Alfred Rehder (1863–1949) of the Arnold Arboretum Herbarium, Massachusetts, USA, author of the *Manual of Cultivated Trees and Shrubs*
Rehmannia for Joseph Rehmann (1788–1831), St Petersburg physician
reichenbachianus -a -um for Heinrich Gottlieb Ludwig Reichenbach (1793–1879), of the Dresden Botanic Garden, Germany
Reineckia for J. Reinecke, German cultivator of tropical plants
Reinwardtia, reinwardtii for Caspar Georg Carl Reinwardt (1773–1854), Director of Leiden Botanic Garden
relaxatus -a -um loose, open, *relaxato, relaxare, relaxavi, relaxatum*
relictus -a -um remnant, primitive, left behind, relict, *relinquo, relinquere, relevi, relictum*
religiosus -a -um sacred, venerated, of religious rites, *religiosus* (the Buddha is reputed to have received enlightenment beneath the bo or peepul tree, *Ficus religiosa*)
remediorus -a -um of medicine, of cures, remedial, *remedium, remedi*
remoratus -a -um hindering, delaying, *remoror, remorari, remoratus*
remotiflorus -a -um having scattered flowers, *remotus-florum*
remotifolius -a -um having remote or scattered leaves, *remotus-folium*
remotus -a -um set aside; distant, secluded, scattered, *removeo, removere, removi, remotum* (e.g. the flowers on the stalk)
remulcus -a -um drooping, *remulceo, remulcere, remulsi* (literally, *remulcum*, a tow-rope)
Remusatia for Abel Remusat (1785–1832), physician and student of the orient
renarius -a -um, reniformis -is -e kidney-shaped, reniform, *renes* a kidney
Renealmia for Paul Reneaulme (1560–1624), author of *Specimen historia plantarum*
renghas from a Malayan vernacular name
renifolius -a -um having kidney-shaped leaves, (*renes, renum*)*-folium*
reniformis -is -e kidney-shaped, (*renes, renum*)*-forma*
repandens, repandus -a -um with a slightly wavy margin, repand, *repandus*
repens creeping, *repo, repere, repsi, reptum* (stoloniferous)
replicatus -a -um double-pleated, doubled down, *replico, replicare*
reptans crawling along, creeping and rooting, present participle of *repto, reptare*
reptatus -a -um crawling along, creeping and rooting, *repto, reptare*
repullulans re-sprouting, present participle of *re-*(*pullulo, pullulare*)
Requienia, requienii for Esprit Requien (1788–1851), student of the floras of S France and Corsica
resectus -a -um shredded, cut off, *re-*(*seco, secare, secui, sectum*)
Reseda Healer, *resedo* (the name in Pliny refers to its use in treating bruises) (***Resedaceae***)
resedi- *Reseda-*
resediflorus -s -um with flowers similar to *Reseda*
resedifolius -a -um with *Reseda*-like foliage, *Reseda-folium*
resiliens recoiling, springing back, rebounding, present participle of *resilio, resilire, resilui*
resinaceus -a -um, resinosus -a -um resinous, producing resin, *resina, resinae*

resinifer -era -erum bearing resin, (*resina, resinae*)-*fero*
resplendens shining brightly, shining out, *resplendeo, resplendere*
restibilis -is -e perennial, able to return, *resto, restare, restiti*; able to stay behind, *restito, restitare*
Restio Rope-maker, *restio, restionis* (use of the fibrous leaves)
restitutus -a -um returning, renewing, restored, *restituo, restituere, restitui, restitutum*
restrictus -a -um confined, sparing, checked, *restringo, restringere, restrinxi, restrictum*
resupinatus -a -um inverted, resupinate, *resupino, resupinare* (e.g. those orchids with twisted ovaries)
ret-, reti- net-, *rete, retis*
retama from the Spanish for *Genista*, from the bushland of S Spain
retatus -a -um netted, net-like, *rete, retis*
retectus -a -um lacking a cover, revealed, open, *retego, retegere, retexi, retectum*
reticosus -a -um net-veined, *rete, retis*
reticulatus -a -um reticulate, conspicuously net-veined, netted, *rete, retis*
reticulosus -a -um somewhat netted with veins, *rete, retis*
Retinispora (*Rerinospora*) a former generic synonym for *Thuja* later used to describe coniferous 'seedling-forms' (produced from juvenile material) that retain (*retineo, retinere, retinui, retentum*) the juvenile foliage characteristic (cf. the process and nature of *Hedera arborescens*)
retinodes tenacious looking, (*retineo, retinere, retinui, retentum*)-*odes*
retirugus -a -um with prominent netted ridges (the sporulating surface of *Leptoglossum retirugum*)
retortus -a -um twisted or turned back, *retorqueo, retorquere, retorsi, retortum*
retractus -a -um drawn backwards; revised; hesitant, *retraho, retrahere, retraxi, retractum*
retro- back-, behind-, backwards-, *retro*
retroflexus -a -um turned backwards or downwards, *retro-*(*flecto, flectere, flexi, flexum*)
retrofractus -a -um stubborn, turned aside, refracted, *retro-*(*frango, frangere, fregi, fractum*)
retrorsus -a -um curved backwards and downwards, *retrorsum*
retrospiralis -is -e with a downwards spiral growth, *retro-*(*spira, spirae*)
retroversus -a -um turned back, *retro-*(*verto, vertere, versi, versum*)
retusiusculus -a -um slightly or shallowly knotched, diminutive of *retusus*
retusus -a -um blunt with a shallow notch at the tip, *retusus* (e.g. leaves; see Fig.7f), retuse
reventus -a -um returning, coming back, *revenio, revenire, reveni, reventum*
reversus -a -um reversed, upside down, resupinate, *revertor, reverti, reversus*
revirescens re-greening, *re-*(*viresco, virescere*) (produces a late growth and flowering if dead-headed)
revolutus -a -um rolled back, rolled out and under (e.g. leaf margin), revolute, *revolvo, revolvere, revolvi, revolutum*
rex king, *rex, regis*
Reynoutria (≡ *Polygonum*)
rhabarbarum foreign rha, ραβαρβαρον (the root of *Rheum officinalis* came from China via the Volga (ρα) and became the 'Volga drug of the foreigners', ρηα-βαρβαρικος, or rhu-barb)
rhabdo- rod-like, rod-, ραβδος, ραβδο-
Rhabdophaga Rod-eater, ραβδος-φαγος (gall mite on leaf margins and twig ends)
rhabdospermus -a -um having rod-shaped seeds, ραβδος-σπερμα
Rhabdothamnus Rod-bush ραβδο-θαμνος (for its entangled habit)
rhabdotus -a -um stiped, with a stipe (stalk), ραβδος
rhache-, -rhachis -is -e backbone-, rachis-, -rachis, ραχις
Rhachicallis Beautiful-rachis, ραχις-καλος

rhacodes slashed-looking, ραχιζω-ωδης
rhadinus -a -um slender, tender, ραδινος
rhaeticus -a -um of the *Raeti* people, from the Central, or Rhaetian Alps of the Swiss–Austrian and Swiss–Italian border
rhaga-, -rhagius -a -um -fissured, -torn, -rent, ραγας
Rhagadiolus Divided, diminutive of ραγας (the inner achenes are caducous but the outer elongate and persist)
rhago- berried, ραξ, ραγος
Rhagodia Berried, ραγος-ωδης
Rhamnella Little-*Rhamnus*, feminine diminutive suffix
rhamnifolius -a -um having leaves similar to those of *Rhamnus, Rhamnus-folium*
rhamnoides resembling *Rhamnus, Rhamnus-oides*
Rhamnus an ancient name, ραμνος, for various prickly shrubs (*rhamnos* in Pliny) (Rhamnus was a town famed for its statue of Nemesis) (***Rhamnaceae***)
rhamphiphyllus -a -um with short, thorn-tipped leafy spurs, botanical Latin from *Rhamnus* and φυλλον
rhaphamistrum *vide raphanistrum*
rhaphi-, rhaphio- needle-like-, needle-, ραφη, ραφις, ραφιδος, ραφιδο-
rhaphidacanthus *vide raphidacanthus*
Rhaphidophora Needle-carrying, ραφιδο-φορα
Rhaphiolepis Needle-scaled, ραφις-λεπις
Rhaphionacme Apically-needled, ραφις-ακμη
rhaphiophyllus -a -um having needle-like leaves, ραφις-φυλλον
Rhaphithamnus Needle-shrub, ραφις-θαμνος (some are thorny)
Rhapis (*Raphis*) Needle, ραφις (the leaf spines)
rhaponticus -a -um rha from the Black Sea area, our vegetable rhubarb (*Rhaponticum*, ρα of Dioscorides, with *pontus*, ≡ *Centaurea*)
rhapto- stitched-, ραπτω, ραπτος, ραπτο-
Rhaptopetalum Seamed-petals (the valvate corolla)
Rhazya for Abu Bekr-er-Rasi (ninth/tenth-century), Arabian physician and writer on medicine
rheithrophilus -a -um liking streams, ρειθρον-φιλος
Rhektophyllum Rent-leaved, ρηκτο-φυλλον (the mature leaves are pinnatisect and perforated)
rhenanus -a -um, rheni- from the environs of the River Rhine (*Rhenus, Rhenanus*)
rheophilus -a -um liking a flow (of water), ρεω-φιλος (rheotropic plants)
rheophyticus -a -um stream-plant, ρεω-φυτον
Rheum Greek name, ρηον in Galen, ρα in Dioscorides, from a Persian name, rewend, for the medicinal roots (rhubarb derives from ρα and βαρβαρος)
Rhexia Rupture, ρηξιο (growth between paired, sessile leaves)
rhin-, rhino- nose-, ρις, ρινος, ριν-
Rhinacanthus Nose-*Acanthus*, ρινος-ακανθος (the flower-shape)
rhinanthoides resembling *Rhinanthus, Rhinanthus-oides*
Rhinanthus Nose-flower, ριν-ανθος (the upper lip of the corolla)
rhinocerotis -i -e of the rhinoceros, ρηινοκερος (a preferred food plant)
rhipi- fan-shaped-, ριπις, ριπιδος, ριπιδο-
Rhipidopteris Fan-leaved, ριπιδος-πτερυξ
rhipo- reed-mat-, wicker-hurdle-, wand-, staff-, ριψ, ριπος
Rhipsalis Wickerwork-like, ριψ (the slender twining stems)
rhiz-, rhizo-, -rhizus root-, -rooted, ριζα
Rhizanthemum Root-flower, ριζ-ανθεμιον (Malaysian parasitic plant)
rhizanthus -a -um flowering from the root, ριζα-ανθος
Rhizobium Root-liver, ριζα-βιοω (root-nodule causing bacteria on legumes)
rhizocephalus -a -um head of roots, many adventitious roots from a large tap root, ριζα-κεφαλη
Rhizoctonia Root-killer, ριζα-κτονος (deuteromycete damping-off and root-rot fungus)

rhizomatus -a -um with rhizomes, having dorsiventral, over- or under-ground, rooting stems, ριζα, ριζωμα
rhizomatosus -a -um with many rhizomes, ριζωμα (as distinct from bulbs)
Rhizophora Carried-on-roots, ριζα-φορα (the long-arched prop-roots) (***Rhizophoraceae***)
rhizophyllus -a -um root-leaved, ριζα-φυλλον (the leaves form marginal roots)
rhizopodius -a -um rooting from the stalk-base; the root-like base or mycelium of fungi, ριζα-(πουσ, ποδος)
-rhizus -a -um -rooted, -root, ριζα
rhod-, rhodo- rose-, rosy-, red-, ροδον, ροδο-
rhodandrus -a -um with red stamens, ροδον-ανηρ
rhodanicus -a -um from the environs of the River Rhone (*Rhodanus*)
Rhodanthe Red-flower, ροδ-ανθος (≡ *Helipterum*)
Rhodanthemum Red-flower, ροδ-ανθεμιον
rhodantherus -a -um with red stamens, ροδ-ανθηρος
rhodanthus -a -um rose-flowered, ροδ-ανθος
rhodensis -is -e, rhodius -a -um from the Aegean island of Rhodes
rhodinsulanus -a -um from Rhode Island, USA
Rhodiola Little-rose, diminutive of ροδον (the rose-fragrant rootstock was *radix rhodiae*)
rhodiolus -a -um like a small rose, *Rhodiola*-like
rhodius -a -um from Rhodes, Greece (*Rhodos, Rhodius*)
Rhodochiton Red-cloak, ροδο-χιτων (the large calyx)
rhododendri of *Rhododendron*, living on *Rhododendron* (*Exobasidium*, basidiomycete fungal gall)
Rhododendron (*um*) Rose-tree, ροδο-δενδρον (an ancient Greek name used for *Nerium oleander*)
Rhodohypoxis Red-*Hypoxis*, ροδο-υπο-οξις
Rhodoleia Like-a-thornless-rose, ροδο-λειος (the flower and smooth stem)
rhodopaeus -a -um, rhodopensis -is -e from Rhodope (Rodopi) mountains, Bulgaria–Greece
Rhodophiala Red-bowled, ροδο-φιαλη (the corolla)
rhodophthalmus -a -um red-eyed, ροδον-οφθαλμος
rhodopis -is -e resembling a rose, ροδον-οπις
rhodopolius -a -um rosy greyish-white, ροδον-πολιος
Rhodora, rhodora Rose-like, ροδον (≡ *Rhododendron*)
Rhodothamnus Rose-shrub, ροδο-θαμνος (the flower colours)
Rhodotypos (*-us*) Rose-type, ροδο-τυπος (floral resemblance)
rhodoxanthus -a -um red and yellow, ροδο-ξανθος
-rhoea -stream, -flow, ρεω (the streaming, ροος, sap or exudate)
rhoeas the old generic name, μεκων ροιας, of the field poppy, *Papaver rhoeas* (the pomegranate, ροια, flower colour)
Rhoeo Flowing, ρεω (etymology uncertain but could refer to the mucilaginous sap)
Rhoicissus Pomegranate-coloured-ivy, ροια-κισσος
rhoifolius -a -um with pomegranate-like leaves, botanical Latin from ροια and *folium*
rhombeus -a -um, rhomboides shaped like a rhombus, ρομβοειδης
rhombi-, rhombicus -a -um, rhomboidalis -is -e, rhomboidosus -a -um diamond-shaped, turbot-shaped, rhombic, ρομβος, *rhombus, rhombi*
Rhombifolium, rhombifolius -a -um with rhombus-shaped leaves, *rhombus-folium*
Rhombiphyllum Rhomboid-leaf , ρομβο-φυλλον (*Rhombiphyllum rhomboideum* almost a tautonym)
rhoophilus -a -um liking creeks or streams, ροος-φιλος
rhopal-, rhopalo- club-, cudgel-, ροπαλον
rhopalanthus -a -um having club-shaped flowers, ροπαλον-ανθος
rhopalocarpus -a -um having club-shaped fruits, ροπαλον-καρπος
rhopalophyllus -a -um with club-shaped leaves, ροπαλον-φυλλον

Rhopalostachya Cudgel-spike, ροπαλον-σταχυον (the strobilar head of sporangia of clubmoss, ≡ *Lycopodium*)
Rhopalostylis Club-shaped-style, ροπαλον-στυλος
-rhops,-rops -bush, -underwood, -shrub, ρωψ
rhumicus -a -um from the island of Rum (Rhum), W Scotland, or the River Rhume area, W Germany
Rhus from an ancient Greek name for a sumach (Arabic, summaq, for the red dye from *Rhus coriaria*)
rhynch-, rhyncho- beak-, ρυγχος, ρυγχο-
Rhynchanthus Beak-flower, ρυγχος-ανθος (the protruding, keeled filament)
Rhynchelytrum (on) Beaked-sheath, ρυγχος-ελυτρον (the shape of the glumes)
Rhynchocoris Beaked-helmet, ρυγχος-κορις (≡ *Rhinanthus*)
rhynchophyllus -a -um having beaked leaves, ρυγχος-φυλλον
rhynchophysus -a -um with a beaked inflated utricle (bellows), ρυγχος-φυσα
Rhynchosia Beak, ρυγχος (the shape of the keel petals)
Rhynchosinapis Beaked-*Sinapis*, botanical Latin from ρυγξος and *Sinapis*
Rhynchospora Beaked-seed, ρυγχος-σπορος (the achene has a distinct beak)
Rhytachne Wrinkled-chaff, ρυτις-αχνη (the rugosity of the lower glumes)
rhyti-, rhytido- wrinkled-, ρυτις, ρυτιδωμα, ρυτιδος, ρυτιδο-
rhytidophylloides resembling (*Viburnum*) *rhytidophyllum, rhytidophyllum-oides*
Rhytidophyllum, rhytidophyllus -a -um Untidy-leaved, with wrinkled leaves, ρυτιδο-φυλλον
rhyzo-, -rhyzus -a -um root-, -rooted, ριζα
Ribes from the Persian, ribas, for the acid-tasting *Rheum ribes*
ribesifolius -a -um, ribifolius -a -um with *Ribes*-like foliage, *Ribes-folium*
ribis -is -e of black currant, living on *Ribes* (*Eriophyes*, acarine gall mite)
richardsonii for Sir John Richardson (1787–1865), companion of Sir John Franklin
Richiea, richeus -a -um for Colonel A. Riche (d. 1791), French naturalist who died during the search for La Peyrouse
ricinifolius -a -um having leaves resembling those of *Ricinus*
ricinocarpus -a -um with *Ricinus*-like fruits, castor-oil-like fruited, botanical Latin from *Ricinus* and καρπος
Ricinodendron *Ricinus*-like-tree, botanical Latin from *Ricinus* and δενδρον (a similarity of the foliage)
Ricinus Tick, *ricinus* (the appearance of the caruncled and coloured seeds)
rifanus -a -um of the N African Rif tribe (*riphaei, riphaeorum*)
rigens stiffening, rigid, *rigeo, rigere*
rigensis -is -e from Riga, Latvia, on the Baltic
rigescens adopting a stiff texture, hardening, present participle of *rigesco, rigescere, rigui*
rigidifolius -a -um stiff-leaved, erect-leaved, *rigidus-folium*
rigidulus -a -um quite stiff, diminutive of *rigidus*
rigidus -a -um stiff, inflexible, *rigidus*
rimarinus -a -um bordered, having a border; of the coast, modern Latin *rimarus*, from Old English, rima, Norse, rimi,
rimicolus -a -um inhabiting cracks or crevices, *rimae-colo*
rimosus -a -um with a cracked surface, furrowed, *rimosus*
ringens with a two-lipped mouth, gaping, *ringens*
ringo from the Japanese vernacular name for *Malus ringo*
Rinorea from a Guyanese vernacular name
riparius -a -um of the banks of streams and rivers, *ripa, ripae*
ripensis -is -e of riversides and stream banks, *ripa, ripae*
Ripogonum Flexible-shoot, ριπος γωνως (growth habit)
ritro a S European name for *Echinops ritro*
ritualis -is -e of ceremonials, belonging to rituals, *ritus, ritus* (*Kniphofia ritualis*)
rivalis -is -e of brooksides and streamlets, *rivus, rivi*
Rivea for Auguste de la Rive, Swiss physiologist

Rivina, riviniana for August Quirinus Rivinus (1652–1722), Professor of Botany at Leipzig
rivularis -is -e, rivulatus -a -um waterside, of the rivers, diminutive of *rivus*
rivulosus -a -um with sinuate marking or grooves, *rivulus*
-rix feminine suffix for masculine nouns ending in *-or*
rizhensis-is -e from Rize, Turkey
robbiae for Mary Anne Robb (1829–1912), who reputedly smuggled *Euphorbia amygdaloides* ssp. *robbiae*, Mrs Robb's bonnet, from Turkey, in a hatbox
robertianus -a -um of Robert, *herba roberti, herba sancti ruperti* (which Robert or Rupert, saint or goblin, is uncertain)
Robinia for Jean Robin (1550–1629) and Vesparian Robin (1579–1600), herbalists and gardeners to Henry VI of France
robur oak timber, strong, hard, *robur, roboris*
robustior more stout or robust, comparative of *robustus*
robustus -a -um of oak, *robur*; strong-growing, robust, *robustus*
Rochea for François de la Roche (1782–1814), French botanical writer
rockii for Joseph Francis Charles (Karl) Rock (1884–1962), American collector in China, Tibet, India, Europe, America and Hawaii
Rodgersia for Rear Admiral John Rodgers (1812–82), expedition commander of the US Navy
rodigazianus -a -um for Señor Rodigas, who collected in S America
rodo- rose, rosy-, red-, ροδον
Rodriguezia for Emanuel Rodriguez, eighteenth century Spanish physician and botanist
Rodrigueziella for João Barbosa Rodrigues (1842–1909), Brazilian botanist
rodriguesii from the island of Rodrigues, Indian Ocean
roebelinii for W. Roebelin, Swiss collector in the Philippines for Sanders
Roegneria for a royal gardener of that name at Oreanda, who assisted K. Koch
Roella for W. Roell, Professor of Anatomy at Amsterdam in the eighteenth century
Roemeria (*Romeria*) for Johann Jacob Römer (1763–1819), Swiss botanist and editor of *Magazin für die Botanik*
roezlii for Benedikt Roezl (1824–85), Austrian collector in Central America.
Rogersia, rogersii for Charles Gilbert Rogers (1864–1937) of the Indian Forestry Service
Rohdea for Michael Rhode, Bremen physician and botanist
romanicus -a -um from Romania
romanus -a -um of Rome, *Roma, Romae*; Roman, *romanus*
Romanzoffia, romanzoffianus -a -um for Prince Nicholas Romanzoff, Russian sponsor of the 1816 expedition around the world
Romneya for Reverend Thomas Romney Robinson (1792–1882), astronomer of Armagh, Ireland
Romulea for Romulus, founder of Rome
Rondeletia for Guillaume Rondelet (1507–66), French physician and aquatic biologist
roribaccus -a -um dewberry, *roridus-baca*
Roridula Dewy, diminutive of *roridus*, (glandular hairy) (***Roridulaceae***)
roridus -a -um apparently with minutely blistered surface, bedewed, *ros, roris*
Rorippa from the old Saxon name, rorippen
rorippifolius -a -um having leaves resembling *Rorippa, Rorippa-folium*
rorulentus -a -um covered in dew, *ros, roris*
Rosa the Latin name, *rosa, rosae* for various roses (***Rosaceae***)
rosaceus a -um looking or coloured like a rose, *rosa, rosae*
rosa-del-monte rose of the mountain, botanical Latin from the Italian
rosae-, rosi- rose-like, rose-coloured, *rosa, rosae*
rosae of roses, living on *rosa* (*Diplolepis*, hymenopteran gall wasp)
rosalbus -a -um white-rose, *rosa-albus*
rosaricus -a -um from El Rosoria, Baja California, NW Mexico
rosarius -a -um of the rose garden, *rosa*

rosarus -a -um of roses, living on roses, *rosa* (symbionts, parasites and saprophytes)
rosasinensis -is -e Chinese-rose, eastern-rose, *rosa-sinensis*
Roscoea for William Roscoe (1753–1831) founder of the Liverpool Botanic Garden
roseatus -a -um flushed rose-pink, *rosa, rosae*
rosellus -a -um bunched like a rose, rosette-like, diminutive of *rosa, rosae*
roseoalbus -a -um white flushed with pink, *rosa-albus*
roseocampanulatus -a -um with little rose-pink bells, diminutive of *Rosa-campanae* (shaped flowers)
roseofractus -a -um broken red, *frango, frangere, fregi, fractum*
roseolus -a -um pink or pinkish, *rosa, rosae*
roseopictus -a -um with rose pink spotting, *rosa-pictus*
roseotinctus -a -um rose-imbued, rose-coloured, *rosa-tinctus*
roseus -a -um rose-like, rose-coloured, *rosa, rosae*
rosiflorus -a -um rose-flowered, *rosa-florum*
rosifolius -a -um, rosaefolius -a -um rose-leaved, *rosa-folium*
rosmarini- *Rosmarinus-*, rosemary-
rosmarinifolius -a -um with leaves resembling *Rosmarinus, Rosmarinus-folium*
rosmariniformis -is -e with a habit like *Rosmarinus, Rosmarinus-forma*
Rosmarinus Sea-dew, Pliny's name, *ros maris, ros-marinus* for a plant of dewy places (it became Mary's rose, or rosemary, in English)
rossii for either Rear Admiral Sir John Ross (1777–1856), arctic navigator, or Herman Ross (1862–1942), German botanist
rossicus -a -um from Russia, Russian, (*Rossica*) (each *Boschniaka rossica* plant is estimated to produce 333 000 seeds!)
rostellatus -a -um with a small beak, beaked, diminutive of *rostrum*
rostochiensis -is -e from Rostock, NE Germany
rostratus -a -um narrowed to a point, with a long straight hard point, beaked, rostrate, *rostratus* (*columna-rostrata* was a column to commemorate a naval victory)
rostrevor *Eucryphia rostrevor* was raised at Rostrevor, Co. Down, Ireland
rostri-, rostris -is -e, rostrus -a -um nose-, beak-like, *rostrum, rostri*
rostriflorus -a -um with nose-shaped flowers, *rostri-florum*
rostripetalus -a -um having beak-like tips on the petals, botanical Latin from *rostrum* and πεταλον
Rosularia Little-rose-like, diminutive from *rosula* (the leaf rosettes)
rosularis -is -e, rosulatus -a -um with leaf rosettes, *rosula*
rotang an Indian vernacular name for a rattan vine
rotatus -a -um flat and circular, wheel-shaped, *rota, rotae*
rothomagensis -is -e from Rouen, France (*Rothomagus, Rotomagus*)
rotulus -a -um like a small wheel, diminutive of *rota, rotae*
rotundatus -a -um becoming rounded, *rotundo, rotundare*
rotundi- rounded in outline or at the apex, spherical, *rotundus, rotundi-*
rotundifolius -a -um having rounded leaves, *rotundus-folium*
rotundilobus -a -um having rounded (leaf) lobes, *rotundus-lobus*
rotundisepalus -a -um having rounded sepals, botanical Latin from *rotundus* and σκεπη
rotundus -a -um plump, round, circular or spherical, *rotundus*
Roxburghia, roxburghii for Dr William Roxburgh (1751–1815), Director of Calcutta Botanic Garden
Royena for Adrian van Royen (1704–99), Professor of Botany at Leiden
Roystonea for General Roy Stone (1836–1905), American soldier
-rrhagus -a -um bursting, ρραγ, stem of ρηγνυμι
-rrhizus -a -um -rooted, ριζα
rubellinus -a -um somewhat reddish, double diminutive of *ruber*
rubellus -a -um reddish, diminutive of *ruber*
rubens blushed with red, ruddy, blushing, *rubeo, rubere*
ruber -bra -brum, rubis -is -e, rubri-, rubro- red, *ruber, rubra, rubrum; ruber, rubri*
ruberrimus -a -um very red, superative of *ruber*

rubescens, rubidus -a -um turning red, reddening, blushing, present participle of *rubesco, rubescere, rubescui*
rubi of brambles, living on *Rubus* (symbionts, parasites and saprophytes)
Rubia Red, *ruber* (the name in Pliny for madder) (***Rubiaceae***)
rubicundus -a -um ruddy-complexioned, reddened, reddish, *rubeo, rubere; rubicundus*
rubidus -a -um reddish, *rubidus*
rubiflorus -a -um having flowers resembling *Rubus, Rubus-florum*
rubifolius -a -um with leaves similar to those of a *Rubus, Rubus-folium*
rubiginosus -a -um rusty-red, brownish-red, *robigo, robiginis*
rubra-euchlora reddish (*Tilia*) *euchlora*
rubricaulis -is -e with reddish stems, *ruber-caulis*
rubriflorus -a -um red-flowered, *ruber-florum*
rubrifolius -a -um red-leaved, *ruber-folium*
rubrimaris -is -e of the Red Sea, between the Nubian and Syrian tectonic plates (called the Red Sea because of colouring effect of the decay products from the periodic intense blooms of *Trichodesmium erythraeum*)
rubripes red-stalked, *ruber-pes*
rubrocyaneus-a -um red and blue, anthocyanosed, *ruber-cyaneus*
rubroglaucus -a -um glaucous-red coloured, *ruber-glaucus*
rubromarginatus -a -um with red margins, *ruber-*(*margo, marginis*) (to the leaves or petals)
rubromucronatus -a -um with red mucronate apices, *ruber-*(*mucro, mucronis*) (to leaves or petals)
rubrostylus -a -um red-styled, *ruber-stilus*
rubrotinctus -a -um imbued with red, blushed, *ruber-tinctus*
rubrum *vide ruber*
Rubus, rubus the ancient Latin name, *rubus*, for brambles, bramble-like
Rudbeckia for Linnaeus' mentor Olaus (Olof) Rudbeck (1630–1702) and his son Olof Rudbeck (1660–1740) both professors of botany at Uppsala
rudentus -a -um creaking; cabled, rope-like, *rudens, rudentis*
ruderalis -is -e of waste places, of rubbish tips, *rudus, ruderis*
rudis -is -e untilled, rough, wild, coarse, *rudus, ruderis*
rudiusculus -a -um wildish, *rudus, ruderis*, with diminutive *-usculus*
Ruellia for John de la Ruelle of Soissons, author of *De natura plantarum* (1536)
rufescens, rufidus -a -um being reddish, turning red, *rufus-essentia*
rufi- red-, reddish-, *rufus*
rufibarbus -a -um red-bearded, *rufus-*(*barba, barbae*)
rufidulus -a -um somewhat rusty-red, diminutive of *rufus*
rufinervis -is -e with red veins, red-nerved, *rufus-nerva*
rufinus -a -um red, *rufus*
rufo-ferrugineus -a -um reddish-brown, chestnut coloured, *rufus-ferrugineus*
rufomicans shining reddish-brown, present participle of *rufus-*(*mico, micare*)
rufo-olivaceus -a -um reddish-olive coloured, *rufus-oliva*
rufotomentosus -a -um with red tomentose hair, *rufus-tomentum*
rufus -a -um, -rufus rusty (-haired), pale- or reddish-brown, red, *rufus* (reds in general)
rugosus -a -um wrinkled, rugose, *rugo, rugare* (e.g. leaf or fruit surfaces)
rugus -a -um having wrinkles or creases, *ruga, rugae*
rugulosus -a um somewhat wrinkled, with small wrinkles, diminutive of *rugosus*
Rulac an Adansonian name (≡ *Acer negundo*)
rumelianus -a -um, rumelicus -a -um from Roumelia, SE Europe
Rumex a name, *rumex, rumicis*, in Pliny for sorrel
rumici- dock-like-, *Rumex*
rumicifolius -a -um with dock-like leaves, *Rumex-folium*
ruminatus -a -um thoroughly mingled, as if chewed, *rumino, ruminare*
rumphii for Georg Everhard Rumpf (1627–1702), Dutch author of *Herbarium Amboinense*

runcinatus -a -um with sharp retrorse teeth (leaf margins), saw-toothed with the fine tips pointing to the base, runcinate, *runcina, runciae; runcinatus*
ruparus -a -um not neat, dirty, ρυπαρος
rupester -tris -tre, rupicola of rock, *rupes*, living in rocky places, *rupes-colo*
rupi-, rupri- of rocks-, of rocky places-, *rupes* rock
Rupicapnos Rock-fumitary, *rupes-capnos* (habitat)
rupicola growing on rocks, *rupes-colo*
rupicoloides resembling *rupicola*
rupifragus -a -um growing in rock crevices; rock-cracking, *rupes-(frango, frangere, fregi, fractum)*
ruppellii for Wilhelm Peter Eduard Rüppell (1794–1884), German naturalist and explorer in N Africa
Ruppia (*Ruppa*) for Heinrich Bernhard Ruppius (1688–1719), German botanist (***Ruppiaceae***)
rupti- interrupted-, broken-, *rumpo, rumpere, rupi, ruptum* to burst or tear
ruralis -is -e of country places, rural, *rus, ruris*
rurivagus -a -um of country roads, country wandering, *ruris-vagus*
Ruschia for Ernst Rusch (1867–1957), S African farmer
rusci- box holly-like, butcher's-broom-like, resembling *Ruscus*
ruscifolius -a -um with leaves resembling the cladodes of *Ruscus, Ruscus-folium*
Ruscus an old Latin name, *ruscum*, for a prickly plant
ruso-, rysso- wrinkled, ρυσος, ρυσσος
Ruspolia for Prince Eugenio Ruspoli (1866–93), explorer, killed by an elephant in Somalia
russatus -a -um reddened, russet, *russus*
Russelia for Dr Alexander Russel FRS (1715–68), author of *Natural History of Aleppo*
russellianus -a -um for either the Dukes of Bedford (Russells), or Mr Russell of Falkirk, or James Russell of Sunningdale Nursery, or George Russell of York (1857–1951), *Lupinus* breeder
russicus -a -um Russian, from Russia, *Rossica*
russocoriaceus -a -um red-leathery, *russus-coriaceus*
russotinctus -a -um red-tinged, *russus-tinctus*
Russula Reddish, *russus* (some have a red cap)
rusticanus -a -um, rusticus -a -um of the countryside, clownish, rustic, *rus, ruris* (*Apium rusticum* is fool's parsley)
Ruta Unpleasantness, the ancient Greek name for rue, ρυτη (***Rutaceae***)
ruta-baga from the Swedish name, rotbagge, ram's root, for swede
rutaecarpus -a -um with fruits similar to *Ruta*, ρυτη-καρπος
ruta-muraria rue-of-the-wall, *ruta-(murus, muri)* a name used in Brunfels
rutgersensis -is -e from Rutgers, State University of New Jersey, USA
ruthenicus -a -um from Ruthenia, Carpathian region of E Europe
rutifolius -a -um with leaves resembling those of *Ruta, Ruta-folium*
rutilans glowing red, turning red, present participle of *rutilo, rutilare*
rutilus -a -um auburn, deep bright glowing red, orange, or golden-yellow, *rutilus*
Ruttya for Dr John Rutty (1697–1775), Irish naturalist
Ruyschia, ruyschianus -a -um for Frederick Ruysch (1638–1731), Professor of Botany at Amsterdam
Ryssopteris Wrinkled-wing, ρυσσος-πτερυξ (the wing on the fruit)
rytidi-, rytido- wrinkled-, ρυτις, ρυτιδος, ρυτιδο-
rytidocarpus -a -um wrinkled-fruit, ρυτιδο-καρπος
rytidophyllus -a -um with wrinkled leaves, ρυτιδο-φυλλον
Rytigynia Wrinkled-ovary, ρυτις-γυνη
ryukyensis -is -e from the Ryukyu islands, off S Japan

sabahanus -a -um from Sabah (N Borneo), E Malaysia
Sabal possibly from a S American vernacular name

sabaneticus -a -um from Santiago Rodrigues (Sabaneta), NW Dominican Republic
sabatius -a -um from Capo di Noli (*Sabathia*), Riviera di Ponente, Liguria, Italy
sabaudus -a -um from Savoy (*Sabaudia*), SE France (*Brassica sabauda* was an old name for savoy cabbage)
Sabbatia for Liberato Sabbati (1714–*c*.79), Italian botanist, author of *Synopsis plantarum*
sabbatius -a -um from Savona, NW Italy, of the Sabbath, *sabbatius*
sabdariffa from a W Indian vernacular name for *Hibiscus sabdariffa*
Sabia from its Bengali vernacular name, sabja-lat (***Sabiaceae***)
Sabicea the Guyanese vernacular name, sabisubi, for *Sabicea aspera*
Sabina, sabina from the Latin name, *herba sabina*, of the people of central Italy, the *Sabini* (for savin, ≡ *Juniperus sabina*, which Pliny described as an abortifacient, later reflected in its use in gin)
Sabinea, sabinianus -a -um for Joseph Sabine (1770–1837) secretary of the Horticultural Society of London
sabrinae from the environs of the River Severn (*Sabrina*)
sabuletorus -a -um of the sands (deserts), of sandy soil communities, *suburra, saburrae*
sabulicolus -a -um, sabulus -a -um living in sandy places, sand-dweller, *sabulum-colo*
sabulosus -a -um sandy, full of sand, of sandy ground, *sabulum*
sacc- sac-, pouch-, σακος, σακκος, σακκο-, σακκ-
saccaticupulus -a -um with a pouch-like cupule, *saccus-cupula*
saccatus -a -um bag-shaped, pouched, saccate, σακκος, *saccus*
saccharatus -a -um with a scattered white coating, sweet-tasting, sugared, *saccharum*
sacchariferus -a -um sugar-producing, bearing sugar, *saccharum-fero*
sacchariflorus -a -um sugar-cane-flowered, *Saccharum-Florum*
saccharinus -a -um, saccharus -a -um sweet, sugary, *saccharum*
sacchariolens sweet-perfumed, with a sweet smell, *saccharum-olens* (*oleo, olere. olui*)
Saccharodendron Sugar-tree, σακχαρον-δενδρον
saccharoides looking like sugar cane, σακχαρο-οειδης
saccharophorus -a -um producing sugar, with a sweet juice, σακχαρο-φορα
saccharosus -a -um having a plentiful sugary sap, *saccharum*
Saccharum, saccharum Sugar, σακχαρον, *saccharum* (for the extract from the solid stem), Arabic, soukar
saccifer -era -erum having a hollowed part, pouch-bearing, bag-bearing, σακκος, σακκο-, σακος, σακο-
Sacciolepis Bag-like-scaled, σακκο-λεπις
sacculatus -a -um having small pocket-like structures, diminutive of *saccus*
sachalinensis -is -e from Sakhalin Island, E Russia
Sacoglottis Pouch-tongue, σακο-γλωττα (the anthers dehisce through basal pouch-like extensions)
sacra sacred, *sacer, sacri* (highly valued frankincense, *Boswellia sacra*)
sacrorum of sacred places, of temples, sacred, *sacer, sacri* (former ritual use; Rome's *Via Sacra* contained most of its temples)
sacrosanctus -a -um held in reverence, sacred, *sacrosanctus*
sadoinsularis -is -e from Sado island, W of Honshu, Sea of Japan
saepium of hedges, *saepes, saepis*
saetabensis -is -e from the area of Játiva (Xátiva) or Alzira (*Saetabis* or *Saetabicula*), Valencia, E Spain
saevus -a -um fierce, cruel, *saevus* (an inappropriate name for the field blewit)
saffroliferus -a -um smelling of true saffron, bearing saffron fragrance, botanical Latin from Arabic, za faran, *oleo* and *fero*
sagatus -a -um cloaked, as if having a soldier's cloak, *sagatus*
sagenarius -a -um of net-fishing or fishing nets, σαγηνη
Sagina Fodder, *sagina* (the virtue of a formerly included species, spurrey)
saginatus -a -um well-fed, stuffed, fattened, *sagino, saginare*
saginoides resembling *Sagina*, *Sagina-oides*

sagittalis -is -e, sagittatus -a -um, sagitti- (*saggitatus -a -um*) arrow-shaped, sagittate, *sagitta, sagittae* (see Fig. 6c)
Sagittaria, sagittarius -a -um Arrowhead, *sagitta*, (*herba sagittaria*, the shape of the leaf-blades re-emphasized in *Sagittaria sagittifolia*)
sagittifolius -a -um with arrow-shaped leaf-blades, *sagitta-folium*
sagu yielding the large starch grains, from Malayan vernacular name, sagu, for the sago palm
saguntinus -a -um from Sagunto, N of Valencia, E Spain
Sagus from the Malayan vernacular name, sagu
sahelicus -a -um from the semi-arid Sahel zone separating the Sahara desert from the more humid savannas to the south
sahyadricus -a -um from the Sahyadri mountains, Maharashtra, India
Saintpaulia for Baron Walter von Saint Paul-Illaire (1860–1910), Berliner who discovered *Saintpaulia ionantha*
sakalavarus -a -um of the Sakalava people of the western savannas of Madagascar
salax lustful, salacious, *salax*
Salaxis an unexplained name by Salisbury (who perhaps lusted, *salax, salacis*, after it)
saldanhensis -is -e from Saldanha Bay, SW S Africa
salebrosus -a -um rough, *salebra, salebrae*
salicarius -a -um, salicinus -a -um willow-like, resembling *Salix*
salice-, salici- willow-like, willow-, *Salix*
salicetorum of willow thickets, *Salix-etorum*
salicifolius -a -um with willow-like leaves, *Salix-folium*
salicinus -a -um resembling willow, of willows, *Salix* (looking like, or growing with or on detritus from willows)
salicolus -a -um inhabiting saline soils, (*sal, salis*)*-colo*
Salicornia Salt-horn, *sal-cornu* (refers to the habitat and the form of the shoot-joints)
salicornioides resembling *Salicornia, Salicornia-oides*
salictorus -a -um of willow habitats, *Salix*
saliens projecting forward, present participle of *salio, salire, salui, saltum*
salignus -a -um of willow-like appearance, willowy, resembling *Salix*
salinus -a -um of saline habitats, halophytic, *salsus*
salisburgensis -is -e from Salzburg (*Salisburgia*) Austria, or Salisbury, England
Salisburia for Richard Anthony Salisbury (1761–1829) English botanist and founder member of the RHS in 1804 (≡ *Ginkgo biloba*)
Salix the Latin name for willows, cognate with sallow (***Salicaceae***)
salmanticus -a -um from Salamanca (*Salmintica*), W Spain
Salmia, salmii for Prince Joseph Maria Franz Anton Hubert Ignaz Salm-Reifferscheid-Dyck (1773–1861), writer on succulent plants
salmoneus -a -um salmon-coloured, pink with a touch of yellow (in mythology, the son of Aeolus, punished for imitating lightning)
salomonis of Solomon (medieval Latin, *sigillum salomonis*, for Solomon's seal, *Polygonatum multiflorum*)
salonitanus -a -um from the environs of the Bay of Salona, Gulf of Corinth, Greece
salpi- tube-, trumpet-, σαλπιγξ, σαλπι-
Salpichroa Tube-of-skin, σαλπι-χροα (the form of the flower)
Salpiglossis Trumpet-tongue, σαλπι-γλωσσα (the shape of the style)
salpingophorus -a -um trumpet-bearing, (σαλπιγξ, σαλπιγγος)-φορος
Salsola Salt, *salsus* (Cesalpino's name for the taste and the habitat)
salsoloides resembling *Salsola, Salsola-oides*
salsuginosus -a -um of salt-marshes, of habitats inundated by salt water, *salsus*
salsus -a -um witty; living in saline habitats, *salsus*
saltatorius -a -um dancing, *saltatorius* (also of discontinuous evolutionary strides)
saltatrix (suggestive of) a female dancer, *saltatrix, saltatricis*
saltensis -is -e from the province of Salta, NW Argentina

saltitans jumping (twitching of the heat-sensitive larva of *Cydia saltitans* in the seed of the Mexican jumping bean *Sebastiana* causes it to jump)
saltuarius -a -um, saltuensis -is -e of woodland meadows, *saltus*
saltuum of glades, woodlands or ravines, *saltus*
saluensis -is -e from the Salween river (Nu Jiang river), China
salutaris -is -e healing, beneficial, wholesome, *salutaris*
salvador from El Salvador, Central America
salvatoris -is -e from Mount San Salvatore, Ticino canton, S Switzerland
salvi-, salviae, salvii- sage-like-, resembling *Salvia*
Salvia Healer, *salveo, salvere*, the old Latin name for sage with medicinal properties (cognates are old French, saulje, sauge and our sage)
salviaefolius -a -um, salvifolius -a -um, salviifolius -a -um sage-leaved, *Salvia-folium*
Salvinia for Professor Antonio Maria Salvini (1633–1722), botanist and Greek scholar of Florence, Italy (***Salviniaceae***)
salviodorus -a -um sage-scented, *Salvia-odorus*
salzmannii for Philipp Salzmann (1781–1851), of Montpellier, who collected in Brazil, Spain, N Africa and S France
saman, Samanea from a S American name, zamang, for the rain tree, *Pithecolobium saman*
samarkandensis -is -e from Samarqand, Uzbekistan
samaroideus -a -um with samara-like fruits, (*samara, samera*)*-oides*
sambac from the Arabic name, zambac, for *Jasminum sambac*
sambuci-, sambucinus -a -um elder-like, resembling *Sambucus*
sambuci growing on elder material, *Sambucus*
sambucifolius -a -um with leaves similar to those of *Sambucus, Sambucus-folium*
Sambucus from the Latin name for the elder tree (*sambuca* was a harp; Gilbert Carter suggests a similarity between the many epicormic shoots and the strings of the σαμβυκη)
samius -a -um from the isle of Samos, Greece
Samolus a name in Pliny, or from a Celtic Druidic name, sal mos (pig food)
Sanchezia for Joseph Sanchez, Professor of Botany at Cadiz
sanctae-rosae holy-rose, *sanctus-rosa*
sancti-johannis for Saint Ivan Rilski (the hermit John of Rila), patron saint of the largest Bulgarian monastery in Rila (St John's wort)
sanctus -a -um holy, sacred, chaste, past participle of *sanctio, sancire, sanxi, sanctum*
sanderae, sanderianus -a -um, sanderi from Henry Frederick Conrad Sander (1847–1920) and family, nurserymen of St Albans and Bruges, importers of many new plants
Sandersonia, sandersonii for John Sanderson (1820–81), Hon. Secretary of the Horticultural Society of Natal
sandwicensis -is -e, sandwicensius -a -um from the Sandwich Islands
sanguinalis -is -e, sanguineus -a -um blood-red, bloody, *sanguis, sanguinis*
Sanguinaria Blood, *sanguis, sanguinis*, (the copious crimson sap)
sanguineolentus -a -um, sanguinolentus -a -um bleeding, bloody-looking, smelling of blood, *sanguis-*(*olens, olentis*)
sanguiniflorus -a -um having blood-red flowers, *sanguis-florum*
Sanguisorba, sanguisorbae Blood-stauncher, *sanguis-*(*sorbeo, sorbere, sorbui*) (has styptic property)
Sanicula Little-healer, *sano, sanare, sanavi, sanatum* (the medicinal property of sanicle)
saniculiformis -is -e looking like *Sanicula* in habit
saniosus -a -um like diseased blood or venom, *sanies*
Sanseveria for Prince Raimond de Sansgrio of Sanseviero (1710–71), Swedish botanist
sansibaricus -a -um from the Zanzibar islands (former Zanzibar Protectorate), Indian Ocean, Tanzania
santalinus -a -um sandal-wood or its resin, santalin, *santalum*

Santalum from the Persian, shandul, for the sandal-wood tree (***Santalaceae***)
santiago of Santiago, Spain, or Chile, or Panama
Santolina Holy-flax, (*sancio, sancire, sanxi, sanctum*)-*linum*
Sanvitalia for the San Vitali (Sanvitali) family of Parma
sap-, sapon- sap-, sweet-tasting-, soapy- (*sapa*, plant-juice, *sapo*, soap)
sapidus -a -um pleasant-tasted, flavoursome, savoury, *sapidus*
sapientium of the wise, of man, *sapiens, sapientis* (implies superiority compared with *troglodytarum*)
Sapindus, sapindus -a -um Indian-soap, contraction of *sapo-indicus* (from its use) (***Sapindaceae***)
Sapium Soapy, *sapo* (refers to the sticky sap)
saponaceus -a -um, saponarius -a -um lather-forming, soapy, *sapo*
Saponaria, saponarius -a -um Soap-like, *sapo, saponis* (lather-forming soapwort)
Sapota, sapota former generic name from the Mexican name, cochil-zapotl, for chicle-tree; see also *zapota* and *Achras* (***Sapotaceae***)
sappan from a Malayan vernacular name, sepang, for *Caesalpinia sappan*
sapphirinus -a -um sapphire-blue, via French, safir, from σαπφειρος
saprio-, sapro- rotten-, σαπρος, σαπρο-
Saprolegnia Putrid-edges, σαπρο-λεγνον
saprophyticus -a -um saprophage, feeding on dead material, σαπρος-φυτον
Saraca from an Asian Indian native vernacular name
saracenicus -a -um, sarracenicus -a -um of the Saracens, *Saraceni* (all Muslim peoples were called σαρακενοι by the Greeks)
Saracha for Isidore Saracha (1733–1803), Spanish Benedictine monk who sent plants to the Madrid Royal Gardens
sarachoides resembling *Saracha, Saracha-oides*
sarc-, sarco- fleshy-, σαρξ, σαρκος, σαρκο-
sarcanthus -a -um with fleshy flowers, σαρκο-ανθος
Sarcobatus Fleshy(-leaved)-thorn-bush, σαρκο-βατος
Sarcodon Fleshy, σαρκωδης
Sarcocapnos Fleshy-fumitory, σαρκο-καπνος
sarcocaulis -is -e soft-stemmed, fleshy-stemmed, σαρκο-καυλος
Sarcocephalus Fleshy-head, σαρκο-κεφαλη (the head of fruits)
Sarcococca Fleshy-berry, σαρκο-κοκκος
sarcoides, sarcodes flesh-like, σαρξ-οειδης, σαρξ-ωδες
Sarcophrynium Fleshy-*Phrynium*, σαρκο-φρυνος (the fleshy fruits)
sarcophyllus -a -um fleshy-leaved, σαρκο-φυλλον
Sarcorhynchus Fleshy-beak, σαρκο-ρυγχος (the swollen spur)
Sarcoscypha Fleshy-goblet, σαρκος-σκυφος (the saprophytic elf-cup fungus)
Sarcosperma Fleshy-seed, σαρκος-σπερμα
Sarcostemma Fleshy-crown, σαρκο-στημων (the coronna)
sardensis -is -e from Lydian Izmir (Sart, *Sardis*) Smyrna, Turkey
sardonius -a -um with the colouration of sardonyx, σαρδονυξ (brownish-red with white)
sardosus -a -um, sardous -a -um from Sardinia, Sardinian (*herba sardoa* was one of the poisons for which Sardinia was famous)
Sargassum from a Portuguese word, sargaço, of unknown meaning
sargentianus -a -um for Professor C. S. Sargent (*vide infra*)
Sargentodoxa Sargent's-glory, for Professor Charles Sprague Sargent (1841–1927), founder and director of Arnold Arboretum, Massachusetts, USA (***Sargentodaxaceae***)
saribus -a -um from the Maluku vernacular name, sariboe
sarisophorus -a -um, sarissophorus -a -um carrying long lanceolate leaves, (*sarisa, sarisae*)-*fero* (literally a Macedonian long lance)
sarisus -a -um, sarissus -a -um long and lanceolate, lance-like, *sarisa, sarisae*
sarmaticus -a -um of the Sarmatians, who inhabited an area now occupied by Ukraine, Belarus and SE Russia

sarmentaceus -a -um, sarmentosus -a -um with long slender stolons or runners, *sarmentum* brushwood
sarmentus -a -um twiggy, like brushwood, *sarmentum, sarmenti*
sarniensis -is -e, sarnius -a -um from Guernsey (*Sarnia*), Channel Isles
saro- broom-like-, σαροω (to sweep)
sarothamni of broom, living on *Sarothamnus* (*Asphondylia*, dipteran gall midge)
Sarothamnus Broom-shrub, σαροω-θαμνος
Sarracenia for Dr Michel Sarrazan (d. 1734), who introduced *Sarracenia purpurea* from Quebec (***Sarraceniaceae***)
sarrachoides from a Brazilian name for another solanaceous genus named for Isidore Saracha (1733–1803), a Benedictine monk who sent plants to Madrid's Royal Gardens, *Saracha-oides*
sarsaparilla from the Mexican-Spanish, zarza-parilla, prickly little vine
sartorii for Andria del Sarto (1486–1531); of tailors, *sartor, sartoris*
Sasa the Japanese name for certain dwarf bamboos
Sasaella Little-dwarf-bamboo, feminine diminutive of *Sasa*
Sasamorpha *Sasa*-shaped, botanical Latin from *Sasa* and μορφη
sasanqua from the Japanese name for the tea-oil-producing *Camellia*
Sassafras from the Spanish name, salsafras, for its medicinal use in breaking bladder and kidney stones (cognate with *Saxifraga*)
satanas the devil's, Satan's, σαταν, σατανος (the devil's *Boletus*)
Satanocrater Satan's-bowl, σατανος-κρατηρ; or Satan's-sin
satanoides resembling (*Boletus*) *satanas*
sathro- humus-, decayed-, perishable-, σαθρος, σαθρο-
satis -is -e tolerable, enough, *satis, sat*; some suggest rather better than adequate
sativus -a -um planted, cultivated, not wild, sown, *sero, serere, sevi, satum*
sativus-atrocaeruleus -a -um cultivated dark blue, *sativus-*(*ater, atri*)*-caeruleus*
satsumanus -a -um from Satsuma peninsula, Kyushu, SW Japan
saturativirens green as grass, full-deep-green, (*satio, satiare, satiavi, satiatum*)*-virens*
saturatus -a -um of intense, full or mixed colouring, *satio, satiare, satiavi, satiatum*
Satureia, Satureja the Latin name, *satureia, satureiorum* in Pliny for a culinary herb, from the Arabic, sattur, savory
satureioides, saturejoides resembling *Satureia, Satureia-oides*
satyrioides resembling *Satyrium, Satyrium-oides*
Satyrium Dioscorides' name for an orchid (in mythology, Satyrion was a drunken woodland god)
Saundersia for William Wilson Saunders FRS (1809–79), of London
saundersiae, saundersii for Mrs Kathleen Saunders (1824–1901), botanical artist in S Africa
saur-, sauro- lizard-like-, lizard-, σαυρα, σαυρος, σαυρο-
Saurauia (*Saurauja*) for Fr. J von Saurau (1760–1832), Italian botanist
saurocephalus -a -um lizard-headed, σαυρο-κεφαλη (colouration)
Sauromatum Lizard, σαυρος (the inner surface of the spathe suggests lizard skin)
Saururus Lizard-tail, σαυρο-ουρα (***Saururaceae***)
saurus -a -um of lizards, σαυρα, σαυρος (σαυροχωρεω, fruit or seed dispersal by lizards)
Saussurea for Horace Bénédict de Saussure (1740–99), Swiss geologist who coined the name geology for his studies in the Alps, and author of *Voyages dans les Alpes*. His son was Nicolas Théodore de Saussure (1767–1845), who confirmed Hale's discoveries on photosynthesis of carbon dioxide
Sauvagesia for François Boissier de Sauvages (1706–67), Professor of Botany at Montpellier
savaganus -a -um of the wild woods, from Latin *silvaticus*, via French, sauvage
savannarus -a um of savannas, from Taino, zavana; or from Savannah, Georgia, USA
savin from Pliny's name, *herba Sabina*, Sabine herb, for *Juniperus sabina*, which was used to procure abortions

saxa-, saxi- rock-, rocks-, *saxum, saxi*
saxatilis -is -e living in rocky places, of the rocks, *saxatilis*
Saxegothaea for Prince Francis Albert Augustus Charles Emanuel of Saxe-Coburg-Gotha (1817–61), Queen Victoria's consort
saxicolus -a -um rock-dwelling, *saxum-colo*
Saxifraga Stone-breaker, *saxum-frango* (living in rock cracks, had the signature for medicinal use in treating gall-, bladder- and kidney-stones) (***Saxifragaceae***)
saximontanus -a -um of screes, of the Rocky Mountains, *saxi-montanus*
saxorus -a -um, saxosus -a -um of rocky or stony places, *saxum*; rocky, stony, *saxosus*
sazensoo the Japanese vernacular name for *Arisaema sazensoo*
scaber -ra -rum coarse, rough, scabrid (like sandpaper), *scaber, scabri*
scaberrimus -a -um the roughest, coarsest, superlative of *scaber*
scaberulus -a -um roughish, somewhat rough, diminutive of *scaber, scabri*
Scabiosa Itch, *scabies, scabiem, scabie* (signature of scurvy involucre, as of medicinal use as a treatment for the disease)
scabiosae of knapweed, living on *Centaurea scabiosa* (*Isocolus*, hymenopteran gall wasp)
scabiosifolius -a -um with leaves resembling those of *Scabiosa, Scabiosa-folium*
scabrellus -a -um somewhat scabrid, diminutive of *scaber, scabri*
scabri- rough-, scabrid-, *scaber, scabri*
scabridoglandulosus -a -um with a scabrid glandular surface, *scaber-glandis*
scabridus -a -um having a rough surface to the touch, *scaber, scabri*
scabrifolius -a -um with rough or scabrid leaf surfaces, *scabri-folium*
scabrilinguis -is -e having a rough tongue, *scabri-lingua*
scabriscapus -a -um with a rough scape, *scabri-scapus*
scabriusculus -a -um somewhat scabrid, diminutive of *scaber, scabri*
scabrosus -a -um rather rough, *scaber, scabri*
scabrus -a -um with a rough surface to the touch, *scaber, scabri*
Scadoxus Shade-glory, σκια-δοξα (the parasol-like flower-heads, ≡ *Haemanthus*)
Scaevola Left-handed, *scaevus* (the one-sided corolla lobes imitate a hand (*Scaevola aemula*, imitating the famed jurist Scaevola Gaius Mucius (507 BC), who convinced the Etruscan ruler, Porsena, of his nobility by burning off his right hand in the altar fire, and thus caused Porsena to make peace with Rome)
scalariformis -is -e with ladder-like form or markings, ladder-like, *scalae, scalarum*
scalaris -is -e with ladder-like markings, *scalae, scalarum*
Scalesia for Scales, the Galapagos naturalist (Santiaga daisy tree)
scalpellatus -a -um knife-like, cutting, *scalpellum*
scalptratus -a -um, scalpturatus -a -um engraved, *scalpo, scalpere, scalpsi, scalptum*
scammonius -a -um purging, σκαμμονια, *scammonea, scammoneae* (purging bindweed, *radix scammoniae* or *Convolvulus scammonia*, cognate with scammony)
scandens climbing, present participle of *scando, scandere*
scandicus -a -um from Scandia, Scandinavian, or Schonen (*Scania*), Sweden
scandinavicus -a -um from Scandinavia
Scandix ancient name, σκανδιξ, for shepherd's needle
scapeosus -a -um, scapiosus -a -um becoming well-scaped, *scapus* (*Primula scapiosa* scape elongates in fruit)
scaphi-, scapho-, scaphy- boat-shaped-, bowl-shaped-, σκαφη, σκαφος, σκαφις, σκαφι-
scaphiglossus -a -um with a dish- or boat-shaped tongue or labellum, σκαφη-γλωσσα
scaphoides, scaphoideus -a -um boat-like, boat-shaped, σκαφοειδης
Scaphopetalum Boat-shaped-petal, σκαφη-πεταλον
scapi-, scapio-, -scapus -a -um clear-stemmed-, scapose-, *scapus, scapi*
scapiferus -a -um, scapiger -era -erum scape-bearing, *scapus-fero*
scapiflorus -a -um with stalked flowers, *scapus-florum*
scapoideus -a -um scape-like, *scapus-oides*
scaposus -a -um with scapes or leafless flowering stems, *scapus, scapi*
-scapus -a -um -peduncled, -stalked, -scaped, -scapose, *scapus, scapi*

scardicus -a -um from the Scardus mountains, S'ar Planina (*Scardia*), Serbia/ Macedonia
scariola (*serriola*) endive-like, of salads, diminutive of *seris*
scariosus -a -um shrivelled, thin, not green, membranous, stiff, scarious, late Latin from *scaria*, a thorny shrub
scarlatinus -a -um brightly coloured; scarlet, medieval Latin *scarlata*, via French, escarlate
scat-, scato- dung-, σκορ, σκατ
scatophagus -a -um dung-feeding, σκατοφαγος
scaturicolus -a -um living near springs or geysers, *scaturigines-colo*
scaturiginsosus -a -um overflowing, gushing, very full, *scaturio, scaturire*
sceleratus -a -um hard, σκληρος; pernicious, vicious, wicked, *sceleratus* (*Ranunculus sceleratus* sap causes ulceration)
Scenedesmus Living-in-strings, σκηνη-δεσμος (morphology, living in bundles)
-scepes -covering, σκεπη
sceptrodes, sceptrus -a -um sceptre-, staff- or wand-like, *sceptrum, sceptri*
schafta a Caspian area vernacular name for *Silene schafta*
schantungensis -is -e from Shantung (Shandong) province of E China
scharffianus -a -um, sharffii for Carl Scharff, who collected on Santa Catherina Island, Brazil *c.* 1888
Schedonorus Near-the-margin, σχεδον-ορος (insertion of the awn, ≡ *Bromus*)
Schefflera for J. C. Scheffler of Danzig
scherzerianus -a -um for M. Scherzer (1821–1903), who found *Anthurium scherzerianum* in Guatemala
Scheuchzeria for the brothers Johann Jakob Scheuchzer (1672–1733) and Johannes Scheuchzer (1684–1738), Professor of Botany at Zurich (***Scheuchzeriaceae***)
schiedianus -a -um. schiedii for Christian J. W. Schiede (1798–1836), traveller in Mexico with Deppe
schillerianus -a -um for Herr Schiller, orchid-growing Consul in Hamburg
Schima etymology unclear; some derive as σχισμα a division, some consider an Arabic origin (σχημα means outward appearance)
schinseng from the Chinese name
schinifolius -a -um with leaves resembling those of *Schinus, Schinus-folium*
Schinus from the Greek name, σχινος, for another mastic-producing plant (*Pistacia*)
schipkaensis -is -e from Schipka Pass, Bulgaria (site of Süleyman Pasha's infamous battle against the Russians)
Schisandra Divided-man, σχιζα-ανδρος (the cleft anthers of the type species) (***Schisandraceae***)
schist-, schismo- divided-, cut-, cleft-, σχιστος, σχισμα
schist-, schisto easily split-, stone-, *schistos* (those that split into layers, schists)
schistaceus -a -um slate-coloured, splitting, σχιστος
schistocalyx with a split calyx, σχιστος-καλυξ
Schistostegia Divided-cover, σχιστος-στεγη (the calyptra)
schistosus -a -um slate-coloured, σχιστος
Schivereckia for S. B. Schivereck from Innsbruck, professor at Lemberg *c.* 1782–1805
schiz-, schizo- cut-, divided-, split-, σχιζειν, σχιζα, σχιζη, σχιζω, σχιζ-
Schizachyrium Split-chaff, σχιζ-αχυρον (the bifid fertile lemmas)
Schizaea Cut, σχιζα, σχιζω (the incised fan-shaped fronds) (***Schizaeaceae***)
Schizandra Split-stamens, σχιζ-ανηρ
Schizanthus Divided-flower, σχιζ-ανθος (the fringed lobes of the corolla in the poor man's orchid)
schizo- fringed, irregularly incised or split, σχιζω
schizocheilus -a -um with an incised lip, σχιζω-χειλος
Schizolegnia Fringed-border, σχιζω-λεγνον
Schizolobium Fringed-lobes, σχιζω-λοβος (the corolla lobes)

Schizomeria, schizomerus -a -um splitting into parts, σχιζω-μερις
Schizonotos Cut-surface, σχιζω-νοτος (≡ *Sorbaria*)
Schizopetalon, schizopetalus -a -um with deeply cut petals, σχιζω-πεταλον
Schizophragma Cleft-wall, σχιζω-φραγμα (the fragmenting capsule walls)
schizophyllus -a -um with split or incised leaves, σχιζω-φυλλον
Schizostachyum Cut-spike, σχιζω-σταχυς
Schizostylis Divided-style, σχιζω-στυλος (the three elongate arms)
-schizus -a -um -cut, -divided, σχιζα
Schkuhria, schkuhrii for Christian Schkuhr (1741–1811), German botanist at Wittenberg
schlechtendalii for either D. K. L. von Schlechtendal (1767–1842), of Xanten, or Diederich F. L. von Schlechtendal (1794–1866), Professor of Botany at Halle
schlechteri for Friedrich Richard Rudolf Schlechter (1872–1925), orchid expert of Berlin
Schlumbergera for Frederic Schlumberger, a Belgian horticulturalist and field botanist
schneideri, schneideriana for J. Christian Schneider
schoen-, schoeno- rush-like, resembling *Schoenus*
Schoenoplectus Rush-plait, σχοινος-πλεκω, σχοινοπλεκτος
schoenoprasus -a -um rush-like leek, σχοινος-πρασον (the leaves)
Schoenus the old name, σχοινος, for rush-like plants
scholaris -is -e of the school, of leisure, of peace, σχολη (*Alstonia scholaris*, dita-bark's tonic properties and writing-board wood)
schomburgkii for Sir Robert Hermann Schomburgk (1804–65), who found *Victoria regia* in S America
Schotia for Richard van der Schott, who accompanied Jacquin in America
schottii for Dr Arthur Schott, who collected in Arizona in 1855 (*Agave schottii*)
schraderianus -a -um for Heinrich Adolf Schrader (1767–1826), who monographed *Verbascum*, or Carl Schrader (1852–1930), astronomer and traveller
Schrankia for Franz von Paula von Schrank (1747–1835), German botanist
Schrebera, schreberi for Johann Christian Daniel von Schreber (1739–1810), a correspondent of Linnaeus
schroederae, schroederianus -a -um for Baroness and Baron Henry von Schröder (1825–1910), orchid growers, or R. Schröder, head gardener at the Agricultural Institute near Moscow
schubertii for Gotthilf Heinrich von Schubert (1780–1860), Austrian physician and traveller in Egypt and Palestine
Schultesia, schultasianus -a -um for Josef August Schultes (1773–1831), Austrian botanical writer
schumannii for Karl Moritz Schumann (1851–1904), of the Berlin Botanical Museum
Schumanniophyton Schumann's-plant, botanical Latin from Schumann and φυτον
schwantesii for Dr Gustav Schwantes, botanist of Kiel, Germany
schweinfurthii for Dr George Angust Schweinfurth (1836–1925), collector in central Africa
Schwenkia for J. T. Schwenk (1619–1671), Professor of Medicine at Jena
scia-, sciadi-, sciado-, scio- overhanging-; shadow-, ghost-, canopy-; umbelled-, σκιαζω, σκια, σκιας, σκιη, σκιαδος, σκιαδο- (used botanically as a suffix meaning umbel)
Sciadanthus Shade-flower, σκιαδος-ανθος
sciadius -a -um shade, canopied, of shade or a canopy, σκιαδος, σκιαδειον
sciadophorus -a -um bearing shade, σκιαδος-φορος
sciadophylloides large-leaved like *Sciadophyllum, Sciadophyllum-oides*
Sciadophyllum Canopy-of-leaves, σκιαδο-φυλλον
Sciadopitys Parasol-pine, σκιαδο-πιτυς (the leaves are crowded at the branch ends)
Sciaphila, sciaphilus -a -um Shade-loving, σκια-φιλος (saprophytic)
Scilla the ancient Greek name, σκιλλα, Latin, *scilla, squilla*, for the squill, *Urginea maritima*

scillifolius -a -um *Scilla*-leaved, *Scilla-folium*
scilloides squill-like, *Scilla-oides*
scilloniensis -is -e from the Scilly Isles (late Latin *Scillonian*)
Scindapsus an ancient Greek name, σκινδαψος, for an ivy-like plant
scintillans gleaming, twinkling, sparkling, *scintillo, scintillare*
scintillula like a small spark, feminine diminutive of *scintilla, scintillae*
sciophilus -a -um shade-loving, σκια-φιλος
scipioniformis -is -e of staff-like habit, *scipionis-forma*
scipionus -a -um wand-like, σκιπων, *scipio, scipionis* (*Calamus scipionum* is used for Malaca canes)
scirpinus -a -um resembling a reed or rush, *Scirpus*
Scirpoides, scerpoides, scirpoideus -a -um rush-like, *Scirpus*-like
Scirpus (*Sirpus*) the old name, *Scirpus*, for a rush-like plant
scissilis -is -e, scissus -a -um splitting easily, split, *scindo, scindere, scidi, scissum*
scitulus -a -um neat, pretty, smart, *scitulus*
scitus -a -um fine, smart, *scitus*
sciuroides curved and bushy, squirrel-tail-like, σκιουρος-οειδης
sciurus -a -um (looking like) a squirrel's tail, σκιουρος
Sclarea, sclarea Clear, medieval Latin, *sclarea* (an old generic name for a *Salvia*, clary, used for eye lotions)
scleracanthus -a -um with hard thorns or spines, σκληρος-ακανθος
Scleranthus Hard-flower, σκληρος-ανθος (texture of the perianth)
scleratus -a -um hardened, σκληρος
Scleria Hard, σκληρος (the hard-coated achenes)
sclero- hard-, σκληρος, σχληρο-
Sclerocarpus, sclerocarpus -a -um Hard-fruit, σκληρο-καρπος
Sclerochitom Hard-coat, σκληρος-χιτων
Sclerochloa Hard-grass, σκληρος-χλοη (≡ *Puccinellia*)
Scleroderma, sclerodermus -a -um Hard-skin, σκληρος-δερμα
scleroneurus -a -um with (prominent) hard veins, σκληρος-(νευρα, νευρον)
sclerophyllus -a -um with hard leaves, leathery-leaved, σκληρος-φυλλον
Scleropoa Hard-pasturage, σκληρος-ποα (≡ *Catapodium*)
Sclerosperma, sclerospermus -a -um Hard-seed, σκληρος-σπερμα (hard albumen of ripe seed)
scleroxylon having hard timber, σκληρος-ξυλον
scobi-, scobiformis -is -e sawdust-like, *scobis*
scobinatus -a -um, scobinus -a -um rough as though rasped, rasp-like, *scobina*
scobinicaulis -is -e with stems appearing to be covered in sawdust or shavings, *scobina-caulis*
scole-, scolo- vermiform-, worm-, σκωληξ
scolecinus -a -um worm-like, σκωληξ
scolio- curved-, bent-, σκολιος
Scoliopus Curved-stem, σκολιο-πους (the rhizomes)
scolopax of the woodcock (shared habitat with *Scolopax rusticola*)
Scolopendrium, scolopendrium Dioscorides' name, σκολοπενδριον, for the hart's tongue fern compares the numerous sori to the legs of a millipede, σκολοπενδρα
scolymoides resembling *Scolymus, Scolymus-oides*
Scolymus the ancient Greek name, σκολυμος, for the artichoke, *Scolymus hispanicus*, and its edible root
scolytus -a -um tortuous, σκολιος (wandering channels of the elm-bark beetle, *Scolytus destructor*)
scopa- broom, *scopae, scoparum*
Scoparia, scoparius -a -um, scopellatus -a -um broom-like, *scopae, scoparum* (use for making besoms)
-scopius -a -um -looking, -watching, σκοπη, σκοπια, σκοπιαζω
Scopolia for Giovani Antonio Scopoli (1723–88), professor at Pavia and writer on plants

scopulinus -a -um twiggy, broom-like, *scopae, scoparum*
scopulorum of cliffs and rock faces, *scopulus, scopuli* (in zoology, *scopula* a tuft of hairs)
scopulosus -a -um like a bristly brush, *scopae, scoparum*
scorbiculatus -a -um with a scurfy texture (*scorbutus* scurvy)
scordiifolius -a -um with leaves resembling those of *Scordium*
scordioides resembling *Scordium*, σκορδιον-οειδης
Scordium, scordium Dioscorides' name, σκορδιον, for a plant with the smell of garlic, σκορδον
Scorodonia an old generic name, σκοροδον, σκορδον, for garlic
scorodonifolius -a -um with leaves resembling those of *Scorodonia*
scorodoprasum (*scordoprasum*) a name, σκορδοπρασον, used by Dioscorides for a plant with intermediate features between garlic, σκορδον, and leek, πρασον
scorpioidalis -is -e, scorpioideus -a -um coiled like the tail of a scorpion, σκορπιος-οειδης (e.g. the axis of an inflorescence)
scorpioides (*scorpoides*) curved like a scorpion's tail (see Fig. 3), σκορπιος-οειδης
Scorpiurus Scorpion-tail, σκορπιος-ουρα (Dioscorides' name, for the coiled fruit of *Scorpiurus sulcata*)
scorteus -a -um leathery, *scorteus*
Scorzonera derivation uncertain but generally thought to refer to use as an antifebrile in snakebite (Italian, scorzone, for the snake *Elaphe longissima*)
scorzonerifolius -a -um with leaves resembling those of *Scorzonera*
scot-, scoto- of the dark-, darkness-, σκοτος, σκοτο-
scoticus -a -um from Scotland, Scottish, modern Latin *scottia*
scotinus -a -um dusky, dark, σκοταιος, σκοτεινος
scotophilus -a -um dark-loving, σκοτο-φιλος (e.g. subterraneous chemotrophic organisms)
scotostictus -a -um with dark spots, dark-punctate, σκοτο-στικτος
Scottellia for George Francis Scott-Elliot (1862–1934), boundary commissioner and plant collector in Sierra Leone, 1891–2
Scottia, scottianus -a -um for Munro Briggs Scott (1889–1917) of Kew, or Robert Scott (1757–1808), Professor of Botany at Dublin
scotticus -a -um from Scotland, Scottish, *scottia*
scouleri for Dr John Scouler (1804–71), Professor of Zoology at Dublin, and collector with David Douglas in NW America
scriblitifolius -a -um scroll-like leaved, with leaves appearing to have letters written upon them, *scribo-littera-folium*
scrinaceus -a -um with lidded-box-like fruits, *scrinium* (as in *Lecythis*) (literally a book-box or letter-case)
scriptus -a -um, -scriptus -a -um marked with lines which suggest writing, *scribo, scribere, scripsi, scriptum*
scrobiculatus -a -um with small depressions or grooves, pitted, *scrobis, scropis*
Scrophularia Scrophula, *scrofa, scrophae*; breeding sows were said to be prone to this glandular disease (signature of the glands on the corolla); many plants were used to treat *scrophulae*, the 'King's disease' (***Scrophulariaceae***)
scrophularifolius -a -um with leaves resembling those of *Scrophularia*
scrotiformis -is -e shaped like a small double bag, pouch-shaped, *scrotum-forma*
scruposus -a -um unstable, of jagged stone habitats, *scrupus, scrupi*
scrupulicolus -a -um living on sharp rocks, (*scrupus, scrupi*)*-colo*
sculptus -a -um carved, *sculpto, sculpere, sculpsi, sculptum*
Scurrula Little-dandy, diminutive feminine of *scurra, scurrae*
scutatus -a -um with a small round shield or buckler, *scutum* (leaves)
Scutellaria Dish, *scutella* (the depression of the fruiting calyx)
scutellarioides resembling *Scutellaria, Scutellaria-oides*
scutellaris -is -e, scutellatus -a -um platter-like, bowl-shaped, *scutella, scutellae*
scutellifolius -a -um with small dish-shaped (knob-like) leaves
scutelliformis -is -e with the shape of small dishes, *scutella-forma*

Scutellinia Small-bowl, *scutella, scutellae*
scutiformis -is -e buckler-shaped, (*scutum, scuti*)-*forma*
Scyphanthus Goblet-flower, σκυφος-ανθος
scyphiferus -a -um bearing goblet or wine-cup (-shaped structures), *scyphus-fero*
scypho-, -scyphus -a -um wine-cup-, beaker-, goblet-, *scyphus*, σκυφος
scyphocalyx with a goblet-shaped calyx, σκυφος-καλυξ
Scyphocephalium Goblet-headed, σκυφος-κεφαλη (the inflorescences contain up to three heads each of numerous flowers)
scyt-, scyto- leathery-, σκυτηνος, σκυτινος, σκυτος, σκυτο-
Scytanthus Leathery-flowered, σκυτος-ανθος (part of the adaptation to attract coprozoic pollinators)
Scytonema Thong-like, σκυτος-νημα (leathery filaments)
scytophyllus -a -um leathery-leaved, σκυτος-φυλλον
se- apart-, without-, out-
Seaforthia for Francis Humberston Mackenzie, Lord Seaforth (1754–1815), patron of botany (≡ *Ptychosperma*)
sebaceus -a -um, sebifer -era -erum tallow-bearing, producing wax, *sebaceus*
Sebaea for Albert Seba (1665–1736), apothecary and author of Amsterdam
sebiferus -a -um providing tallow, wax-bearing, *sebum-fero*
sebosus -a -um full of wax, *sebum, sebi*
Secale the Latin name, *secale*, for a grain like rye (not cognate with sea-kale, *Crambe maritima*)
secalinus -a -um rye-like, resembling *Secale*
Secamone from the Arabic, squa mona
sechellarus -a -um from the Seychelles, Indian Ocean
Sechium from the W Indian vernacular, chacha
seclusus -a -um hidden, isolated, secluded, *seclusus*
sectilis -is -e as though cut into portions, for cutting, *sectilis*
-sectus -a -um, -sect cut to the base, -divided, -partite, *seco, secare, secui, sectum*
secundatus -a -um following behind, one-sided, secund, *secundum* (all the florets are disposed to one side)
secundi-, secundus -a -um turned-, secund, one-sided, *secundum* (as when flowers are all to one side of an inflorescence)
secundiflorus -a -um with the flowers all facing one direction, secund-flowered, *secundum-florum*
secundirameus -a -um with secund branching, *secundum-ramus*
Securidaca Axe-like, *securis* (from the shape of the winged fruits)
securifolius -a -um having leaves shaped like axe-heads, *securis- folium*
securiger -era -erum axe-bearing, *securis-gero* (the shape of some organ)
Securinega Axe-refuser, *securis*-(*nego, negare, negavi, negatum*) (the hardness of the timber of some species)
sedi- stonecrop-like, *Sedum*
sedifolius -a -um with leaves resembling *Sedum, Sedum-folium*
sediformis -is -e with the habit of a stonecrop, *Sedum-forma*
sedoides (*sedioides*) resembling *Sedum, Sedum-oides*
Sedum a name, *sedo*, in Pliny (refers to the plant's 'sitting' on rocks etc in the case of cushion species)
seemannii for Berthold Carl Seemann (1825–71), German collector in tropical America and Pacific Islands
segetalis -is -e, segetus -a -um of the cornfields, growing amongst crops, *seges, segetis*
segregatus -a -um dissociated, a component separated from a super-species, *segrego, segregare, segregavi, segregatum*
seguieri, seguierianus -a -um for Jean François Seguier (1703–84), botanist of Nîmes
seguinis -is -e harming themselves, σε-γυιοω (*Dieffenbachia*, dumb-canes)
seiro- rope-like, rope, σειρα
sejugus -a -um with six leaflets, *sex-iugum*
sejuntus -a -um separated, solitary, *se*-(*iungo, iungere, iunxi, iunctum*)

sekukuniensis -is -e from the Transvaal area once ruled by the Pedi chief Sekhukhune, SE Africa (Sekukuniland)
Selaginella a diminutive of *Selago* (see below) (***Selaginellaceae***)
selaginoides clubmoss-like, *Selaginella-oides*
Selago the name in Pliny for *Lycopodium*, from the Celtic name for the Druidic collection of *Juniperus sabina*, verbal noun with feminine suffix from *seligo, seligere, selegi, selectum* to choose or select (***Selaginaceae***)
seleni-, seleno- moon-, σεληνη
Selenicereus Moon-*Cereus*, botanical Latin from σεληνη and *Cereus* (night flowering)
selenites with a lunar or moon-like appearance, σεληνη
selensis -is -e moonshine, σεληνη; from Sela, Yunnan, China
Selinum the name, σελινον, in Homer for a celery-like plant with lustrous petals (relates etymologically with *Silaum* and *Silaus*, and several derivations have been suggested)
sellaeformis -is -e, selliformis -is -e with both sides hanging down, saddle-shaped, *sella-forma* (e.g. of leaves)
Selliera for Natale Sellier, French engraver for Cavannilles
selligerus -a -um saddled, saddle-bearing, chaired, *sella-gero*
selloanus -a -um, selloi, sellovianus -a -um, sellowii, for Friedrich Sellow (Sello) (1789–1831), German botanist from Potsdam, collector in Brazil
selskianus -a -um for Ilarion Segiewitsch Selskey (1808–61), of the Russian Geographical Society, Irkutsk
semecarpifolius -a -um half-*Carpinus*-leaved (the second leaf form of *Quercus semecarpifolia* is undulate or entire)
Semele for the daughter of Kadmos, mother of Bacchus
semenovii, semenowii for Peter Petrowitsch Semenow-Tian-Shansky (1827–1914), Russian traveller
semestris -is -e half-yearly, of a half year, *semestris*
semi- half-, *semi-*
semialatus -a -um half-winged, *semi-alatus*
Semiaquilegia Half-*Aquilegia* (corolla differs in being saccate, not spurred)
Semiarundinaria Half-*Arundinaria* (some treat as *Arundinaria*)
semiatratus -a -um half in mourning, *semi-atratus* (flower colouration)
semibarbatus -a -um half-bearded, *semi-barbatus*
semibulbosus -a -um somewhat (half) bulbous, *semi-bulbosus*
semicastratus -a -um cut halfway around, *semi-(castro, castrare)*
semiclausus -a -um half-closed, *semi-clausus* (not fully open)
semicordatus -a -um cordate or heart-shaped on one side only, *semi-cordatus*
semicylindricus -a -um half-terete, botanical Latin from *semi* and κυλινδρος
semidecandrus -a -um with (about) five stamens, botanical Latin from *semi* and δεκα-ανηρ (*Tibouchina semidecandra* has ten stamens but five have yellow anthers and the other five form a self-coloured platform for visiting pollinators)
semideciduus -a -um half-deciduous, retaining some leaves all year, *semi-deciduus*
semidentatus -a -um half-dentate, *semi-dentatus*
semiexsertus -a -um half-exserted, *semi-exsertus* (stamens or style)
semiglobatus -a -tm half-globe-shaped, *semi-globosus* (pileus of dung roundhead fungus)
semilanceatus -a -um somewhat (half) lance-shaped, *semi-lanceatus* (the sharply umboed pileus of liberty cap fungus)
semiliberus -a -um half-separated, *semi-liberatus* (the cap is not wholly united to the stipe)
semilunatus -a -um half-moon-shaped, *semi-lunatus*
semiorbicularis -is -e hemispherical, half-round, *semi-orbiculatus*
semiovatus -a -um shaped like the narrow end of an egg, *semi-ovatus* (the pileus)
semipersistens half-persistent, *semi-persistens*
semipileatus -a -um somewhat felt-cap-like, *semi-pileatus* (fruiting body texture)
semipinnatus -a -um half-pinnate, *semi-pinnatus* (leaves)

semiplenus -a -um half-doubled, *semi-plenus*
semisanguineus -a -um somewhat bloody in colour, *semi-sanguineus* (less so than *sanguineus*)
semisectus -a -um cut halfway (to the base), *semi-sectus*
semiteres half-cylindrical, *semi-teres* (in cross-section)
semitomentosus -a -um half-tomentose, *semi-tomentosus* (*Viburnum*)
semiverticillatus -a -um half-verticillate, half-whorled, *semi-verticillatus* (grass panicles)
semivivus -a -um having a long dormant period, *semi-vivus*
semnoides resembling (*Rhododendron*) *semnum* (≡ *praestans*)
semocordatus -a -um half-heart-shaped, *semi-cordatus*
semotus -a -um remote; distinct, past participle of *semoveo, semovere, semovi, semotum*
semper- always-, ever-, *semper*
semperaureus -a -um continuously golden, *semper-aureus*
semperflorens ever-flowering, with a long flowering season, present participle from *semper-(floreo, florere, florui)*
sempervirens always green, *semper-(vireo, virere, virui)*
sempervirens-sibirica (*Saxifraga*) *sempervirens* from Siberia
sempervivoides, sempervivus -a -um houseleek-like, *Sempervivum-oides*
Sempervivum, sempervivus -a -um Always-alive, never-die, always living, *semper-vivus*
senarius -a -um composed of six (parts), six-partite, *senarius* (literally a trimeter)
sendaicus -a -um from Sendai, Japan
Senebiera for Jean Senebier (1742–1809), Swiss physiologist
Senecio Old-man, *senex, senis* (the name in Pliny refers to the grey hairiness as soon as fruiting commences)
senecioides (*senecoides*) groundsel-like, *Senecio-oides*
senecionis -is -e of ragworts, living on *Senecio* (symbionts, parasites and saprophytes)
senega of the American Seneca Indians (*Polygonum senega* used as treatment for rattlesnake-bite, Seneca snake-root)
senegalensis -is -e from Senegal, W Africa
senescens ageing, turning hoary with whitish hairs, *senesco, senescere, senui*
senetti a Plant Variety Rights registered name used to market *Pericallis*
seni- six-, six-each-, *seni, senorum*
senifolius -a -um six-leafleted, *seni-folium*
senilis -is -e aged, like an old person, grey-haired, *senilis* (*senex* an old man)
Senna, senna from the Arabic name, sana, for the laxative leaves and pods
sensibilis -is -e, sensitivus -a -um sensitive to a stimulus, irritable, *sentio, sentire, sensi, sensum*
sensu a term used, between the species name and its authority, to indicate that the name is used in the sense used by that named author
senticosus -a -um thorny, full of thorns, *sentis, sentis*
sentis -is -e briar-like, thorny, *sentis*
seorsus -a -um with its own beginning; apart, distinct, different, *se-orsum*
sepal-, -sepalus -a -um sepal-, -sepalled, σκεπη
sepiaceus -a -um dark-clear-brown, sepia coloured (σηπια, *sepia*, cuttle-fish)
sepiarius -a -um, sepius -a -um growing in hedges, of hedges, *sepes, saepes*
sepikanus -a -um from the environs of the Sepik river, Papua New Guinea
sepincolus -a -um hedge-dweller, inhabitant of hedges, *sepes-colo, saepes-colo*
sepius -a -um sepia-coloured, σηπια, *sepia*, a cuttle-fish
sept-, septem- seven-, *septem*
septalis -is -e of September, of the seventh month of the Roman year, *septem* (flowering or fruiting)
septangulus -a -um seven-edged, seven-angled, *septem-angulus*
septi-, septatus -a -um having partitions, septate, *septum, septa; septi-*

septemfidus -a -um with seven divisions, seven-cut, *septem-fidus*
septemlobus -a -um with seven lobes, *septem-lobus*
septifragus -a -um having a capsule whose valves break away from the partitions, *septem-(frango, frangere, fregi, fractum)*
septentrionalis -is -e of the north, of northern areas, *septentrionalis*
septupli- sevenfold-, *septuplus*
septuplinervius -a -um seven-nerved from the base, *septuplus-(nervus, nervi)*
sepulcralis -is -e (sepulchralis) of funerals, of graveyards, of tombs, *sepulcrum, sepulcri*
Sepultaria Buried, *sepelio, sepelire, sepelivi, sepultum*
sepultus -a -um buried, *sepultus*
sequax pursuing, following; trailing, *sequax, sequacis*
Sequoia for the N American Indian half-breed, George Gist (Sequoyah) (1770–1843) who invented the Cherokee alphabet
Sequoiadendron *Sequoia*-tree (resemblance in size)
Serapias, serapias for the Egyptian deity, Serapis (name, σεραπιας, used in Dioscorides for an orchid, ≡ *Cephalanthera* and *Epipactis*)
serbicus -a -um from Serbia
serenanus -a -um from La Serena, Coquimbo, N Chile
Serenoa for Sereno Watson (1826–92), American botanist
sergipensis -is -e from Sergipe, Brazil
seri-, serici-, sericans, sericeus -a -um silky, silky-hairy, *seres* (sometimes implying Chinese)
serialis -is -e, seriatus -a -um with transverse or longitudinal rows, *series, seriem*
sericanthus -a -um with silky flowers, σηρικος-ανθος
sericatus -a -um silken, σηρικος, *serica, sericorum*
sericellus -a -um minutely silky, diminutive of *serica, sericorum*
sericeovillosus -a -um with long silky hair, *serica-villosus*
sericeus -a -um with silky hair, like silk, σηρικος
sericifer -era -erum silk-bearing, *serica-fero*
sericifolius -a -um silky-leaved, *serica-folium*
Sericocarpus Silky-fruit, σηρικος-καρπος
sericofer -era -erum silk-bearing, *serica-fero*
sericophyllus -a -um with silky leaf-surfaces, σηρικος-φυλλον
Sericotheca Silken-case, σηρικος-θηκη (the pericarp) (≡ *Holodiscus*)
sericus -a -um silken, silky; from China (*Seres*)
Seringia, seringeana for Nicholas Charles Seringe (1776–1858), Director of the Botanic Garden at Lyon
Seriola Little-jar, *seriola, seriolae* (the amphora-like shaped tips of the scape branches) (≡ *Hypochaeris*)
Seriphidium from the Aegean island of Seriphos
-seris -potherb, σηρις, σεριδος
Serissa from the Indian vernacular name for *Serissa foetida*
serissimus -a -um silkiest, superlative of *sericeus*
Serjania for Paul Serjeant, French priest and botanist
serjaniaefolius -a -um with leaves resembling those of *Serjania*
serotinus -a -um autumnal, of late season, late, *sero, serius* (flowering or fruiting)
serpens, serpentarius -a -um, serpentinus -a -um creeping, serpentine, *serpo, serpere, serpsi, serptum*
serpentilinguus -a -um snake-tongue-like, *(serpens, serpentis)-linguus*
serpentini, serpentinicus -a -um of (growing on) serpentine rocks, late Latin *serpentinus*
Serpula Slow-spreader, *serpo, serpere, serpsi, serptum* (dry rot fungus)
serpyllaceus -a -um resembling thyme, *serpyllum*
serpyllifolius -a -um thyme-leaved, *serpyllum-folium*
serpyllum from an ancient name for thyme, (σ)ερπυλλος, *serpyllum* (ερπω, ερπυζω creeping)

serra, serra-, -serras saw, saw-like-, serrate-, deeply cut, *serratus*
Serrafalcus for Domenico Lo Faso Pietrasanta, Duke of Serrafalco, archaeologist (≡ *Bromus*)
serrarius -a -um saw-like, *serra, serrae* (leaf margins)
serratifolius -a -um with markedly serrate leaves, *serratus-folium*
serratipetalus -a -um with toothed petals, botanical Latin from *serratus* and πεταλον
serratodigitatus -a -um having digitate leaves with serrate leaflets, *serratus-digitatus*
Serratula, serratula Saw-tooth. *serra, serrae* (the name in Pliny for betony, re-used by Gerard because saw-wort, *Serratula tinctoria*, has saw-toothed leaves)
serratus -a -um edged with forward pointing teeth, serrate, *serra, serrae* (see Fig. 4d)
serriolus -a -um in ranks, of salad, diminutive of *seris* (from an old name for chicory)
serrulatus -a -um edged with small teeth, finely serrate, serrulate, diminutive of *serratus*
sertatus -a -um, sertulatus -a -um of garlands, garlanded, *sero, serere, sertum*
sertuliferus -a -um bearing garlands, (*sertum, serti*)*-fero* (of flowers)
sesameus -a -um sesame-like, *Sesamum*
sesamoides resembling *Sesamum*, σησαμον-οειδης, *Sesamum-oides*
Sesamum Hippocrates' name, σησαμον, from the Semitic name, simsim, cognate with sesame
Sesbania from the Arabic name for *Sesbania sesban*
Seseli the ancient Greek name, σεσελι, σεσελις
Sesleria for Leonardo Sesler (d. 1785), naturalist and physician of Venice
sesleriiformis -is -e with the habit of *Sesleria*
sesqui- one-and-one-half-, *sesqui*
sesquiorgyalis -is -e being about nine feet, or one and a half fathoms (high or long), botanical Latin from *sesqui* and οργυια
sesquipedalis -is -e about 18 inches long, the length of a foot and a half, *sesqui-pedalis*
sesquitertius -a -um of four to three, *sesquitertius* (the sexes in androgynous catkin structures)
sessili-, sessilis -is -e attached without a distinct stalk, sessile, sitting on, *sessilis*
sessiliflorus -a -um without distinct stalks to the flowers, *sessilis-florum*
sessilifolius -a -um leaves without petioles, sessile-leaved, *sessilis-folium*
sessilioides appearing sessile, *sessilis-oides*
Sesuvium etymology uncertain
seta-, setaceus -a -um, (saetaceus), seti- bristly, with bristles or stiff hairs, *seta, setae, saeta, saetae*
Setaria Bristly, *seta, setae* (most have hairs subtending the spikelets)
setchuenensis -is -e, setschwanensis -is -e from Sichuan (Szechwan, Setchuan) province, China
Setcreasea derivation obscure
seti- bristle-, bristly-, *saeta, saetae, seta, setae*
seticaulis -is -e having a bristly stalk, *seta-caulis*
setifer -era -erum, setiger -era -erum, seti- bearing bristles, bristly, *seta-fero*
setifolius -a -um with bristly-surfaced leaves, *seta-folium*
setipodus -a -um with bristly stems, botanical Latin from *seta* and ποδιον
setispinus -a -um bristle-spined, *seta-spina*
setosus -a -um covered with bristles or stiff hairs, *seta, setae*
setuliformis is -e thread-like, with minute bristles, *setula-forma*
setulosus -a -um with fine bristles, diminutive of *setosus*
-setus -a -um -bristled, *seta, setae*
Severinia for M. A. Severino (1580–1656), anatomist from Naples
severus -a -um terrible, strict, severe, *severus*
sex- six-, *sex*
sexangularis -is -e, sexangulus -a -um six-angled, *sex-angulus* (stems)
sexflorus -a -um six-flowered, with six-flowered racemes, *sex-florum*
sexstylosus -a -um with six styles, *sex-stilus*

sextupli- six fold-, six-partite, medieval Latin *sextuplus*
seyal an Arabic vernacular name for *Acacia seyal* timber (shittim wood)
shallon from the Chinook Indian name, kl-kwa-sha-la, for *Gaultheria shallon*
sharonensis -is -e from Sharon, Connecticut, USA
shastensis -is -e from the Shasta area of the Cascade range, N California
shawianus -a -um, shawii for Walter Robert Shaw, botanist of Illinois and Manilla *c.* 1871–3
shensianus -a -um from Shaanxi or Shanxi provinces of N China
Shepherdia for John Shepherd (1764–1836), curator of Liverpool University Botanic Garden
Sherardia for William Sherard (1659–1728) and his brother James Sherard
Sherbournia for Margaret Dorothea Sherbourn (1791–1846), who was the first to flower *Sherbournia foliosa* in England
sheriffii for Major George Sherriff (1898–1967), collector in Tibet and Bhutan
Shibataea for Keita Shibata (1878–1949), Japanese botanist
shirasawanus -a -um from a Japanese name
shirensis -is -e from the environs of the Shire river which overflows from Lake Nyassa, Malawi
shittim the biblical name for the wood of *Acacia nilotica*, from which the ark was to be built
Shortia, shortii for Dr Charles W. Short (1794–1863), botanist of Kentucky, USA
Shoshonea of the Shoshone or Shoshoni indigenous American people, or from their territory
shuttleworthii for Edward Shuttleworth (1829–1909), who collected for Bull's nursery at Chiswick
siamensis -is -e, siameus -a -um, siamicus -a -um from Thailand (formerly Siam)
Sibbaldia for Robert Sibbald (1643–1720), Professor of Medicine at Edinburgh and author of *Scotia illustrata*
Sibbaldiopsis similar to *Sibbaldia*, botanical Latin from *Sibbaldia* and ωψις
Sibaraea, sibiraeus -a -um Siberian (the provenance of *Sibaraea altaiensis*)
sibericus -a -um from Siberia, Siberian
Sibthorpia, sibthorpianus -a -um, sibthorpii for Professor Humphrey Sibthorp (1713–97), of Oxford, and his son John (1758–96), English botanist
siccatus -a -um, siccus -a -um of dry places, dried out, *sicco, siccare, siccavi, siccatum*
sicerarius -a -um (for holding) strong drink, σικερα (*Lagenaria siceraria*, bottle gourd)
siculi- dagger-shaped-, *sica, sicae*
siculiformis -is -e shaped like a small dagger, *sicula-forma*
siculus -a -um from Sicily, Sicilian (*Sicilia*)
sicyoides from Sicyon, Peloponnese
Sicyos a name, σικυος, used by Theophrastus for a cucumber
Sida from a Greek name, σιδε, used by Theophrastus for a water-lily and a pomegranate tree
Sidalcea Like-*Sida*-and-*Alcea*
sidereus -a -um iron-hard, of iron-like nature, σιδηρειος; σιδηρος, σιδηρο-
Sideritis the Greek name, σιδεριτος, for plants used on wounds caused by iron weapons
siderophloius -a -um iron-hard-barked, σιδηρος-φλοιος
siderophyllus -a -um with iron-hard or rusty haired leaves, σιδηρος-φυλλον
siderostichus -a -um with rigid rows, σιδηρος-στιχος (of spikelets)
Sideroxylon, sideroxylon Iron-wood, σιδηρος-ξυλον (the hard timber of the miraculous berry)
sidoides resembling *Sida, Sida-oides*
sieberi for Franz Wilhelm Sieber (1785–1844), of Prague, who travelled widely in the tropics
sieboldiana, sieboldii for Philipp Franz von Siebold (1796–1866), German physician and plant collector in Japan
Sieglingia for Professor Siegling, botanist of Erfurt

siehei for Walther Siehe (1859–1928) of Berlin
sierrae of sierras, of jagged mountain chains, via Spanish from *serra*
Sigesbeckia (*Siegesbeckia, Sigesbekia*) for Johann Georg Siegesbeck (1686–1755), physicist and botanist, Director of the Botanic Garden at St Petersburg
sigillatus -a -um with the surface marked with seal-like impressions, *sigillatus*
sigma-, sigmato- S-shaped, σιγμα, σιγματος
sigmoideus -a -um S-shaped, σιγμα-οειδης
signatus -a -um well-marked, designated, signed, *signo, signare, signavi, signatum*
sikangensis-is -e from Si Kiang (Xijiang), China
sikkimensis -is -e from Sikkim, Indian Himalayas
sikokianus -a -um from Shokoku island, Japan
silaifolia with narrow leaves as in pepper saxifrage, *Silaum silaus*
Silaum meaning uncertain; from Sila forest area of S Italy? (see *Selinum*)
silaus an old generic name, *silaus*, in Pliny used for pepper saxifrage
Silene Theophrastus' name for another catchfly, *Viscaria* (others derive it from Bacchus' companion, Silenos, or from σαιλον, saliva)
sileni- Silene-like
silesiacus -a -um from Silesia, SW Poland
siliceus -a -um of sand, growing on sand, *silex, silicis* (silicate)
silicicolus -a -um growing on siliceous soils, *silicis-colo*
siliculosus -a -um having broad pods or capsules from which the two valves fall and leave a false membrane (*replum*) with the seeds, *silicula*
siliquastrum (*siliquastris*) from the old Latin name for a pod-bearing tree, cylindric-podded, *siliqua-astrum*
siliquosus -a -um having elongate pods or capsules as the last, *siliqua*
siliquus -a -um, -siliquus -a -um, siliqui- podded, with pods, *siliqua*
sillamontanus -a -um from Cerro de La Silla, S America
Silphium an ancient Greek name, σιλφιον, for a resinous plant, reapplied to the N American compass-plant, *Silphium laciniatum*
silvaticus -a -um of woodlands, of the woods, *silva, silvae*
silvester -tris -tre woodland, wild, *silva, silvae*
silvicolus -a -um woodland dwelling, *silva-colo*
silvigaudens rejoicing the woodland, present participle from *silva-*(*gaudeo, gaudere, gavisus*)
Silybum Dioscorides' name, σιλυβον (for a thistle-like plant)
Simaba from the Guyanese vernacular name
Simarouba (*Simaruba*) from the Carib name for bitter damson (***Simaroubaceae***)
simensis -is -e from Arabia, *Simenia*, Middle-Eastern
Simethis after the Oread nymph, Simaethis
simiarus -a -um monkey-like, liked by monkeys, *simia, simiae* (flowers)
simili-, similis -is -e resembling other species, like, the same, similar, *similis*
similiflorus -a -um having the flowers all alike, *similis-florum* (e.g. in an umbel)
simius -a flat-nosed, σιμος; of the ape, *simius, simii*, or monkey, *simia, simiae* (flower-shape or implying inferiority)
Simmondsia, simmondsii for Arthur Simmonds (1892–1968), Secretary to the RHS 1956–62
simonii for Gabriel Eugène Simon (b. 1829), French Consul and collector in the East
simonsii for Reverend Jelinger Simons (1778–1851), of Leyton, Essex
simonsianus -a -um for Dr J. C. Simons, who collected in Assam *c.* 1895
simorrhinus -a -um monkey-snouted, σιμια-ρινος (spathe)
simplex undivided, entire, single, *simplex, simplicis*
simplicaulis -is -e, simplicicaulis -is -e with an unbranched stem, *simplicis-caulis*
simplice-, simplici- undivided, simple, *simplicis*
simplicifolius -a -um with undivided leaves, *simplici-folium*
simplicior undivided, *simplex, simplicis*
simplicissimus -a -um the least divided, superlative of *simplex*

simpliciusculus -a -um somewhat undivided, diminutive of *simplex*
simsii for John Sims (1749–1831), editor of *Botanical Magazine*
simulans, simulatus -a -um similar, resembling, imitating, present participle of *simulo, simulare, simulavi, simulatum*
simulatrix imitator (feminine form of *simulator*, *Salix* being feminine)
sinaicus -a -um from Sinai, Egypt
sinaloensis -is -e from Sinaloa state, NW Mexico
Sinapis the old name, σιναπι, used by Theophrastus for mustard, *sinapi, sinapis* (Celtic, nap, for cabbage-like plants)
Sinapistrum Wild-mustard, *Sinapis-istrum*
sinapizans becoming mustard like, *Sinapis* (smelling of radish)
Sinarundinaria Chinese-*Arundinaria, sino-Arundinaria*
sindicus -a -um from Sind province, Indus valley, Pakistan
sinensis -is -e (*chinensis -is -e*) from China, Chinese
singularis -is -e unusual, singular, unique, extraordinary, *singularis*
sinicus -a -um, sino- of China, Chinese (*Sinica*)
sinistrorsus -a -um turned to the left, *sinister, sinistri*, twining clockwise upwards as seen from above, sinistral,
Sinningia for Wilhelm Sinning (1794–1874), head gardener at Bonn University
sino- Chinese-, Chinese form of-, *sinica, sino-*
Sinobambusa Chinese-*Bambusa*
Sinocalycanthus Chinese-*Calycanthus*
sinofalconeri Chinese form of (*Rhododendron*) *falconeri*
Sinofranchetia for Adrien Franchet (1834–1900), French botanist who described many Chinese plants
sinograndis -is -e Chinese form of (*Rhododendron*) *grande*
Sinomenium Chinese-moonseed, μηνη, (the curved stone of the fruit)
sino-ornatus -a -um the Chinese form of *ornatus -a -um*
sinoplantaginea Chinese (*Primula*) *plantaginea* (≡ *P. nivalis*)
sinopurpurea Chinese (*Primula*) *purpurea* (≡ *P. macrophylla*)
Sinowilsonia, sinowilsonii for E. H. Wilson (1876–1930), introducer of Chinese plants
sinuatus -a -um, sinuosus -a -um, sinuus -a -um with a wavy margin, sinuate, winding, waved, *sinuo, sinuare, sinuavi, sinuatm* (see Fig. 4c)
siphiliticus -a -um see *syphiliticus -a -um*
sipho-, -siphon tubular-, -pipe, -tube, σιφον
Siphocampylus (*Siphocampylos*) Curved-tube, σιφον-καμπυλος (the corolla tube)
siphonanthus -a -um with pipe-like flowers, σιφον-ανθος
sisalanus -a -um from Sisal, Yucatan, Mexico (the fibre of *Agave sisalana* was exported from the port of Sisal)
sisarus -a -um Dioscorides' name for a plant with an edible root
siskiyouensis -is -e from the Siskiyou mountains, Oregon and California, USA
Sison a name, σισων, used by Dioscorides
sissoo a Bengali vernacular name for sisso tree (*Dalbergia sissoo*)
sisymbrifolius -a -um with leaves resembling those of *Sisymbrium, Sisymbrium-folium*
sisymbrii of hedge mustard, living on *Sisymbrium* (*Dasyneura*, dipteran gall midge)
Sisymbrium ancient Greek name, σισυμβριον, συσυμβρον (for various plants)
Sisyrinchium, sisyrinchium Pig-snout, Theophrastus' name, σισυριγχιον, for an iris (συς-ρυγχος, pig's snout, they dig for the sweet tubers)
Sitanion Food-grain, σιτος
sitchensis -is -e from Sitka, Baranof island, Alaska (*Picea sitchensis*, Sitka spruce)
sitiacus -a -um from the environs of Sitias Bay, NE Crete
Sitolobium Wheat-lobes, σιτος-λοβος (the shape of the pinnae, ≡ *Dicksonia*)
Sium an old Greek name, σιον, for water plants (Celtic, sin, water)
sivasicus -a -um from Megalopolis-Sebasteia (Sivas), central Turkey
skapho- see *scapho-*

Skimmia from a Japanese name, miyami shikimi
skio- see *scia-*, *scio-*, σκια
skiophilus -a -um shade-loving, σκια-φιλος
skole-, *scolo-* vermiform-, worm-, σκωληξ
skolecosporus -a -um with elongate, worm-like spores, σχωληξ-σπορα
skolio- see *scolio*, σκολιος, σκολιο-
skoto- see *scot*, *scoto-*, σκοτος, σκοτο-
skotophilus -a -um dark-loving, living in darkness, σκοτο-φιλος
Sloanea for Sir Hans Sloane (1660–1753), Irish physician and botanist, founder of the British Museum and Chelsea Physic Garden, President of the Royal Society
smaragdiflorus -a -um emerald-green flowered, σμαραγδος, emerald
smaragdinus of emerald, emerald-green, σμαραγδινος
Smeathmannia, *smeathmannii* for Henry Smeathman (1742–86), who collected plants in Sierra Leone in 1771–2, and proposed the settlement of freed slaves in Freetown
smilaci- *Smilax*-like
smilacifolius -a -um having leaves resembling those of *Smilax*, *Smilax-folium*
Smilacina diminutive of *Smilax*
Smilax from an ancient Greek name, σμιλαξ (σμιλη a scraper, for the prickly stems) (***Smilacaceae***)
Smithia for James Edward Smith (1759–1828), writer on the Greek flora and founder of the Linnean Society
smithiae, *smithianus -a -um*, *smithii* for James Edward Smith (*vide supra*), or Professor Christen Smith (1785–1816), Norwegian botanist (*Aeonium smithii*), or John Smith (1798–1888), gardener at Kew, or Joannes Jacobus Smith (1867–1947), Dutch specialist on Indonesian plants
Smithiantha for Matilda Smith (1854–1926), botanical artist at Kew
Smithicodonia the composite name for hybrids between *Smithiantha* and *Eucodonia*
Smyrnium Myrrh-fragrant, σμυρνιον, σμυρνα-ιον (the fragrance)
soboliferus -era -erum bearing soboles, producing vigorous shoots from the stem at ground level, bearing offspring, (*suboles*, *subolis*)-*fero*
socialis -is -e, *sociatus -a -um*, *sociarus -a -um* in pure stands, dominant, growing in colonies, *socies*
socotranus -a -um from Socotra island, Indian Ocean, Yemen
Socratea for the Greek philosopher Socrates (469–399 BC)
soda alkaline, medieval Latin from Arabic, suwwad (the calcined ash of *Salsola kali*)
sodiferus -a -um bringing pleasure, *sodes-fero*
sodomeus -a -um from the Dead Sea area (Sodom)
soja a vernacular name for the seeds of *Glycine soja*, soya beans
sol-, *solis -is -e* sun-, of the sun, *sol*, *solis*
solan-, *solani-* potato-, *Solanum*-like-
solanaceus -a -um potato-like, resembling *Solanum*
solanantherus -a -um with potato-flower-like stamens, *Solanum-anthera* (positioning)
Solandra, *solanderi* for Daniel Carlsson Solander FRS (1736–82), Swede who accompanied Sir Joseph Banks and Captain James Cook
Solanum Comforter, *solor*, *solavi*, *solatus* (an ancient Latin name, *solanum*, in Pliny) (***Solanaceae***)
solaris -is -e of sunny habitats, of the sun, *solaris*
Soldanella, *soldanella* Little-coins, from the Italian, soldo, for a small coin, diminutive of *soldo* (the leaves)
soldanelloides resembling *Soldanella*
soleae- sandal-, *solea*
Soleirolia for Captain Joseph François Soleirol (1791–1863), collector of Corsican plants
solen-, *soleno-* box-, tube-, σωλην, σωληνος, σωληνο-

Solena Tubular, σωλην (a wrong interpretation of the anther structure)
Solenangis Tube-vessel, σωλην-αγγειον
Solenomelus Theophrastus' name, σωληνομελος (for the tubular limb of the perianth)
Solenopsis Tube-like, σωλην-οψις (the very long white tubular flowers of the poisonous *S. longiflora*, ≡ *Isotoma longiflora*)
Solenostemon Tube-stamens, σωληνος-στεμον (their united filaments)
-solens -tubed, -tubular, σωλην
Solidago Uniter, from *solido, solidare*, verbal noun with feminine suffix (Brunfels' name for its use as a healing medicine)
Solidaster the composite name for hybrids between *Solidago* and *Aster*
solidifolius -a -um entire-leaved, *solidi-folium*
solidus -a -um a coin; complete, entire, solid, dense, not hollow, *solidus*
solitarius -a -um the only species (of a monotypic genus); with individuals growing in extreme isolation, solitary, lonely
Sollya for Richard Horsman Solly (1778–1858), plant anatomist
solonis for Solon (630–560 BC), the famed Athenian statesman and law-giver
solstitialis -is -e (*solsistialis*) of midsummer, *solstitium* (flowering about 11 July, St Barnabas' Day)
solutus -a -um loosened, independent, free, undone, *solvo, solvere, solvi, solutum*
soma-, -somus -a -um- -bodied, σωμα
somaliensis -is -e from Somalia, E Africa
somnians asleep, sleeping, present participle of *somnio, somniare* (dormant buds)
somnifer -era -erum sleep-inducing, sleep-bearing, *somnus-fero*
sonchi- *Sonchus*-like-
sonchifolius -a -um with *Sonchus*-like leaves, *Sonchus-folium*
Sonchus the Greek name, σογχος, σογκος (for a thistle)
songaricus -a -um from Dzungaria, Dzhungarsky Ala-Tau (Songaria), Kazakhstan/China
Sonneratia for Pierre Sonnerat (1749–1841), who collected in several areas of the tropics (***Sonneratiaceae***)
sonorus -a -um from the Sonoran desert, Baja California, N America
sophera an Arabian name for a pea-flowered tree, like *Sophora*
sophia knowledge, craft, wisdom, σοφια, (the use of flixweed, in treating dysentery or flux, *Sophia chirurgorum*, Sophia of the craft of surgeons)
Sophora from an Arabic name, sophera, for a pea-flowered tree
sophro- discreet-, modest-, σωφρων
Sophronitis Modesty, σωφρων (the small flowers)
soporificus -a -um sleep-bringing, soporific, *soporo, soporare*
Sopubia from an Indian vernacular name
Sorbaria Mountain-ash-like, *Sorbus* (from the form of the leaves)
Sorbaronia the composite name for hybrids between *Sorbus* and *Aronia*
sorbifolius -a -um with *Sorbus*-like leaves, *Sorbus-folium*
Sorbopyrus hybrids between *Sorbus* and *Pyrus*
Sorbus the ancient Latin name, *sorbum*, for the fruit of the service tree (cognate with sorb and service)
sordidus -a -um neglected, dirty-looking, *sordidus*
sorediatus -a -um heaped, mounded, σωρος (with patches of loose cells)
sorediiferus -a -um bearing soredia, σωρος-φερω (on lichens)
Sorghastrum Wild-*Sorghum*, *Sorghum-astrum*
Sorghum from the Italian name, sorgho (medieval Latin *sorgum*)
soriferus -a -um bearing sori, σωρος-φερω (on ferns)
soro- a heap-, a head-, σωρος (compound structures like pineapples, breadfruits and mulberries, of fruits and receptacles, are called *soroses*)
sorophorus -a -um bearing sori, σωρος-φερω
sororis -is -e, sororius -a -um very closely related, sisterly, *soror, sororis*
Soulangia for Etienne Soulange-Bodin (*vide infra*) (≡ *Phylica*)

soulangiana, soulangii for Etienne Soulange-Bodin (1774–1846), French horticulturalist
Sowerbaea for James Edward Sowerby (1787–1871), author of *English Botany* and artist
spadiceus -a -um chestnut-brown, having a spadix, date-coloured, σπαδιξ, σπαδικος
spadicigerus -a -um bearing spadices, botanical Latin from σπαδιξ and *gero* (*spadix-gero*, bearing chestnut brown)
span- few-, sparse-, σπανιος, σπανο-
spananthus -a -um having few flowers, sparsely flowered, σπανος-ανθος
spanioclemus -a -um with few shoots, σπανιος-κλημα
Sparaxis Torn-one, σπαρασσο (the lacerated spathaceous bracts)
sparganifolius -a -um having leaves resembling *Sparganium*, *Sparganium-folium*
Sparganium Dioscorides' name, σπαργανιον (σπαργανον was a swaddling band) (***Sparganiaceae***)
Sparrmannia for Dr Anders Sparrmann (1748–1820), Swede who was on Cook's second voyage
sparsi- scattered, *spargo, spargere, sparsi, sparsum*
sparsiflorus -a -um botanical Latin to imply having scattered or few flowers, *sparsus-florum*
sparsisorus -a -um having few sori, *sparsus-sorus*
sparsus -a -um sparse, scattered, dispersed, few, *spargo, spargere, sparsi, sparsum*
Spartina old name, σπαρτον, for various plants used for making ropes
Spartium Binding or Broom, σπαρτιον, diminutive of σπαρτον (former uses for binding and sweeping)
spartum esparto-providing, rope, σπαρτον (fibre-producing grasses)
spath-, spathi-, spatho-, spathus -a -um spathulate-, spathe-, sheath, *spatha*, σπαθη
spathaceus -a -um with a spathe-like structure, *spatha* (bracts or calyx)
Spathicarpa Spathed-fruit, σπαθη-καρπος (adnation to the spathe)
spathiflorus -a -um with flowers having spathe-like bracts, *spatha-florum*
Spathiphyllum Leafy-spathe, σπαθη-φυλλον
Spathodea Spathe-like, σπαθη-οιδα (the calyx)
spathularis -is -e spoon-shaped, diminutive of *spatha* (leaf or other organ)
spathulatus -a -um, spathuli- shaped like a spoon, diminutive of *spatha* (see Fig. 6b)
spathulifolius -a -um, spathulaefolius -a -um with broadsword-shaped leaves, *spatha-folium*
spatulae-, -spatulatus spoon-, -spatulate, from *spathula*, diminutive of *spatha*
spatiosus -a -um spacious, wide, large, ample, scattered, σπαρτος, *spatiosus*
speciosissimus -a -um most handsome or showy, superlative of *speciosus*
speciosus -a -um showy, handsome, semblance, good-looking, *speciosus*
spectabilis -is -e admirable, spectacular, good-looking, *specto, spectare, spectavi spectatum*
spectatissimus -a -um the most spectacular, superlative of *spectabilis*
specuicolus -a -um cave-dwelling, inhabiting caves, *specus, specus*
Specularia Mirror, *speculum* (*Specularia speculum*, Venus' looking glass, ≡ *Legousia hybrida*)
specularius -a -um, speculum shining, mirror-like, *speculum, speculi*
speculatrix watcher, feminine form of *speculator*, from *speculor, speculare, speculatus* (*Iris speculatrix*) (however, σπεκουλατωρ executioner)
speculatus -a -um shining, as if with mirrors, *speculum*
speculum-veneris Venus' looking-glass, *speculum-*(*Venus, Veneris*)
speir- wreathed, twisted-, coiled, *spira*, σπειρα
Speirantha Twisted-flower, or Wreathed-flowers, σπειρα-ανθος
speiranthus -a -um with twisted flowers, σπειρα-ανθος
speirea scattered, spreading, σπειρω
speirostachyus -a -um twisted or spiralled spikes, σπειρα-σταχυς
spelta medieval Latin from an old Saxon name, spelta, for a bearded wheat with two-grained spikelets

speluncae, speluncarum (*spelunchae*) of caves, cave-dwelling, *spelunca, speluncae*
speluncatus -a -um, speluncosus -a -um cavitied, full of holes, *spelunca, speluncae*
sperabilis -is -e desirable, to be hoped for, *spero, sperare, speravi, speratum* (*sperata* a bride!)
sperabiloides resembling *Rhododendron sperabile*, (*Rhododendron*) *sperabile-oides*
Spergula Scatterer, *dispergo, dispergere, dispersi, dispersum* (l'Obel's name refers to the discharge of the seeds)
Spergularia Resembling-*Spergula*
sperm-, spermato-, -spermus -a -um seed-, -seed, -seeded, σπερμα, σπερματο
Spermatophtyta Seed-plants, σπερματο-φυτον
sphacelatus -a -um necrotic, scorched, gangrened, σφακελος
sphacioticus -a -um damaged-looking, of gangrened appearance, σφακελος
sphaer-, sphaero- globular-, spherical-, ball-, σφαιρα, σφαιρο-
Sphaeralcea Spherical-*Alcea*, σφαιρα-αλκαια (the shape of the fruit)
sphaerandrus -a -um with stamens held in an orb-like manner, σφαιρα-ανδρος
Sphaeranthus, sphaeranthus -a -um Globe-flowered, σφαιρα-ανθος
sphaericus -a -um globe-like, *sphaera, sphaerae* (flower-shape)
sphaeroblastus -a -um with spherical growth, spherical bud, σφαιρο-βλαστος
sphaerocarpus -a -um orbicular-fruited, with spherical fruits, σφαιρο-καρπος
sphaerocephalon, sphaerocephalus -a -um round-headed, σφαιρο-κεφαλη
Sphaerocodon Spherical-bell, σφαιρο-κωδων (the corolla)
Sphaerosmeria Globose-parted, σφαιρο-μερος
sphaerospermus -a -um with spherical seeds, σφαιρο-σπερμα (achenes)
sphaerostachyus -a -um with a rounded spike or flower-head
Sphaerostemma Spherical-crown, σφαιρο-στεμμα (≡ *Schizandra*)
sphaerulus -a -um somewhat rounded, diminutive of *sphaera, sphaerae*
sphagnicolus -a -um living in *Sphagnum* communities, *Sphagnum-colo*
Sphagnum Latinized by Pliny from the Greek, σφαγνος (for a moss on trees) and re-used by Dillenius for bog-moss (***Sphagnaceae***)
sphegiferus -a -um bearing wasps, σφηξ-φορος
sphegodes resembling wasps, (σφηξ, σφηκος)-ωδες (flower shape)
sphenantherus -a -um with club- or wedge-shaped anthers, σφην-ανθηρος
spheno- wedge-shaped-, σφην, σφηνος, σφηνο-
Sphenoclea Wedge-shaped-cup, σφηνος-χλαινα (the calyx) (***Sphenocleaceae***)
Sphenopteris Wedge-fern, σφηνος-πτερυξ
Sphenopus Wedge-shaped-stalk, σφηνος-πους
Sphenotoma Cut-into-wedges, σφηνος-τομη
sphericus -a -um globular, spherical, σφαιρα
Sphinctrinus, sphinctrinus -a -um Close-together, σφιγγειν, (the apothecia); resembling *Sphinctrinus*
sphondylius -a -um rounded, σφονδυλιον (σπονδυλος spinning whorl or disc)
spica, spicati- with an elongate inflorescence of sessile flowers, spiked, spicate
spicant spikenard, spike, ear, tufted, *spica, spicae* (Bauhin equates the origin with *indica spica*, or spikenard; some derive it from an ancient German name)
spicatus -a -um, spicifer -era -erum with a spicate inflorescence, *spicatus* (see Fig. 2a)
spica-venti ear of the wind, tuft of the wind, *spica-ventus* (application not clear)
spiciferus -a -um, spicigerus -a -um bearing spicate inflorescences, *spica-*(*fero* or *gero*)
spiciformis -is -e shaped into a spiked inflorescence, *spica-forma*
spiculi- spicule-, dart-, sting-, small-thorn-, *spiculum*
spiculifolius -a -um with thorn-like leaves, spicule-leaved, *spiculum-folium*
spiculosus -a -um spiked, *spica, spicae* (shape or surface structures)
Spigelia for Adrian van der Spiegel (1578–1625), Professor of Anatomy at Padua
Spilanthes Stained-flower, σπιλος-ανθος (receptacular marks of some species)
spilo- stained-, σπιλοω, σπιλος
spilofolius -a -um spotted-leaved, botanical Latin from σπιλος and *folium* (*Spilographa alternata*, rose-hip fly)
spilotus -a -um stained-looking, of stained appearance, σπιλος

spina-christi Christ's thorn, Old English Latin (*spina, spinae*)-*christus* (χριστος)
Spinacia Prickly-one, from the Arabic, isbanakh (the fruit walls of spinach, *Spinacia oleracea*)
spinalbus -a -um with white spines, *spina-albus* (on leaf margins)
spinatus -a -um having spines, becoming spiny, *spina, spinae*
spinescens becoming spiny, *spina-essentia*
spineus -a -um spiny, with spines, *spina, spinae*
spinidens with prickly teeth, *spina-dens*
spinifer -era -erum, spinifex prickly, bearing spines, *spina-fero*
spinosissimus -a -um most spiny, superlative of *spinosus*
spinosus -a -um spiny, with spines, *spina, spinae*
Spinovitis Spiny-vine, *spina-vitis*
spinulifer -era -erum, spinulosus -a -um bearing small spines, diminutive of *spina*
spir- twisted-, coiled-, σπειρα
Spiraea Garland, σπειρα (Theophrastus' name for a plant used for making garlands)
spiralis -is -e, spiratus -a -um twisted, spiral, *spira, spirae*
spiralisepalus -a -um having twisted sepals, *spira-sepalus*
Spiranthes Twisted, σπειρα-ανθος (the inflorescence)
spirellus -a -um small-coiled, σπειρα
spiro- twisted-, coiled-, σπειρα
Spirodela Obvious-spiral, σπειρα-δηλος (the mode of budding daughter thalli of the colony, ≡ *Lemna*)
Spironema Twisted-thread, σπειρα-νημα (the spiral stamens)
spissifolius -a -um with compact foliage, densely leaved, *spissus-folium*
spissus -a -um compact, dense, *spissus* (growth habit)
spithamaeus -a -um a span (almost eight inches, 20 cm), σπιθαμη, modern Latin *spithamaeus* (a short span in height or length)
splendens gleaming, striking, present participle of *splendeo, splendere*
splendentior more splendid or striking, comparative of *splendens*
splendidissimus -a -um most splendid, superlative of *splendens*
splendidus -a -um strikingly fine, *splendidus*
Spodiopogon Grey-beard, σποδια-πωγων (the hairy inflorescence)
spodo- dust, ashes, ash-grey, σποδια, σποδος, σποδο-
spodochrus -a -um ashen, greyish-skinned, σποδο-χοως
Spondianthus *Spondias*-flowered, *Spondias-anthus*
Spondias Theophrastus' name refers to the plum-like fruit
spondioides resembling *Spondias, Spondias-oides*
spongiopsis spongy-looking, σπογγος-οψις, *spongia-opsis*
spongiosipes spongy-stalked, σπογγος-πεσ (the felt-like tomentum on the stipe)
spongiosus -a -um spongy, σπογγος, *spongia*
sponhemicus -a -um from Sponheim (*Sponhemium*), Rhine, Germany
spontaneus -a -um natural, independent, by chance, modern Latin *spontaneus* (*sua sponte*, of one's own accord)
sporadicus -a -um, sporadus -a -um scattered, widely dispersed, σπορας
sporo-, sporo- spore-, seed-, σπορα, σπορος
Sporobolus Seed-ejector, σπορα-(βολω , βαλλειν) (the seed emerges from a mucilaginous coat)
-sporus -a -um -seed, -seeded, -spored, -σπορος
Spraguea, spraguei for Isaac Sprague (1811–95), American botanical illustrator
Sprekelia for Dr Johann Heinrich von Sprekelsen (1691–1764), who wrote on *Yucca Draconis foliis*
Sprengelia for Christian Konrad Sprengel (1750–1816), Brandenburg writer on the fertilization of flowers
sprengelii for Kurt Polykarp Joachim Sprengel (1766–1833), Professor of Halle
sprengeri for Karl (Carlo) Sprenger (1846–1917), German nurseryman in Vomero, Italy
spretus -a -um (*sprettus*) despised, spurned, *sperno, spernere, sprevi, spretum*

spumarius -a -um foamy, frothing, *spuma, spumae*
spumescens becoming frothy, of frothy appearance, *spumesco, spumescere*
spumeus -a -um foaming, frothy, *spuma, spumae*
spumosus -a -um with a frothy or foamy appearance, *spuma, spumae*
spurcatus -a -um fouled, nasty, filthy, *spurcus*
spurius -a -um false, bastard, *spurius*
squalens, squalidus -a -um untidy, dingy, squalid, *squaleo, squalere, sqalui*
squamarius -a -um, squamosus -a -um scale-clad, covered with scales, with scale-like leaves, *squama, squamae*
squamatus -a -um with small scale-like leaves or bracts (*squamae*), squamate
squameus -a -um scaly, *squama, squamae*
squamigerus -era -erum scale-bearing, *squama-gero*
squamosorodicosus -a -um with gnawed scales, with irregular-shaped scales, *squamosus-*(*rodo, rodere, rosi, rosum*)
squarrosus -a -um rough, *squarrosus* (when closely overlapping leaves have protruding tips or sharp edges, cognate with *squamosus*)
squarrulosus -a -um somewhat rough or scurfy, diminutive of *squarrosus*
squillus -a -um shrimp-like; squill-like, *squilla, squillae*
stabilis -is -e firm, lasting, not changeable, *stabilis*
stachy- spike-like-, σταχυς, resembling *Stachys*
Stachygynandrum Spiked-female-male, σταχυς-γυνη-ανδρος (≡ *Lycopodium pro parte*)
stachyoides resembling Stachys, σταχυς-οειδης
-stachyon , -stachys, stachyus -a -um -spiked, narrowly-panicled, σταχυς
Stachys Spike, σταχυς (the Greek name used by Dioscorides for several dead-nettles)
Stachytarpheta Thick-spike, σταχυς-ταρφυς (the densely flowered spike)
Stachyurus Spiked-tail, σταχυς-ουρα (the shape of the inflorescence) (***Stachyuraceae***)
Stackhousia for John Stackhouse (1740–1819), British botanist (***Stackhousiaceae***)
stagnalis -is -e of pools, *stagno, stagnare*
stagninus -a -um of swampy or boggy ground, *stagnum*
stamineus -a -um with prominent or many stamens, *stamineus* (*stamen* a filament)
staminodiosus -a -um with many sterile stamens or staminodes, modern Latin from *stamen, staminis*
staminosus -a -um the stamens being a marked feature of the flowers, *stamen, staminis*
standishii for John Standish (1814–75), of Standish and Noble nurseries at Sunningdale, Berkshire
Stangeria for William Stanger (d. 1854), Surveyor General of Natal
Stanhopea for Philip Henry, Fourth Earl of Stanhope (1781–1855), President of the Medico-Botanical Society
Stanleya, stanleyi for Edward, Lord Stanley, Thirteenth Earl of Derby (1775–1851), of Knowsley, ornithologist and lover of natural history art
stans self-supporting, upright, erect, standing, *sto, stare, steti, statum*
Stapelia named by Linnaeus for Johannes Bodaeus von Stapel (d. 1631), Dutch physician of Amsterdam
stapeliiformis -is -e with the habit of *Stapelia, Stapelia-forma*
stapfianus -a -um, stapfii for Dr Otto Stapf (1857–1933), Keeper of the Herbarium at Kew
staphisagrius -a -um like wild grapes, σταφυλη-αγριος
Staphylea Cluster, σταφυλη (Pliny refers a name, *staphylodendron*, to the bunched flowers; used by Linnaeus for the nature of the inflorescence) (***Staphyleaceae***)
-staphylos -raceme, -bunch, σταφυλη (as of grapes)
stasophilus -a -um living in stagnant water, loving stagnant waters, στασο-φιλος
-states -standing, -placed, στατος
Statice Astringent, στατικος (Dioscorides' name, στατικη, for the *Limonium* of gardeners) (≡ *Limonium*)

staticifolius -a -um with leaves resembling those of *Statice*
Stauntonia, stauntonii for Sir George Leonard Staunton (1737–1801), Irish traveller in China
stauro- palisade-; cross-shaped-, crosswise-, cruciform-, σταυρος, σταυρο-
Staurogyne Cross-shaped-ovary, σταυρος-γυνη
stauropetalus -a -um cruciform-petalled, with petals forming a cross σταυρος-πεταλον
staurosporus -a -um with cross-shaped (quadrangular to stellate) spores, σταυρο-σπορος
stearnii for Professor William Thomas Stearn (1911–2001), author of numerous botanical works, of the British Museum (Natural History)
stegano-, stego-, -stegia, stegno- covered-over-, roofed-, -cover, στεγανος, στεγη, στεγνος, στεγνο-, στεγος-, στεγο-
Stegnogramma Covered-in-lines, στεγνος-γραμμα
steiro- barren-, στειρος
Steirodiscus Barren-disc, στειρο-δισκος (*Steirodiscus euryopoides*)
Steironema Barren-threads, στειρο-νημα (the staminodes)
Stellaria, stellaria Star, *stella* (the appearance of stitchwort flowers)
stellaris -is -e, stellatus -a -um with spreading rays, stellate, star-like, *stella*
Stellera, stellerianus -a -um for Georg Wilhelm Steller (1709–46), German who collected in Russia
stelliger -era -erum star-bearing, *stella-gero*
stellipilus -a -um with stellate hairs, *stella-pilus*
stellulatus -a -um small-starred, with small star-like flowers, diminutive of *stellatus*
-stema, -stima -crowned, -wreathed, στεμμα (botanically for a flower-head)
-stemon -stamen, -stamened, στημων, *stamen, stamenis* (a thread or warp)
sten-, steno- short-, narrow-, στενος, στενο-, στεν-
Stenanthera, stenantherus -a -um Narrow-anthers, στεν-ανθηρος
Stenanthium Narrow-panicled-one, στεν-ανθινος
stenarthus -a -um having narrow or short internodes, στεν-αρθρον
stenaulus -a -um narrow-tubed, στεν-αυλος
stenobotryus -a -um having slender or narrow bunches, στενο-βοτρυς (of fruits)
Stenocarpus Narrow-fruit, στενο-καρπος (the flattened follicular fruits)
stenocarpus -a -um with narrow fruits, στενο-καρπος
stenocephalus -a -um with a narrow head, στενο-κεφαλη (of flowers)
Stenochlaena Narrow-cloak, στενο-(χ)λαινα (sporangia cover the entire surface of the linear fertile pinnae)
stenodes compressed-looking, στενος-ωδες
Stenoglottis Narrow-tongue, στενο-γλωττα (the lip)
stenolepis -is -e with narrow scales, στενο-λεπις (glumes or bracts)
Stenoloma Narrow-fringe, στενο-λωμα (the narrow indusium)
stenomeres with slender parts, στενο-μερος
Stenomeson (*Stenomesson*) Narrow-in-the-middle, στενο-μεσον (μεσσον) (the shape of the corolla tube)
stenopetalus -a -um narrow-petalled, στενο-πεταλον
stenophyllus -a -um narrow-leaved, στενο-φυλλον
stenopterus -a -um narrowly winged, with a narrow wing, στενο-πτερον (to the stem)
stenosiphon with a narrow tube, στενο-σιφον (of the corolla)
stenostachyus -a -um having narrow spike-like inflorescences, στενο-σταχυς
Stenotaphrum Shallow-depression, στενο-ταφρος (the florets are recessed into cavities in the flowering rachis)
stenothyrsus -a -um with narrow thyrsoid inflorescences, στενο-θυρσος
Stenotus Narrow-lobed, στενο-ωτος (corolla segments)
stephan-, stephano- crowned-, crown-, wreathed-, στεφανη; στεφανος, στεφανο-
Stephanandra Male-crown, στεφαν-ανδρος (the arrangement of the persistent stamens)

Stephania for Christian Friedrich Stephan (1757–1814), German botanist who worked in Moscow and St Petersburg
Stephanotis Crowned-ear, στεφαν-ωτος (the auricled staminal crown); used by the Greeks for plants suitable for making chaplets or crowns, στεφανος
-stephanus -a -um -crowned, στεφανος, στεφανιτης
stepporus -a -um, stepposus -a -um of the steppes, modern Latin from Russian, step
stercorarius -a -um growing on or smelling of dung, *stercus, stercoris*
Sterculia Dung, *stercus* (the evil-smelling flowers of some species) (Sterculius was a heathen god) (***Sterculiaceae***)
sterculiaceus -a -um like *Sterculia*, with a foul smell, *stercus*
stereo- solid-, stiff, constant, στερεος, στερεο-
stereophyllus -a -um with stiff leaves, στερεο-φυλλον
Stereospermum Solid-seed, στερεος-σπερμα
Stereum Stiff, στερεος (the leathery brackets of the fruiting stage)
sterilis -is -e infertile, barren, sterile, *sterilis*
Sternbergia, sternbergii for Count Kaspar Moritz von Sternberg of Prague (1761–1838), Czech author of *Revisio saxifragarum* (1810)
sternianus -a -um for Colonel Sir Frederick Claude Stern (1884–1967), horticultural pioneer of Highdown, Worthing (varietal names 'Highdown' and *highdownensis*)
stevenagensis -is -e from Stevenage, England
stevenianus -a -um, stevenii for Christian von Steven (1781–1863), Finnish Director of Nikita Botanic Garden, Crimea
stevensonii for Sir William Stevenson, Governor of Mauritius, 1857–63
Stevia for Dr Peter James Esteve (d. 1566), Professor of Botany at Valencia
Stewartia, stewartii for John Stewart (1713–92), Third Earl of Bute and patron of botany
-stichus -a -um -ranked, -rowed, στιξ, στιχος
stict-, sticto- ,-stictus -a -um punctured-, -spotted, στικτος, στικτο-
Stictocardia Spotted-interior, στικτος-καρδια (the corolla)
stictocarpus -a -um with spotted fruits, στικτος-καρπος
stictophyllus -a -um with spotted leaves, στικτος-φυλλον
stigma- spot-, point-, stigma-, στιγμα
Stigmaphyllon Leaf-stigma, στιγμα-φυλλον (the stigma is slightly flattened)
stigmaticus -a -um having a conspicuous stigma, *stigma, stigmatis*
stigmosus -a -um spotted, marked, *stigma, stigmatis*
-stigmus -a -um -spotted, -dotted, -marked, *stigma, stigmatis*
Stilbe Shining, στιλβω
stillatus -a -um drop-like, *stillo, stillare, stillavi, stillatum* (gelatinous fruiting stage)
stimulans tormenting, exciting, present participle of *stimulo, stimulare, stimulavi, stimulatum*
stimulosus -a -um pole-like, goad-like, *stimulus, stimuli*
Stipa Tow, στυππειον, *stupa, stuposus* (Greeks used the feathery inflorescences, like hemp, for caulking and plugging)
Stipagrostis *Stipa*-like field grass, στυππειον-αγρωστις, *Stipa-Agrostis*
stipellatus -a -um with stipels, diminutive of stipula (in addition to stipules)
stipitatoglanduosus -a -um having stalked glands, *stipitatus-glandulosus*
stipitatus -a -um with a stipe or stalk or trunk, *stipes, stipitis*
stipticus -a -um styptic, στυπτικος (στυπτηρια, alum)
stipulaceus -a -um, stipularis -is -e, stipulatus -a -um, stipulosus -a -um with conspicuous stipules, *stipula, stipulae* (literally a blade or stalk or stubble)
stiriacus -a -um, styriacus -a -um from Steyr (Styria), Austria
stoechas Dioscorides' name for a lavender growing on the Iles d'Hyeres, Toulon, which were called 'Stoichades'
Stokesia for Dr Jonathan Stokes (1755–1831), who worked with Withering on his arrangement of plants
stolonifer -era -erum spreading by stolons, with stems rooting at the nodes, *stolonis fero*

stoloniflorus -a -um flowering on creeping stems, *stolonis-flores*
Stomatium Mouth, στομα (from the pairs of toothed leaves)
-stomus -a -um- mouthed, στομα
Storax see *Styrax*
stracheyi for General Sir Richard Strachey FRS (1817–1908), collector in the Himalayas, 1846–9
stragulatus -a -um, stragulus -a -um carpeting, covering, choking, *strangulo, strangulare, strangulavi, strangulatum*
stramine-, stramineus -a -um straw-coloured, *stramen, straminis*
stramonii- *Stramonium*-like-
stramonium from a name, στρυχνος μανικος, used by Theophrastus for the thorn apple, *Datura stramonium*, possibly from Tartar, turman
strangulatus -a -um constricting, strangling; with irregular constrictions, *strangulo, strangulare, strangulavi, strangulatum*
Stranvaesia for William Thomas Horner Fox-Strangways (1795–1865), Earl of Ilchester and botanist
strateumaticus -a -um forming an army, forming groups, στρατευμα, στρατευματος
Stratiotes Soldier, στρατιωτης (Dioscorides' name, στρατιωτης ποταμιος, for an Egyptian water plant with sword-shaped leaves)
Stravinia the composite name for hybrids between *Stranvaesia* and *Photinia*, when held to be distinct genera
Streblochaete Twisted-hair, στρεβλοω-χαιτη (the long awns of the lemmas)
Strelitzia for Charlotte, Duchess of Mecklenburg-Strelitz (1744–1818), wife of George III (bird of paradise flowers) (***Strelitziaceae***)
strepens rustling, rattling, present participle of *strepo, strepere, strepui*
Strephonema Twisted-threads, στρεφω-νημα (the stamens)
strepsi-, strept-, strepto- twisted-, coiled-, στρεπτος, στρεπτο-
Streptocarpus Twisted-fruit, στρεπτος-καρπος (the fruits contort as they mature)
Streptochaeta Twisted-bristles, στρεπτος-χαιτη (the twined awns)
Streptogyne Twisted-ovary, στρεπτος-γυνη (the tangled, spinulose stigmas)
Streptolophus Twisted-crest, στρεπτος-λοφος (the fruit's bristles)
Streptomyces Coiled-fungus, στρεπτος μυκης (filamentous bacteria forming mycelia with branching strands of spores)
streptophyllus -a -um with spirally twisted leaf blades, στρεπτο-φυλλον
Streptopus Twisted-stalk, στρεπτος-πους
Streptosolen Twisted-tube, στρεπτος-σωλην (the corolla tube is spirally twisted below the expanded part)
striatellus -a -um, striatulus -a -um somewhat marked with parallel lines, grooves or ridges, *striata, striatae* (literally a scallop)
striatiflorus -a -um with striped flowers, *striatus-florum*
striatus -a -um ridged, striped, with a rippled or lined surface, modern Latin from *stria*, a furrow
stribrnyi for Vaclav Stribrny (b. 1853), botanist of Lidice, Prague
stricti-, stricto- straight, strict, *stringo, stringere, strinxi, strictum*
strictiforme drawn together, of erect habit, *strictus-forma*
strictior straighter, more erect, comparative of *strictus*
strictipes with a straight stalk, *strictus-pes*
strictissimus -a -um the most erect, superlative of *strictus*
strictus -a -um erect, close, stiff, *stringo, stringere, strinxi, strictum*
Striga Swathe, *strigosus* thin (most are rigidly erect)
strigatus -a -um straight, rigid, *Striga*-like
strigilifolius -a -um having rigid, bristle-like leaves, *strigilis-folium*
strigillosus -a -um with short rigid bristles, *strigil, strigilis*
strigosus -a -um scraggy, thin, lank; with rigid hairs or bristles, strigose, *strigosus*
strigulosus -a -um somewhat strigose, diminutive of *strigosus*
striolatus -a -um faintly striped, finely lined, diminutive of *stria*
strobiformis -is -e of a conical habit, botanical Latin from στροβιλος and *forma*

strobil-, *strobili-* cone-bearing, cone-, στροβιλος
strobilaceus -a -um cone-like, cone-shaped, στροβιλος
Strobilanthes Cone-flower, στροβιλος-ανθος (the dense inflorescence)
strobilifer -era -erum bearing cones, botanical Latin from στροβιλος and *fero*
strobus an ancient name for an incense-bearing tree (*Pinus strobus* has large seed-cones that were burnt at festivals, στροβιλος a fir cone) (στροφη, στρεφω to whirl or spin, for the whorled structure of a strobilus)
Stromanthe Bedded-flower, στρωμα-ανθος (the form of the inflorescence)
Strombocactus Top-cactus, στρομβος-κακτος (the rootstock)
strombuli- top-like-, snail-shell-like-, *strombus*, whirligig, στρομβος
strombuliferus -a -um snail-like, bearing spirals, *strombus-fero* (as with the fruits of some *Medicago* species)
strongyl-, *strongylo-* round-, rounded-, στρογγυλος, στρογγυλο-
strongylanthus -a -um having rounded flowers, στρογγυλο-ανθος
strongylophyllus -a -um with rounded leaves, στρογγυλο-φυλλον
Strophanthus Twisted-flower, στροφη-ανθος (the elongate lobes of the corolla)
strophio- turned-over, turning-, στροφη, στροφος, στροφο-
strophiolatus -a -um having a distinctive caruncle on the seed, στροφη
Strumaria Tumour, στρωμα (the style is expanded in the middle)
strumarius -a -um, strumosus -a -um cushion-like, swollen, *struma, strumae* (signature for use in treatment of swollen necks)
strumi- cushion-like-swelling-, wen-, goitre-like-, στρωμα
strupi- thong-like-, strapped-, *stroppus*
Struthiola Sparrow, *strutheus* (the seeds are somewhat like the bird's beak)
Struthiopteris Ostrich-fern, στρουθιον-πτερυς, the shuttlecock fern (στρουθιον or στρουθος was a sparrow or an ostrich, now called *Struthio camelus*)
struthius -a -um feathery, ostrich-like, στρουθος
Strychnos Linnaeus reapplied Theophrastus' name for poisonous solanaceous plants, στρυχηνος, (modern Latin *strychnos*)
Stuartia see *Stewartia*
stuartianus -a -um for Stuart Henry Low, father (1826–90) or son (*c*. 1863–1952), both nurserymen of Clapton, orchid growers
studiosorus -a -um of specialists, *studeo, studere, studui*
stupeus -a -um, stuppeus -a -um woolly, rough-tufted, *stupa; stuppa*
stuposus -a -um, stuposus -a -um shaggy with matted tufts of long hairs, tousled, tow-like, *stupa; stuppa* (*stuposus*)
stygius -a -um of the underworld, Stygian, *Styx, Stygis, Stygos; stygius* (*Globularia stygia* spreads by subterranean stolons); growing in stagnant or foul water
stylaris -is -e, stylosus -a -um with a conspicuous or prominent style, στυλος, *stilus*
Stylidium Column, στυλος (the united styles and stamens) (***Stylidiaceae***)
stylo-, *-stylus -a -um* style-, -styled, στυλος, στυλο-
Stylochiton Covered-pillar, στυλος-χιτων (the spathe enfolds the elongate male portion of the spadix)
styloflexus -a -um having a curved style, *stilus-(flecto, flectere, flexi, flexum)*
Stylomecon Styled-poppy, στυλος-μηκων (the ovary has a distinct style)
Stylophorum Style-bearing, στυλος-φορα (distinctively styled)
stylosus -a -um with a prominent or persistent styles, *stylosus*
Styphelia Hard or Solid, στυφελος (the leaf texture)
styphelioides resembling *Styphelia*, στυφελος-οειδης
Styphnolobium Rough-pod, στυφελος-λοβος
stypticus -a -um astringent, styptic, στυπτικος
styraciflUus -a -um flowing with gum, *storax-fluo*
Styrax ancient Greek name, στυραξ, Latin, *storax* (for *Styrax officinalis*, storax gum tree and its resin) (***Styracaceae***)
styrido- cruciform-, σταυρος
Suaeda from the Arabic, suwed-mullah, for *Suaeda baccata*

suaveolens sweet-scented, *suavis*-(*olens, olentis*)
suavis -is -e sweet, agreeable, pleasant, delightful, *suavis*
suavissimus -a -um most sweetly scented, superlative of *suavis*
sub-, suc-, suf-, sug- below-, under-, approaching-, nearly-, just-, less than-, usually-
subacaulis -is -e almost without a stem, *sub-a-caulis*
subadpressus -a -um slightly appressed, *sub-adpressus*
subaequalis -is -e with almost equal parts, *sub-aequalis* (perianth lobes)
subalpinus -a -um subalpine, growing below the snow-line, *sub-alpinus*
subarcticus from the far north but not arctic, *sub-arcticus*
subaridus -a -um requiring dry conditions, from sub-arid habitats, *sub-aridus*
subaspersus -a -um slightly rough-surfaced, *sub-aspersus*
subaxillaris -is -e from below a node, *sub*-(*axilla, axillae*) (developmental feature)
subbalaustinus -a -um *vide balaustinus -a -um*
subbiflorus -a -um mostly two-flowered, resembling *biflorus*, *sub-bi-florum* (a comparative relationship)
subblandus -a -um slightly pleasing, *sub-blandus*
subcaeruleus -a -um slightly blue, *sub-caeruleus*
subcanescens somewhat grey-haired, *sub-canus-essentia*
subcaninus -a -um slightly thorny, more or less wild, *sub-canina*
subcanus -a -um slightly grey, *sub-canus*
subcapitatus -a -um with loose heads, *sub*-(*caput, capitis*)
subcarinatus -a -um almost keeled, *sub-carinata*
subcaulis -is -e with a very short stem, *sub-caulis*
subclausus -a -um almost enclosed, *sub-clausus*
subcoeruleus -a -um slightly blue, *sub-coeruleus*
subcollinus -a -um of the lower foothills, *sub-collis*
subcompressus -a -um slightly compressed from side to side, *sub-compressus*
subconcolor mostly of one colour, *sub*-(*concolor, concoloris*)
subcostatus -a -um slightly veined or ribbed, *sub-costatus*
subcrenatus -a -um slightly crenate, *sub-crenatus*
subcrenulatus -a -um very slightly crenate, *sub-crenatulus*
subcuneatus -a -um slightly wedge-shaped, *sub-cuneatus*
subcyaneus -a -um almost corn-flower blue, *sub-cyaneus*
subdentatus -a -um slightly toothed, *sub-dentata*
subdiaphenus -a -um semi-transparent, *sub-diaphenus*
subdulcis -is -e less than sweet, *sub-dulcis* (mild, turning bitter)
suber corky, *suber, suberis* (the ancient Latin name for the cork oak, *Quercus suber*)
suberatus -a -um corky or becoming corky, *suber*
suberectus -a -um growing at an angle, not quite upright, *sub-erectus*
suberosus -a -um slightly bitten, *sub-erosus*; corky, *suber*
subfalcatus -a -um slightly sickle-shaped, *sub-falcatus*
subflabellatus -a -um slightly fan-shaped, *sub-flabellatus*
subfulgens brightish, not quite shining, *sub*-(*fulgeo, fulgere, fulsi*)
subglandulosus -a -um slightly glandular, *sub-glandulosus*
subglobisporus -a -um with almost globose spores, botanical Latin from *sub-globus* and σπορα
subhirsutus -a -um almost hirsute, hairy, *sub-hirsutus*
subhirtellus -a -um very slightly hairy, *sub-hirtellus*
subincarnatus -a -um almost flesh-like, *sub-carnatus*
subintegrus -a -um almost entire, *sub-integrus*
subinundatus -a -um marginally amphibious, from the edges of wet habitats, *sub*-(*inundo, inundare, inundavi, inundatum*)
sublanceolatus -a -um almost lanceolate, *sub-lanceolatus*
sublateritius -a -um almost brick red, *sub-lateritius*
sublatus -a -um elated, lofty, *sub-latus*
sublignosus -a -um slightly woody, *sub-lignosus*

sublobatus -a -um slightly lobed, *sub-lobatus*
sublustris -is -e faintly luminous, glimmering, almost shining, *sub-(lustro, lustrare, lustravi, lustratum)*
submammillaris -is -e slightly mammillate, *sub-mammillaris*
submersus -a -um underwater, submerged, cf. *demersus*
submollis -is -e slightly soft, *sub-mollis*
submontanus -a -um sub-montane, from foothills, *sub-montanus*
subnodulosus -a -um slightly noduled, *sub-nodulosus*
suboliferus -era -erum bearing offspring, *suboles, subolis* (see *soboliferus -era -erum*)
suboppositus -a -um almost opposite, *sub-oppositus* (leaved)
subovalis -is -e, subovatus -a -um almost ovate, slightly ovate, *sub-ovalis*
subpeltatus -a -um slightly peltate, not attached at the edge, *sub-peltatus*
subpraestans almost excellent, quite distinguished, *sub-praestans*
subrepens slightly creeping, *sub-repens*
subrigidus -a -um slightly stiff, *sub-rigidus*
subrotundus -a -um almost round, slightly rounded, *sub-rotundus*
subsericeus -a -um slightly silky, *sub-sericeus*
subserratus -a -um slightly toothed, *sub-serratus*
subsessilis -is -e very short-stalked, almost sessile, *sub-sessilis*
subsimilis -is -e slightly resembling, *sub-similis*
subsimplex mostly undivided, *sub-simplex*
subspicatus -a -um somewhat spike-like, *sub-spicatus*
subspinosus -a -um slightly thorny, *sub-spinosus*
subterraneus -a -um below ground, underground, *sub-(terra, terrae)*
subtiliflorus -a -um having delicate flowers, *subtilis-florum*
subtilis -is -e fine, slender, delicate, *subtilis*
subtilissimus -a -um (*subtillissimus -a -um*) the most delicate, superlative of *subtilis*
subtomentosus -a -um slightly tomentose, *sub-tomentosum*
subtriphyllus -a -um mostly three-leaved, botanical Latin from *sub-tri* and φυλλον
subtropicus -a -um from the borders of the tropics, subtropical; half-hardy, modern Latin, via Middle English, from *sub* and τροπη
subturbinatus -a -um almost conical or top-shaped, *sub-turbinatus*
Subularia Awl, *subula, subulae* (the leaf shape)
subulatus -a -um awl-shaped, *subula, subulae*
subuli- shaped-like-an-awl-, *subula, subulae*
subulifolius -a -um with long-pointed, awl-like leaves, *subula-folium*
subulosus -a -um somewhat awl-shaped, diminutive of *subula*
subumbellatus -a -um slightly umbelled, *sub-umbellatus*
subumbrans lightly shading, *sub-(umbro, umbrare)*
subuniflorus -a -um mostly single flowered, *sub-uni-florum*
suburbanus -a -um around habitations, *sub-urbanus* (literally, near the city)
subverticillatus -a -um slightly whorled or disc-like, *sub-verticillatus*
subvestitus -a -um slightly clothed, *sub-vestitus* (with an indumentum)
subvillosus -a -um slightly shaggy-haired, *sub-villosus*
succedaneus -a -um following, *succedo, succedere, successi, successum*
succiferus -a -um sappy, producing sap, (*sucus, succus, suci*)-*fero*
succinctus -a -um armed, ready, *succingo, succingere, succinxi, succinctum*
succineus -a -um, sucineus -a -um of amber, *succinum* (the colour)
succiniferus -a -um bearing amber, *succinum* (*Pinus succinifera*)
succiruber -era -erum, succirubrus -a -um with reddish-amber sap, *succus*
Succisa Cut-off, past participle of *succido, succidere, succidi* (the rhizome of *Succisa pratensis*)
succisus -a -um (*succissus*) cut off from below, abruptly ended, *succido, succidere, succidi*
succosus -a -um full of sap, sappy, *sucus, suci; succus, succi*
succotrinus -a -um from Socotra, Indian Ocean

Succowia for Georg Adolph Suckow (1751–1813), professor at Heidelberg
succubus -a -um lying upon, with a lower distichous leaf overlain by the next upper leaf on the same side of the stem, late Latin *succubo, succubare* (classically for a female nocturnal demon, see *incubus*)
succulentus -a -um fleshy, soft, juicy, succulent, full of sap, *succus* (*sucus*)
sucidus -a -um, sucosus -a -um sappy, juicy, *sucidus* (full of juice, *sucus, suci*)
sudanensis -is -e from the Sudan, Sudanese
sudeticus -a -um from the Südetenland of Czechoslovakia and Poland
suecicus -a -um from Sweden (*Suecia*), Swedish
suendermannii for F. Sündermann (b. 1864), alpine nurseryman of Lindau
suffocatus -a -um suffocating, *suffoco, suffocare* (the flower heads of *Trifolium suffocatum* turn to the ground)
suffrutescens slightly or becoming shrubby, *sub-frutex-essentia*
suffruticosus -a -um somewhat shrubby at the base, soft-wooded and growing yearly from ground level, *sub-frutex*
suffruticulosus -a -um small shrub growing annually from the base, diminutive of *suffruticosus*
suffultus -a -um supported, propped-up, *suffulcio, suffulcire, suffulsi, suffultum*
suffusus -a -um tinged, coloured, blushed, *suffundo, suffundere, suffudi, suffusum*
Suillus Of-pigs, συς, συος, *sus, suis*
suionum of the Swedes (*Sviones*)
sulcatus -a -um furrowed, grooved, sulcate, *sulcus, sulci*
sulfureus -a -um, sulphureus -a -um pale-yellow, sulphur-yellow, *sulfur, sulphur*
sultana a synonym for Japanese plum (*Prunus salicina* × *simonsii*)
sultani for the Sultan of Zanzibar, Seyyid Bagrash ibn Seyyid Said (d. 1888)
sumatranus -a -um from Sumatra, Indonesia
sumulus -a -um excellent, *summus*
sundaicus -a -um from Soenda (Sunda) Island, Java
super-,supra- above-, over-, *super-*
superbiens becoming superb, present participle of *superbio, superbire*
superbissimus -a -um the most magnificent, superlative of *superbus*
superbus -a -um arrogant, proud, magnificent, superb, *superbus*
superciliaris -is -e eyebrow-like, with eyebrows, with hairs above, *super-ciliaris*
superfluus -a -um overflowing, *super-(fluo, fluere, fluxi, fluxum)*
supernatans living on the surface of water, *super-natans*
supinus -a -um lying flat, extended, supine, *supino, supinare, supinavi, supinatum*
supra- above-, on-the-surface-of-, *supra-*
supra-axillaris -is -e from above the axillary bud, *supra-axillaris*
supracanus -a -um grey-haired towards the top, more grey-haired, *supra-canus*
supradecompositus -a -um growing on decaying matter, *supra-de-compositus*
suprafolius -a -um growing on a leaf, *supra-folium*
supranubius -a -um of very high mountains, from above the clouds, *supra-(nubes, nubis)*
surattensis -is -e, suratanus -a -um from Surat, SE Gujarat state, India
surculosus -a -um shooting, suckering, freely producing young shoots, *surculus, surculi*
surinamensis -is -e from Suriname, S America (formerly Dutch Guiana)
surrectus -a -um not quite upright or erect, rising, leaning, *sub-rectus*
surrepens sprawling, creeping, *subrepo, subrepere, subrepsi, subreptum*
sursum- high-up-, forwards-and-upwards-, upwards, *sursum*
susianus -a -um from Shush (Susa), Iran
suspendus -a -um, suspensus -a -um hung up, hanging, pendent, suspended, *suspendo, suspendere, suspensi, suspensum*
susquehannae from Susquehanna, Philadelphia, USA
sutchuenensis -is -e (*sutchuensis -is -e*) from Sichuan (Sutchuan), W China
Sutera, suteri for Johann Rudolf Suter (1766–1827), Professor of Botany at Berne, author of *Flora Helvetica* (1802)

Sutherlandia for James Sutherland (*c.* 1639–1719), Superintendent of Edinburgh Botanic Garden and botanical writer
sutherlandii for Dr Peter Cormac Sutherland (1822–1900), Surveyor General in Natal
Swainsonia for Isaac Swainson (1746–1812), plant grower of Twickenham
Swartzia, swartzianus -a -um for Olof Peter Swartz (1760–1818), Professor of Botany at Stockholm
swatensis -is -e from Swat, Pakistan
swazicus -a -um from Swaziland, S Africa
sweginzowii for Nicholas A. Zvegintzov (1848–1920), Russian Governor of Latvia
Swertia for Imanuel Swert of Harlem, writer of *Florilegium* (1620)
Swietenia for Gerard van Swieten (1700–72), Dutch botanist and writer
sy-, syl-, sym-, syn-, syr-, sys- with-, together with-, united-, joined-
Syagrus Wild-boar, συαγρος (in Pliny this was a kind of date palm, Martius transferred it to a *Cocos*-like American genus)
sycamorus fig-fruited, συκη-μορεα (of the fig, *Ficus sycamorus*)
sychno- many-times-, frequent-, συχνος, συχνο-
Sychnosepalum Many-sepalled, συχνος-σκεπη
syco-, sycon- fig-like-fruit-, fig-, συκη, συκεη, συκον
Sycomorus, sycomorus -a -um Fig-mulberry, συκη-μορεα (the biblical fig, *Sycomorus antiquorum*, is *Ficus sycamorus*); Gerard re-used it for *Acer pseudoplatanus* because of a similarity of leaf shape, and its pleasant shade
Sycoparrotia the composite name for hybrids between *Sycopsis* and *Parrotia*
Sycopsis Fig-resembler, συκη-οψις (looks like some shrubby *Ficus*)
sylhetensis -is -e from Sylhet, NE Bangladesh
sylvaticus -a -um, sylvester -tris -tre wild, of woods or forests, sylvan, *sylva, sylvae*
sylvicolus -a -um inhabiting woods, *sylvae-colo*
sym- united, συμ- (with suffixes starting with b, m, or p)
symbiotic living with, συμ-βιοω
symmetricus -a -um actinomorphic, regular, συμ-μετρος
symmixis marriage, mixing together, συμ-μιξις
symphertos united, συμφερτος
symphiandrus -a -um having fused stamens, συμφερτος-ανηρ
sympho-, symphy- growing-together-, συμφυω-
Symphonia Harmonious, συμφωνος (they are united with five groups of three linear anthers alternating with the stigmatic lobes)
Symphoria Brought-together, or Useful, συμ-φερω
Symphoricarpos (us) Clustered-fruits, συμ-φερω-καρπος (the bunched berries)
Symphyandra United-stamens, συμφυω-ανδρος
symphytifolius -a -um with leaves resembling those of *Symphytum*
Symphytum Grow-together-plant, συμ-φυτον (Dioscorides' name, συμφυτον, for healing plants, including comfrey, *conferva* of Pliny)
symplo- braided-together-, united-, conjoined-, συμ-πλοκος
Symplocarpus Connected-fruits, συμπλοκη-καρπος (the compound fruit of skunk cabbage)
Symplocos United, συμπλοκη (the connected stamens) (***Symplocaceae***)
sympodialis -is -e with sympodial growth, from succeeding lateral buds, συμ-ποδος
syn- together-, συν-, συγ-, συ- (for suffixes not starting with b, m or p)
Synadenium Joined-glands, συν-αδην (the united involucral glands)
syncarpus -a -um with a compound ovary, aggregate-fruited, συν-καρπος
synciccus -a -um with *Cicca*, with flowers of different sexes in the same inflorescence (*ciccus* pomegranate pip)
Syneilesis Pressed-together, συμ-ειλεω; Folded-together, συμ-ειλυω
Syngonanthus Fused-female-flower, συν-γυνη-ανθος (fusion of the petals of the female flowers)
Syngonium United ovaries, συν-γυνη

Synsepalum, synsepalus -a -um United-sepals, συν-σκεπη (the tubular lower half of the calyx)
Synthyris Collected-little-doors, συν-(θυρις, θυριδος) (the capsule valves)
syphiliticus -a -um of syphilis, first named as a character in Girolamo Fracastoro's poem *Syphilis, sive morbus Gallicus* (1530) or 'Syphilis, the French disease' (*Lobelia syphilitica* was used to treat the disease)
syriacus -a -um from Syria, Syrian
Syringa Pipe, συριγξ, συριγγος (formerly for *Philadelphus* but re-applied by Dodoens use of the hollow stems to make flutes) (the nymph Syringa was changed into a reed)
syringanthus -a -um lilac-flowered, *Syringa*-flowered, συριγγος-ανθος
syringiflorus -a -um lilac-flowered, *Syringa-flora*
Syringodea Pipe-like, συριγγος-ωδης (the tubular part of the perianth)
syrticolus -a -um of sand banks, living on sand banks, *syrtis-colo*
systylius -a -um calyptrate, with a lid or cap, συν-στυλος (on moss capsules)
systylus -a -um with the styles joined together, συν-στυλος (*Rosa systyla*)
syzigachne with scissor-like glumes, συ-ζυγος-αχνη
Syzygium Paired, συ-ζυγος (from the form of branching and opposite leaves; formerly applied to *Calyptranthus*)
szechuanicus -a -um, szetschuanicus -a -um, szechwanensis -is -e from Szechwan, W China
szovitsianus -a -um, szovitsii for Jihann Nepomuk Szovitz (d. 1830), Hungarian collector in the Caucasus and Armenia

tabacicomus -a -um with a tobacco-coloured head or hair, *tabacum-(coma, comae)*
tabacinus -a -um tobacco-coloured, pale-brown, (*Nicotiana tabacum*) *tabacinus*
tabacum (tabaccum) from the Mexican-Spanish, Carib vernacular name, tabaco, for the pipe used for smoking the leaves of *Nicotiana tabacum*
Tabebuia from a Brazilian vernacular name
Tabernaemontana, tabernaemontanus -a -um, tabernaemontani for Jacob Theodore Mueller von Bergzabern of Heidelberg (1520–90), physician and herbalist (his Latinization of Bergzabern)
tabescens decaying, melting, wasting away, *tabesco, tabescere, tabescui*
tabidus -a -um wasting away, melting or decaying, *tabeo, tabere*
tabulaeformis -is -e, tabuliformis -is -e flat and circular, plate-like, *tabulae-forma*
tabulaemontanus -a -um, tabulamontanus -a -um from Table Mountain, S Africa, or from Tafelberg, Surinam *tabulae-montanus*
tabularis -is -e, tabuli- table-flat, flattened, *tabula, tabulae*
tabulatus -a -um layer upon layer, storied, *tabulatum, tabulati*
tacamahaca from an Aztec vernacular name for the resin from *Populus tacamahaca*, and *Calophyllum inophyllum*
Tacazzea from the environs of the Takazze river, Ethiopia
tacazzeanus -a -um from the Takazze river, Ethiopia
Tacca from a Malayan vernacular name, taka, for arrowroot (***Taccaceae***)
Taccarum *Tacca-arum* (implies intermediate looks but not hybridity)
taccifolius -a -um with leaves like *Tacca, Tacca-folium*
Tachia etymology uncertain (Speed, ταχος)
Tachiadenus *Tachia*-glanded, botanical Latin from *Tachia* and αδηνος (the circle of glands around the ovary)
Tacinga an anagram of Catinga, the Brazilian semi-arid scrubland vegetation type
tactilis -is -e sensitive to touch, *tango, tangere, tetigi, tactum*
tacubayensis -is -e from Tacubaya, south of Mexico City, Mexico
taediger -era -erum, taedifer -era -erum torch-bearing, *taeda, taedae, taedifer*
taediosus -a -um loathsome, *taedet, taedere, taeduit, taesum*
taedus -a -um an ancient name, *taeda*, for resinous pine cones used for torches
taegetus -a -um from Mount Taygetos, Greece

taenialis -is -e, taenianus -a -um shaped (segmented) like a tapeworm, *Taenia*-like, ribbon, ταινια
Taeniatherum Ribbon-bristled-one, ταινια-(αθηρ, αθηρος) (the awns of medusa's head grass)
Taeniophyllum Ribbon-leaf, ταινια-φυλλον
Taeniopsis Ribbon-like, ταινια-οψις (the frond shape, ≡ *Vittara*)
Taeniopteris Ribbon-fern, ταινια-πτερυξ
Taeniorrhiza Ribbon-root, ταινια-ριζα
taeniosus -a -um with a ribbon, bearing a head band, ταινιοω
Taenitis Ribbon-like, ταινια (frond shape)
tagal a Philippine vernacular name, tagal, for the native mangrove, *Ceriops*
Tagetes for Tages, Etruscan god of the underworld and grandson of Jupiter
tagliabuana for the brothers Tagliabe
taheitensis -is -e from Tahiti, Pacific Ocean
taimyrensis -is -e from the most northern Taimyr peninsular, N Siberia, Russia
Tainia Ribbon, ταινια (the labellum)
taipeicus -a -um from Taipei Shan, Shensi, China
taitensis -is -e from the Taita hills, Rift highlands, Kenya
Taiwania, taiwanensis -is -e, taiwanianus -a -um from Formosa (Taiwan), Formosan
taiwanicola inhabiting Taiwan, botanical Latin from Taiwan (Formosa) and *colo*
takesimanus -a -um, takesimensis -is -e from the Takeshima Islands, Japan Sea
Takhtajania for Armen Leonovich Takhtajan (b. 1910), Russian systematic botanist of St Petersburgh's Komarov Institute, author of *Diversity and Classification of Flowering Plants* (1997)
Talamancalia from the Talamanca mountains of Costa Rica
talamancanus -a -um from the Talamanca mountains of Costa Rica
talasicus -a -um winged, enduring, wretched, audacious, ταλας (Telasius was the Roman god of weddings)
Talauma from a W Indian vernacular name
Talinum, Talinopsis from an African vernacular name; some derive it from θαλια, a green branch, for its verdure
Talisia from the Guyanese vernacular name, toulichi
tamarici-, tamarisci- *Tamarix*-like-
tamarindi- tamarind-like, *tamarindus*
Tamarindus Indian-date, from the Arabic, tamr-hindi, Hindustan-date
Tamarix the late Latin name, *tamariscus*, for the Spanish area of the River Tambo (*Tamaris*); others derive it from the Hebrew, tamar, for a palm tree (***Tamaricaceae***)
Tamaulipa from Tamaulipas, Mexico
tamnifolius -a -um bryony-leaved, *Tamus-folium* (*Tamnus* of Pliny)
Tamonea from the Guyanese vernacular name, tamone
tampicanus -a -um from Tampico, NE Mexico
tamukeyama the Japanese name for a cultivar 'Crimson Queen'
Tamus from the name, *taminia uva*, in Pliny for a kind of vine
tana- long-, τανaος, τανυ-
tanacet-, tanaciti- tansy-like-, *Tanacetum-*
tanacetifolius -a -um with leaves resembling those of *Tanacetum, Tanacetum-folium*
Tanacetopsis Resembling-*Tanacetum, Tanacetum-opsis*
Tanacetum from the Latin, *tanazita*, and cognate with αθανασια (tansy was placed amongst the winding sheets of the dead to repel vermin)
tanaiticus -a -um from the region of the River Tanais (Don) in Sarmatia
Tanakaea, tanakae for Yoshio Tanaka (1836–1916), Japanese botanist and entomologist (*Useful Plants of Japan*)
tananicus -a -um from the river Don (*Tanais*), Russia
tanastylus -a -um having a long, excerted style, τανaος-στυλος
tangerinus -a -um deep orange-red, tangerine-coloured; from Tangier, Morocco (Tanger)
tangshen from Tang shan province of China, vernacular for bastard ginseng

tanguticus -a -um of the Tangut tribe of Gansu, the former Hsi Hsia (Xi Xia) kingdom of W China and NE Tibet, Tibetan
Tanquania from the Tanqua Karoo and National Park area, S Africa
tapein-, tapeino- humble-, modest-, level-, insignificant-, ταπεινος
Tapeinanthus, tapeinanthus -a -um Low-flower, ταπεινος-ανθος (refers to the small stature, *Amaryllidaceae*)
Tapeinia Insignificant, ταπεινος
Tapeinidium Modest-sized, ταπεινοω
Tapeinochilus Modest-lip, ταπεινος-χειλος (refers to the small labellum)
Tapeinosperma Small-seed, ταπεινος-σπερμα
Tapeinostemon Short-stamened-one, ταπεινος-στημων
tapesi-, tapeti- carpet-like-, carpet-, ταπης, ταπητος, ταπις, ταπιδος
tapetiformis -is -e resembling a carpet, (*tapeta, tapetae*)*-forma*
tapetis -is -e carpeting, ταπητος, *tapeta, tapetae*
tapetoides, tapetodes carpet-like, ταπητος-οειδης, ταπητος-ωδης, *tapeta-oides*
Taphrina Depression, ταφρος (in which the naked asci are produced)
taphro-, -taphrum ditch-, -depression, ταφρος, ταφρο-
Taphrospermum Depressed-seed, ταφρος-σπερμα
Tapinanthus Modest-flower, ταπεινος-ανθος (corolla lobes small and erect, *Loranthaceae*)
tapiriceps shaped like a tapir's head, *Tapirus-caput* (*Catasetum tapiriceps*)
Tapirira from the Tupi vernacular name, tapyra, for *Tapirira guianensis*
tapirorum of the tapirs, *Tapirus*, from Tupi, tapyra (their food plant)
Tapiscia an anagram of *Pistacia*
taprobanensis -is -e from Sri Lanka (Ceylon) (*Taprobane*)
Tapura from the vernacular name in Guiana
tarantinus -a -um from Taranto (*Tarentum*) province, S Italy
taraxaci- dandelion-like-, *Taraxacum-*
taraxacifolius -a -um with leaves resembling those of a dandelion
taraxacoides resembling *Taraxacum*, dandelion-like, *Taraxacum-oides*
Taraxacum Disturber (from the Arabic name, tarakhshagog, or talkhchakok, for a bitter herb)
tarayensis -is -e from the Tarai, or Terai, 'moist land' region of N India–S Nepal
Tarchonanthus Tarragon-flowered-one, botanical Latin from Arabic, tarkhon, and ανθος
tardans slow, late, retarded, *tardo, tardare, tardavi, tardatum*
tardi-, tardus -a -um slow, reluctant, late, *tardus*
tardiflorus -a -um reluctant- or late-flowering, (*tardo, tardare, tardavi, tardatum*)*-florum*
tardissimus -a -um the latest or slowest, superlative of *tardus*
tardivus -a -um late to appear, slow-growing, *tardus*
Tarenna from the Sri Lankan vernacular name, tarana
tarentinus -a -um from Taranto (*Tarentum*), S Italy
tarokoensis -is -e from the T'ai-lu-ko (Taroko) gorge, central Taiwan
taronensis -is -e from the valley of the Taron, a headstream of the Irrawaddy river, Burma
tartareus -a -um, tartrus -a -um infernal, of the underworld, *tartarus*; with a loose crumbling surface
tartaricus -a -um, tataricus -a -um from Tartary, central Asia
tasmanicus -a -um from Tasmania (a name given for Abel Janszoon Tasman (1603–59), the Dutch navigator)
tataricus -a -um from Tartary (*Tataria*), Russia–Mongolia, or Tatar Strait area off Sakhalin Island
tatarinovii, tatarinowii for Alexander Tatarinov (1817–86), author of a catalogue of Chinese drugs
tatsiensis -is -e from Kangding (Tatsien-lu), W Sichuan, China
tatula from an old name for a *Datura*

tauri from the Taurus range, Asia Minor, Turkey
tauricensis -is -e, tauricus -a -um from the Crimea (*Tauria*)
tauricola inhabiting the Crimea (*Tauric Chersonese*)
tauricus -a -um of the Crimea (*Tauric Chersonese*)
taurinus -a -um from Turin, Italy (*Augusta Taurinorum*), or of bulls, *tauri*
tauscheri for Gyula Tauscher (1832–82), Hungarian physician
Tauschia for Ignaz Friedrich Tausch (1793–1848), naturalist and director of the Duke of Canal de Malabaillas' garden, Prague
Tavaresia for Joachim da Silva Tavares (1866–1931), Portuguese cleric and botanist (some attribute it for Jose Tavares de Macedo)
Taverniera for J. B. Tavernier (1605–89), traveller in the Levant
tax-, taxi-, taxo- orderly-, order-, τασσω, τασσειν, ταξις
taxi-, taxi yew-like-, resembling *Taxus*, living on *Taxus* (symbionts, parasites and saprophytes)
taxicolus -a -um living with or on yew, *Taxus-colo*
taxifolius -a -um having leaves resembling those of *Taxus, Taxus-folium*
Taxillus Little-dice, *taxillus*
taxodioides resembling *Taxodium, Taxodium-oides*
Taxodium Yew-like, resembling *Taxus*, botanical Latin from *Taxus* and ωδης (***Taxodiaceae***)
taxoides resembling yew, *Taxus-oides*
taxonomy orderly law, classification, ταξις-νομος
taxophilus -a -um yew-loving, botanical Latin from *Taxus* and φιλος (and also other material)
Taxus the ancient Latin name, *taxus*, for yew of Dioscorides, poisonous in all parts except the fleshy aril; ταχον a bow (***Taxaceae***)
taygetes, taygeteus -a -um from the Taíygetos mountain range, S Greece, (*Taygete* was a Pleiad)
tazettus -a -um little cup, ταζετα (the corona of *Narcissus tazetta*)
Tchihatchewia, tchihatchewii for Count Pierre A. de Tchihatchef (1812–90), Russian-born traveller and writer
technicus -a -um special, technical, artistic, τεχνικος
Tecoma, Tecomaria from the Mexican name, tecomaxochitl
Tecomanthe *Tecoma*-flowered, *Tecoma-anthus*
Tecomella feminine diminutive from *Tecoma*
Tecophilaea for Tecofila Billiotti-Colla, botanical illustrator, daughter of Professor Luigi Colla of Turin (***Tecophilaeaceae***)
Tectaria Roofed, *tectum* (the complete indusium)
tectificus -a -um sheltering, forming a roof, *tectum-(facio, facere, feci, factum)*
Tectiphiala Bowl-roofed, botanical Latin from *tectum* and φιαλη
Tectona from the Tamil name, tekka, for teak
tectorius -a -um, tectorus -a -um of a plasterer; of rooftops, growing on rooftops, of the tiles, *tectum, tecti*
tectus -a -um with a thin covering, hidden, tectate, *tego, tegere, texi, tectum*
Teedia for J. G. Teede, German botanist and traveller
Teesdalia for Robert Teesdale (*c.* 1740–1804), Yorkshire botanist and author of a Flora of Castle Howard
tef the Arabic name for *Eragrostis abyssinica* (tef grass)
tegens covering, hiding, concealing, protecting, *tego, tegere, texi, tectum*
tegetiformis -is -e mat-like, forming mats, *tegetis-forma*
tegetus -a -um mat-like, *teges, tegetis*
tegmentosus -a -um roof-like, covering, *tegimen, tegiminis*
tegulaneus -a -um overlapping, like tiles, *tegula, tegulae*
tegumentus -a -um covered (e.g. indusiate), *tego, tegere, texi, tectum*
Teijsmanniodendron for Johannes Elias Teijsmann (Teysmann) (1808–82), Dutch botanist at the Bogor Buitenzorg gardens, Java, botanical Latin from Teijsmann and δενδρον

teino- elongate, stretch, τεινω, τονος
tel-, tele- far-, far-off-, afar-, τηλου, τηλε, τηλ-; complete, perfect, full in number, τελειος
Telanthophora Bearing-perfect-flowers, τελειος-ανθος-φορα
teledapos far-rending, τηλε-δαπτω (*Rubus teledapos*)
Telekia for Samuel Teleki de Szek (1739–1822), Chancellor of Transylvania and patron of botany
telephiifolius -a -um having leaves resembling those of *Telephium*
telephioides resembling *Sedum telephium, telephium-oides*
Telephium, telephium Distant-lover, τηλε-φιλος (a Greek name, τηλεφιλον, for a plant thought to be capable of indicating reciprocated affection)
teleuto-, telio- terminal-, completion-, accomplishment, an end-, τελευτη
Telfairia for Dr Charles Telfair (1778–1838), Irish surgeon and botanist
Teliostachya Full-flowered-spike, τελειος-σταχυς
Telipogon Bearded-end, τελε-πωγων (the column)
Tellima an anagram of *Mitella*
tellimoides resembling *Tellima, Tellima-oides*
telmataia, telmateius -a -um of swamps or marshes, of muddy water, τελμα, τελματος
Telopea Seen-at-a-distance, τηλωπος (the conspicuous crimson flowers)
telopeus -a -um conspicuous, seen from afar, τηλωπο
Telosma Fragrant-from-afar, τηλ-οσμη (many have strong perfumes)
temenius -a -um of sacred precincts or holy places, τεμενιος
Temnocalyx Severed-calyx, τεμνω-καλυξ (the truncate, rim-like calyx)
Templetonia for John Templeton (1766–1825), Irish botanist, founder of the Belfast Society of Natural History and Philosophy
temulentus -a -um (*temulum, temulus*) drunken, intoxicating, nodding irregularly, *temulentus* (toxic seed of ryegrass)
tenacellus -a -um somewhat tenacious, diminutive of *tenax, tenacis*
tenacissimus -a -um most tenacious, superlative of *tenax* (esparto-grass)
tenago- swamp-, shallow-water-, τεναγος
Tenagocharis Shallow-water-pleasure, τεναγος-χαρις
Tenaris from a S African vernacular name
tenax gripping, stubborn, firm, persistent, tenacious, *tenax, tenacis*
tenebrosus -a -um somewhat tender, dark, gloomy, of shade, *tenebrae* darkness
tenelliflorus -a -um having dainty flowers, diminutive from *teneri-florum*
tenellus -a -um delicate, tender, diminutive of *tener*
tenens enduring, persisting, present participle of *teneo, tenere, tenui*
teneri-, tener -era -erum, tenerus -a -um soft, tender, delicate, *tener, teneri*
tenerifa, teneriffae, from Tenerife, Atlantic Ocean
tenerrimus -a -um quite or most soft, *tener, teneri*
tennesseensis -is -e from Tennessee, USA
tenorei, tenoreanus -a -um for Michele Tenore (1780–1861), Professor of Botany at Naples
tentaculatus -a -um, tentaculosus -a -um with sensitive glandular hairs, modern Latin *tentaculum*
tenthridiniferus -a -um insect-bearing, harbouring sawflies, *Tenthridinoidea-fero* (*Ophrys* flower form)
tenui-, tenuis -is -e persisting, tenacious, slender, thin, fine, *tenuo, tenuare, tenuavi, tenuatum*
tenuicaulis -is -e with slender stems, *tenuis-caulis*
tenuidolius -a -um shaped like a narrow wine-jar, *tenuis-dolium*
tenuiflorus -a -um with slender (tubular) flowers, *tenuis-florum*
tenuifolius -a -um slender-leaved, with narrow leaves, *tenuis-folium*
tenuior more slender, *tenuis*
tenuipes slender-stemmed, *tenuis-pes*
tenuiramis -is -e with slender branches, *tenuis-(ramus, rami)*

tenuisectus -a -um with slender divisions, *tenuis-(seco, secare, secui, sectum)* (of the leaves)
tenuisiliquus -a -um with slender pods, *tenuis-(siliqua, siliquae)*
tenuispinus -a -um with slender spines, *tenuis-(spina, spinae)*
tenuistachyus -a -um narrowly panicled or spicate, botanical Latin from *tenuis* and σταχυς
teocote the Mexican vernacular name for the twisted-leaved or Aztec pine
tephro-, tephrus -a -um ash-grey-, ashen, τεφρα, τεφροω, τεφρο-
tephrodes ashen-coloured, τεφρο-ωδες
tephropeplus -a -um ash-grey robed, τεφρο-πεπλος
tephrosanthos ashen-flowered, τεφρο-ανθος
Tephroseris Ashen-potherb, τεφρο-(σερις, σεριδος)
Tephrosia Ashen, τεφροω (the leaf colour)
tephrotrichus -a -um with ash-grey hairs, τεφρο-τριχος
Tepualia from Tepual, Chile
Tepuia, Tepuianthus from the high table-lands, tepuis, of Guiana and Venezuela
ter- three-times, triple-, thrice-, *ter*
terato- prodigious-, monstrous-, τερας, τερατος, τερατο-
Teratophyllum Prodigious-leaf, (τερας, τερατος)-φυλλον
terebinthi- *Pistacia*-like-, turpentine-
terebinthifolius -a -um with leaves like those of *Pistacia*, *Pistacia terebinthus-folium*
terebinthinus a former name, τερεβινθινος, for Chian turpentine tree, *Pistacia terebinthus*.
terebinthus turpentine, τερεβινθος (turpentine was first produced from *Pistacia terebinthus*)
teres -etis -ete, tereti- quill-like, rounded, cylindrical, terete, *teres, teretis*
tereticornis -is -e with cylindrical horns, *teretis-cornu*
teretifolius -a -um with terete leaves, cylindrical-leaved, *teretis-folium*
teretiusculus -a -um somewhat smoothly rounded, weakly rounded, diminutive of *teres, teretis*
tergeminus -a -um three-twins, three-paired, *ter-geminus*
tergi- at the back-, *tergum, tergi, tergo-*
Terminalia Terminal, *terminus* (the leaves are frequently crowded at the ends of the branches)
Terminaliopsis Resembling-*Terminalia*, botanical Latin from *Terminalia* and οψις
terminalis -is -e terminal, *termino, terminare, terminavi, terminatum* (the flower on the end of the stem)
ternatanus -a -um, ternateus -a -um from Ternate, Maluku (Moluccus) Islands, Indonesia
ternatipartitus -a -um with (floral) parts in threes, *ternatus-(partitio, partitionis)*
ternatus -a -um, ternati-, terni- with parts in threes, ternate, *terni, ternatus* (see Fig. 5e)
terniflorus -a -um with flowers in threes, *terni-florum*
ternifolius -a -um three-leaved, with leaves in threes, *terni-folium*
Ternstroemia (Ternströmia) for Christopher Ternström (1703–54), Swedish cleric and naturalist in China
Terpsichore Dancing-delightfully, τερψις-χορος; in mythology, Terpsichore was the muse of lyric poetry and dance
terracinus -a -um from Terracina (*Tarracina*), Latina province, S Italy
terra-novae from Newfoundland, *terra-nova*
terra-reginae from Queensland, Australia, of the name
terrester -ris -tre, terrestris -is -e growing on the ground, *terrestris* (not epiphytic or aquatic)
terreus -a -um earthen, earth-coloured, *terra, terrae*
terricolor earth-coloured, *(terra, terrae)-color*
tersi- neat-, clean-, *tergeo, tergere, tersi, tersum*
Tersonia Clean, *tersus* (leafless stemmed)

tertio- third-, *tertius*
tescus -a -um of wild or waste areas or desert, *tesqua, tesquorum* (*tesca*)
tesquicolus -a -um of waste land, of desert land, *tesqua-colo*
tesquorus -a -um of waste lands and deserts, *tesquorum*
tessellatus -a -um (tesselatus) chequered, mosaic-like, tessellated, with small (rectangular) areas of colour, *tessella, tessellae*
testaceus -a -um brownish-yellow, terracotta, brick-coloured, *testa, testae*
testicularis -is -e, testiculatus -a -um tubercled, having some testicle-shaped structure, *testiculus, testiculi* (e.g. a tuber or fruit)
Testudinaria Tortoise-like, *testudineus* (the outer layer of the root, ≡ *Dioscorea*)
testudinarium resembling tortoise shell, *testudineus*
teter -era -erum having a foul smell, hideous, repulsive, *teter; taeter, taetri*
tetra- square-, four-, τετρα
Tetracarpidium Four-carpelled-one, τετρα-καρπιδιον, *tetra-carpidium*, diminutive of καρπος, *carpum* (ovaries of the *Euphorbiaceae* are predominantly tricarpellary)
tetracarpus -a -um fruiting as four carpels, τετρα-καρπος
Tetracentron Four-spurs, τετρα-κεντρον (the spur-like appendages of the fruit) (***Tetracentraceae***)
Tetracera Four-horns, τετρα-κερας (the lobed fruiting capsule)
Tetraclinis Four-ranked-leaves, τετρα-κλινη
Tetracme Four-points, τετρα-ακμη (the shape of the fruit)
Tetracoccus Four-berried, τετρα-κοκκος (the four-lobed fruit)
Tetractomia Four-cornered, τετρα-τομη (the fruit)
tetradactylus -a -um four-fingered, with four elongate lobes, τετρα-δακτυλος
Tetradenia Four-glands, τετρα-αδην
Tetradium Four-partite, τετραδιον (a group of four soldiers) (tetramerous floral structure)
Tetradymia Fourfold, τετραδυμος (the fourfold flower-heads and their involucral bracts)
Tetragastris Four-bellies, τετρα-γαστηρ (the fruit shape)
Tetragonia Four-angled, τετρα-γωνια (the shape of the fruit)
Tetragonolobus, tetragonolbus -a -um Quadrangular-pod, τετρα-γωνια-λοβος (the fruit)
Tetragonotheca Tetragonal-case, τετρα-γωνια-θηκη (the pseudo-nuts)
tetragonus -a -um four-angled, square, τετρα-γωνια
tetrahit four-times, τετρα (tetraploid)
tetralix a name, τετραλιξ, used by Theophrastus for the cross-leaved state when the leaves are arranged in whorls of four
Tetrameles Four-apple, τετρα-μελεα (the tetramerous flowers)
Tetrameranthus Four-partite-flowered, τετρα-μερος-ανθος
Tetramerista Four-divisions, τετρα-μεριστης (the flowers)
Tetramerium Four-partite, τετρα-μερος (flowers)
tetramerus -a -um with the (floral) parts in fours, four-partite, τετρα-μερος
Tetramicra Four-small(-parts), τετρα-μικρος (the four-partite anther)
tetrandrus -a -um with four stamens, four-anthered, τετρα-ανηρ
Tetranema Four-stamens, τετρα-νημα (fifth absent as in *Penstemon*)
Tetraneuris Four-nerved, τετρα-νευρα (four-veined or ribbed)
Tetrapanax Four-partite-*Panax*, τετρα-παν-ακεσις (four-partite floral structure of Chinese rice paper tree)
tetrapetalus -a -um with four petals, τετρα-πεταλον
tetraphyllus -a -um four-leaved, with leaves in fours, τετρα-φυλλον
Tetrapleura Four-ribbed, τετρα-πλευρα (the angular legume)
tetraplus -a -um fourfold, τετρα(-πλοος,-πλους) (e.g. ranks of leaves)
Tetrapogon Four-bearded, τετρα-πωγων (the awns of the lemmas)
Tetrapteris (Tetrapterys), tetrapterus -a -um Four-winged, τετρα-πτερον (stems, seeds or fruits)
tetraquetrus -a -um sharply four-angled, *tetra-quadra* (*Arenaria tetrequetra*)

Tetrardisia Four-partite-*Ardisia* (the floral structure)
Tetraria Four-partite, τετρα
Tetrarrhena Four-stamened, τετρα-αρρην (grasses typically have three)
tetrasepalus -a -um having four sepals, τετρα-σκεπη
tetraspermus -a -um four-seeded, τετρα-σπερμα
Tetraspis Four-shields, τετρα-ασπις
tetrastachyus -a -um with four spikes, τετρα-σταχυς
tetrastichus -a -um with four rows, four-ranked, τετρα-στιχος (spikelets)
Tetrastigma Four-stigmas, τετρα-στιγμα (the four-lobed stigma)
Tetratheca Four-cells, τετρα-θηκη (the anthers)
Tetrazygia Four-yoked (partite), τετρα-ζυγος (the tetramerous flowers)
Tetrorchidium Four-small-testicles, τετρα-ορχις, diminutive Latinized suffix (the anther-lobes)
teucrioides resembling *Teucrium*, τευκριον-οειδης
Teucrium, teucrium Dioscorides' name, τευκριον, perhaps for Teucer, hero and first King of Troy
teuscheri for R. Teuscher (1827–84), planter in Java
Teuscheria for Heinrich Teuscher (1891–1984), German botanist
teutliopsis resembling a small beet, τευτλιον-οψις
texanus -a -um, texensis -is -e from Texas, USA, Texan
textilis -is -e used for weaving, woven, *textilis* (fibres or leaflets)
teydeus -a -um from Pico de Teide, Tenerife, Canary Islands
Teysmannia for E. J. Teysmann (1808–82), Dutch gardener
Thaia from Thailand (Thai, formerly Siam)
thalami- bedchamber-, receptacle-, θαλαμος,
Thalassia Of-sea-water, θαλασσα (marine aquatic turtle grass)
Thalassicola Sea-dweller, botanical Latin, *thalassa-colo* (planktonic)
thalassicus -a -um sea-green, θαλασσα, *thalassicus*
thalassinus -a -um of the sea, sea-green, θαλασσα, *thalassicus*
Thalassodendron Sea-tree, θαλασσα-δενδρον (seagrass)
Thalia, thalianus -a -um for Johannes Thal (1542–83), German botanist, author of *Sylva Hercynia* (1588) (Thalia, from Thale, was the eighth muse, presiding over comedy and idyllic poetry; she was also one of the three Graces and patroness of festive meetings)
thalictrifolius -a -um with leaves similar to those of *Thalictrum, Thalictrum-folium*
thalictroides resembling *Thalictrum, Thalictrum-oides*
Thalictrum a name, θαλικτρον, used by Dioscorides for another plant, θαλλω to grow green
thalidi-, thallo thallus-, θαλος, θαλλος (a vegetative body without differentiation into stem and leaves)
thamn-, thamno-, -thamnus -a -um -shrub-like, -shrubby, θαμνος, θαμνο-
Thamnea Shrub, θαμνος (the habit)
Thamnobryum Bushy-moss, θαμνο-βρυον (Derbyshire feather-moss)
Thamnocalamus Bushy-reed, θαμνο-καλαμος
Thamnocharis Pleasing-shrub, θαμνο-χαριεις
Thamnochortus Pasture-shrub, θαμνο-χορτος
Thamnosma Odorous-shrub, θαμνο-οσμη (Turpentine smelling)
thanatophorus -a -um bearing death, deadly poisonous, θανατος-φορος
thapsi similar to *Thapsia*
Thapsia ancient Greek name, θαψια, used by Theophrastus for a poisonous plant
thapsiformis -is -e resembling *Thapsus* (*Verbascum thapsus*), *Thapsus-forma*
Thapsus, thapsus from the island of Thapsos, an old generic name, θαψος, for *Cotinus coggygria* (Thapsus(os) was a N African town and site of a victory by Caesar)
thasius -a -um from the wooded Aegean island of Thasos, Greece
Thaumasianthus Extraordinary-flower, θαυμασι-ανθος; θαμβεω to astonish
Thaumastochloa Strange-grass, θαυμαζω-χλοη

Thaumatocaryon Strange-nut, θαυματο-καρυον
Thaumatococcus Amazing-berry, θαυματο-κοκκος (crimson with shining black seeds, on forest floor, the aril containing a powerful sweetening protein)
thaumus -a -um extraordinary, strange, marvellous, θαυμα, θαυμασιος, θαυμασι-; θαυμαζω, θαυματο-
Thea the Latinized Chinese name, T'e (***Theaceae***)
thebaicus -a -um from the ancient area of Thebes (*Thebais*), Egypt (doum palm)
thebanus -a -um from Thebes (*Thivai*), Boeotia department of Greece
-theca, theco-, -thecus -a -um box-, -chambered, -cased, θηκη
Thecacoris Helmet-celled, θηκη-κορυς (the anthers)
theciferus -a -um bearing chambers, θηκη-φορος
-thecius -a -um -cased, -chambered, θηκη
Thecocarpus Cased-fruit, θηκη-καρπος (the fruit is not schizocarpic)
Thecostele Hollow-pillar, θηκη-στηλη (the nectary or the gynostegium)
theezans tea-like, resembling *Thea*
theifer -era -erum tea-bearing, *Thea-fero*
theifolius -a -um with leaves like *Camellia sinensis*, tea-leaved, *Thea-folium*
theio- brimstone-, smoky-, θειον, θειο-
theioglossus -a -um sulphur-tongued, smoke-tongued, θειο-γλωσσα
theionanthus -a -um smoke-flowered, with haze-like inflorescences, θειον-ανθος
thele-, thelo-, thely- prolific, female-, nipple-, θηλυς, θηλυ-
thelegonus -a -um with nipple-like irregularities along the angles, θηλυς-γωνια
thelephorus -a -um covered in nipple-like prominences, θηλυ-φορα
Thelepogon Nippled-beard, θηλυ-πωγων (the tuberculate lower glumes)
Thelesperma Nippled-seed, θηλυ-σπερμα (the papillose achenes)
Thelethylax Papillate-pouch, θηλυ-θυλακος
Thelocactus Nipple-cactus, θηλυ-κακτος (the protuberances on the stem ribs)
Thelycrania Female-cornelian-cherry, θηλυς-κρανεια (the name, θηλυκρανεια, used by Theophrastus)
Thelygonum (Theligonum) Girl-begetter, θηλυ-γυνη (claimed by Pliny to determine the conception of girl offspring) (***Thelygoniaceae***)
Thelymitra Hooded-woman, θηλυ-μιτρα (the shape of apex of the column of woman's cap orchid)
Thelypodiopsis Resembling-*Thelypodium*, θηλυ-ποδιον-οψις
Thelypodium Stalked-female, θηλυ-ποδιον (the gynophore)
Thelypteris (*Thelipteris*) Female-fern, θηλυς-πτερις (Theophrastus' name, θηλυπτερις, for a fern) (***Thelypteridaceae***)
Themeda from an Arabic vernacular name, tha emed
Themistoclesia for Themistocles (528–462 BC), Athenian statesman and soldier
Thenardia for Louis Jaques Thenard (1777–1857), French biochemist (Thenard's blue porcelain pigment)
Theobroma God's-food, θεος-βρωμα (Aztec, cacahuatl, cocoa bean)
Theodorea for Theodoro M. F. P. da Silva, Brazilian statesman (≡ *Rodriguesiella*)
theoides resembling tea-plant, *Thea-oides*
Theophrasta, theophrastii for Theophrastus (370–285 BC), Greek scholar and father of botany (***Theophrastaceae***)
therei- summer-, summer-time-, θερεια, θερει-
Thereianthus Summer-flowering, θερεια-ανθος
theriacus -a -um (theriophonus, for *theriophobus?)* antidote, θηριακη, *theriaca* (theriacs are antidotes to poisons and bites of wild beasts, θηρ, θεριον,θερος; cognate with the original meaning of treacle)
thermalis -is -e of warm springs, θερμη
thermophilus -a -um liking warmth, θερμη-φιλος
Thermopsis Lupin-like, θερμος-οψις
thero- harvest-, summer-, θερος
Theropogon Summer-beard, θερος-πωγων (*Convallaria*-like flowered)
Therorhodion Summer-rose, θερος-ροδον (floral comparison)

Thesium a name in Pliny for a bulbous plant (root parasites)
Thespesia Divine, θεσπεσιος (commonly cultured round temples)
thessalicus -a -um, thessalus -a -um from Thessaly (Thessalía), N Greece
thessalonicus -a -um from Thessaloniki (Thessalonica), Greece
Thevetia for Andre Thévet (1502–92), French traveller in Brazil and Guiana
thianschanicus -a -um, thianshanicus -a -um from Tien Shan (Tian Shan), central Asia
Thibaudia for J. M. B. Thibault de Chanvalon (1725–88), Secretary of the Linnean Society of Paris
thibetanus -a -um, thibeticus -a -um from Tibet (see *tibetanus*)
thigmo- touch-, θιγμα, θιγγανω, θιγμο-
thinicolus -a -um shifting-sand-dwelling, θινος-*colo*
thino- dune-, sand-, strand-, θις, θινος, θινο-
thinophilus -a -um sand-dune-loving, θινο-φιλος
Thinopyrum Strand-wheat, θινο-πυρος
thirsi- contracted-panicle-, θυρσος (see Fig. 3)
thirsiflorus -a -um with flowers in thyrses, thyrsoid, botanical Latin from θυρσος and *florum*
Thistletonia, thistletonii, Thiseltonia, thiseltonii for Sir William Turner Thiselton-Dyer (1843–1928), Director of Kew (1885–1905)
Thladiantha Eunuch-flower, θλαδιας-ανθος (female flowers have aborted stamens)
Thlaspi the name, θλασπις, used by Hippocrates for cress with seeds which, when crushed, θλαω, were used as a condiment
tholiformis -is -e rotunda-like, vaulted, domed, θολος, *tholus*
Thomasia, thomasii for either Pierre Thomas and his son Abraham Thomas (1788–1859), collectors of Swiss plants, or Graham Stuart Thomas OBE (1909–2003), plantsman and writer on roses, or Dr David John Thomas (1813–71), physician and botanist in New South Wales
Thompsonella for Charles Henry Thompson (1870–1931), American botanist
thomsonae for the wife of Reverend W. C. Thomson in Old Calabar, W Africa (*Clerodendron thomsonae*)
Thomsonia for Anthony Todd Thomson (1778–1849), professor at University College London
thomsonianus -a -um, thomsonii for either Anthony Todd Thomson (1778–1849), professor at University College London, or Thomas Thomson (1817–78), Superintendent of Calcutta Botanic Garden and contributor to Hooker's *Flora of India*, or George Thomson (1819–78), Missionary in W Africa, or Joseph T. Thomson (1858–95), French naval officer and writer on W African orchids
Thonningia for Peter Thonning (1775–1848), Danish doctor on the 1799–1803 W African expedition
thora etymology uncertain; of corruption, of death, φθορα (a medieval name for a poisonous buttercup)
Thouinia, Thouinidium for André Thouin (1747–1824), French botanist, Curator of the Jardin des Plantes, Paris
thracicus -a -um from the ancient region of Thrace (*Thracia*), modern Balkans
thrasi-, thrasy- enduring-, stout-, bold-, θαρσος, θρασος, θρασυ-
Thrasia Bold, θαρσος, θρασος
Thrasyopsis Resembling-*Thrasia*, θρασια-οψις
thrausto- brittle, θραυστος
Thrinax, -thrinax Fan, θριναξ (the flabellate leaves of fan palms)
Thrincia Capping, θριγκος (the toothed scales of the outer pappus of *Leontodon*)
thripticos effeminate, θρυπτιχος
-thrix -hair, -haired, θριξ, τριχος, τριχη
Thrixospermum Hair-seeded, θριξ-σπερμα (distinctive seed morphology)
Thryallis the name used by Theophrastus for *Verbascum*
Thryptomene Feeble-courage, θρυπτω-μενος (the lowly nature of the first species described)

Thuarsea, thouarsii for Louis Marie Aubert du Petit Thours (1758–1831), French writer on African orchids
Thuidium *Thuja*-like
Thuja (Thuya) Theophrastus' name, θυια, for a resinous, fragrant-wooded tree, θυον (θυος incense, used during worship)
Thujopsis (Thuyopsis) Resembling-*Thuja*, θυι-οψις
Thunbergia for Carl Per (Karl Pehr, *Caroli Petri*) Thunberg (1743–1822), Swedish physician and professor at Uppsala, who travelled in Africa, Japan and Batavia
Thunia for Count Franz A. Graf von Thun Hohenstein of Tetschin (1786–1873), Bohemian orchidologist
Thuranthos Incense-flower, θυς-ανθος (the fragrance)
thurifer -era -erum, turifer -era -erum incense-bearing, frankincense-producing, (*tus, turis*)-*fero*
thuringiacus -a -um from mid-Germany (*Thuringia*)
thusculus -a -um from Tusculum, near Rome; incense, *tus, turis*
Thuspeinanta an anagram of *Tapeinanthus*
thuyioides, thyoides *Thuja*-like, *Thuja-oides*
thylac-, thylaci-, thylaco- pouched-, θυλακος, θυλακιον
thylaciochilus -a -um the corolla having a pouch-like lip, θυλακιον-χειλος
Thylacopteris Pouched-fern, θυλακος-πτερον
Thylacospermum Pouched-seed, θυλακος-σπερμα
thymbra an ancient Latin name, *thymbra, thymbrae,* in Pliny for a savoury, thyme-like plant
thymbriphyrestus -a -um mixed with thyme, botanical Latin from *thymbra* and φυρω
Thymelaea Thyme-olive, θυμος-ελεια (the leaves and fruit), Dioscorides' name, θυμελαια, for *Daphne gnidium* (***Thymeliaceae***)
thymifolius -a -um Thyme-leaved, *thymum-folium*
thymoides Thyme-like, *Thymus-oides*
thymoliferus -a -um thymol-bearing, botanical Latin from θυμος and *-ol* (oil) and *fero*
Thymophylla Thyme-leaf, θυμος-φυλλον (foliar fragrance, ≡ *Dyssodia*)
Thymus Theophrastus' name, θυμος, for a plant used in sacrifices, θυω to burn incense, θυοω to perfume
thyoides *Thuja*-like, θυια-οειδης
thyri-, thyro- entrance-, door-, θυρα; small door-, θυρις, θυριδος
Thyridachne Little-door-chaff, (θυρις, θυριδος)-αχνη (the lower lemma)
Thyridolepis Little-door-scale, (θυρις, θυριδος)-λεπις (the lower glume)
Thyrocarpus Door-fruited-one, θυρα-καρπος
Thyroma Shield-growths, θυρεος-σωμα (the glands)
thyrs-, thyrsi-, -thyrsos baccic staff; wreath, contracted-panicle-, θυρσος, *thyrsus* (see Fig. 3d)
Thyrsacanthus Thyrsoid-*Acanthus*, θυρσος-ακανθα (≡ *Odontonema*)
Thyrsanotus Fringed, θυρσανωτος (the members of the inner perianth)
Thyrsanthemum, thyrsanthus -a -um Wand-flowered-one, with flowers borne in a thyrse, θυρσος-ανθεμιον
thyrsiflorus -a -um with thyrsoid inflorescences, *thyrsus-florum*
thyrsoideus -a -um with a pyramidal panicle-, thyrsoid, θυρσος-οειδης (see Fig. 3d)
Thyrsopteris Thyrsoid-fern, θυρσος-πτερον (the sori are in racemose bunches)
Thyrsostachys Wand-spiked-one, θυρσος-σταχυς
thysano-, thysanoto- fringed-, θυσανωτος, θυσανο-
Thysanocarpus Fringed-fruit, θυσανο-καρπος (the fringed capsule)
Thysanoglossa Fringed-tongue, θυσανο-γλωσσα (the labellum)
Thysanolaena Fringed-cloak, θυσανο-χλαινα (the upper lemma)
Thysanotus Fringed, θυσανωτος (the inner perianth); Fringed-ear, θυσανο-οτος
Tianshaniella, tianschanicus -a -um from Tien Shan (Tian Shan), central Asia
Tiarella Little-turban, τιαρα (the shape of the capsules)
tiarelloides resembling Tiarella, τιαρα-οειδης
tibae from Tivoli (*Tibur*); of flutes, *tibia, tibiae*

tibesticus -a -um from the Tibesti Massif, Chad–Libya–Niger, central Sahara region
tibetanus -a -um, tibeticus -a -um from Tibet, Tibetan
tibicinis piper's or flute-player's, *tibicina, tibicinae*
tibicinus -a -um hollow-reed-like, flute-like, *tibia, tibiae*
Tibouchina from a Guyanese vernacular name
Tibouchinopsis Resembling-*Tibouchina, Tibouchina-opsis*
Ticodendron Little-tree, botanical Latin from the Spanish vernacular diminutive, tico, and δενδρον
Ticoglossum Small-tongue, botanical Latin from a Spanish vernacular diminutive suffix, tico, and γλωσσα
Ticorea from a Guyanese vernacular name for *Ticorea foetida*
ticus -a -um fecund, fruitful, bringing forth, τικτω
Tidestromia for Ivar Frederick Tidestrøm (1864–1956), Swedish-American botanist
Tieghemella for Phillippe Eduard Leon van Tieghem (1839–1914), French botanist, professor at Paris
tienschanicus -a -um from Tien Shan, central Asia
tigerstedtii for Carl Gustaf Tigerstedt (1886–1957), Finnish dendrologist
tiglius -a -um the apothecaries' name, *grana tiglia*, for the purgative croton oil (tigline) from *Croton tiglium*
Tigridia, tigridus -a -um Tiger, *tigris* (the colour marking of the perianth)
tigrinus -a -um tiger-toothed, striped, spotted, *tigris*
tigurinus -a -um from Zurich (*Turicum, Tigurinus*)
tikunorus -a -um of the Amazonian Tukuna (Tikuna) people
Tilia Wing (the ancient Latin name, *Tilia*, for the linden tree, ελατη) (***Tiliaceae***)
Tiliacora from a Bengali vernacular name
tiliae-, tiliaceus -a -um lime-like, resembling *Tilia*
tiliarus -a -um of lime (*Contarinia*, a dipteran gall midge on *Tilia*)
tiliifolius -a -um, tilliaefolius -a -um having leaves resembling those of *Tilia, Tilia-folium*
Tillaea for Michelangelo Tilli (1655–1740), Professor of Botany at Pisa (≡ *Crassula*)
tillaea from a former generic name, *Tillaea*
Tillandsia for Elias Erici Tillands (Tillandz, Tillander) (1640–93), Finnish botanist and Professor of Medicine at Abo and Uppsala
timetius -a -um precious, honoured, valued, τιμηεις, τιμιος
Timonius from a Malayan vernacular name
Tinantia for François Auguste Tinant (1803–58), of Luxembourg
tinctorius -a -um used for dyeing, *tingo, tingere, tinxi, tinctum*
tinctorum of the dyers, *tinctor, tinctoris*
tinctus -a -um coloured, dipped, imbued, *tingo, tingere, tinxi, tinctum*
Tinea for Vincenze Tineo (1791–1856), Sicilian Professor of Botany at Palermo
tingens dyeing, staining, present participle of *tingo, tingere, tinxi, tinctum*
tingitanus -a -um from Tangiers, Morocco (*Tingis, Tingitanus*)
tini- *Tinus*-like (*Viburnum*-like)
Tinnea for the three Dutch ladies, Henriette Tinne, her daughter Alexandria Tinne, and her sister Adrienne van Calellen, on the 1863 Nile Expedition
tinus the old Latin name, *tinus, tini*, for *Laurustinus* (≡ *Viburnum*)
Tipuana from the Tipuani valley, Bolivia (provenance of pride of Bolivia, *Tipuana tipu*)
Tipularia Cranefly orchid, *tippula, tippulae* water-spider (tenuous perianth segments)
tipuliformis -is -e resembling a cranefly or tipulid (*tippula, tipula*)
tipuloides cranefly- or tipulid-like, *Tipula-oides*
Tiquilia a S American vernacular name for 'crinklemat' desert plants
tirolensis -is -e from the Tyrol, Austria, Tyrolean
tirucalli a Malayan vernacular name for a latex
titan gigantic, τιτανικος, for the race of giant gods of mythology, *Titan, Titanis*
titania exceptional; for Titania, Shakespeare's Queen of the Fairies

titano- chalk-, lime-, τιτανος
Titanopsis Sun-like, τιταν-οψις (resemblance of the flower to a small sun)
Titanotrichum Lime-hair, τιτανος-τριχος (has lime secreting hairs)
titanotus -a -um chalky-eared, τιτανος-ωτος
titanus -a -um of the Titans, massive, gigantic, very large, τιτανικος
Tithonia after Tithonus of Greek mythology, son of Laomedon, brother of Priam and favourite of the goddess Aurora
tithymaloides spurge-like, τιθυμαλλος-οειδης
Tithymalus (*Tithymallus*) an ancient name, τιθυμαλλος, used for plants with latex, spurges
-tmemus -a -um -cut, -τμημα
Tmesipteris Separate-wing, (τμηγω, τμημα, τμησις)-πτερον
tmoleus -a -um from the Tmolus mountain, Lydia, Turkey
Tobagoa from Tobago, W Indies
tobira the Japanese vernacular name, tobera, for the timber of *Pittosporun tobira*
toco- offspring-, τοκος
Tococa the vernacular name, tococo, from Guiana
Tocoyena from the vernacular name in Guiana
todayensis -is -e from Todaya, Mindanao, Philippines
Toddalia from the Malabar vernacular, kaka-toddali, for lopez-root
Todea for Henrich Julius Tode (1733–97), German cleric and mycologist, author of *Fungi Mecklenbergensis* (1790)
Toechima Wall-covered, τοιχος-ειμα
tofaceus -a -um of tufa, tufa-coloured, gritty, *tofus, tofi*
Tofieldia for Thomas Tofield (1730–79), Yorkshire naturalist
togatus -a -um robed, gowned, *togatus*
togoicus -a -um from Togo, W Africa
tokyoensis -is -e from Tokyo, Japan
Tolmiea, tolmiei for Dr William Fraser Tolmie (1812–86), Scottish surgeon of the Hudson Bay Company
tolminsis -is -e from Tolmin, former Yugoslavia
tolonensis -is -e from Toulon (*Tolona*), France
Tolpis an Adansonian name of uncertain derivation
toluiferus -a -um producing balsam of tolu, *tolu-fero* (*Myroxylon toluifera*)
Tolumnia etymology uncertain (Tolumnius was a soothsayer in Virgil's *Aeneid*)
Tolypella Little-forceful-one, τολυπευω (compact growth habit of *Tolypella prolifera*)
tolypephorus -a -um wound up into a ball, bearing convolutions, τολυπε-φορος
tommasinianus -a -um, tommasinii for Muzio Spirito de Tommasini (1794–1879), Italian botanist
tomentellus -a -um somewhat hairy, diminutive from *tomentum*
tomentosus -a -um thickly matted with hairs, *tomentum* (padding)
tomi-, -tomus -a -um cutting-, -cut, -incised, τομη
-tonae -elongated, stretched, τονος (gerund form of τεινω, to stretch)
Tonestus an anagram of *Stenotus*
tongaensis -is -e, tongensis -is -e from Tonga, SW Pacific
tonkinensis -is -e from the area of the Gulf of Tonkin, Vietnam
tonsus -a -um shaven, sheared, shorn, *tondeo, tondere, totondi, tonsum*
Toona, toona from an Indian vernacular name for *Cedrella toona* (*Toona ciliata*)
tophaceus -a -um *vide tofaceus*
topiarius -a -um of ornamental gardens, *topiarius*
topo- place-, locality-, τοπος (refers to a definable population grouping)
tordylioides similar to *Tordylium*
Tordylium the name, τορδυλιον, used by Dioscorides for an umbellifer
Torenia for Reverend Olof Torén (1718–53), chaplain in India, Surat and China to the Swedish East India Company
torfosus -a -um growing in bogs, *torfosus*
Torilis a meaningless name by Adanson

toringo a Japanese vernacular name for a *Malus*
toringoides toringo-like, *toringo-oides*
Tormentilla, tormentillus -a -um an ancient Latin name, anguish, torment (the powdered rhizome of *Potentilla erecta*, tormentil, was used to treat diarrhoea etc.)
torminalis -is -e of colic, *tormina, torminum* (used medicinally to relieve colic)
torminosus -a -um (subject to) causing colic, *tormina, torminum*
tornatus -a -um rounded off, turned, *torno, tornare, tornavi, tornatum* (the coiled legumes)
torosus -a -um cylindrical with regular constrictions, *torosus* (literally, muscular)
torquatus -a -um with a (chain-like) collar, necklaced, *torques, torquis*
Torreya, torreyanus -a -um for Dr John Torrey (1796–1873), American botanist, significant contributor to the *Flora of North America*
Torreyochloa Torrey's-grass (*vide supra*)
Torricellia for Evangelista Torricelli (1608–47), Italian physicist and microscopist, inventor of the mercury barometer
torridus -a -um frost-bitten, dried up, of very hot places, *torridus*
torti-, tortilis -is -e, tortus -a -um twisted, *torqueo, torquere, torsi, tortum*
tortilipetalus -a -um having twisted petals, botanical Latin from *tortilis* and πεταλον
Tortula Twisted, *tortus* (the 32 spirally twisted teeth of the peristome)
Torulinium Tuft-like, *torulus, toruli*
tortuosus -a -um with complicate or winding stem growth, possessive of *tortus*
tortus -a -um complicated, meandering, winding, *tortuosus* (irregularly twisted stems)
torulosus -a -um muscular, swollen or thickened at intervals, tufted, possessive of *torulus, toruli*
torus -a -um ornamental, mounded, bulging, knotted, *torus, tori* (fruits)
torvus -a -um fierce, harsh, sharp, *torvus*
tosaensis -is -e from the area around Tosa Bay, Japan
totarus -a -um from a New Zealand Moari name, totara
Tournefortia for Joseph Pitton de Tournefort (1656–1708), renaissance plant systematist, author of *Institutiones rei herbariae* (1710)
Tourrettia for Marc Antoine Louis Claret de la Tourrett (1729–93), French naturalist and writer
tovarensis -is -e from the Sierras Tovar, Venezuelan Andes
Tovaria for Simon de Tovar, Spanish physician and botanist
Townsendia, townsendii for David Townsend (1787–1858), Pennsylvanian botanist, USA
Townsonia for William Townson (1850–1926), English botanist in New Zealand
toxi-, toxicarius -a -um, toxic, *toxicum*, containing a poisonous principle (τοξικον-φαρμακον, poison for an arrow)
Toxicodendron, toxicodendron Poison-tree, τοξικον-δενδρον (≡ *Rhus, Anacardiaceae*)
Toxicodendrum Poison-tree, τοξικο-δενδρον (≡ *Hyaenanche*, υαινα-αγχω, *Euphorbiaceae*)
toxicus -a -um poisonous, *toxicum*
toxifera -a -um poisonous, poison-bearing, *toxicum-fero*
toxispermus -a -um having poisonous seeds, τοξικον-σπερμα
Toxocarpus Poison-fruit, τοξικον-καρπος
toza from a S African native name
Tozzia for L. Tozzi (1663–1717), Italian botanist
trabeculatus -a -um cross-barred, diminutive from *trabs, trabis*, beam
Trachelium, trachelium Neck, τραχηλος (old name for a plant used for throat infections, *Campanula trachelium*, throatwort)
trachelo- neck-, τραχηλος, τραχηλο-
Trachelospermum Necked-seed, τραχηλος-σπερμα
trachy- shaggy-, rough-, τραχυς, τραχυ-, -τραχεια
Trachyandra Rough-stamens, τραχυς-ανηρ
trachyanthus -a -um having shaggy flowers, τραχυς-ανθος

Trachycalymma Rough-covering, τραχυς-καλυμμα
Trachycarpus Rough-fruit, τραχυς-καρπος
Trachymene Rough-membrane, τραχυς-μενινξ (ridged fruit wall, ≡ *Didiscus*)
trachyodon short-toothed, rough-toothed, τραχυς-οδων
Trachyphrynium Rough-*Phrynium* (the rough-coated fruit)
trachyphyllus -a -um rough-leaved, τραχυς-φυλλον
trachypodus -a -um rough-stalked, τραχυς-ποδος
Trachypogon Shaggy-bearded, τραχυς-πωγων
Trachypteris Rough-fern, τραχυς-πτερον
Trachyspermum Rough-seed, τραχυς-σπερμα
Trachystemon Rough-stamen, τραχυς-στεμων (hairy filaments)
Tracyina for Joseph Prince Tracy (1879–1953), Californian botanist
Tradescantia for Old John Tradescant (1567–1638), gardener to Charles I, and his son John Tradescant (1608–62), both travellers and collectors (their collection, Tradescant's Ark, was the basis of the Ashmolean Museum, Oxford)
tragacantha yielding gum-tragacanth (from a Greek plant name, τραγακανθα, goat-thorn, *Astragalus tragacantha*), also used for a section of *Astragalus*
tragacanthoides resembling *Tragacantha, Tragacantha-oides*
Traganum Of-goats, τραγος (grazing pastures)
Tragia for Hieronymus Tragus (Jerome Bock) (1498–1554), German Lutheran physician and herbalist, author of *Kreuter Buch* (1539)
trago- goat-, τραγος, τραγο-
tragoctanus -a -um goat's-bane, τραγος-κτονος
tragophyllus -a -um with leaves resembling those of *Tragus*, τραγος-φυλλον
Tragopogon Goat-beard, τραγος-πωγων (Theophrastus' name, τραγοπωγων, refers to the pappus of the fruit)
tragopogonis -is -e of goat-beard, living on *Tragopogon* (*Aulacidea*, hymenopteran gall wasp)
Tragopyrum Goat-wheat, τραγος-πυρος (≡ *Atraphaxis*)
Tragus, tragus Goat, τραγος (Dioscorides' name for the plants they eat; burr grass)
traillianus -a -um for James William Helenus Trail (1851–1919), Professor of Botany at Aberdeen
trajectilis -is -e, trajectus -a -um bridging, passing over, *traicio, traicere, traieci, traiectum* (separation of anther loculi by the connective)
tranquebaricus -a -um from Tranquebar, Tamil Nadu state, India
tranquillans calming, present participle of *tranquillo, tranquillare*
trans- through-, beyond-, across-, *trans*
transalpinus -a -um crossing the Alps, *trans-(alpes, alpium)*
transbaikalicus -a -um from the Siberian region E of Lake Baikal (Baykal)
transcaspius -a -um from the Transcaspian region of the former USSR (now Turkmenistan, SW Kazakhstan, W Uzbekistan)
transcaucasicus -a -um from the region of the former USSR called Transcaucasia (now Georgia, Armenia, Azerbaijan)
transens latticed, intertwined, *transenna, transennae*
transhyrcanus -a -um beyond the ancient region of Hyrcania, *trans-hyrcania* (east of the Caspian Sea)
transiens surpassing, passing-over, *transeo, transire, transii, transitum*
transitorius -a -um between, intermediate, transitory, *transitus*
translucens almost transparent, allowing some light to pass, present participle of *transluceo, translucere*
transmontanus -a -um across or beyond the mountains, *trans-montanus*
transmutatus -a -um not constant, shifting or changing, *transmuto, transmutare*
transparens shining-through, permitting light penetration, present participle from *trans-(pareo, parere, parui, paritun)*
transsilvanicus -a -um, transsylvanicus -a -um from Transylvania, Romania
transvaalensis -is -e from Transvaal, S Africa
transversus -a -um athwart, across, collateral, transverse, *transversus*

transwallianus -a -um from Pembroke, S Wales (*Transwallia*, beyond Wales)
Trapa from *calcitrapa*, a four-spiked weapon used in battle to maim cavalry horses' hooves (for the horned fruit of water chestnut) (***Trapaceae***)
Trapella diminutive of *Trapa*
trapeziformis -is -e lozenge-shaped, trapezoid, Latin from τραπεζιον (pinnae)
trapezioides lozenge-shaped, shaped like a deformed square, trapezoid, τραπεζιον-οειδης
Traubia for Hamilton Paul Traube (1890–1983), American botanist
Traunsteinera, traunsteineri for Joseph Traunsteiner (1798–1850), Austrian pharmacist of Kitzbühel, student of the Tyrolean flora
Trautvetteria, trautvetteri for Ernst Rudolf von Trautvetter (1809–89), Prussian botanist, Director of St Petersburg Botanic Garden
travancorensis -is -e, travancoricus -a -um from Travancore, S India
traversii for William Thomas Locke Travers (1819–1903), Irish botanist in New Zealand
treculeanus -a -um *Treculia*-like
Treculia for Auguste Adolphe Lucien Trécul (1818–96), French botanist and explorer
Trema Aperture, τρημα (the pitting of the testa)
Tremacron Large-aperture, τρημα-μακρον (anther dehiscence)
Tremandra Shaking-man, (*tremor, tremere, tremui*)-*andrus* (the versatile stamens) (***Tremandraceae***)
Trematolobelia Holed-*Lobelia*, botanical Latin from τρημα and *Lobelia* (the capsule)
Tremella Quiverer, τρεμω, *tremo, tremere, tremui* (jelly-like mass)
tremellosus -a -um trembling, diminutive of *tremulus* (the gelatinous fruiting body)
tremuloides aspen-like, resembling *Populus tremula, tremula-oides*
tremulus -a -um trembling, shaking, quivering, *tremulus*
trepidus -a -um restless, trembling, *trepido, trepidare, trepidavi, trepidatum*
Trevesia for the family Treves de Bonfigli of Padua
Trevoa for Señor Trevo, Spanish botanist
Trevoria for Sir James Trevor Lawrence (1831–1913), English orchidologist, President of RHS
Trewia for Christoph Jakob Trew (1695–1769), German physician, botanist and explorer
tri- three-, τρεις, τρια, τρι-, *tres-*
triacanthophorus -a -um bearing spines in threes, triple-thorned, τρια-ακανθος-φορα
triacanthos, triacanthus -a -um three-spined, τρι-ακανθα
triactinus -a -um three-rayed, three-splendoured, τρι-(ακτις, ακτινος)
triadelphus -a -um with the stamens grouped into three bundles, τρια-αδελφος
Triadenum Three-glanded-one, τρια-αδην (staminodes)
triandrus -a -um three-stamened, having three stamens in the flower, τρι-ανηρ
triangulari-, triangularis -is -e, triangulatus -a -um three-angled, triangular, *tri-angulus, anguli* (leaves)
triangularivalvis -is -e with triangular valves, *tri-angulus-valvae* (of the fruit or fruiting head)
triangulus -a -um triangular, *tri-angulus* (leaflets)
Triantha, trianthus -a -um Three-flowered, τρι-ανθος
Trianthema Triple-flowered, τρεις-ανθεμιον
triaristus -a -um with three awns, *tri-arista*
Trias Three-partite, τρεις, τρια (the arrangement of the glassy green flowers)
Tribulus Three-lobes, τριβολος (the shape of the fruit)
Tricalysia Three-coverings, τρι-καλυξ (corolla, calyx and cupular bracteoles)
tricamarus -a -um having three vaults, three-chambered, τρι-καμαρα
Tricardia Three-hearts, τρια-καρδια (the sepals)
tricarinatus -a -um with three keels or ridges, *tri-(carina, carinae)*
tricarpellatus -a -um having fruits of three carpels, τρι-καρπος, *tri-carpellum*
triceps having three heads, *tri-ceps*

Triceratorrhynchus Three-horned-nose, τρια-κερατος-ρυγχος (the rostellum)
Trichantha Hair-flower, (θριξ, τριχος)-ανθος (the fine pedicels)
trichanthus -a -um with hairy flowers, τριχος-ανθος
Trichilia Three-partite, τριχα- (the three-celled ovary)
Trichinium Hair-covered, τριχινος (literally, of hair)
Trichloris Triple-*Chloris* (three-awned)
tricho-, -trichus -a -um, trich- hair-like-, -hairy, θριξ, τριχος, τριχη
trichocarpus -a -um with a hairy ovary, τριχος-καρπος
Trichocaulon, trichocaulon Hairy-stem, τριχος-καυλος
Trichocentrum Hair-spur, τριχος-κεντρον (the spur of the labellum)
Trichocereus Hairy-*Cereus*, τριχος-*Cereus* (the areoles)
Trichoceros Hairy-horns, τριχος-κερας (lateral elongate processes on the column)
Trichocladus, trichocladus -a -um Hairy-branched, τριχος-κλαδος
Trichocolea Hairy-sheath, τριχος-κολεος (to the gametophyte)
Trichocoronis Hairy-crowned, botanical Latin from τριχος and *corona*
trichodes of hairy appearance, τριχος-ωδης
Trichodesma Hair-bound, τριχος-δεσμα (the anthers are intermingled with hairs)
Trichodiadema Bristled-circlet, τριχος-διαδημα (the fringing bristles of the perianth parts)
trichodium hair-like, τριχος-ωδης
trichoglossus -a -um hairy-tongued, τριχος-γλωσσα
Trichoglottis Hairy-tongue, τριχος-γλωττα (the labellum)
Trichogonia Hairy-ridged, τριχος-γωνια (the pseudo-nuts)
trichoides hair-like, τριχος-οειδης
Tricholaena Hairy-cloak, τριχος-(χ)λαινα (the hair-covered spikelets)
tricholepis -is -e hairy-scaled, τριχος-λεπις (small, densely pruinose leaves)
Trichomanes, trichomanes Hair-scarcity, τριχος-μανος (Theophrastus' name for maidenhair spleenwort) (the protrusive soral axes)
trichomanifolius -a -um maidenhair-leaved, *Trichomanes-folium*
trichomanoides maidenhair-like, *Trichomanes-oides*
Trichonema Hair-threads, τριχος-νημα (the anthers, ≡ *Romulea*)
Trichoneura Hair-nerved, τριχος-νευρα (the excurrent awn point of the lemmas)
Trichoon Hairy-egg, τριχος-ωον (*Phragmites*)
Trichopetalum Hairy-petalled-one, τριχος-πεταλον
Trichophorum, trichophorus -a -um Hair-carrier, τριχος-φορα (perianth bristles)
trichophyllus -a -um with hair-like leaves, τριχος-φυλλον
Trichopilia Hairy-cap, τριχος-πιλος (a small felted cap covers the anther)
Trichopteryx Hair-winged, τριχος-πτερυξ (the upper lemmas have a tuft of hair near each margin)
Trichoptilium Hair-winged-one, τριχος-πτιλον (the receptacular scales)
Trichopus Hairy-stalked-one, τριχος-πους (***Trichpodaceae***)
trichorhizus -a -um hair-rooted, with fine roots, τριχος-ριζα
Trichosacme Hair-point, τριχος-ακμη (the corolla has a hairy appendage)
Trichosalpinx Hairy-trumpet, τριχος-σαλπιγξ (the sheaths)
Trichosandra Bristly-anthers, τριχος-ανηρ
Trichosanthes Hair-flower, τριχος-ανθος (the fringed corolla lobes of serpent cucumber)
trichosanthus -a -um with hairy flowers, τριχος-ανθος
Trichoscypha Hairy-cup, τριχος-σκυφος
Tichospermum, trichospermus -a -um Hairy-seeded, τριχος-σπερμα
Trichostachys Slender-spike, τριχος-σταχυς
Trichostema Slender-wreath, τριχος-στεμον
Trichostigma Hairy-stigma, τριχος-στιγμα
trichostomus -a -um hairy-mouthed, with a hairy-throated flower
trichotocephalus -a -um three-forked-headed, τριχος-τομη-κεφαλη (trichotomous inflorescences)
trichotomus -a -um three-forked, triple-branched, τριχος-τομη

Trichotosia Hairy, τριχωμα, τριχωτο (the indumentum)
tricoccus -a -um three-seeded, three-berried, τρι-κοκκος
tricolor three-coloured, *tri-color*
tricornis -is -e, tricornutus -a -um with three horns, *tri-cornus*
tricostatus -a -um with three ridges, three-ribbed, *tri-costatus*
Tricuspidaria Triple-tooth, *tri-cuspidis* (the petals) (≡ *Crinodendron*)
tricuspidatus -a -um with three teeth, three-toothed, *tri-cuspidis* (petals)
tricuspis -is -e with three points, *tri-*(*cuspis, cuspidis*)
Tricyrtis Three-domes, τρι-κυρτος (the nectaries on the bases of the three outer tepals)
Tridactyle Three-fingered, τρι-δακτυλος (the lobes of the lip)
tridactylites, tridactylus -a -um three-fingered, τρι-δακτυλος-ιτης
Tridax Three-toothed, θριδαξ, θριδακος (Theophrastus' name for a lettuce; ligulate florets are often three-fid)
Tridens, tridens Three-toothed, three-pronged, *tri-*(*dens, dentis*)
tridentatus -a -um, tridentinus -a -um three-toothed, three-pronged, *tri-*(*dens, dentis*)
Tridesmostemon Three-banded-stamen, τρι-δεσμος-στεμων
triduus -a -um three days, *triduum, tridui* (the flower)
triennialis -is -e, triennis -is -e lasting for three years, *triennium* (plant duration)
Trientalis, trientalis -is -e Four-inches, a third of twelve, a third of a foot in length, *triens, trientis* (Cordus', *herba trientalis*, signifying the stature of wintergreen)
trifasciatus -a -um three-banded, *tri-fasciatus*
trifidus -a -um divided into three, three-cleft, *trifidus*
triflorus -a -um three-flowered, *tri-flora*
trifoliatus -a -um trifoliate, having three leaflets, *tri-folium* (see Fig. 8d)
trifolii of clover, living on *Trifolium* (*Dasyneura*, dipteran gall midge)
trifoliobinatus having two trifoliate leaflets, *tri-folium-binatus*
trifoliolatus -a -um having three-leaflets, *tri-foliola*
Trifolium Trefoil, *tri-folium* (the name in Pliny for trifoliate plants)
trifolius -a -um with three leaflets, trifoliate, *tri-folium*
trifurcatus -a -um three-forked, divided into three equal parts, *tri-*(*furca, furcae*)
trigintipetalus -a -um having (about) thirty petals, *triginta-petalum*
triglans three-nutted-fruits, containing three nuts, *tri-glans*
triglochidiatus -a -um with three-barbed bristles, botanical Latin from τρι-γλωχις, *tri-glochidium*
Triglochin Three-angled, τρι-γλωχις (the fruits)
triglumis -is -e having three glumes, *tri-gluma*
Trigonachras Three-angled-pear, botanical Latin from τρι-γωνια and *acras*
Trigonella Triangle, feminine diminutive of τρι-γωνια (the perianth of fenugreek seen from the front)
Trigonia Three-angled, τρια-γωνια (the fruit)
Trigonidium Triangular-form, τρι-γωνον-ειδος
trigoniflorus -a -um having petals forming a triangle, botanical Latin from τρι-γωνον and *florum*
Trigonobalanus Three-angled-nut, τρι-γωνια-βαλανος (the fruit)
Trigonophyllum Three-cornered-leaf, τρι-γωνια-φυλλον
trigoniophyllus -a -um having triangular leaves, τρι-γωνια-φυλλον
Trigonospermum Three-angled-seed, τρι-γωνια-σπερμα
Trigonostemon Three-edged-stamen, τρι-γωνια-στεμων
Trigonotis Triangular-eared, τρι-γωνον-ωτος (bracts)
trigonus -a -um three-angled, with three flat faces and angles, τρι-γωνια
trigynus -a -um with a three-partite ovary, τρι-γυνη
trijugus -a -um three-yoked, united in threes, *tri-*(*iugum, iugi*)
trilamellatus -a -um with three thin plates or scales, three-layered, *tri-lamellatus*
Trilepis Three-scaled, τρι-λεπις (the perianth scales)
Trilidium *Trilisa*-like
trilineatus -a -um marked with three lines, *tri-lineatus*

Trilisa Tripled, τριλιξ (the pappus structure)
Trillidium Little-*Trillium* (diminutive suffix, floral resemblance)
Trillium Triple-lily, τρι-λιριον (the parts are conspicuously in threes, lily-like) (***Trilliaceae***)
trilobatus -a -um, trilobus -a -um three-lobed, τρι-λοβος
trilocularis -is -e with a three-chambered ovary or fruit, *tri-loculus*
trilophus -a -um with three crests, τρι-λοφια
trimaculatus -a -um marked with three spots, three-spotted, *tri-*(*macula, maculae*)
trimerus -a -um with a series of three parts, τρι-μερος (floral organs)
trimestris of three months, maturing in three months, *tri-mensis* (*Lavatera trimestris*) (cf. semester, from German, from Latin *semestris*)
Trimeza Three-big(-ones), τρι- μειζων (comparative of μεγας; for the larger outer tepals)
Trimorpha Three-forms, τρι-μορφη (capitulae)
trimus -a -um lasting three years, three years old, *trimus*
trinervis -is -e, trinervius -a -um three-nerved, *tri-nervus* (three-veined leaves)
trineus -a -um three fold, *trinus* (floral structure)
Trinia for Karl Bernhard Friehher von Trinius (1778–1844), German physician and botanist in St Petersburg, Russia
triniifolius -a -um having leaves resembling those of *Trinia*
Triniochloa Trinius'-grass, botanical Latin from Trinius and χλοη
trinus -a -um in threes, *trini, trinorum*
Triodanis etymology uncertain
Triodia Three-toothed, τρεις-οδους (the apex of the lemmas, ≡ *Sieglingia*)
Triolena Three-armed, τρεις-ολενε (the appendages of the stamens)
trionus -a -um three-clawed, τρεις-ονυξ
trionychon three-clawed, τρεις-ονυχος (the bract and paired bracteoles)
Triopteris Three-winged, τρεις-πτερον (the three-winged samaras)
triornithophorus -a -um bearing three birds, the flower-heads, τρεις-(ορνις, ορνιθος)-φορα (*Linaria* flower-heads – often has four flowers)
Triosteum Three-bones, τρεις-οστεον (the three bony nutlets)
tripartitus -a -um divided into three segments, *tripartitus, tripertitus*
Tripetaleia Three-petals, τρεις-πεταλον (the tripartite floral arrangement)
tripetaloides similar to *Tripetaleia*, τρεις-πεταλον-οειδης, *Tripetaleia-oides*
tripetalus -a -um three-petalled, τρεις-πεταλον
Triphasia Triple, τριφασιος (floral parts)
Triphora Bearing-three(-lobes), τρεις-φορα (the labellum)
triphyllos three-leaved, with three leaflets, τρεις-φυλλον
triphyllus -a -um three-leaved, τρια-φυλλον
Triphyophyllum Growing-three-leaf(-forms), τρεις-φυη-φυλλον (one with two apical hooks, one entire and one filamentous with *Drosera*-like glands) (*Dioncophyllum* produces only two; ***Dioncophyllaceae***)
Triphysaria Three-bladdered-one, τρεις-φυσα (pouches of lower lip)
tripinnatus -a -um having thrice-pinnate leaves, *tri-pinnatus*
Triplachne Triple-scaled, τριπλους-αχνη
Triplaris In-threes, *triplaris* (the floral parts)
Triplasis Three-times-more, τριπλασιος (the lemmas have an awn and two subulate lobes)
Tripleurospermum Three-ribbed-seed, τρεις-πλευρον-σπερμα (the achene has three ribs)
tripli-, triplo- triple-, threefold-, τριπλους
triplinervis -is -e with three veins (leaves)
Triplochiton Three-coverings, τριπλους-χιτων (the flowers have a series of petaloid staminodes within the staminal ring, forming the third layer)
Triplochlamys Three-covers, τριπλους-χλαμυς, τριπλο-
Triplophyllum Triple-leaved-one, τριπλους-φυλλον (primary frond division)
Triplotaxis Three-ranked, τριπλους-ταξισ (the involucral bracts)

Tripogandra Three-hairy-stamens, τρεις-πωγων-(ανηρ, ανδρος) (bearded filaments)
Tripogon Three-beards, τρεις-πωγων (the tufts at the base of the lemma veins)
Tripolium, tripolium Theophrastus' name, τριπολιον, for a plant with three times the strength of *Teucrium polium*
Tripsacum Three-fragments, τρεις-ψακας (disarticulation of the fruiting head)
tripteranthus -a -um having three-winged flowers, τρι-πτερυξ-ανθος
Tripteris, tripteris -is -e Three-winged, τρεις-πτερυξ (the seed; various structures)
Tripterygium Three-wings, τρεις-πτεριγιον (the three-winged fruits)
Tripterocalyx Three-winged-calyx, τρεις-πτερο-καλυξ
Tripterospermum Three-winged-seeded-one, τρεις-πτερο-σπερμα
Tripterygium Triple-winged, τρεις-πτερον (the fruits)
Triptilion Three-wings, τρι-πτιλον (the pappus divisions)
tripyrenus -a -um having a fruit of three pyrenes, τρι-πυρην, *tri-pyrena* (stones of drupes)
triqueter, triquetrus -a -um Sicilian; three-cornered, three-edged, three-angled, *triquetrus* (stems)
triquetrifolius -a -um having three-angled leaves, *triquetrus-folium*
triquinatus -a -um divided into three and then into five lobes, with three groups of five, *tri-quinatus*
Triraphis Three-needled-one, τρι-ραφις (lemmas have an awn and two awn-like excurrent lateral nerves)
Trisepalum Three-sepalled-one, τρεις-σκεπη
triserialis -is -e with series or successions of three (structures), *tri-(series, seriem)*
triserratus -a -um triple-toothed, *tri-serratus*
Trisetaria Three-awned, *tri-saeta*
Trisetella Three-small-hairs, *tri-saetella* (feminine diminutive suffix from *saeta*) (the apices of the sepals)
Trisetum Three-awns, *tri-saeta*
Trismeria Three-partite, τρεις-μερις
trispermus -a -um three-seeded, τρεις-σπερμα
Tristachya Three-spikeleted, τρεις-σταχυς (the spikelets are in triads along a narrow raceme or panicle)
Tristagma Three-droppers, τρι-σταγμα (the septal nectaries of the ovary)
Tristania for Marquis Jules Marie Claude de Tristan (1776–1861), French botanist
tristaniicarpus -a -um with *Tristania*-like fruits
Tristaniopsis resembling *Tristania, Tristania-opsis*
Tristellateia Three-starred, τρι-στελλα (the stellate shape of the three fruits)
tristis -is -e bitter, sad, gloomy, dull-coloured, melancholy, *tristis*
trisulcus -a -um, trisulcatus -a -um three-grooved, three-grooved, *tri-(sulcus, sulci)*
Triteleia Triplicate, τρι-τελειος (the flower parts are in threes)
Triteleiopsis Resembling *Triteleia*, τρι-τελειος-οψις
triternatus -a -um three times in threes, with three trifoliate leaflets, *tri-ternatus* (division of the leaves)
Trithrinax Triple-fan, τρι-θριναξ (the leaves)
triticeus -a -um wheat-like, *triticum*
tritici of wheat, living on *Triticum* (symbionts, parasites and saprophytes)
Triticum the classical name, *triticum*, for threshing grain, *tero, tritum*
tritifolius -a -um with polished or rubbing leaves, *tritus-folium*
Tritonia Weathercock, τριτον (the variable disposition of the stamens; Triton was a minor sea god and son of Neptune)
Tritoniopsis Resembling-*Tritonia*
tritus -a -um in common use, past participle of *tero, terere, triti, tritum*
Triumfetta for Giovani Battista Trionfetti (1658–1708), Italian botanist
triumphans exultant, triumphal, celebrating, *triumpho, triumphare, triumphavi, triumphatum*
triumvirati of three commissioners, *triumvir, triumviri* (a group of three senior citizens, *triumvir*, ran a Roman town, suggesting decoration such as mayoral regalia)

Triuris Three-tailed-one, τρεις-ουρα (the subulate extensions of the connectives) (***Triuridaceae***)
trivialis -is -e common, ordinary, wayside, of crossroads, *trivium*
trixago *Trixis*-like, *trixis* with feminine suffix
-trix, -tricis suffix indicating possessive of a feminine plural noun (e.g. *histrico* theatre, *histrix* theatrical, *histricis* of theatricals)
Trixis Triple, τριξος (three-angled fruits)
Trizeuxis Three-yoked, τρι-ζευξις (the three united perianth segments)
Trochetia for René Joachim Henri du Trochet (1771–1847), French plant physiologist who elucidated the phenomenon of osmosis
Trochetiopsis *Trochetia*-like, resembling *Trochetia*, *Trochetia-opsis*
trocho- wheel-like-, hooped-, wheel-, τροχος, τροχο-
Trochocarpa Wheel-fruit, τροχος-καρπος (the radial cells of the fruit)
Trochodendron Wheel-tree, τροχος-δενδρον (the radially spreading stamens) (***Trochodendraceae***)
Trochomeria Part-of-a-wheel, τροχος-μερος (the male flower's radiating filiform petals on their long pedicels)
trochopteranthemus -a -um with flowers resembling electric fans, winged-wheel-flowered, botanical Latin from τροχος-πτερον-ανθεμιον
troglodytarum of cave-dwellers, apes or monkeys (Linnaeus' *Musa troglodytarum* implied inferiority or unsuitability for man; cf. *sapientium* and *paradisiaca*)
troglodytes hole-dwelling, τρωγλο-δυτης (*Troglodytes troglodytes* a wren)
Troglophyton Cave-plant, τρωγλο-φυτον (habitat)
trojanus -a -um from Troy, Trojan
Trollius Closed-in-flower (Gesner's name, *flos trollius*, from the Swiss-German name, trollblume, for Gerard's globe flower)
troodi from Mount Trudos, Cyprus
Tropaeolum Trophy, τροπαιον, *tropaeum* (the gardener's *Nasturtium* was likened by Linnaeus to the routed losers' shields and helmets displayed in the manner of the Greeks after victories in battle) (***Tropaeolaceae***)
Trophaeastrum Somewhat like a trophy, botanical Latin from *tropaeum* with *astrum*
Trophis Food, τροφη (eaten by cattle)
-trophus -a -um -nourished, τροφις, τροφοεις
tropicalis -is -e, tropicus -a -um of the tropics, tropical, τροπη, τροπικος, *tropicus*
Tropidia Little-keel, diminutive of τροπις
Tropidocarpum Keeled-fruit, botanical Latin from τροπις and καρπος (each of the two halves of the capsule has a keel)
tropicus -a -um of the tropics, tropical, between the sun's turning points, or latitudes called Cancer and Capricorn, via Middle English from τροπη, τροπικος
-tropis -keeled, τροπις
-tropus -a -um -turning, τροπη
Trudelia for Nikolaus Trudel, Swiss orchidologist
trullatus -a -um shaped like a bricklayer's trowel, *trulla, trullae*
trulliferus -a -um bearing (leaves) shaped like a scoop or trowel, (*trulla, trullae*)-*fero*
trullifolius -a -um with trowel-shaped leaves (*trulla, trullae*)-*folium*
trullus -a -um, trullis -is -e ladle-shaped, scoop-shaped, *trulla, trullae*
truncatulus -a -um, truncatus -a -um cut off, blunt-ended, *trunco, truncare, truncavi, truncatum* (the apex of a leaf) (see Fig. 7d)
truncicolus -a -um tree-trunk dweller, (*truncus, trunci*)-*colo*
trunciflorus -a -um having truncated corollas, *trunco-florum*
truncorus -a -um of tree-stumps, *truncus, trunci*
Trymalium Eye-of-the-needle, τρυμαλια (the perforations of the capsule)
tsangpoensis -is -e from the area of the Brahmaputra, or Tsang Po river, Tibet
tschonoskii for Chonosuka Sugawa Tschonoski, who collected for Maximonwicz in Japan *c.* 1873
tsinghaicus -a -um from Ching-hai (Tsinghai), Tibetan highlands, NW China

Tsingia from the Madagascan vernacular name, tsingy, for the limestone areas that are the generic provenance
tsintauensis -is -e from Tsingtau, Kwangsi Chuan, China
Tsuga from the Japanese vernacular name for the hemlock cedar
tsugetorus -a -um of *Tsuga* associations
tsugifolius -a -um with *Tsuga*-like leaves, *Tsuga-folium*
tsu-shimensis -is -e from Tsu-shima, Japan
Tsusiophyllum *Tsusia*-leaved, botanical Latin from the Japanese, tsutsuji, and φυλλον (*Tsusia* is a sectional name in *Rhododendron*)
Tsutsutsi from a Japanese vernacular name, tsutsuji (the name of a section of *Rhododendron*)
tuan a Chinese vernacular name for *Tilia tuan*
tubaeflorus -a -um with trumpet-shaped flowers, *tubae-flora*
tubaestylus -a -um hollow-styled, (*tuba, tubae*)-*stilus*
Tubaria Trumpet-shaped, *tuba, tubae*
tubatus -a -um trumpet-shaped, *tuba, tubae*
Tuber Truffle, *tuber, tuberis*, a tumour, swelling, lump, or truffle (used botanically for an anatomically more or less spheroid organ)
Tuberaria Tuber, *tuber, tuberis* (rootstock of *Tuberaria vulgaris*)
tuberculatus -a -um, tuberculosus -a -um knobbly, warted, warty, tuberculate, diminutive of *tuber* (the surface texture)
tuberculiflorus -a -um having tuberculate flower surfaces, *tuberculatus*
tubergenianus -a -um, tubergenii for Messrs C. G. van Tubergen of Haarlem, Holland, est. 1868, bulb importers and growers
tuberiferus -a -um bearing tubers, *tuber-fero*
Tuberolabium Swollen-lip, *tuber-labium* (the swellings on the labium)
Tyberostylis Swollen-style, botanical Latin from *tuber* and στυλος
tuberosus -a -um swollen, tuberous, comparative of *tuber*
tubi- tube-, pipe-, *tubus, tubi*
tubifer -era -erum, tubulosus -a -um tubular, bearing tubular structures, *tubi-fero*
tubiflorus -a -um with trumpet-shaped flowers, *tubi-florum*
tubiformis -is -e tube-shaped, tubular, *tubi-forma*
Tubilabium Tubular-lip, *tubi-labium*
tubispathus -a -um with a tube-forming spathe, *tubi-spatha*
tubuliformis -is -e narrowly tube-shaped, diminutive of *tubus*, with *forma*
tubulosus -a -um large-tubular, comparative of *tubus*
Tuctoria an anagram of the grass genus *Orcuttia*
tucumaniensis -is -e, tucumanus -a -um from the Tucuman province of Argentina, Argentinian
tuguriorus -a -um of hut-dwellers, *tugurium, tuguri*
tuitans guarding, *tueor, tueri, tuitus; tutus* (of leaves that adopt a sleep-position)
tul- warted-, τυλος
tulbaghensis -is -e from Tulbagh, an early S African township, named for Rijk Tulbagh
Tulbaghia for Rijk Tulbagh (1699–1771), one-time Dutch Governor of the Cape of Good Hope
Tulipa original seed sent by Ogier Gheselin de Busbecq (1522–92), Viennese Ambassador to Suliman the Magnificent, described as tulipan, from the Persian name, dulbend or thoulyban, for a turban
tulipi- tulip-, *Tulipa*-like-
tulipiferus -a -um tulip-bearing, having tulip-like flowers, *Tulipa-fero*
tumacabus -a -um from the environs of Tumaco Island and Tumaco Bay, SW Colombia
Tumamoca from the area of Tumamoc Hill, Arizona, USA
tumefaciens causing swellings, *tume-*(*facio, facere, feci, factum*) (*Agromonas* stem galls on swedes)
tumescens becoming inflated, swelling, tumescent, *tumesco, tumescere, tumescui*

tumidi-, tumidus -a -um swollen, tumid, *tumidus*
tumidicarpus -a -um with swollen fruits, *tumidus-carpus*
tumidinodus -a -um with swollen nodes, *tumidus-nodus*
tumidissinodus -a -um with very tumid or swollen-noded, superlative of *tumidus-nodus*
tumulorum of burial mounds, of tumuli, *tumulus, tumuli*
tunbrigensis -is -e from Tonbridge, Kent
tunguraguae from the area of the Tungurahua volcano, Ecuador
Tunica Undergarment, *tunica* (the bracts below the calyx)
tunicatus -a -um coated, having a covering or tunic, tunicate, *tunica*
tuolumnensis -is -e from Tuolumne river and county, California, USA
Tupa, tupa from a Chilean vernacular name (≡ *Lobelia*)
tupelo swamp-tree, ιτω-οπιλωα (≡ *Nyssa sylvatica*)
tupi-, tupis-, tupus -a -um mallet-like-, *tupi*
Tupidanthus Mallet-flowered-one, *tupis-anthus* (flower-bud form)
Tupistra Mallet, τυπις (the shape of the stigmatic head)
Turbina Whirl, *turbo, turbinis* (the perianth)
turbinatus -a -um, turbiniformis -is -e conical, top-shaped, turbinate, *turbineus*
turbith untidy, crowded, τυρβη
turcicus -a -um from Turkey, Turkish (after the founding ruler Kemal Atatürk)
turcomanicus -a -um, turcumaniensis -is -e from Turkestan, central Asia
turcorus -a -um of the Turks, Turkish (after the founding ruler Kemal Atatürk)
turczaninovii, turczaninowii for Nickolai Stepanovitch Turczaninov (1796–1864), Russian author of *Flora Baicolensis-Dahurica*
turfaceus -a -um, turfosus -a -um growing in bogs, modern Latin *torfaceus, turfaceus*
Turgenia for Alexander Turgenev, Chancellery Director to Prince Gollintzin, in Russia; some derive as *turgeo*, to swell
turgescens becoming distended, becoming turgid, *turgesco, turgescere*
turgidus -a -um bombastic, swollen, inflated, turgid, *turgidus*
turgiphalliformis -is -e erect-phallus-shaped, via seventeenth late Latin *turgidus-phallus-forma*
turio- sucker-, scaly-shoot-, *turio, turionis*
turioniferus -a -um throwing up scaly suckers from ground level, *turionis-fero*
turkestanicus -a -um from Turkestan, S Kazakhstan
turkmenorus -a -um of the Turcoman people, Turkmenistan
Turnera for Reverend William Turner (1508–68), Tudor botanist of Wells, author of *A New Herbal* (1568) (***Turneraceae***)
Turpinia for Pierre Jean François Turpin (1775–1840), French botanical artist
turpis -is -e ugly, deformed, repulsive, *turpis*
Turraea for Georgio della Turra (1607–88), Professor of Botany at Padua
Turraeanthus *Turraea*-flowered, *Turraea-anthus*
Turricula Turreted, diminutive of *turris, turris*
turriculatus -a -um like a high turret or steeple, *turris, turris*
Turrita, turrita, Turritis Tower, towering, *turris, turris* (tower mustard)
turritus -a -um towering, tower-shaped, straight stemmed, *turritus*
Tussacia for F. Richard de Tussac (1751–1837), French botanist, author of *Flora of the Antilles* (1808)
tussilagineus -a -um resembling a small *Tussilago, Tussilago*-like
Tussilago Coughwort, *tussis* with feminine suffix (medicinal use of leaves for treatment of coughs, onomatopoeic)
Tutcheria for William James Tutcher (1867–1920), of the Hong Kong Forestry and Botany Department
tutelatus -a -um, tutelus -a -um protector, guardian, charm, *tutela, tutelae* (*tutamentum*)
Tweedia, tweedii for James Tweed (1775–1862), of Glasgow, who collected for Kew in Argentina
tycho- by chance-, τυχη

tyermannii for John Simpson Tyermann from Cornwall (*c*. 1830–89), Superintendent at Liverpool Botanic Garden *c*. 1871
Tylanthera Knob-like-anther, τυλη-ανθερος
tylicolor dark-grey, coloured like a woodlouse, modern Latin
tylo- knob-, callus-, swelling-, τυλη
Tylocodon deceptive anagram of *Cotyledon* (not swollen-bell)
tylodes knobbly, callosed, τυλη-ωδες
Tylophora Callus-bearing, τυλη-φορα (for the pollen masses)
Tylostigma Knob-like-stigma, τυλη-στιγμα
tympani- drum-, *tympanum (typanum), tympani*
tymphresteus -a -um from Mount Tymphrestos (Timfrestos, Tymfristos), Greece
Typha a Greek name, τυφη, used by Theophrastus for various plants (***Typhaceae***)
typhinus -a -um, typhoides bulrush-like, *Typha-oides*, relating to fever
typhofolius -a -um with leaves resembling those of *Typha*
Typhonium from an ancient Greek name; some derive as τυφωνος, a hurricane
Typhonodorum Gift-of-the-storm, τυφωνος-δωρον
typicus -a -um, -typus the type, typical, τυπος
-typus -original pattern, -figure, *typus, typi*
tyrianthinus -a -um Purple-flowered, botanical Latin from *Tyrus* and ανθινος, scarlet-purple-coloured (Tyrian purple, the scarlet dye from mollusc shells, πορφυρος)
tyrius -a -um royal purple, Tyrian purple, *Tyrus*
Tytonia for A. Tyton, a patron of botany (≡ *Hydrocera*)
tytthanthus -a -um small-flowered, τιτθος-ανθος (*Alchemilla tytthantha*)
tzumu a Chinese vernacular name for *Sassafras tzumu*, of which one plant was grown in Britain, in 1900, and lost when transplanted to Kew

Uapaca from the Madagascar vernacular name
ubatubanus -a -um from the environs of Ubatuba, SE Brazil
uber -is -e breast, teat; rich, luxuriant, full, fruitful, *uber, uberis*
uberiformis -is -e formed like a breast or udder, *uberis-forma*
uberrimus -a -um very fruitful, superlative of *uber*
Ubochea an anagram of *Bouchea*
ucranicus -a -um from the Ukraine, Ukrainian
udensis -is -e from the River Uda or the Uden district of Siberia
udisilvestris -is -e of damp woodland undergrowth, *udus-sylvestris*
Udora of water, υδωρ (the habitat) (≡ *Elodea*)
uduensis -is -e from Udu, New Guinea
udus -a -um wet, damp, *udus*
ugandae, ugandensis -is -e from Uganda, E Africa
Ugni, ugni from a Chilean vernacular name, uñi, for *Ugni molinae* (*Myrtus ugni*)
-ugo -having (a feminine suffix in generic names)
ugoensis -is -e from Mount Ugo, Luzon, Philippines
ulcerosus -a -um full of sores, knotty, lumpy, *ulcerosus*
Uleiorchis for Ernst Heinrich Georg Ule (1854–1915), German botanist and plant collector in S America
-ulentus -a -um -abundant, -full, -being (comparative suffix)
Uleophytum Ule's-plant, botanical Latin from Ule and φυτον (*vide supra*)
Ulex an ancient Latin name in Pliny for a thorny shrub
ulicifolius -a -um with *Ulex*-like leaves, *Ulex-folium*
ulicinus -a -um, ulicoides resembling *Ulex, Ulex-oides*
uliginosus -a -um marshy, of swamps or marshes, *uligo, uliginis*
Ullucus the Ecuadorian vernacular name, olloco, for the edible tubers of *Ullucus tuberosus*
-ullus -a -um -smaller, -lesser
Ulmaria, ulmaria Elm-like, *Ulmus* (Gesner's name refers to the appearance of the leaves)

ulmariae of meadow sweet, living on *Filipendula ulmaria* (*Dasyneura*, dipteran gall midge)
ulmarius -a -um growing with or on elm debris, *Ulmus*
ulmifolius -a -um elm-leaved, *Ulmus*-leaved, *Ulmus-folium*
ulmi-, ulmi elm-like, of elms, living on *Ulmus* (symbionts, parasites and saprophytes)
ulmoides elm-like, *Ulmus-oides*
Ulmus the ancient Latin name, *Ulmus*, for elms, Celtic, ulm (***Ulmaceae***)
ulo- shaggy-, pernicious, ουλος
ulophyllus -a -um shaggy-leaved, ουλο-φυλλον
-ulosus -a -um minutely-, somewhat-
Ulothrix Shaggy-hair, ουλος-θριξ (the coarse filaments of this green alga)
ultonius -a -um from Ulster, modern Latin
ultra- beyond-, more than-, *ultra*
-ulus -a -um -tending to, -having somewhat
Ulva Sedge, *ulva, ulvae* (sea-lettuce grows in watery habitat)
ulvaceus -a -um resembling the green alga *Ulva*, sea-lettuce
umbellaris -is -e, umbellatus -a -um with the branches of the inflorescence all rising from the same point, umbellate, *umbella, umbellae* (literally, a parasol) (see Fig. 2e)
umbelli- umbel-, *umbella, umbellae*
umbellifer -era -erum umbel-bearing, shade carrying, *umbella-fero, umbra-fero*
umbelliformis -is -e umbel shaped, *umbella-forma*
Umbellularia Little-umbel, diminutive of *umbella* (the inflorescences)
umbellulatus -a -um umbelled, *umbella, umbellae*
umbilicatus -a -um, umbilicus -a -um navelled, with a navel, *umbilicus, umbilici*
Umbilicus Navel, *umbilicus* (e.g. the depression in the leaf surface above the peltate insertion of the petiole)
umbo- knob-like-, *umbo, umbonis*
umbonatus -a -um with a raised central boss or knob, *umbo, umbonis*
umbracul- umbrella-like-, shading, *umbra, umbrae*
umbraculiferus -a -um shade-giving, arbour-bearing, bearing parasols (e.g. large leaves), *(umbraculum, umbraculi)-fero*
umbrarus -a -um, umbrinus -a -um umber-coloured, the colour of raw umber, *umbra* (Italian, terra di ombra, earth of shade)
umbraticolus -a -um occupying shaded habitats, *umbra-colo* (literally, idler or lounger)
umbrophilus -a -um shade-loving, botanical Latin from *umbra* and φιλος
umbrosus -a -um growing in shade, shade-loving, *umbra*
umidus -a -um damp, dank, moist, *umidus*
un- one-, single-; not-
unalascheensis -is -e, unalaschensis -is -e, unalaschkensis -is -e from Unalaska, largest of the Aleutian islands
Uncaria Hook, *uncus, unci* (shrubs climbing with hooked inflorescence peduncles)
uncatus -a -um, uncus with hooks, hook-like, hooked, *uncatus, uncus*
unci- hook-, *uncus, unci*
uncialis -is -e an inch in length, *uncia, unciae*
Uncifera Hook-bearer, *unci-fero*
uncifolius -a -um hook-leaved, *unci-folium* (retrorse marginal teeth)
uncinatus -a -um with hooks, barbed, *uncinatus* (see Fig. 8e)
uncinellus -a -um with small hooks, diminutive of *uncus*
Uncinia Much-hooked, *uncinatus* (the sharply reflexed apex of the extended spikelet axis) (see Fig. 8e)
uncipes with a hooked stalk, *unci-pes*
unctuosus -a -um with a smooth shiny surface, fatty, greasy, *uncatus*
undatus -a -um, undosus -a -um not flat, billowy, undulate, waved, *unda, undae*
undulatifolius -a -um with wavy leaf surfaces, *unda-folium*
undulatus -a -um wavy, *undo, undare, undavi*

undulifolius -a -um wavy-leaved, *unda-folium*
unedo the Latin name for the *Arbutus* tree and its fruit, meaning 'I eat one'
Ungeria for Franz Joseph Andreas Nicolaus Unger (1800–70), Austrian biologist of Vienna University, author of *Genera et species plantarum fossilium* (1850)
Ungernia for Baron Franz von Ungern-Sternberg (1800–68), of Dorpat (Tartu), Estonia
Ungnadia (*Ugnandia*) for Baron David von Ungnad, sixteenth-century Austrian diplomat in Constantinople *c.* 1576–82, who introduced horse-chestnut to Vienna
ungui- half an inch-; clawed-, *unguis, unguis*
unguicularis -is -e, unguiculatus -a -um with a small claw, *unguis,* or stalk (e.g. the petals)
unguilobatus -a -um, unguilobus -a -um with claw-like lobes, *unguis-lobus* (the leaf margins)
unguinosus -a -um slimy, greasy, *unguen, unguinis*
unguipetalus -a -um the petals having a distinct claw, *unguis-petalum*
unguis-cati with recurved thorns, cat's-clawed, *unguis-catus* (late Latin)
ungulatus -a -um clawed, *unguis*
Ungulipetalum Clawed-petal, botanical Latin from *unguis* and πεταλον
uni-, unio- one-, single-, *unus, uni-*
unibracteatus -a -um having one bract, *uni-bracteatus*
unicanaliculatus -a -um single-channelled, *uni-canalis*
unicapsularis -is -e the fruit being a single capsule, *uni-capsula*
unicolor single-coloured flowers, *uni-color*
unicus -a -um solitary, unique, *unicus*
uniflorus -a -um one-flowered, *uni-florum*
unifoliatus -a -um with a single leaflet, *uni-foliatus*
Unifolium, unifolius -a -um One-leaf, *uni-folium* (≡ *Maianthemum*); having a single leaf
unifurcatus -a -um divided only once, *uni-(furca, furcae)*
unigemmatus -a -um having a single bud, *uni-gemmatus*
uniglumis -is -e with one glume, *uni-glumis*
unilateralis -is -e one-sided, unilateral, *uni-(latus, lateris)*
unilocularis -is -e with a one-chambered ovary, *uni-loculus*
Uniola an ancient Latin plant name, *unio, unionis* a single large pearl (application uncertain)
unioloides resembling *Uniola, Uniola-oides* (American sea oats)
uniseriatus -a -um of a single row, *uni-series*
unisiliquosus -a -um with a single siliqua, *uni-siliquosus* (fruit)
unitus -a -um joined, united, *uniter*
Unonopsis Resembling-*Unona* (Annonaceae, ≡ *Xylopicriopsis*)
uplandicus -a -um from Uppland, Sweden
uporo Fijian vernacular name of cannibal tomato (*Solanum anthropophagorum*)
Upuna a vernacular name, upun batu, for *Upuna borneensis*, Borneo ironwood
uragogus -a -um diuretic, *urina*
uralensis -is -e from the Ural mountains, Russia
uralum from vernacular name, urala swa, for *Hypericum uralum*
Urania Heavenly, ουρανιος (elegant palms)
uranthus -a -um with tailed flowers, ουρ-ανθος
Uraria Tailed-one, ουρα (the long, bracteate racemes)
uratus -a -um tailed, with a tail, ουρα
Urbananthus Urban's-flower, botanical Latin from Urban and ανθος (*vide infra*)
Urbania, urbanii for Ignaz (Ignatz) Urban (1848–1931), of Berlin Botanical Museum, who made significant contributions to the botany of tropical America
Urbanodendron Urban's-tree, botanical Latin from Urban and δενδρον (*vide supra*)
urbanus -a -um, urbicus -a -um of the town, urban, *urbs, urbis*
Urceola Urn-shaped (flower), feminine of *urceolus*
urceolaris -is -e, urceolatus -a -um pitcher-shaped, urn-shaped, *urceolus, urcioli*

Urceolina Urn-like-one, diminutive of *urceolus* (the flower shape)
Uredo Blight, *uredo* (from the scorched appearance of infected host plants, *uro* burn)
Urelytrum Tailed-cover, ουρ-ελυτρον (the long-awned lower glumes)
Urena from the Malabar vernacular name, aramina, for the fibre plant *Urena lobata*
urenissimus -a -um most burning, most fiery, most stinging, superlative of *urens*
urens acrid, stinging, burning, *uro, urere, ussi, ustum*
Urera Burning, *uro, urere, ussi, ustum* (cow itch)
Urginea from the Algerian type locality, the area of the Beni Urgin tribe
urnalis -is -e, urnulus -a -um resembling a small urn (diminutive of *urna, urnae*)
urnigerus -a -um urn-bearing, *urna-gero*
uro-, -urus -a -um tail-, -tailed, ουρα
Urobotrya Tailed-cluster, ουρα-βοτρυς (the pendulous yellow fruits)
Urochloa Tailed-grass, ουρα-χλοη (the racemose inflorescence)
Urochondra Tailed-grain, ουρα-χονδρος (the caryopsis)
Urogentias Tailed-gentian, botanical Latin from ουρα and *Gentiana*
Uromyrtus Tailed-myrtle, botanical Latin from ουρα and *Myrtus*
Urophyllum, urophyllus -a -um Tail-leaved, ουρα-φυλλον (the drip tip)
Uroskinnera for George Ure Skinner (1804–67), English merchant and collector of Central American plants
Urospatha Tailed-spathe, ουρα-σπαθη (projection on the spathe)
Urospermum Tailed-seed, ουρα-σπερμα (the beaked achenes)
urseolatus -a -um crowded, hemmed-in, *urgeo, urgere, ursi*
Ursinia for Johannes Heinrich Ursinus (1608–66), German cleric and botanist of Regensburg, author of *Arboretum biblicum* (1663)
ursinus -a -um, ursorus -a -um bear-like, *ursus, ursi, ursa, ursae* (the smell), northern (under the *Ursa Major* constellation)
urtic-, urticae- nettle-, *Urtica-*
Urtica Sting, *uro, urere, ussi, ustum* (the Latin name, *urtica, urticae*) (***Urticaceae***)
urticae of nettles, living on *Urtica* (symbionts, parasites and saprophytes)
urticifolius -a -um nettle-leaved, with leaves resembling *Urtica, Urtica-folium*
uruguayensis -is -e from Uruguay, SE coast of S America
urumiensis -is -e from Orumiyeh, by Lake Urmia, Iran
urundeuva a S American vernacular name for a hardwood timber
Urvillea for Jules Sébastien César Dumont d'Urville (1790–1884), French circum-navigator (eponymous marine alga, New Zealand island and river)
usambarensis -is -e from the Usambara mountains of NE Tanzania
-usculus -a -um -ish, -somewhat (a diminutive suffix)
usitatissimus -a -um most useful, superlative of *usitatus*
usitatus -a -um everyday, ordinary, useful, *usitatus*
Usnea a name of uncertain meaning by Adanson
usneoides resembling *Usnea, Usnea-oides* (hanging in long threads)
ussuriensis -is -e from the environs of the Ussuri river (Wu-su-li chiang) of the China–Siberian border
ustalis -is -e of a glowing colour, *uro, urere, ussi, ustum*
ustaloides of scorched appearance, *uro, urere, ussi, ustum*
ustulatus -a -um frosted, scorched-looking, *uro, urere, ussi, ustum*
ustulescens becoming scorched or dried-out-looking, present participle of *uro, urere, ussi, ustum*
ustus -a -um parched, *uro, urere, ussi, ustum*
usuriensis -is -e, ussuriensis -is -e from the Ussuri river area, between China and Siberia
utahensis -is -e from Utah, USA
utilis -is -e serviceable, useful, *utilis*
utilissimus -a -um the most useful, superlative of *utilis*
utri- bottle-, bag-, *uter, utris*

Utricularia Little-womb, *utriculus*, diminutive of *uterus* (the underwater traps of the bladderwort)
utricularis -is -e, utriculatus -a -um with utricles, bladder-like, *utriculus*
utriculosus -a -um bladder-like, inflated, *utriculus*
utriformis -is -e bag-shaped, *uter-forma*
utrigerus -era -erum bearing bladders, *uter-gero*
-utus -a -um -having
uva-crispa curly-bunch, *uva-crispus*, botanical Latin via old French, grozelle, and German, kraus, crisped or curled)
Uvaria, uvaria from an old generic name (clustered fruits, like a bunch of grapes, *uva, uvae*)
Uvariastrum Somewhat-like-*Uvaria, Uvaria-ad-instar*
uvariifolius -a -um with leaves resembling those of *Uvaria*
Uvariodendron *Uvaria*-like-tree, botanical Latin from *Uvaria and* δενδρον
Uvariopsis Like-*Uvaria, Uvaria-opsis*
uva-ursi bear's-berry, *uva-ursus* (Latin equivalent of the Greek-derived name, *Arctostaphylos*)
uva-vulpis fox's-berry, *uva-vulpes*
uvidus -a -um drunken; wet, damp, moist, *uvidus*
uvifer -era -erum fruiting in clusters, grape-bearing, *uva-fero*
uviformis -is -e in a clustered mass, like a cluster of swarming bees, *uva-forma*
Uvularia Palate, *uvula* (the signature for medicinal use, either from the hanging flowers or from the fruits) (***Uvulariaceae***)

Vaccaria Cow-fodder (d'Aléchamps' name from *vacca*, a cow)
vaccarus -a -um of cow pastures, *vacca*
vaccini-, vaccinioides bilberry-like, resembling *Vaccinium*, υακινθος-οειδης
vacciniaceus -a -um bilberry-like, *Vaccinium*
vaccinii-, vaccinii of ericaceous plants, living on *Vaccinium* (symbionts, parasites and saprophytes)
vacciniiflorus -a -um with *Vaccinium*-like flowers
vacciniifolius -a -um with *Vaccinium*-like leaves
Vaccinium a Latin name of great antiquity with no clear meaning (may be cognate with *Hyacinthus*, for the dark blue, υακινθος, colour of the fruit); various derivations have been suggested (***Vacciniaceae***)
vaccinus -a -um the colour of a red cow, of cows, *vacca, vaccae*
vacillans variable, swinging, versatile, *vacillo, vacillare*
vagans of wide distribution, wandering, *vagor, vagari, vagatus*
Vagaria Wandering, *vagor, vagari, vagatus* (first identified after it turned up in Paris, apparently from the Levant)
vagensis -is -e from the River Wye (*Vaga*)
vaginans, vaginatus -a -um having a sheath, *vagina*, sheathed (as the stems of grasses by the leaf-sheaths)
vaginervis -is -e with veins arranged in no apparent order, *vagans-vervus*
vagus -a -um uncertain, varying, inconstant, fickle, wandering, *vagus*
Vahlia, vahlii for Martin Hendriksen Vahl (1749–1804), Norwegian botanist, Director of Copenhagen Botanic Garden, author of *Symbolae botanicae* (1794)
Valantia, vaillantii, valantia for Sebastien Vaillant (*Valantius*) (1669–1722), French physician and botanist
valdensis -is -e, valdensius -a -um from Vaud, Switzerland (*Valdia*)
Valdivia, valdivianus -a -um, valdiviensis -is -e from Valdivia, Chile
valentinus -a -um from Valencia, Spain (*Valentia*)
valerandi for Dourez Valerand, sixteenth-century botanist
Valeriana Health, *valeo, valere, valui, valitum* (from a medieval name for valerian's medicinal use, or for *Valerius*, who first used it medicinally, or from the Roman province of *Valeria*, W Hungary) (***Valerianaceae***)

Valerianella diminutive of the name *Valeriana*
Valerioanthus Valerian-flowered (≡ *Peltanthera*-flowered)
valesiacus -a -um see *vallesiacus*
validus -a -um, validi- well-developed, strong, *validus*
Vallariopsis Resembling-*Vallaris, Vallaris-opsis*
Vallaris Pallisade, *vallus* (its use for fence-making in Java)
Vallea for Felice Valle (d. 1747), Italian botanist, author of *Flora Corsicae* (1761)
vallerandii for Eugene Vallerand
Vallesia, vallesia for Francisco Vallesio (*Vallesius*) (d. 1592), Spanish physician to Philip II of Spain
vallesiacus -a -um, vallesianus -a -um of valleys, *valles, vallis*; from Vallais (*Vallesia, Wallis*), Switzerland
vallicolus -a -um of ramparts, living in valleys, *vallis-colo*
vallis-gratiae from the valley of grace, *vallis-*(*gratia, gratiae*)
vallis-mariae from the Mariental valley, Namibia, botanical Latin from *vallis* and Mariental
vallis-mortae from Death Valley, California, USA, *vallis-*(*mors, mortis*)
Vallisneria for Antonio Vallisnieri de Vallisnera (1661–1730), Italian physician and naturalist, professor at Padua
Vallota for Pierre Vallot (1594–1671), French botanist and garden writer
valperedisiacus -a -um from the environs of Valparaiso, Chile
valvatus -a -um having valvate dehiscence, *valvae, valvarum* (literally, with folding doors)
valverdensis -is -e from Valverde, Hierro, Canary Islands
valvulatus -a -um articulated, jointed, diminutive of *valvae*
vampirus -a -um of the vampire, emphasises the significance of the generic name of the bizarre orchid, *Dracula vampira*
Vancouveria, vancouverianus -a -um for Captain George Vancouver (1758–98), English navigator on the *Discovery*'s exploration of NW coastal America
Vanda from the Sanskrit name
vandasii for Dr Karel Vandas (1861–1923), Professor of Botany at Brno (Brünn)
Vandopsis Resembling-*Vanda, Vanda-opsis*
Vangueria from the Madagascan vernacular name, voa-vanguer or vavangue
Vangueriella feminine diminutive from *Vangueria*
Vanguerioposis Looking-like-*Vangueria, Vangueria-opsis*
Vanhouttea, vanhouttei for Louis Benoit van Houtte (1810–76), Belgian nurseryman, author of *Hortus Vanhoutteanus* (1846)
Vanieria for Jaques de Vanier (1664–1739), French Jesuit and author of *Praedum rusticum*
vanikorensis -is -e from Vanikoro, Santa-Cruz Islands
Vanilla Little-sheath, diminutive of Spanish *vaina*, Latin, *vagina* (from the Spanish name, vainilla, for a small sheath, describing the fruit)
Vanzilia for Dorothy van Zijl
Vargasiella for Julio Cesar Vargas Calderón (1907–60), Peruvian botanist and orchidologist
vari-, varii- differing, changing, diverse, varying, *vario, variare, variavi, variatum, varii-, vario-*
variabilis -is -e variable, not constant, *variabilis*
variabillimus -a -um very variable, comparative of *variabilis*
varians changing, varying, present participle of *vario*
variatus -a -um several, various, *variatio, variationis*
varicosus -a -um with dilated veins or filaments, varicose, *varicosus*
variegatus -a -um irregularly coloured, blotched, variegated, *variegatus*
variicolor of several colours, changing colour, *varii-color*
variifolius -a -um variable-leaved, botanical Latin *varii-folium*
variolaris -is -e, variolatus -a -um pock-marked, pitted, from modern Latin, *variola*, for a smallpox pustule

variolosus -a -um smallpox-like, very pock-marked, with large dimples, *variola*
variopictus -a -um diversely spotted, with coloured spotting, *vario-(pingo, pingere, pinxi, pictum)*
varius -a -um coloured, spotted, variable, changing, fickle, variegated, *varius*
vartani from Vartan, Sweden
vas-, vasi- duct- (tube), vessel- (container), *vas, vasis; vasa*
Vascellum Little-bowl, diminutive of *vasculum* (the fruiting body becomes bowl-shaped)
vasconicus -a -um from the Basque country of Spain, *Vasco*
vascularis -is -e possessing vessels, botanical Latin from *vasculum* (of the conductive tissue)
vasculosus -a -um having large vessels or thick cell walls, *vasculosus*
vasculum a small vessel, *vasculum* (also used for the container used by field botanists)
Vaseyanthus for George Vasey (1822–93) American physician and botanist at the US Department of Agriculture
Vaseyochloa Vasey's-grass (*vide supra*)
vastatrix denuder, devastator, ravager, feminine form of *vastator*, from *vasto, vastare, vastavi, vastatum*
vastus -a -um empty, desolate, very large, vast, *vastus*
Vateria for Abraham Vater (1684–1751), German physician and botanist
Vatica Soothsayer, *vates, vatis* (*strychnos, herba vatica*, has sundry uses from heightening the senses to killing)
Vauanthes V-flower, botanical Latin from *vau* and ανθος (the V-shaped marks on the petals)
vaupesanus -a -um from Vaupés department, SE Colombia
Vauquelinia for Nicolas Louis Vauquelin (1763–1829), French chemist who discovered the element chromium (atomic number 24)
Vavaea from the W Pacific Philippine island group called Vava'u
Vavilovia for Nikolai Ivanovitch Vavilov (1887–1943), Russian geneticist and plant breeder, victimized by the autocrat T. D. Lysenko
vectensis -is -e from the Isle of Wight (*Vectis insula*)
vedrariensis -is -e, vedrarius -a -um from Verrières, Paris, France
vegetus -a -um spritely, growing strongly or quickly, vigorous, *vegetus*
Veitchia for James Veitch junior (1815–69) and his son John Gould Veitch (1839–70), nurserymen of Chelsea
veitchianus -a -um, veitchii, veitchiorum for Messrs Veitch, nurserymen of Exeter and Chelsea, est. by John Veitch (1725–1839) and responsible for many plant introductions
velaris -is -e, velatus -a -um concealed, veiling, veiled, *velo, velare, velavi, velatum*
velebiticus -a -um from the Velebit mountains, Croatia
Vella from the Celtic name, velar, for cress
Velleia (Velleja) for Major Thomas Velley (1748–1806), phycologist
Vellereophyton Fleecy-plant, *vellus, velleris* (the woolly indumentum)
vellereus -a -um densely long-haired, fleecy (*vellus*, a fleece)
Vellozia for José Mariano de la Conceicão Velloso (Veloso, Vellozo) (1742–1811), Capuchin monk and botanist who edited Vandelli's works on Brazil (***Velloziaceae***)
Velloziella for Joaquim Velloso de Miranda (1733–1815), Portuguese botanist and collector in S America (≡ *Digitalis, pro parte*)
velosus -a -um veiled, mycological Latin, *velatus*
velox swift, rapid-growing, *velox, velocis*
Veltheimia for August Ferdinand Graf von Veltheim (1741–1808), German patron of botany
veluti- down-like-, velvety, from French, velouté
velutinellus -a -um finely velvety, diminutive of *velutinus*
velutinosus -a -um, velutinus -a -um with a soft silky down-like covering, velvety, *velutinus* from French velouté

velutipes with a velvety stalk, botanical Latin from French velouté and *pes*
venator hunting-pink coloured, of the hunter, *venator, venatoris* (the flowers of *Rhododendron venator* are 'hunting-pink')
veneficiorus -a -um, veneficus -a -um of sorcerers or poisoners, *veneficium, veneficii*
Venegasia for Miguel Venegas (1680–1764), Mexican Jesuit, author of *A Natural and Civil History of California* (1759)
venenatus -a -um poisonous, magic, *veneno, venenare; venenatus*
veneniferus -a -um poison-bearing, *venenum-fero*
venenosus -a -um very poisonous, comparative of *venenatus*
veneris -is -e revered, past participle of *veneror, venerari, veneratus*
venetus -a -um of Venice, Venetian (*Veneti, Venetus*)
Venidium Veined, *vena, venae* (the decurrent leaf base on the stem)
venosus -a -um conspicuously veined, comparative of *vena, venae*
Ventenata for Etienne Pierre Ventenat (1757–1808), French librarian, writer and botanist
Ventilago Of-the-wind, *ventilo, ventilare* with feminine suffix (wind dispersal of the winged mericarps)
ventilator flabellate, fan-shaped, winnower-like; juggler, *ventilator*
ventosus -a -um like the wind, fickle, *ventosus*
ventri- belly-, *venter*
ventricosus -a -um bellied out below, distended to one side, expanded, ventricose, *venter*
Ventricularia Small-bellied, *ventriculus* (depression on the labellum)
ventriculosus -a -um slightly bellied, diminutive of *ventriculus*
ventriosoporus -a -um having pot-bellied spores, botanical Latin from *ventricosus* and σπορα
venulosus -a -um with fine veins, *venula*, finely veined, diminutive of *vena*
venustulus -a -um quite charming (diminutive of *venustus*)
venustus -a -um graceful, beautiful, charming, *venustus*
Vepris Brambly, *vepre, vepris* (≡ *Todalia*)
veratrifolius -a -um false-hellebore-leaved, with leaves like *Veratrum, Veratrum-folius*
Veratrum Truly-black, *vere-(ater, atris)* (for the roots of false-hellebore, the Latin name, *veratrum*, for a hellebore)
verbanensis -is -e from the area of Lake Maggiore (*Lacus Verbanus*), Italy
verbasci- mullein-like, resembling *Verbascum*
verbascifolius -a -um mullein-leaved, with leaves resembling *Verbascum, Verbascum-folium*
Verbascum a name, *barbascum*, in Pliny (for the bearded stamens)
Verbena Sacred-bough, from the Latin name, *verbena, verbenae*, for the leafy twigs carried by priests, used in wreaths for Druidic ritual, and in medicine (used by Virgil and Pliny for vervain, *Verbena officinalis*), Celtic, ferfain (***Verbenaceae***)
verbenaca, verbeni- from a name in Pliny, vervain-like
verbeniflos *Verbana*-flowered
Verbesina *Verbena*-like (resembles some species)
verecundus -a -um bashful, modest, shy, *verecundus*
veris of spring, *ver, veris* (flowering time), genuine, true, standard
verlotiorum (verlotorum) for the brothers Verloti, who introduced *Artemesia verlotiorum* from China and noted its relation to *A. vulgaris*
vermi- worm-like-, worm-, grub-, *vermis*
vermicularis -is -e worm-like, grub-like, *vermis*
vermiculatus -a -um inlaid or marked with wavy lines, *vermiculatus* (cognate, via French and Old English, with vermilion, from *vermis, vermiculus*)
vernalis -is -e of spring, vernal, *ver, veris* (flowering time)
verniciferus -era -erum producing varnish, *vernix-fero*
vernicifluus -a -um from which flows a varnish, *vernix-(fluo, fluere, fluxi, fluxum)*
vernicosus -a -um glossy, varnished, *vernicosus*

vernix varnish, Old French, vernis, from medieval Latin *vernix*, for fragrant resin
Vernonanthera *Vernonia*-flowering, botanical Latin from *Vernonia* and ανθερος
Vernonia for William Vernon (1680–1711), English botanist and collector in N America
vernonoides *Vernonia*-like, *Vernonia-oides*
vernus -a -um of the spring, *ver, veris*
veronic-, veronici- *Veronica*-like
Veronica Fuchs' name, for Saint Veronica who wiped the sweat from Christ's face, may be cognate with *Betonica* and *Vettonica*; various derivations have been suggested (Arabic viru-niku; and, as patron saint of photography, Latin *vere-icon*, true image)
Veronicastrum Somewhat resembling *Veronica, Veronica-ad-instar*
veronici of speedwell, living on *Veronica* (*Eriophyes veronici*, acarine gall mite)
Verpa Circumcised, *verpus* (meaning of feminine case not clear)
verrucarius -a -um of warts, *verruca, verrucae* (*nasturtium verrucarium* was the Englished version of the German name, Warzen Kress, for *Coronopus squamatus*, wart-cress)
verrucosus -a -um with a warty surface, verrucose, warted, *verruca, verrucae*
Verrucularia Warty, diminutive of *verruca*
verruculosus -a -um somewhat warty, *verruca-ulosus*
versatilis -is -e revolving, versatile, *versatilis*
Verschaffeltia, verschaffeltii for Ambrose Colletto Alexandre Verschaffelt (1825–86), Belgian horticulturalist writer on *Camellia*
versi- several-, turning-, changing-, *verso* (*verto, vertere, verti, versum*)
versicolor varying or changeable in colour, *verto-color*
versipellis -is -e werewolf-like, changing skin, changing appearance, *versipellis*
verticill-, verticilli- with whorls of-, whorled-, arranged into a disc-, *verticillus*
verticillaris -is -e, verticillatus -a -um having whorls (several leaves or flowers all arising at the same level on the stem), verticillate, *verticillus*
verticillaster with whorls of flowers, *verticillus-aster*
verticilliflorus -a -um having whorls of flowers, *verticillus-florum*
Verticordia for the personification of the goddess Venus as goddess of chastity, *Venus Verticordae*, (*verto, vertere, verti, versum*)*-corda*
veruculatus -a -um cylindric and somewhat pointed, like a small pike, *veruculatus*
verucundus -a -um true, shy, modest, *verecundus*
verus -a -um true, genuine, *verus*
verutus -a -um shaped like a javelin, armed with javelin-like structures, *verutum*
vescus -a -um small, feeble, undernourished, *vescus*; edible, *vescor, vesci*
Vesicaria Bladder-like, *vesica* (the inflated fruit)
vesicarius -a -um inflated, bladder-like, *vesica*
vesicatorius -a -um blistering, *vesicula* (sap causing a local allergic reaction)
vesiculiferus -a -um covered with blister-like irregularities, *vesicula-fero*
vesiculosus -a -um inflated, composed of little blisters, vesiculous, *vesicula, vesiculae*
vespertilionis -is -e, vespertilis -is -e bat-like, *vespertilio* (with two large lobes)
vespa resembling a wasp, *vespa, vespae*
vespertinus -a -um of the evening, *vesper, vesperis* (evening-flowering)
vestae for Vesta, a Roman goddess of the household (Vestal Virgins)
vestalis -is -e white, chaste, *vestalis*
Vestia for Lorenz Chrysanth von der Vest (1776–1840), Austro-German physician and botanist, Professor of Chemistry at Graz
vestiarius -a -um well clothed, *vestio, vestire, vestii, vestitum* (with foliage)
vestitus -a -um covered, clothed, *vestio, vestire, vestii, vestitum* (with hairs)
Vetiveria Latinized English version of S Indian name, vettiveru, for khus-khus grass
Vetrix Osier, from Italian, vetrice, for an osier
vetulus -a -um somewhat old or wizened, *vetulus*

vexans annoying, wounding, present participle of *vexo, vexare, vexavi, vexatum*
Vexatorella Troublesome, feminine diminutive of *vexator, vexatoris*
Vexillabium Pennant-lipped, *vexillum-labium* (like long flags)
vexillaris -is -e, vexillarius -a -um with a standard, *vexillum, vexilli* (as the large 'sail' petal of a pea-flower)
vialis -is -e, viarum ruderal, of the wayside, *via, viae; vialis*
viaticus -a -um travelling allowance, of journeys, *viaticus*
viatoris is -e of the roadways, of travellers, *viator, viatoris*
vibecinus -a -um wealed, *vibix, vibicis* (the pale striae on the drying pilea)
viberni of wayfaring tree, living on *Viburnum* (*Eriophyes*, acarine gall mite)
viburnifolius -a -um having leaves like *Viburnum*
viburnoides *Viburnum*-like, *Viburnum-oides*
Viburnum the Latin name for the wayfaring tree, *Viburnum lanata*
vicarius -a -um proxy, substitute, *vicarius*
Vicia Binder, the Latin name in Pliny for a vetch, *vincio, vincire, vixi, victum* to bind
viciae-, vicii-, vicioides bound, vetch-like-, resembling *Vicia, Vicia-oides*
viciifolius -a -um vetch-leaved, *Vicia-folium*
vicinus -a -um neighbouring, kindred, *vicinis* (closely allied taxon)
Victoria, victoriae for Queen Victoria (Alexandrina Victoria) (1819–1901)
victoriae-mariae for Queen Mary (Victoria Mary Augusta Louise Olga Pauline Claudine Agnes of Teck) (1867–1953)
victorialis -is -e victorious, *victoria, victoriae* (protecting, *Allium victorialis* was worn as a protective talisman by Bohemian miners)
victoria-reginae for Queen Alexandrina Victoria (1819–1901)
vidalianus -a -um, vidalii for either Captain Vidal RN, who collected in the Azores in 1842, or Sebastian Vidal de Soler (1842–89), Spanish botanist in the Philppines
viduiflorus -a -um bereft of flowers, (*viduo, viduare*)-*florum; viduus-florum*
vietnamensis -is -e from Vietnam
vietus -a -um shrivelled, *vietus*
vigilis -is -e awake, watching, *vigil, vigilis*
Vigna for Domenico Vigna (d. 1647), Professor of Botany at Pisa
Vignea for Gislain François de la Vigne (d. 1805), Professor of Botany at Charkow (≡ *Carex*)
Viguiera for L. G. Alexander Viguier (1790–1867), French physician and botanist
vilis -is -e common, of little value, *vilis*
Villadia for Manuel Villada (b. 1841), Mexican scientist
Villanova for Thomas M. Villanova (1757–1802), botany professor at Valencia
Villarsia, villarsii for Dominique Villars (1745–1814), professor at Grenoble, France
villi- fleecy, shaggy, *villus, villi*
villicaulis -is -e with a shaggy stem, *villi-caulis*
villiferus -ero -erum carrying shaggy hairs, clad in a shaggy coat, *villi-fero*
villiflorus -a -um having shaggy flowers, *villus-florum*
villipes with a long-haired stalk, *villi-pes*
villosipes with a very hairy stalk, *villosus-pes*
villosulus -a -um slightly hairy, finely villous (diminutive of *villosus*)
villosus -a -um with long rough hairs, shaggy, villous, *villosus*
Vilmorinia, vilmorinianus -a -um, vilmorinii for Pierre Philippe André Levêque de Vilmorin (1776–1862), or the French nurserymen Vilmorin-Andrieux
viminalis -is -e, vimineus -a -um with long slender shoots suitable for wicker or basketwork, of osiers, osier-like, with pliant twigs, *vimen, viminis*
Viminaria Twiggy, *vimem, viminis* (broom-like habit)
vinaceus -a -um of the vine, the colour of wine, *vinum*
Vinca Bond, *vinculum* (Pliny's name, *vinca pervinca*, refers to the use of periwinkle in wreaths; *pervinco, pervincere, pervici, pervictum* to overcome)
vincentinus -a -um from the environs of the San Vincenti, El Salvador
Vincetoxicopsis resembling *Vincetoxicum, vinco-toxicum-opsis*

Vincetoxicum Poison-beater, *vinco-toxicum* (the supposed antidotal property of *Vincetoxicum officinale* to snakebite)
vinci-, vincoides periwinkle-like, resembling *Vinca, Vinca-oides*
vinciflorus -a -um *Vinca*-flowered, periwinkle-flowered, *Vinca-florum*
vinculans binding, fettering, present participle of *vincio, vincire, vixi, vinctum; vinculum*, a fetter or prison
vindobonensis -is -e from Vienna (*Vindobona*), Viennese
vinealis -is -e of vines and the vineyard, growing in vineyards, *vinea, vineae*
vinicolor wine-red, (*vinum, vini*)-*color*
vinifer -era -erum wine-producing, wine-bearing, *vini-fero*
vinosus -a -um wine-red, wine-like, *vinum, vini*
Viola the Latin name, *viola, violae*, applied to several fragrant plants (the equivalent of the Greek name, ιον) (***Violaceae***)
violaceolineatus -a -um marked with violet lines, *violaceus-lineatus*
violaceus -a -um violet-coloured, *violaceus*
violae of or upon *Viola* species (*Urocystis*, smut fungus)
violarius -a -um dyer of violet; of the violet-bed, violet-coloured, *violarius*
violascens turning violet, *viola, violae*
violeipes with a violet-coloured stalk, *viola-pes*
violiflorus -a -um with violet-like flowers, violet flowered, *viola-florum*
violoides Viola-like, *Viola-oides*
Viorna, viornus -a -um Road decoration, *via-*(*orno, ornare, ornavi, ornatum*), from the French name for traveller's joy, *Clematis vitalba*
viperatus -a -um viper-like, *vipera, viperae* (markings)
viperinus -a -um snake's, serpent's, *viperinus*
Virecta Grassy-sward, *virecta, virectorum*
Virectaria Resembling-*Virecta*
virens, -virens green, -flourishing, -vigorous, *vireo, virere, virui*
virescens light-green, turning green, *viresco, virescere*
virgatus -a -um twiggy, with straight slender twigs, *virga, virgae*
virgaurea, virga-aurea rod-of-gold, golden-rod, *virga-aureus* (golden rod of Turner)
Virgilia for Publius Vergilius Mato (70–19 BC), Roman epic poet known as Virgil, author of the unfinished *Aeneid*
virginalis -is -e, virgineus -a -um maidenly, virginal, purest white, *virginalis*
virginianus -a -um, virginiensis -is -e from Virginia, USA, Virginian
virginicus -a -um from the Virgin Islands, Virginian
virgulatus -a -um twiggy, wand-like, *virgula*
virgultorum of thickets, *virgulta, virgultorum*
virgunculus -a -um of little girls, *virguncula, virgunculae*
viridescens greenness, becoming green, turning green, *viridis-essentia*
viridi-, viridis -is -e, viridus -a -um youthful, fresh-green, *viridis*
viridicatus -a -um entirely green, *viridis*
viridiflavus -a -um greenish yellow, green with yellow, *viridis-flavus*
viridiflorus -a -um green-flowered, *viridis-florum*
viridifolius -a -um green-leaved, *viridis-folium*
viridifrons having green fronds, *viridis-*(*frons, frondis*)
viridifuscus -a -um greenish brown, green with brown, *viridis-fuscus*
viridiglaucescens with a greenish white bloom, *viridis-glaucus-essentia*
viridior becoming green, greening, *viridior, viridiare*
viridissimus -a -um greenest, very green, superlative of *viridis*
viridistriatus -a -um with green stripes, *viridis-stria*
Viridivia commemorative play on words for P. J. Greenway (*viridis-via*) (1897–1980), systematist of the East African Agricultural Research Station, Dar es Salaam
viridulus -a -um greenish, diminutive of *viridis*
virosus -a -um slimy, rank, poisonous, with an unpleasant smell, *virosus*
Viscaria Bird-lime, *viscum* (the sticky stems of German catchfly)

viscatus -a -um clammy, *visco, viscare*
viscidi-, viscidus -a -um sticky, clammy, viscid, *visco, viscare*
viscidiflorus -a -um having sticky flowers, *viscidus-florum*
viscidifolius -a -um having sticky leaves, *viscidus-folium*
viscidulus -a -um slightly sticky, somewhat viscid (diminutive of *viscidus*)
viscosepalus -a -um with sticky sepals, *visco-sepala*
viscosissimus -a -um stickiest, very sticky, superlative of *viscosus*
viscosus -a -um sticky, viscid, *visco, viscare*
Viscum the ancient Latin name, *viscum*, for mistletoe or the birdlime from its berries (Aristotle knew that the mistle thrush, *Turdus viscivorus*, excreted seeds onto apple trees, hence mistle twigs or mistletoe)
Vismia for M. de Visme, a Portuguese merchant
visnaga the Spanish vernacular name for *Ammi*; some derive as old Norse, visna, withering
Visnea for Giraldo Visne, Portuguese botanist
vistulensis -is -e from the environs of the River Vistula
vitaceus -a -um vine-like, resembling *Vitis*
vitalba vine-of-white, *vitis-alba* (old generic name for the appearance of fruiting *Clematis vitalba*)
Vitaliana, vitalianus -a -um for Vitaliano Donati (1717–62), professor at Turin
Vitellaria Egg-yolk-coloured, *vitellus* (*V. mammosa* is the marmalade tree) (≡ *Butyrospermum*)
vitellinus -a -um dull reddish yellow, the colour of egg-yolk, *vitellus*
Vitex an ancient name used in Pliny possibly for chaste tree, *Vitex agnus-castus*
viti-, vitoides vine-like, resembling *Vitis, Vitis-oides*
Viticella, viticellus -a -um Small vine, diminutive of *vitis* (≡ *Clematis pro parte*)
viticenus -a -um, viticoides *Vitex*-like, *Vitex-oides*
Viticola, viticolus -a -um Vine-parasite, inhabiting the vine, *Vitis-colo*
viticulosus -a -um sarmentose; producing tendrils, vine-like, diminutive from *Vitis*
vitiensis -is -e from the Fijian islands (Viti Levu)
vitifolius -a -um vine-leaved, with leaves resembling those of *Vitis, Vitis-folium*
vitigineus -a -um growing as a vine, *vitigenus*
vitilis -is -e with unpigmented areas (medically called *vitiligo* or *leucoderma*)
Vitis the Latin name, *vitis, vitis*, for the grapevine (***Vitaceae***)
vitis-idaea Theophrastus' name, αμπελος παρα Ιδης, for the vine of Mount Ida or Idaea, Greece
vitreus -a -um, vitricus -a -um glassy, vitreous, *vitrum, vitri*
vitrinus -a -um of woad; of glass, *vitrum, vitri*
Vittadinia for Carlo Vittadini (1800–65), Italian physician and mycologist
vittae- banded-, filleted-, ribboned-, *vitta, vittae*
Vittaria Ribbon, *vitta, vittae* (for the shape of the fronds) (***Vittariaceae***)
vittarioides *Vittaria*-like, *Vittaria-oides*
vittatus -a -um striped lengthwise, banded longitudinally, *vitta, vittae*
vittiformis -is -e band-like, *vitta-forma*
vittiger -era -erum, vittigerus -a -um bearing lengthwise bands or stripes, (*vitta, vittae*)*-gero*
vivax long-lived, *vivax, vivacis* (flowering for a long time)
vividus -a -um lively, vivid, *vividus*
viviparus -a -um producing plantlets (often in place of flowers, or from bulbils, or as precocious germination on the parent plant) viviparous, *viviparus*
vivus -a -um enlivened, long-lasting. natural, *vivus*
vix- difficult-, hardly-, *vix*
Voacanga the Madagascan vernacular name
Voandzeia from the Madagascan name, voandzou, for the underground bean
Voanioala from the Madagascan name for this rare palm
Vogelia, vogelii for Christian Benedict Vogel (1745–1825), professor at Altdorf, or J. R. Theodor Vogel (1812–1841) of the 1841 Niger expedition

volgaricus -a -um, volgensis -is -e from the River Volga, Russia
Volkameria, volkameri, volkamerianus -a -um for Johann Georg Volkamer (1662–1744), writer on the flora of Nuremberg
volubilis -is -e spinning; entwining, enveloping, *volubilis*
volucris -s -e resembling a small winged insect, *volucris*
volutaris -is -e, volutus -a -um with rolled leaves, rolled, *voluto, volutare*
Volvariella Small-volva (the bag-like remnant of the veil, on the stipe)
Volvox Turner or Roller, *volvo, volvere, volvi, volvutum* (locomotion)
vomeformis -is -e, vomiformis -is -e shaped like ploughshares, *vomer, vomis*
vomeraceus -a -um ploughshare-like, *vomer, vomis* (flower shape)
vomerculus -a -um like a small ploughshare, *vomer, vomis* (diminutive of *vomer*)
vomerensis -is -e from Vomero, Naples, Italy
vomitorius -a -um causing regurgitation, of vomiting, emetic, *vomo, vomere, vomui, vomitum*
Vonitra the Madagascan vernacular name for *Vonitra fibrosa*
-vorus -a -um devouring, eating, *voro, vorare, voravi, voratum*
Vossia commemorative attribution uncertain
Vouacapoua a S American vernacular name, wacapou, for *Vouacapoua americana*
Voyria a French Guianan vernacular name for the ghost plant
Vriesia for Willem Hendrik de Vriese (1806–62), Dutch physician and botanist, professor at Leiden
Vrydagzynea for Theodore Daniel Vrydag Zynen, Dutch pharmacist
vulcanicolus -a -um living on volcanic soils, *vulcanus-colo*
vulcanicus -a -um, vulcanorus -a -um fiery, of volcanoes or volcanic soils, for *Vulcan* the god of fire
vulgaris -is -e, vulgatus -a -um usual, of the crowd, common, vulgar
vulnerans wounding, present participle of *vulnero, vulnerare, vulneravi, vulneratum*
vulnerarius -a -um of wounds, *vulnus, vulneris* (wound-healing property) (kidney vetch, *Anthyllis vulneraria*, was commended by Lyte for renal problems)
vulnerus -a -um marked, wounded, *vulnus* a wound
vulparia fox-bane (*vulpes* the fox) (Turner used an earlier Latin, *lycoctonum*, to produce the name, wolf's-bane, for *Aconitum vulparia*)
Vulpia, Vulpiella for Johann Samuel Vulpius (1760–1846), German botanist and pharmacist of Pforzheim
vulpinoideus -a -um *Vulpia*-like (fescue-like)
vulpinus -a -um fox-like, of the fox, *vulpes*, (colouration, shape of an inflorescence, inferiority)
vulvaria cleft, of the vulva, *volva, volvae; vulva, vulvae* (Durante's name refers to the smell of *Chenopodium vulvaria*)
Vvedenskya, Vvedenskyella for Aleksandr Ivanovich Vvedensky (1904–41), Russian reformist cleric and philosopher

Wachendorfia for Evert Jacob van Wachendorf (1702–58), Dutch physician, Professor of Botany at Utrecht
wagenerianus -a -um for M. Wagener (1813–60), German collector in tropical America
Wahlenbergia for Georg Wahlenberg (1780–1851), Professor of Botany at Uppsala and author of *Flora Lapponica* (1812)
wakefieldii for Reverend Thomas Wakefield (1836–1901), collector in E tropical Africa
Walafrida for Walafrid Strabo (808–849), Benedictine Abbot at Reichenau, author of *Liber de cultura hortorum* (830)
Waldsteinia, waldsteinii For Count Franz de Paula Adam Waldstein-Wartenburg (1759–1823), Austrian botanist and writer
Walkera for Dr Richard Walker (1679–1764), founder of the Cambridge Botanic Garden

walkeri, walkerianus -a -um for General George Warren Walker (d. 1844), collector in India and Ceylon
Wallaceodendron Wallace's-tree, botanical Latin from Wallace and δενδρον, for Alfred Russell Wallace (1823–1913), English naturalist, extensive author and evolutionist believing in natural selection before Darwin
Wallichia, wallichianus -a -um, wallichii for Nathaniel (Nathan Wolff) Wallich (1786–1854), Danish botanist and author, Curator of Calcutta Botanic Garden
wallisii for Gustav Wallis (1830–78), collector in the Andes for the William Bull nursery in Chelsea
Waltheria for Augustin Friedrich Walther (1688–1746), German physician and botanist
wandoensis -is -e from the area of the Wando river, S California, USA
Warburgia for Otto Warburg (1859–1938), German botanist, professor at Berlin Humboldt University
wardii for Frank Kingdon-Ward (1885–1958), collector of E Asian plants, and for Dr Nathaniel Bagshaw Ward (1791–1868), inventor of the Wardian case
warianus -a -um from the area of the Waria river, Papua New Guinea
warleyensis -is -e of Warley Place, Essex, home of Miss Ellen Ann Willmott (1858–1934), who developed the garden without regard to financial cost and produced many new cultivars there
Warmingia, warmingianus -a -um for Professor Johannes Eugenius Bülow Warming (1841–1924), Danish ecological botanist, professor at København (Copenhagen)
Warrea, Warreella, Warreopsis for Frederick Warre, nineteenth-century, British orchidologist
Warszewiczella, warszewiczianus -a -um (*warscewiczianus*), *warszewiczii* (*warscewiczii*) for Józef von Warszewicz (1812–66), Polish orchid collector for Messrs van Houtte and Inspector at Krakow Botanic Garden (Orchidaceae)
Warszewiczia (*Warscewiczia*) for Józef von Warszewicz (1812–66), *vide supra* (*Rubiaceae*)
Wasabia the Japanese vernacular name for *Wasabia wasabi* (Japanese horseradish)
Washingtonia for George Washington (1732–1799), first American President
washingtonianus -s -um, washingtonensis -is -e from Washington, USA
watereri for the Waterer Nursery, Bagshot, Surrey
Waterhousea for Frederick George Waterhouse (1815–98), Australian botanist
watermaliensis -is -e from Watermal, Belgium
Watsonia, watsonianus -a -um, Watsonium for Sir William Watson (1715–87), English student of sciences and extensive author
watsonii for William Watson (1858–1925), Curator of the Royal Botanic Gardens, Kew
wattianus -a -um, wattii for Sir George Watt (1851–1930), writer on Indian plants
webbianus -a -um, webbii for either Philip Barker Webb (1793–1854), author of *Histoire naturelle des Isles Canaries*, or Captain W. S. Webb, an associate of Wallich and collector in the central Himalayas *c.* 1810
Weberbauera for August Weberbauer (1871–1948), German botanist in Peru, author of *Die Pflanzenwelt der Peruanischen Anden* (1924)
Websteria for George W. Webster (1833–1914), American botanist
weddellianus -a -um, weddellii for Dr Hugh Algernon Weddell (1819–77), botanist and traveller
Wedelia for George Wolfgang Wedel (1645–1721), German physician, professor at Jena
Weigela (*Weigelia*) for Christian Ehrenfried von Weigel (1748– 1831), German physician, Professor of Botany at Griefswald
Weinmannia for Johann Wilhelm Weinmann of Ratisbon (1683–1741), author of *Phytanthoza iconographica* (1737–45)
weinmannianus -a -um for Johann A. Weinman (1782–1858), Director of St Petersburg Botanic Garden

Weldenia for Franz Ludwig Frieherr von Welden (1780–1853), alpine naturalist in the Austrian army

Wellingtonia for Sir Arther Wellesley (1769–1852), Duke of Wellington, (≡ *Sequoia*)

Wellstedia for Lieutenant J. R. Wellsted (1805–42), Belgian with the East India Company's *Palinurus* survey of NE Africa (***Wellstedtiaceae***)

Welwitchia, welwitschii for Dr Friedrich Martin Welwitsch (1806–72), Austrian physician, naturalist and traveller, director of Lisbon Botanic Garden (***Welwitschiaceae***)

Wendlandia, wendlandianus -a -um, wendlandii for Johann Christoph Wendland (1755–1828), his son Heinrich Ludolph Wendland (1729–1869) and his grandson Herman Wendland (1825– 1903), successive Curators of Herrenhausen Botanic Garden

Werneria for Abraham Gottleib Werner (1749–1817), German geologist

wernerifolius -a -um with foliage resembling that of (*Senecio*) *werneri*

Westringia for Johan Peter Westring (1753–1833), botanist and lichenologist, physician to the senile King Charles XIII of Sweden

Wetria an anagram of *Trewia*

Wettsteiniola for Richard Wettstein Ritter von Westerheim (1863–1931), Austrian systematist and palaeontologist

wherryi for Thomas Theodore Wherry (1885–1982), professor at Philadelphia

Whipplea, whipplei for Lieutenant Amiel Weeks Whipple (1818–63), pioneer on the transcontinental American Railway, Pacific Ocean survey (1853–4)

whiteanus -a -um for W. H. White (*c*. 1859–1942), gardener to Sir Trevor Lawrence at Dorking

whitei for either A. S. White of Findiaweni, Natal, or Gilbert White (1720–93), or Cyril Tenison White (1890–1950), Queensland botanist

Whitfieldia for Thomas Whitfield, collector in Sierra Leone and Gambia, 1843–8

Whitlavia for F. Whitlaw, Irish botanist (≡ *Phacelia*)

whittallii for Edward Whittall (1851–1917), collector in Turkey

wichuraianus -a -um, wichurianus -a -um, wichurii, wichurae for Max Ernst Wichura (1817–66), German botanist

Widdringtonia for Captain Samuel Edward Widdrington (1787–1856), English botanist and explorer

Wiesneria for Julius Ritter von Wiesner (1838–1916), Czech- born Austrian botanist and explorer

Wigandia for Johannes Wigand (1523–87), Prussian Bishop of the N European region of Pomerania (Poland–Germany), and botanical writer

Wightia, wightii for Dr Robert Wight (1796–1872), Scottish physician and botanist, Superintendent of Madras Botanic Garden, author of *Icones plantarum Indiae orientalis* (1840)

Wikstroemia for Johan Emanuel Wikström (1789–1856), Swedish botanist

Wilbrandia for Johann Bernhard Wilbrand (1779–1846), German physician and botanist, professor at Geissen

wildpretii for Wolfredo Wildpret de la Torre (b. 1933) of the Orotava Botanic Garden, Tenerife

Wilkesia for Lieutenant Charles Wilkes (1798–1877), American explorer of Pacific islands and NW American coasts

Willdenowia for Carl Ludwig von Willdenow (1765–1812), German physician and naturalist, Director of Berlin Botanical Garden, author of *Flora Berolinensis prodromus* (1787)

williamsii for a number of plant collectors, plant introducers, breeders and gardeners, of whom Robert Statham Williams (1859–1945) was an American plant collector in the Philippines, Benjamin Samuel Williams (1824–1890) was an orchidologist (author of the *Orchid-grower's Manual*), Percival Dacre Williams (1865–1935) created the garden at Lanarth in Cornwall, also his cousin John Charles Williams (1861–1939) of Caerhays Castle, Dr A. H. Williams, President of the National Rose Society (1933–4), and Louis Otto Williams (b. 1908), an American botanist

Willkommia for Heinrich Moritz Willkomm (1821–95), German botanist and explorer of Spain
willmottiae, willmottianus -a -um for Miss Ellen Anne Willmott (1858–1934), of Warley Place, Essex, gardening devotee and plant introducer
wilsoniae for Mrs Ernest Henry Wilson (*vide infra*)
wilsonianus -a -um, wilsonii for several Wilsons, of whom George Ferguson Wilson (1822–1902) established the Wisley wild garden, and Dr Ernest Henry Wilson (1876–1931) collected for Messrs Veitch of Chelsea and later became Director of the Arnold Arboretum, Massachusetts
winitii for Phya Winit Wanadorn, Thai collector in Thailand (Siam) *c.* 1924
winteri for either Ferdinand Winter (1835–88), of Eiffel, or E. L. Winter, Commissioner of Kumaon, N India *c.* 1908
Winteria for Captain Winter, who sailed with Francis Drake (*Drimys winteri*) (***Winteraceae***)
wintonensis -is -e from Winchester (*Venta*), or Winton, Somerset or Australia
wisleyensis -is -e from the RHS garden, Wisley, Surrey
Wisteria (Wistaria) for Caspar Wistar (1761–1818), American anatomist of Pennsylvania University
Witheringia for William Withering (1741–99), English physician and naturalist who wrote on the use of *Digitalis* for dropsy and other conditions
witmannianus -a -um, witmannii for Herr Witmann, who collected in Caucasus–Taurus *c.* 1840
witotorus -a -um from the area of the Huitoto peoples, N Peru / S Colombia
wittebergansis -is -e from Witteberg, Cape Province, S Africa, famed for its palaeobotanic record
Wittrockia, wittrockianus -a -um for Professor Veit Brecher Wittrock (1839–1914), of Stockholm, Swedish Director of Hortus Bergianus and author of *Morphologisk biologiska och systematiska studier öfver Viola tricolor och hennes närmäre anförvandter*
Wittsteinia for George Christian Wittstein (1810–87), writer on plant names and chemistry
wockeanus -a -um, wockei for Erich Wocke (1863–1941), who founded the Alpenpflanzengarten at Oliva, Danzig, author of *Die Kultupraxis der Alpenpflanzen*
Wodyetia from an Australian aboriginal name for *Wodyetia bifurcata* (foxtail palm)
woerlitzensis -is -e from Wörlitz, Anhalt-Dessau, Germany
Wolffia for Johann Friedrich Wolff (1778–1806), German doctor and author on *Lemna*
wolffii for Herman Wolff (1866–1929), veterinary surgeon and botanist
wolfianus -a -um, wolfii for either Ferdinand Otto Wolf (1838–1906), Professor of Botany at Sitten, or Franz Theodor Wolf (1841–1921), German geologist and botanist, who monographed *Potentilla*
Wolffiella a diminutive of *Wolffia*
wolgaricus -a -um from the region of the River Volga, Russia
Wollemi from Wollemi Canyon, Blue Mountains, New South Wales, Australia (*Wollemi nobilis*, the latest 'living fossil', found 1996)
woodii for John Medley Wood (1827–1915), Curator of Durban Botanic Garden
Woodsia for Joseph Woods (1776–1864), English architect, botanist and author of *The Tourist's Flora* (1852)
Woodwardia for Thomas Jenkinson Woodward of Suffolk (1745–1820) ('one of the best English botanists' – Sir J. E. Smith)
Woollsia for William Woolls (1814–93), English cleric and botanist in Australia
Worsleya for Arthington Worsley (1861–1944), English civil engineer and botanist in Brazil
Wrightia for Dr William Wright (1740–1827), Scottish physician and botanist, who found *Cinchona jamaicensis*
wrightianus -a -um for Charles Henry Wright (1864–1941), of the Herbarium at Kew

wrightii for Charles Wright (1811–85), from Connecticut, collector in Cuba
Wulfenia, Wulfeniopsis, wulfenianus -a -um, wulfenii for Franz Xavier Freiherr von Wulfen (1728–1805), Austrian Jesuit and naturalist, writer on plants
Wulffia for Johann Christoph Wulff (d. 1767), German physician and botanist, author of *Flora Borussica*
Wullschlaegelia for Heinrich Rudolf Wullschlaegel (1805–64), Russian theologian and orchidologist
Wurdackanthus for John Julius Wurdack (1921–98), American systematic botanist at the Smithsonian Institute
Wurmbea for Friedrich van Wurmb, eighteenth-century Dutch naturalist and Secretary of the Batavian Academy of Sciences
wutaiensis -is -e from Mount Wutai, Shanxi province, China
Wyethia for Nathaniel Jarvis Wyeth (1802–56), Boston fur-trader and plant collector
wyomingensis -is -e from an Indian vernacular name meaning 'the land of vast plains'; either from Wyoming state or one of the Wyoming counties, USA
wytaiensis -is -e from Wutai, Shanxi province, China

xalapensis -is -e from Xalapa (Jalapa), Mexico (see *jalapa*)
xanth-, xanthi-, xantho- yellow-, ξανθος, ξανθο-, ξανθ-; *xanthus, xantho-*
xanthacanthus -a -um yellow-thorned, ξανθ-ακανθα
xanthellus -a -um pale yellow, diminutive from ξανθος
Xantheranthemum Yellow-*Eranthemum*, ξανθ-εραω-ανθεμιον (or Lovely-yellow-flower)
xanthifolius -a -um yellow-leaved, *xanthus-folium*
xanthinoides resembling (*Rosa*) *xanthina*
xanthinus -a -um yellow, ξανθος
Xanthisma Of-yellow, ξανθισμα (star of Texas)
Xanthium Dioscorides' name, ξανθιον, for cocklebur, from which a yellow, ξανθος, hair dye was made
xanthocalyx with a yellow calyx, ξανθο-καλυξ
xanthocarpus -a -um yellow-fruited, ξανθο-καρπος
xanthocephalus -a -um yellow-headed, ξανθο-κεφαλη
Xanthoceras Yellow-horn, ξανθο-κερας (the glandular processes on the disc)
Xanthocercis Yellow-staff, ξανθο-κερκις (bark colour)
xanthochlorus -a -um yellow-green, yellow with green, ξανθο-χλωρος
xanthochrous -a -um yellow coloured, yellow-skinned, ξανθο-χροα
xanthochymus -a -um with a yellow exudate or sap, botanical Latin from ξανθο and *chymus*
xanthocodon yellow bell, ξανθο-κωδον (flowers)
Xanthocyparis Yellow-*Cyperus*, ξανθο-κυπειρος (Vietnamese)
xanyhodermus -a -um with a yellow skin, ξανθο-δερμα
xanthomelas dark-yellow, ξανθο-μελας
Xanthomyrtus Yellow-myrtle, *xantho-Myrtus*
xanthoneurus -a -um with yellow veins, ξανθο-νευρα
Xanthopappus Yellow-pappus, *xantho-pappus*
Xanthophyllum Yellow-leaved-one, ξανθο-φυλλον (age-related colouration)
xanthophloeus -a -um with yellow bark, ξανθο-φλοιος (fever wood, *Cinchona*)
Xanthorhiza (Xanthorrhiza) Yellow-root, ξανθο-ριζα (roots yield a yellow dye)
Xanthorrhoea (Xanthorrhaea) Yellow-flow, ξανθο-ρεω (the yellow sap of the grass tree) (***Xanthoxorrhoeaceae***)
Xanthosia Yellow, ξανθος (the covering of yellow down)
Xanthosoma Yellow-body, ξανθο-σωμα (some have yellow stem tissues in rhizomes)
xanthospilus -a -um yellow-spotted, ξανθο-σπιλος
Xanthostemon Yellow-stamened-one, ξανθο-στημων
xanthostephanus -a -um with a yellow crown, ξανθο-στεφανος (flower heads)

xanthoxyloides resembling *Xanthoxylum*, ξανθο-ξυλον-οειδης
Xanthoxylum (-on), xanthoxylon, xanthoxylum Yellow-wooded, ξανθο-ξυλον (the timber of the toothache tree)
xen-, xenico-, xeno- foreign-, unnatural-, strange-, ξενος
xenanthus -a -um having unusual or strange flowers, ξενος-ανθος
xenogenus -a -um of strange birth, of uncertain ancestry, ξενος-γενος
xer-, xero- dry-, ξερος, ξερο-, ξερ-; ξηρος, ξηρο-, ξηρ-; ξερα dry land
xerampelinus -a -um clothed with dark colours, *xerampelinae, xerampelinarum*
Xeranthemum Dry-flower, ξερος-ανθεμιον (immortelle)
Xerochloa Dry-grass, ξερος-χλοη
xerographicus -a -um with dry markings, with chalk marks, ξερο-γραφις
Xerolirion Dry-lily, ξερος-λειριον
Xeronema Dry-thread, ξερος-νημα (the persistent filaments)
xerophilus -a -um drought-loving, living in dry places, ξηρο-φιλος
Xerophyllum Dry-leaf, ξερος-φυλλον (*Xerophyllum* × *tenax*, elk grass)
Xerophyta Dry-plant, ξερος-φυτον
xerophyticus -a -um, xerophyton drought plant, ξερος-φυτον
Xerorchis Dry-orchid, ξερος-ορχις
Xerosicyos Dry-gourd, ξερος-σικυος
Xerospermum Dry-seeded-one, ξερος-σπερμα
Xerothamnella Little-dry-bush, botanical Latin diminutive from ξερος-θαμνος
Xerotia Of-dry-land, ξερα
xestophyllus -a -um having polished leaves, ξεω-φυλλον
Ximenia for Francisco Ximenez (Ximenes), Spanish monk and naturalist who wrote on Mexican plants in 1615
xiphi-, xipho- sword-, ξιφος (elongate and with an acute apex)
Xiphidium Dagger, ξιφιδιον (the leaf shape)
xiphioides, xiphoides *Xiphium*-like, *Xiphium-oides*, sword-like, shaped like a sword, ξιφος-οειδης
Xiphion (um), xiphium Sword, ξιφος (old generic name, ξιφιον, from the Greek name for a cornflag or *Gladiolus*)
xiphochilus -a -um with a sword-shaped lip, ξιφος-χειλος
xiphophyllus -a -um with sword-shaped leaves, ξιφος-φυλλον
Xiphopteris Sword-fern, ξιφος-πτερυξ (the firm, mostly simple fronds)
Xolisma an uncertain name for a genus containing species with such vernacular names as 'maleberry', 'fetterbush' and 'staggerbush'
xyl-, xylo-, -xylon, -xylum woody-, -wooded, -timbered, wood-, ξυλον, ξυλο-, ξυλ-
Xylanthemum Woody-flowered-one, ξυλ-ανθεμιον (perianth texture)
Xylaria Belonging-to-wood, ξυλον (timber-rotting fungi)
Xylia Wood, ξυλον (an ironwood)
Xylobium Wood-life, ξυλον-βιος (epiphytic)
xylocanthus -a -um, xylonacanthus -a -um woody-thorned, ξυλον-ακανθος
xylocarpus -a -um woody-fruited, ξυλον-καρπος
xylophilus -a -um wood-loving, ξυλον-φιλος (of wood attacking fungi)
xylophyllus -a -um having hard foliage, tough-leaved, ξυλον-φυλλον
Xylopia Bitter-wood, ξυλον-πικρια (the Greek for such wood)
xylopicron bitter-wooded, ξυλον-πικρος
xylopodus -a -um with woody stalks, ξυλον-ποδος
xylorrhizus -a -um woody-rooted, ξυλον-ριζα
Xylosma Fragrant-wooded, ξυλον-οσμη
xylosteoides resembling *Xylosteon, Xylosteon-oides*
Xylosteon an Adansonian name, ≡ *Lonicera, pro parte*
xylosteum hard-wooded, ξυλον-οστεον (wood-bone)
xyridiformis -is -e razor-sharp, resembling *Xyris* in habit, *Xyris-forma*
Xyris Greek name, ξυρις, used by Dioscorides for *Iris foetidissima;* ξυρον a razor (***Xyridaceae***)
Xysmalobium Fragmented-lobes, ξυσμα-λοβος (divisions of the corona)

Yabea for Yoshitaba Yabe (1876–1931), Japanese botanist
yakushimanus -a -um, yakusimanus -a -um, yakusimensis -is -e from the island of Yakushima, S of Japan
yamatensis -is -e from Yamato, Honshu, Japan
yanthinus -a -um bluish-purple, violet (see *ianthinus -a -um*)
yargongensis -is -e from the Yar Gong gorge, Tibet
yebrudii of the Yebrud, Syria
yedoensis -is -e from Tokyo (known as Edo, Yedo or Jedo before 1868), Japan
yemensis -is -e from Al Yaman (the Yemen), Arabia
yesoensis -is -e, yezoensis -is -e from Hokkaido island (Yezo, Yesso, Jezo, Jesso), Japan
yoco a S American vernacular name for *Paulinia yoco*
yosemitensis -is -e from the Yosemite valley, California, USA
youngianus -a -um for Messrs Young, nurserymen on the Milford estate, Epsom, until 1862
ypsilo- lofty, steep, stately, proud, υψηλος, on high, υψι-
Ypsilopus Erect-stalk, υψηλο-πους (the caudicles; some derive as uppercase upsilon-shaped, Y)
ypsilostylus -a -um proud-styled, υψηλος-στυλος
yucatanensis -is -e from Yucatán state, SW Mexico
Yucca from a Carib name, yuca, for cassava (*Manihot*), for its enlarged roots, incorrectly applied by Gerard
yuccifolius -a -um with *Yucca*-like leaves
yuccoides resembling *Yucca*
yukonensis -is -e from the Yukon, Alaska/Canada
yulan from the Chinese name for *Magnolia denudata*
yungasensis -is -e from the tropical 'warm lands' of Bolivia, which the Aymara call Yungas
yungningensis -is -e from Nan-ning (Yung-ning), China
yunnanensis -is -e from Yunnan, China
yuraguanus -a -um from Yuraguana, Cuba
Yutajea from Yutaje, Guayana Venezuelan Highland
Yvesia for Alfred Marie Saint-Yves (1855–1933), French agrostologist

za- most-, much-, many-, very-, ζα-
Zabelia, zabelii for Hermann Zabel (1832–1912), German dendrologist
zabucajo the vernacular name for sapucaua nuts, *Lecythis zabucajo*
zacatecasensis -is -e from the state of Zacatecas, Mexico
Zacateza an anagram of *Tacazzea*
zagricus -a -um from the Zagros mountains of E Iran
zalaccus -a -um very resinous, botanical Latin from ζα and *lacca*
zaleucus -a -um very white, vivid-white, ζα-λευκος
zalil from an Afghan vernacular name for *Delphinium zalil*
Zaluzianskya for Adam Zalusiansky von Zalusian (1558–1613), Bohemian physician and botanist, author of *Methodus herbariae*
zaman as *saman*, from a S American vernacular name
zambac as *sambac*, from an Arabic vernacular name
zambesiacus -a -um, zambesinus -a -um from the Zambezi river area, SE Africa
Zamia Parched-one, *zamia* (a name in Pliny refers to the sterile appearance of the staminiferous cones; αζω dried up)
zamii- resembling *Zamia*
Zamioculcas *Zamia*-like-*Culcas* (this Zanzibar aroid has *Zamia*-like leaves, with 6 to 8 pairs of alternate pinnae)
zanguebarius -a -um from Zanzibar, E Africa (Zanguebar)
Zannichellia for Giovani Garolamo Zannichelli (1662–1729), Venetian physician, chemist and botanist (***Zannichelliaceae***)

Zanonia for Giacoma Zanoni (1615–82), Italian botanist, author of *Istoria botanica* (1615)
Zantedeschia for Giovani Zantedeschi (1773–1846), Italian doctor and botanist
zantho- yellow, ξανθος, ξανθ0-
Zanthorhiza (Zanthorrhiza) Yellow-root, ξανθο-ριζα (see *Xanthorhiza*)
Zanthoxylum Yellow-wood, ξανθο-ζυλον (toothache tree)
zanzibarensis -is -e, zanzibaricus -a -um from Zanzibar, E Africa
zapellito the Brazilian vernacular name for *Cucurbita zapellito*
zaplutus -a -um very rich, highly treasured, very powerful, ζα-πλουτος
zapota from the Mexican name, cochil-zapotl, for the chicle tree, *Achras sapota*
Zapoteca commemorating the Zapotec civilization of Oaxaca, Mexico
Zataria from an Arabian vernacular name, za atar, for *Zataria multiflora*
Zauschneria for Johann Baptist Joseph Zauschner (1737–99), professor at Prague
zawadskii for Alexander (Jan Antoni) Zawadski (1798–1868), of Brno, present-day Czech Republic
zazil (zalil) from an Afghan name for a *Delphinium*
Zea from the Greek name for another cereal, possibly for spelt, ζεια (ζαω to live)
Zebrina, zebrinus -a -um from the Portuguese for a wild ass; the modern meaning is striped with different colours, zebra-striped
zedoaria an Indian vernacular name, zedoari, for fruits of *Curcuma zedoaria*
Zehneria for Joseph Zehner, German plant illustrator
zelandicus -a -um from New Zealand
Zelkova from the Caucasian name, tselkwa, for *Zelkova carpinifolia*
zenii for Chien P'ie, Chinese botanist
Zenkeria, Zenkerella for Johann Carl Zenker (1799–1837), German botanist
Zenobia an ancient Greek name, for Septimia Zenobis, Znwbya Bat Zabbai (d. AD 274) Queen of Palmyra (Roman colony in what is now Syria)
zeo- joined with-, ζευ- (plus a former name, e.g. *Zeobromus, Zeocriton*)
Zephyra, zephyrius -a -um Of the west, ζεφυρος, (Chilean), or western (for Indonesian plants, flowering or fruiting during the monsoon season)
Zephyranthes West-wind-flower, ζεφυρος-ανθος (introduced from America)
Zerna a Greek name, ζερνα (for the *Cyperus*-like spikelets)
zerumbet an Indian vernacular name for a plant
zetlandicus -a -um from the Shetland Isles, Scotland (*Zetlandia*)
Zeugites Paired, ζευγος (the spikelets)
Zeuxine Yoked, ζευξις (the arrangement of the perianth members)
Zeyheria for Johann Michael Zeyher (1770–1843),German horticulturalist
zeylanicus -a -um from Sri Lanka (Ceylon), Singhalese (Zeylona) (*Taprobane*)
Zexmenia for Francisco Ximenez (*c.* 1615), Spanish monk and botanical writer (anagrammatic name)
zibethinus -a -um of the civet (the foul-smelling fruits of *Durio zibethina* are used to trap the Asiatic civet (*Vivera zibetha*)
Zieria for John Zier (d. 1796), Polish botanist
Zigadenus (Zygadenus) Yoked-glands, ζυγος-αδηv (paired glands at the perianth base)
zigomeris -is -e *vide zygomeris*
zimapani from Zimapan, Mexico
Zimmermannia, zimmermannii for Albrecht W. P. Zimmermann (1860–1931), German botanist
Zimmermanniopsis Resembling-*Zimmermannia*, botanical Latin from *Zimmermannia* and οψις
Zingeria for Basil Zinger (1836–1907), Russian botanist
Zingiber the Greek name, ζιγγιβερις, from a Sanskrit name, singabera, or shrigavera, for the spice, possibly from an Indian or oriental source, inchi (a root), cognate with ginger (***Zingiberaceae***)
Zinnia for Johann Gottfried Zinn (1727–59), German professor of pharmacology and director of the botanic garden at Göttingen

zionis -is -a from Zion National Park, SW Utah, USA
Zizania an ancient Greek name, ζιζανιον (darnel, for a wild plant)
Zizaniopsis Resembling-*Zizania*
zizanoides resembling Canadian wild rice, ζιζανιον-οειδης
Zizia, zizii for John Baptist Ziz (1779–1829), botanist of Mainz
Ziziphora Resembling-*Zizyphus*
Zizyphus ancient Greek name, ζιζυφον, for *Zizyphus jujuba* (from the Arabic, zizouf or zizafun, for *Z. lotus*)
zizyphus resembling *Zizyphus*
Zoisia, Zoysia, zoysii for Karl von Zoys (Zois) (1756–1800), Austrian botanist
zombensis -is -e from the area of Zomba, Shire Highlands, Malawi
Zombia a Haitian vernacular name for the palm *Zombia antillarum*
zonalis -is -e girdled with distinct bands or concentric zones of colour, *zona, zonae*
zonarius -a -um belt-like, restricted to a narrow zone, *zona, zonae*
zonatus -a -um with zones of colour markings, *zona, zonae*
zooctonus -a -um poisonous, ζωο-κτονος (creature-killing)
Zornia for Johannes Zorn (1739–99), a German botanist, author of *Icones plantarum medicinallium* (1779–84)
zoster- girdle-, ζωστηρ-
Zostera Ribbon (Theophrastus' name, ζωστηρ, for a marine plant) (***Zosteraceae***)
zoutpansbergensis -is -e from the salt-pan-mountain area, Soutpansberg, northern Transvaal, S Africa
Zoysia for Karl von Zoys (1756–1800), Austrian botanist and collector
zuluensis -is -e from Zululand (KwaZulu-Natal), S Africa
zumi a Japanese name
Zygia Paired, ζυγος
zygis, zyge yoke-like, ζυγος (paired flowers)
Zygnema Paired-thread, ζυγος-νημα (at conjugation)
zygo- paired-, balanced-, yoked-, ζευγος; ζυγον, ζυγος, ζυγο- (the goddess of marriage, Hera, was also known as Juno Zygia)
Zygocactus Jointed-stem-cactus, ζυγο-κακτος
Zygochloa Yoke-grass, ζυγος-χλοη
Zygogynum Joined-ovary, ζυγος-γυνη
zygomeris -is -e with twinned parts, ζυγος-μερις
zygomorphus -a -um bilateral, of balanced form, ζυγο-μορφη
Zygopetalum Yoked-petals, ζυγος-πεταλον (two united basally to the column)
Zygophlebia Joined-veins, ζυγος-(φλεψ, φλεβος)
Zygophyllum, zygophyllus -a -um Yoked-leaves, ζυγος-φυλλον (some bean capers have conspicuously paired leaves) (***Zygophyllaceae***)
Zygosepalum Yoked-sepals, ζυγος-σκεπη (≡ *Menadenium*)
Zygostates Placed-yoke, ζυγος-στατος (the lateral extensions to the base of the column)
Zygotritonia Bilateral-*Tritonia*, ζυγος-τριτον (the upper perianth lobe is hooded and the lower four are recurved; *Tritonia* is actinomorphic)
zymo- of fermentation, ζυμοω, ζυμο-, to leaven, ζυμη

Figures

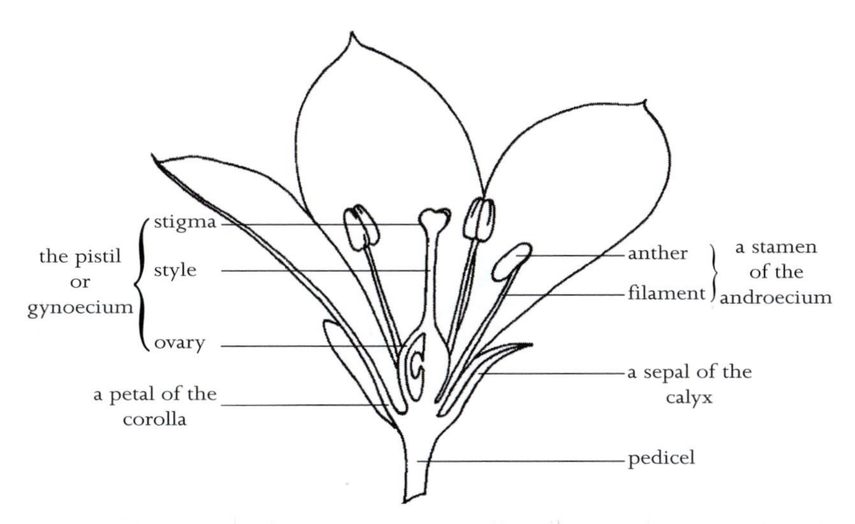

Figure 1. The parts of a flower, as seen in a stylized flower cut vertically in half.

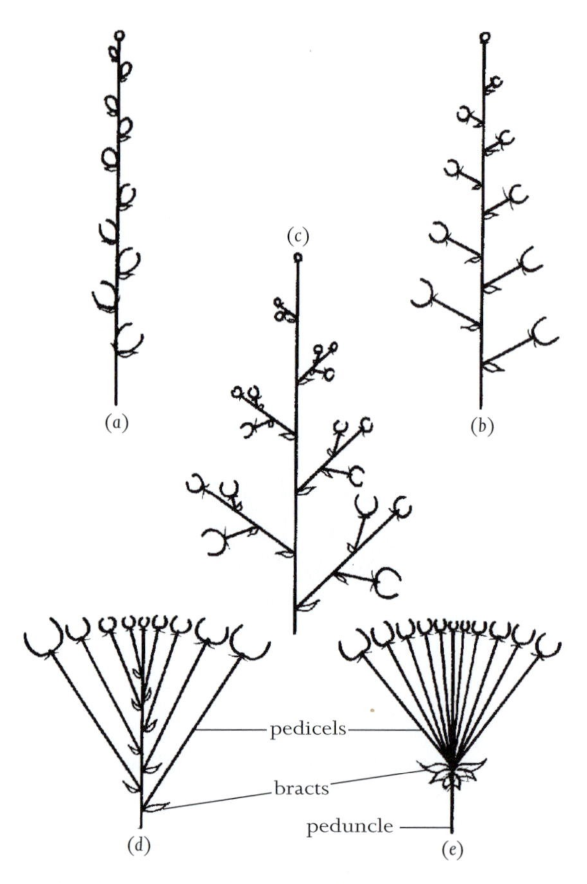

Figure 2. Types of inflorescence which provide specific epithets:
(*a*) A spike (e.g. *Actaea spicata* L. and *Phyteuma spicatum* L.).
(*b*) A raceme (e.g. *Bromus racemosus* L. and *Sambucus racemosa* L.).
(*c*) A panicle (e.g. *Carex paniculata* L. and *Centaurea paniculata* L.).
(*d*) A corymb (e.g. *Silene corymbifera* Bertol. and *Teucrium corymbosum* R. Br.).
(*e*) An umbel (e.g. *Holosteum umbellatum* L. and *Butomus umbelatus* L.).
In these inflorescences the oldest flowers are attached towards the base and the youngest towards the apex.

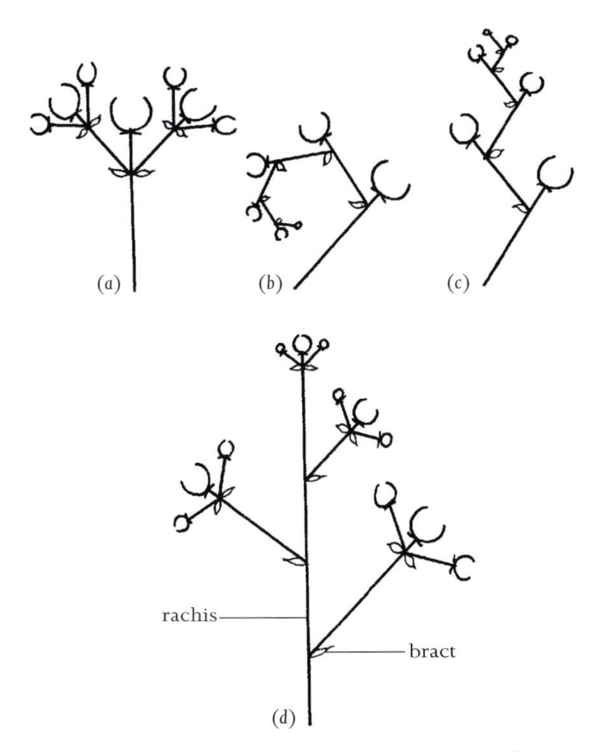

Figure 3. Types of inflorescence which provide specific epithets:
(*a*), (*b*) and (*c*) are cymes (e.g. *Saxifraga cymosa* Waldts. & Kit.).
(*b*) may have the three-dimensional form of a screw, or bostryx.
(*c*) may be coiled, or scorpioid (e.g. *Myosotis scorpioides* L.).
(*d*) is a raceme of cymes, or a thyrse (e.g. *Ceanothus thyrsiflorus* Eschw.).
In these inflorescences the oldest flower terminates the axis and younger ones are axillary to and below it.

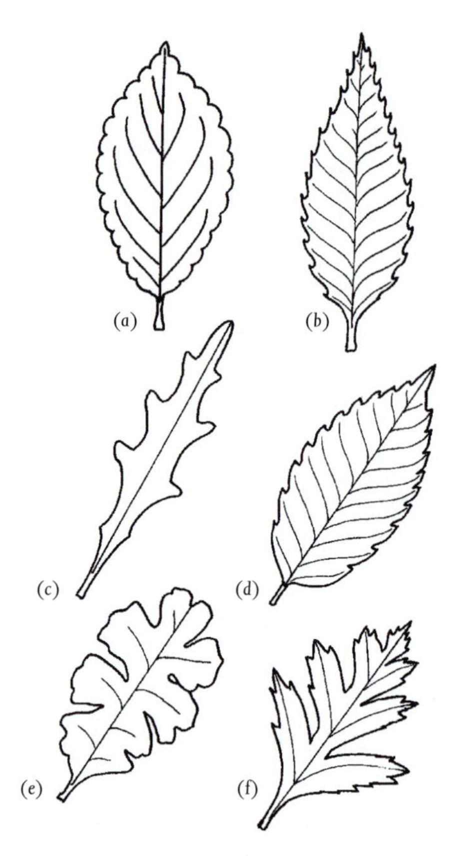

Figure 4. Leaf-margin features which provide specific epithets:
(*a*) Crenate (scalloped as in *Ardisia crenata* Sims).
(*b*) Dentate (toothed as in *Castanea dentata* Borkh.). This term has been used for a range of marginal tooth shapes.
(*c*) Sinuate (wavy as in *Matthiola sinuata* (L.) R. Br.). This refers to 'in and out' waved margins, not 'up and down' or undulate waved margins.
(*d*) Serrate (saw-toothed as in *Zelkova serrata* (Thunb.) Makino).
(*e*) Lobate (lobed, as in *Quercus lobata* Née).
(*f*) Laciniate (cut into angular segments as in *Crataegus laciniata* Ucria).

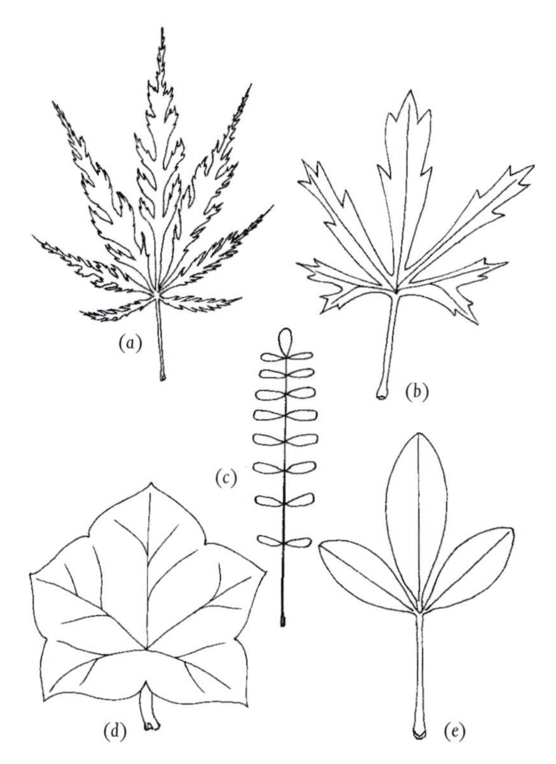

Figure 5. Some leaf shapes which provide specific epithets:
(*a*) Palmate (e.g. *Acer palmatus* Thunb. 'Dissectum'. As this maple's leaves mature, the secondary division of the leaf-lobes passes through incised-, *incisum*, to torn-, *laciniatum*, to dissected-, *dissectum*, lobed, from one central point.
(*b*) Pedate (e.g. *Callirhoe pedata* Gray). This is distinguished from palmate by having the lower, side lobes themselves divided.
(*c*) Pinnate (e.g. *Ornithopus pinnatus* Druce). When the lobes are more or less strictly paired it is called paripinnate, when there is an odd terminal leaflet it is called imparipinnate, and when the lobing does not extend to the central leaf-stalk it is called pinnatifid.
(*d*) Peltate (e.g. *Pelargonium peltatum* (L.) Ait.) has the leaf-stalk attached on the lower surface, not at the edge.
(*e*) Ternate (e.g. *Choisya ternata* H. B. K.) In other ternate leaves the three divisions may be further divided, ternately, palmately, or pinnately.

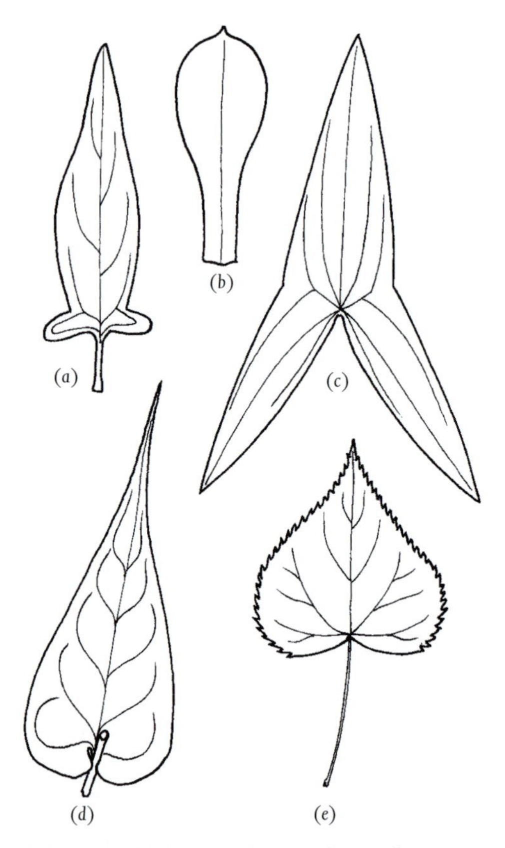

Figure 6. More leaf shapes which provide specific epithets:
(*a*) Hastate (e.g. *Scutellaria hastifolia* L.), with auricled leaf-base.
(*b*) Spathulate (e.g. *Sedum spathulifolium* Hook.).
(*c*) Sagittate (e.g. *Sagittaria sagittifolia* L.), with pointed and divergent auricles.
(*d*) Amplexicaul (e.g. *Polygonum amplexicaule* D. Don), with the basal lobes of the leaf clasping the stem.
(*e*) Cordate (e.g. *Tilia cordata* Mill.), heart-shaped.

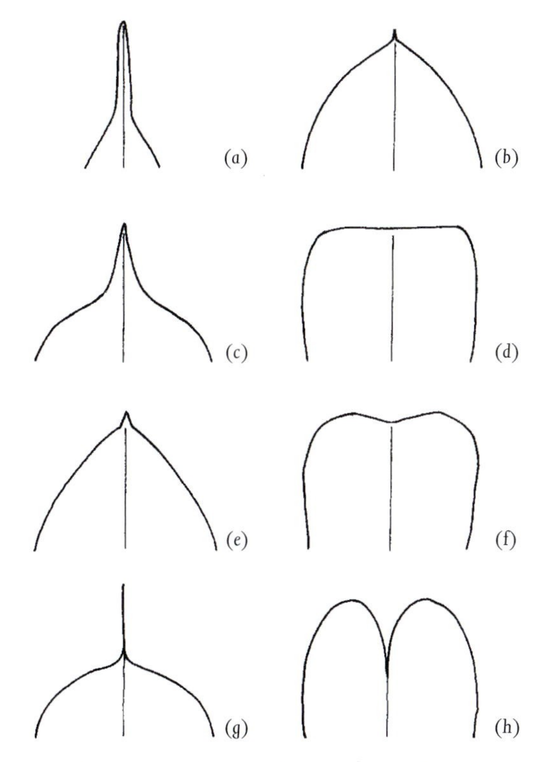

Figure 7. Leaf-apex shapes which provide specific epithets:
(*a*) Caudate (e.g. *Ornithogalum caudatum* Jacq.), with a tail.
(*b*) Mucronate (e.g. *Erigeron mucronatus* DC.), with a hard tooth.
(*c*) Acuminate (e.g. *Magnolia acuminata* L.), pointed abruptly.
(*d*) Truncate (e.g. *Zygocactus truncatus* K.Schum.), bluntly foreshortened.
(*e*) Apiculate (e.g. *Braunsia apiculata* Schw.), with a short broad point.
(*f*) Retuse (e.g. *Daphne retusa* Hemsl.), shallowly indented.
(*g*) Aristate (e.g. *Berberis aristata* DC.), with a hair-like tip, not always restricted to describing the leaf-apex.
(*h*) Emarginate (e.g. *Limonium emarginatum* (Willd.) O. Kuntze), with a deep mid-line indentation.

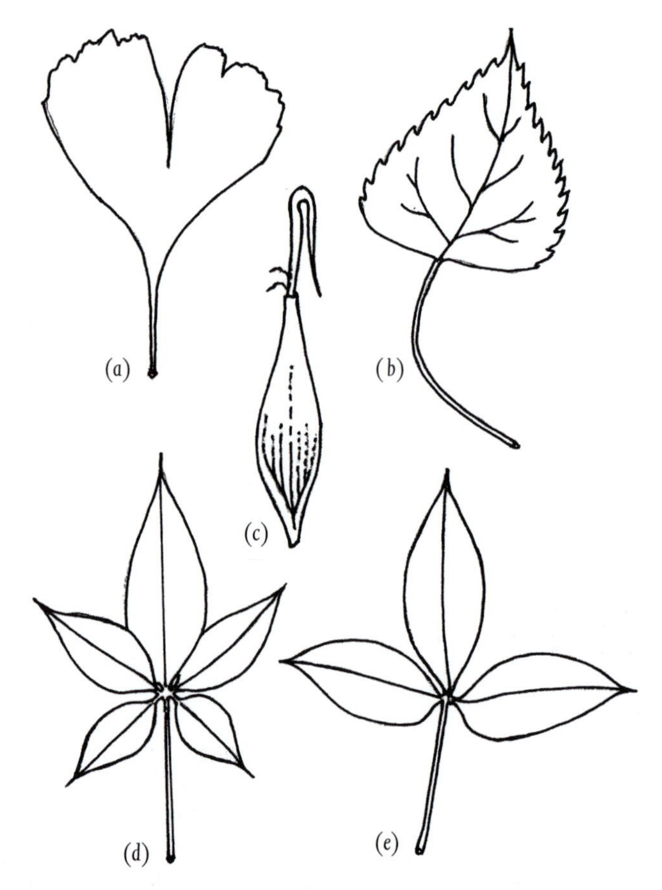

Figure 8. Other shapes which provide specific epithets:
(*a*) Bilobed (e.g. *Ginkgo biloba*), deeply cleft into two lobes.
(*b*) Deltoid (e.g. *Populus deltoides*), triangular, almost an equilateral triangle.
(*c*) Uncinate (e.g. the spikelet axis of *Uncinia uncinata*), formed into a hook.
(*d*) Digitate (e.g. *Adansonia digitata*), with five lobes arising from the apex of the petiole.
(*e*) Trifoliate (e.g. *Ptelea trifoliata*), with three leaflets.

Bibliography

Adanson, M. 1763–4 *Familles des plantes*. Paris.
Albertus Magnus 1478 *Liber aggregations seu liber secretorum Alberti magni de virtutibus herbarum*. Johann de Anunciata de Augusta.
Bagust, H. 2001 *Plant Names: Common & Botanical*. Oxford: Helicon.
Bailey, L. H. 1949 *Manual of Cultivated Plants*. New York: Macmillan.
Bateson, W. 1909 *Mendel's Principles of Heredity*. Cambridge.
Bauhin, C. 1620 *Prodromus theatri botanici*. Frankfurt.
Bauhin, C. 1623 *Pinax theatri botanici*. Basel.
Boerhaave, H. 1710 & 1720 *Index plantarum* . . . Leiden.
Brickell, C. D. *et al*. 1980 *International Code of Nomenclature for Cultivated Plants*. In *Regnum Vegetabile*, Vol. 104. Deventer.
Brickell, C. D. *et al*. 2004 *International Code of Nomenclature for Cultivated Plants – Code International pour la Nomenclature des Plantes Cultivées*, 7th edn. *Acta Horticultuae*, 647. Leuven: International Society for Horticultural Science.
Britten, J. & Holland, R. 1886 *A Dictionary of English Plant Names*. London: English Dialect Society.
Brummitt, R. K. 1992 *Vascular Plant Families and Genera*. London: Royal Botanic Gardens, Kew.
Brunfels, O. 1530–6 *Herbarium vivae eicones* . . . Strasbourg.
Caesalpino, A. 1583 *De plantis libri xvi*. Florence.
Camp, W. H., Rickett, H. W. & Weatherby, C. A. (eds.) 1947 Rochester Code. *Brittonia* **6**(1), 1–120. Massachusetts: Chronica Botanica.
Candolle, A. de 1867 *Lois de la nomenclature botanique*. Paris: H. Georg.
Candolle, A. P. de 1813 *Théorie élémentaire de la botanique*. Paris.
Chittenden, F. J. (ed.) 1951 *Royal Horticultural Society Dictionary of Gardening*, Vols. 1–4, and Supplements 1956 and (ed. P. M. Synge) 1969. Oxford: Oxford University Press.
Cordus, V. 1561–3 *Annotationes in pedacii Dioscorides* . . . Strasbourg.
Correns, C. 1900 G. Mendel's Regel über das Verhalten der Nach-kommenschaft der Rassenbastarde. *Berichte* **18**, 158.
Cube, J. von 1485 *German Herbarius*. Mainz.
Darwin, C. R. 1859 *The Origin of Species By Means of Natural Selection*. London: J. Murray.
Dioscorides, Pedanius *De materia medica*. John Goodyer translation of 1655 (ed. R. T. Gunther). Oxford: Oxford University Press.
Dodoens, R. 1583 *Stirpium historiae pemptades*. Antwerp.
l'Ecluse, C. 1583 *Stirpium nomenclator pannonicus*. Német-Hjvár.
Farr, E. R. *et al*. 1979–86. Index Nominum Genericorum and Supplement 1. In *Regnum Vegetabile*, Vols. 100, 101, 102, 113, The Hague.
Fernald, M. L. 1950 *Gray's Manual of Botany*. New York: American Book Company.
Fuchs, L. 1542 *De historia stirpium* . . . Basel.
Gilbert-Carter, H. 1964 *Glossary of the British Flora*, 3rd edn. Cambridge: Cambridge University Press.
Greatwood, J., Hunt, P. F., Cribb, P. J. & Stewart, J. (eds.) 1993 *The Handbook on Orchid Nomenclature and Registration*, 4th edn. London: International Orchid Commission.
Green, M. L. 1927 The history of plant nomenclature. *Kew Bulletin*, 403–15.
Greuter, W. *et al*. 2000 *International Code of Botanical Nomenclature (St. Louis Code)*. Königstein, Germany: Koeltz.

Grew, N. 1682 *The Anatomy of Plants*. London.
Grigson, G. 1975 *An Englishman's Flora*. St Albans: Hart Davis.
Ivimey-Cook, R. B. 1974 *Succulents: a Glossary of Terms and Descriptions*. Oxford: National Cactus and Succulent Society.
Jackson, B. D. 1960 *A Glossary of Botanic Terms*, 4th edn. London: Duckworth.
Jeffrey, C. 1978 *Biological Nomenclature*, 2nd edn. London: Edward Arnold.
Johnson, A. T. & Smith, H. A. 1958 *Plant Names Simplified*. Feltham.
Jung, J. 1662 Doxoscopiae; 1679 Isagoge phytoscopica. In *Opuscula botanica-physica*. Coburg 1747
Jussieu, A. L. 1789 *De genera plantarum*. Paris.
Linnaeus, C. 1735 *Systema naturae*. Leiden.
Linnaeus, C. 1738 *Classes plantarum*. Leiden.
Linnaeus, C. 1751 *Philosophia botanica*. Stockholm & Amsterdam.
Linnaeus, C. 1753 *Species plantarum*. Stockholm.
Linnaeus, C. 1754 *Genera plantarum*, 5th edn. Stockholm.
Linnaeus, C. 1759 *Systema naturae*, 10th edn. Stockholm.
Linnaeus, C. 1762–3 *Species plantarum*, 2nd edn. Stockholm.
Linnaeus, C. 1764 *Genera plantarum*, 6th edn. Stockholm.
Lyte, Henry 1578 *Niewe Herball, or Historie of Plantes*.
MacLeod, R. D. 1952 *Key to the Names of British Plants*. London: Pitman & Sons.
Malpighi, M. 1671 . . . *anatomie plantarum* . . . London (1675–9).
Matthioli, Pierandrea A. G. 1572 *Commentarii in sex libros Pedanii Dioscoridis*.
Mendel, G. J. 1866 *Versuche uber Planzenhybriden*. Brno.
Mentzel, C. M. 1682 *Index nominum plantarum multilinguis* (*universalis*). Berlin.
Morison, R. 1672 *Plantarum umbelliferum distributio nova*. Oxford.
Morison, R. 1680 *Plantarum historia universalis* . . . Oxford.
l'Obel, M. 1576 *Plantarum seu stirpium historia* . . . Antwerp.
Pankhurst, A. 1992 *Who Does Your Garden Grow*? Colchester: Earl's Eye.
Paracelsus (Bombast von Hohenheim) 1570 *Dispensatory and Chirurgery . . . Faithfully Englished by W. D.* London 1656.
Parkinson, J. 1629 *Paradisi in sole paradisus terrestris*. London: H. Lownes & R. Young (reprinted by Methuen, 1904).
Pliny the Elder (Gains Plinius Secundus) (AD 23–79) Thirty-seven books of *Historia naturalis*, of which 16 deal with plants.
Plowden, C. C. 1970 *A Manual of Plant Names*. London: Allen & Unwin.
Porta, Giambatista della (Johannes Baptista) 1588 *Phytognomica*. Naples.
Prior, R. C. A. 1879 *On the Popular Names of British Plants*. 3rd edn. London.
Rauh, W. 1979 *Bromeliads*. English translation by P. Temple. Dorset: Blandford Press.
Ray, J. 1682 *Methodus plantarum*. London.
Ray, J. 1686–1704 *Historia plantarum*. London.
Rivinus, A. Q. 1690 *Introductio generalis in rem herbariam*. Leipzig.
Schultes, R. E. & Pease, A. D. 1963 *Generic Names of Orchids: Their Origin and Meaning*. London: Academic Press.
Smith, A. W. 1972 *A Gardener's Dictionary of Plant Names* (revised and enlarged by W. T. Stearn). London: Cassell.
Sprague, T. A. 1950 The evolution of botanical taxonomy from Theophrastus to Linnaeus. In *Lectures on the Development of Taxonomy*. Linnean Society of London.
Stafleu, F. A. & Cowan, R. S. 1976 – Taxonomic literature, edn 2. In *Regnum Vegetabile* Vol. 94 (1976),Vol. 98 (1979),Vol. 105 (1981),Vol. 110 (1983),Vol. 112 (1985),Vol. 115 (1986) . . . Utrecht.
Stafleu, F. A. *et al.* (eds.) 1983 International Code of Botanical Nomenclature. In *Regnum Vegetabile*, Vol. 111. Utrecht.
Stearn, W. T. 1992 *Botanical Latin*, 4th edn Newton Abbot: David & Charles.
Stearn, W. T. 1992 *Stearn's Dictionary of Plant Names for Gardeners*, 2th edn London: Cassell.

Styles, B. T. (ed.) 1986 *Infraspecific Classification of Wild and Cultivated Plants*. Oxford: The Systematics Association, Special Vol. 29.
Sutton, W. S. 1902 On the morphology of the chromosome group in *Brachystola magna*. *Biological Bulletin*, **4**, 24–39.
Theophrastus 1483 *De causis plantarum lib vi*. Bartholomaeum Confalonerium de Salodio.
Tournefort, J. P. de 1694 *Elemens de botanique*. Paris.
Tournefort, J. P. de 1700 *Institutiones rei herbariae*. Paris.
Trehane, R. P. 1995 *International Code of Nomenclature for Cultivated Plants*. Wimborne, Dorset: Quarterjack.
Tschermak, E. 1900 Über künstliche Kreuzung bei *Pisum sativum*. *Biologisches Zentralblatt*, **20**, 593–5.
Turner, W. 1538 *Libellus de re herbaria*.
Turner, W. 1548 *The Names of Herbes* (The Ray Society, Vol. 145, London, 1965).
Turner, W. 1551–68 *A New Herbal*. London & Cologne.
UPOV 1985 *International Convention for the Protection of New Varieties of Plants*. Texts of 1961, 1972, 1978. UPOV publication 293E. Geneva.
de Vries, H. 1900 Sur la loi de disjonction des hybrides. *Comptes Rendues*, 130, 845–7. Paris. (Das Spaltungsgesetz der Bastarde; Vorlaufige Mitteilung. *Ber. Dtsch. Bot. Ges*. **18**, 83–90, 1900.)
Willis, J. C. 1955 *A Dictionary of Flowering Plants and Ferns*, 6th edn. Cambridge: Cambridge University Press.
Wilmott, A. J. 1950 Systematic botany from Linnaeus to Darwin. In *Lectures on the Development of Taxonomy*. Linnean Society of London.
Zimmer, G. F. 1949 *A Popular Dictionary of Botanical Names and Terms*. London: Routledge & Kegan-Paul.

Index